Snakes of the World

Snakes of the World

A Supplement

Jeff Boundy

CRC Press
Taylor & Francis Group
Boca Raton London New York

CRC Press is an imprint of the
Taylor & Francis Group, an **informa** business

Dedication

To Ken Williams, 1934–2017

Ken is my academic brother, not only in a titular sense by both of us receiving our doctoral degrees from Douglas Rossman, but via our passion for maintaining a complete compendium of literature toward a summary of all the world's snakes, living and extinct.

Contents

Introduction

"We found only a few species missing from the book."

–Uetz, Davison & Ellis, 2014, from their review of Wallach, Williams, & Boundy, 2014.

That quote from Uetz et al. may be the highest acclaim that we could have wished for a work documenting (almost) 3783 snakes with 860,000+ words. The present work is meant to be used as a companion to that first taxonomic summary of all the world's snakes, both living and fossil, *Snakes of the World: A Catalogue of Living and Extinct Species*, by Van Wallach, Ken Williams, and Jeff Boundy. Williams and Wallach conceived the world snake project during the early 1980s, based on shared interests in producing a taxonomic summary of snakes that was geographically and temporally complete. Boundy, independently working on a similar summary, joined that project in the mid-2000s, and in 2012 a decision was made to use the end of 2012 as a cutoff date for new material. We finalized the manuscript during 2013, Van already having secured CRC Press as publisher through Chuck Crumly. The book was published in April 2014, but Ken had already been caching literature since the first days of 2013. From his seamless efforts, Ken moved forward to working on two documents: new species accounts of novel taxa with new data for existing species, and a new literature cited. Once the "WWB" book was published, he offered that we three begin work on an addendum. Van declined to participate, and I began sending Ken PDFs and literature citations for a rapidly growing pair of manuscripts that the two of us planned to eventually publish.

Ken's work remained on pace through initial diagnoses and treatments for cancers during 2014–2016, but incarcerations in medical facilities during much of 2017 resulted in little additional work to the manuscript pair, and Ken passed away at home in November of that year. Within a month, his widow, Viola, blessed me with access to his computer from which I retrieved the last versions, compared them to the latest he had forwarded to me, and I proceeded to "catch up" with new literature. As with the Wallach, Williams, and Boundy book, Chuck Crumly offered a publishing outlet, and we agreed on a cutoff date of end of 2019.

Aside from the literature document, Ken had begun sections on papers dealing with valid vs. invalid names, papers discussing museum collections, species accounts for species for which there were additions or taxonomic changes, and a summary of geographical literature. To these I added a summary of papers that cover snake classification at the suprageneric level, and a list of genera in each family or subfamily. Because the alphabetic arrangement by genus is continued from the original book, this latter list is offered as a cross-reference for individuals wishing to locate each genus in particular families. Ken was creating accounts for those species for which there were taxonomic changes and/or pertinent biological or distributional data. I have added all recognized species to provide an updated list of all snake species. In comparison, up to the end of 2012 there were 651 genera and 3783 species of snakes, and at the end of 2019 there are 685 genera and 4182 species. Within those totals, the number of extinct (fossil) genera increased from 112 to 138, and species increased from 274 to 314.

Aside from a half-dozen papers, all of the publications cited in the literature are in JB's possession. That fact limits the geographical sources section of this work due to financial constraints of buying the many new books that have been recently published.

ACKNOWLEDGEMENTS

I thank Ken's widow, Viola ("Vi") Williams, for her help in sorrowful times. Van Wallach, Bill Lamar, and Alex Pyron provided helpful comments and hard-to-get literature, and I thank everyone who sent PDFs. Lastly, I thank Peter Uetz for creating and maintaining his Reptile Database, with which I was able to compare our respective compilations.

Novelties

Our objective is not to introduce new names or combinations within this work, but in the following cases there is either sufficient evidence to relocate a taxon to another genus, or a need, based on the genera and their content that we recognize, to do so. The following represent new combinations or changes to existing nomenclature:

Bothrops lojanus to *Bothrocophias* (Viperidae)
Lytorhynchus levitoni to *Rhynchocalamus* (Colubridae)
Rhinophis grandis vs. *R. saffragamus* (Uropeltidae)
Sinonatrix yapingi to *Trimerodytes* (Natricidae)
Trimeresurus gunaleni to *Parias* (Viperidae)
Trimeresurus yingjiangensis to *Popeia* (Viperidae)
Sinovipera sichuanensis to *Viridovipera* (Viperidae)

Nomenclatural Matters

CHARINIDAE

Quintero-A. & Shear (2016) requested the International Code of Zoological Nomenclature to suppress the family group name Charinidae Gray due to its homonymy with Charinidae Quintero-A. (for *Charinus*, an arthropod). Savage & Crother (2017) recommended modification of the family group name to Charinaidae, and Pyron (2017) requested to conserve the family-group name Charinidae Gray by modifying the arthropod version to Charinusidae. This Case, 3688, remains open.

CALAMARIIDAE, PAREATIDAE, XENODERMATIDAE

Savage (2015) provides the rationale for the correct spelling of the family names for the type genera *Calamaria*, *Pareas*, and *Xenodermus*, as Calamariidae, Pareidae, and Xenodermidae.

RAYMOND HOSER PUBLICATIONS

In Wallach et al. (2014), we considered names published by Raymond Hoser in his journal *Australasian Journal of Biology* to be illegitimate (*nomina illegitima*), and we listed them as such under each taxon to which they pertained. Our rationale, that their naming violated the spirit of the International Code of Zoological Nomenclature, would be confirmed or nullified if ruled upon by the ICZN. Hoser (2013) challenged that his journal, and therefore his names, is legitimate, and submitted a request for a ruling on the matter to the ICZN (Case 3601). His specific case was the priority of his name *Spracklandus* (2009) over *Afronaja* Wallach, Wüster & Broadley, which was published six months later.

Rhodin (2015), with 69 co-signees, argued that Hoser's megalomania was initiating a dual taxonomy of names that, whether properly published or not, would be ignored by most researchers. They requested that the issues of *Australasian Journal of Herpetology* published to that point be placed on the Official List of Rejected Works in Zoology. H. Kaiser (2013) proposed a method of screening and standardizing nomenclatural works, targeting *AJH*, and H. Kaiser et al. (2013a) argued further for a standardized procedure for publishing taxonomic novelties.

Hoser published several rebuttals (e.g., 2015a, b) to criticisms, but the arguments have ceased for several years pending decision of the ICZN. Case 3601 remains open as of February 2020. Rather than produce a long list of names proposed by Hoser since 2012, we here refer to the *Australasian Journal of Herpetology* by issue number, and provide taxa covered in each:

16: *Acrantophis, Anilios, Anomochilus, Atropoides, Boiga, Candoia, Chrysopelea, Cylindrophis, Eryx, Furina,* Hydrophiinae, *Malayopython, Pseudonaja*
17: *Calliophis,* Colubroidea families, *Dendrelaphis, Gerrhopilus, Liophidium, Liopholidophis, Suta,* Tropidophiidae, Ungaliophiinae, Uropeltidae, Viperidae tribes
19: *Bitis, Bothrops, Causus,* Crotalinae
20: *Atractaspis, Atractus, Boulengerina, Chamaelycus, Dendrelaphis, Drysdalia, Gongylosoma, Hapsidophrys, Hemachatus, Liopeltis, Lycophidion, Mopanveldophis*
21: *Dasypeltis*
22: *Acrochordus*
23: *Acanthophis, Aparallactus, Aspidomorphus, Cacophis, Chrysopelea, Cryptophis, Lycodon, Malayopython, Oligodon, Suta, Xerotyphlops*
30: *Demansia, Liasis, Vipera*
31: *Acanthophis, Achalinus, Candoia, Denisonia, Loveridgelaps, Salomonelaps, Tropidechis, Xenodermus*
33: *Aplopeltura, Boiga, Crotalus, Emydocephalus, Hoplocephalus, Malayopython, Montivipera, Oligodon,* Typhlopidae
35: *Boiga, Telescopus*
36: *Bungarus, Calloselasma, Morelia*
37: *Crotalus, Elachistodon, Suta*
40: *Vermicella*

Papers Covering Museum Collections and/or Types

Uetz et al. (2019) provide background from a world perspective on type material in collections, and a link to their file of the type material for all extant reptile species.

Savage & McDiarmid (2017) evaluate all of Georgio Jan's *nomina*, including nomenclatural availability and type material.

Vanzolini & Myers (2015) illustrate and discuss the Wied type material in the American Museum of Natural History collection.

The following works cover museum collections:

Andreone & Gavetti (2010, Museo Regionale di Scienze Naturali di Torino).

Bauer & Wahlgren (2013, Linck Collection, Naturalenkabinett Waldenburg).

Ceríaco et al. (2014b, Museu de História Natural da Universidade do Porto).

Céspedez, J. & Alvarez, B.B. (2005, Universidad Nacional del Nordeste, Corrientes, Argentina [UNNEC])

Conradie et al. (2019, Port Elizabeth Museum).

Doria (2010, Museo Civico di Storia Naturale "G. Doria").

Dotsenko (2003, National Museum of Natural History, Ukrainian Academy of Sciences).

Džukić et al. (2017, Institute for Biological Research "Siniša Stanković," University of Belgrade).

Ferraro & Williams (2006, Museo de La Plata).

Flores-Villela et al. (2016, Museo de Historia Natural Alfredo Dugès).

Ganesh & Asokan (2010, Government Museum, Chennai).

García-Díez & González-Fernández (2013, Museo Nacional de Ciencias Naturales).

Gonzales & Montaño-F (2005, Museo de Historia Natural "Noel Kempff Mercado").

Guo et al. (2012a, 2016a, Chengdu Institute of Biology).

Hamdan et al. (2013, Museu de Zoologia, Universidade Federal da Bahia).

Kaya & Özuluğ (2017, Zoology Museum, Istanbul University).

Mekinić et al. (2015, Natural History Museum in Split).

Mlíkovský et al. (2011, National Museum, Prague).

Moskvitin & Kuranova (2006, Zoological Museum, Tomsk State University).

Nistri (2010, Museo di Storia Naturale, University of Florence).

Prudente et al. (2019, Museu Paraense Emílio Goeldi).

Scali (2010, Museo Civico di Storia Naturale di Milano).

Scrocchi & Kretzschmar (2017, Fundación Miguel Lillo).

Serna-Botero & Ramírez-Castaño (2017, Museo de Historia Natural de la Universidad de Caldas).

Y. Werner & Shacham (2010, Hebrew University of Jerusalem).

Classification

The following papers discuss snake classification in a broad phylogenetic sense, and/or at the suprageneric level.

BROAD/BASAL

Conrad (2008) presents alternative phylogenies and morphological diagnoses of the Squamata, including Serpentes, from the Ophidia basal node through to Macrostomata.

Wiens et al. (2012) produce a phylogeny using 44 nuclear genes and representatives of then-recognized families, except Xenophidiidae. They find that the Scolecophidia are not monophyletic.

Pyron et al. (2013) produce a phylogeny using numerous genes of 1262 snake species, establishing a new and subsequently consistent arrangement of snake families and subfamilies.

Palci et al. (2013b) use new osteological data to place hind-limbed Cretaceous snakes in a phylogeny in context with other stem-group snakes.

Scanferla et al. (2013) produce a phylogeny of families, and primitive and/or monotypic genera to determine placement of †*Kataria*.

K.T. Smith (2013) discusses classification and familial assignments of some fossil taxa as follows: (1) †*Eoanilius europae* is probably a booid; (2) †*Calamagras weigeli* belongs to the Ungaliophiinae; (3) †*Paraplatyspondylia batesi* is probably a species of †*Platyspondylia*; (4) †*Ogmophis compactus* belongs to the Loxocemidae.

Hsiang et al. (2015) discuss snake origins, and present a revised phylogeny.

Reeder et al. (2015) use morphological and molecular data to produce a phylogeny of Squamata that places snakes as sister taxon to the Mosasauria. Anomalepidiae are paraphyletic with respect to Typhlopidae and Leptotyphlopidae, making the Scolecophidia non-monophyletic. These families are sisters to remaining snakes, followed by *Dinilysia* and *Najash*, with remaining snakes forming two clades.

Figueroa et al. (2016) present a ten gene-phylogeny using 1652 snake species, resulting in a modified resolution of suprageneric snake classification.

Streicher & Wiens (2016) use 1.4 million base pairs from nuclear genes to produce a phylogeny using one or two representatives from snake families. The phylogenies reveal (1) a non-monophyletic Scolecophidia, (2) an association between Aniliidae and Tropidophiidae, (3) and a clade containing *Casarea*, *Loxocemus*, and *Xenopeltis* with the Pythonidae.

Zheng & Wiens (2016) produce a combination phylogeny based on numerous genes and taxa that is relatively similar to the one produced by Wiens et al. (2012).

Weinell & Brown (2018) use DNA sequence data to produce a phylogeny of the Lamprophiidae that resolves Philippine taxa (*Cyclocorus*, *Hologerrhum*, *Myersophis*, *Oxyrhabdium*) into a subfamily-level clade that they name Cyclocorinae. *Buhoma* remains as the sister taxon to the remainder of the family aside from the Prosymninae. *Micrelaps* resolves as the sister taxon to *Psammodynastes* plus the Lamprophiinae, and *Aparallactus* forms a clade with *Atractaspis* and *Homoroselaps* to the exclusion of other Aparallactinae. Due to this latter arrangement, the authors suggest including the Aparallactinae within the Atractaspidinae.

Garberoglio et al. (2019) produce a morphology-based phylogeny that emphasizes relationships of certain pre-Serpentes snake taxa.

R.O. Goméz et al. (2019) produce a skeletal-based phylogeny of early fossil snakes, focusing on the Madtsoiidae.

MADTSOIIDAE

Rio & Mannion (2018) produce a phylogeny of the Madtsoiidae based on vertebral characters.

"SCOLECOPHIDIA"

Hedges et al. (2014) construct a molecular-based phylogeny of Typhlopidae, from which they erect four subfamilies, and new genera. They include data for 92 named and 60 yet-to-be described species.

Pyron & Wallach (2014) produce a DNA sequence-based phylogeny of the Typhlopidae using 95 species, which conform to four subfamilies, and a revised arrangement of genera. Their taxonomic conclusions are backed by data from internal and external morphology of nearly all recognized species.

Nagy et al. (2015) add genes and taxa to the gene matrices studied by Hedges et al. (2014) and Pyron & Wallach (2014), concluding that *Antillotyphlops*, *Cubatyphlops*, and *Sundatyphlops* are diagnosable, and that all native Madagascar species are assignable to *Madatyphlops*.

Miralles et al. (2018) use 14 nuclear genes to examine relationships of the Scolecophidia. The resulting phylogeny reveals the Scolecophidia to be paraphyletic, with anomalepids *Liotyphlops* and *Typhlophis* as sister taxa to the Alethinophidia.

UROPELTIDAE

Pyron et al. (2016c) provide a revised phylogeny and a taxonomic summary of the Uropeltidae. *Pseudotyphlops* and Sri Lankan *Uropeltis* are included in *Rhinophis*.

Cyriac & Kodandaramaiah (2017) use DNA sequence data to produce a phylogeny of 32 described plus some undescribed species from all genera of Uropeltidae, which resolve among four clades. (1) *Brachyophidium* is nested within *Teretrurus*, and they recommend synonymy of the two genera. (2) All Sri Lankan genera and species (radiation) are nested within Indian *Rhinophis*, rendering *Rhinophis* and *Uropeltis* non-monophyletic as currently recognized. (3) Indian *Uropeltis* includes multiple species.

PYTHONIDAE AND BOIDAE

Pyron et al. (2014b) critique recent Boidae phylogenies, and propose elevating all subfamilies to family level: Sanziniidae, Charinidae, Erycidae, Candoiidae, and Boidae s.s. They provide a morphological and phylogenetic definition for each, as well as list extant species for each.

Reynolds et al. (2014b) present a DNA sequence-based phylogeny of most species of Pythonidae and Boidae. The pythonids arranged into seven clades that correspond to the genera recognized herein, except for those subsequently argued as valid by Barker et al. (2015), who recommend continued recognition of *Apodora* and *Leiopython* (subsumed within *Liasis* and *Bothrochilus*, respectively, by Reynolds et al.). In addition, Reynolds et al. recover *Morelia viridis* as sister taxon to *Antaresia*, but they do not recommend reassignment of *viridis*. The boids are arranged into eight clades plus *Calabaria*. The four innermost clades correspond to the Boinae, and the other four correspond to the other subfamilies recognized by Pyron et al. (2014b), with Charininae and Ungaliophiinae branching within one of those clades.

Barker et al. (2015) review the taxonomy of the Pythonidae, and continue to recognize *Apodora* and *Leiopython* based on their distinctive morphologies.

Barker et al. (2018) review the Pythonidae of Asia, the East Indies, and New Guinea.

Reynolds & Henderson (2018) present a synopsis of the living genera, species, and subspecies of the Superfamily Booidea. For each taxon, they provide a subheading of taxonomy, type specimen, distribution, and conservation status.

COLUBROIDEA

Pyron et al. (2014a) attempt to resolve several difficult nodes in basal Caenophidea and Colubroidea. They conclude that monophyly of both groups remains uncertain. Other conclusions are that (1) Homalopsidae is sister taxon to Lamprophiidae, Elapidae, and Colubridae, (2) relationships between Lamprophiidae subfamilies are poorly resolved, (3) three clades are evident within a broad Colubridae (a: Sibynophiinae, Grayiinae, Calamariinae and Colubrinae, b: Pseudoxenodontinae and Dipsadinae, c: Natricinae).

Zaher et al. (2019) produce a DNA sequence-based phylogeny of over 1200 caenophidian species using up to 15 genes. They provide a morphological and temporal basis for their stem clades. The arrangement in Elapidae is stepwise, although the Australasian and marine taxa are recognized as the subfamily Hydrophiinae. Within the Dipsadidae are two speciose clades: Dipsadinae and Xenodontinae. There are some basal taxa that are assigned to the Carphophiinae, and several others that are paraphyletic relative to the named subfamilies (listed as subfamily "unnamed" below; e.g., *Diadophis*, *Farancia*, *Heterodon*, and *Thermophis*).

HOMALOPSIDAE

J.C. Murphy & Voris (2014) provide a synopsis of the Homalopsidae, including keys to the genera and species.

Quah et al. (2018c) produce an mtDNA sequence-based phylogeny of Homalopsidae that supports previous phylogenies, but includes *Raclitia*, which recovers as sister taxon to *Erpeton*.

PAREIDAE

Deepak et al. (2018) use data from six genes to evaluate the phylogenetic relationship of *Xylophis*. The three species form a sister clade to the Pareatidae, and are allocated to a new subfamily, Xylophiinae, within the pareatids. The latter clade contains a clade containing *Asthenodipsas*, and another containing *Pareas*, to which *Aplopeltura* is the sister taxon.

DIPSADIDAE

Pyron et al. (2015) present a revised phylogeny of the Dipsadinae. They define the tribes Nothopsini (*Nothopsis*) and Diaphorolepidini (*Diaphorolepis*, *Emmochliophis*, *Synophis*). They (2016a) present a revised phylogeny of the species of the Diaphorolepidini.

Arteaga et al. (2018) present a DNA sequence-based phylogeny of the Dipsadini, which places *Sibynomorphus* within *Dipsas*.

COLUBRIDAE

X. Chen et al. (2014) present a phylogeny for *Gonyophis* and related taxa, and recommend combining *Gonyophis*, *Rhadinophis*, and *Rhynchophis* with *Gonyosoma*.

L.D. Wilson & Mata-Silva (2015) present a synopsis of the *Tantilla* clade of the Sonorini (*Geagras*, *Scolecophis*, *Tantilla*, *Tantillita*).

Mirza et al. (2016) provide a DNA-sequence based phylogeny of numerous Old World Colubrini that includes an "arid clade" comprising *Bamanophis*, *Dolichophis*, *Eirenis s.l.*, *Hemerophis*, *Hemorrhois*, *Hierophis*, *Lytorhynchus*, *Macroprotodon*, *Mopanveldophis*, *Orientocoluber*, *Platyceps*, *Spalerosophis*, and *Wallaceophis*, to the exclusion of African (*Thelotornis*, *Thrasops*) and Asian (*Coelognathus*) forest taxa. Mirza & Patel (2018) add *Rhynchocalamus* and *Wallophis* to the arid clade.

X. Chen et al. (2017) present a revised phylogeny of the "ratsnakes" (Coronellini plus Lampropeltini). (1) The placement of *Elaphe zoigeensis* and *Orthriophis taeniurus* leave the respective genera paraphyletic, which the authors resolve by placing *Orthriophis* within an expanded *Elaphe*. (2) They return *Zamenis scalaris* to the monotypic *Rhinechis*, though it remains sister taxon to remaining *Zamenis*. (3) The species of *Lampropeltis* are segregated into two clades: the blotched/cross-banded species, and the ringed, tricolor species. (4) In order to recognize the Lampropeltini as distinct from Old World species, the authors revive Coronellini for both Old and New World species, and recognize five subtribes: Coronellina, Elapheina, Euprepiophina, Oreocryptophina, and Lampropeltina.

Dahn et al. (2018) use sequence data from seven genes to produce a phylogeny of the Lampropeltini. Species content of previously established genera are confirmed, and evidence of multiple species are confirmed or discovered for *Arizona*, *Cemophora*, *Pseudelaphe*, *Rhinocheilus*, and *Senticolis*.

Cox et al. (2018) present a phylogeny of taxa of Sonorini related to *Sonora*. They resolve five clades, and elect to refer these to subgenera, which requires reducing *Chilomeniscus* and *Chionactis* to subgenera.

A.K. Mallik et al. (2019) use a variety of genes to produce a phylogeny of the Ahaetullinae, which partitions two clades: (1) *Dryophiops* (*Proahaetulla* (*Ahaetulla*)), and (2) *Chrysopelea* and *Dendrelaphis*. The 11 species evaluated for the latter form two clades. They also provide a key to the five genera.

Pseudoxyrhophiidae

Ruane et al. (2015) produce a phylogeny sampled from most genera of Malagasy pseudoxyrhophiids that sorts them among two clades that correspond to maxillary tooth morphology.

Atractaspididae

Portillo et al. (2018) provide a phylogeny of the Aparallactinae, but without *Brachyophis* or *Hypoptophis* material. Aparallactines are monophyletic minus *Micrelaps*. There are three primary clades: (1) *Aparallactus*, (2) *Chilorhinophis* as sister taxon to *Polemon*, and (3) *Macrelaps-Amblyodipsas-Xenocalamus*. Within the latter, *Xenocalamus* are within *Amblyodipsas*, and are hypothesized to bear longer snouts for occurrence in arid areas. Also, *Amblyodipsas concolor* lies without other *Amblyodipsas* species as sister taxon to *Macrelaps*. The authors recommend against taxonomic changes until additional taxa are analyzed. They also found *Micrelaps* to be unrelated to either the Atractaspidinae or Aparallactinae.

Portillo et al. (2019a) provide a phylogeny of the Atractaspidinae, *Atractaspis* and *Homoroselaps*, confirming the monophyly of both genera as sister taxa.

Natricidae

McVay & Carstens (2013) produce a phylogeny of existing genera of the Natricid tribe Thamnophiini using DNA sequence data. Two primary clades each contain two clades: (1) *Storeria* plus *Clonophis, Virginia, Seminatrix, Regina alleni, R. rigida*, (2) *Adelophis, Thamnophis* plus *Nerodia, Tropidoclonion, Regina graham, R. septemvittata*. Due to paraphyly, the two species of *Virginia* are recognized as separate genera (*Haldea* resurrected for *V. striatula*), and due to paraphyly and monophyly, *Seminatrix* and the two *Regina* of the first clade are included in a resurrected *Liodytes*.

Takeuchi et al. (2018) analyze Asian natricids using DNA sequence data. Several clades are resolved: (1) *Opisthotropis* is sister taxon to European and New World taxa, (2) Sri Lankan endemics *Aspidura* and *Haplocercus*, and (3) a speciose clade in which *Macropisthodon rudis* (=*Pseudagkistrodon*), *Trachischium* and *Hebius* for a sister clade to remaining species. In the speciose group, the structure has the African genera *Afronatrix* and *Natriciteres* as a sister taxon, followed by a clade containing *Atretium* and the *Xenochrophis piscator* species group (=*Fowlea*). The remainder of the clade contains *Balanophis*, other *Xenochrophis*, and other *Macropisthodon* mingled among *Rhabdophis*. The authors recommend combining the latter clade into a broader *Rhabdophis*.

Giri et al. (2019a) produce a DNA sequence-based phylogeny for some natricids, with a focus on Asian taxa. Their phylogeny shows the close relationship of *Trimerodytes, Smithophis*, and *Opisthotropis*, the distinction of the nuchal-gland clade apart from *Rhabdophis*, and the paraphyly of *Atretium yunnanensis* with *Xenochrophis* s.s.

Elapidae

Sanders et al. (2013) provide a phylogeny of viviparous marine Elapidae (i.e., sea snakes) using DNA sequence data from combined mt- and nDNA from 39 species and 15 genera. Taxa partition into two primary clades: *Emydocephalus* and *Aipysurus*, and a clade containing the remaining taxa. The semi-aquatic and microcephalic genera *Ephalophis, Hydrelaps, Microcephalophis*, and *Parahydrophis* are sister taxa to a group referred to as the core *Hydrophis* group. Due to the paraphyly of this group, in which classically recognized genera are scattered among subclades and hierarchies with *Hydrophis*, s.s., Sanders et al. elect to recognize this group as an expanded *Hydrophis*, s.l. See also Ukuwela et al. (2016) for supporting results.

M.S.Y. Lee et al. (2016a) use multiple-gene DNA sequence data to produce a phylogeny of more than half of known species of Elapidae. *Calliophis* are sister taxa to all remaining species, which form three primary clades: (1) *Sinomicrurus* and New World species (*Micruroides* and *Micrurus*), (2) *Hemibungarus, Dendroaspis*, and the "cobras" (*Aspidelaps, Hemachatus, Naja* s. l., *Ophiophagus*, and *Walterinnesia*), (3) *Bungarus* plus the Australopapuan genera and all marine species.

J.L. Strickland et al. (2016) present a phylogeny of the Hydrophiinae, based on mt- and nDNA sequence data, which does not recommend taxonomic change.

VIPERIDAE

Stümpel & Joger (2009) summarize the viperid genera *Daboia*, *Macrovipera* and *Montivipera* in SW Asia.

Guo et al. (2015) produce an mtDNA-based phylogeny for numerous South Asian pit vipers, which supports recognition of recently recognized genera: *Craspedocephalus, Parias, Popeia, Trimeresurus* and *Viridovipera*, with *Himalayophis* and *Sinovipera* as monotypic.

Alencar et al. (2016) used large DNA datasets to evaluate relationships of roughly 80% of known species of Viperidae.

Their results include a monophyletic *Bothrops* that includes the genera *Bothriopsis, Bothropoides, Bothrops, Rhinocerophis,* with *B. pictus* as sister taxon to the remaining species. Retention of the four genera would require a new genus for *B. pictus*.

Timms et al. (2019) evaluate mtDNA gene sequences for most species related to *Bothrops*, and recover phylogenies that support four clades (*Bothriopsis, Bothropoides, Bothrops,* and *Rhinocerophis*), but would require a new genus for the sister arrangement of *Bothrops lojanus*, or *B. lojanus* and *B. pictus*.

Classification and List of Genera

The following classification is intended to associate snake genera with their respective families and subfamilies. In the case of fossil taxa, the placement is often subjective, and based on suggested relationships in original descriptions. The non-alethinophidian classification is based in part on Garberoglio et al. (2019), Gomez et al. (2019), Hsiang et al. (2015), Miralles (2018), Onary et al. (2017), and Reeder et al. (2015). The "henophidian" relationships are based in part on Pyron et al. (2014b), Reynolds et al. (2014b), Streicher & Wiens (2016), Zheng and Wiens (2016). The colubroidean classification follows Zaher et al. (2019), but adds the Cyclocorinae of Weinell & Brown (2018) as a family to match the hierarchical levels of Zaher et al.

OPHIDIA
Incertae sedis
†*Coniophis*, †*Diablophis*, †*Eophis*, †*Herensugea*, †*Najash*, †*Norisophis*, †*Parviraptor*, †*Portugalophis*, †*Seismophis*, †*Tetrapodophis*

†DINILYSIIDAE
†*Dinilysia*

SERPENTES
Incertae sedis
†*Lunaophis*, †*Tuscahomaophis*, †*Vectophis*, †*Xiaophis*

†ANOMALOPHIIDAE
†*Anomalophis*, †*Krebsophis*, †*Russellophis*

†LAPPARENTOPHIIDAE
†*Lapparentophis*, †*Pouitella*

†MADTSOIIDAE
†*Adinophis*, †*Alamitophis*, †*Eomadtsoia*, †*Gigantophis*, †*Madtsoia*, †*Menarana*, †*Nanowana*, †*Nidophis*, †*Patagoniophis*, †*Platyspondylophis*, †*Rionegrophis*, †*Sanajeh*, †*Wonambi*, †*Yurlunggur*

†NIGEROPHIIDAE
†*Amananulam*, †*Indophis*, †*Kelyophis*, †*Nigerophis*, †*Nubianophis*, †*Woutersophis*

†PALAEOPHIIDAE
†Archaeophiinae
†*Archaeophis*
†Palaeophiinae
†*Palaeophis*, †*Pterosphenus*

†SIMOLIOPHIIDAE
†*Eupodophis*, †*Haasiophis*, †*Mesophis*, †*Pachyophis*, †*Pachyrhachis*, †*Simoliophis*

"SCOLECOPHIDIA"
ANOMALEPIDIDAE
Anomalepis, Helminthophis, Liotyphlops, Typhlophis

GERRHOPILIDAE
Cathetorhinus, Gerrhopilus

TYPHLOPIDAE
Afrotyphlopinae
Afrotyphlops, Letheobia, Rhinotyphlops
Asiatyphlopinae
Acutotyphlops, Anilios, Argyrophis, Cyclotyphlops, Grypotyphlops, Indotyphlops, Malayotyphlops, Ramphotyphlops, Sundatyphlops, Xerotyphlops
Madatyphlopinae
Madatyphlops
Typhlopinae
Amerotyphlops, Antillotyphlops, Cubatyphlops, Typhlops

XENOTYPHLOPIDAE
Xenotyphlops

LEPTOTYPHLOPIDAE
Leptotyphlopinae
Epacrophis, Leptotyphlops, Myriopholis, Namibiana
Epictinae
Epictia, Habrophallos, Mitophis, Rena, Rhinoguinea, Rhinoleptus, Siagonodon, Tetracheilostoma, Tricheilostoma, Trilepida

ALETHINOPHIDEA
Incerta Sedis
†*Cerberophis*, †*Goinophis*

ANOMOCHILIDAE
Anomochilus

CYLINDROPHIIDAE
Cylindrophis, †*Michauxophis*

ANILIIDAE
Anilius, †*Australophis*, †*Colombophis*, †*Eoanilius*, †*Hoffstetterella*

UROPELTIDAE
Melanophidium, Platyplectrurus, Plectrurus, Pseudoplectrurus, Rhinophis, Teretrurus, Uropeltis

MACROSTOMATA
BOOIDEA
Incertae Sedis
†*Anilioides*, †*Bavarioboa*, †*Botrophis*, †*Cadurcoboa*, †*Calamagras*, †*Cheilophis*, †*Chubutophis*, †*Conantophis*, †*Coprophis*, †*Daunophis*, †*Dawsonophis*, †*Gaimanophis*,

†Geringophis, †Hechtophis, †Helagras, †Hordleophis, †Huberophis, †Itaboraiophis, †Kataria, †Lithophis, †Ogmophis, †Ophidioniscus, †Palaelaphis, †Paleryx, †Paulacoutophis, †Plesiotortrix, †Pollackophis, †Pterygoboa, †Rageophis, †Sanjuanophis, †Tachyophis, †Tallahattaophis, †Titanoboa, †Totlandophis, †Tregophis †Waincophis

LOXOCEMIDAE
Loxocemus

XENOPELTIDAE
Cryptophidion, Xenopeltis

PYTHONIDAE
Antaresia, Aspidites, Bothrochilus, Leiopython, Liasis, Malayopython, Morelia, †Palaeopython, Python, Simalia

BOIDAE
Boa, Chilabothrus, Corallus, Epicrates, Eunectes, †Rukwanyoka

CALABARIIDAE
Calabaria

CANDOIIDAE
Candoia

CHARINIDAE
Charininae
Charina, Lichanura, †Paraepicrates
Ungaliophiinae
†Dunnophis, Exiliboa, †Paraplatyspondylia, †Paraungaliophis, †Platyspondylia, Ungaliophis

ERYCIDAE
†Albaneryx, †Bransateryx, †Cadurceryx, Eryx

SANZINIIDAE
Acrantophis, Sanzinia

TROPIDOPHIIDAE
†Boavus, †Falseryx, †Messelophis, †Rottophis, †Szyndlaria, Trachyboa, Tropidophis

BOLYERIIDAE
Bolyeria, Casarea

XENOPHIDIIDAE
Xenophidion

CAENOPHIDIA
ACROCHORDIDAE
Acrochordus

COLUBROIDEA
Incerta Sedis
†Ameiseophis, Blythia, †Dakotaophis, †Floridaophis, †Headonophis, Micrelaps, Montaspis †Nebraskophis,

†Procerophis, †Renenutet, Scaptophis, Tetralepis, †Texasophis, †Thaumastophis

HOMALOPSIDAE
Bitia, Brachyorrhos, Calamophis, Cantoria, Cerberus, Dieurostus, Djokoiskandarus, Enhydris, Erpeton, Ferania, Fordonia, Gerarda, Gyiophis, Heurnia, Homalophis, Homalopsis, Hypsiscopus, Karnsophis, Kualatahan, Mintonophis, Miralia, Myron, Myrrophis, Phytolopsis, Pseudoferania, Raclitia, Subsessor, Sumatranus

PAREIDAE
Pareinae
Aplopeltura, Asthenodipsas, Pareas
Xylophiinae
Xylophis

PSEUDOXENODONTIDAE
Plagiopholis, Pseudexenodon

XENODERMIDAE
Achalinus, Fimbrios, Parafimbrios, Stoliczkia, Xenodermus

DIPSADIDAE
Subfamily unnamed
Cercophis, Diadophis, †Dryinoides, Enuliophis, Enulius, Farancia, Heterodon, Lioheterophis, †Paleofarancia, †Paleoheterodon, Stichophanes, Thermophis
Carphophiinae
Carphophis, Contia
Dipsadinae
Adelphicos, Amastridium, Atractus, Cenaspis, Chersodromus, Coniophanes, Cryophis, Diaphorolepis, Dipsas, Emmochliophis, Geophis, Hydromorphus, Hypsiglena, Imantodes, Leptodeira, Ninia, Nothopsis, Omoadiphas, Plesiodipsas, Pliocercus, Pseudoleptodeira, Rhadinaea, Rhadinella, Rhadinophanes, Sibon, Synophis, Tantalophis, Tretanorhinus, Trimetopon, Tropidodipsas, Urotheca
Xenodontinae
Alsophis, Amnesteophis, Apostolepis, Arcanumophis, Arrhyton, Boiruna, Borikenophis, Caateboia, Calamodontophis, Caraiba, Chapinophis, Clelia, Conophis, Coronelaps, Crisantophis, Cubophis, Ditaxodon, Drepanoides, Echinanthera, Elapomorphus, Erythrolamprus, Eutrachelophis, Gomesophis, Haitiophis, Helicops, Hydrodynastes, Hydrops, Hypsirhynchus, Ialtris, Lygophis, Magliophis, Manolepis, Mussurana, Ocyophis, Oxyrhopus, Paraphimophis, Parapostolepis, Phalotris, Philodryas, Phimophis, Pseudalsophis, Pseudoboa, Pseudoeryx, Pseudotomodon, Psomophis, Ptychophis, Rhachidelus, Rodriguesophis, Saphenophis, Schwartzophis, Siphlophis, Sordellina, Tachymenis, Taeniophallus, Thamnodynastes, Tomodon, Tropidodryas, Uromacer, Xenodon, Xenopholis

CALAMARIIDAE
Calamaria, Calamorhabdium, Collorhabdium, Etheridgeum, Macrocalamus, Pseudorabdion, Rabdion

GRAYIIDAE
Grayia

SIBYNOPHIIDAE
Scaphiodontophis, Sibynophis

COLUBRIDAE
Incertae Sedis
†Chotaophis, †Gansophis, †Proptychophis, †Sardophis, †Sivaophis

Ahaetullinae
Ahaetulla, Chrysopelea, Dendrelaphis, Dryophiops, Proahaetulla

Colubrinae
Aeluroglena, Aprosdokedophis, Archelaphe, Argyrogena, Arizona, Bamanophis, Bogertophis, Boiga, Cemophora, Chironius, Coelognathus, Coluber, Colubroelaps, Conopsis, Coronella, Crotaphopeltis, Cyclophiops, Dasypeltis, Dendrophidion, Dipsadoboa, Dispholidus, Dolichophis, Drymarchon, Drymobius, Drymoluber, Dryocalamus, Eirenis, Elaphe, Euprepiophis, Ficimia, Geagras, Gongylosoma, Gonyosoma, Gyalopion, Hapsidophrys, Hemerophis, Hemorrhois, Hierophis, †Hispanophis, Lampropeltis, Leptodrymus, Leptophis, Liopeltis, Lycodon, Lytorhynchus, Macroprotodon, Masticophis, Mastigodryas, Meizodon, †Miocoluber, Mopanveldophis, Muhtarophis, Oligodon, Oocatochus, Opheodrys, Oreocalamus, Oreocryptophis, Orientocoluber, Orthriophis, Oxybelis, Palusophis, Pantherophis, †Paracoluber, †Paraoxybelis, †Paraxenophis, †Periergophis, Philothamnus, Phrynonax, Phyllorhynchus, Pituophis, Platyceps, Pseudelaphe, †Pseudocemophora, Pseudoficimia, Ptyas, Rhamnophis, Rhinobothryum, Rhinocheilus, Rhynchocalamus, Salvadora, Scaphiophis, Scolecophis, Senticolis, Simophis, Sonora, Spalerosophis, Spilotes, Stegonotus, Stenorrhina, Symphimus, Sympholis, Tantilla, Tantillita, †Tauntonophis, Telescopus, Thelotornis, Thrasops, Toxicodryas, Trimorphodon, Wallaceophis, Wallophis, Xenelaphis, Xyelodontophis, Zamenis, †Zelceophis, †Zilantophis

CYCLOCORIDAE
Cyclocorus, Hologerrhum, Myersophis, Oxyrhabdium

LAMPROPHIIDAE
Boaedon, Bothrolycus, Bothrophthalmus, Chamaelycus, Dendrolycus, Gonionotophis, Gracililima, Hormonotus, Inyoka, Lamprophis, Limaformosa, Lycodonomorphus, Lycophidion, Mehelya, Pseudoboodon

PSEUDOXYRHOPHIIDAE
Alluaudina, Amplorhinus, Brygophis, Compsophis, Ditypophis, Dromicodryas, Duberria, Elapotinus,

Heteroliodon, Ithycyphus, Langaha, Leioheterodon, Liophidium, Liopholidophis, Lycodryas, Madagascarophis, Micropisthodon, Pararhadinaea, Parastenophis, Phisalixella, Pseudoxyrhophus, Thamnosophis

PSEUDASPIDIDAE
Psammodynastes, Pseudaspis, Pythonodipsas

PROSYMNIDAE
Prosymna

PSAMMOPHIIDAE
Dipsina, Hemirhagerrhis, Kladirostratus, Malpolon, Mimophis, Psammophis, Psammophylax, Rhagerhis, Rhamphiophis, Taphrometopon

ATRACTASPIDIDAE
Aparallactinae
Amblyodipsas, Aparallactus, Brachyophis, Chilorhinophis, Hypoptophis, Macrelaps, Poecilopholis, Polemon, Xenocalamus

Atractaspidinae
Atractaspis, Homoroselaps

NATRICIDAE
Adelophis, Afronatrix, Amphiesma, Amphiesmoides, Anoplohydrus, Aspidura, Atretium, Ceratophallus, Clonophis, Elapoidis, Fowlea, Haldea, Hebius, Helophis, Herpetoreas, Hydrablabes, Hydraethiops, Iguanognathus (?), Isanophis, Limnophis, Liodytes, Lycognathophis, †Micronatrix, †Mionatrix, Natriciteres, Natrix, †Neonatrix, Nerodia, Opisthotropis, †Palaeonatrix, Paratapinophis, Pseudagkistrodon, Regina, Rhabdophis, Rhabdops, Smithophis, Storeria, Thamnophis, Trachischium, Trimerodytes, Tropidoclonion, Tropidonophis, Virginia, Xenochrophis

ELAPIDAE
Acanthophis, Afronaja, Aipysurus, Antaioserpens, Aspidelaps, Aspidomorphus, Austrelaps, Boulengerina, Brachyurophis, Buhoma, Bungarus, Cacophis, Calliophis, Cryptophis, Demansia, Dendroaspis, Denisonia, Drysdalia, Echiopsis, Elapognathus, Elapsoidea, Emydocephalus, Ephalophis, Furina, Hemachatus, Hemiaspis, Hemibungarus, Hoplocephalus, Hydrelaps, Hydrophis, †Incongruelaps, Kolpophis, Laticauda, Loveridgelaps, Microcephalophis, Micropechis, Micruroides, Micrurus, Naja, Notechis, Ogmodon, Ophiophagus, Oxyuranus, Parahydrophis, Parapistocalamus, Parasuta, Paroplocephalus, Pseudechis, Pseudohaje, Pseudolaticauda, Pseudonaja, Rhinoplocephalus, Salomonelaps, Simoselaps, Sinomicrurus, Suta, Thalassophis, Toxicocalamus, Tropidechis, Uraeus, Vermicella, Walterinnesia

VIPERIDAE

Incerta Sedis
†*Laophis*

Azemiopinae
Azemiops

Viperinae
Atheris, Bitis, Causus, Cerastes, Daboia, Echis, Eristicophis, Macrovipera, Montatheris, Montivipera, Proatheris, Pseudocerastes, Vipera

Crotalinae
Agkistrodon, Atropoides, Bothriechis, Bothrocophias, Bothrops, Calloselasma, Cerrophidion, Craspedocephalus, Crotalus, Deinagkistrodon, Garthius, Gloydius, Himalayophis, Hypnale, Lachesis, Mixcoatlus, Ophryacus, Ovophis, Parias, Peltopelor, Popeia, Porthidium, Protobothrops, Sistrurus, Trimeresurus, Tropidolaemus, Viridovipera

Species Accounts by Genus

ACANTHOPHIS Daudin, 1803 (Elapidae)

Comments: Maddock et al. (2015) produce a phylogeny of populations from seven of the species, based on DNA sequence data. Papuan (*A. laevis*) populations are sister to Australian populations, which form two clades of three species each.

Acanthophis antarcticus (Shaw & Nodder, 1802 *in* 1789–1813)

Acanthophis ceramensis (Günther, 1863)

Distribution: Add Indonesia (Ambon, Bisa, Haruku, Nusa Laut, Obi, Saparua, Selaru, Yamdena), Lang (2013, as *Acanthophis* spp.).

Acanthophis cryptamydros Maddock, Ellis, Doughty, Smith & Wüster, 2015. Zootaxa 4007(3): 308–312, figs. 5–8.

Holotype: WAM 174083, a 482 mm male (R. Ellis, G. Bourke & R. Barrett, 8 March 2014).

Type locality: "1 km north-west of Theda Station homestead, Western Australia (14°46′59.10″S, 126°29′22.02″E)," Australia.

Distribution: NW Australia (Kimberley region of Western Australia, including Bigge, Boongaree, Koolan, Molena, and Wulalam Islands).

Acanthophis hawkei Wells & Wellington, 1985

Acanthophis laevis Macleay, 1877

Distribution: Add Papua New Guinea (Manam Island), Clegg & Jocque (2016).

Acanthophis praelongus Ramsay, 1877

Acanthophis pyrrhus Boulenger, 1898

Acanthophis rugosus Loveridge, 1948

Acanthophis wellsi Hoser, 1998

ACHALINUS W.C.H. Peters, 1869 (Xenodermidae)

Achalinus ater Bourret, 1937

Distribution: Add Vietnam (Bac Kan), Teynié et al. (2015).

Achalinus emilyae Ziegler, Nguyen, Pham, Nguyen, Pham, Schingen, Nguyen & Le, 2019b. Zootaxa 4590(2): 260–265, figs. 8–10.

Holotype: IEBR 4465, a 415+ mm female (M. van Schingen & D.K.T. Pham, 5 May 2016).

Type locality: "Dong Son-Ky Thuong Nature Reserve (21°10.15′N, 107°9.58′E, at an elevation of 348.5 m above sea level), Hoanh Bo District, Quang Ninh Province, Vietnam."

Distribution: N Vietnam (Bac Giang, Quang Ninh), 349 m. Known from two specimens.

Achalinus formosanus Boulenger, 1908

Achalinus hainanus Huang *in* Hu, Zhao & Huang, 1975

Achalinus jinggangensis (Zong & Ma, 1983)

Distribution: Add China (Hunan), J. Wang et al. (2019).

Achalinus juliani Ziegler, Nguyen, Pham, Nguyen, Pham, Schingen, Nguyen & Le, 2019b. Zootaxa 4590(2): 253–255, figs. 2–5.

Holotype: IEBR A2018.8, a 355 mm male (C.T. Pham, 14 June 2014).

Type locality: "forest of Duc Quang Commune (22°43.084′N, 106°39.653′E, at an elevation of 477 m above sea level), Ha Lang District, Cao Bang Province, northern Vietnam."

Distribution: N Vietnam (Cao Bang), 477–1590 m.

Achalinus meiguensis Hu & Zhao, 1966

Holotype: Corrected to CIB 008091, Guo et al. (2012a).

Distribution: Add China (Yunnan), Ziegler et al. (2019b).

Achalinus niger Maki, 1931

Achalinus rufescens Boulenger, 1888

Distribution: Add China (Guihou), Ziegler et al. (2019b).

Achalinus spinalis W.C.H. Peters, 1869

Distribution: Add China (Guizhou, Hunan), J. Wang et al. (2019); China (Chekiang, Yunnan), Ziegler et al. (2019b); Japan (Ryukyu Is.: Inokawa, Kametoku, Koshiki), Ziegler et al. (2019b).

Achalinus timi Ziegler, Nguyen, Pham, Nguyen, Pham, Schingen, Nguyen & Le, 2019b. Zootaxa 4590(2): 258–260, figs. 6, 7.

Holotype: IEBR A2018.10, a 178 mm male (A.V. Pham, 12 May 2014).

Type locality: "forest within the Copia Nature Reserve, near Hua Ty Village (21°20.105′N, 103°35.860′E), Co Ma Commune, Thuan Chau District, Son La Province, northern Vietnam,… at an elevation of ca. 1470 m above sea level."

Distribution: N Vietnam (Son La), 1470 m. Known only from the holotype.

Achalinus werneri Van Denburgh, 1912

Achalinus yunkaiensis J. Wang, Li & Wang *in* J. Wang et al., 2019. Zootaxa 4674(4), 474–478, figs. 2, 3.

Holotype: SYS R001903, a 343 mm male (J. Wang & H. Chen, 10 April 2018).

Type locality: "Dawuling Forestry Station (22.27580°N, 111.19524°lE; 1500 m a.s.l.), Maoming City, Guangdong Province, China."

Distribution: S China (Guangdong), 900–1600 m.

ACRANTOPHIS Jan, 1860 *in* Jan & Sordelli, 1860–1866 (Sanziniidae)

Acrantophis dumerili Jan, 1860 *in* Jan & Sordelli, 1860–1866

Comments: mtDNA analysis by Orozco-Terwengel et al. (2008) suggests the possibility of a new species that morphologically resembles *A. dumerili*, but is paraphyletic with

A. madagascariensis. Mezzasalma et al. (2019) describe the chromosomes.

Acrantophis madagascariensis (A.M.C. Duméril & Bibron, 1844)

ACROCHORDUS Hornstedt, 1787 (Acrochordidae)

Comments: Sanders et al. (2010) evaluate molecular and morphological data for the three living species, finding *A. javanicus* to be the sister taxon to the other two. They provide a table of morphological characters for each species, and discuss divergence dating, with †*A. dehmi*. The results confirm that *Acrochordus* forms a sister group to Colubroidea within Caenophidia.

Acrochordus arafurae McDowell, 1979
†**Acrochordus dehmi** Hoffstetter, 1964
Acrochordus granulatus (Schneider, 1799)

Distribution: Add West Malaysia (Johor), Voris (2015); West Malaysia (Selangor), Voris (2017); Philippines (Calauit, Guimaras, Negros, Siquijor), Leviton et al. (2018); Indonesia (Halmahera), Lang (2013); Papua New Guinea (Manus Island), Clegg & Jocque (2016).

Acrochordus javanicus Hornstedt, 1787

ACUTOTYPHLOPS Wallach, 1995 (Typhlopidae: Asiatyphlopinae)

Comments: Hedges et al. (2014), Pyron & Wallach (2014) provide generic diagnoses.

Acutotyphlops banaorum Wallach, R. Brown, Diesmos & Gee, 2007
Acutotyphlops infralabialis (Waite, 1918)
Acutotyphlops kunuaensis Wallach, 1995
Acutotyphlops solomonis (H.W. Parker, 1939)
Acutotyphlops subocularis (Waite, 1897)

ADELOPHIS Dugès *in* Cope, 1879 (Natricidae)

Comments: This genus remains paraphyletic within *Thamnophis*, Zaher et al. (2019).

Adelophis copei Dugès *in* Cope, 1879
Adelophis foxi Rossman & Blaney, 1968

ADELPHICOS Jan, 1862 (Dipsadidae: Dipsadinae)

Comments: Smith et al. (2001) recommend recognition of *A. sargii* and *A. visoninum* as species based on discrete morphological differences. Previously LaDuc (1996) had proposed recognition of *sargii* and *newmanorum* in an unpublished thesis, which Smith et al. did not reference. All three taxa are variously recognized as distinct species (with increasing frequency), and we refer to them as such herein.

Adelphicos daryi Campbell & Ford, 1982
Adelphicos ibarrorum Campbell & Brodie, 1988
Adelphicos latifasciatum Lynch & H.M. Smith, 1966
Adelphicos newmanorum Taylor, 1950. Univ. Kansas Sci. Bull. 33(11): 443, 445, pl. 4.

Holotype: LSUMZ 204, a 310 mm male (M. Newman, 7 May 1947).

Type locality: "Xilitla region, San Luis Potosí," Mexico.

Distribution: NE Mexico (Nuevo Leon, E Queretaro, E San Luis Potosi, S Tamaulipas), 680–1605 m.

Adelphicos nigrilatum H.M. Smith, 1942

Adelphicos quadrivirgatum Jan, 1862

Synonyms: Remove *Rhegnops visoninus* Cope, 1866, *Rhegnops sargii* J.G. Fischer, 1885, *Adelphicos newmanorum* Taylor, 1950.

Distribution: Limited to S Mexico (Chiapas, Oaxaca, E Tabasco, Veracruz).

Adelphicos sargii J.G. Fischer, 1885. Jahrb. Hamburg. Wiss. Anst. 2: 92–93. (*Rhegnops sargii*)

Lectotype: BMNH 1946.1.6.28, an adult female (Sarg).

Type locality: "Guatemala"; restricted to Volcan Zunil, Suchitepequez, by Smith & Talor (1950).

Distribution: Mexico (SE Chiapas), S Guatemala, 500–1550 m.

Adelphicos veraepacis Stuart, 1941

Adelphicos visoninum Cope, 1866. Proc. Acad. Nat. Sci. Philadelphia 8: 128–129. (*Rhegnops visoninus*)

Holotype: USNM 24899, female (B. Parsons).

Type locality: "Honduras"; corrected to Belize by Cochran (1961).

Distribution: SE Mexico (N Chiapas), N Guatemala, Brelize, N Honduras.

†**ADINOPHIS** Pritchard, McCartney, Krause & Kley, 2014. J. Vert. Paleontol. 34(5): 1081. (†Madtsoiidae)

Type species: †*Adinophis fisaka* Pritchard, McCartney, Krause & Kley, 2014 by original designation.

Distribution: Upper Cretaceous of Madagascar.

†**Adinophis fisaka** Pritchard, McCartney, Krause & Kley, 2014. J. Vert. Paleontol. 34(5): 1081–1083, fig. 1.

Holotype: UA 9941 (orig. FMNH PR 2577), a posterior trunk vertebra.

Type locality: "Locality MAD93-73, Berivotra Study Area, Mahajanga Basin, northwestern Madagascar."

Distribution: Upper Cretaceous (Maastrichtian) of NW Madagascar (Mahajanga).

AELUROGLENA Boulenger, 1898 (Colubridae: Colubrinae)

Aeluroglena cucullata Boulenger, 1898

AFRONAJA Wallach, Wüster, & Broadley, 2009 (Elapidae)

Synonyms: Remove *Palaeonaja* Hoffstetter, 1939.

Fossil records: Add middle Miocene (Orleanian) of Germany, Čerňanský et al. (2017, as *Naja* sp.); Miocene/Pliocene transition (late Turolian) of Greece, Georgialis et al. (2019, as *Naja* sp.).

Comments: See under *Naja*.

†**Afronaja antiqua** (Rage, 1976)

Fossil records: Add late Miocene (Vallesian) of Morocco, Blain et al. (2013a).

Afronaja ashei (Wüster & Broadley, 2007)
Afronaja crawshayi (Günther, 1894)
† **Afronaja depereti** (Hoffstetter, 1939)
† **Afronaja iberica** (Szyndlar, 1985)

Afronaja katiensis (Angel, 1922)

Distribution: Add Senegal (Fatick, Kolda, Matam), Mané & Trape (2017); Benin, Trape & Baldé (2014).

Afronaja mossambica (W.C.H. Peters, 1854)

Distribution: Add Angola (Cauando-Cubango), Conradie et al. (2016b).

Afronaja nigricincta (Bogert, 1940)

Afronaja nigricollis (J.T. Reinhardt, 1843)

Distribution: Add Mauritania (Gorgol, Nouakchott), Padial (2006); Senegal (Louga, Sedhiou, Ziguinchor), Mané & Trape (2017); Liberia, Trape & Baldé (2014); Mali (Mopti, S. Tombouctou), Trape & Mané (2017); Niger (Tillaberi), Trape & Mané (2015); Uganda (Kaabong), D. Hughes et al. (2017b); Angola (Benguela, Bie, Cuando Cubango, Cuanza Norte, Cuanza Sul, Malanje, Zaire), M.P. Marques et al. (2018).

Afronaja nubiae (Wüster & Broadley, 2003)

Afronaja pallida (Boulenger, 1896)

† *Afronaja robusta* (Meylan, 1987)

AFRONATRIX Rossman & Eberle, 1977 (Natricidae)

Afronatrix anoscopus (Cope, 1861)

Distribution: Add Liberia (Grand Gedeh), Rödel & Glos (2019).

AFROTYPHLOPS Broadley & Wallach, 2009 (Typhlopidae: Afrotyphlopinae)

Synonyms: Add *Megatyphlops* Broadley & Wallach, 2009.

Comments: Hedges et al. (2014), Pyron & Wallach (2014) provide generic diagnoses.

Afrotyphlops angolensis (Bocage, 1866)

Distribution: Add Gabon (Moyen-Ogooue), Pauwels et al. (2017a); Angola (Cuanza Norte, Moxico), M.P. Marques et al. (2018).

Afrotyphlops anomalus (Bocage, 1873)

Distribution: Add Angola (Cabinda), M.P. Marques et al. (2018).

Afrotyphlops bibronii (A. Smith, 1846 *in* 1838–1849)

Afrotyphlops blanfordii (Boulenger, 1889)

Afrotyphlops brevis (Scortecci, 1929)

Afrotyphlops chirioi Trape, 2019. Bull. Soc. Herpétol. France 169: 28–32, figs. 1–4.

Holotype: MNHN 2006.0536, a 154 mm specimen of unknown sex (L. Chirio, 29 January 1996).

Type locality: "Berbérati (04°15′N/15°47′E) en République centrafricaine."

Distribution: SW Central African Republic (Mambere-Kadei, Ombella-M'Poko). Known from two specimens.

Afrotyphlops congestus (A.M.C. Duméril & Bibron, 1844)

Distribution: Add Gabon (Ogooue-Lolo), Pauwels et al. (2016a).

Afrotyphlops decorosus (Buchholz & Peters *in* W.C.H. Peters, 1875)

Comments: Hedges et al. (2014) transfer *decorosus* from *Letheobia* to *Afrotyphlops* based on its morphological similarity to *A. obtusus*, which is genetically allied to *Afrotyphlops*.

Afrotyphlops elegans (W.C.H. Peters, 1868)

Afrotyphlops fornasinii (Bianconi, 1849)

Afrotyphlops gierrai (Mocquard, 1897)

Distribution: Add Tanzania (Nguru Mtns., Morogoro), Menegon et al. (2008).

Afrotyphlops kaimosae (Loveridge, 1935)

Afrotyphlops liberiensis (Hallowell, 1848)

Comments: Trape & Baldé (2014) note that preliminary molecular data support recognition of this species as distinct from *A. punctatus*.

Afrotyphlops lineolatus (Jan, 1864 *in* Jan & Sordelli, 1860–1866)

Distribution: Add Mali (Koulikoro, Sikasso), Trape & Mané (2017); Angola (Cuanza Norte), Branch (2018); Angola (Bengo, Cuanza Norte, Cuanza Sul, Lunda Sul, Malanje, Zaire), M.P. Marques et al. (2018).

Afrotyphlops mucruso (W.C.H. Peters, 1854)

Distribution: Add Angola (Lunda Sul), M.P. Marques et al. (2018); Angola (Moxico), Branch (2018).

Afrotyphlops nanus (Broadley & Wallach, 2009)

Afrotyphlops nigrocandidus (Broadley & Wallach, 2000)

Distribution: Low elevation of 633 m, Lyakurwa (2017).

Afrotyphlops obtusus (W.C.H. Peters, 1865)

Comments: The DNA sequence-based phylogenies that Hedges et al. (2014) and Pyron & Wallach (2014) produce show that *obtusus* is the sister taxon to *Afrotyphlops*, and those authors transfer it to *Afrotyphlops* from *Letheobia*.

Afrotyphlops punctatus (Leach *in* Bowdich, 1819)

Distribution: Add Mali (Koulikoro), Trape & Mané (2017). Upper elevation of 1900 m, Ineich et al. (2015).

Afrotyphlops rondoensis (Loveridge, 1942)

Afrotyphlops rouxestevae Trape, 2019. Bull. Soc. Herpétol. France 169: 32–33, figs. 5–7.

Holotype: SMF 16639, a 525 mm female, collector unknown.

Type locality: "Douala (04°03′N/09°42′E) au Cameroun."

Distribution: SW Cameroon (Littoral). Known only from the holotype.

Afrotyphlops schlegelii (Bianconi, 1849)

Distribution: Add Angola (Huila), M.P. Marques et al. (2018); Angola (Namibe), Ceríaco et al. (2016).

Afrotyphlops schmidti (Laurent, 1956)

Afrotyphlops steinhausi (F. Werner, 1909)

Afrotyphlops tanganicanus (Laurent, 1964)

Afrotyphlops usambaricus (Laurent, 1964)

AGKISTRODON Palisot de Beauvois, 1799 (Viperidae: Crotalinae)

Comments: Porras et al. (2013) revise the Mesoamerican species, and Burbrink & Guiher (2015) revise the United States species. Porras et al. (2013) were unable to assign the following populations to species: Mexico (inland Chiapas, C Veracruz [=*Agkistrodon bilineatus lemosespinali* H.M. Smith & Chiszar, 2001]), C Guatemala (Baja Verapaz, Huehuetenango).

Agkistrodon bilineatus (Günther, 1863)

Synonyms: Remove *Agkistrodon bilineatus russeolus* Gloyd, 1972 and *Agkistrodon bilineatus howardgloydi* Conant, 1984.

Distribution: Pacific side of Mexico (Chiapas, ext. SW Chihuahua, Colima, Guerrero, Jalisco including Maria Madre Island, Michoacan, Morelos, Nayarit, S Oaxaca, Sinaloa, S Sonora), Guatemala (Escuintla, Quezaltenango, Santa Rosa), El Salvador (Cuscatlan, La Libertad, Sonsonate), ext W Honduras (Copan). Arenas-Monroy & Ahumada-Carrillo (2015) confirm the presence in central Jalisco, Mexico.

Comments: Porras et al. (2013) elevate *howardgloydi* and *russeolus* to species level.

Agkistrodon conanti Gloyd, 1972. Proc. Biol. Soc. Washington 82(1):226–230, fig. 6. (*Agkistrodon piscivorus conanti*)

Holotype: USNM 165962, a young adult male (R.P. Elliott, J. Wariner & P. Pinnel, 16 July 1966).

Type locality: "'at edge of Rochelle-Cross Creek Road, about 7 miles southeast' of Gainesville, Alachua County, Florida," USA.

Distribution: SE USA (SE Alabama, Florida, S Georgia).

Comments: Elevated to species, based on DNA sequence data, by Burbrink & Guiher (2015). See comments under *A. piscivorus*.

Agkistrodon contortrix (Linnaeus, 1766)

Synonyms: Remove *Agkistrodon mokasen laticinctus* Gloyd & Conant, 1934, *Agkistrodon mokeson pictigaster* Gloyd & Conant, 1943.

Distribution: Remove western part of range (C and W Texas, C Oklahoma, N Mexico).

Agkistrodon howardgloydi Conant, 1984. Proc. Biol. Soc. Washington 97(1): 135–140, fig. 1. (*Agkistrodon bilineatus howardgloydi*)

Holotype: AMNH 125525 a 746 mm male (L.W. Porras & J.C. Rindfleish, 8 August 1982).

Type locality: "0.8 km north of Mirador el Cañon del Tigre, Parque Nacional Santa Rosa, Provinia Guanacaste, Costa Rica."

Distribution: S Honduras (Choluteca, Valle), W Nicaragua (Granada, Leon, Managua, Masaya), NW Costa Rica (Guanacaste), NSL-600 m. McCranie & Gutsche (2016) add Honduras (Valle: Isla Zacate Grande) to the distribution. Add Honduras (El Paraíso), McCranie et al. (2014b); Nicaragua (Leon), Sunyer et al. (2014).

Comments: Porras et al. (2013) elevate this taxon to species level on the basis of morphology, mtDNA sequence, and geography.

Agkistrodon laticinctus Gloyd & Conant 1934. Occas. Pap. Mus. Zool. Univ. Michigan (283):2–5, figs. 1, 2. (*Agkistrodon mokasen laticinctus*)

Synonyms: *Agkistrodon mokeson pictigaster* Gloyd & Conant, 1943.

Holotype: UMMZ 75599, a 793 mm male (W.A. Bevan & R.F. Harvey, Oct. 1933).

Type locality: "26 miles northwest of San Antonio, Bexar County, Texas," USA.

Distribution: NC Mexico (NE Chihuahua, N Coahuila), SC USA (C Oklahoma, W & C Texas).

Comments: Elevated to species, based on DNA sequence data, by Burbrink & Guiher (2015).

Agkistrodon piscivorus (Lacépède, 1789)

Synonyms: Remove *Agkistrodon piscivorus conanti* Gloyd, 1972.

Distribution: Remove most of Florida, SE Alabama, and S Georgia from the range, to that of *A. conanti*.

Comments: J.L. Strickland et al. (2014) use molecular data to detect two clades, and hypothesize that Florida and Texas were glacial refugia. Their data suggest extensive, subsequent gene flow in the geographic hiatus.

Agkistrodon russeolus Gloyd, 1972a. Proc. Biol. Soc. Washington 84(40): 327–330. (*Agkistrodon bilineatus russeolus*)

Holotype, KU 70905, an 850 mm male (J.B. Tulecke, 20 July 1962).

Type locality: "11.7 km. north of Pisté, Yucatán, México."

Distribution: Mexico (Campeche, Quintana Roo, Yucatán), N Belize (Belize, Orange Walk), 20 m. Add Mexico (Tabasco), Charruau et al. (2014).

Comments: Porras et al. (2013) elevate this taxon to species level on the basis of morphology, mtDNA sequence data, and geography.

Agkistrodon taylori Burger & Robertson, 1951

AHAETULLA Link, 1807 (Colubridae: Ahaetullinae)

Comments: Mohapatra et al. (2017) provide a key to the species. Deepak et al. (2019) provide an mtDNA sequence-based phylogeny of *Ahaetulla* species that has *A. laudankia* nested within *A. nasuta*.

Ahaetulla anomala Annandale, 1906. Mem. Asiatic Soc. Bengal 1(10): 196. (*Dryophis mycterizans anomalus*)

Holotype: ZSI 15463, a 956 mm male (G. Cummusky).

Type locality: "Rámanád" district, Tamil Nadu, India; modified to "Santragachi, Howrah district, West Bengal, India" by Mohapatra et al. (2017).

Distribution: E India (Jharkhand, Orissa, Tamil Nadu, West Bengal) and S Bangladesh (Khuina).

Comments: Mohapatra et al. (2017) revalidate *D. m. anomalus* based on morphological differences, and phylogenetic placement using DNA-sequence data.

Ahaeutulla dispar (Günther, 1864)

Ahaetulla fasciolata (J.G. Fischer, 1885)

Ahaetulla fronticincta (Günther, 1858)

Distribution: Add Myanmar (Tanintharyi), Mulcahy et al. (2018).

Ahaetulla laudankia Deepak, Narayanan, Sarkar, Dutta & Mohapatra, 2019. J. Nat. Hist. 53: 501–513, figs. 3–5.

Holotype: ZSI CZRC-6403, a 1237 mm female (M.V. Nair & S.K. Dutta, 15 June v2010).

Type locality: "Bangriposi, Mayurbhanj district, Odisha state (22.142167N, 86.520025E)," India.

Distribution: C India (Maharashtra, Odisha, Rajasthan).

Ahaetulla mycterizans (Linnaeus, 1758)

Distribution: Add Myanmar (Tanintharyi), J.L. Lee et al. (2015).

Ahaetulla nasutus (Lacépède, 1789)

Synonyms: Remove *Dryophis mycterizans anomalus* Annandale, 1906.

Distribution: Add Nepal (Chitwan, Nawalparasi), Pandey (2012); Nepal (Parsa), Bhattarai et al. (2018b); India (Arunachal Pradesh), Purkayastha (2018); Myanmar (Mandalay), Platt et al. (2018).

Ahaetulla perroteti (A.M.C. Duméril, Bibron & Duméril, 1854)

Ahaetulla prasina (F. Boie, 1827)

Distribution: Add Bhutan (Zhemgang), Tshewang & Letro (2018); Thailand (Rayong, and Koh Chang I.), Chan-ard and Makchai (2011); Cambodia (Siem Reap), Geissler et al. (2019); Vietnam (Hai Phong: Cat Ba Island), T.Q. Nguyen et al. (2011); Vietnam (Quang Ngai), Nemes et al. (2013); West Malaysia (Pulau Singa Besar), B.L. Lim et al. (2010); West Malaysia (Pulau Tioman), K.K.P. Lim & Lim (1999); Indonesia (Anambi, Bangka, Batu, Bawean, Kangean, Karimunjawa, Mentawai, Natuna, Nias, Panaitan, Sanana, Sangihe, Sebuku, Simeulue, Sula, Ternate, We Islands), Lang (2013); Philippines (Batan, Bohol, Calauit, Camiguin Norte, Camiguin Sur, Leyte, Marinduque, Masbate, Mindoro, Sabtang Is.), Leviton et al. (2018); Philippines (Cebu I.), Supsup et al. (2016); Philippines (Tablas I.), Siler et al. (2012).

Ahaetulla pulverulenta (A.M.C. Duméril, Bibron & Duméril, 1854)

Distribution: Add Bangladesh (Rajshahi Division, Naogaon District), Ahmad et al. (2015a); Bangladesh (Bagerhat District, Khulna), Neumann-Denzau & Denzau (2010). Remove India (Gujarat), Patel & Vyas (2019).

AIPYSURUS Lacépède, 1804 (Elapidae)

Aipysurus apraefrontalis M.A. Smith, 1926

Distribution: Add coastal NW Western Australia, Sanders et al. (2015).

Comments: Sanders et al. (2015) use mtDNA sequence and morphological data to confirm that coastal populations are resident, rather than vagrants, from the Timor Sea reefs, where it apparently no longer occurs. They provide a revised description.

Aipysurus duboisii Bavay, 1869

Aipysurus eydouxii (Bibron *in* Gray, 1849)

Distribution: Rasmussen et al. (2014) exclude Australia from the geographic range.

Aipysurus foliosquama M.A. Smith, 1926

Distribution: Add coastal NW Western Australia, Sanders et al. (2015).

Comments: As with *A. apraefrontalis*, Sanders et al. (2015) use mtDNA sequence and morphological data to confirm that coastal populations are resident, rather than vagrants, from the Timor Sea reefs, where it apparently no longer occurs. They provide a revised description.

Aipysurus fuscus (Tschudi, 1837)

Aipysurus laevis Lacépède, 1804

Aipysurus mosaicus Sanders, Rasmussen, Elmberg, Mumpuni, Guinea, Blias, Lee & Fry, 2012

Aipysurus pooleorum L.A. Smith, 1974

Aipysurus tenuis Lönnberg & Andersson, 1913

†*ALAMITOPHIS* Albino, 1986 (†Madtsoiidae)

†*Alamitophis argentinus* Albino, 1986

†*Alamitophis elongatus* Albino, 1994

†*Alamitophis tingamarra* Scanlon, 2005

†*ALBANERYX* Hoffstetter & Rage, 1972 (Erycidae)

†*Albaneryx depereti* Hoffstetter & Rage, 1972

†*Albaneryx volynicus* Zerova, 1989

Distribution: Add late Middle Miocene (Astaracian) of NE Kazakhstan, Ivanov et al. (2019, as cf. *volynicus*); late Middle Miocene (Astaracian or Vallesian) of Hungary, Venczel (2011, as *A.* cf. *volynicus*).

ALLUAUDINA Mocquard, 1894 (Pseudoxyrhophiidae)

Alluaudina bellyi Mocquard, 1894

Alluaudina mocquardi Angel, 1939

ALSOPHIS Fitzinger, 1843 (Dipsadidae: Xenodontinae)

Comments: Bochaton et al. (2019) describe and illustrate vertebrae of an undescribed *Alsophis* species from the Late Pleistocene of Marie-Galante, Guadeloupe Islands.

Alsophis antiguae H.W. Parker, 1933

Synonyms: Remove *Alsophis antiguae sajdaki* Henderson, 1990, left there by Wallach et al. (2014) in error.

Alsophis antillensis (Schlegel, 1837)

Fossil records: Late Pleistocene of Marie-Galante, Guadeloupe Islands, Bochaton et al. (2019).

Alsophis danforthi Cochran, 1938

Alsophis manselli H.W. Parker, 1933

Alsophis rijgersmaei Cope, 1870

Distribution: Add St. Barthelemy (Ilet Tortue), Questel (2011).

Alsophis rufiventris (A.M.C. Duméril, Bibron & Duméril, 1854)

Alsophis sajdaki Henderson, 1990

Alsophis sanctonum T. Barbour, 1915

Alsophis sibonius Cope, 1879

†*AMANANULAM* McCartney, Roberts, Tapanila & O'Leary, 2018 (Nigerophiidae)

Type species: *Amananulam sanogoi* McCartney, Roberts, Tapanila & O'Leary, 2018, by monotypy.

Distribution: Palaeocene of Mali.

†*Amananulam sanogoi* McCartney, Roberts, Tapanila & O'Leary, 2018. Acta Palaeontol. Polon. 63(2): 214–215, fig. 6.
Holotype: CNRST-SUNY 462, a mid-trunk vertebra.
Type locality: "Locality Mali-19, in Northeastern Mali."
Distribution: Palaeocene of NE Mali.

AMASTRIDIUM Cope, 1861 (Dipsadidae: Dipsadinae)
Amastridium sapperi (F. Werner, 1903)
Distribution: Hansen & Vermilya (2015) report a second record for Queretaro, Mexico, and an upper elevation of 1635 m.
Amastridium veliferum Cope, 1861
Distribution: Add Panama (Panama, Panama Oeste), Ray & Ruback (2015);

AMBLYODIPSAS W.C.H. Peters, 1857 (Atractaspididae: Aparallactinae)
Comments: See *Xenocalamus*. Portillo et al. (2018) produce a phylogeny using most of the species, which had *A. concolor* as sister taxon to *Macrelaps*. They recommend against a taxonomic change until additional species can be studied.
Amblyodipsas concolor (A. Smith, 1849 *in* 1838–1849)
Amblyodipsas dimidiata (Günther, 1888)
Amblyodipsas katangensis Witte & Laurent, 1942
Amblyodipsas microphthalma (Bianconi, 1852)
Amblyodipsas polylepis (Bocage, 1873)
Distribution: Add Angola (Bengo), M.P. Marques et al. (2018); Angola (Moxico), Portillo et al. (2018).
Amblyodipsas rodhaini (Witte, 1930)
Amblyodipsas teitana Broadley, 1971
Amblyodipsas unicolor (J. T. Reinhardt, 1843)
Distribution: Add Guinea (Kankan), Portillo et al. (2018); Ivory Coast (Yamoussoukro), Portillo et al. (2018); Mali (Sikasso), Trape & Mané (2017); Chad (Logone Oriental), Portillo et al. (2018); South Sudan (Eastern Equatoria), Ullenbruch & Böhme (2017).
Amblyodipsas ventrimaculata (Roux, 1907)
Distribution: Add Angola (Cuando Cubango, Huila), Branch (2018); Angola (Moxico), Portillo et al. (2018); South Africa (Limpopo), Portillo et al. (2018).

†*AMEISEOPHIS* Holman, 1976 (Colubroidea: incerta sedis)
†*Ameiseophis robinsoni* Holman, 1976

AMEROTYPHLOPS Hedges, Marion, Lipp, Marin & Vidal, 2014. Caribbean Herpetol. 49: 43. (Typhlopidae: Typhlopinae)
Type species: *Typhlops brongersmianus* Vanzolini, 1976 by original designation.
Distribution: S Middle America and much of South America east of the Andes.
Comments: Hedges et al. (2014), Pyron & Wallach (2014) provide generic diagnoses. Graboski et al. (2019) use multiple genes to produce a phylogeny of eight of the species of *Amerotyphlops*.

Amerotyphlops amoipira (Rodrigues & Junca, 2002)
Distribution: Add Brazil (Alagoas, Rio Grande do Norte), P. Brito & Freire (2012); Brazil (N. Maranhao), Graboski et al. (2019); Brazil (Sergipe), Graboski et al. (2015). Upper elevation of 629 m, Guedes et al. (2014).
Amerotyphlops arenensis Graboski, Filho, Silva, Prudente & Zaher, 2015. Zootaxa 3920(3): 444–447, figs. 1, 2.
Holotype: MZUSP 20042, a 191 mm male (G.P. Filho, 19 Nov. 2008).
Type locality: "the Reserva Ecológica Mata do Pau Ferro (06° 58'12" S, 35° 42'15" W; ca. 600 m), municipality of Areia, state of Paraíba, Brazil."
Distribution: Brazil (Paraiba), 400–600 m. Add Brazil (Bahia, Minas Gerais), Graboski et al. (2019); Brazil (Pernambuco), Roberto et al. (2018).
Amerotyphlops bongersmianus (Vanzolini, 1976)
Distribution: Add Brazil (Alagoas, Maranhão, Santa Catarina, Tocantins), Graboski et al. (2015); Brazil (Parana), Dainesi et al. (2019); Brazil (Piauí), Dal Vechio et al. (2013); Brazil (Sao Paulo: Moela Island), Abegg et al. (2019); Paraguay (Boqueron), Cabral & Weiler (2014). Upper elevation of 847 m, Guedes et al. (2014).
Amerotyphlops costaricensis (Jiménez & Savage, 1963)
Amerotyphlops lehneri (Roux, 1926)
Amerotyphlops microstomus (Cope, 1866)
Amerotyphlops minuisquamus (Dixon & Hendricks, 1979)
Distribution: Add Guyana (Potaro-Siparuni), Graboski et al. (2015).
Amerotyphlops paucisquamus (Dixon & Hendricks, 1979)
Distribution: Add Brazil (Paraiba), R. França et al. (2012); Brazil (Alagoas, Ceara, Maranhão, Rio Grande do Norte), Graboski et al. (2015).
Amerotyphlops reticulatus (Linnaeus, 1758)
Distribution: Add Brazil (Ceara), Guedes et al. (2014); Brazil (Maranhão, Roraima), Graboski et al. (2015).
Comments: Caicedo-Portilla (2011) summarizes data for this species in Colombia.
Amerotyphlops stadelmani (K.P. Schmidt, 1936)
Amerotyphlops tasymicris (J.P.R. Thomas, 1974)
Amerotyphlops tenuis (Salvin, 1860)
Distribution: Add Mexico (Chiapas), A.G. Clause et al. (2016).
Amerotyphlops trinitatus (Richmond, 1965)
Amerotyphlops tycherus (Townsend, L.D. Wilson, Ketzler & Luque-Montes, 2008)
Distribution: Add Honduras (Olancho), and low elevation of 1400 m, McCranie & Valdés-Orellano (2012).
Amerotyphlops yonenagae (Rodrigues, 1991)
Distribution: Guedes et al. (2014) give the elevation of the type locality as 475 m.

AMNESTEOPHIS C.W. Myers, 2011 (Dipsadidae: Xenodontinae)

Amnesteophis melanauchen (Jan, 1863)

AMPHIESMA A.M.C. Duméril, Bibron & Duméril, 1854 (Natricidae)

Synonyms: Remove *Herpetoreas* Günther, 1860, *Hebius* Thompson, 1913, *Paranatrix* Mahendra, 1984.

Comments: Guo et al. (2014) present a DNA sequence-based phylogeny that shows *Amphiesma stolatum* to be paraphyletic with all other species of *Amphiesma*, the latter of which they partition into the resurrected genera *Hebius* and *Herpetoreas*. Guo et al. present a generic diagnosis, and state that some species currently in *Hebius* may belong in *Amphiesma*.

Amphiesma stolatum (Linnaeus, 1758)

Distribution: Add Nepal (Nawalparasi), Pandey et al. (2018); India (Gujarat), Parmar & Tank (2019); India (Madhya Pradesh), Manhas et al. (2018a); Sri Lanka (Sabaragamuwa), Peabotuwage et al. (2012); Thailand (Chanthaburi), Chan-ard et al. (2011); Cambodia (Siem Reap), Geissler et al. (2019); China (Nan Ao Island), Qing et al. (2015); Vietnam (Hai Phong: Cat Ba Island), T.Q. Nguyen et al. (2011).

AMPHIESMOIDES Malnate, 1961 (Natricidae)

Amphiesmoides ornaticeps (F. Werner, 1924)

AMPLORHINUS A. Smith 1847 *in* 1838–1849 (Pseudoxyrhophiidae)

Amplorhinus multimaculatus A. Smith 1847 *in* 1838–1849

†ANILIOIDES Auffenberg, 1963 (Booidea: incerta sedis)

†*Anilioides minuatus* Auffenberg, 1963

†*Anilioides nebraskensis* Holman, 1976

ANILIOS Gray, 1845 (Typhlopidae: Asiatyphlopinae)

Comments: J. Marin et al. (2013) produce phylogenies using mt- and nDNA sequence data for 27 species. Some species are polyphyletic and/or paraphyletic, and the authors believe that the number of species in the genus is underestimated by half. Hedges et al. (2014), Pyron & Wallach (2014) provide generic diagnoses. Shea (2015) discusses the diagnostics of the genus *Anilios* as currently comprised, and concludes that it is diagnosable solely by molecular data. See under *Sundatyphlops*.

Anilios affinis (Boulenger, 1889)

Anilios ammodytes (Montague, 1914)

Anilios aspina (Couper, Covacevich & S. Wilson, 1998)

Distribution: Range extension 470 km NW of previous known sites, Vanderduys (2013).

Comments: Vanderduys (2013) describes a third known specimen.

Anilios australis Gray, 1845

Anilios batillus (Waite, 1894)

Anilios bicolor (P. Schmidt *in* W.C.H. Peters, 1858)

Anilios bituberculatus (W.C.H. Peters, 1863)

Anilios broomi (Boulenger, 1898)

Anilios centralis (Storr, 1984)

Anilios chamodracaena (Ingram & Covacevich, 1993)

Anilios diversus (Waite, 1894)

Anilios endoterus (Waite, 1918)

Anilios erycinus (F. Werner, 1901)

Anilios fossor Shea, 2015. Zootaxa 4033(1): 104–109, figs. 1–3.

Holotype: NTM R14324, a 245 mm female (J.R. Cole, 15 October 1989).

Type locality: "Glen Annie, Ruby Gap Nature Park, Northern Territory, Australia." Probable coordinates are "23°28′S 134°58′E, at an altitude of *ca* 500 m."

Distribution: Central Australia (Northern Territory), 500 m. Known only from the holotype.

Anilios ganei (Aplin, 1998)

Anilios grypus (Waite, 1918)

Anilios guentheri (W.C.H. Peters, 1865)

Anilios hamatus (Storr, 1981)

Anilios howi (Storr, 1983)

Anilios insperatus Venchi, S. Wilson & Borsboom, 2015. Zootaxa 3990(2): 273–275, figs.1–2.

Holotype: QM J54987, a 97 mm specimen (A. Borsboom, 19 May 1992).

Type locality: "Warrill View, Department of Primary Industries Animal Genetic Centre Farm (2749′ S, 15237′ E)," Queensland, Australia.

Distribution: Australia (SE Queensland). Known only from the holotype.

Anilios kimberleyensis (Storr, 1981)

Anilios leptosoma (Robb, 1972)

Comments: Ellis et al. (2017) provide a revised description, color photograph of the holotype, and map of known localities.

Anilios leucoproctus (Boulenger, 1889)

Anilios ligatus (W.C.H. Peters, 1879)

Distribution: McNab et al. (2014) discuss new records for Cape York Peninsula, Queensland, Australia, and provide a map of known localities.

Anilios longissimus (Aplin, 1998)

Anilios margaretae (Storr, 1981)

Anilios micromma (Storr, 1981)

Anilios minimus (Kinghorn, 1929)

Anilios nema (Shea & Horner, 1996)

Anilios nigrescens Gray, 1845

Anilios nigricauda (Boulenger, 1895)

Anilios nigroterminatus (H.W. Parker, 1931)

Anilios obtusifrons Ellis, Doughty, Donnellan, Marin & Vidal, 2017. Zootaxa 4323(1): 15–19, figs. 3, 6, 7.

Holotype: WAM R146400, a 204 mm specimen, stated to be both male and female in the original description (D. Algaba & B. Maryan, 2001).

Type locality: "23 km south of Kalbarri (27°55′19″S; 114°09′48″E), Western Australia," Australia.

Distribution: Australia (extreme WC Western Australia).

Anilios pilbarensis (Aplin & Donnellan, 1993)

Anilios pinguis (Waite, 1897)

Anilios proximus (Waite, 1893)

Anilios robertsi (Couper, Covacevich & S. Wilson, 1998)

Anilios silvia (Ingram & Covacevich, 1993)

Anilios splendidus (Aplin, 1998)

Anilios systenos Ellis, Doughty, Donnellan, Marin & Vidal, 2017. Zootaxa 4323(1): 12–15, figs. 3, 5.

Holotype: WAM 114892, a 268 mm male (T. Backshall, 1992).

Type locality: "15 km east of Geraldton (28°46′S; 114°37′E), Western Australia," Australia.

Distribution: Australia (extreme WC Western Australia).

Anilios torresianus Boulenger, 1889. Ann. Mag. Nat. Hist, series 6, 4(23): 362. (*Typhlops torresianus*)

Syntypes: BMNH 1946.1.11.48–49, a halfgrown specimen and 400 mm adult (S. Macfarlane).

Type locality: "Murray Island, Torres Straits," Queensland, Australia.

Distribution: S Papua New Guinea and NE Australia.

Comments: Hedges et al. (2014) resurrect this taxon for non-Sunda populations, which are referred to *Sundatyphlops polygrammicus*.

Anilios tovelli (Loveridge, 1945)

Anilios troglodytes (Storr, 1981)

Anilios unguirostris (W.C.H. Peters, 1867)

Anilios waitii (Boulenger, 1895)

Anilios wiedii (W.C.H. Peters, 1867)

Anilios yampiensis (Storr, 1981)

Anilios yirrikalae (Kinghorn, 1942)

Anilios zonula Ellis, 2016. Herpetologica 72(3): 273–276, figs. 1–2.

Holotype: WAM 171667, a 187 mm female (V. Kessner, 20 May 2009).

Type locality: "Storr Island, Western Australia, Australia, (15°57′8.71″S, 124°33′49.54″E;…)."

Distribution: NW Australia (Augustus and Storr Islands, Kimberley region of Western Australia).

ANILIUS Oken, 1816 (Aniliidae)

Anilius scytale (Linnaeus, 1758)

Distribution: Add Brazil (Acre), Bernarde et al. (2011b).

Comments: Natera-Mumaw et al. (2015) elevate *A. s. phelpsorum* to species based on morphological differences in Venezuelan specimens, but do not provide a morphological or geographical definition of a unique taxon.

ANOMALEPIS Jan, 1860 *in* Jan & Sordelli, 1860–1866 (Anomalepididae)

Anomalepis aspinosus Taylor, 1939

Distribution: Add Peru (Amazonas), González-Carvajal et al. (2018).

Anomalepis colombia Marx, 1953

Distribution: Add Colombia (Antioquia, Quindio), González-Carvajal et al. (2018) and Vanegas-Guerrero et al. (2019). Elevation 599–1700 m.

Comments: F.J.M. Santos & Reis (2019) provide a revised description and photographs of the holotype. Vanegas-Guerrero et al. (2019) report and describe second and third known specimens.

Anomalepis flavapices J.A. Peters, 1957

Distribution: Add Peru (Amazonas), Vanegas-Guerrero et al. (2019).

Anomalepis mexicana Jan, 1860 *in* Jan & Sordelli, 1860–1866

Distribution: Add Costa Rica (Puntarenas, San Jose), elevation to 1028 m, Stickel et al. (2017); Panama (Bocas del Toro), González-Carvajal et al. (2018), Panama (Panama Oeste), Ray & Ruback (2015); Colombia (Bolívar), González-Carvajal et al. (2018).

†ANOMALOPHIS Auffenberg, 1959 (†Anomalophiidae)

†Anomalophis bolcensis (Massalongo, 1859)

ANOMOCHILUS Berg, 1901 (Anomochilidae)

Anomochilus leonardi M.A. Smith, 1940

Distribution: Low elevation of 70 m, Yaakob (2003).

Anomochilus montana I. Das, Lakim, Lim & Hui, 2008

Anomochilus weberi (Lidth de Jeude *in* Weber, 1890)

ANOPLOHYDRUS F. Werner, 1909 (Natricidae)

Anoplohydrus aemulans F. Werner, 1909

ANTAIOSERPENS Wells & Wellington, 1985 (Elapidae)

Comments: Couper et al. (2016) revise the genus as two genetically and morphologically distinct species.

Antaioserpens albiceps Boulenger, 1898. Ann. Mag. Nat. Hist., Series 7, 2(11): 414. (*Pseudelaps albiceps*)

Synonyms: *Denisonia rostralis* De Vis, 1911, *Rhynchelaps fuscicollis* Lönnberg & Andersson, 1913.

Holotype: BMNH 1946.1.17.50, a 160 mm specimen (R. Broom).

Type locality: "Port Douglas," Queensland, Australia.

Distribution: NE Australia (NE Queensland)

Antaioserpens warro (De Vis, 1884)

Synonyms: Remove *Pseudelaps albiceps*, Boulenger, 1898, *Denisonia rostralis* De Vis, 1911, *Rhynchelaps fuscicollis* Lönnberg & Andersson, 1913.

Distribution: Limited to SC Queensland, Australia.

ANTARESIA Wells & Wellington, 1984 (Pythonidae)

Antaresia childreni (Gray, 1842)

Antaresia maculosa (W.C.H. Peters, 1873)

Antaresia perthensis (Stull, 1932)

Antaresia stimsoni (L.A. Smith, 1985)

ANTILLOTYPHLOPS Hedges, Marion, Lipp, Marin & Vidal, 2014. Caribbean Herpetol. 49: 44. (Typhlopidae: Typhlopinae)

Type species: *Typhlops hypomethes* Hedges & J.P.R. Thomas, 1991 by original designation.

Distribution: Antillean West Indies.

Comments: The validity of *Antillotyphlops* is affirmed by Nagy et al. (2015).

Antillotyphlops annae (Breuil, 1999)

Antillotyphlops catapontus (J.P.R. Thomas, 1966)

Antillotyphlops dominicanus (Stejneger, 1904)

Antillotyphlops geotomus (J.P.R. Thomas, 1966)

Antillotyphlops granti (Ruthven & Gaige, 1935)

Antillotyphlops guadaloupensis (Richmond, 1966)

Antillotyphlops hypomethes (Hedges & J.P.R. Thomas, 1991)

Antillotyphlops monastus (J.P.R. Thomas, 1966)

Antillotyphlops monensis (K.P. Schmidt, 1926)

Antillotyphlops naugus (J.P.R. Thomas, 1966)

Antillotyphlops platycephalus (A.M.C. Duméril & Bibron, 1844)

Distribution: Add Turks and Caicos Islands (Big Ambergris Cay), Reynolds & Niemiller (2010a).

Antillotyphlops richardii (A.M.C. Duméril & Bibron, 1844)

APARALLACTUS A. Smith, 1849 *in* 1838–1849 (Atractaspididae: Aparallactinae)

Comments: Portillo et al. (2018) analyze seven of the species using DNA sequence data, and conclude that *A. capensis*, *A. lunulatus* and *A. modestus* are species complexes, with *A. guentheri* nested within *A. capensis*.

Aparallactus capensis A. Smith, 1849 *in* 1838–1849

Distribution: Add Rwanda, Portillo et al. (2018); Mozambique (Cabo Delgado), Portillo et al. (2018); Angola (Cuando Cubango), Branch (2018).

Aparallactus guentheri Boulenger, 1895

Distribution: Add Tanzania (Morogoro), Menegon et al. (2008).

Aparallactus jacksonii (Günther, 1888)

Distribution: Add Tanzania (Morogoro), Menegon et al. (2008).

Aparallactus lineatus (W.C.H. Peters, 1870)

Aparallactus lunulatus (W.C.H. Peters, 1854)

Distribution: Add Chad (Logone Oriental), Portillo et al. (2018).

Aparallactus modestus (Günther, 1859)

Distribution: Add Guinea (Faranah), Portillo et al. (2018); Gabon (Ogooue-Lolo), Pauwels et al. (2016a).

Aparallactus moeruensis Witte & Laurent, 1953

Aparallactus niger Boulenger, 1897

Aparallactus nigriceps (W.C.H. Peters, 1854)

Aparallactus turneri Loveridge, 1935

Aparallactus werneri Boulenger, 1895

APLOPELTURA A.M.C. Duméril, 1853 (Pareidae: Pareinae)

Aplopeltura boa (H. Boie, 1828)

Distribution: Add West Malaysia (Penang), Quah et al. (2013); West Malaysia (Pulau Singa Besar), B.L. Lim et al. (2010); West Malaysia (Terengganu), Sumarli et al. (2015); Philippines (Bohol, Leyte, Samar Is.), Leviton et al. (2018); Philippines (Dinagat), Sanguila et al. (2016); Philippines (Luzon), Sy & Binaday (2016).

APOSTOLEPIS Cope, 1862 (Dipsadidae: Xenodontinae)

Comments: Nogueira et al. (2012) provide a key to the species groups: *dimidiata*, *dorbignyi*, *flavotorquata*, *longicaudata*, *nigroterminata* (alternatively *nigrolineata* group). L.A. Martins & Lema (2015) discuss specimens of eight species from southwestern Brazil, from which they create an *A. borellii* species group. Lema (2016) provides a key to species of the *A. dimidiata* group. Cabral et al. (2017) present a diagnostic synopsis for all species of *Apostolepis*. Lema et al. (2017) summarize the *A. nigroterminata* group. Entiauspe-Neto et al. (2019) present a synoptic list of currently recognized species, and caution that the piecemeal nature of descriptions of taxa from small series may lead to synonymizations of some once a broad review of the genus, using multiple characters, is accomplished.

Apostolepis adhara D.P.F. França, Barbo, Silva, Silva & Zaher, 2018. Zootaxa 4521(4): 540–546, figs. 1–3, 6.

Holotype: CEPB 6554, a 258 mm female (N.J. da Silva and team).

Type locality: "the region surrounding the São Salvador Hydroelectric Power Plant (12°48′18.96″S, 48°13′11.79″W ca. 120 m above sea level…), municipality of São Salvador do Tocantins, state of Tocantins, Brazil."

Distribution: Brazil (S Tocantins), about 120 m. Known from two specimens from the type locality.

Apostolepis albicollaris Lema, 2002

Distribution: Elevational range of 400–1100 m, Nogueira et al. (2012).

Comments: Nogueira et al. (2012) provide a revised description, locality map, and color photograph.

Apostolepis ambinigra (W.C.H. Peters, 1869)

Apostolepis ammodites Ferrarezzi, Barbo & Albuquerque, 2005

Distribution: Add Brazil (Maranhao), F.M. Santos et al. (2018).

Apostolepis arenarius Rodriguez, 1993

Apostolepis assimilis (J.T. Reinhardt, 1861)

Distribution: Add Brazil (Distrito Federal, Mato Grosso, Mato Grosso do Sul, Santa Catarina), Lema and Renner (2011). Low elevation of 254 m, Guedes et al. (2014).

Apostolepis barrioi Lema, 1978. Comunic. Museu Ciências PUCRGS 18/19:30–32, figs. 1–4.

Holotype: MACN 49402, a 390 mm male (G.J. Williner, February 1965).

Type locality: "Rio Ipane, Cororo, Paraguai."

Distribution: SE South America. SE Brazil (Mato Grosso do Sul, Minas Gerais, Sao Paulo), E Paraguay (Amambay, Concepcion, San Pedro).

Comments: Revalidated from the synonymy of *A. dimidiata* by Cabral et al. (2017), who provide a revised species description. Entiauspe-Neto et al. (2019) argue that it is synonymous with *A. dimidiata*, being just a sympatric extreme within color pattern variation in the latter.

Apostolepis borellii Peracca, 1904

Distribution: Add Brazil (Mato Grosso), L.A. Martins & Lema (2015, as cf. *A. borellii*).

Comments: L.A. Martins & Lema (2015) revalidate *A. borellii* on the basis of morphological distinctiveness from *A. nigroterminata*. Lema & Renner (2015) provide a species account with color photographs and distribution map.

Apostolepis breviceps Harvey, Gonzales & Scrocchi, 2001

Apostolepis cearensis Gomés, 1915

Distribution: Add Brazil (Maranhao), F.M. Santos et al. (2018); Brazil (Pernambuco), Pedrosa et al. (2014); Brazil (Tocantins), Zamprogno et al. (1998) and Lema & Renner (2007).

Comments: Zamprogno et al. (1998) provide a description, color photograph and locality map.

Apostolepis cerradoensis Lema, 2003

Apostolepis christineae Lema, 2002

Distribution: Add Bolivia (Santa Cruz), and elevation range of 103–496 m, Entiauspe-Neto & Lema (2015); Brazil (Mato Grosso do Sul), Ferreira et al. (2017, as *A. cf christineae*).

Apostolepis dimidiata (Jan, 1862)

Synonyms: Remove *Apostolepis barrioi* Lema, 1978.

Distribution: Add Brazil (Distrito Federal), Lema and Renner (2012, as *A. cf. dimidiata*); Brazil (Mato Grosso), L.A. Martins & Lema (2015); Brazil (Rio de Janeiro), Cabral et al. (2017); Brazil (Santa Catarina), D.P.F. França et al. (2018); Paraguay (Alto Parana), Cabral et al. (2017); Paraguay (Amambay, Central, Guaira), Cabral & Weiler (2014).

Apostolepis dorbignyi (Schlegel, 1837)

Apostolepis flavotorquata (A.M.C. Duméril, Bibron & Duméril, 1854)

Distribution: Add Brazil (Minas Gerais), Silviera (2014); Brazil (SE Para), D.P.F. França et al. (2018); Brazil (Sao Paulo), Nogueira et al. (2012).

Apostolepis freitasi Lema, 2004

Comments: Lema & Renner (2005, 2007) describe new specimens (formerly known only from the holotype), and provide a revised description of the species.

Apostolepis gaboi Rodrigues, 1993

Distribution: Elevation range of 395–405 m, Guedes et al. (2014).

Comments: Guedes et al. (2018) provide a revised description, color photographs, and distribution map based on the study of over thirty additional specimens.

Apostolepis goiasensis Prado, 1943

Neotype: MCP 9192, designated by Lema (2015), not described.

Neotype locality: "Uberlândia, Minas Gerais, Cerrado region of Central Brazil."

Distribution: Add Brazil (Mato Grosso), Loebmann & Lema (2012); Brazil (Tocantins), Lema and Renner (2012).

Comments: Loebmann & Lema (2012) provide new morphological data and a map of known localities. Lema (2015) describes and illustrates the holotype, now destroyed, and designates a neotype.

Apostolepis intermedia Koslowski, 1898

Distribution: Add Brazil (Mato Grosso), D.P.F. França et al. (2018); Paraguay (San Pedro), and upper elevation of 204 m, Entiauspe et al. (2014).

Apostolepis kikoi F.M. Santos, Entiauspe-Neto, Araújo, Souza, Lema, Strüssmann & Albuquerque, 2018. Zoologia 35(e26742):2–6, figs. 1–5.

Holotype: MCP 12096, a 288 mm female (Faunal Rescue Team, 2000).

Type locality: "Manso multi-use reservoir and hydroelectrical power plant – locally known as APM Manso – constructed at the confluence of the rivers Manso and Casca, Chapado dos Guimarães (15°27'39"S, 55°45'00"W; 811 m.a.s.l.), Mato Grosso, Brazil."

Distribution: Brazil (Mato Grosso), 811 m. Known only from the type locality.

Apostolepis lineata Cope, 1887

Apostolepis longicaudata Gomés *in* Amaral, 1921

Distribution: Add Brazil (Bahia), D.P.F. França et al. (2018); Brazil (Para), F.M. Santos et al. (2018); Brazil (Paraiba), R. França et al. (2012, as *A. cf. longicaudata*).

Apostolepis mariae Borges-Nojosa, Lima, Bezerra & Harris, 2017. Rev. Nordest. Zool. 10(2): 78–82, figs. 1, 2.

Holotype: CHUFC 3131, a 603 mm female (D.M. Borges-Nojosa, 23 November 1997).

Type locality: "Maciço de Baturité, Sitio Olho d'água dos Tangarás, Pacoti Municipality, State of Ceará, Brazil (04°14'13,7"S/38°54'58,6"W…)."

Distribution: NE Brazil (Ceara).

Apostolepis multicincta Harvey, 1999

Distribution: Add Argentina (Jujuy), F.B. Gallardo et al. (2017); Bolivia (Cochabamba), Mendoza-Miranda & Muñoz-S. (2017).

Apostolepis nelsonjorgei Lema & Renner, 2004

Distribution: Add Brazil (Mato Grosso, Para), D.P.F. França et al. (2018).

Apostolepis niceforoi Amaral, 1935

Apostolepis nigrolineata (W.C.H. Peters, 1869)

Synonyms: Remove *Apostolepis quinquelineata* Boulenger, 1896, *Apostolepis rondoni* Amaral, 1925.

Comments: Lema et al. (2017) remove *A. quinquelineata* and *A. rondoni* from the synonymy based on morphological differences. They present a revised description and diagnosis, color photographs, and a locality map.

Apostolepis nigroterminata Boulenger, 1896

Distribution: Add Brazil (Acre), F.M. Santos et al. (2018); Peru (Cuzco, Madre de Dios), Lema & Renner (2015); Peru (Puno), Llanqui et al. (2019).

Comments: Lema & Renner (2015) provide a species account with color photographs and distribution map.

Apostolepis phillipsae Harvey, 1999

Distribution: Add Brazil (Mato Grosso), and upper elevation of 247 m, Colli et al. (2019); Brazil (Mato Grosso do Sul), L.A. Martins & Lema (2015, as *A.* aff. *phillipsi*).

Comments: Michels & Bauer (2004) corrected the matronym to *phillipsae*. Colli et al. (2019) describe an additional specimen, and provide a revised species diagnosis.

Apostolepis pymi Boulenger, 1903

Distribution: Add Brazil (Goias), Lema and Renner (2012).

Apostolepis quinquelineata Boulenger, 1896. Cat. Snakes Brit. Mus. 3: 235, plate 10.

Holotype: BMNH 1946.1.9.59, a 165 mm male (J. Quelch).

Type locality: "Demerara," Guyana.

Distribution: Guyana, N Brazil (Amapa, Amazonas, Roraima).

Comments: Recognized as a species distinct from *A. nigrolineata* by Nogueira et al. (2012), Lema et al. (2017).

Apostolepis quirogai Giraudo & Scrocchi, 1998

Apostolepis roncadori Lema 2016. Caderno Pesquisa, Biol. 28(1): 2–7, fig. 1.

Holotype: BMNH 1972.429, a 252 mm male (Xavantina-Cachimbo Expedition).

Type locality: "Serra do Roncador, Mato Grosso, Brazil (12°51′ S 51°46′ W), at 700 m on the sea level."

Distribution: C Brazil (E Mato Grosso), 700 m. Known only from the holotype.

Apostolepis rondoni Amaral, 1925. Commissão Linhas Telegr. Estrat. Matto Grosso Amazonas (84): 25.

Holotype: MNRJ.

Type locality: no locality given; stated as Rondon, N Mato Grosso by Amaral (1977).

Distribution: W Brazil (SW Amazonas, W Mato Grosso, Rondonia).

Comments: Recognized as a species distinct from *A. nigrolineata* by Nogueira et al. (2012), Lema et al. (2017).

Apostolepis serrana Lema & Renner, 2006

Apostolepis striata Lema, 2004

Apostolepis tenuis Ruthven, 1927

Apostolepis tertulianobeui Lema, 2004. Acta Biol. Leopold. 26(1): [155].

Synonyms: *Apostolepis parassimilis* Lema & Renner, 2011.

Holotype: MCN 8535, a 398 mm male.

Type locality: Uberlândia, Minas Gerais, Brazil.

Distribution: EC Brazil (Bahia, Goias, Minas Gerais, Sao Paulo).

Comments: H.C. Costa & Bérnils (2015) report that *A. tertulianobeui* and *A. parassimilis* are based on the same holotype. They reiterate diagnostic characters between *A.*

tertulianobeau and *A. assimilis*, but without studying the latter, they make no decision regarding the validity of the former. A member of the *assimilis* group. Descriptive paper not seen; data based on Uetz et al. (2019 et seq.).

Apostolepis thalesdelemai Borges-Nojosa, Lima, Bezerra & Harris, 2017. Rev. Nordest. Zool. 10(2): 82–85, fig. 3.

Holotype: CHUFC 2341, a 647 mm female (D.M. Borges-Nojosa, 4 June 1999).

Type locality: "Planalto da Ibiapaba, Murimbeca locality, Ubajara Municipality, State of Ceará, Brazil (03°49′14,3″S/40°54′16,8″W…)."

Distribution: NE Brazil (Ceara).

Apostolepis underwoodi Lema & P. Campbell, 2017. Res. Rev.: J. Zool. Sci. 5(1): 20–27, figs. 5–13.

Holotype: BMNH 1927.8.1.180, a 382 mm female.

Type locality: "Buena Vista, Santa Cruz Department, Bolivia (17°27′32″S, 63°39′33″W), 398 m OLS."

Distribution: C Bolivia (Santa Cruz).

Apostolepis vittata (Cope, 1887)

Distribution: Lower elevational range of approximately 500 m, Lema & Renner (2004).

Comments: Lema & Renner (2004) describe two additional specimens from Mato Grosso, Brazil.

APROSDOKETOPHIS Wallach, Lanza & Nistri, 2010 (Colubridae: Colubrinae)

Aprosdoketophis androoeoi Wallach, Lanza & Nistri, 2010

ARCANUMOPHIS Smaga, Ttito & Catenazzi, 2019. Zootaxa 4671(1): 132. (Dipsadidae: Xenodontinae)

Type species: *Liophis problematicus* Myers, 1986, by monotypy.

Distribution: WC South America (Peru).

Comments: Smaga et al. (2019) use two mtDNA genes to produce a phylogeny that has *L. problematicus* as sister taxon to other Xenodontines (*Erythrolamprus*, *Lygophis* and *Xenodon*).

Arcanumophis problematicus (C.W. Myers, 1986)

Comments: Smaga et al. (2019) describe and illustrate a second specimen, collected at 1960 m.

†*ARCHAEOPHIS* Massalongo, 1859 (†Palaeophiidae: Archaeophiinae)

†*Archaeophis proavus* Massalongo, 1859
†*Archaeophis turkmenicus* Tatarinov, 1963

ARCHELAPHE Schulz, Böhme & Tillack, 2011 (Colubridae: Colubrinae)

Archelaphe bella (Stanley, 1917)

ARGYROGENA F. Werner, 1924 (Colubridae: Colubrinae)

Comments: S. Das et al. (2019) examine relationships of *A. fasciolata* using cranial morphology and mtDNA sequence data, both of which indicate it is a sister taxon to *Platyceps*.

Argyrogena fasciolata (Shaw, 1802)

Distribution: Sagadevan et al. (2019) document records from the SE coast of India.

Argyrogena vittacaudata (Blyth, 1854)

ARGYROPHIS Gray, 1845 (Typhlopidae: Asiatyphlopinae)

Synonyms: *Asiatyphlops* Hedges, Marion, Lipp, Marin & Vidal, 2014.

Type species: *Argyrophis bicolor* Gray 1845, designated by Stejneger (1907).

Distribution: SE Asia from N Pakistan to Sumatra and SE China.

Comments: Hedges et al. (2014), Pyron & Wallach (2014) provide generic diagnoses.

Argyrophis bothriorhynchus (Günther, 1864)

Argyrophis diardii (Schlegel, 1839 *in* 1837–1844)

Distribution: Add Nepal (Nawalparasi), Pandey (2012).

Argyrophis fuscus (A.M.C. Duméril & A.H.A. Duméril, 1851)

Argyrophis giadinhensis (Bourret, 1937)

Argyrophis hypsobothrius (F. Werner, 1917)

Comments: Hedges et al. (2014) assign *hypsobothrius* to *Indotyphlops*, but Pyron & Wallach (2014) argue that it is morphologically assignable to *Argyrophis*.

Argyrophis klemmeri (Taylor, 1962)

Argyrophis koshunensis (Oshima, 1916)

Argyrophis muelleri (Schlegel, 1839 *in* 1837–1844)

Distribution: Add West Malaysia (Pahang), Zakaria et al. (2014); West Malaysia (Perak), Hurzaid et al. (2013).

Argyrophis oatesii (Boulenger, 1890)

Argyrophis roxaneae (Wallach, 2001)

Argyrophis siamensis (Günther, 1864)

Argyrophis trangensis (Taylor, 1962)

ARIZONA Kennicott *in* Baird, 1859 (Colubridae: Colubrinae)

Comments: E.A. Myers et al. (2017a) provide a DNA-sequence based phylogeny with four primary clades: (1) Baja California (=*A. pacata*), (2) western, (3) Chihuahuan Desert, (4) Great Plains, the last three currently referred to *A. elegans*.

Arizona elegans Kennicott *in* Baird, 1859

Comments: Molecular data studied by E.A. Myers et al. (2017a) and Dahn et al. (2018) indicate the existence of a western and eastern clade.

Arizona pacata Klauber, 1946

Comments: Molecular data studied by E.A. Myers et al. (2017a) and Dahn et al. (2018) confirm the validity as a species.

†***Arizona voorhiesi*** Parmley & Holman, 1995

ARRHYTON Günther, 1858 (Dipsadidae: Xenodontinae)

Arrhyton ainictum Schwartz & Garrido, 1981

Arrhyton dolichura F. Werner, 1909

Arrhyton procerum Hedges & Garrido, 1992

Arrhyton redimitum (Cope, 1862)

Arrhyton supernum Hedges & Garrido, 1992

Arrhyton taeniatum Günther, 1858

Distribution: Add Cuba (Camaguey: Sabinal Key), A. Gonzalez & Iturriaga (2014); Cuba (Mayabeque, Villa Clara), Amaro-Valdés & Morell-Savall (2017).

Arrhyton tanyplectum Schwartz & Garrido, 1981

Arrhyton vittatum (Gundlach *in* W.C.H. Peters, 1861)

ASPIDELAPS Fitzinger, 1843 (Elapidae)

Aspidelaps lubricus (Laurenti, 1768)

Aspidelaps scutatus (A. Smith, 1849 *in* 1838–1849)

ASPIDITES W.C.H. Peters, 1877 (Pythonidae)

Aspidites melanocephalus (Krefft, 1864)

Aspidites ramsayi (Macleay, 1882)

ASPIDOMORPHUS Fitzinger, 1843 (Elapidae)

Aspidomorphus lineaticollis (F. Werner, 1903)

Aspidomorphus muellerii (Schlegel, 1837)

Distribution: Add Indonesia (Gag Island), Lang (2013); Papua New Guinea (Kairiru, New Britain and New Ireland Islands), Clegg & Jocque (2016); Papua New Guinea (Oro), O'Shea et al. (2018b, suppl.).

Aspidomorphus schlegelii (Günther, 1872)

Distribution: Add Papua New Guinea (East Sepik), Kraus (2013); Papua New Guinea (Sandaun), O'Shea et al. (2018b, suppl.). Elevation range NSL-326 m, O'Shea et al. (2018b, suppl.).

ASPIDURA Wagler, 1830 (Natricidae)

Aspidura brachyorrhos (F. Boie, 1827)

Aspidura ceylonensis (Günther, 1858)

Aspidura copii Günther, 1864

Aspidura deraniyagalae Gans & Fetcho, 1982

Aspidura desilvai Wickramasinghe, Bandara, Vidanapathirana & Wickramasinghe, 2019. Zootaxa 4559(2): 266–273, figs. 2–4, 6, 7.

Holotype: NMSL-NH 2019.01.02, a 193 mm male (L.J.M. Wickramasinghe & D.R. Vidanapathirana, 7 July 2018).

Type locality: "Riverstone, Knuckles, Matale District, Central Province, Sri Lanka (07°31′39″ N, 80°44′01″ E, elevation 1420 m)."

Distribution: Sri Lanka (Central), 995–1700 m.

Aspidura drummondhayi Boulenger, 1904

Aspidura guentheri Ferguson, 1876

Aspidura ravanai Wickramasinghe, L.J.M., Vidanapathirana, Kandambi, Pyron & Wickramasinghe, 2017. Zootaxa 4347(2): 277, 289–290, fig. 3–5.

Holotype: NMSL-NH 2017.19.01, a 293 mm male (L.J.M. Wickramasinghe, D.R. Vidanapathirana, S.C. Ariyarathne, A.W.A. Chanaka, D. Kandambi & M.D.G. Rajeev, 5 February 2010).

Type locality: "Sri Pada upper region, Ratnapura district, Sabaragamuwa province, Sri Lanka, (06°48′28″N, 080°29′23″E, elevation 1,680 m a.s.l.)."

Distribution: S Sri Lanka (Sabraragamuwa), 1650–2000 m.

Aspidura trachyprocta Cope, 1860

Distribution: Add Sri Lanka (Sabaragamuwa), Wickramasinghe et al. (2019).

ASTHENODIPSAS W.C.H. Peters, 1864 (Pareidae: Pareinae)

Comments: Quah et al. (2019a) review populations related to *A. laevis* on Borneo.

Asthenodipsas jamilinaisi Quah, Grismer, Lim, Anuar & Imbun, 2019a. Zootaxa 4646(3): 508–515, figs. 4, 5.

Holotype: SP 4076, a 456 mm male (P.Y. Imbun et al., 14 May 1991).

Type locality: "Mount Trusmadi, Tambunan, Sabah, East Malaysia (estimated: N 5.552776, E 116.516667, 2612 m a.s.l.)."

Distribution: East Malaysia (Sabah), 1668–2612 m.

Asthenodipsas laevis (H. Boie *in* F. Boie, 1827)

Asthenodipsas lasgalenensis Loredo, Wood, Quah, Anuar, Greer, Ahmad & Grismer, 2013. Zootaxa 3664(4): 514–517, figs. 2, 3.

Holotype: LSUHC 8869, a 529 mm male (L.L. Grismer, P.L. Wood, J.L. Grismer & C.K. Onn, 2 March 2008).

Type locality: "Bukit Larut, Perak, Malaysia (4° 44.596′ N, 100° 45.537′ E; 1184 m)."

Distribution: Highlands of West Malaysia (Pahang, Perak), above 800 to at least 2050 m.

Comments: See under *A. vertebralis*.

Asthenodipsas malaccana W.C.H. Peters, 1864

Asthenodipsas steubingi Quah, Grismer, Lim, Anuar & Imbun, 2019a. Zootaxa 4646(3): 504–506, figs. 2, 3.

Holotype: SP 4679, a 619 mm female (P.Y. Imbun, F.T.Y. Yu & Safric, 25 March 2009).

Type locality: "Minduk Sirung (Alab-Mahua trail), Crocker Range Park, Sabah, East Malaysia (estimated N 5.823240, E 116.347238, 1859 m a.s.l.)."

Distribution: E. Malaysia (Sabah), 971–1859 m.

Asthenodipsas tropidonota Lidth de Jeude 1923. Zool. Meded. 7: 243. (*Amblycephalus tropidonotus*)

Lectotype: RMNH 4902b, a 693 mm male (P.J. van Houten), designated by Grossmann & Tillack (2003).

Type locality: "Tapanoeli in Sumatra."

Distribution: Indonesia (W Sumatra: Lampung, Panjang).

Comments: See under *A. vertebralis*.

Asthenodipsas vertebralis (Boulenger, 1900)

Synonyms: Remove *Amblycephalus tropidonotus* Lidth de Jeude, 1923.

Distribution: Populations from Sumatra are now referable to *A. tropidonotus* according to Loredo et al. (2013).

Comments: Molecular and morphological data studied by Loredo et al. (2013) show that *A. vertebralis* is comprised of three species: *A. lasgalenensis*, *A. tropidonotus* and *A. vertebralis*.

ATHERIS Cope, 1862 (Viperidae: Viperinae)

Comments: Menegon et al. (2014) use mtDNA sequence data to create a phylogeny of the East African species plus several western species. The analysis confirms the monotypy of *Atheris*, and recovers four clades within the eastern species that correspond to tectonic and climatic events.

Atheris acuminata Broadley, 1998

Atheris anisolepis Mocquard, 1887

Atheris barbouri Loveridge, 1930

Atheris broadleyi D. Lawson, 1999

Distribution: Upper elevation of 1398 m, Ineich et al. (2015).

Atheris ceratophora F. Werner, 1896

Comments: Using mtDNA sequence data, Menegon et al. (2014) identify possible cryptic species within *A. ceratophora*.

Atheris chlorechis (Pel, 1851)

Distribution: Add Nigeria, Trape & Baldé (2014).

Atheris desaixi Ashe, 1968

Atheris hirsuta Ernst & Rödel, 2002

Distribution: Add Liberia (Nimba), and upper elevation of 585 m (Penner et al. (2013).

Comments: Penner et al. (2013) provide a description of a second known specimen.

Atheris hispida Laurent, 1955

Distribution: Add Uganda (Central), Groen et al. (2019).

Atheris katangensis Witte, 1953

Atheris mabuensis Branch & Bayliss, 2009

Atheris matildae Menegon, Davenport & Howell, 2011

Atheris nitschei Tornier, 1902

Atheris rungweensis Bogert, 1940

Comments: Using mtDNA sequence data, Menegon et al. (2014) identify possible cryptic species within *A. rungweensis*.

Atheris squamigera (Hallowell, 1855)

Distribution: Add Ghana, Trape & Baldé (2014); South Sudan (Central Equatoria), Ullenbruch & Böhme (2017); Angola (Bengo), Branch (2018); Angola (Cuanza Norte), M.P. Marques et al. (2018).

Atheris subocularis J.G. Fischer, 1888

ATRACTASPIS A. Smith, 1849 *in* 1838–1849 (Atractaspididae: Atractaspidinae)

Comments: Portillo et al. (2019a) analyze DNA sequence data for 14 of the species. They confirm that there are two clades, characterized by venom gland size and geographic distribution (Middle East and N Africa vs. Subsaharan).

Atractaspis andersonii Boulenger, 1905

Distribution: Add Yemen (Lahi), Portillo et al. (2019a).

Atractaspis aterrima Günther, 1863

Distribution: Add Mali (Koulikoro), Trape & Mané (2017).

Atractaspis battersbyi Witte, 1959

Atractaspis bibronii A. Smith, 1849 *in* 1838–1849

Distribution: Add Angola (Benguela), Branch (2018).

Atractaspis boulengeri Mocquard, 1897

Distribution: Add Gabon (Ogooue-Lolo), Portillo et al. (2019a); Congo (Niari), Portillo et al. (2019a).

Atractaspis branchi Rödel, Kucharzewski, Mahlow, Chirio, Pauwels, Carlino, Sambolah & Glos, 2019. Zoosyst. Evol. 95(1): 108–114, figs. 1–4.

Holotype: ZMB 88529, a 284 mm female (M.-O. Rödel, G. Sambola & J. Glos, 6 April 2018).

Type locality: "Liberia, Lofa region, Foya Forest, 08°01′16.2″N, 010°25′31.4″W, 317 m a.s.l."

Distribution: S Guinea (Nzerekore), N Liberia (Lofa), 317–486 m.

Atractaspis congica W.C.H. Peters, 1877

Distribution: Add Angola (Benguela, Cuanza Norte, Cuanza Sul, Huila, Uige, Zaire), Branch (2018); Angola (Bie, Cabinda, Huambo, Lunda Sul, Malanje), M.P. Marques et al. (2018); Angola (Luanda), Portillo et al. (2019a).

Atractaspis corpulenta (Hallowell, 1854)

Distribution: Add Congo (Niari), Portillo et al. (2018).

Atractaspis dahomeyensis Bocage, 1887

Distribution: Add Mali (Koulikoro), and upper elevation of 386 m, Trape & Mané (2017); Chad (Logone Oriental), Portillo et al. (2019a).

Atractaspis duerdeni Gough, 1907

Distribution: Add South Africa (Northern Cape), Portillo et al. (2019a).

Atractaspis engaddensis Haas, 1950

Holotype: Y. Werner & Shacham (2010) report that the un-numbered holotype has been lost from HUJ.

Distribution: Add Saudi Arabia (Ar-Riyad), Al-Sadoon (1989).

Atractaspis engdahli Lönnberg & Andersson, 1913

Atractaspis fallax W.C.H. Peters, 1867

Atractaspis irregularis (J.T. Reinhardt, 1843)

Distribution: Add Guinea (Faranah), Portillo et al. (2019a).

Atractaspis leleupi Laurent, 1950

Atractaspis leucomelas Boulenger, 1895

Atractaspis magrettii Scortecci, 1928

Atractaspis microlepidota Günther, 1866

Atractaspis micropholis Günther, 1872

Distribution: Add Mauritania (Hodh Ech Chargui), Padial (2006); Chad, Portillo et al. (2019a).

Atractaspis phillipsi Barbour, 1913

Atractaspis reticulata Sjöstedt, 1896

Distribution: Add Nigeria (Plateau), Rödel et al. (2019), Cameroon (Central), Rödel et al. (2019); Central African Republic (Sangha-Mbaere), Rödel et al. (2019), Equatorial Guinea, Rödel et al. (2019); Gabon (Estuaire), Rödel et al. (2019); Angola (Cuanza Norte), Rödel et al. (2019). Upper elevation of 1800 m, Rödel et al. (2019). Rödel et al. (2019) list and map known localities.

Comments: Rödel et al. (2019) redescribe the holotype of *A. reticulata*.

Atractaspis scorteccii Parker, 1949

Atractaspis watsonii Boulenger, 1908

Distribution: Add Benin, Trape & Baldé (2014); Niger (Diffa, Tillaberi, Zinder), Trape & Mané (2015). Low elevation of 41 m, Trape & Mané (2017).

ATRACTUS Wagler, 1828 (Dipsadidae: Dipsadinae)

Comments: Passos et al. (2013c) provide a key to *Atractus* known from the Guiana Shield. Arteaga et al. (2017) provide a phylogeny for 22 mostly Ecuadorian species using mtDNA sequence data. Passos et al. (2018a) provide a key to species of the *A. collaris* group. Melo-Sampaio et al. (2019) present a DNA sequence-based phylogeny of species from the Guiana Shield region. See under *Geophis*.

Atractus aboiporu Melo-Sampaio, Passos, Fouquet, Prudente & Torres-Carvajal, 2019. Syst. Biodiv. 17(3): 223–225, figs. 10, 11.

Holotype: MPEG 25796, a 298 mm female (U. Galatti, D. Silvano & B. Pimenta, 9 November 2000).

Type locality: "Serra do Navio, Amapá, Brazil."

Distribution: N Brazil (Amapa).

Atractus acheronius Passos, Rivas-Fuenmayor & Barrio-Amorgós, 2009

Atractus albuquerquei Cunha & Nascimento, 1983

Distribution: Add Brazil (Amazonas, Minas Gerais), and upper elevation of 943 m, B.T.M. do Alburquerque et al. (2017); Brazil (Tocantins), Passos et al. (2019b). B.T.M. do Alburquerque et al. (2017) map known localities.

Atractus alphoneshogei Cunha & Nascimento, 1983

Distribution: Remove Colombia, and add elevation range NSL-50 m, Passos et al. (2018a).

Comments: In the *A. collaris* species group, Passos et al. (2013d; 2018a). Passos et al. (2018a) provide a revised description, drawing of the holotype, and a locality map.

Atractus altagratiae Passos & Fernandes, 2008

Atractus alytogrammus Köhler & Kieckbusch, 2014. Zootaxa 3872(3):292–295, figs. 1–3.

Holotype: SMF 88371, a 332 mm male (F. Medem, 9 January 1957).

Type locality: "the Serrania de la Lindosa (2.46782°, −72.73155°), south of the municipality of San José del Guaviare, department of Guaviare, Colombia.

Distribution: Colombia (Guaviare). Add Colombia (Meta), 209 m, Angarita-Sierra (2019).

Comments: Angarita-Sierra (2019) describes the external morphology and hemipenes of a second specimen.

Atractus andinus Prado, 1944

Atractus apophis Passos & Lynch, 2010

Atractus atlas Passos, Scanferla, Melo-Sampaio, Brito & Almendáriz, 2019a. Anais Acad. Brasil. Ciéncias 91(supl. 1): 3–8, figs. 2–4.

Holotype: MEPN 14203, a 926+ mm female (A. Almendáriz, J. Brito, J. Hurtado & J. Puchaicela, 26 July 2011).

Type locality: Rio Blanco (03°55′2.08″S 78°30′9.81″W, ca. 1850 m above sea level…), Paquisha, municipality of Paquisha, province of Zamora-Chinchipe, Ecuador."

Distribution: SE Ecuador (Morona-Santiago, Zamora-Chinchipe), 1700–2100 m.

Atractus atratus Passos & Lynch, 2010

Atractus attenuatus C.W. Myers & Schargel, 2006

Atractus avernus Passos, Chiesse, Torres-Carvajal & Savage, 2009

Atractus ayeush Esqueda-González, 2011. Herpetotropicos 6(1–2): 36–37, figs. 1, 2.

Holotype: ULABG 5461, a 218 mm female (L.F. Esqueda-G., 27 December 2002).

Type locality: "the stream 'Quebrada La Concepción', sector 5, San Felipe, near Finca El Cocorucho, approx. 10°36′95″N and 69°29′62″W. Altitude 1050 m.a.s.l., Municipio Urdaneta, Lara State, Venezuela."

Distribution: NW Venezuela (Lara), 1050 m. Known only from the holotype.

Atractus badius (H. Boie *in* F. Boie, 1827)

Distribution: Add Brazil (Amapa), Schargel et al. (2013).

Atractus biseriatus Prado, 1941

Atractus bocki F. Werner, 1909

Atractus bocourti Boulenger, 1894

Atractus boimirim Passos, Prudente & Lynch, 2016a. Herpetol. Monogr. 30, 6–11, figs. 5–9.

Holotype: MPEG 17908, a 306 mm male (N. Jorge da Silva and team, between 14 November 1988 and 29 March 1989).

Type locality: "Jamari River…, Vila Cachoeira de Samuel (08°45′S, 63°27′W, ca. 100 m asl), municipality of Porto Velho, state of Rondônia, Brazil."

Distribution: Central Brazil (Amazonas, Pará, Rondonia), 30–145 m.

Comment: In the *Atractus pantostictus* group.

Atractus boulengerii Peracca, 1896

Atractus caete Passos, Fernandes, Bérnils & Moura-Leite, 2010

Atractus careolepis Köhler & Kieckbusch, 2014. Zootaxa 3872(3): 295–297, figs. 4–6.

Holotype: SMF 68413, a 207 mm male (M. Henning & F. Klaaßen, 1 August 1970).

Type locality: "the Punta de Betin (approximately 11.2522°, −74.2197°, 15 m asl), municipality of Santa Marta, province of Magdalena, Colombia."

Distribution: Colombia (Magdalena). Known only from the holotype.

Atractus carrioni H.W. Parker, 1930

Distribution: Add Peru (Huancabamba, Piura), 1500–2600 m, Passos et al. (2013b).

Comments: Passos et al. (2013b) present a revised diagnosis and morphological variation in the species. In the *A. roulei* species group, Passos et al. (2013b).

Atractus caxiuana Prudente & Santos-Costa, 2006

Distribution: Add Colombia (Vaupes) and Brazil (Rondonia), Passos et al. (2013d).

Comments: Passos et al. (2013d) provide a revised diagnosis and species account. In the *A. collaris* species group, Passos et al. (2013d; 2018a).

Atractus cerberus Arteaga, Mebert, Valencia, Cisneros-Heredia, Peñafiel, Reyes-Puig, Vieira-Fernandes & Guayasamin, 2017. ZooKeys 661:105–108, fig. 5.

Holotype: MZUTI 4330, a 235 mm male (J.L. Vieira-Fernandez and A. Arteaga, 6 November 2015).

Type locality: "Pacoche, province of Manabí, Ecuador (S1.06664, W80.88123; 280 m)."

Distribution: Ecuador (Manabi). Known only from two specimens from the type locality.

Comments: In the *A. iridescens* group according to Arteaga et al. (2017).

Atractus charitoae Silva-Haad, 2004

Atractus chthonius Passos & Lynch, 2010

Atractus clarki Dunn & Bailey, 1939

Atractus collaris Peracca, 1897

Synonyms: Add *Leptocalamus limitaneus* (Amaral, 1935).

Distribution: Add Colombia (Caqueta, Guainia, Vaupes), Peru (Ucayali), Brazil (Amazonas), and elevation range of 100–700 m, Passos et al. (2018a).

Comments: In the *A. collaris* species group, Passos et al. (2013d; 2018a). Passos et al. (2018a) synonymize *A. limitaneus* due to morphological overlap, and provide a revised description, color photographs, and a locality map for *A. collaris*. They also note that the holotype of *L. limitaneus* was not destroyed by fire.

Atractus crassicaudatus (A.M.C. Duméril, Bibron & Duméril, 1854)

Distribution: Add Colombia (Meta), Paternina & Capera-M. (2017).

Comments: Paternina & Capera-M. (2017) provide a species account for Colombia.

Atractus dapsilis Melo-Sampaio, Passos, Fouquet, Prudente & Torres-Carvajal, 2019. Syst. Biodiv. 17(3): 216–219, figs. 3–5.

Holotype: MNRJ 14914, a 354 mm male (E.G. Pereira & team, 1 February 2007).

Type locality: "Platô Teófilo, Flona Saracá-Taquera (1°42′51.6″S, 56°24′34.0″W), alt. 97 m asl, Oriximiná, Pará, Brazil."

Distribution: N Brazil (E Amazonas, W Para), 41–180 m.

Atractus darienensis C.W. Myers, 2003

Atractus depressiocellus C.W. Myers, 2003

Distribution: Add Colombia (Chocó), 100 m, Echavarría-Rentería et al. (2015).

Atractus duboisi (Boulenger, 1880)

Atractus duidensis Roze, 1961

Comment: Passos et al. (2013c) present a redescription and photographs of the type material.

Atractus dunni Savage, 1955

Distribution: Add Ecuador (Carchi, Imbabura, Santo Domingo, elevation 1688–2286 m), Arteaga et al. (2017).

Atractus echidna Passos, Mueses-Cisneros, Lynch & Fernandes, 2009

Atractus ecuadorensis Savage, 1955

Distribution: Elevation 1507 m, Arteaga et al. (2017).

Atractus edioi N.J. Silva, Rodrigues-Silva, Ribeiro, Souza & Souza, 2005

Atractus elaps (Günther, 1858)

Distribution: Add Ecuador (Pastaza), Meneses-Pelayo & Passos (2019); Peru (Huanuco, San Martin), Meneses-Pelayo & Passos (2019).

Atractus emigdioi Gonzáles-Sponga, 1971

Atractus emmeli (Boettger, 1888)

Synonyms: Add *Atractus boettgeri* Boulenger, 1896, *Atractus balzani* Boulenger, 1898, *Atractus taeniatus* Griffin, 1916, *Atractus paravertebralis* Henle & Ehrl, 1991.

Distribution: Add Bolivia (Cochabamba, Pando, Santa Cruz), Peru (Huanuco), and Brazil (Acre, Rondonia), Passos et al. (2018b); Low elevation of 90 m, Passos et al. (2019b).

Comments: Based on overlapping morphology, Passos et al. (2018b) synonymize *A. balzani* with *A. emmeli*. Passos et al. (2019b) provide a revised description, locality map, and color photographs. They also synonymize *A. boettgeri*, *A. paravertebralis* and *A. taeniatus* based on morphology.

Atractus eriki Esqueda, La Marca & Bazo, 2005

Distribution: Add Venezuela (Zulia), and elevation range of 900–1200 m, Natera-Mumaw et al. (2015)

Atractus erythromelas Boulenger, 1903

Distribution: Add Venezuela (Tachira), Meneses-Pelayo & Passos (2019). Upper elevation of 2200 m, Natera-Mumaw (2015).

Atractus esepe Arteaga, Mebert, Valencia, Cisneros-Heredia, Peñafiel, Reyes-Puig, Vieira-Fernandes & Guayasamin, 2017. ZooKeys 661:108–110, fig. 6.

Holotype: MZUTI 3758, a 285 mm male (A. Arteaga, 12 Sept. 2014).

Type locality: "Caimito, Esmeraldas Province, Ecuador (N0.69620, W80.090472; 102 m)."

Distribution: Ecuador (Esmeraldas). Known only from two specimens from the type locality.

Comments: In the *A. iridescens* group according to Arteaga et al. (2017).

Atractus favae (Filippi, 1840)

Atractus flammigerus (H. Boie *in* F. Boie, 1827)

Distribution: Revised to E Guiana Shield: E Suriname (Paramaribo, Sipaliwini), French Guiana (Cayenne, Coast, Saint-Laurent du Maroni), Brazil (Amapa), NSL-600 m, Passos et al. (2017).

Comments: Passos et al. (2017) provide a revised diagnosis and description, color photographs, and a locality map. They note that numerous records for this species are based on mis-identifications with *A. snethlageae* and *A. torquatus*.

Atractus franciscopaivai Silva-Haad, 2004

Comments: The specific epithet was misspelled in Wallach et al. (2014).

Atractus francoi Passos, Fernandes, Bérnils & Moura-Leite, 2010

Atractus fuliginosus (Hallowell, 1845)

Distribution: Add Venezuela (Anzoategui, Aragua, Barinas, Carabobo, Cojedes, Miranda, Portuguesa, Vargas, Yaracuy),

Atractus gaigeae Savage, 1955

Distribution: Add Ecuador (Napo, Orellana, Pastaza, Sucumbios), and lower elevation of 200 m, Passos et al. (2018a). Upper elevation 972 m, Arteaga et al. (2017).

Comments: In the *A. collaris* species group, Passos et al. (2013d; 2018a). Passos et al. (2018a) provide a revised description, color photographs, and a locality map.

Atractus gigas C.W. Myers & Schargel, 2006

Distribution: Add Ecuador (Carchi), Passos et al. (2019a); Ecuador (Santo Domingo), Arteaga et al. (2017).

Atractus guentheri (Wucherer, 1861)

Atractus guerreroi C.W. Myers & Donnelly, 2008

Atractus heliobelluomini Silva-Haad, 2004

Atractus heyeri Esqueda-González & McDiarmid *in* Natera-Mumaw et al., 2015. Atlas Serp. Venezuela: 407–411, fig. 1.

Holotype: USNM 247764, a 497 mm female (J.B. Heppner, August 1981).

Type locality: "17 Km SE Parque Nacional Yacambú, estado Lara, aproximadamente 1510 m de altitude, coordenades geográficas 9°42′26″N y 69°34′34″O."

Distribution: NW Venezuela (Lara), 1510 m. Known only from the holotype.

Atractus hoogmoedi Prudente & Passos, 2010

Comments: In the *A. collaris* species group, Passos et al. (2013d; 2018a).

Atractus hostilitractus C.W. Myers, 2003

Atractus imperfectus C.W. Myers, 2003

Atractus indistinctus Prado, 1940

Distribution: Add Colombia (Cesar), Meneses-Pelayo & Passos (2019); Venezuela (Merida, Tachira), Meneses-Pelayo & Passos (2019).

Atractus insipidus Roze, 1961

Distribution: Add Brazil (Rondônia), Bernarde et al. (2012b). Elevation corrected to 952 m, Passos et al. (2013c). The type locality straddles the Venezuela-Brazil border.

Comments: Passos et al. (2013c) presented a redescription and photograph of the holotype.

Atractus iridescens Peracca, 1896

Distribution: Add Colombia (Cauca), and upper elevation of 1251 m, Vera-Pérez et al. (2018);

Ecuador (Carchi, Esmeraldas, Pichincha), Arteaga et al. (2017).

Atractus lancinii Roze, 1961

Atractus lasallei Amaral, 1931

Atractus latifrons (Günther, 1868)

Distribution: Add Bolivia (Beni), Brazil (Roraima), Colombia (Vaupes), Almeida et al. (2014).

Comments: Almeida et al. (2014) provide a redescription of the holotype, and describe variation within this widespread species.

Atractus lehmanni Boettger, 1898

Distribution: Low elevation of 1748 m, Arteaga et al. (2017).

Atractus longimaculatus Prado, 1940

Atractus loveridgei Amaral, 1930

Atractus macondo Passos, Lynch & Fernandes, 2008

Atractus maculatus (Günther, 1858)

Distribution: Add: Brazil (Pernambuco), Abegg et al. (2017b); Brazil (Sergipe), Guedes et al. (2014, as *A.* aff. *maculatus*).

Atractus major Boulenger, 1894

Synonyms: Add *Atractus arangoi* Prado, 1940.

Distribution: Add Ecuador (Morona-Santiago), Almendáriz et al. (2014); Ecuador (Orellana, Sucumbios, Zamora-Chinchipe), Arteaga et al. (2017); Ecuador (Pastaza), Schargel et al. (2013); Peru (Cuzco, Ucayali), Schargel et al. (2013); Bolivia (La Paz), Passos et al. (2018b); Colombia (Caqueta, Putumayo), Schargel et al. (2013); Venezuela (Bolivar), Schargel et al. (2013); Brazil (Acre), Bernarde et al. (2011b); Brazil (Rondônia), Bernarde et al. (2012b).

Comments: Schargel et al. (2013) find that the holotype of *A. arangoi* (which is extant, not destroyed as erroneously stated by Wallach et al., 2014) agrees in all respects with *A. major*. They provide a revised diagnosis and color photographs of *A. major*.

Atractus manizalesensis Prado, 1940

Distribution: Add Colombia (Risaralda), and increase elevation range to 1500–2160 m, Rojas-Morales et al. (2017). They provide evidence that specimens from Cauca are not *A. manizalesensis*.

Comments: Rojas-Morales et al. (2017) provide a species account, with description, locality map and color photographs. The species epithet is misspelled in Wallach et al. (2014).

Atractus mariselae Lancini, 1969

Atractus marthae Meneses-Pelayo & Passos, 2019. Copeia 107(2): 251–256, figs. 1–4.

Holotype: UIS R3027, a 306 mm male (E. Meneses-Pelayo, 23 October 2014).

Type locality: "Colombia, department of Santander, municipality of Santa Bárbara, Vereda Esparta, 07°01′5.38″N, 72°53′43.04″W, ca. 2400 m above sea level."

Distribution: Colombia (Santander), 2220–2400 m.

Atractus matthewi Markezich & Barrio-Amorgós, 2004

Atractus medusa Passos, Mueses-Cisneros, Lynch & Fernandes, 2009

Distribution: Add Ecuador (Esmeraldas), 55 m, Cisneros-Heredia & Romero (2015).

Atractus melanogaster F. Werner, 1916

Distribution: Add Colombia (Quindio), Quintero-Ángel et al. (2012, as cf *A. melanogaster*).

Atractus melas Boulenger, 1908

Atractus meridensis Esqueda & La Marca, 2005

Atractus michelae Esqueda & La Marca, 2005

Atractus microrhynchus (Cope, 1868)

Distribution: Add Ecuador (Loja, Manabi, elevation to 1545 m), Arteaga et al. (2017).

Atractus mijaresi Esqueda & La Marca, 2005

Atractus modestus Boulenger, 1894

Distribution: Add Ecuador (Carchi, El Oro, low elevation of 1019 m), Arteaga et al. (2017).

Atractus multicinctus (Jan, 1865 *in* Jan & Sordelli, 1860–1866)

Distribution: Add Ecuador (Esmeraldas), Arteaga et al. (2017).

Atractus multidentatus Passos, Rivas-Fuenmayor & Barrio-Amorgós, 2009

Atractus nasutus Passos, Arredondo, Fernandes & Lynch, 2009

Atractus natans Hoogmoed & Prudente, 2003.

Distribution: Add Brazil (Marajó Island, Pará), G.M. Rodrigues et al. (2015).

Atractus nicefori Amaral, 1930

Atractus nigricaudus K.P. Schmidt & Walker, 1943

Atractus nigriventris Amaral, 1933

Atractus obesus Marx, 1960

Distribution: Add Colombia (Cauca), and elevation range of 1300–2800 m, Vera-Pérez et al. (2018).

Atractus obtusirostris F. Werner, 1916

Atractus occidentalis Savage, 1955

Distribution: Add Ecuador (Cotopaxi, Santo Domingo, elevation to 1985 m), Arteaga et al. (2017).

Atractus occipitoalbus (Jan, 1862)

Distribution: Add Colombia (Antioquia) and upper elevation of 1280 m, C.M. Marín et al. (2017); Ecuador (Napo, Pastaza, Sucumbios), Amendáriz & Orcés (2004); Bolivia (Cochabamba), Quinteros-Muñoz (2013).

Atractus ochrosetrus Esqueda & La Marca, 2005

Atractus oculotemporalis Amaral, 1932

Atractus orcesi Savage, 1955

Atractus paisa Passos, Arredondo, Fernandes & Lynch, 2009

Atractus pamplonensis Amaral, 1935

Atractus pantostictus Fernanes & Puorto, 1994

Atractus paraguayensis F. Werner, 1924

Type locality: Restricted to Bernal Cué, 25°16′,19.16S, 57°15′24.83W, Cordillera, Paraguay by Cabral and Cacciali (2018).

Distribution: Add Brazil (Rio Grande do Sul, Santa Catarina), Entiauspe-Neto & Abegg (2013, who also map all records).

Comments: In the *A. paraguayensis* species group, Passos et al. (2013a).

Atractus paucidens (Mocquard *in* Despax, 1910)

Distribution: Add Ecuador (El Oro, Santa Elena, Santo Domingo, elevation 72–1409 m), Arteaga et al. (2017).

Atractus pauciscutatus K.P. Schmidt & Walker, 1943

Atractus peruvianus (Jan, 1862)

Atractus poeppigi (Jan, 1862)

Atractus potschi Fernandes, 1995

Distribution: Add Brazil (Pernambuco), Roberto et al. (2017).

Comments: Passos et al. (2016b) expand on the diagnosis and morphological variation based on additional specimens. In the *A. paraguayensis* species group, Passos et al. (2013a).

Atractus punctiventris Amaral, 1933

Holotype: MLS 254, incorrectly stated as destroyed by Wallach et al. (2014).

Distribution: Add Colombia (Boyacá, Casanare), Passos et al. (2016).

Comments: Passos et al. (2016a) present a redescription, and illustrate the holotype, and assign it to the *A. flammigerus* group.

Atractus pyroni Arteaga, Mebert, Valencia, Cisneros-Heredia, Peñafiel, Reyes-Puig, Vieira-Fernandes & Guayasamin, 2017. ZooKeys 661: 111–112, fig. 7.

Holotype: MZUTI 5107, a 477 mm female (J.L. Vieira-Fernandez and C. Durán, 23 May 2016).

Type locality: "between Balzapamba and Bilován, province of Bolívar, Ecuador (S1.83601, W79.13322; 2026 m)."

Distribution: Ecuador (Bolivar). Known only from the holotype.

Comments: In the *A. roulei* group according to Arteaga et al. (2017).

Atractus resplendens F. Werner, 1901

Distribution: Elevation to 1962 m, Arteaga et al. (2017).

Atractus reticulatus (Boulenger, 1885)

Atractus riveroi Roze, 1961

Distribution: Add Brazil (Roraima), Fraga et al. (2017); 980–1800 m.

Comments: Passos et al. (2013c) present a redescription and photographs of the type material. Fraga et al. (2017) report on the geographic variation and distribution on the basis of additional specimens.

Atractus ronnie Passos, Fernandes & Borges-Nojosa, 2007

Comments: Ferreira-Silva et al. (2019) provide morphological data on numerous additional specimens.

Atractus roulei Mocquard *in* Despax, 1910

Distribution: Upper elevation to 3029 m, Arteaga et al. (2017).

Comment: In the *A. roulei* species group, Passos et al. (2013b).

Atractus sanctaemartae Dunn, 1946

Atractus sanguineus Prado, 1944

Atractus savagei Salazar-Valenzuela, Torres-Carvajal & Passos, 2014. Herpetologica 70(3), 351–357, figs. 1–4.

Holotype: QCAZ 8713, a 377 mm female (O. Torres-Carvajal, S. Aldás-Alarcó & E.E. Tapia, 25 February 2009).

Type locality: "surroundings of Chilmá Bajo on the way to Tres Marías waterfall (0°51′53.82″N, 78°2′59.23″W…; 2071 m above sea level [asl], Tulcán County, Carchi Province, Ecuador."

Distribution: Extreme NC Ecuador (Carchi), 2071–2420 m.

Comment: In the *Atractus paucidens* group.

Atractus schach (H. Boie *in* F. Boie, 1827)

Distribution: Limited to the Guiana Shield region of Guyana, French Guiana and Suriname, see Comments below. The following Brazilian records of "*Atractus schach*" represent other species, but we do not know which: add Brazil (Marajó Island), Moreira-Rodrigues et al. (2015); Brazil (Acre), Bernarde et al. (2011b); Brazil (Rondônia), Bernarde et al. (2012b).

Comments: Melo-Sampaio et al. (2019) present a revised diagnosis and description, and restrict the geographic range.

Atractus serranus Amaral, 1930

Atractus snethlageae Cunha & Nascimento, 1983

Distribution: Add Ecuador (Morona-Santiago, to 835 m), Arteaga et al. (2017); Ecuador (Orellana, Pastaza, Sucumbios, Tungurahua), Schargel et al. (2013); Peru (Amazonas, Cuzco, Loreto, Madre de Dios, Ucayali), Schargel et al. (2013). Upper elevation is 1800 m, Schargel et al. (2013).

Comments: Schargel et al. (2013) provide a revised description and color photographs.

Atractus spinalis Passos, Junior, Recoder, Sena, Dal Vechio, H. Pinto, Mendonça, Cassimiro & Rodrigues, 2013a. Pap. Avulsos Zool. 53(6): 77–83, figs. 1–4.

Holotype: MZUSP 20760, a 274 mm male (M.T. Rodrigues, M.T. Junior, R.S. Recoder, M.A. de Sena, F. Dal Vechio, H.B. de A. Pinto, S.H.S.T. Mendonça & J. Cassimiro, 26 March 2011).

Type locality: "Alto do Palácio (19°15′35.5″S, 43°31′55.2″W; 1357 m above sea level…), Parque Nacional da Serra do Cipó, municipality of Morro do Pilar, state of Minas Gerais, Southeastern Brazil."

Distribution: SE Brazil (SE Minas Gerais), 1354–1430 m.

Comments: In the *A. paraguayensis* species group, Passos et al. (2013a).

Atractus steyermarki Roze, 1958

Synonyms: Add *Atractus guerreroi* C.W. Myers & Donnelly, 2008.

Distribution: Elevation range 500–2244 m, Passos et al. (2013c).

Comment: Passos et al. (2013c) present a redescription and photographs.

Atractus stygius Passos, Azevedo, Nogueira, Fernandes & Sawaya 2019. Herpetol. Monogr. 33: 17–20, fig. 14.

Holotype: MNRJ 26734, a 319+ mm male (Universidade Federal de Mato Grosso team, 2009).

Type locality: "Bocaiúva (12°29′50″S, 57°52′30″W; ca. 312 m), Craveri River, municipality of Brasnorte, state of Mato Grosso, Brazil."

Distribution: C Brazil (Mato Grosso), 200–500 m.

Atractus sururucu Prudente & Passos, 2008

Comments: Passos et al. (2013c) present a redescription and photographs of the holotype. In the *A. collaris* species group, Passos et al. (2013d; 2018a).

Atractus tamaensis Esqueda & La Marca, 2005

Atractus tamessari Kok, 2006

Distribution: 500–2200 m.

Comments: Passos et al. (2013c) present a redescription and photographs.

Atractus taphorni Schargel & García-Pérez, 2002

Atractus tartarus Passos, Prudente & Lynch, 2016a. Herpetol. Monographs 30: 11–16, figs. 10–14.

Holotype: MNRJ 16511, a 372 mm male (R.S. Bérnils, H. Wogel, & P.S. Abe, 7 February 2008).

Type locality: "Vila Palestina (04°40′S, 47°56′W, ca. 200 m asl), municipality of Rondon do Pará, state of Pará, Brazil."

Distribution: NC Brazil (Maranhao, Pará), 50–400 m.

Comment: In the *Atractus flammigerus* group.

Atractus thalesdelemai Passos, Fernandes & Zanella, 2005

Atractus titanicus Passos, Arredondo, Fernandes & Lynch, 2009

Distribution: Add Colombia (Quindio, Risaralda), and upper elevation of 2656 m, Vanegas-Guerrero et al. (2014).

Atractus torquatus (A.M.C. Duméril, Bibron & Duméril, 1854)

Atractus touzeti Schargel, Lamar, Passos, Valencia, Cisneros-Heredia & Campbell, 2013. Zootaxa 3721(5): 466–469, figs. 6, 7.

Holotype: FHGO 517, an 1195 mm female (P. Pearman, 16 August 1992).

Type locality: "Cosanga-Archidona road (00°37′S, 77°48′W; 2200 m) in the Cordillera de Los Guacamayos, Province of Napo, Ecuador."

Distribution: Ecuador (Napo), 2200 m. Add Ecuador (Pastaza, 1355 m), Arteaga et al. (2017).

Atractus trefauti Melo-Sampaio, Passos, Fouquet, Prudente & Torres-Carvajal, 2019. Syst. Biodiv. 17(3): 220–223, figs. 6, 7, 9.

Holotype: MNRJ 26709, a 266 mm male (A. Fouquet, E. Courtois & M. Dewynter, 18 December 2012).

Type locality: "Route de l'Est N2, Roura, French Guiana, (4°29′19.7″N, 52°21′01.4W; 43 m asl)."

Distribution: French Guiana and N Brazil (Amapa, N Para), 43 m.

Atractus trihedrurus Amaral, 1926

Atractus trilineatus Wagler, 1828

Distribution: Add Guyana (Upper Takutu-Upper Essequibo), J.C. Murphy et al. (2019b). Upper elevation of 698 m, Arteaga et al. (2017).

Comments: J.C. Murphy et al. (2019b) produce a DNA sequence based phylogeny of various *A. trilineatus* populations in comparison with thirty other *Atractus* species. *A. trilineatus* is sister taxon to the remaining species, and shows little genetic differentiation between populations from Trinidad and Tobago, and those from the Guianas.

Atractus trivittatus Amaral, 1933

Atractus turikensis Barros-Blanco, 2000

Distribution: Add Colombia (Cesar), and upper elevation of 2540 m, Montes-Correa et al. (2017).

Atractus typhon Passos, Mueses-Cisneros, Lynch & Fernandes, 2009

Distribution: Add Ecuador (Esmeraldas), and low elevation of 63 m, Arteaga et al. (2017).

Atractus univittatus (Jan, 1862)

Atractus variegatus Prado, 1942

Atractus ventrimaculatus Boulenger, 1905

Atractus vertebrolineatus Prado, 1940

Atractus vertebralis Boulenger, 1904

Atractus vittatus Boulenger, 1894

Atractus wagleri Prado, 1945

Atractus werneri Peracca, 1912

Atractus zebrinus (Jan, 1862)

Atractus zidoki Gasc & Rodrigues, 1979

Comments: In the *A. collaris* species group, Passos et al. (2013d; 2018a).

ATRETIUM Cope, 1861 (Natricidae)

Comments: See under *Xenochrophis*.

Atretium schistosum (Daudin, 1803)

Atretium yunnanensis Anderson, 1879

ATROPOIDES Werman *in* Campbell & Brodie, 1992 (Viperidae: Crotalinae)

Atropoides indomitus E.N. Smith & Ferrari-Castro, 2008

Distribution: Add Honduras (El Paraiso), McCranie et al. (2013a); Honduras (Choluteca, Francisco Morazan), and upper elevation of 1681 m, Solís et al. (2017b); Medina-Flores et al. (2016, range extension).

Comment: McCranie et al. (2013a) support its distinctness from *A. occiduus*.

Atropoides mexicanus (A.M.C. Duméril, Bibron & Duméril, 1854)

Distribution: Add Nicaragua (Atlantico Norte, Nueva Segovia), Sunyer et al. (2014).

Atropoides nummifer (Rüppell, 1845)

Atropoides occiduus (Hoge, 1966)

Distribution: Mexico (Chiapas, range extension), García-Padilla (2015).

Atropoides olmec (Pérez-Higareda, H.M. Smith & Juliá-Zertuche, 1985)

Atropoides picadoi (Dunn, 1939)

Comments: Alencar et al. (2016), using four mtDNA genes, resolve *A. picadoi* as the sister taxon to *Cerrophidion* + all other *Atropoides*, as had previous studies. The phylogeny of Zaher et al. (2019) has *A. picadoi*, instead, as the sister taxon to *Cerrophidion* + *Porthidium*. Its ambiguous placement renders any of these genera paraphyletic without placing *picadoi* in its own genus.

†*AUSTRALOPHIS* Gómez, Báez & Rougier, 2008 (Aniliidae)

†*Australophis anilioides* Gómez, Báez & Rougier, 2008

AUSTRELAPS Worrell, 1963 (Elapidae)

Austrelaps labialis (Jan, 1859)
Austrelaps ramsayi (Krefft, 1864)
Austrelaps superbus (Günther, 1858)

AZEMIOPS Boulenger, 1888 (Viperidae: Azemiopinae)

Comments: Orlov et al. (2013) review the systematics of *Azemiops*.

Azemiops feae Boulenger, 1888

Distribution: Myanmar (Tanintharyi), China (Anhui, Fujian, Gansu, Guangxi, Guizhou, Hunan, Jiangxi, Shaanxi, Sichuan, SE Xixang, Yunnan, Zhejiang), Vietnam (Lai Chau, Lao Cai), 700–2800 m.

Comments: Orlov et al. (2013) provide a species account with a revised diagnosis and color photographs.

Azemiops kharini Orlov, Ryabov & T.T. Nguyen, 2013. Russian J. Jerpetol. 20(2): 119–125, figs. 18–27.

Synonyms: *Azemiops albocephala* Orlov, Ryabov & T.T. Nguyen, 2013 (*nomen nudum*).

Holotype: ZISP 26028, a 759 mm male (N.L. Orlov & S.A. Ryabov, 12 June 2003).

Type locality: "Tam Dao Mountain, Tam Dao Village, Vinh Phuc Province, Vietnam; 900 m a.s.l."

Distribution: N Vietnam (Bac Thai, Cao Bang, Lang Son, Vinh Phuc), 600–1800 m.

BAMANOPHIS Schätti & Trape, 2008 (Colubridae: Colubrinae)

Bamanophis dorri (Lataste, 1888)

†*BAVARIOBOA* Szyndlar & Schleich, 1993 (Booidea: incerta sedis)

†*Bavarioboa bachensis* Szyndlar & Rage, 2003
†*Bavarioboa crocheti* Szyndlar & Rage, 2003
†*Bavarioboa hermi* Szyndlar & Schleich, 1993
†*Bavarioboa herrlingensis* Szyndlar & Rage, 2003
†*Bavarioboa minuta* Szyndlar & Rage, 2003
†*Bavarioboa ultima* Szyndlar & Rage, 2003
†*Bavarioboa vaylatsae* Szyndlar & Rage, 2003

BITIA Gray, 1842d (Homalopsidae)

Comments: J.C. Murphy & Voris (2014) provide a generic diagnosis.

Bitia hydroides Gray, 1842

Distribution: Add Borneo, J.C. Murphy & Voris (2014).

Comments: J.C. Murphy & Voris (2014) provide a description and illustrations.

BITIS Gray, 1842 (Viperidae: Viperinae)

Fossil records: Add upper Pliocene of Tanzania, Rage & Bailon (2011, as *Bitis* n. sp. or *B. olduvaiensis*).

Bitis albanica Hewitt, 1937

Bitis arietans (Merrem, 1820)

Distribution: Add Mauritania (Gorgol), Padial (2006); Gabon (Nyanga), Pauwels et al. (2017c); Angola (Bengo, Benguela, Bie, Cuanza Norte, Cunene, Huambo, Huila, Lunda Sul, Malanje, Zaire), M.P. Marques et al. (2018); Angola (Cauando-Cubango), Conradie et al. (2016b); Angola (Namibe), Ceríaco et al. (2016). Crochet et al. (2015) report a southward range extension to the central coast of Western Sahara.

Comments: Alshammari (2011) evaluates variation in RNA sequence data between Arabian populations. Barlow et al. (2013), using mt and nuclear DNA sequence data, discover a number of clades in southern Africa that corresponded with hypothetical refugia of the late Pliocene.

Bitis armata (A. Smith, 1826)

Holotype: Number corrected to PEM R6769 by Conradie et al. (2019).

Bitis atropos (Linnaeus, 1758)

Bitis caudalis (A. Smith, 1839 *in* 1838–1849)

Distribution: Add Angola (Benguela, Huila, Luanda), M.P. Marques et al. (2018).

Bitis cornuta (Daudin, 1803)

Bitis gabonica (A.M.C. Duméril, Bibron & Duméril, 1854)

Distribution: Add Benin, Trape & Baldé (2014); Nigeria (Gongola), Nneji et al. (2019); Gabon (Moyen-Ogooue, Nyanga), Pauwels et al. (2017c); Angola (Benguela, Malanje), P.S. Oliveira et al. (2016); Angola (Cabinda), Branch (2018); Angola (Cuanza Norte, Zaire), M.P. Marques et al. (2018); Tanzania (Morogoro), Menegon et al. (2008), Lyakurwa (2017).

Bitis harenna Gower, Wade, Spawls, Böhme, Buechley, Sykes & Colston, 2016b. Zootaxa 4093(1): 43–50, figs.1–7.

Holotype: ZMUC R68255, a 655 mm female (deposited by S.J. Birket-Smith, date of collection unknown, but possibly 1966 or 1967).

Type locality: "Dodola (from maps: 6.98° N, 39.18° E c. 2,400 m elevation), Oromia Region, Ethiopia."

Distribution: Ethiopia (Oromia), 2400 m. Known only from the holotype and a photo voucher.

Comments: The holotype was previously identified as *B. parviocula* by Böhme (1977) as well as others. The describers consider it a member of the subgenus *Macrocerastes*.

Bitis heraldica (Bocage, 1889)

Distribution: Add Angola (Benguela), and upper elevation of 2302 m, F.M.P. Gonçalves et al. (2019, authors map known localities); Angola (Cuanza Sul, Zaire), M.P. Marques et al. (2018).

Bitis inornata (A. Smith, 1838 *in* 1838–1849)

Bitis nasicornis (Shaw & Nodder, 1792 *in* 1789–1813)

Distribution: Add Sierra Leone & Togo, Trape & Baldé (2014); Angola (Bengo, Cuanza Sul, Uige), Branch (2018); Angola (Cuanza Norte), M.P. Marques et al. (2018).

†*Bitis olduvaiensis* Rage, 1973

Bitis parviocula Böhme, 1977

Comments: Gower et al. (2016b) expand on the description through photographs of non-collected specimens.

Bitis peringueyi (Boulenger, 1888)

Distribution: Add Angola (Namibe), M.P. Marques et al. (2018).

Bitis rhinoceros (Schlegel, 1855)

Bitis rubida Branch, 1997

Bitis schneideri (Boettger, 1886)

Bitis worthingtoni Parker, 1932

Bitis xeropaga Haacke, 1975

BLYTHIA Theobald, 1868 (Colubroidea: incerta sedis)

Blythia hmuifang Vogel, Lalremsanga & Vanlalhrima, 2017. Zootaxa 4276(4), 570–576, figs. 1–6.

Holotype: MZMU 714, a 175 mm male (H.T. Lalremsanga & Z. Fanai, 14 April 2015).

Type locality: "Hmuifang Community forest, Aizawl District, Mizoram, India, 23°27'13.5''N; 92°45'09.5''E, elevation 1,442 m asl."

Distribution: NE India (Mizoram), 1442–1458 m.

Blythia reticulata Blyth, 1854

Distribution: Add India (Mizoram) and Bangladesh (Chittagong), Vogel et al. (2017).

BOA Linnaeus, 1758 (Boidae)

Synonyms: Remove †*Pseudoepicrates* Auffenberg, 1963.

Comments: Hynková et al. (2009) produce an mtDNA-based phylogeny for mainland populations that contains a cis- and a trans-Andean clade. Reynolds et al. (2014b) evaluate DNA sequence-data for five subspecies, also resulting in two clades, and they formally recognize *B. imperator* as a species for the Middle America populations. Card et al. (2016) use mtDNA sequence-data that support the mainland species split advised by Hynková et al. and Reynolds et al., and establish a third species, *B. sigma*, for western Mexico. We recognize *B. nebulosa* and *B. orophias* as species following Henderson & Powell (2009) and Reynolds & Henderson (2018), although previously suggested by Price & Russo (1991).

†*Boa blanchardensis* Bochaton & Bailon, 2018. J. Vert. Paleontol. 38(1462829): 3–9, figs. 3–6.

Holotype: MEC-A-18.1.1.1, a parabasisphenoid.

Type locality: "'layer 10' of Blanchard Cave," Marie-Galante Island, French Guadeloupe.

Distribution: Late Pleistocene (34,000–15,000 ya) of Marie-Galante Island, French Guadeloupe.

Boa constrictor Linnaeus, 1758

Synonyms: Remove *Boa orophias* Linnaeus, 1758, *Boa eques* Eydoux & Souleyet, 1841, *Boa diviniloquax mexicana* Jan, 1863, *Boa ortonii* Cope, 1878, *Boa constrictor isthmica* Garman, 1883, *Epicrates sabogae* Barbour, 1906, *Constrictor constrictor sigma* H.M. Smith, 1943, †*Neurodromicus barbouri* Vanzolini, 1952, †*Neurodromicus stanolseni* Vanzolini, 1952, *Constrictor constrictor nebulosa* Lazell, 1964, *Boa constrictor longicauda* Price & Russo, 1991.

Distribution: Remove Central America, cis-Andean Colombia, Ecuador and Peru, and Lesser Antilles. Add Aruba (introduced), Quick et al. (2005); Colombia (Cesar), Medina-Rangel (2011); Colombia (Norte de Santander), Armesto et al. (2011); Colombia (Santander), E. Ramos and Meza-Joya (2018); Ecuador (Manabi), Almendáriz et al. (2012); Brazil (Acre), Bernarde et al. (2011b); Brazil (Amapa), Sanches et al. (2018); Brazil (Ceara), Mesquita et al. (2013); Brazil (Espirito Santo), Silva-Soares et al. (2011); Brazil (Maranhão), J.P. Miranda et al. (2012); Brazil (Paraíba), R. França et al. (2012); Brazil (Piauí), Dal Vechio et al. (2013); Brazil (Rio Grande do Norte), Sales et al. (2009); Brazil (Rio de Janeiro: Ilha Grande), C.F.D. Rocha et al. (2018); Brazil (Tocantins), Dal Vechio et al. (2016); Paraguay (Presidente Hayes), P. Smith & Atkinson (2017); Paraguay (San Pedro), P. Smith et al. (2016); Argentina (Catamarca, Formosa, Jujuy, Santa Fe, Tucuman), Waller et al. (2012); Argentina (La Rioja), Kass et al. (2018); Argentina (San Juan), T.A. Martínez et al. (2015).

Fossil records: Add late Pleistocene (Lujanian) of Argentina (Albino & Brizuela, 2014).

Comments: Bushar et al. (2015) used DNA sequence data to suggest that the invasive population in Aruba probably originated from escaped or released snakes from northern South America. Reynolds & Henderson (2018) recognize four subspecies.

Boa imperator Daudin, 1803. Hist. Nat. Rept. 5: 150–152.

Synonyms: *Boa eques* Eydoux & Souleyet, 1841, *Boa diviniloquax mexicana* Jan, 1863, *Boa ortonii* Cope, 1878, *Boa constrictor isthmica* Garman, 1883, *Epicrates sabogae* Barbour, 1906, *Boa constrictor longicauda* Price & Russo, 1991.

Syntypes: lost from MNHN.

Type locality: "Mexique, à Carthagène [Bolívar, Colombia], et dans le royaume de Choco [Colombia]"; restricted to Colombian Choco by Dunn and Saxe (1950:161) and "Córdoba," Veracruz, México by Smith and Taylor (1950:347).

Distribution: E Mexico (Campeche, Chiapas, Hidalgo, E Oaxaca, Puebla, Queretaro, Quintana Roo, San Luis Potosi, Tamaulipas, Veracruz, Yucatan), Belize, Guatemala, Honduras, El Salvador, Nicaragua, Panama, W Colombia (Antioquia, Cauca, Choco, Valle del Cauca), W Ecuador (Manabi), and NW Peru (Cajamarca, La Libertad, Lambayeque, Piura, Tumbes). Add Mexico (Yucatan: Isla Contoy), Noguez & Ramírez-Bautista (2008); Honduras (Islas Exposición, Garrobo, Inglesera, El Pacar, Sirena, del Tigre, Zacate Grande), McCranie & Gutsche (2016); Panama (Panama Oeste), Ray & Ruback (2015); Colombia (Gorgona I.), Urbina-Cardona et al. (2008); Colombia (Valle del Cauca: Isla Palma), Giraldo et al. (2014).

Comments: Suárez-Atilano et al. (2014) produce a phylogeny from genetic data that recognizes two monophyletic lineages: one in Central America, the Yucatan Peninsula, and western Gulf of Mexico states, and one on the Pacific coast of Mexico (here recognized as *B. sigma*), with secondary contact between the two lineages in Oaxaca. Reynolds & Henderson (2018) recognize two subspecies. See under *Boa*.

Boa nebulosa Lazell, 1964. Bull. Mus. Comp. Zool. 132(3): 264–266, figs. 4, 5. (*Constrictor constrictor nebulosa*)

Holotype: MCZ R65493, a 1438 mm female (J.D. Lazell, 30 July 1959).

Type locality: "Woodford Hill, Dominica."

Distribution: Dominica, West Indies.

Comments: Recognized as a species distinct from *Boa constrictor* by Henderson & Powell (2009) and Reynolds & Henderson (2018).

Boa orophias Linnaeus, 1758. Syst. Nat. 10: 215.

Holotype: An unnumbered specimen in Museum de Geer collection within the Royal Museum of Stockholm.

Type locality: "Habitat … "; restricted to Praslin, St. Lucia by Lazell (1964).

Distribution: St. Lucia, West Indies.

Comments: Recognized as a species distinct from *Boa constrictor* by Henderson & Powell (2009) and Reynolds & Henderson (2018).

Boa sigma H.M. Smith, 1943. Proc. U. S. Natl. Mus. 93(3169): 411–412. (*Constrictor constrictor sigma*)

Holotype: USNM 46484, a skinned female of approximately 200 cm (E.W. Nelson & E.A. Goldman, 12 May 1897).

Type locality: "María Madre Island, Tres Marías Islands," Nayarit, Mexico.

Distribution: W Mexico (SW Chihuahua, Colima, SW Durango, Guerrero, Jalisco, Michoacan, Morelos, Nayarit, W Oaxaca, Sinaloa, Sonora). Add Mexico (Jalisco: San Pancho I.), Hernández-Salinas et al. (2014); Mexico (Zacatecas), Bañuelos-Alamillo et al. (2016). Upper elevation of 1420 m, Jacobs (2018).

Comments: See under *Boa*.

BOAEDON A.M.C. Duméril, Bibron & Duméril, 1854 (Lamprophiidae)

Comments: Greenbaum et al. (2015) present a DNA sequence-based phylogeny of populations of six Central African species. Trape & Mediannikov (2016) present an mtDNA-based phylogeny for Central and West African species.

Boaedon abyssinicus (Mocquard, 1906)

Boaedon angolensis Bocage, 1895. Herpétol. Angola Congo: 80. (*Boodon lineatus angolensis*)

Syntypes: originally in MBL, most, if not all, destroyed by fire.

Type locality: "Loanda…; Novo Redondo…; Benguella, Dumbe, Capangombe," Angola.

Distribution: Angola (Benguela, Cuando Cubango, Cuanza Norte, Cunene, Huila, Moxico, Namibe).

Comments: Marques et al. (2018) recognize *angolensis* as a species, and state that ongoing studies will define its taxonomic status and geographic distribution.

Boaedon arabicus H.W. Parker, 1930

Boaedon capensis A.M.C. Duméril, Bibron & Duméril, 1854

Comments: Trape & Mediannikov (2016) list *Boaedon mentalis* as a species apart from *B. capensis*, but without comment.

Boaedon erlangeri Sternfeld, 1908

Boaedon fuliginosus (H. Boie *in* F. Boie, 1827)

Synonyms: Remove *Alopecion variegatum* Bocage, 1867.

Distribution: Add Morocco (Tiznit), Barnstein et al. (2012); Mauritania (Nouakchott), Padial (2006); Mali (Kayes, Koulikoro, Mopti, Seguo, Sikasso), Trape & Mané (2017); Tanzania (Morogoro), Menegon et al. (2008), Lyakurwa (2017); Niger (Diffa, Tahoua, Tillaberi, Zinder), Trape & Mané (2015); Angola (Bengo, Cuando Cubango, Lunda Sul), M.P. Marques et al. (2018, as *B. fuliginosus* complex). See comments.

Comments: *Boaedon fuliginosus* is polyphyletic according to the DNA-based phylogeny that Greenbaum et al. (2015) present. Trape & Mediannikov (2016) consider *B. fuliginosus* to be limited to West Africa, describing at least two new species from populations from Cameroon, Chad, Central African Republic and Congo. However, they do not indicate the eastern limits of the range, nor do they discuss East African populations. See comments under *B. variegatus*.

Boaedon geometricus (Schlegel, 1837)

Boaedon lineatus A.M.C. Duméril, Bibron & Duméril, 1854

Synonyms: Remove *Boodon lineatus angolensis* Bocage, 1895.

Distribution: Add Liberia (Lofa), Rödel & Glos (2019); Mali (Segou), Trape & Mané (2017). See comments.

Comments: Trape & Mediannikov (2016) consider *B. lineatus* to be limited to West Africa, describing at least two new species from eastern populations. However, they do not indicate the eastern limits of the range. See comments under *B. angolensis*.

Boaedon littoralis Trape *in* Trape & Mediannikov, 2016. Bull. Soc. Herpétol. France 159: 100–105, figs. 29–32.

Holotype: MNHN 1964.11, a 1016 mm female (A. Stauch, September 1963).

Type locality: "Pointe-Noire (République Populaire de Congo)."

Distribution: WC Africa. W Congo (Kouilou) and S Gabon (Nyanga).

Boaedon longilineatus Trape *in* Trape & Mediannikov, 2016. Bull. Soc. Herpétol. France 159: 94–99, figs. 9, 25, 27, 28.

Holotype: MNHN R2015.0093 (orig. IRD 2521.N), a 913 mm female (villager, 18 November 2015).

Type locality: "Fieng-bac (09°51′01″N, 15°04′38″E; altitude 326 m) près de Fianga (Tchad, région du Mayo-Kebbi Est)," Chad.

Distribution: WC Africa. Chad (Batha, Mayo-Kebbi Est, Mayo-Kebbi Ouest) and Cameroon (Far North, North), 326–366 m.

Boaedon maculatus H.W. Parker, 1932

Boaedon olivaceus (A.H.A. Duméril, 1856)

Distribution: Add Gabon (Estuaire), Pauwels et al. (2017a); DR Congo (Kasai Occidental), Greenbaum et al. (2015).

Boaedon paralineatus Trape & Mediannikov, 2016. Bull. Soc. Herpétol. France 159: 88–94, figs. 8, 21–24.

Holotype: MNHN R2015.0085 (orig. IRD 2069.N), a 748 mm male (villager, 22 May 2015).

Type locality: "Baïbokoum (Tchad, région du Logone Oriental) devant sa maison (07°44′57″N, 15°41′28″E, altitude 491 m)."

Distribution: C Africa. S Chad (Logone Oriental), Cameroon and Central African Republic (Bangui, Haut-Mbomou, Haute-Kotto, Ombella-Mpoko, Ouham-Pende), 491 m. Add Nigeria (Gongola), Nneji et al. (2019).

Boaedon perisilvestris Trape & Mediannikov, 2016. Bull. Soc. Herpétol. France 159: 82–88, figs. 5, 17–20.

Holotype: MNHN R2015.0083 (orig. IRD TR.4232), a 436 mm male (J.-F. Trape, 15 December 2013).

Type locality: "l'ancien centre ORSTOM de Brazzaville près de l'ancien bâtiment de la trypanosomiase (04°16′39″S, 15°14′20″E; altitude 398 m)," Congo Brazzaville.

Distribution: WC Africa. Cameroon (Adamaoua, Centre), S Chad (Logone Oriental), Central African Republic (Bangui, Haut-Mbomou, Haute-Kotto, Ouham-Pende, Vakaga), Gabon (Haut-Ogooue), Congo (Brazzaville, Cuvette, Kouilou, Lekoumou, Plateaux, Pool), 398 m. Add Nigeria (Gongola), Nneji et al. (2019); W Congo-Zaire (Equateur), Greenbaum et al. (2017).

Boaedon radfordi Greenbaum, Portillo, Jackson & Kusamba, 2015. African J. Herpetol. 64(1): 24–29, figs. 3, 4.

Holotype: UTEP 20995 (orig. EBG 2422), a 553 mm male (SVL) (C. Kusamba, W.M. Moninga, M.M. Aristote & E. Greenbaum, 7 July 2009).

Type locality: "Shatuma-Abis village, Lendu Plateau, Orientale Province, D[emocratic]R[epublic of the]C[ongo] (2.01315° N, 30.84066° E, 2000 m)."

Distribution: Democratic Republic of the Congo (Nord Kivu, Oriente, Sud Kivu), Uganda (Southern, Western), 640–2362 m.

Boaedon subflavus Trape *in* Trape & Mediannikov, 2016. Bull. Soc. Herpétol. France 159: 75–82, figs. 6–8, 13–16.

Holotype: MNHN R2015.0066, a 643 mm male (villager, 29 May 2015).

Type locality: "village de Kumao (07°36′50″N, 15°36′45″E; altitude 553 m)…situé à 17 km au sud-ouest de Baibokoum (Tchad, region du Logone Oriental)."

Distribution: C Africa. Cameroon (North), Chad (Batha, Guera, Chari-Baguirmi, Logone Occidental, Logone Oriental, Mayo-Kebbi Est, Mayo-Kebbi Ouest, Ouaddai), Central African Republic (Vakaga), WC Sudan. 553 m.

Boaedon upembae (Laurent, 1954)

Comments: Greenbaum et al. (2015) transfer *Lycodonomorphus upembae* to *Boaedon* based on phylogenetic placement using DNA sequence data.

Boaedon variegatus Bocage, 1867. J. Sci. Math. Phys. Nat., Series 1, 1: 227. (*Alopecion variegatum*)

Syntypes: MBL (n = 4), (J. Anchieta), MBL (n = 1), (J.A. Botelho), all probably destroyed by fire.

Type locality: "Benguella e Dombe…..Novo Redondo," Angola.

Distribution: W Angola (Benguela, Cuanza Sul, Huambo, Luanda, Namibe).

Comments: Marques et al. (2018) recognize *variegatus* as a species, and state that ongoing studies will define its taxonomic status and geographic distribution

Boaedon virgatus (Hallowell, 1854)

Distribution: Upper elevation of 1772 m, Ineich et al. (2015).

†*BOAVUS* Marsh, 1871 (Tropidophiidae)

†*Boavus affinis* Brattstrom, 1955
†*Boavus brevis* Marsh, 1871
†*Boavus idelmani* Gilmore, 1938
†*Boavus occidentalis* Marsh, 1871

BOGERTOPHIS Dowling & Price, 1988 (Colubridae: Colubrinae)

Bogertophis rosaliae (Mocquard, 1899)
Bogertophis subocularis (A.E. Brown, 1902)

BOIGA Fitzinger, 1826 (Colubridae: Colubrinae)

Synonyms: Add *Elachistodon* J.T. Reinhardt, 1863.

Boiga andamanensis (Wall, 1909)

Boiga angulata (W.C.H. Peters, 1861)

Distribution: Add Philippines (Bohol I.), Leviton et al. (2018).

Comments: Leviton et al. (2018) believe *B. angulata* is likely a color variant of *B. drapiezii*.

Boiga barnesii (Günther, 1869)

Boiga beddomei (Wall, 1909)

Distribution: Add India (Gujarat), Patel & Vyas (2019); Sri Lanka (Sabaragamuwa), Peabotuwage et al. (2012).

Boiga bengkuluensis Orlov, Kudryavtsev, Ryabov & Shumakov, 2003

Boiga bourreti Tillack, Ziegler & Quyet, 2004

Boiga ceylonensis (Günther, 1858)

Distribution: Add Sri Lanka (Sabaragamuwa), Peabotuwage et al. (2012).

Boiga cyanea (Bibron *in* A.M.C. Duméril, Bibron & Duméril, 1854)

Distribution: Add Thailand (Rayong, and Koh Chang I.), Chan-ard and Makchai (2011); Cambodia (Siem Reap), Geissler et al. (2019); Vietnam (Dien Bien), Dung et al. (2014); West Malaysia (Pulau Singa Besar), B.L. Lim et al. (2010).

Boiga cynodon (H. Boie *in* F. Boie, 1827)

Distribution: Add Philippines (Bohol, Calayan, Camiguin Norte, Carabao, Inampulugan, Lubang, Negros, Pan de Azucar, Siquijor Is.), Leviton et al. (2018); Philippines (Tablas I.), Siler et al. (2012). Remove India (Gujarat), Patel & Vyas (2019).

Boiga dendrophila (F. Boie, 1827)

Distribution: Add Myanmar (Tanintharyi), J.L. Lee et al. (2015); West Malaysia (Kedah), Shahriza et al. (2013); West Malaysia (Pulau Singa Besar), B.L. Lim et al. (2010), West Malaysia (Terengganu), Sumarli et al. (2015); Philippines (Calayan, Siargao Is.), Leviton et al. (2018).

Boiga dightoni (Boulenger, 1894)

Boiga drapiezii (H. Boie *in* F. Boie, 1827)

Distribution: Add Myanmar (Tanintharyi), J.L. Lee et al. (2015); West Malaysia (Pahang), Zakaria et al. (2014); West Malaysia (Terengganu), Sumarli et al. (2015); Philippines (Luzon Is., Sorsogon Prov., 373 m), Binaday & Lobos (2016); Philippines (Mindanao I.), Leviton et al. (2018).

Boiga flaviviridis Vogel & Ganesh, 2013. Zootaxa 3637(2): 159–161, figs. 1, 2.

Holotype: BMNH 1911.9.8.4, an 890+ mm female (presented by F. Wall, 1911).

Type locality: "Berhampore, state of Orissa, India…, 19°18′ 57″ N 84°47′ 38″ E, 31 m asl."

Distribution: E India (Andhra Pradesh, incl. Sriharikota Is., Orissa), 11–680 m. Add India (Chattisgarh), Khandal et al. (2016a); India (Maharashtra), Sharma et al. (2016).

Boiga forsteni (A.M.C. Duméril, Bibron & Duméril, 1854)

Distribution: Add India (Jharkhand), Chaudhuri et al. (2017). Vamdev et al. (2019) map the distribution in India.

Boiga fusca (Gray, 1842)

Boiga gocool (Gray, 1835 *in* Gray & Hardwicke, 1830–1835)

Boiga guangxiensis Wen, 1998

Distribution: Add China (Yunnan), Hodges (2018); Vietnam (Binh Phuoc), Neang et al. (2017); Vietnam (Hai Duong), Ryabov & Orlov (2010); Vietnam (Hai Phong: Cat Ba Island), T.Q. Nguyen et al. (2011); Vietnam (Quang Ngai), Nemes et al. (2013); Cambodia (Mondulkiri), Neang et al. (2017).

Comments: The specific epithet was misspelled in Wallach et al. (2014).

Boiga hoeseli Ramadhan, Iskandar & Subasri, 2010

Boiga irregularis (Merrem *in* Bechstein, 1802)

Distribution: Add Indonesia (Gorong, Haruku, Saparua, Sula, Ternate, Tidore Islands), Lang (2013); Indonesia (Talaud Island, introduction), A. Koch et al. (2009); Papua New Guinea (Ambitle, Manam, Rambutyo and Tarawai Islands), Clegg & Jocque (2016); Papua New Guinea (Oro), O'Shea et al. (2018b, suppl.).

Boiga jaspidea (A.M.C. Duméril, Bibron & Duméril, 1854)

Distribution: Add West Malaysia (Terengganu), Sumarli et al. (2015).

Boiga kraepelini Stejneger, 1902

Distribution: Add Vietnam (Ha Giang, Ninh Binh, Phu Tho, Quang Binh, Quang Ninh, Thanh Hoa), Ananjeva et al. (2015).

Comment: Ananjeva et al. (2015) model habitat to predict the potential distribution.

Boiga multifasciata (Blyth, 1860)

Distribution: Add Bhutan (Punakha) and lower elevation of 1200 m, Koirala et al. (2016); Bhutan (Trongsa, Wangdue Phodrang), Tshewang & Letro, 2018).

Boiga multomaculata (F. Boie, 1827)

Distribution: Add Myanmar (Tanintharyi), Mulcahy et al. (2018); China (Nan Ao Island), Qing et al. (2015); Cambodia (Siem Reap), Geissler et al. (2019); low elevation to 700 m, Chettri et al. (2010).

Boiga nigriceps (Günther, 1863c)

Distribution: Add West Malaysia (Terengganu), Sumarli et al. (2015); Indonesia (Bali), Vink & Shonleben (2015).

Boiga nuchalis (Günther, 1875)

Boiga ochracea (Theobald *in* Günther, 1868)

Distribution: Add Bhutan (Trongsa, Wangdue Phodrang), Tshewang & Letro, 2018).

Boiga philippina (W.C.H. Peters, 1867)

Distribution: Add Philippines (Babuyan Claro I.), Leviton et al. (2018).

Boiga quincunciata (Wall, 1908)

Distribution: Add India (West Bengal), and upper elevation of 744 m, Ghosh & Mukherjee (2019).

Boiga ranawanei Samarawickrama, Samarawickrama Wijesena & Orlov, 2005. Russ. J. Herpetol. 12(3): 216–219, figs. 1–5, 7, 11, 12, 14.

Holotype: CMS, number not given, an 899 mm male (V.A.M.P.K. Samarawickrama, 20 October 2004).

Type locality: "Bulawaththa, Gannoruwa Forest, Kandy District, Central Province, Sri Lanka…07°17′16″ N 80°35′36″ E, altitude 640 m a.s.l."

Distribution: C Sri Lanka.

Comments: Vogel & Ganesh (2013) tentatively recognize *ranawanei* for comparative purposes.

Boiga saengsomi Nutaphand, 1985

Boiga schultzei Taylor, 1923

Comments: Leviton et al. (2018) believe *B. schultzei* is likely a color variant of *B. drapiezii*.

Boiga siamensis Nutaphand, 1971

Distribution: Add Nepal (Chitwan), Pandey et al. (2018); Myanmar (Tanintharyi), Mulcahy et al. (2018); Cambodia (Siem Reap), Geissler et al. (2019).

Boiga tanahjampeana Orlov & Ryabov, 2002

Boiga thackerayi Giri, Deepak, Captain, Pawar & Tillack, 2019. J. Bombay Nat. Hist. Soc. 116: 3–7, figs. 1–4.

Holotype: BNHS 2371, an 1163 mm male (T. Thackeray & S. Pawar, 27 July 2016).

Type locality: "near Humbarli, Koyna, Satara district, Maharashtra, India."

Distribution: WC India (Maharashtra). Known only from the type locality.

Boiga trigonata (Schneider *in* Bechstein, 1802)

Distribution: Add Iran (North Khorasan, Semnan, South Khorasan), Safaei-Mahroo et al. (2015); Nepal (Nawalparasi), Pandey et al. (2018).

Boiga wallachi I. Das, 1998
Boiga walli M.A. Smith, 1943

Boiga westermanni (J.T. Reinhardt, 1863)

Distribution: Add India (Gujarat, Madhya Pradesh), Vyas (2013); India (Telangana), Dandge & Tiple (2016, who map known localities); India (Punjab, Uttarakhand), Sharma (2014); India (Rajasthan), Khandal et al. (2016b). Mohan et al. (2018) provide a locality map.

Comments: Mohan et al. (2018) using DNA sequence-data, find *Elachistodon westermanni* to resolve within *Boiga*, as sister taxon to other Indian species. They transfer *E. westermanni* to *Boiga*.

BOIRUNA Zaher, 1996 (Dipsadidae: Xenodontinae)

Boiruna maculata (Boulenger, 1896)

Distribution: Add Brazil (Parana), Souza-Filho et al. (2015); Brazil (Rio Grande do Sul), M.B. dos Santos et al. (2012); Argentina (San Juan), Laspiur & Nenda (2018). Upper elevation of 1898 m, Quinteros-Muñoz (2015).

Boiruna sertaneja Zaher, 1996

Distribution: Add Brazil (Maranhao, Rio Grande do Norte, Sergipe), Guedes et al. (2014). Upper elevation of 1000 m, Guedes et al. (2014). P. Brito & Gonçalves (2012) document known and new records.

BOLYERIA Gray, 1842 (Bolyeriidae)

Bolyeria multocarinata (H. Boie *in* F. Boie, 1827)

BORIKENOPHIS Hedges & Vidal *in* Hedges, Couloux & Vidal, 2009 (Dipsadidae: Xenodontinae)

Borikenophis portoricensis (J.T. Reinhardt & Lütken, 1862)

Synonyms: Remove *Alsophis portoricensis prymnus* Schwartz, 1966.

Distribution: Add USA (Puerto Rico: Culebrita I.), Torres-Santana (2010).

Borikenophis prymnus Schwartz, 1966. Stud. Fauna Curaçao Caribb. Islands 23(90): 194–198, figs. 45, 53. (*Alsophis portoricensis prymnus*)

Holotype: MCZ 77226, an 811+ mm male (D.C. Leber & A. Schwartz, 13 May 1961).

Type locality: "Caja de Muertos, Puerto Rico," USA.

Distribution: S Puerto Rico, USA.

Comments: Grazziotin et al. (2012) and Zaher et al. (2019) recognize *prymnus* as a species distinct from *B. portoricensis*.

Borikenophis sancticrucis (Cope, 1862)
Borikenophis variegatus (K.P. Schmidt, 1926)

BOTHRIECHIS W.C.H. Peters, 1859 (Viperidae: Crotalinae)

Comments: Townsend et al. (2013b) provide an mtDNA-based phylogeny for ten of the species. They consider specimens from the Sierra de Sulaco, Yoro, Honduras to be unassignable to recognized species. Mason et al. (2019) produce a multi-gene phylogeny of the eleven species, which they use to determine the biogeographical patterns of their evolution.

Bothriechis aurifer (Salvin, 1860)
Bothriechis bicolor (Bocourt, 1868)

Bothriechis guifarroi Townsend, Medina-Flores, L.D. Wilson, Jadin & Austin, 2013b. ZooKeys 298: 85–91, figs. 2, 3, 5, 6.

Holotype: UTA R60303, a 734 mm male (E. Aguilar, A. Contreras, L. Gray, L.A. Herrera-B., M. Medina-Flores, A. Portillo, A. Stubbs & J.H. Townsend, 25 July 2010).

Type locality: "La Liberacíon…, 15.5302°N, 87.2939°W (lDD), 1,015 m elevation, Refugio de Vida Silvestre Texíguat, Departamento de Atlántida, Honduras."

Distribution: Honduras (Atlantida), 1015–1450 m.

Bothriechis lateralis W.C.H. Peters, 1862

Distribution: Add Panama (Panama Oeste), Ray & Ruback (2015).

Bothriechis marchi (Barbour & Loveridge, 1929)

Type locality: Initially stated as probably "El Oro, Municipio de Quimistán in the Sierra del Espíritu Santo to the northwest of the town of Quimistán," L.D. Wilson & McCranie (1992).

Distribution: Add Guatemala (Zacapa), Townsend et al. (2013b). Remove Honduras (Atlantida) populations to *B. guifarroi*, Townsend et al. (2013b).

Bothriechis nigroviridis W.C.H. Peters, 1859

Distribution: Add Panama (Bocas del Toro), Doan et al. (2016).

Comments: Doan et al. (2016) discuss geographic variation.

Bothriechis nubestris Doan, Mason, Castoe, Sasa & Parkinson, 2016. Zootaxa 4138(2): 277–278, figs. 4, 5, 7.

Holotype, UTA R9637, a 911 mm female (P. Seigfried, October 1973).

Type locality: "San Isidro de El General, Province of San José, Costa Rica; approximately 3000 m."

Distribution: Costa Rica (Cartago, Limón, San José), 2400–3000+ m.

Comments: *B. nubestris* occupies northern and central parts of the Cordillera de Talamanca, populations formerly assigned to *B. nigroviridis*.

Bothriechis rowleyi (Bogert, 1968)

Distribution: Grünwald et al. (2016a) provide additional records from Chiapas of this uncommon snake.

Bothriechis schlegelii (Berthold, 1845)

Distribution: Add Mexico (Oaxaca), Wylie & Grünwald (2016); Nicaragua (Boaco), Obando & Sunyer (2016b); Panama (Panama Oeste), Ray & Ruback (2015); Colombia (Cesar, Chocó, Córdoba, Cundinamarca, Huila, Magdalena, Meta, Nariño, Risaralda, Valle del Cauca), and increase elevation to 2970 m, C. Gómez & Buitrago-González (2017); Colombia (Quindio), Quintero-Ángel et al. (2012); Ecuador (Azuay, Canar, Cotopaxi, Imbabura, Los Rios, Loja, Santo Domingo), Valencia et al. (2016).

Bothriechis supraciliaris (Taylor, 1954)

Bothriechis thalassinus Campbell & E.N. Smith, 2000

Distribution: Add Guatemala (Zacapa), Townsend et al. (2013b).

BOTHROCHILUS Fitzinger, 1843 (Pythonidae)

Comments: See under *Leiopython*. Barker et al. (2015) reject the synonymy of *Leiopython* with *Bothrocheilus*, as proposed by Reynolds et al. (2014) as a conservative arrangement, largely based on observed morphological differences, and until genetic material is analyzed for four of the species of *Leiopython*.

Bothrochilus boa (Schlegel, 1837)

BOTHROCOPHIAS Gutberlet & Campbell, 2001 (Viperidae: Crotalinae)

Bothrocophias andianus Amaral, 1923

Comments: Carrasco et al. (2019), Timms et al. (2019) include *B. andianus* within *Bothrocophias*.

Bothrocophias campbelli (Freire-Lascano, 1991)

Distribution: Add Ecuador (Esmaraldas, Santo Domingo), Valencia et al. (2016).

Bothrocophias colombianus (Rendahl & Vestergren, 1940)

Bothrocophias hyoprora (Amaral, 1935)

Distribution: Add Brazil (Mato Grosso, Pará), Carvalho et al. (2013). Upper elevation of 1200 m, Valencia et al. (2016). L.S. Oliveira et al. (2018) map known localities.

Comment: Carvalho et al. (2013) provide range records and describe hemipenial morphology and chromosomes.

Bothrocophias lojana (Parker, 1930) **new combination**

Distribution: Add Peru (Cajamarca), Carrasco et al. (2016); Ecuador (Azuay, El Oro), and elevation range of 600–2900 m, Valencia et al. (2016).

Comments: Alencar et al. (2016), Timms et al. (2019) and Zaher et al. (2019) all resolve *Bothrops lojanus* within *Bothrocophias* based DNA sequence data.

Bothrocophias microphthalmus (Cope, 1875)

Distribution: Add Brazil (Rondônia), Bernarde et al. (2012b); Peru (Cajamarca, Huanuco, San Martin), Carrasco et al. (2016); Peru (Puno), Llanqui et al. (2019, as *B.* cf. *microphthalmus*). Low elevation of 720 m, Valecia et al. (2016).

Bothrocophias myersi Gutberlet & Campbell, 2001

BOTHROLYCUS Günther, 1874 (Lamprophiidae)

Bothrolycus ater Günther, 1874

Distribution: Upper elevation of 1900 m, Ineich et al. (2015).

BOTHROPHTHALMUS W.C.H. Peters, 1863 (Lamprophiidae)

Bothrophthalmus brunneus Günther, 1863

Bothrophthalmus lineatus (W.C.H. Peters, 1863)

BOTHROPS Wagler *in* Spix, 1824 (Viperidae: Crotalinae)

Synonyms: *Bothriopsis* W.C.H. Peters, 1861, *Rhinocerophis* Garman, 1881, *Bothropoides* Fenwick, Gutberlet, Evans & Parkinson, 2009.

Distribution: Middle and South America.

Comments: Alencar et al. (2016) resolve a monophyletic *Bothrops* that includes clades for the genera *Bothriopsis*, *Bothropoides*, *Bothrops*, *Rhinocerophis*, with *B. pictus* as sister taxon to the included clades. Retention of the four genera would require a new genus for *B. pictus*. A similar result is obtained by Timms et al. (2019) and Zaher et al. (2019) who both recover *B. lojanus* within *Bothrocophias*. We combine the four monophyletic clades, with *B. pictus*, within *Bothrops* as recently recognized, and export *lojanus* to *Bothrocophias*. Carrasco et al. (2019) present a DNA-sequence based phylogeny of the *Bothropoides* clade.

Bothrops alcatraz (O.A.V. Marques, Martins & Sazima, 2002)

Bothrops alternatus A.M.C. Duméril, Bibron & Duméril, 1854

Distribution: Add Brazil (Distrito Federal), Carrasco et al. (2019); Brazil (Mato Grosso do Sul), Ferreira et al. (2017); Brazil (Marinheiros I., Rio Grande do Sul), Quintela et al. (2011).

Fossil record: Late Pleistocene of Argentina (Albino & Brizuela, 2014).

Bothrops ammodytoides Leybold, 1873

Bothrops asper (Garman, 1884)

Distribution: Add Mexico (Puebla), Saldarriaga-Córdoba et al. (2009); Belize (Cayo), Saldarriaga-Córdoba et al. (2009); Guatemala (Alta Verapaz, Escuintla, Izabal, Quetzaltenango), Saldarriaga-Córdoba et al. (2009); Honduras (Copan, Toro), Saldarriaga-Córdoba et al. (2009); Nicaragua (Rivas), Sunyer et al. (2014) and Martínez-Fonseca (2016b); Nicaragua (Zelaya), Saldarriaga-Córdoba et al. (2009); Panama (Panama, Panama Oeste), Ray & Ruback (2015); Colombia (Antioquia, Arauco, Atlantico, Bolivar, Boyaca, Caldas, Cauca, Cesar, Chocó, Magdalena, Nariño, Santander, Sucre, Tolima, Valle del Cauca), Díaz-Ricaurte et al. (2018a); Colombia (Córdoba), Carvajal-Cogollo et al. (2007); Colombia (Norte de Santander), Armesto et al. (2011); Ecuador (Manabi), Almendáriz et al. (2012), Ecuador (Azuay, Bolivar, Carchi, Cotopaxi, El Oro, Esmeraldas, Guayas, Loja, Los Rios, Manabí), Cisneros-Heredia & Touzet (2004); Ecuador (Canar, Imbabura, Santa Elena, Santo Domingo), Valencia et al. (2016). Farr & Lazcano

(2017) detail records for Tamaulipas, Mexico. Díaz-Ricaurte et al. (2018a) detail records for Colombia.

Comments: Saldarriaga-Córdoba et al. (2009) find that morphological variation partition both geographically and/or ecologically. A comparison with mtDNA sequence data finds no correspondence with the morphological data. The mtDNA data partition into two clades that diverge northward and southward in Costa Rica.

Bothrops atrox (Linnaeus, 1758)

Distribution: Add Colombia (Amazonas, Arauca, Caqueta, Casanare, Cundinamarca, Guainia, Guaviare, Meta, Putumayo, Vaupes, Vichada), and upper elevation of 1826 m, Quiñones-Betancourt et al. (2018); Ecuador (Sucumbios), Valencia et al. (2016); Peru (Cusco, Pasco), Carrasco et al. (2019); Peru (Huanuco), Schlüter et al. (2004); Brazil (Ceara), Guedes et al. (2014); Brazil (Mato Grosso), Carrasco et al. (2019). Quiñones-Betancourt et al. (2018) detail records for Colombia.

Comments: The synonym *Trigonocephalus colombiensis* Hallowell, 1845 is found in the phylogenies of Alencar et al. (2016) and Zaher et al. (2019) as a revived *Bothrops colombiensis*. The authors do not provide a rationale for recognizing *colombiensis*. Bill Lamar (pers. comm.) states that the syntypes of *colombiensis* are bleached to being patternless, and cannot be arguably associated with a taxonomically distinct population.

Bothrops ayerbei Folleco-Fernández, 2010

Comments: Ramírez-Chaves & Solari (2014) argue that *B. ayerbei* is not a valid name because it was published in an online version of a journal prior to 2011, which is not a valid publication according to the International Code of Zoological Nomenclature.

Bothrops barnetti Parker, 1938

Comment: Based on a single mtDNA gene, Alencar et al. (2016) resolve *Bothrops barnetti* within the *Bothropoides* clade.

Bothrops bilineatus (Wied-Neuwied, 1821)

Holotype: AMNH R4006, a 585 mm male (Prince M. du Wied), according to Vanzolini & Myers (2015).

Distribution: Add Ecuador (Sucumbios), Valencia et al. (2016); Peru (Cusco), and upper elevation of 2000 m, Venegas et al. (2019); Peru (Huanuco), Schlüter et al. (2004); Peru (Amazonas, San Martin), Carrasco et al. (2019); Peru (Puno), Llanqui et al. (2019); Brazil (Acre, Amapá, Amazonas, Roraima), Bernarde et al. (2011a); Brazil (Minas Gerais), Feio & Caramaschi (2002); Brazil (Pernambuco), Roberto et al. (2018). Upper elevation of 1060 m, Moraes et al. (2017).

Bothrops brazili Hoge, 1954

Distribution: Add Peru (Cusco, Ucayali), Carrasco et al. (2019); Ecuador (Orellana, Pastaza), and upper elevation of 1000 m, Valencia et al. (2016).

Bothrops caribbaeus (Garman, 1887)
Bothrops chloromelas (Boulenger, 1912)

Bothrops cotiara (Gomes, 1913)

Bothrops diporus Cope, 1862

Distribution: Add Bolivia (Chuquisaca, Santa Cruz), Carrasco et al. (2019); Bolivia (Tarija), Ocampo & Fernandez (2014); Paraguay (Caaguazu, Cordillera), Carrasco et al. (2019). Found in the following departments of Paraguay: Alto Paraguay, Boqueron, Caazapa, Central, Itapua, Paraguari and Presidente Hayes according to Cabral & Weiler (2014).

Bothrops erythromelas Amaral, 1923

Bothrops fonsecai Hoge & Belluomini, 1959

Bothrops insularis (Amaral, 1922)

Bothrops isabelae Sandner-Montilla, 1979

Bothrops itapetiningae (Boulenger, 1907)

Distribution: Add Brazil (Distrito Federal), Carrasco et al. (2019).

Comments: Leão et al. (2014) discuss morphology and variation.

Bothrops jararaca (Wied-Neuwied, 1824)

Distribution: Add Brazil (Rio de Janeiro: Ilha Grande), C.F.D. Rocha et al. (2018); Paraguay (Canindeyu), Cabral & Weiler (2014). Upper elevation of 1027 m, Guedes et al. (2014).

Bothrops jararacussu Lacerda, 1884

Bothrops jonathani Harvey, 1994

Bothrops lanceolatus (Bonnaterre, 1790)

Bothrops leucurus Wagler *in* Spix, 1824

Distribution: Add Brazil (Paraíba), R. França et al. (2012). Elevation confirmed to 900 m, Guedes et al. (2014).

Bothrops lutzi (Miranda-Ribeiro, 1915)

Distribution: Upper elevation of 847 m, Guedes et al. (2014). Moura et al. (2013b) map known localities.

Bothrops marajoensis Hoge, 1966

Bothrops marmoratus V. Silva & Rodrigues, 2008

Comment: Based on two mtDNA genes, Alencar et al. (2016) resolve *Bothrops marmoratus* within the *Bothropoides* clade, and using additional genes, Carrasco et al. (2019) obtain the same result.

Bothrops mattogrossensis Amaral, 1925

Distribution: Add Bolivia (Potosi), Rivas et al. (2018); Paraguay (Caaguazu, Central, Concepcion, Itapua, Neembucu, Paraguari), Cabral & Weiler (2014).

Bothrops medusa (Sternfeld, 1920)

Bothrops monsignifer Timms, Chaparro, Venegas, Salazar-Valenzuela, Scrocchi, Cuevas, Leynaud & Carrasco, 2019. Zootaxa 4656(1): 102–113, figs. 1, 5–9.

Holotype: MNK 5556, a 1280 mm female (locals, 11 March 2017).

Type locality: "13 km southwest to Refugio Los Volcanes (18°11′51.10″S, 63°40′5.95″W; 1658 m above sea level…), Cuevas Ecological Center, province of Florida, department of Santa Cruz, Bolivia."

Distribution: SE Peru (Puno), W Bolivia (La Paz, Santa Cruz), 891–2133 m.

Bothrops moojeni Hoge, 1966

Bothrops muriciensis Ferrarezzi & Freire, 2001

Bothrops neuwiedi Wagler *in* Spix, 1824

Distribution: Add Brazil (Tocantins), Dal Vechio et al. (2016). F.G.R. França et al. (2012) map the distribution.

Bothrops oligolepis (F. Werner, 1901)

Distribution: Add Peru (Pasco), Carrasco et al. (2019).

Bothrops osbornei Freire-Lascano, 1991

Distribution: Add Ecuador (Santo Domingo), and low elevation of 400, Valencia et al. (2016).

Bothrops otavioi Barro, Grazziotin, Sazima, Martins & Sawaya, 2012

Bothrops pauloensis Amaral, 1925

Distribution: Add Brazil (Tocantins), Dal Vechio et al. (2016).

Bothrops pictus (Tschudi, 1845)

Comments: Based on mtDNA sequence data, Alencar et al. (2016) resolve *B. pictus* as the sister taxon to *Bothrops* sensu lato, which would require a new genus for *B. pictus* in order to recognize *Bothriopsis*, *Bothropoides*, *Bothrops* and *Rhinocerophis*.

Bothrops pirajai Amaral, 1923

Distribution: elevation modified to 88–835 m, M.A. Freitas et al. (2014a).

Comments: M.A. Freitas et al. (2014a) present a description and distribution map.

Bothrops pubescens (Cope, 1870)

Distribution: Add Argentina (Corrientes, Entre Rios, Misiones), Carrasco et al. (2019, as *B.* aff. *pubescens*); Brazil (Parana, Santa Catarina), Carrasco et al. (2019, as *B.* aff. *pubescens*).

Bothrops pulchra (W.C.H. Peters, 1862)

Distribution: Add Ecuador (Pichincha, Tungurahua, Zamora-Chinchipe), Valencia et al. (2016).

Bothrops punctatus (García, 1896)

Distribution: Add Ecuador (Carchi, Santo Domingo), Valencia et al. (2016); Ecuador (Pichincha), Arteaga et al. (2013). Low elevation of 0 m, Vera-Pérez et al. (2018).

Comments: Ospina-L. (2017) provides a species account for Colombia.

Bothrops rhombeatus García, 1896

Comments: Ramírez-Chaves & Solari (2014) argue that *B. rhombeatus* is a *nomen dubium* because there is no known type material, and the original description does not exclude it from being *B. asper*. They recommend assigning a neotype to fix the name to populations as currently recognized.

Bothrops roedingeri Mertens *in* Titschack, 1942

Bothrops sanctaecrucis Hoge, 1966

Distribution: Add Bolivia (La Paz) and upper elevation of 1935 m, Miranda-Calle & Aguilar-Kirigin (2011).

Bothrops sazimai Barbo, Gasparini, Almeida, Zaher, Grazziotin, Gusmão, Ferrarini & Sawaya, 2016. Zootaxa 4097(4): 516–521, figs. 2–5.

Holotype: MZUSP 22228, a 738 mm male ("our team," 14 May 2013).

Type locality: "Ilha dos Franceses (20°55'36"S, 40°45'15"W), municipality of Itapemirim, Itaoca beach, state of Espírito Santo, Brazil."

Distribution: Brazil (Ilha dos Franceses, Espírito Santo).

Bothrops sonene Carrasco, Grazziotin, Farfán, Koch, Ochoa, Scrocchi, Leynaud & Chaparro, 2019. Zootaxa 4565(3): 310–314, figs. 4–6. (*Bothrops sonene*)

Holotype: MUBI 12319, a 1073 mm female (J.A. Ochoa, 23 April 2013).

Type locality: "Pampa Juliaca (12°57'46.79"S, 68°55'22.61"W; 206 m above sea level…), Pampas del Heath (Bahuaja-Sonene National Park), district of Tambopata, province of Tambopata, department of Madre de Dios, Peru."

Distribution: SE Peru (Madre de Dios), 206–210 m.

Bothrops taeniatus Wagler *in* Spix, 1824

Distribution: Add Ecuador (Chimborazo, Pichincha, Sucumbios), Valencia et al. (2016); Peru (Cusco, Junin), Carrasco et al. (2019); Brazil (Acre), J.R.D. Souza et al. (2013).

Bothrops venezuelensis Sandner-Montilla, 1952

†BOTROPHIS Rochebrune, 1880 (Booidea: incerta sedis)

†Botrophis gaudryi Rochebrune, 1880

BOULENGERINA Dollo, 1886 (Elapidae)

Comments: Wüster et al. (2018) produce a DNA sequence-based phylogeny, which, combined with morphological data, resolve four species within *B. melanoleuca*. See under *Naja*.

Boulengerina annulata (Buchholz & Peters *in* W.C.H. Peters, 1876)

Distribution: Add Angola (Lunda Norte), Branch (2018).

Boulengerina christyi Boulenger, 1904

Boulengerina guineensis Broadley, Trape, Chirio, Ineich & Wüster *in* Wüster et al., 2018. Zootaxa 4455(1): 79–81, fig. 5. (*Naja* [*Boulengerina*] *guineensis*)

Holotype: MNHN 1921.0485, a 2220 mm male (P. Chabanaud, 1919–1921).

Type locality: "N'Zébéla, Macenta Perfecture, Nzérékoré region of forested southeastern Guinea (8° 05'N, 9° 05'W), elev. 490 m."

Distribution: W Africa. At least Guinea (Nzerekore), Guinea Bissau, Sierra Leone (Eastern, Northern, Southern), Liberia (Bong, Grand Gedeh, Montserrado, Margibi, Nimba), Ivory Coast, Ghana (Eastern, Volta, Western), W Togo.

Boulengerina melanoleuca (Hallowell, 1857)

Synonyms: Remove *Naja melanoleuca subfulva* Laurent, 1955), *Naja melanoleuca aurata* Stucki-Stirn, 1979.

Distribution: C Africa. At least SE Ghana, S Benin (Kouffo), Nigeria (Ibadan), Cameroon (Central, Est, Sud-Ouest), Central African Republic, Gabon (Estuaire, Haut-Ogooue, Moyen-Ogooue, Ngounie, Ogooue-Ivindo, Ogooue-Maritime, Woleu-Ntem), Congo (Likouala), Democratic Republic of

Congo (Orientale, South Kivu), Angola (Benguela, Cabinda, Cuanza Norte, Cuanza Sul, Huambo, Huila, Lunda Norte, Malanje, Namibe). Add Nigeria (Gongola), Nneji et al. (2019); Equatorial Guinea, Gabon (Nyanga), Pauwels et al. (2017a); Angola (Namibe), Ceríaco et al. (2016). Eliminate island of São Tomé from the range, Ceríaco et al. (2017). Eliminate bulk of far West Africa, plus most of eastern and southern Africa, Wüster et al. (2018).

Comments: Wüster et al. (2018) restrict the definition and range after splitting portions into other species: *B. guineensis*, *B. savannula*, *B. subfulva*. They provide a revised diagnosis, color illustrations, and a map of localities.

Boulengerina multifasciata (F. Werner, 1902)

Boulengerina peroescobari Ceríaco, Marques, Schmiz & Bauer, 2017. Zootaxa 4324(1): 130–132, fig. 4. (*Naja* [*Boulengerina*] *peroescobari*)

Holotype: MUHNAC/MB03-001065, a 2080 mm male (L. Ceríaco, M. Marques & A.C. Sousa, 24 February 2016).

Type locality: "the vicinity of Praia Inhame (0.028636° N, 6.523203° E, WGS-84; 17 m above sea level), São Tomé, Republic of São Tomé e Príncipe."

Distribution: Republic of São Tomé e Príncipe (São Tomé), 17–281 m.

Boulengerina savannula Broadley, Trape, Chirio & Wüster *in* Wüster et al., 2018. Zootaxa 4455(1): 81–83, fig. 7. (*Naja* [*Boulengerina*] *savannula*)

Holotype: MNHN 2018.0002, a 2191 mm male (T. Chirio, 31 March 2007).

Type locality: "Niénié, W Biosphere Reserve, Benin, 11.05920 °N, 2.20488 °E, elevation 272 m."

Distribution: W Africa. At least Senegal (Dakar, Fatick, Kolda, Tambacounda, Thies), Gambia, Guinea (Kindia), Ivory Coast, Ghana (Eastern), E Togo, Benin (Atakora), SW Mali (Kayes, Koulikoro, Sikasso), Niger (Niamey), Chad (Logone Oriental), Nigeria (Kaduna), N Cameroon (Extreme Nord). Add Senegal (Kedougou), Monasterio et al. (2016); Senegal (Ziguinchor), Mané & Trape (2017); Niger (Niamey), Trape & Mané (2015).

Boulengerina subfulva Laurent, 1955. Rev. Zool.-Bot. Afric. 51(1–2): 132–135. (*Naja melanoleuca subfulva*)

Synonyms: *Naja melanoleuca aurata* Stucki-Stirn, 1979.

Holotype: MRAC 17514, female (Van der Borght, May 1953).

Type locality: "Lwiro), 1850 m, Terr. de Kabare, Kivu," Democratic Republic of the Congo.

Distribution: C and S Africa. At least SE Nigeria (Cross Rivers), Cameroon (Nord-Ouest), Chad (Lac), Congo (Likouala), Central African Republic (Haut-Mbomou, Ombella M'Poko), South Sudan (Bahr Al Ghazal, Eastern Equatoria), W Ethiopia (Gemu Gofa, Illubabor, Kefa, Shoa), S Somalia (Jubbada Hoose), Kenya (Central, Coast, Eastern, Nyanza, Rift Valley, Western), Uganda (Central, Eastern, Northern, Western, and Lake Victoria Islands of Busi, Massambwa, Sanga and Sese), E Democratic Republic of Congo (Katanga, Maniema, Oriental, South Kivu), Rwanda, Burundi, Tanzania (Arusha,

Kigoma, Kilimanjaro, Lindi, Mtwara, Mwanza, Shinyanga, Tanga, and Islands of Kagera, Mafia, Zanzibar), Mozambique (Gaza, Inhambane, Manica, Maputo, Sofala, Tete, Zambezia, and Islands of Bazaruto, Inhaca, Portuguesa), Malawi (Central, Northern, Southern), Zambia (Copperbelt, Central, Eastern, Luapula, Northern, North-Western), Angola (Cuanza Norte, Cuanza Sul, Malanje), E Zimbabwe (Manicaland), South Africa (KwaZulu-Natal).

Comments: Wüster et al. (2018) recognize this species as distinct from *B. melanoleuca* based on morphological and genetic uniqueness. They provide a diagnosis, color photographs and map of localities.

BRACHYOPHIS Mocquard, 1888 (Atractaspididae: Aparallactinae)

Brachyophis cornii Scortecci, 1932
Brachyophis krameri Lanza, 1966
Brachyophis revoili Mocquard, 1888

BRACHYORRHOS Kuhl *in* Schlegel, 1826 (Homalopsidae)

Comments: J.C. Murphy & Voris (2014) provide a generic diagnosis and key to the species.

Brachyorrhos albus (Linnaeus, 1758)

Comments: J.C. Murphy & Voris (2014) provide a description and illustrations.

Brachyorrhos gastrotaenius (Bleeker, 1860)

Type locality: Ambon is in error, the correct locality being Buru, J.C. Murphy and Voris (2014).

Distribution: limited to the island of Buru, Indonesia, J.C. Murphy and Voris (2014).

Comments: J.C. Murphy & Voris (2014) provide a description and illustrations of the holotype.

Brachyorrhos raffrayi (Sauvage, 1879)

Comments: J.C. Murphy & Voris (2014) provide a description and illustrations of the holotype.

Brachyorrhos wallacei J.C. Murphy, Mumpuni, Lang, Gower & Sanders, 2012

Comments: J.C. Murphy & Voris (2014) provide a description and illustrations of the holotype.

BRACHYUROPHIS Günther, 1863 (Elapidae)

Brachyurophis approximans (Glauert, 1954)
Brachyurophis australis (Krefft, 1864)

Brachyurophis campbelli Kinghorn, 1929. Rec. Australian Mus. 17(4): 191, fig. 2. (*Rhynchoelaps campbelli*)

Synonyms: *Rhynchelaps woodjonesii* Thomson, 1934.

Holotype: AM R9387, a 140 mm specimen (W.D. Campbell, December 1928).

Type locality: "Almaden, Queensland," Australia.

Distribution: NE Australia (N Queensland).

Comments: Cogger (2014) recognizes *B. campbelli* as a species distinct from *B. roperi* and *B. semifasciatus*, stating that there is no convincing evidence to recognize them as conspecific.

Brachyurophis fasciolatus (Günther, 1872)

Brachyurophis incinctus (Storr, 1968)

Distribution: northward range extension in Queensland, Australia, Schembri & Jolly (2017).

Brachyurophis morrisi (Horner, 1998)

Brachyurophis roperi (Kinghorn, 1931)

Synonyms: Remove *Brachyurophis campbelli* Kinghorn, 1929, *Rhynchelaps woodjonesii* Thomson, 1934.

Distribution: Limited to NC Australia (N Northern Territory, N Western Australia).

Comments: See under *B. campbelli*.

Brachyurophis semifasciatus Günther, 1863

†*BRANSATERYX* Hoffstetter & Rage, 1972 (Erycidae)

†*Bransateryx vireti* Hoffstetter & Rage, 1972

BRYGOPHIS Domergue & Bour, 1988
(Pseudoxyrhophiidae)

Brygophis coulangesi (Domergue, 1988)

BUHOMA Ziegler, Vences, Glaw & Böhme, 1997 (Elapidae)

Comments: Zaher et al. (2019), using DNA sequence data, resolve *Buhoma* and Indian *Calliophis* species in a novel position as sister taxa to all Elapidae.

Buhoma depressiceps (F. Werner, 1897)

Buhoma procterae (Loveridge, 1922)

Buhoma vauerocegae (Tornier, 1902)

BUNGARUS Daudin, 1803 (Elapidae)

Bungarus andamanensis Biswas & Sanyal, 1978

Bungarus bungaroides (Cantor, 1839)

Bungarus caeruleus (Schneider, 1801)

Distribution: Add Nepal (Parsa), Bhattarai et al. (2018b); Bhutan (Zhemgang), Tshewang & Letro (2018); Bangladesh (Barisal, Chittagong, Dhaka, Khulna, Mymensingh, Rajshahi, Rangpur, Sylhet), Ahsan & Rahman (2017); India (Goa), Chowdhury & Chaudhuri (2017); India (Gujarat), and low elevation of 37 m, Vyas (2014); India (Nagaland), Yanthungbeni et al. (2018); India (Odisha), Debata (2017).

Bungarus candidus (Linnaeus, 1758)

Distribution: Add Thailand (Chanthaburi), Chan-ard et al. (2011); Cambodia (Ratanikiri), S.N. Nguyen et al. (2017a); Cambodia (Siem Reap), Geissler et al. (2019); Vietnam (Binh Thuan, Quang Ngai), S.N. Nguyen et al. (2017a).

Comments: S.N. Nguyen et al. (2017a) examine several *Bungarus* specimens from Vietnam that are morphologically similar to *B. magnimaculatus*, but based on DNA sequence-data are referable to *B. candidus*.

Bungarus ceylonicus Günther, 1864

Bungarus fasciatus (Schneider, 1801)

Distribution: Add India (Chhattisgarh, Jharkhand), Srinivasulu et al. (2009); Nepal (Nawalparasi), Pandey (2012); Bangladesh (Barisal, Chittagong, Dhaka, Khulna, Mymensingh, Rajshahi, Rangpur, Sylhet), Ahsan & Rahman (2017); Myanmar (Tanintharyi), Platt et al. (2018).

Bungarus flaviceps J.T. Reinhardt, 1843

Distribution: Add West Malaysia (Kedah), Shahriza et al. (2013); West Malaysia (Pahang), Zakaria et al. (2014).

Comments: Zaher et al. (2019) resolve *B. flaviceps* as sister taxon to *Sinomicrurus*, *Micruroides* and *Micrurus*, in a separate subfamily from other *Bungarus* species.

Bungarus lividus Cantor, 1839

Distribution: Add Nepal (Parsa), Bhattarai et al. (2018b); Bhutan (Trongsa), Tshewang & Letro, 2018); Bangladesh (Mymensingh, Rangpur), Ahsan & Rahman (2017).

Bungarus magnimaculatus Wall & Evans, 1901

Distribution: Add Myanmar (Magway), S.N. Nguyen e al. (2017a).

Bungarus multicinctus Blyth, 1860

Distribution: Add Vietnam (Hai Phong: Cat Ba Island), T.Q. Nguyen et al. (2011).

Bungarus niger Wall, 1908

Distribution: Add Bangladesh (Chittagong, Dhaka, Khulna, Mymensingh, Sylhet), Ahsan & Rahman (2017).

Bungarus persicus Abtin, Nilson, Mobaraki, Hosseini & Dehgannejhad, 2014. Russian J. Herpetol. 21(4): 244–247, figs. 2, 3, 5–10.

Holotype: ZMGU 3121, a 630 mm female (E. Abtin, 29 September 2013).

Type locality: "north of Sarbaz, Baluchistan, Iran."

Distribution: SE Iran (Sistan and Baluchistan) and SW Pakistan (Baluchistan).

Bungarus sindanus Boulenger, 1897

Distribution: Add Afghanistan (Nangahar, Paktika), and upper elevation of 1200 m, Wagner et al. (2016b).

Bungarus slowinskii Kuch, Kizirian, Truong, Lawson, Donnelly & Krebs, 2005

Distribution: Add Thailand (Nan), Smits & Hauser (2019).

Bungarus walli Wall, 1907

Distribution: Add Bangladesh (Barisal, Dhaka, Khulna, Rajshahi, Rangpur), Ahsan & Rahman (2017).

Bungarus wanghaotingi Pope, 1928

CAAETEBOIA Zaher, Grazziotin, Cadle, R.W. Murphy, Moura-Leite & Bonatto, 2009 (Dipsadidae: Xenodontinae)

Caaeteboia amarali (Wettstein, 1930)

CACOPHIS Günther, 1863 (Elapidae)

Cacophis churchilli Wells & Wellington, 1985

Cacophis harriettae Krefft, 1869

Cacophis krefftii Günther, 1863

Cacophis squamulosus (A.M.C. Duméril, Bibron & Duméril, 1854)

†*CADURCERYX* Hoffstetter & Rage, 1972 (Erycidae)
†*Cadurceryx filholi* Hoffstetter & Rage, 1972

†*CADURCOBOA* Rage, 1978 (Booidea: incerta sedis)
†*Cadurcoboa insolita* Rage, 1978

CALABARIA Gray, 1858 (Calabariidae)
Calabaria reinhardtii (Schlegel, 1848)
Distribution: Add Sierra Leone, Trape & Baldé (2014); Gabon (Woleu-Ntem), Pauwels et al. (2017b); Angola (Cabinda), M.P. Marques et al. (2018).

†*CALAMAGRAS* Cope, 1873 (Booidea: incerta sedis)
Comments: K.T. Smith (2013) concludes that *C. weigeli* belongs to the Ungaliophiidae, and recommends re-examination of the type species, *C. murivorus*, in order to determine whether or not *C. weigeli* warrants a new genus.

†*Calamagras angulatus* Cope, 1873
†*Calamagras floridanus* Auffenberg, 1963
†*Calamagras gallicus* Rage, 1977
†*Calamagras murivorus* Cope, 1873
†*Calamagras platyspondyla* Holman, 1976
†*Calamagras primus* Hecht in McGrew, 1959
†*Calamagras turkestanicus* Danilov & Averianov, 1999

†*Calamagras weigeli* Holman, 1972
Distribution: Add late Eocene (Chadronian) of USA (North Dakota), K.T. Smith (2013).

CALAMARIA H. Boie *in* F. Boie, 1827 (Calamariidae)
Calamaria abramovi Orlov, 2009
Calamaria abstrusa Inger & Marx, 1965
Calamaria acutirostris Boulenger, 1896

Calamaria albiventer (Gray, 1835 *in* Gray & Hardwicke, 1830–1835)
Distribution: Add Malaysia (Malacca), Quah et al. (2018a).
Comments: Quah et al. (2018a) describe a recently collected specimen, and discuss previous records.

Calamaria alidae Boulenger, 1920

Calamaria andersoni Yang & Zheng, 2018. Copeia 106(3): 486–488, figs. 2, 3.
Holotype: SYS R001699, a 351 mm male (J.-H. Yang, 6 May 2016).
Type locality: "China, Yunnan Province, Yingjiang County, Tongbiguan Town, 24°32′53.30″N, 97°35′59.10″E, approx. 1520 m above sea level."
Distribution: SE China (W Yunnan), 1520 m. Known only from the holotype.

Calamaria apraeocularis M.A. Smith, 1927
Calamaria banggaiensis Koch, Arida, McGuire, Iskandar & Böhme, 2009
Calamaria battersbyi Inder & Marx, 1965
Calamaria bicolor A.M.C. Duméril, Bibron & Duméril, 1854
Calamaria bitorques W.C.H. Peters, 1872
Calamaria boesemani Inger & Marx, 1965

Calamaria borneensis Bleeker, 1860
Calamaria brongersmai Inger & Marx, 1965
Calamaria buchi Marx & Inger, 1955
Calamaria butonensis Howard & Gillespie, 2007
Calamaria ceramensis Rooij, 1913
Calamaria concolor Orlov, Nguyen, Nguyen, Ananjeva & Ho, 2010
Calamaria crassa Lidth de Jeude, 1922
Calamaria curta Boulenger, 1896
Calamaria doederleini Gough, 1902

Calamaria dominici Ziegler, Tran, Babb, Jones, Moler, Van Devender & Nguyen, 2019d. Rev. Suisse Zool. 126 (1): 17–23, figs. 1–4.
Holotype: IEBR A2018.1, a 421 mm female (A.V. Tran, 28 May 2017).
Type locality: "Ta Dung Nature Reserve, Dak Nong Province, Central Highlands, Vietnam, at an elevation of 1240 m asl."
Distribution: C Vietnam (Dak Nong), 1240 m. Known only from the holotype.

Calamaria eiselti Inger & Marx, 1965
Calamaria everetti Boulenger, 1893
Calamaria forcarti Inger & Marx, 1965
Calamaria gervaisii A.M.C. Duméril, Bibron & Duméril, 1854
Distribution: Add Philippines (Camiguin Norte, Masbate Is.), Leviton et al. (2018); Philippines (Carabao I.), Siler et al. (2012).

Calamaria gialaiensis Ziegler, Nguyen & Nguyen, 2009
Calamaria gimlettii Boulenger, 1905
Calamaria grabowskyi J.G. Fischer, 1885
Calamaria gracillima Günther, 1872
Calamaria griswoldi Loveridge, 1938
Calamaria hilleniusi Inger & Marx, 1965
Calamaria ingeri Grismer, Kaiser & Yaakob, 2004
Calamaria javanica Boulenger, 1891
Calamaria joloensis Taylor, 1922
Calamaria lateralis Mocquard, 1890
Calamaria lautensis Rooij, 1917
Calamaria leucogaster Bleeker, 1860
Calamaria linnaei H. Boie *in* F. Boie, 1827
Calamaria longirostris Howard & Gillespie, 2007
Calamaria lovii Boulenger, 1887

Calamaria lumbricoidea H. Boie *in* F. Boie, 1827
Distribution: Add Malaysia (Penang), Quah et al. (2019b); Philippines (Biliran I.), Leviton et al. (2018); Philippines (Camiguin), Sanguila et al. (2016).

Calamaria lumholtzii Andersson, 1923
Calamaria margaritophora Bleeker, 1860
Calamaria mecheli Schenkel, 1901
Calamaria melanota Jan, 1865 in Jan & Sordelli, 1860–1866
Calamaria modesta A.M.C. Duméril, Bibron & Duméril, 1854
Calamaria muelleri Boulenger, 1896
Calamaria nuchalis Boulenger, 1896
Calamaria palavanensis Inger & Marx, 1965

Calamaria pavimentata A.M.C. Duméril, Bibron & Duméril, 1854

Distribution: Add India (Mizoram), Vogel et al. (2017); Thailand (Nakhon Si Thammarat), Pauwels & Chan-ard (2016); Vietnam (Bac Giang), Hecht et al. (2013); Vietnam (Quang Ngai), Nemes et al. (2013).

Calamaria prakkei Lidth de Jeude, 1893

Calamaria rebentischi Bleeker, 1860

Calamaria sangi Q.T. Nguyen, Koch & Ziegler, 2010

Calamaria schlegeli A.M.C. Duméril, Bibron & Duméril, 1854

Calamaria schmidti Marx & Inger, 1955

Calamaria septentrionalis Boulenger, 1890

Distribution: Add China (Nan Ao Island), Qing et al. (2015); Vietnam (Bac Giang), Hecht et al. (2013).

Calamaria suluensis Taylor, 1922

Calamaria sumatrana Edeling, 1870

Calamaria thanhi Ziegler & Le, 2005

Calamaria ulmeri Scackett, 1940

Calamaria virgulata H. Boie in F. Boie, 1827

Calamaria yunnanensis Chernov, 1962

CALAMODONTOPHIS Amaral, 1963 (Dipsadidae: Xenodontinae)

Calamodontophis paucidens (Amaral, 1935)

Calamodontophis ronaldoi Franco, Carvalho & Lema, 2006

CALAMOPHIS A.B. Meyer, 1874 (Homalopsidae)

Comments: J.C. Murphy & Voris (2014) provide a generic diagnosis and key to the species.

Calamophis jobiensis A.B. Meyer, 1874

Comments: J.C. Murphy & Voris (2014) provide a description.

Calamophis katesandersae J.C. Murphy, 2012

Comments: J.C. Murphy & Voris (2014) provide a description and illustrations of the holotype.

Calamophis ruuddelangi J.C. Murphy, 2012

Comments: J.C. Murphy & Voris (2014) provide a description and illustrations of the holotype.

Calamophis sharonbrooksae J.C. Murphy, 2012

Distribution: Add Papua (Soron Selatan Division), elevation ~260 m, O'Shea & H. Kaiser (2016).

Comments: J.C. Murphy & Voris (2014) provide a description and illustrations of the holotype. O'Shea & H. Kaiser (2016) describe a third specimen.

CALAMORHABDIUM Boettger, 1898 (Calamariidae)

Calamorhabdium acuticeps Ahl, 1933

Calamorhabdium kuekenthali Boettger, 1898

CALLIOPHIS Gray, 1835 *in* Gray & Hardwicke, 1830–1835 (Elapidae)

Comments: Leviton et al. (2014) revise the species of the Philippines. Brown et al. (2018) present an mtDNA sequence-based phylogeny of southeast Asian Elapidae that has *Calliophis* comprising several clades that share a most recent common ancestor with the remaining Australasian elapids. On the other hand, Zaher et al. (2019) have Indian species of *Calliophis* as sister clade to a paraphyletic *C. bivirgatus* plus all other Elapidae *Calliophis* appears to be polyphyletic.

Calliophis beddomei M.A. Smith, 1943

Distribution: Low elevation of 420 m, Jins et al. (2014).

Comments: Ganesh & Ramanujam (2014) describe a new specimen from type locality.

Calliophis bibroni (Jan, 1863)

Comments: Shankar & Ganesh (2009) summarize known records and add geographical and morphological data for additional specimens.

Calliophis bilineatus W.C.H. Peters, 1881. Monatsb. Preuss. Akad. Wiss. Berlin 1881: 109. (*Callophis bilineatus*)

Holotype: ZMB 10004, a 325 mm specimen (O. Koch)

Type locality: "Insel Palawan," Philippines.

Distribution: Philippines (Balabac, Busuanga, Calauit, Culion, Palawan, Sangat Islands), NSL-305 m.

Comments: See under *C. intestinalis*.

Calliophis bivirgatus (F.Boie, 1827)

Distribution: Add West Malaysia (Terengganu), Sumarli et al. (2015).

Calliophis castoe E.N. Smith, Ogale, Deepak & Giri, 2012

Calliophis gracilis Gray, 1835 *in* Gray & Hardwicke, 1830–1835

Distribution: Add West Malaysia (Pahang), Zakaria et al. (2014).

Calliophis haematoetron E.N. Smith, Manamendra-Arachchi & Somaweera, 2008

Calliophis intestinalis (Laurenti, 1768)

Synonyms: Remove *Callophis intestinalis philippina* Günther, 1864, *Callophis bilineatus* W.C.H. Peters, 1881, *Callophis intestinalis suluensis* Steindachner, 1891.

Distribution: Remove the Philippines.

Comments: Leviton et al. (2014, 2018) recognize three Philippine populations as distinct species: C. bilineatus, C. philippina and C. suluensis.

Calliophis maculiceps (Günther, 1858)

Distribution: Add Thailand (Ranyong, Koh Chang I.), Chan-ard and Makchai (2011); Cambodia (Siem Reap), Geissler et al. (2019); West Malaysia (Pulau Singa Besar), B.L. Lim et al. (2010).

Calliophis melanurus (Shaw, 1802)

Distribution: Add India (Chhattsghar), Deshmukh et al. (2018).

Calliophis nigrescens Günther, 1862

Calliophis philippina Günther, 1864. Rept. British India: 349. (*Callophis intestinalis philippina*)

Holotype: BMNH 1946.1.19.22, a 430 mm male (H. Cuming).

Type locality: "Philippine Islands."

Distribution: Philippines (Bohol, Camiguin Sur, Dinagat, Mindanao, Samar Is.), 5–1433 m.

Comments: See under *C. intestinalis*.

Calliophis salitan R.M. Brown, Smart, Leviton & E.N. Smith, 2018. Herpetologica 74(1):93–99, figs. 1, 3–7.

Holotype: PNM 9844 (original number RMB 8291), a 997 mm male (J.B. Fernandez, 28 July 2007).

Type locality: "Mt. Cambinlia, sitio Cambinlia (Sudlon), Barangay Santiago, Municipality Loreto, Dinagat Islands Province, Dinagat Island, Mindanao P[leistoce]A[ggregate] I[sland]C[omplex], Philippines, 195 m elevation (10.3436833°N, 125.6181167°E."

Distribution: Dinagat Island, Philippines, 195 m. Known only from the holotype.

Calliophis suluensis Steindachner, 1891. Sber. Akad. Wiss. Wien, Math.-Nat. Klasse 100(1): 295–296. (*Callophis intestinalis suluensis*)

Syntypes: NMW 27199.1-7 (Meyerink, 1885), 27199.8-16 (Meyerink, 1890).

Type locality: "Sulu-Inseln," Philippines.

Distribution: Philippine Islands (Jolo, Siasi Is.), 31 m.

Comments: See under *C. intestinalis*.

CALLOSELASMA Cope, 1860 (Viperidae: Crotalinae)

Calloselasma rhodostoma (Kuhl, 1824)

Distribution: Add Cambodia (Siem Reap), Geissler et al. (2019).

CANDOIA Gray, 1842 (Candoiidae)

Candoia aspera (Günther, 1877)

Distribution: Add Papua New Guinea (Blupblup, Boisa, Garove, Kairiru, Manam, Mussau and Wallis Islands), Clegg & Jocque (2016); Papua New Guinea (Oro), O'Shea et al. (2018b, suppl.).

Comments: Reynolds & Henderson (2018) recognize two subspecies.

Candoia bibroni (A.M.C. Duméril & Bibron, 1844)

Candoia carinata (Schneider, 1801)

Distribution: Add Indonesia (Boano, Gorong, Neira, Yamdena Islands), Lang (2015); Papua New Guinea (Northern), Kraus (2013); (Blupblup, Boisa, Kranket, Laing, Tatawai, Umboi and Wallis Islands), Clegg & Jocque (2016).

Comments: Reynolds & Henderson (2018) recognize two subspecies.

Candoia paulsoni (Stull, 1956)

Distribution: Add Indonesia (Tidore Island), Lang (2013); Papua New Guinea (Crown, Long and New Ireland Islands), Clegg & Jocque (2016); Papua New Guinea (Oro), O'Shea et al. (2018b, suppl.). The records for NE Sulawesi are questioned by A. Koch et al. (2009).

Comments: Reynolds & Henderson (2018) recognize six subspecies.

Candoia superciliosa (Günther, 1863)

Comments: Reynolds & Henderson (2018) recognize two subspecies.

CANTORIA Girard, 1858 (Homalopsidae)

Comments: J.C. Murphy & Voris (2014) provide a generic diagnosis. *Cantoria annulata* was removed to *Djokoiskandarus* by J.C. Murphy, 2011.

Cantoria violacea Girard, 1858

Distribution: Add Myanmar (Tanintharyi), Mulcahy et al. (2018). Remove New Guinea.

Comments: J.C. Murphy & Voris (2014) provide a description.

CARAIBA Zaher, Grazziotin, Cadle, R.W.Murphy, Moura-Leite & Bonatto, 2009 (Dipsadidae: Xenodontinae)

Synonyms: Remove *Haitiophis* Hedges & Vidal *in* Hedges et al. (2009).

Comments: Using DNA sequence-data, Krysko et al. (2015) and Zaher et al. (2019) revive *Haitiophis* for *Caraiba anomala*, despite the sister taxa relationship of *anomala* and *C. andreae*, and the resulting monophyletic genera.

Caraiba andreae (J.T. Reinhardt & Lütken, 1862)

CARPHOPHIS Gervais *in* d'Orbigny, 1843 (Dipsadidae: Carphophiinae)

Carphophis amoenus (Say, 1825)

Carphophis vermis (Kennicott, 1859)

CASAREA Gray, 1842 (Bolyeriidae)

Casarea dussumieri (Schlegel, 1837)

CATHETORHINUS A.M.C. Duméril & Bibron, 1844 (Gerrhopilidae)

Distribution: Unknown, but Timor, Indonesia seems likely, Pyron & Wallach (2014).

Comments: Pyron & Wallach (2014) transfer *Cathetorhinus* to the Gerrhopilidae, based on shared features of head scalation. They provide a generic diagnosis.

Cathetorhinus melanocephalus A.M.C. Duméril & Bibron, 1844

CAUSUS Wagler, 1830 (Viperidae: Viperinae)

Causus bilineatus Boulenger, 1905

Distribution: Add Angola (Bengo), Branch (2018); Angola (Huila), M.P. Marques et al. (2018).

Causus defilippii (Jan, 1863)

Causus lichtensteinii (Jan, 1859)

Causus maculatus (Hallowell, 1842)

Distribution: Add Niger (Tillaberi), Trape & Mané (2015), Gabon (Moyen-Ogooue), Pauwels et al. (2017a); Angola (Bengo), Branch (2018); Angola (Cuanza Norte), M.P. Marques et al. (2018).

Causus rasmusseni Broadley, 2014. Arnoldia Zimbabwe 10:343, 348, figs. 1–8.

Holotype: NMZB 10704, a 685 mm female (purchased from local resident, 8 October 1990).

Type locality: "Ikelenge, Mwililunga District, north-western Zambia."

Distribution: Angola (Cauando-Cubango), Zambia (North-Western), 1224 m.

Comment: Related to *C. rhombeatus* (Bradley, 2014). Conradie et al. (2016b) discuss the validity of *C. rasmusseni* on the basis of a new specimen from Angola.

Causus resimus (W.C.H. Peters, 1862)

Distribution: Add Angola (Bengo, Luanda), M.P. Marques et al. (2018); Angola (Cabinda), Branch (2018).

Causus rhombeatus (Lichtenstein, 1823)

Distribution: Add Zambia (Muchinga), Broadley (2014); Angola (Bengo, Bengeula, Bie, Cabinda ?, Cuando Cubango, Cuanza Norte, Cuanza Sul, Huambo, Huila, Luanda, Lunda Sul, Malanje, Zaire), M.P. Marques et al. (2018); Angola (Namibe), Ceríaco et al. (2016).

CEMOPHORA Cope, 1860 (Colubridae: Colubrinae)

Comment: Weinell & Austin (2017) present a DNA sequence-based phylogeny.

Cemophora coccinea (Blumenbach, 1788)

Cemophora lineri Williams, Brown & L.D. Wilson, 1966. Texas J. Sci. 18(1): 85–87, fig. 1. (*Cemophora coccinea lineri*)

Holotype: AMNH 75307, a 579+ mm female (E.A.Liner & R. Whitten, 29 June 1963).

Type locality: "34.5 miles S Riviera, Kenedy County, Texas," USA.

Distribution: Extreme SC USA (S Texas).

Comment: Weinell & Austin (2017) elevate this taxon to species.

CENASPIS Campbell, E.N. Smith & Hall, 2018. J. Herpetol. 52(4): 460. (Dipsadidae: Dipsadinae)

Type species: *Cenaspis aenigma* Campbell, Smith & Hall, 2018 by monotypy.

Distribution: SW Mexico.

Cenaspis aenigma Campbell, E.N. Smith & Hall, 2018. J. Herpetol. 52(4): 460–462, figs. 1–3.

Holotype: UTA R10544, a 258 mm male (J. Ornelas-Martínez, 6 July 1976).

Type locality: "'La Loma,' located some 20–25 km (by road) W-NW of Rizo de Oro (sometimes known as Nueva Tenochtitlán), Chiapas, Mexico."

Distribution: SW Mexico (Chiapas).

CERASTES Laurenti, 1768 (Viperidae: Viperinae)

Comments: Al-Fares (2014) presents a natural history overview, with an emphasis on *C. gasperettii*.

Cerastes boehmei Wagner & Wilms, 2010

Cerastes cerastes (Linnaeus, 1758)

Distribution: Add Morocco (Tata, Tiznit), Damas-Moreira et al. (2014); Mauritania (Nouakchott, Tagant), Padial (2006); Algeria (El Oued), Mouane et al. (2014); Algeria (Tindouf), Donaire et al. (2000), Libya (Al Wahat, Jabal al Akhdar, Jafara, Jufra, Marj, Misratah, Murqub, Nuqat al Khams, Tripoli, Wadi al Hayaa, Wadi al Shath), Bauer et al. (2017); Libya (Sabha, Tarabulus), Frynta et al. (2000), Tunisia (Medenine), Frynta et al. (2000); Mali (Kidal), Trape & Mané (2017). Crochet et al. (2015) report a specimen from SW Western Sahara.

Comments: Alshammari (2011) evaluates variation in RNA sequence data between Arabian populations.

Cerastes gasperettii Leviton & S. Anderson, 1967

Distribution: Add Saudi Arabia (Ha'il), Alshammari et al. (2017); Iran (Ilam), Safaei-Mahroo et al. (2015).

Comments: Alshammari (2011) evaluates variation in RNA sequence data between Arabian populations, which corroborates recognition of *C. gasperettii* as a species distinct from *C. cerastes*.

Cerastes vipera (Linnaeus, 1758)

Distribution: Add Mauritania (Nouakchott), Padial (2006); Algeria (El Oued), Mouane et al. (2014); Algeria (Tindouf), Donaire et al. (2000); Libya (Al Wahat, Butnan, Ghat, Kufrah, Marj, Misratah, Murzuq, Nalut, Tripoli, Zawiyah), Bauer et al. (2017); Mali (Kidal), Trape & Mané (2017); Sudan (Northern), Abukashawa et al. (2018). Crochet et al. (2015) report specimens from intermediate areas of known localities within Western Sahara.

CERATOPHALLUS Cope, 1893 (Natricidae)

Comments: Based on DNA sequence data in Takeuchi et al. (2018), *C. vittatus* is paraphyletic with respect to some *Xenochrophis* species and *Rhabdophis conspicillatus*.

Ceratophallus vittatus (Linnaeus, 1758)

Distriution: Introduced into Puerto Rico, USA, Herrera-Montes et al. (2015).

†*CERBEROPHIS* Longrich, Bhullar & Gauthier, 2012 (Alethinophidia: incerta sedis)

†*Cerberophis robustus* Longrich, Bhullar & Gauthier, 2012

CERBERUS Cuvier, 1829 (Homalopsidae)

Comments: J.C. Murphy & Voris (2014) provide a generic diagnosis.

Cerberus australis (Gray, 1842)

Comments: J.C. Murphy & Voris (2014) provide a description and illustrations.

Cerberus dunsoni J.C. Murphy, Voris & Karns, 2012

Comments: J.C. Murphy & Voris (2014) provide a description and illustrations.

Cerberus microlepis Boulenger, 1896

Comments: J.C. Murphy & Voris (2014) provide a description and illustrations.

Cerberus rynchops (Schneider, 1799)

Distribution: Add Myanmar (Tanintharyi), Mulcahy et al. (2018).

Comments: J.C. Murphy & Voris (2014) provide a description and illustrations. Vyas et al. (2013) comment on the distribution at the western edge of the range.

Cerberus schneiderii (Schlegel, 1837)

Distribution: Add Thailand (Nakhon Si Thammarat), Pauwels & Sumontha (2016); West Malaysia (Pulau Singa Besar), B.L. Lim et al. (2010); Indonesia (Bawean, Madura Islands), Lang (2013); Indonesia (Salibabu Island), A. Koch et al. (2009); Philippines (Masbate, Siquijor Is.), Leviton et al. (2018); Philippines (Sibuyan I., Tablas I.), Siler et al. (2012).

Comments: J.C. Murphy & Voris (2014) provide a description and illustrations of the lectotype.

CERCOPHIS Fitzinger, 1843 (Dipsadidae: subfamily unnamed)

Synonyms: Add *Uromacerina* Amaral, 1930

Comments: Hoogmoed et al. (2019) find *Cercophis auratus* and *Uromacerina ricardinii* to represent the same taxon.

Cercophis auratus (Schlegel, 1837)

Synonyms: Add *Uromacer ricardinii* Peracca, 1897.

Distribution: Suriname (Paramaribo), E Brazil (Bahia, Espirito Santo, Minas Gerais, Para, Parana, Rio de Janeiro, Rio Grande do Sul, Santa Catarina, Sao Paulo), NSL–850 m.

Comments: Hoogmoed et al. (2019) provide a revised description, color photographs and locality map.

CERROPHIDION Campbell & Lamar, 1992 (Viperidae: Crotalinae)

Cerrophidion godmani (Günther, 1863)
Cerrophidion petlalcalensis López-Luna, Vogt & Torre-Loranca, 1999
Cerrophidion sasai Jadin, Townsend, Castoe & Campbell, 2012
Cerrophidion tzotzilorum (Campbell, 1985)
Cerrophidion wilsoni Jadin, Townsend, Castoe & Campbell, 2012

Distribution: Add Nicaragua (Matagalpa, Nueva Segovia), Fernández et al. (2017); Nicaragua (Jinotega, Nueva Segovia), Sunyer et al. (2017).

CHAMAELYCUS Boulenger, 1919 (Lamprophiidae)

Chamaelycus christyi Boulenger, 1919
Chamaelycus fasciatus (Günther, 1858)
Chamaelycus parkeri (Angel, 1934)
Chamaelycus werneri (Mocquard, 1902)

CHAPINOPHIS Campbell & E.N. Smith, 1998 (Dipsadidae: Xenodontinae)

Chapinophis xanthocheilus Campbell & E.N. Smith, 1998

Comments: Villatoro-Castañeda & Ariano-Sánchez (2017) describe and illustrate a fourth known specimen.

CHARINA Gray, 1849 (Charinidae: Charininae)

Charina bottae (Blainville, 1835)
†***Charina prebottae*** Brattstrom, 1958
Charina umbratica Klauber, 1943

†***CHEILOPHIS*** Gilmore, 1938 (Booidea: incerta sedis)

†***Cheilophis huerfanoensis*** Gilmore, 1938

CHERSODROMUS J.T. Reinhardt, 1861 (Dipsadidae: Dipsadinae)

Comments: Canseco-Márquez et al. (2018) revise the genus.

Chersodromus australis Canseco-Márquez, Ramírez-González and Campbell, 2018. Zootaxa 4399(2): 9–11, figs. 6, 7.

Holotype: MZFC 17618, a 220 mm female (E.P. Ramos, 27 June 1995).

Type locality: "San Isidro La Gringa (17°04.591′ N, 94°03.844′ W; 350 m asl), Municipality of Santa Maria Chimalapa, Oaxaca, Mexico."

Distribution: SE Mexico (NE Oaxaca) at 350 m. Known only from the holotype.

Chersodromus liebmanni J.T. Reinhardt, 1861

Distribution: Add Mexico (SE Puebla) and lower elevation of 1000 m, Canseco-Márquez et al. (2018).

Comments: Canseco-Márquez et al. (2018) provide a species account, locality map and color photograph.

Chersodromus nigrum Canseco-Márquez, Ramírez-González and Campbell, 2018. Zootaxa 4399(2): 12, 15, figs. 8, 9.

Holotype: MZFC 17619, a 334 mm male (I.S. Zavaleta, 28 May 2005).

Type locality: "Xucayucan (19°53′47.9″ N, 97°28′43.7″ W; 1493 m asl), Municipality of Tlatlauquitepec, Puebla, Mexico."

Distribution: EC Mexico (N Puebla), 1493–1560 m.

Chersodromus rubriventris (Taylor, 1949)

Distribution: Add Mexico (N Hidalgo), and upper elevation of 1650 m, Canseco-Márquez et al. (2018).

Comments: Canseco-Márquez et al. (2018) provide a species account, locality map and color photograph.

CHILABOTHRUS A.M.C. Duméril & Bibron, 1844. Erpétol. Gén. 6: 562. (Boidae)

Synonyms: *Homalochilus* J.G. Fischer, 1856, *Piesigaster* Seoane, 1880, †*Pseudoepicrates* Auffenberg, 1963, *Boella* H.M. Smith & Chiszar, 1992.

Type species: *Boa inornata* Reinhardt 1843 by monotypy.

Distribution: West Indies.

Comments: Reynolds et al. (2013) use sequence data from multiple genes to produce a phylogeny of West Indian boidae. Due to paraphyly of mainland *Epicrates* with *Eunectes*, West Indian species are assigned to their own genus, for which the name *Chilabothrus* is available.

Chilabothrus angulifer Bibron, 1840 *in* Ramón de la Sagra, 1838–1843

Chilabothrus argentum Reynolds, Puente-Rolón, Geneva, Avila-Rodriguez & Herrmann, 2016. Breviora (549): 14–16, figs. 1, 3–6.

Holotype: MCZ R193527, an 1142 mm female (Reynolds, Puente-Rolón, Geneva, Avila Rodriguez & Herrmann, 21 October 2015).

Type locality: "the Conception Island Bank, Bahamas."

Distribution: Bahamas (Conception Island Bank), NSL.

Chilabothrus chrysogaster (Cope, 1871)

Synonyms: Remove *Epicrates chrysogaster schwartzi* Buden, 1975.

Distribution: Add Turks and Caicos Islands (Gibbs Cay), Reynolds & Niemiller (2010b). Remove Aklins and Crooked Islands, Reynolds et al. (2018a).

Comments: See under *C. schwartzi*. Reynolds & Henderson (2018) recognize two subspecies.

Chilabothrus exsul (Netting & Goin, 1944)

Distribution: Add Bahamas (Sandy Cay), Krysko et al. (2013).

Chilabothrus fordii (Günther, 1861)

Comments: Reynolds & Henderson (2018) recognize three subspecies.

Chilabothrus gracilis J.G. Fischer, 1888

Comments: Reynolds & Henderson (2018) recognize two subspecies.

Chilabothrus granti Stull, 1933. Occas. Pap. Mus. Zool. Univ. Michigan (267): 1–2. (*Epicrates inornatus granti*)

Holotype: MCZ 33847, a male (C. Grant).

Type locality: "Tortola Island," British Virgin Islands.

Distribution: All of range of *C. monensis* except Mona Islands.

Comments: Rodríguez-Robles et al. (2015) use mtDNA sequence-data to formally recognize *C. m. granti* and *C. m. monensis* as separate species.

Chilabothrus inornatus (J.T. Reinhardt, 1843)

Chilabothrus monensis (Zenneck, 1898)

Synonyms: Remove *Epicrates inornatus granti* Stull, 1933.

Distribution: Restricted to Puerto Rico (Mona Is.), NSL.

Comments: See under *C. granti*.

Chilabothrus schwartzi Buden, 1975. Herpetologica 31(2): 173–174. (*Epicrates chrysogaster schwartzi*)

Holotype: LSUMZ 27500, a 960 mm female (D.W. Buden, 27 April 1972).

Type locality: "the settlement of Delectable Bay, Acklins Island, Bahama Islands."

Distribution: Bahamas (Aklins I., Crooked I.).

Comments: Reynolds et al. (2018a) use mtDNA sequence data to determine relationships among *Chilabothrus* taxa, which resolve *schwartzi* as sister taxon to *C. argentum*, rather than as a population within *C. chrysogaster*. They provide a revised description of *C. schwartzi*.

†***Chilabothrus stanolseni*** Vanzolini, 1952. J. Paleontology 26(3): 455, figs. 6–11. (*Neurodromicus stanolseni*)

Synonyms: †*Neurodromicus barbouri* Vanzolini, 1952.

Holotype: MCZ 1977, part, an anterior precloacal vertebra.

Type locality: "Thomas Farm, Gilchrist County, Florida," USA.

Distribution: Early Miocene (Hemingfordian) of Florida, USA.

Comments: Onary & Hsieu (2018) review the type material of *Neurodromicus stanolseni* and its synonym, *N. barbouri*, and reassign all to *Chilabothrus*.

Chilabothrus striatus (J.G. Fisher, 1856)

Distribution: Reynolds et al. (2014a) document a specimen from Vieques Island, Puerto Rico, USA, but suspect that it may have been released there.

Comments: Reynolds & Henderson (2018) recognize three subspecies.

Chilabothrus strigilatus (Cope, 1862)

Distribution: Add Bahamas (Great Ragged Island), Reynolds & Puente-Rólon (2016).

Comments: DNA sequence data used by Reynolds et al. (2013) support the recognition of two species within *C. striatus*, and they resurrect *H. strigilatus* for populations on the Bahamas. Reynolds & Henderson (2018) recognize five subspecies.

Chilabothrus subflavus (Stejneger, 1901)

CHILORHINOPHIS F. Werner, 1907 (Atractaspididae: Aparallactinae)

Chilorhinophis butleri F. Werner, 1907
Chilorhinophis gerardi Boulenger, 1913

CHIRONIUS Fitzinger, 1826 (Colubridae: Colubrinae)

Comments: Hamdan et al. (2017) use DNA sequnce data from 16 species of *Chironius* to produce a phylogeny. The monophyly of the genus is supported, and the studied species are allocated among four primary clades. Torres-Carvajal et al. (2019) perform a similar analysis using 20 species, resulting in a more hierarchical pattern.

Chironius bicarinatus (Wied-Neuwied, 1820)

Distribution: Add Brazil (Ceara), Roberto & Loebmann (2016); Brazil (Paraíba), Guedes et al. (2014); Brazil (Mato Grosso do Sul), Ferreira et al. (2017); Brazil (Sao Paulo: Ilha Anchieta), Cicchi et al. (2009).

Chironius brazili Hamdan & Fernandes, 2015. Zootaxa 4012(1):107–113, figs. 7–9.

Holotype: MNRJ 17480, a 1381 mm male (A.C.A. Lopes, October 2008).

Type locality: "at RPPN Santuário do Caraça (20°05′, 43°29′29″W, 1262 m asl), municipality of Catas Altas, state of Minas Gerais, Brazil."

Distribution: Brazil (Cerrado biome of Distrito Federal, Goias, Minas Gerais, Rio Grande do Sul, São Paulo), 70–1360 m. Abegg et al. (2016) confirm the occurrence in Rio Grande do Sul.

Chironius carinatus (Linnaeus, 1758)

Distribution: Add St. Vincent and The Grenadines (St. Vincent), P. Araújo et al. (2019); Colombia (Cesar), Medina-Rangel (2011); Colombia (Huila), Moreno-Arias & Quintero-Corzo (2015); Colombia (Norte de Santander), Armesto et al. (2011); Venezuela (Portuguesa), Seijas et al. (2013); Venezuela (Aragua, Tachira), P. Araújo et al. (2019); Ecuador (Pastza) P. Araújo et al. (2019); Peru (Huanuco), Schlüter et al. (2004); Guyana (East Berbice-Corentyne, Upper Demerara-Berbice), P. Araújo et al. (2019); Suriname (Brokopondo, Wanica), P. Araújo et al. (2019); Brazil (Piaui), Guedes et al. (2014); Brazil (Acre), Bernarde et al. (2011b); Brazil (Amapa, Ceara, Paraiba, Rondonia, Tocantins), P. Araújo et al. (2019). P. Araújo et al. (2019) list and map known localities.

Chironius challenger Kok, 2010

Chironius cochranae Hoge & Romano, 1969. Mem. Inst. Butantan 35: 93–95, figs. 1, 2.

Holotype: USNM 158103, a 1972+ mm male (E. Dente, 19 July 1965).

Type locality: "Agua Preta Utinga (near Belem), State Pará, Brazil."

Distribution: NC South America. E Venezuela, Guyana (Barima-Waini, Cuyuni-Mazaruni, Essequibo Islands-West Demerara, Potaro-Siparuni, Upper Demerara-Berbice, Upper Takutu-UpperEssequibo), Suriname (Brokopondo, Marowijne, Nickerie, Sipaliwini), French Guiana (Cayenne), N Brazil (Amapa ?, NE Amazonas, Maranhao, N Para), to 676 m.

Comments: Torres-Carvajal et al. (2019) recognize this taxon (as *Chironius* cf. *cochranae*) for Guianan populations of *C. multiventris*.

Chironius diamantina Fernandes & Hamdan, 2014. Zootaxa 3881(6): 564–570, figs. 1–3.

Holotype: MZUFBA 1657, a 1018 mm female (collector unknown, November 2005).

Type locality: "municipality of Morro de Chapéu (11° 33′ 9″S, 41° 9′ 27″W, about 1000 m above sea level…), oriental zone of Chapada Diamantina, state of Bahia, Brazil."

Distribution: Brazil (central Bahia), 1000–1310 m.

Comments: Allied to *C. flavolineatus* according to Fernandes & Hamden (2014).

Chironius exoletus (Linnaeus, 1758)

Distribution: Add Panama (Panama Oeste), Ray & Ruback (2015); Ecuador (Carchi, Orellana, Santo Domingo de los Tsachilas, Sucumbios, Zamora-Chinchipe), Torres-Carvajal et al. (2019); Brazil (Ceara, Paraíba); Guedes et al. (2014); Brazil (Acre), D.P.F. França et al. (2017); Brazil (Mato Grosso do Sul), Ferreira et al. (2017); Brazil (Paraíba), R. França et al. (2012); Brazil (Piauí), Dal Vechio et al. (2013); Brazil (Rondônia), Bernarde et al. (2012b).

Comments: Hamdan et al. (2017), using DNA-sequence data, determine that there are likely a northern and southern species within *C. exoletus*, and Torres-Carvajal et al. (2019) identify three putative genetic species within. Sudré et al. (2017) use morphological characters to conclude that *C. pyrrhopogon* is correctly placed as a synonym of *C. exoletus*, and they could find no geographic pattern for the morphotype within *C. exoletus*. They also conclude based on morphological variation that the latter is a species complex.

Chironius flavolineatus (Jan, 1863)

Lectotype: Hamdan et al. (2014) report the discovery of one of Jan's syntypes that was supposedly destroyed during WW II. The newly discovered lectotype, MSNM R2729, is the same specimen upon which the lectotype illustration of Dixon et al. was based.

Distribution: Add Brazil (Espirito Santo), Silva-Soares et al. (2011); Brazil (Maranhão), J.P. Miranda et al. (2012); Brazil (Distrito Federal, Rio Grande do Norte, Sergipe), Hamdan et al. (2014); Brazil (Piauí), Dal Vechio et al. (2013). Hamdan & Fernandes (2015) provide a locality map, and state the elevational range as 3–1171 m.

Comment: Hamdan et al. (2014) redescribe the lectotype. Hamdan & Fernandes (2015) present a revised diagnosis and description, including that of the hemipenes and skull. Hamdan et al. (2017), using DNA-sequence data, recovered two inland populations as distinct, unnamed species.

Chironius flavopictus (F. Werner, 1909)

Distribution: Add Panama (Los Santos), Saenger & Ray (2016); Panama (Panama Oeste), Ray & Ruback (2015); Ecuador (Los Rios, Pichincha, Santo Domingo de Los Tsachilas), Torres-Carvajal et al. (2019).

Chironius foveatus Bailey, 1955

Distribution: Add Brazil (Rio de Janeiro: Ilha Grande), C.F.D. Rocha et al. (2018).

Chironius fuscus (Linnaeus, 1758)

Distribution: Add Ecuador (Orellana, Sucumbios, Zamora-Chinchipe), Torres-Carvajal et al. (2019); Peru (Huanuco), Schlüter et al. (2004); Peru (San Martin), Torres-Carvajal et al. (2019); Brazil (Amapa, Mato Grosso, Rondonia), L.P. Nascimento et al. (2013); Brazil (Mato Grosso do Sul), Ferreira et al. (2017); Brazil (Santa Catarina), and lower elevation to n.s.l., G.A.S. Filho et al. (2012).

Comments: Torres-Carvajal et al. (2019) identify four putative genetic species within *C. fuscus*.

Chironius grandisquamis (W.C.H. Peters, 1868)

Distribution: Add Nicaragua (Rivas), Guevara et al. (2015); Panama (Panama Oeste), Ray & Ruback (2015); Colombia (Valle del Cauca: Isla Palma), Giraldo et al. (2014).

Chironius laevicollis (Wied-Neuwied, 1824)

Chironius laurenti Dixon, Wiest & Cei, 1995

Distribution: Add Brazil (Acre), D.B. Miranda et al. (2014), Brazil (Mato Grosso do Sul), Ferreira et al. (2017).

Chironius leucometapus Dixon, Wiest & Cei, 1995

Chironius maculoventris Dixon, Wiest & Cei, 1995

Distribution: Add Paraguay (Alto Paraguay), Cabral & Weiler (2014); Brazil (Rio Grande do Sul), F.J.M. Santos et al. (2015).

Chironius monticola Roze, 1952

Distribution: Add Colombia (Cauca), and upper elevation of 3169 m, Vera-Pérez et al. (2018); Ecuador (Carchi), Torres-Carvajal et al. (2019); Peru (Cajamarca), Koch et al. (2018); Peru (Huancabamba), Torres-Carvajal et al. (2019).

Comments: Torres-Carvajal et al. (2019) identify four putative genetic species within *C. monticola*.

Chironius multiventris K.P. Schmidt & Walker, 1943

Distribution: Add Ecuador (Orellana, Sucumbios), Torres-Carvajal et al. (2019). Remove Guianan region to *C. cochranae*.

Comments: Torres-Carvajal et al. (2019) identify two putative genetic species within *C. multiventris*, with the two being paraphyletic with *C. foveatus* and *C. laurenti*, respectively.

Chironius quadricarinatus (F. Boie, 1827)

Distribution: Add Brazil (Espirito Santo, Parana), Sudré et al. (2017); Paraguay (Cordillera), Weiler & Ortega (2016).

Chironius scurrulus (Wagler *in* Spix, 1824)

Distribution: Add Ecuador (Orellana), Torres-Carvajal et al. (2019); Ecuador (Zamora-Chinchipe), Almendáriz et al. (2014); Peru (Cusco), Torres-Carvajal et al. (2019); Brazil (Acre), Bernarde et al. (2011b); Brazil (Amapa), Torres-Carvajal et al. (2019); Brazil (Rondônia), Bernarde et al. (2012b).

Chironius septentrionalis Dixon, Wiest & Cei, 1995

Distribution: Add Brazil (Roraima), Moraes et al. (2017).

Chironius spixii (Hallowell, 1845)

Distribution: Add Colombia (Santander), E. Ramos and Meza-Joya (2018).

Chironius vincenti (Boulenger, 1891)

†*CHOTAOPHIS* Head, 2005 (Colubridae: incerta sedis)

†*Chotaophis padhriensis* Head, 2005

CHRYSOPELEA H. Boie *in* Schlegel, 1826 (Colubridae: Ahaetullinae)

Chrysopelea ornata (Shaw, 1802)

Distribution: Add Nepal (Nawalparasi), Pandey et al. (2018); Nepal (Parsa), Bhattarai et al. (2018b); India (Tamil Nadu), Melvinselvan & Nibedita (2016); Myanmar (Ayeyarwady), Platt et al. (2018); Thailand (Chanthaburi), Chan-ard et al. (2011); Thailand (Khon Kaen), Nurngsomsri et al. (2014); Thailand (Rayong, and Koh Chang I.), Chan-ard and Makchai (2011); Cambodia (Siem Reap), Geissler et al. (2019); Vietnam (Hai Phong: Cat Ba Island), T.Q. Nguyen et al. (2011); West

Malaysia (Jerejak Island), Quah et al. (2011); West Malaysia (Penang), Quah et al. (2013); West Malaysia (Pulau Tioman), K.K.P. Lim & Lim (1999); Bangladesh (Rajshahi Division, Naogaon District), Ahmad & Alam (2015).

Chrysopelea paradisi H. Boie in F. Boie, 1827

Distribution: Add Thailand (Krabi: Ko Yung and Phi Phi Don Is.), Milto (2014); West Malaysia (Pulau Singa Besar), B.L. Lim et al. (2010); Philippines (Calayan, Dalupiri, Masbate, Siagao Is.), Leviton et al. (2018); Philippines (Tablas I.), Siler et al. (2012); Indonesia (Pulau Bangkuru off Sumatra), Tapley & Muurmans (2016).

Chrysopelea pelias (Linnaeus, 1758)

Chrysopelea rhodopleuron F. Boie, 1827

Distribution: Add Indonesia (Kei Cecil, Yamdena Islands), Lang (2013).

Chrysopelea taprobanica M.A. Smith, 1943

Distribution: Add India (Andhra Pradesh), Guptha et al. (2015); India (Tamil Nadu), and increase elevation range to 657 m, S. Narayanan et al. (2017a, as *C.* cf. *taprobanica*).

†*CHUBUTOPHIS* Albino, 1993 (Booidea: incerta sedis)

†*Chubutophis grandis* Albino, 1993

CLELIA Fitzinger, 1826 (Dipsadidae: Xenodontinae)

Clelia clelia (Daudin, 1803)

Distribution: Add Mexico (S Chiapas), Hernández-Ordóñez et al. (2015); Panama (Panama), Ray & Ruback (2015), Colombia (Amazonas, Arauca, Boyacá, Caldas, Quindío, Sucre, Tolima, Vichada), Díaz-Ricaurte et al. (2018b); Colombia (Huila), Moreno-Arias & Quintero-Corzo (2015); Colombia (Valle del Cauca: Isla Palma), Giraldo et al. (2014); Peru (Cajamarca), Koch et al. (2018); Peru (Puno), Llanqui et al. (2019); Brazil (Acre), Bernarde et al. (2011b).

Comments: Arquilla & Lehtinen (2018) report differences in juvenile color pattern between cis-Andean and trans-Andean plus Central American specimens.

Clelia equatoriana (Amaral, 1924)

Distribution: Add Peru (Cajamarca, Piura), Chávez-Arribasplata et al. (2016). Upper elevation of 2200 m, Vera-Pérez et al. (2018).

Clelia errabunda Underwood, 1993

Clelia hussami Morata, Franco & Sanchez, 2003

Clelia langeri Reichle & Embert, 2005

Clelia plumbea (Wied-Neuwied, 1820)

Distribution: Add Brazil (Piauí, Tocantins), Dal Vechio et al. (2016); Brazil (Rondônia), Bernarde et al. (2012b).

Clelia scytalina (Cope, 1867)

Distribution: Add Mexico (Michoacan), Pérez-Hernandez et al. (2015); Grünwald et al. (2016a, range extension in Jalisco, Mexico).

CLONOPHIS Cope, 1889 (Natricidae)

Clonophis kirtlandi (Kennicott, 1856)

COELOGNATHUS Fitzinger, 1843 (Colubridae: Colubrinae)

Coelognathus enganensis (Vinciguerra, 1892)

Coelognathus erythrurus (A.M.C. Duméril, Bibron & Duméril, 1854)

Distribution: Add Philippines (Barit, Basilan, Batan, Bongao, Calayan, Dalupiri, Inampulugan, Masbate, Siquijor Is.) Leviton et al. (2018); Philippines (Romblon, Tablas I.), Sy & Tan (2015a).

Comments: Leviton et al. (2018) discuss color pattern variation.

Coelognathus flavolineatus (Schlegel, 1837)

Distribution: Add India (Great Nicobar I., Andaman & Nicobars), Harikrishnan et al. (2010, as *Coelognathus* sp.); West Malaysia (Pahang), Zakaria et al. (2014).

Coelognathus helena (Daudin, 1803)

Synonyms: *Coelognathus helena nigriangularis* Mohapatra, Schulz, Helfenberger, Hofmann & Dutta, 2016. Russ. J. Herpetol. 23(2): 132–137, figs. 15–20, 22, 23, 29.

Holotype: BNHS 3470, a 1386 mm female (S.K. Behera & P.P. Mohapatra, November, 2007).

Type locality: "a limestone cave near Gupteswar Temple, Koraput District, Odisha, India, 18°49'17.02" N 82°10'0.04" E, 462 m a.s.l."

Distribution: Add India (Arunachal Pradesh), Purkayastha (2018); India (Chhattisgarh, Jharkhand, Odisha, Uttarakhand), Mohapatra et al. (2016); Nepal (Nawalparasi), Pandey (2012); Nepal (Parsa), Bhattarai et al. (2018b). Lower elevation range is NSL, Mohapatra et al. (2016).

Comments: Mohapatra et al. (2016) summarize geographic variation and delineate three subspecies in Indian populations.

Coelognathus philippinus (Griffin, 1909)

Distribution: Add Philippines (Calauit I.), Leviton et al. (2018).

Coelognathus radiatus (F. Boie, 1827)

Distribution: Add India (Andhra Pradesh, Uttarakhand), Javed et al. (2010); Nepal (Chitwan), Pandey (2012); Nepal (Bara), Pandey et al. (2018); Nepal (Parsa), Bhattarai et al. (2018b); Bhutan (Trongsa, Wangdue Phodrang), Tshewang & Letro, 2018); Myanmar (Sagaing), Platt et al. (2018); Myanmar (Tanintharyi), Mulcahy et al. (2018); Thailand (Chanthaburi), Chan-ard et al. (2011); Thailand (Rayong, and Koh Chang I.), Chan-ard and Makchai (2011); Thailand (Tak), Vogel & Han-Yuen (2010).

Coelognathus subradiatus (Schlegel, 1837)

Distribution: Add Timor-Leste (Ataúro Island), H. Kaiser et al. (2013b).

COLLORHABDIUM Smedley, 1932 (Calamariidae)

Collorhabdium williamsoni Smedley, 1932

†*COLOMBOPHIS* Hoffstetter & Rage, 1977 (Aniliidae)

†*Colombophis portai* Hoffstetter & Rage, 1977
†*Colombophis spinosus* Hsiou, Albino & Ferigolo, 2010

COLUBER Linnaeus, 1758 (Colubridae: Colubrinae)

Comments: See under *Masticophis*.

Coluber constrictor Linnaeus, 1758

Distribution: Historical range in Maine, USA, discussed by Persons & Mays (2017).

COLUBROELAPS Orlov, Kharin, Ananjeva, Nguyen & Nguyen, 2009 (Colubridae: Colubrinae)

Colubroelaps nguyenvansangi Orlov, Kharin, Ananjeva, Nguyen & Nguyen, 2009

COMPSOPHIS Mocquard, 1894 (Pseudoxyrhophiidae)

Compsophis albiventris Mocquard, 1894
Compsophis boulengerii (Peracca, 1892)
Compsophis fatsibe (Mercurio & Andreone, 2005)
Compsophis infralineatus (Günther, 1882)
Compsophis laphystius (Cadle, 1996)
Compsophis vinckei (Domergue, 1988)
Compsophis zeny (Cadle, 1996)

†*CONANTOPHIS* Holman & Harrison, 2000 (Booidea: incerta sedis)

†*Conantophis alachuaensis* Holman & Harrison, 2000

CONIOPHANES Hallowell *in* Cope, 1860 (Dipsadidae: Dipsadinae)

Coniophanes alvarezi Campbell, 1989

Distribution: Upper elevation of 2170 m, García-Padilla & Mata-Silva (2014h).

Coniophanes andresensis Bailey, 1937. Occas. Pap. Mus. Zool. Univ. Michigan (362): 4–5. (*Coniophanes fissidens andresensis*)

Synonyms: *Coniophanes brevifrons* Bailey, 1937.

Holotype: MCZ 31867, a female (J.B. Zetek, 30 April 1931).

Type locality: "San Andres Island, off the coast of Nicaragua," Colombia.

Distribution: Colombia (San Andres Island).

Comments: Recognized as a species distinct from *C. fissidens* by others, including Caicedo-Portillo (2014), who provides a summary of the species.

Coniophanes bipunctatus (Günther, 1858)

Distribution: Add Honduras (Isla Guanaja), Solís et al. (2017a); Nicaragua (Atlantico Norte, Rio San Juan), Sunyer et al. (2014).

Coniophanes dromiciformis (W.C.H. Peters, 1863)

Distribution: Add Ecuador (Manabi), Almendáriz et al. (2012).

Coniophanes fissidens (Günther, 1858)

Synonyms: Remove *Coniophanes brevifrons* Bailey, 1937, *Coniophanes fissidens andresensis* Bailey, 1937.

Distribution: Add Honduras (Santa Barbara), McCranie et al. (2013b); Panama (Panama Oeste), Ray & Ruback (2015). Remove Colombia (San Andres Island) to *C. andresensis*.

Coniophanes imperialis (Baird, 1859)

Distribution: Add Mexico (Guerrero), Palacios-Aguilar & Flores-Villela (2018); Mexico (México, Nuevo León, 360 m), Rangel-Patiño et al. (2015), García-Padilla et al. (2016a); Belize (Ambergris Cay), Pavón-Vázquez (2016b).

Coniophanes joanae C.W. Myers, 1966

Distribution: Add Panama (Panama Oeste), Ray & Ruback (2015).

Coniophanes lateritius Cope, 1862

Distribution: Range extensions and supposed first records for Michoacán, Suazo-Ortuño et al. (2014), and Sonora, Lara-Resendiz et al. (2016), Mexico. Palacios-Aguilar et al. (2018) confirm the record for Guerrero by the collection of two specimens.

Coniophanes longinquus Cadle, 1989

Coniophanes melanocephalus (W.C.H. Peters, 1870)

Distribution: Solano-Zavaleta et al. (2014) discuss known localities, add Mexico (Guerrero) to the range, and report an upper elevation of 1540 m.

Coniophanes meridanus K.P. Schmidt & Andrews, 1936

Coniophanes michoacanensis Flores-Villela & H.M. Smith, 2009

Distribution: Calzada-Arciniega & Palacios-Aguilar (2015) document a third locality, in Guerrero, Mexico.

Coniophanes piceivittis Cope, 1870

Distribution: Add Honduras (El Paraíso, La Paz), McCranie & Gutsche (2013a), McCranie et al. (2014b); Nicaragua (Atlantico Sur: Great Corn Island), Durso & Norberg (2016).

Coniophanes quinquevittatus (A.M.C. Duméril, Bibron & Duméril, 1854)

Coniophanes sarae Ponce-Campos & H.M. Smith, 2001

Coniophanes schmidti Bailey, 1937

Coniophanes taylori Hall, 1951

Distribution: Add Mexico (Oaxaca), Grünwald et al. (2016a). Low elevation of 200 m, Palacios-Aguilar & Flores-Villela (2018).

†*CONIOPHIS* Marsh, 1892 (Ophidia: incerta sedis)

Distribution: Add late Eocene (Chadronian) of USA (North Dakota), K.T. Smith (2013, as *Coniophis* sp., but with some doubt as to genus).

†*Coniophis carinatus* Hecht *in* McGrew, 1959
†*Coniophis cosgriffi* Armstrong-Ziegler, 1978
†*Coniophis dabiebus* Rage & C. Werner, 1999
†*Coniophis platycarinatus* Hecht *in* McGrew, 1959
†*Coniophis precedens* Marsh, 1892

CONOPHIS W.C.H. Peters, 1860 (Dipsadidae: Xenodontinae)

Conophis lineatus (A.M.C. Duméril, Bibron & Duméril, 1854)

Distribution: Add Mexico (Yucatan: Isla Contoy), Noguez & Ramírez-Bautista (2008); Honduras (Isla Exposición, Isla Zacate Grande), McCranie & Gutsche (2016); Nicaragua (Jinotega), Ubeda-Olivas & Sunyer (2015b).

Comments: Schätti & Kucharzewski (2017) discuss the type material.

Conophis morai Pérez-Higareda, López-Luna & H.M. Smith, 2002

Conophis vittatus W.C.H. Peters, 1860

Distribution: Add Mexico (México), Hernández-Gallegos et al. (2014).

CONOPSIS Günther, 1858 (Colubridae: Colubrinae)

Conopsis acuta (Cope *in* Ferrari-Pérez, 1886)
Conopsis amphisticha (H.M. Smith & Laufe, 1945)
Conopsis biserialis Taylor & H.M. Smith, 1942

Distribution: Delete Mexico (Hidalgo), reported in error, Lemos-Espinal & G.R. Smith (2015). Elevation range 1550–2800 m, Palacios-Aguilar & Flores-Villela (2018).

Conopsis lineata (Kennicott *in* Baird, 1859)
Conopsis megalodon (Taylor & H.M. Smith, 1942)
Conopsis nasus (Günther, 1858)

CONTIA Baird & Girard, 1853 (Dipsadidae: Carphophiinae)

Contia longicaudae Feldman & Hoyer, 2010
Contia tenuis (Baird & Girard, 1852)

†*COPROPHIS* Parris & Holman, 1978 (Booidea: incerta sedis)

†*Coprophis dakotaensis* Parris & Holman, 1978

CORALLUS Daudin, 1803 (Boidae)

Comments: Henderson (2015) presents a review of the genus, including a key to the species.

Corallus annulatus (Cope, 1875)

Distribution: Add Nicaragua (Rio San Juan), Sunyer et al. (2014); Colombia (Narino, Valle del Cauca), Pinto-Erazo & Medina-Rangel (2018). Pinto-Erazo & Medina-Rangel (2018) list known localities.

Comments: Henderson (2015) provides a description, color photographs and a distribution map.

Corallus batesii (Gray, 1860)

Comments: Henderson (2015) provides a description, color photographs and a distribution map.

Corallus blombergi (Rendahl & Vestergen, 1941)

Distribution: Add Colombia (Narino) and Ecuador (Azuay), Pinto-Erazo & Medina-Rangel (2018). Pinto-Erazo & Medina-Rangel (2018) list known localities.

Comments: Henderson (2015) provides a description, color photographs and a distribution map.

Corallus caninus (Linnaeus, 1758)

Comments: Henderson (2015) provides a description, color photographs and a distribution map.

Corallus cookii Gray, 1842

Comments: Henderson (2015) provides a description, color photographs and a distribution map.

Corallus cropanii (Hoge, 1954)

Comments: Henderson (2015) provides a description, color photographs and a distribution map.

Corallus grenadensis (Barbour, 1914)

Comments: Henderson (2015) provides a description, color photographs and a distribution map.

Corallus hortulanus (Linnaeus, 1758)

Distribution: Add Ecuador (Orellana, Sucumbios), Yánez-Muñoz et al. (2017); Brazil (Alagoas), Guedes et al. (2014); Brazil (Mato Grosso do Sul), Ferreira et al. (2017); Brazil (Paraiba), Sampaio et al. (2018); Brazil (Piauí), Dal Vechio et al. (2013); Brazil (Rio de Janeiro: Ilha Grande), C.F.D. Rocha et al. (2018); Brazil (Tocantins), Dal Vechio et al. (2016).

Comments: Duarte et al. (2015) report six distinct color pattern morphs, with some geographic associations, in the Amazon basin of Brazil. Henderson (2015) provides a description, color photographs and a distribution map.

†*Corallus priscus* Rage, 2001

Corallus ruschenbergerii (Cope, 1875)

Distribution: Add Colombia (Norte de Santander), Armesto et al. (2011); Venezuela (Aragua), Calcaño & Barrio-Amorós (2017).

Comments: Henderson (2015) provides a description, color photographs and a distribution map.

CORONELAPS Lema & Hofstadler-Deiques, 2010 (Dipsadidae: Xenodontinae)

Coronelaps lepidus (J.T. Reinhardt, 1861)

Distribution: Add Brazil (Ceara), Roberto & Loebmann (2016).

Comments: Lema (2006) redescribes the holotype of *Elapomorphus coronatus*.

CORONELLA Laurenti, 1768 (Colubridae: Colubrinae)

Synonyms: Remove *Wallophis* Werner, 1929.

Distribution: Eliminate SW Asia from the range.

Coronella austriaca Laurenti, 1768

Distribution: Add Sweden (Åland), Galarza et al. (2015); Liechtenstein, Kühnis (2006); Spain (Asturias, Leon), Meijide (1985); Andorra, Orriols & Fernàndez (2003); Croatia (Cres Island), Bonte (2012); Montenegro, Polović & Čađenović (2013); Serbia, Tomović et al. (2014); Slovakia (Prešov), Pančišin & Klembara (2003) and Jablonski (2011); Iran (Alborz, Golestan, Mazandaran, Tehran), Safaei-Mahroo & Ghaffari (2015), Iran (West Azerbaijan), Safaei-Mahroo et al. (2015).

Fossil Records: Add Upper Pliocene/Lower Pleistocene of Spain, Blain (2009); Middle Pleistocene (Ionian) of Italy, Delfino (2004, as cf. *C. austriaca*); Upper Pleistocene (Tarantian) of Spain, Barroso-Ruiz and Bailon (2003); Late Pleistocene (post-Tarantian) of Belgium, Blain et al. (2019).

Comments: X. Santos et al. (2008) use mtDNA sequence data to evaluate history of fragmented populations on the Iberian Peninsula. Sztencel-Jabłonka et al. (2015) use mtDNA and microsatellite data that reveal a deep divergence between populations in eastern and western Poland. Galarza et al. (2015) also use mtDNA and microsatellite sequence data to produce a rangewide phylogeny that reveals western, central, and eastern European clades.

Coronella girondica (Daudin, 1803)

Distribution: Add Spain (Asturias), Meijide (1985); Spain (Isla Meda Gran), Pedrocchi & Pedrocchi-Rius (1994); Morocco (Figuig), Rosado et al. (2016); Morocco (Oujda, Taounate), Mediani et al. (2015b).

Fossil records: Add upper Pliocene (Villanyian) of Spain, Blain (2009); lower Pleistocene (Calabrian) of Spain, Blain et al. (2007).

Comments: X. Santos et al. (2012) produce an mtDNA based phylogeny that shows distinct clades in SE Spain, N Africa, SW Iberia, and the remainder of the Iberian Peninsula eastward to Italy.

†*Coronella miocaenica* Venczel, 1998

CRASPEDOCEPHALUS Kuhl & Hasselt, 1822 (Vipera: Crotalinae)

Comments: See comments under *Trimeresurus*.

Craspedocephalus andalcsensis (David, Vogel, Vijayakumar & Vidal, 2006)

Craspedocephalus borneensis (W.C.H. Peters, 1872)

Craspedocephalus brongersmai (Hoge, 1969)

Craspedocephalus gramineus (G. Shaw, 1802)

Distribution: Chowdhury et al. (2017) confirm the presence in West Bengal, India.

Craspedocephalus malabaricus (Jerdon, 1854)

Craspedocephalus puniceus Kuhl, 1824

Craspedocephalus rubeus (Malhotra, Thorpe, Mrinalini & Stuart, 2011)

Craspedocephalus strigatus (Gray, 1842)

Distribution: Ganesh & Chandramouli (2018) demonstrate that the known range is limited to only the Nilgiris of Kerala and Tamil Nadu at elevations over 1800 m. Published records from other mountain ranges and localities in the Western and Eastern Ghats are in error.

Craspedocephalus trigonocephalus (Donndorff, 1798)

Craspedocephalus wiroti (Klemmer in Trutnau, 1981)

CRISANTOPHIS Villa, 1971 (Dipsadidae: Xenodontinae)

Crisantophis nevermanni Dunn, 1937

CROTALUS Linnaeus, 1758 (Viperidae: Crotalinae)

Comments: Wüster & Bérnils (2011) discuss a suggested partition of *Crotalus* into multiple genera, based on a published phylogeny. Bryson et al. (2014) revise the *C. triseriatus* group. Blair & Sánchez-Ramírez (2016) present a phylogeny of all but several *Crotalus* species, based on sequence data from ten DNA genes. M. Davis et al. (2016) use morphometric and molecular data to revise the *Crotalus viridis* complex. Rubio &

Keyler (2013) discuss the species of the Western USA. Schuett et al. (2016a, b) discuss the species of Arizona, USA.

Crotalus adamanteus Palisot de Beauvois, 1799

Crotalus angelensis Klauber, 1963. Trans. San Diego Nat. Hist. Soc. 13(5): 75–80, figs. 1, 2. (*Crotalus mitchellii angelensis*)

Holotype: SDSNH 51994, a 1331 mm male (R. Moran, 22 March 1963).

Type locality: "about 4 miles southeast of Refugio Bay, at 1500 feet elevation, Isla Ángel de la Guarda, Gulf of California, Mexico (near 29° 29½′ N, 113° 33′ W)."

Distribution: Mexico (Angel de la Guarda I., Baja California).

Comments: Meik et al. (2015), using data from multiple genes, recognize *angelensis* as a species distinct from *C. mitchellii*.

Crotalus aquilus Klauber, 1952

Distribution: Add Mexico (Zacatecas), Carbajal-Márquez et al. (2015).

Comment: Bryson et al. (2014) present relationships as a part of the *C. triseriatus* complex.

Crotalus armstrongi J.A. Campbell, 1978. Trans. Kansas Acad. Sci. 81(4): 365–369, fig. 1. (*Crotalus triseriatus armstrongi*)

Holotype: UTA 6258 (orig. JAC 2874), a 484 mm male (J.A. Campbell & B.L. Armstrong, 23 July 1976).

Type locality: "Rancho San Francisco, 1.5 mi NW Tapalpa, Jalisco, Mexico, elevation 2103 m."

Distribution: WC Mexico (W Jalisco, N Michoacan), 2105 m.

Comment: Elevated to species level by Bryson et al. (2014) as part of the *C. triseriatus* complex. They discuss a putative new species from populations in SE Nayarit.

Crotalus atrox Baird & Girard, 1853

Distribution: Add Mexico (Aguascalientes), Quintero-Díaz & Carbajal-Márquez (2017c); Mexico (Jalisco), Villalobos-Juárez & Sigala-Rodríguez (2019).

Comments: Schield et al. (2015) evaluate population structure, and conclude that there is an eastern and western clade, with admixture from SE Arizona to W Texas. E.A. Myers et al. (2017a) provide a DNA-sequence based phylogeny of USA populations, with three primary clades: one west of the Continental Divide, and two east of it.

Crotalus basiliscus (Cope, 1864)

Distribution: Add Mexico (Aguascalientes), Carbajal-Márquez et al. (2015c). Carbajal-Márquez et al. (2015f) report a second specimen from Zacatecas.

Crotalus campbelli Bryson, Linkem, Dorcas, Lathrop, Jones, Alvarado-Díaz, Grünwald & Murphy, 2014. Zootaxa 3826(3): 486–489, figs. 7, 8.

Holotype, KU 73649 (orig. field number PLC 3216), an adult female (P.L. Clifton, 25 October 1962).

Type locality: "Sierra de Cuale, 9 km N El Teosinte, municipality of Talpa de Allende, state of Jalisco, Mexico."

Distribution: WC Mexico (W Jalisco, and the Sierra de Manantlán in S Jalisco/N Colima), 2009–2515 m. Nayarit was added to the distribution by Luja & Grünwald (2015), and Flores-Guerrero & Sánchez-González (2016) added records for Jalisco.

Comments: A member of the *C. triseriatus* complex.

Crotalus catalinensis Cliff, 1954

Crotalus cerastes Hallowell, 1854

Crotalus cerberus (Coues *in* Wheeler, 1875)

Comments: Douglas et al. (2016) find five geographically separated clades from mtDNA sequence data.

Crotalus concolor Woodbury, 1929

Crotalus culminatus Klauber, 1952

Distribution: Low elevation of 50 m, Palacios-Aguilar & Flores-Villela (2018). Upper elevation of 1734 m, Peralta-Fonseca & García-Padilla (2015).

Crotalus durissus Linnaeus, 1758

Synonyms: Remove *Crotalus horridus unicolor* Lidth de Jeude, 1887, *Crotalus pulvis* Ditmars, 1905.

Distribution: Add Colombia (Magdalena), Rueda-Solano & Castellanos-Barliza (2010); Colombia (Arauca, Boyacá, Cesar, Vichada), Díaz-Ricaurte et al. (2018c); Brazil (Alagoas, Sergipe), and upper elevation of 1100 m, Guedes et al. (2014); Brazil (Distrito Federal), Vanzolini & Calleffo (2002); Brazil (Tocantins), Dal Vechio et al. (2016); Bolivia (Potosi), Rivas et al. (2018). Prigioni et al. (2013) detail known records from Uruguay. Remove Netherlands Antilles (Aruba).

Crotalus enyo (Cope, 1861)

Crotalus ericsmithi Campbell & Flores-Villela, 2008

Crotalus estebanensis Klauber, 1949

Crotalus helleri Meek, 1905

Synonyms: Add *Crotalus viridis caliginis* Klauber, 1949.

Comments: M. Davis et al. (2016) found no mtDNA sequence divergence between the insular *C. caliginis* and mainland *C. helleri*.

Crotalus horridus Linnaeus 1758

Comment: Bushar et al. (2014) evaluated genetic structure between some NE USA populations at a non-taxonomic level.

Crotalus intermedius Troschel *in* J.W. Müller, 1865

Crotalus lannomi Tanner, 1966

Crotalus lepidus (Kennicott, 1861)

Synonyms: Remove *Crotalus lepidus morulus* Klauber, 1952, *Crotalus lepidus castaneus* Julia-Zertuche & Treviño, 1978.

Distribution: Remove Mexico (Tamaulipas, = *C. morulus*).

Comment: Bryson et al. (2014) present relationships as a part of the *C. triseriatus* complex, and remove *C. l. morulus* as a full species.

Crotalus lorenzoensis Radcliff & Maslin, 1975. Copeia 10975(3): 490–492. (*Crotalus ruber lorenzoensis*)

Holotype: SDSNH 46009, an 838 mm male (Univ. Colorado Museum Exped., 1966).

Type locality: "San Lorenzo Sur Island in the Gulf of California, Baja California Norte, Mexico."

Distribution: Mexico (San Lorenzo Sur I., Baja California).

Comments: Grismer (1999) recognizes this taxon as a species based on morphological distinctiveness and allopatry from mainland *C. ruber*.

Crotalus lutosus Klauber, 1930

Synonyms: Add *Crotalus confluentus abyssus* Klauber, 1930.

Comments: M. Davis et al. (2016) found some mtDNA sequence divergence, but no morphometric differences between the *C. abyssus* and *C. lutosus*, and recommended that the former be subsumed within the latter. They act as first revisors in giving nomenclatural priority to *C. lutosus*.

Crotalus mitchellii (Cope, 1861)

Synonyms: Remove *Caudisona pyrrha* Cope, 1867, *Crotalus goldmani* Schmidt, 1922, *Crotalus mitchelli aureus* Kallert, 1927, *Crotalus mitchellii muertensis* Klauber, 1949, *Crotalus mitchellii angelensis* Klauber, 1963.

Distribution: Limited to Mexico (Baja California Sur, and Carmen, Cerralvo, Espiritu Santo, Monserrate, Partida Sur, San Jose, Santa Margarita Islands).

Comments: See under *C. pyrrhus*.

Crotalus molossus Baird & Girard, 1853

Distribution: Upper elevation of 2863 m, Olvera & Badillo (2016a)

Comments: E.A. Myers et al. (2017a) provide a DNA-sequence based phylogeny of mostly USA populations, with three primary clades: one on either side of the Continental Divide (the latter confirming the status of *C. ornatus*), and a third based on a specimen from SE Durango, Mexico.

Crotalus morulus Klauber, 1952. Bull. Zool. Soc. San Diego (26): 52 (footnote) (*Crotalus lepidus morulus*).

Synonyms: *Crotalus lepidus castaneus* Julia-Zertuche &Treviño, 1978.

Holotype: UMMZ 101376, an adult female (F. Harrison, 1950).

Type locality: "10 miles northwest of Gómez Farías on the trail to La Joya de Salas, Tamaulipas, Mexico, at an altitude of about 5300 ft."

Distribution: NE Mexico (Sierra Madre Oriental in SE Coahuila, C Nuevo León and SW Tamaulipas).

Comments: Elevated to species level based on DNA sequence data by Bryson et al. (2014); in the *C. triseriatus* complex.

Crotalus oreganus Holbrook, 1840

Crotalus ornatus Hallowell, 1855

Distribution: Add Mexico (Nuevo León), Nevárez-de los Reyes et al. (2016a). Low elevation of 691 m, García-Padilla et al. (2016).

Comments: See under *C. molossus*.

Crotalus polisi Meik, Schaack, Flores-Villela & Streicher, 2018. J. Nat. Hist. 52(13–16): 15–17, fig. 5.

Holotype: MZFC 26408, a 468 mm female (J.M. Meik, S. Schaack & M.J. Ingrasci, 18 March 2010).

Type locality: "Cabeza de Caballo Island, Municipality de Ensenada, Baja California, Mexico. Coordinates: N 28.971 W 113.479."

Distribution: Mexico (Cabeza de Caballo I., Baja California).

Crotalus polystictus (Cope, 1865)

Crotalus pricei Van Denburgh, 1895

Distribution: Upper elevation of 3305 m, Contreras-Lozano et al. (2011).

Crotalus pusillus Klauber, 1952

Distribution: Add Mexico (N Colima), Bryson et al. (2014).

Comments: Bryson et al. (2014) present relationships as a part of the *C. triseriatus* complex.

Crotalus pyrrhus Cope, 1867. Proc. Acad. Nat. Sci. Philadelphia 18(4): 310. (*Caudisona pyrrha*)

Synonyms: *Crotalus goldmani* Schmidt, 1922, *Crotalus mitchelli aureus* Kallert, 1927, *Crotalus mitchellii muertensis* Klauber, 1949.

Holotype: USNM 6606 Cochran (1961) states that the holotype was collected by E. Coues in 1870, which is after the species was named.

Type locality: "Arizona," USA, more specifically from Canyon Prieto, Yavapai Co., Cochran (1961).

Distribution: SW USA (W Arizona, S California, S Nevada, SW Utah), NW Mexico (Baja California, including El Muerto, Smith Islands).

Comments: Meik et al. (2015), using data from multiple genes, recognize *pyrrhus* as a species distinct from *C. mitchellii*, and find the insular population referred to as *muertensis* to have little or no differentiation from mainland *C. pyrrhus*.

Crotalus ravus Cope, 1865

Distribution: Upper elevation of 3192 m, Palacios-Aguilar & Flores-Villela (2018).

Comments: Bryson et al. (2014) present relationships as a part of the *C. triseriatus* complex.

Crotalus ruber Cope, 1892

Synonyms: Remove *Crotalus ruber lorenzoensis* Radcliff & Maslin, 1975.

Distribution: Remove San Lorenzo Sur Island, Baja California.

Comments: See under *C. lorenzoensis*.

Crotalus scutulatus (Kennicott, 1861)

Distribution: Fernández-Badillo et al. (2016a) confirm the presence of *C. scutulatus* in México, Mexico.

Comments: Cardwell et al. (2013) discuss the type material and its geographic origin. E.A. Myers et al. (2017a) provide a DNA sequence-based phylogeny with three primary clades: those in eastern Mexico, which are sister populations to a pair of clades on either side of the Continental Divide. Schield et al. (2018) obtain a similar phylogeny using additional samples, but a deeper split in southern and eastern Mexican populations that correspond to the Mexican Plateau and to the isolated subspecies *salvini*. They do not recommend taxonomic changes. Watson et al. (2019) use morphology to test the potential for species boundaries, but find that variation is clinal across clades, and they, too, recommend against taxonomic changes.

Crotalus simus Latreille *in* Sonnini & Latreille, 1801a

Distribution: Add Honduras (El Paraíso), McCranie et al. (2014b); Nicaragua (Esteli), Sunyer et al. (2014).

Crotalus stejnegeri Dunn, 1919

Crotalus stephensi Klauber, 1930

Crotalus tancitarensis Alvarado-Díaz & Campbell, 2004

Crotalus thalassoporus Meik, Schaack, Flores-Villela & Streicher, 2018. J. Nat. Hist. 52(13–16): 18–19, fig. 5.

Holotype: MZFC 26410, a 369 mm male (J.M. Meik, S. Schaack & M.J. Ingrasci, 19 March 2010).

Type locality: "Piojo Island, Municipality de Ensenada, Baja California, Mexico. Coordinates: N 29.018 W 113.465."

Distribution: Mexico (Piojo I., Baja California).

Crotalus tigris Kennicott *in* Baird, 1859

Crotalus tlaloci Bryson, Linkem, Dorcas, Lathrop, Jones, Alvarado-Diaz, Grünwald & Murphy, 2014. Zootaxa 3826(3): 483–486, fig. 4.

Holotype: MZFC 3666, an adult female (E. Hernández-García, 20 June 1986).

Type locality: "'Los Llanos' (18°36′N, 99°37′W; 2200–2300 m above sea level…), 10 km by road from Taxco to Tetipac, Sierra de Taxco, municipality of Tetipac, state of Guerrero, Mexico."

Distribution: WC Mexico (NC Guerrero, México, W Morelos, E Michoacán), cloud forest, 2000–2400 m. Upper elevation of 2520 m, Palacios-Aguilar & Flores-Villela (2018).

Crotalus totonacus Gloyd & Kaufield, 1940

Distribution: Add Mexico (Hidalgo) and upper elevation of 1715 m, Frías et al. (2015). Farr et al. (2015) document known records for Nuevo Leon, Mexico.

Comment: Farr et al. (2015) find no morphological data to support recognition of *C. durissus neoleonensis* as a taxon.

Crotalus transversus Taylor, 1944

Crotalus triseriatus (Wagler, 1830)

Synonyms: Remove *Crotalus triseriatus armstrongi* Campbell, 1978.

Distribution: Limited to C Mexico (Mexico, E Michoacan, Morelos, WC Veracruz).

Comment: Bryson et al. (2014) present relationships as a part of the *C. triseriatus* complex.

Crotalus tzabcan Klauber, 1952

Crotalus unicolor Lidth de Jeude, 1887. Notes Leyden Mus. 9: 133–134. (*Crotalus horridus unicolor*)

Synonyms: *Crotalus pulvis* Ditmars, 1905.

Holotype: RMNH 613, a 62 cm specimen (J.R.H. Neervoort van de Poll).

Type locality: "Aruba."

Distribution: Aruba in the Netherlands Antilles.

Comments: The taxon *unicolor* is variously recognized as a subspecies of *C. durissus*, or as a species. Zaher et al. (2019) recognize it as a species that is sister taxon to *C. durissus*.

Crotalus vegrandis Klauber, 1941

Crotalus viridis (Rafinesque-Schmaltz, 1818)

Crotalus willardi Meek, 1906

Comments: Barker, *in* Schuett et al. (2016a), recommends that the five subspecies of *C. willardi* be elevated to full species based on morphological charactcristics of each, and probable underlying genetic differences between the subspecies.

CROTAPHOPELTIS Fitzinger, 1843 (Colubridae: Colubrinae)

Crotaphopeltis barotseensis Broadley, 1968

Distribution: Add Angola (Cuando Cubango), Branch (2018).

Crotaphopeltis braestrupi J.B. Rasmussen, 1985

Crotaphopeltis degeni (Boulenger, 1906)

Crotaphopeltis hippocrepis (J.T. Reinhardt, 1843)

Distribution: Add Mali (Koulikoro, Sikasso), Trape & Mané (2017).

Crotaphopeltis hotamboeia (Laurenti, 1768)

Distribution: Add Mali (Mopti, Segou), Trape & Mané (2017); Niger (Dosso, Tillaberi, Zinder), Trape & Mané (2015); Angola (Bie, Cauando-Cubango), Conradie et al. (2016b); Angola (Namibe), Ceríaco et al. (2016); Angola (Bengo, Benguela, Cabinda, Cuanza Norte, Cuanza Sul, Cunene, Huambo, Huila, Lunda Norte, Malanje, Zaire), M.P. Marques et al. (2018).

Crotaphopeltis tornieri (F. Werner, 1908)

CRYOPHIS Bogert & Duellman, 1963 (Dipsadidae: Dipsadinae)

Cryophis hallbergi Bogert & Duellman, 1963

CRYPTOPHIDION Wallach & Jones, 1994 (Xenopeltidae)

Cryptophidion annamense Wallach & Jones, 1994

CRYPTOPHIS Worrell, 1961. W. Aust. Nat. 8: 26. (Elapidae)

Synonyms: *Unechis* Worrell, 1961.

Type species: *Hoplocephalus pallidiceps* Günther 1858 by original designation.

Distribution: New Guinea and E Australia.

Comments: *Cryptophis* is re-recognized apart from the monotypic *Rhinoplocephalus*, Cogger (2014).

Cryptophis boschmai (Brongersma & Knaap-van Meeuwen, 1964)

Cryptophis incredibilis (Wells & Wellington, 1985)

Cryptophis nigrescens (Günther, 1862)

Cryptophis nigrostriatus (Krefft, 1964)

Cryptophis pallidiceps (Günther, 1858)

CUBATYPHLOPS Hedges, Marion, Lipp, Marin & Vidal, 2014. Caribbean Herpetol. 49: 46. (Typhlopidae: Typhlopinae)

Type species: *Typhlops biminiensis* Richmond, 1955 by original designation.

Distribution: Cuba, Cayman Islands, and the Bahamas.

Comments: The validity of *Cubatyphlops* is affirmed by Nagy et al. (2015).

Cubatyphlops anchaurus (J.P.R. Thomas & Hedges, 2007)
Cubatyphlops anousius (J.P.R. Thomas & Hedges, 2007)
Cubatyphlops arator (J.P.R. Thomas & Hedges, 2007)
Cubatyphlops biminiensis (Richmond, 1955)
Cubatyphlops caymanensis (Sackett, 1940)
Cubatyphlops contorhinus (J.P.R. Thomas & Hedges, 2007)
Cubatyphlops epactius (J.P.R. Thomas, 1968)
Cubatyphlops golyathi (Domínguez & Moreno, 2009)
Cubatyphlops notorachius (J.P.R. Thomas & Hedges, 2007)
Cubatyphlops paradoxus (J.P.R. Thomas, 1968)
Cubatyphlops perimychus (J.P.R. Thomas & Hedges, 2007)
Cubatyphlops satelles (J.P.R. Thomas & Hedges, 2007)

CUBOPHIS Hedges & Vidal *in* Hedges, Couloux & Vidal, 2009 (Dipsadidae: Xenodontinae)

Cubophis brooksi Barbour, 1914. Mem. Mus. Comp. Zool. 44(2): 333. (*Alsophis brooksi*)

Syntypes: MCZ 7893 (n = 2), (G. Nelson, March 1912).

Type locality: "Little Swan Island, Caribbean Sea."

Distribution: Honduras (Swan Is: Little Swan I.).

Comments: McCranie et al. (2017) and Zaher et al. (2019) recognize the Swan Island population as a species distinct from *C. cantherigerus*.

Cubophis cantherigerus (Bibron, 1840 *in* Ramon de la Sagra, 1838–1843)

Synonyms: Remove *Alsophis brooksi* Barbour, 1914.

Distribution: Add Bahamas (Anguilla Cays, Cay Sal Bank), Krysko et al. (2015). Remove Swan Islands, to *C. brooksi*.

Comments: See under *C. brooksi*.

Cubophis caymanus (Garman, 1887)
Cubophis fuscicaudus (Garman, 1888)
Cubophis ruttyi (Grant, 1940)
Cubophis vudii (Cope, 1862)

CYCLOCORUS A.M.C. Duméril, 1853 (Cyclocoridae)

Cyclocorus lineatus (J.T. Reinhardt, 1843)

Distribution: Add Philippines (Calayan, Camiguin Norte, Marinduque Is.), Leviton et al. (2018).

Cyclocorus nuchalis Taylor, 1923

Distribution: Add Philippines (Dinagat), Sanguila et al. (2016).

CYCLOPHIOPS Cope, 1888 (Colubridae: Colubrinae)

Fossil records: Early Pleistocene of Japan (Okinawa), Ikeda et al. (2016, as *Cyclophiops* sp.).

Comments: Figueroa et al. (2016), based on DNA sequence data, synonymize *Cyclophiops* with *Ptyas* due to paraphyly. *Ptyas* species arrange in sister clades, each containing one of the two sampled *Cyclophiops* taxa. The polyphyly could be resolved by resurrecting *Zapyrus* Günther, 1864, for the clade containing *C. multicinctus*.

Cyclophiops doriae Boulenger, 1888

Synonyms: Add *Ablabes hamptoni* Boulenger, 1900.

Distribution: Known only from NE India (Manipur), N Myanmar (Kachin, Mandalay), SE China (SE Yunnan).

Comments: Meetei et al. (2018) re-examined the holotype of *A. hamptoni* and conclude that the characters used to distinguish it are inadequate to separate it from *C. doriae*. They provide a revised description of *C. doriae*, with a locality map and color photographs.

Cyclophiops herminae (Boettger, 1895)

Cyclophiops major (Günther, 1858)

Distribution: Add Vietnam (Ha Giang), Ziegler et al. (2014).

Cyclophiops multicinctus (Roux, 1907)

Distribution: Add Thailand (Nan [?], Uttaradit), Hauser (2018); Laos (Bolikhamxay, Khammouane, Louangphrabang, Saravan), Hauser (2018); Vietnam (Bac Giang), Hecht et al. (2013); Vietnam (Ha Giang), Ziegler et al. (2014); Vietnam (Lam Dong, Lang Son), Hauser (2018); Vietnam (Son La), Pham et al. (2014); upper elevation of 1400 m, Hauser (2018).

Cyclophiops semicarinatus (Hallowell, 1861)

CYCLOPTYPHLOPS Bosch & Ineich, 1994 (Typhlopidae: Asiatyphlopinae)

Comments: Hedges et al. (2014), Pyron & Wallach (2014) provide generic diagnoses.

Cyclotyphlops deharvengi Bosch & Ineich, 1994

CYLINDROPHIS Wagler, 1828 (Cylindrophiidae)

Comments: Amarasinghe et al. (2015b) revise the concept of *C. ruffus* as consisting of four named species. Their arrangement leaves unnamed the populations in China, Laos, Thailand, Cambodia, Malaysia, Sumatra, and most of Borneo. Kieckbusch et al. (2018) partially rectify this situation.

Cylindrophis aruensis Boulenger, 1920

Distribution: Lang (2013) provides evidence that the species does not occur on the Aru Islands, and that the only known locality is Damar Island, Indonesia.

Cylindrophis boulengeri Roux, 1911

Cylindrophis burmanus M.A. Smith, 1943. Fauna Brit. India, Rept. Amphib. 3:97. (*Cylindrophis rufus burmanus*)

Lectotype: BMNH 1940.3.3.1, a 330 mm specimen (F.J. Meggitt), designated by Amarasinghe et al. (2015b).

Type locality: lectotype from "Rangoon, [Rangoon Prov.,] Burma."

Distribution: Burma (Ayeyarwady, Bago, Kachin, Mandalay, Rakhaing, Rangoon, Sagaing, Shan, Taninthayi, Tenasserim, Yangon).

Comments: Recognized as a species by Amarasinghe et al. (2015b), who provide a diagnosis and description. Kieckbusch et al. (2018) provide additional localities.

Cylindrophis engkariensis Stuebing, 1994

Cylindrophis isolepis Boulenger, 1896

Cylindrophis jodiae Amarasinghe, Campbell, Hallermann, Sidik, Supriatna & Ineich, 2015b. Amphibian & Reptile Conservation 9(1): 42–44, figs. 4, 5.

Holotype: MNHN 1911.0196, a 430 mm specimen (P. Eberhardt, prior to 1911).

Type locality: "Annam, Central Vietnam."

Distribution: SE Asia. Burma (Tanintharyi); Thailand (Bangkok, Chiang Mai, Phang Nga, Phetchaburi, Saraburi); Vietnam (An Giang, Ba Ria-Vung Tau, Ca Mau, Can Tho, Da Nang, Dong Nai, Ho Chi Minh City, Kien Giang, Quang Binh, Quang Nam, Quang Tri, Tay Ninh, Thua Thien-Hue, Vinh Phuc), Hong Kong and probably also SE China populations of *Cylindrophis*, Malaysia (Kedah). Add Vietnam (Binh Phuoc), T.Q. Nguyen et al. (2014).

Commenst: May be related to undescribed species in Cambodia and Thailand *fide* Amarasinghe et al. (2015b). Kieckbusch et al. (2016, 2018) provide additional localities.

Cylindrophis lineatus Blanford, 1881

Cylindrophis maculatus (Linnaeus, 1758)

Cylindrophis melanotus Wagler, 1828 *in* 1828–1833

Distribution: Add Indonesia (Sanana I., Moluccas), Kieckbusch et al. (2016).

Cylindrophis opisthorhodus Boulenger, 1897

Cylindrophis osheai Kieckbusch, Mader, H. Kaiser & Mecke, 2018. Zootaxa 4486(3): 237–243, figs. 1–3.

Holotype: RMNH 5460, a 486 mm female (D.S. Hoedt, 1865).

Type locality: "Boano Island, Central Maluku Regency, northern Maluku Province, Indonesia."

Distribution: Indonesia (Maluku: Boano Island).

Cylindrophis ruffus (Laurenti, 1768)

Synonyms: Remove *Scytale scheuchzeri* Merrem, 1820.

Add ***Cylindrophis mirzae*** Amarasinghe, Campbell, Hallermann, Sidik, Supriatna & Ineich 2015b. Amphibian & Reptile Conservation 9(1): 44–45, figs. 6, 7.

Holotype, MNHN 3279, a 429 mm specimen (J.F.T. Eydoux, during the expedition on the vessel La Favorite, 1829–1832).

Type locality: "Singapore."

Distribution: Possibly limited to Indonesia (Java).

Comments: Amarasinghe et al. (2015) restrict *C. ruffus* to Java, and provide a revised diagnosis and description. Kieckbusch et al. (2016) review the taxonomic history. They conclude that *Anguis rufa javanica* Gray, 1849 is a *species inquirenda*, and that *Scytale scheuchzeri* is not a synonym of *Cylindrophis*, but is likely a colubroid snake. They find that characters used to distinguish *C. mirzae* on Singapore are overlapped by values for sympatric and allopatric *C. ruffus*, and that *C. mirzae* is a synonym of the latter.

Cylindrophis subocularis Kieckbusch, Mecke, Hartmann, Ehrmantraut, O'Shea & H. Kaiser, 2016. Zootaxa 4093(1): 11–17, figs. 3–5.

Holotype: RMNH 8785, a 395 mm female (F. Kopstein, February 1937).

Type locality: "Grabag, Purworejo Regency (formerly Koetoardjo), Central Java Provence (Jawa Tengah), Java, Indonesia."

Distribution: Indonesia (SC Java). Known for certain only from the type locality.

Cylindrophis yamdena L.A. Smith & Sidik, 1998

DABOIA Gray, 1842 (Viperidae: Viperinae)

†***Daboia maghrebiana*** (Rage, 1976)

Daboia mauritanica (Gray, 1849)

Synonyms: Add *Vipera lebetina deserti* Anderson, 1892.

Distribution: Add Morocco (Al Hoceima), Mediani et al. (2015b); Morocco (Marrakech), P. Roux & Slimani (1992); Algeria (Tiaret), Ferrer et al. (2016). Also included is the range of the former *D. deserti*: E Tunisia and NW Libya (see comments), to which add Libya (Butnan, Jabal al Gharbi, Jafara, Nalut, Tripoli, Zawiyah), Bauer et al. (2017).

Comments: Martínez-Freiría et al. (2017a) use mtDNA sequence data to examine paleoecological effects on phylogeography of North African *Daboia*. The analysis recovers seven sublineages, six of which occur in western and southern Morocco. The seventh lineage ranges from eastern Morocco to Libya, and subsumes populations previously referred to as *D. deserti*, which they synonymize with *D. mauritanica*.

†***Daboia maxima*** (Szyndlar, 1988)

Daboia palaestinae (F. Werner, 1938)

Distribution: Add Turkey (Hatay), Göçmen et al. (2018).

Comments: Volynchik (2012) finds little microgeographic variation in morphology from different habitats in Israel.

Daboia russelii (Shaw & Niodder, 1797 *in* 1789–1813)

Distribution: Add India (Assam), Sengupta et al. (2019); India (Goa), Trivedi & Desai (2019).

†***Daboia sarmatica*** (Chkhikvadze & Lungu *in* Zerova, Lungu & Chkhikvadze, 1987)

Daboia siamensis (M.A. Smith, 1917)

Distribution: Add Cambodia (Oddar Meanchey, Siem Reap), Neang et al. (2015).

†**DAKOTAOPHIS** Holman, 1976 (Colubroidea: incerta sedis)

†***Dakotaophis greeni*** Holman, 1976

DASYPELTIS Wagler, 1830 (Colubridae: Colubrinae)

Comments: Saleh & Sarhan (2016) provide an mtDNA-based phylogeny of nine species of *Dasypeltis*. Bates & Broadley (2018) revise the taxonomy of species occurring in northeastern Africa and Arabia.

Dasypeltis abyssina (A.M.C. Duméril, Bibron & Duméril, 1854)

Distribution: Elevation range 1800–2450 m, Bates & Broadley (2018).

Comments: Bates & Broadley (2018) provide a species account, with description, color photos, and distribution map. The specific epithet is misspelled in Wallach et al. (2014).

Dasypeltis arabica Bates & Broadley, 2018. Indago 34(1):46–52, figs. 27–29.

Holotype: BMNH 1987.2192, a 714 mm female (M. Al-Safadi).

Type locality: "Sana'a, North Yemen (15°19′N, 44°14′E)."

Distribution: W Yemen, Saudi Arabia (Asir), 1300–2300 m.

Dasypeltis atra Sternfeld, 1913

Distribution: Add Tanzania (Arusha, Kagera, Kigoma), Bates & Broadley (2018).

Comments: Bates & Broadley (2018) provide a species account, with description, color photos, and distribution map.

Dasypeltis bazi Saleh & Sarhan, 2016. Bull. Soc. Herpetol. France 160: 31–36, figs. 3–5.

Holotype: AUZC R09458, a 521 mm female (September 2009).

Type locality: "Abu-Gandir, Faiyum Governorate, Egypt (29°15′12 N, 30°40′35 E)."

Distribution: Egypt (Cairo, Fayoum). Bates & Broadley (2018, as *Dasypeltis* cf. *bazi*) add SE Sudan (Al Bahr Al Ahmar).

Comments: Bates & Broadley (2018) provide a species account, with description, color photos, and distribution map.

Dasypeltis confusa Trape & Mané, 2006

Distribution: Add Ghana, Liberia and Nigeria, Trape & Baldé (2014); Mali (Sikasso), Trape & Mané (2017); South Sudan (Central Equatoria, Eastern Equatoria), Ethiopia (no precise locality), Kenya (Rift Valley, Western), Rwanda and Uganda (Buikwe, Buliisa, Gulu, Lira, Moyo, Serere), Bates & Broadley (2018); Angola (Cuando Cubango, Malanje), Branch (2018). Increase elevation to 2044 m, Ineich et al. (2015).

Comments: Bates & Broadley (2018) provide a species account, with description, color photos, and distribution map.

Dasypeltis crucifera Bates & Broadley, 2018. Indago 34(1):42–46, figs. 23–25.

Holotype: ZMB 7631, a 506 mm female (purchased, latter 1800s).

Type locality: "Bogos, Eritrea." See discussion in Bates & Broadley (2018).

Distribution: Eritrea, 600–1417 m.

Dasypeltis fasciata A. Smith, 1849 *in* 1838–1849

Distribution: Add Sierra Leone, Trape & Baldé (2014); Gabon (Ogooue-Lolo), Pauwels et al. (2016a).

Comments: Bates & Broadley (2018) provide a species account, with description, color photos, and distribution map.

Dasypeltis gansi Trape & Mané, 2006

Distribution: Add Ghana and Ivory Coast, Trape & Baldé (2014); Mali (Segou), Trape & Mané (2017); Niger (Tillaberi, Zinder), Trape & Mané (2015).

Dasypeltis inornata A. Smith, 1849 *in* 1838–1849

Dasypeltis latericia Trape & Mané, 2006

Dasypeltis loveridgei Mertens, 1954. Zool. Anz. 152(9/10): 213–214. (*Dasypeltis scabra loveridgei*)

Holotype: SMF 46642, a 663 mm female (W. Metzler, 29 October 1952).

Type locality: "Farm Finkenstein bei Windhoek, Südwestafrika."

Distribution: Namibia, and perhaps adjacent areas.

Comments: Elevated to species by Bates & Broadley (2018).

Dasypeltis medici (Bianconi, 1859)

Distribution: Add Kenya (Eastern, Nairobi, North Eastern) and Tanzania (Arusha, Dodoma, Manyara), and increase elevation to 2500 m, Bates & Broadley (2018).

Comments: Bates & Broadley (2018) provide a species account, with description, color photos, and distribution map.

Dasypeltis palmarum (Leach *in* Tuckey, 1818)

Dasypeltis parascabra Trape, Mediannikov & Trape, 2012

Dasypeltis sahelensis Trape & Mané, 2006

Distribution: Add Morocco (Tiznit), Crochet et al. (2015); Niger (Tillaberi), Trape & Mané (2015). Upper elevation of 467 m, Trape & Mané (2015). Jiménez-Robles et al. (2017) document additional localities.

Dasypeltis scaber (Linnaeus, 1758)

Synonyms: Remove *Dasypeltis scabra loveridgei* Mertens, 1954.

Distribution: Add South Sudan (Eastern Equatoria), Bates & Broadley (2018); Angola (Benguela, Cabinda, Cuando Cubango, Cuanza Norte, Huambo, Huila, Luanda, Lunda Sul, Malanje, Moxico, Zaire), M.P. Marques et al. (2018). Remove Egypt, Saleh & Sarhan (2016), much or all of Namibia, Bates & Broadley (2018).

Comments: Bates & Broadley (2018) provide a species account, with description, color photos, and distribution map.

Dasypeltis taylori Bates & Broadley, 2018. Indago 34(1):27–29, figs. 13, 14.

Holotype: BMNH 1949.2.2.2, a 202 mm male (Colonel R.H.R. Taylor, 20 April 1932).

Type locality: "Haud [former British Somaliland, 08°20′N, 46°00′E at 2100 ft = 640 m] in northern Somalia."

Distribution: Djibouti and Somalia (Awdal, Sanaag, Togdheer, Woqooyi Galbeed), 640–1370 m.

†***DAUNOPHIS*** Swinton, 1926 (Booidea: incerta sedis)

†***Daunophis langi*** Swinton, 1926

†***DAWSONOPHIS*** Holman, 1979 (Booidea: incerta sedis)

†***Dawsonophis wyomingensis*** Holman, 1979

DEINAGKISTRODON Gloyd, 1979 (Viperidae: Crotalinae)

Deinagkistrodon acutus (Günther, 1888)

Distribution: Add Vietnam (Ha Giang), and upper elevation of 1640 m, Pham et al. (2017); Vietnam (Yen Bai), Le et al. (2018).

DEMANSIA Gray, 1842 (Elapidae)

Demansia angusticeps (Macleay, 1888)
Demansia calodera Storr, 1978
Demansia flagellatio Wells & Wellington, 1985
Demansia olivacea (Gray, 1842)
Demansia papuensis (Macleay, 1877)
Demansia psammophis (Schlegel, 1837)
Demansia quaesitor Shea in Shea & Scanlon, 2007
Demansia reticulata (Gray, 1842)
Demansia rimicola Shea in Shea & Scanlon, 2007
Demansia rufescens Storr, 1978
Demansia shinei Shea in Shea & Scanlon, 2007
Demansia simplex Storr, 1978
Demansia torquata Günther, 1862
Demansia vestigiata (De Vis, 1884)

DENDRELAPHIS Boulenger, 1890 (Colubridae: Ahaetullinae)

Comments: Rooijen et al. (2015) revise the species of Australo-Papua. Wickramasinghe (2016) provides a key to Sri Lanka species

Dendrelaphis andamanensis (J. Anderson, 1871)
Dendrelaphis ashoki Vogel & Rooijen, 2011
Dendrelaphis bifrenalis (Boulenger, 1890)
Dendrelaphis biloreatus Wall, 1908

Dendrelaphis calligastra (Günther, 1867)

Synonyms: Add *Dendrophis salomonis* Günther, 1872. Remove *Dendrophis calligastra keiensis* Mertens, 1926.

Distribution: Distribution includes that of *D. salomonis*: Bougainville and the Solomon Islands. Add Indonesia: Misool and Yapen Islands), Rooijen et al. (2015); Papua New Guinea (Karkar, Kranket, Manam, Manus, Mussau, New Ireland, Tarawai and Wallis Islands), Clegg & Jocque (2016); Papua New Guinea (Darnley Island), Rooijen et al. (2015). Remove Indonesia (Babar Island), Rooijen et al. (2015).

Comments: Rooijen et al. (2015) discuss the syntypes of *Dendrelaphis schlenckeri* Ogilby, 1898. They also synonymize *Dendrophis salomonis*, remove *D. c. keiensis*, and provide a species account with a revised description and color photographs. Additional morphological data are used by Rooijen & Vogel (2016) to confirm the synonymy of *D. salomonis*. They designate, describe and illustrate a neotype (ZMB 24024) for *D. c. distinguendus*, and describe the syntypes of *D. salomonis* as well as provide photographs of one of them.

Dendrelaphis caudolineatus (Gray, 1834 *in* Gray & Hardwicke, 1830–1835)

Distribution: Add Indonesia (Pulau Bangkuru off Sumatra), Tapley & Muurmans (2016). Patel et al. (2019) discuss the veracity of India records.

Dendrelaphis caudolineolatus (Günther, 1869)
Dendrelaphis chairecacos (F. Boie, 1827)

Dendrelaphis cyanochloris (Wall, 1921)

Distribution: Add Bhutan (Trongsa), Tshewang & Letro, 2018); Laos (Champasak), Teynié et al. (2004); West Malaysia (Terengganu), Sumarli et al. (2015).

Dendrelaphis flavescens Gaulke, 1994

Dendrelaphis formosus (F. Boie, 1827)

Distribution: Add Myanmar (Tanintharyi), Mulcahy et al. (2018); West Malaysia (Pulau Singa Besar), B.L. Lim et al. (2010); West Malaysia (Pulau Tioman), K.K.P. Lim & Lim (1999); West Malaysia (Terengganu), Sumarli et al. (2015).

Dendrelaphis fuliginosus Griffin, 1909

Distribution: Add Philippines (Cebu I.), Leviton et al. (2018).

Dendrelaphis gastrostictus (Boulenger, 1894)

Comments: Rooijen et al. (2015) provide a species account with a revised description and color photographs. Rooijen & Vogel (2016) re-examine four syntypes of *D. meeki* and confirm that they resemble *D. gastrostictus*.

Dendrelaphis girii Vogel & Rooijen, 2011
Dendrelaphis grandoculis (Boulenger, 1890)
Dendrelaphis grismeri Vogel & Rooijen, 2008

Dendrelaphis haasi Rooijen & Vogel, 2008

Distribution: Add Myanmar (Tanintharyi), Mulcahy et al. (2018).

Dendrelaphis hollinrakei Lazell, 2002
Dendrelaphis humayuni Tiwari & Biswas, 1973
Dendrelaphis inornatus Boulenger, 1897

Dendrelaphis keiensis Mertens, 1926. Senckenberg. Biol. 6: 277–278. (*Dendrophis calligastra keiensis*)

Holotype: SMF 18662, a 1352 mm female (H. Merton & J. Roux, 29 May 1908).

Type locality: "Dulah, Kei-Inseln," Maluku, Indonesia.

Distribution: E Indonesia (Babar, Kei Dulah and Timor Laut Islands). Add Indonesia (Maluku: Kei Besar, Kei Cecil, Kur Islands), Karin et al. (2018); Indonesia (Dulah, Yamdena Islands), Lang (2013).

Comments: Rooijen et al. (2015) elevate this taxon to species based on distinct morphology from *D. calligastra*. They provide a species account with revised description and color photographs.

Dendrelaphis kopsteini Vogel & Rooijen, 2007
Dendrelaphis levitoni Rooijen & Vogel, 2012

Dendrelaphis lineolatus Hombron & Jacquinot, 1842. Voyage Pole Sud. Zoologie, Atlas: plate 2, fig. 1, and Vol. 3: 20–21. (*Dendrophis lineolata*)

Synonyms: *Dendrophis punctulatus astrostriata* A.B. Meyer, 1874, *Dendrophis punctulatus fasciata* A.B. Meyer, 1874, *Dendrophis elegans* Ogilby, 1891.

Holotype: MNHN 5081, unsexed.

Type locality: "Nouvelle-Guinée."

Distribution: E Indonesia (Papua, and Misool, Salawati and Yapen Islands), Papua New Guinea (the precise distribution is not detailed by Rooijen et al., 2015).

Comments: Rooijen et al. (2015) revalidate this taxon based on morphological distinctiveness from *D. punctulatus*. They designate neotypes for *Dendrophis punctulatus astrostriata* A.B. Meyer, 1874 and *D. p. fasciata* A.B. Meyer, 1874, and provide a species account with a description and color photographs.

Dendrelaphis lorentzii (Lidth de Jeude, 2011)

Distribution: Rooijen et al. (2015) add E Indonesia (Bivak and Salawati Islands, Papua), and New Guinea (Normanby Island, Milne Bay).

Comments: Rooijen et al. (2015) provide a species account with a revised description and color photographs.

Dendrelaphis luzonensis Leviton, 1961

Distribution: Add Philippines (Marinduque I.), Leviton et al. (2018).

Dendrelaphis macrops Günther, 1877. Proc. Zool. Soc. London 1877: 131, fig. 2. (*Dendrophis macrops*)

Synonyms: *Dendrophis breviceps* Macleay, 1877, *Dendrophis papuae* Ogilby, 1891.

Holotype: BMNH 1946.1.23.42, an 1145 mm male (G. Brown).

Type locality: "Duke-of-York Island," East New Britain Distr., Papua New Guinea.

Distribution: E Indonesia (Papua, including Numfor Island), Papua New Guinea (East New Britain, East Sepik, Madang, Western, including Daru and Duke of York Islands). Add Papua New Guinea (Kairiru and Mioko Islands), Clegg & Jocque (2016).

Comments: Rooijen et al. (2015) determine that *Dendrophis macrops* Günther, 1877 has temporal priority over *D. breviceps* Macleay, 1877. They revalidate this species from *D. punctulatus* based on morphological distinctiveness, and provide a species account with a description and color photographs.

Dendrelaphis marenae Vogel & Rooijen, 2008

Distribution: Add Indonesia (Sula Islands), Lang (2013); Philippines (Carabao I., Tablas I.), Siler et al. (2012); Philippines (Cagayan Sulu, Guimaras, Jolo, Marinduque, Siagao Is.), Leviton et al. (2018).

Dendrelaphis modestus Boulenger, 1894

Distribution: Add Indonesia (Bisa, Kasiruta, Tidore Islands), Lang (2013).

Dendrelaphis ngansonensis (Bourret, 1935)

Distribution: Add Cambodia (Pursat), Neang et al. (2015); Vietnam (Dien Bien), Dung et al. (2014); Vietnam (Hai Phong: Cat Ba Island), T.Q. Nguyen et al. (2011); Vietnam (Quang Ngai), Nemes et al. (2013).

Dendrelaphis nigroserratus Vogel, Rooijen & Hauser, 2012

Dendrelaphis oliveri (Taylor, 1950)

Dendrelaphis papuensis Boulenger, 1895

Distribution: Limited to the Trobriand Islands, Milne Bay Province, Papua New Guinea according to Rooijen et al. (2015), who discount mainland New Guinea records.

Comments: Rooijen et al. (2015) provide a species account with a revised description and color photographs.

Dendrelaphis philippinensis (Günther, 1879)

Distribution: Leviton et al. (2018) could find no basis for the inclusion of the following Philippine islands in the distribution: Camiguin, Dinagat, Siquijor, Surigao.

Dendrelaphis pictus (Gmelin, 1789)

Distribution: Add Myanmar (Bago, Rakhine, Sagaing), Platt et al. (2018); Myanmar (Tanintharyi), Mulcahy et al. (2018); Cambodia (Siem Reap), Geissler et al. (2019); Vietnam (Hoa Binh), Nguyen et al. (2018); West Malaysia (Jerejak Island), Quah et al. (2011); West Malaysia (Kedah), Shahriza et al. (2013); West Malaysia (Pulau Singa Besar), B.L. Lim et al. (2010).

Dendrelaphis proarchos (Wall, 1909)

Distribution: Add India (Arunachal Pradesh), Purkayastha (2018).

Dendrelaphis punctulatus (Gray *in* King, 1827)

Synonyms: Add *Dendrophis fuscus* Jan, 1863. Remove *Dendrophis lineolata* Hombron & Jacquinot, 1842, *Dendrophis striolatus* Peters, 1867, *Dendrophis punctulatus astrostriata* A.B. Meyer, 1874, *Dendrophis punctulatus fasciata* A.B. Meyer, 1874, *Dendrophis macrops* Günther, 1877, *Dendrophis breviceps* Macleay, 1877, *Dendrophis elegans* Ogilby, 1891, *Dendrophis papuae* Ogilby, 1891.

Distribution: Restricted to Australia according to Rooijen et al. (2015) through revalidation of other taxa to New Guinea populations.

Comments: Rooijen et al. (2015) discuss the syntypes of *D. bilorealis* Macleay, 1884, and provide a species account with a revised description and color photographs.

Dendrelaphis schokari (Kuhl, 1820)

Distribution: Add Sri Lanka (Sabaragamuwa), Peabotuwage et al. (2012).

Dendrelaphis sinharajensis Wickramasinghe, 2016. Zootaxa 4162(3): 506–510, figs. 2–7.

Holotype: NMSL 2016.06.01, a 995 mm female (L.J.M. Wickramasinghe).

Type locality: "Mideripitiya, Sinharaja Forest, Deniyaya, Matara District, Southern Province (N 06° 21′ 24.72″, E 080° 29′ 21.63″), 285 m," Sri Lanka.

Distribution: Sri Lanka (Southern, Sabaragamuwa), 285 m.

Comments: Wickramasinghe (2016) tentatively assigns *sinharajensis* to *Dendrelaphis*.

Dendrelaphis striatus (Cohn, 1905)

Distribution: Add Myanmar (Tanintharyi), Mulcahy et al. (2018).

Dendrelaphis striolatus W.C.H. Peters, 1867. Monatsb. Preuss. Akad. Wiss. Berlin 1867: 25. (*Dendrophis striolatus*)

Holotype: ZMB 5450, an 884 mm male (collection of C. Semper).

Type locality: "Pelew [=Palau]-Inseln."

Distribution: Palau (Babelthuap, Koror, Ngeaur, Ngerekebesang, Ngerukeuid and Ngermalk Islands).

Comments: Rooijen et al. (2015) revalidate this species from *D. punctulatus* based on morphological distinctiveness, and provide a species account with a description and color photographs.

Dendrelaphis subocularis (Boulenger, 1888)
Dendrelaphis terrificus (W.C.H. Peters, 1872)
Dendrelaphis tristis (Daudin, 1803)

Distribution: Add India (Uttarakhand), Joshi et al. (2019); Nepal (Bara), Pandey et al. (2018); Nepal (Parsa), Bhattarai et al. (2018b).

Dendrelaphis underwoodi Rooijen & Vogel, 2008
Dendrelaphis walli Vogel & Rooijen, 2011

DENDROASPIS Schlegel, 1848 (Elapidae)

Dendroaspis angusticeps (A. Smith, 1849 *in* 1838–1849)
Dendroaspis intermedia Günther, 1865
Dendroaspis jamesoni (Traill, 1843)

Distribution: Add Gabon (Estuaire), Pauwels et al. (2017b); Angola (Bengo, Cuanza Sul), Branch (2018); Angola (Cabinda, Cuanza Norte, Huambo, Luanda, Malanje, Zaire), M.P. Marques et al. (2018). Ceríaco et al. (2018) provide evidence that the species does not occur on São Tomé.

Dendroaspis polylepis Günther, 1864

Distribution: Add Angola (Cauando-Cubango), Conradie et al. (2016b); Angola (Bie, Huila, Lunda Norte, Malanje, Zaire), M.P. Marques et al. (2018); Mozambique (Inhambane), Jacobsen et al. (2010).

Dendroaspis viridis (Hallowell, 1844)

DENDROLYCUS Laurent, 1956 (Lamprophiidae)

Dendrolycus elapoides (Günther, 1874)

DENDROPHIDION Fitzinger, 1843 (Colubridae: Colubrinae)

Dendrophidion apharocybe Cadle, 2012

Distribution: Add Panama (Panama Oeste), Ray & Ruback (2015).

Dendrophidion atlantica Freire, Caramaschi & Gonçalves, 2010

Distribution: Add Brazil (Paraiba), Barbosa et al. (2019); Brazil (Pernambuco), Nacimento & Santos (2016).

Dendrophidion bivittatum (A.M.C. Duméril, Bibron & Duméril, 1854)

Distribution: Add Colombia (Huila), Moreno-Arias & Quintero-Corzo (2015).

Dendrophidion boshelli Dunn, 1944
Dendrophidion brunneum (Günther, 1858)

Distribution: Add Ecuador (Esmeraldas), Torres-Carvajal et al. (2019); Peru (Lambayeque), Venegas (2005).

Dendrophidion clarku Dunn, 1933

Distribution: Panama (Panama Oeste), Ray & Ruback (2015).

Dendrophidion crybelum Cadle, 2012
Dendrophidion dendrophis (Schlegel, 1837)

Distribution: Add Ecuador (Zamora-Chinchipe), Torres-Carvajal et al. (2019); Bolivia (La Paz), Cortez-Fernandez (2005, new for Bolivia); Brazil (Marajo I., Para), G.M.

Rodrigues et al. (2015); Brazil (Rondônia), Bernarde et al. (2012b).

Dendrophidion graciliverpa Cadle, 2012
Dendrophidion nuchale (W.C.H. Peters, 1863)
Dendrophidion paucicarinatum (Cope, 1894)
Dendrophidion percarinatum (Cope, 1894)

Distribution: Add Panama (Panama Oeste), Ray & Ruback (2015); Colombia (Huila), Moreno-Arias & Quintero-Corzo (2015).

Dendrophidion prolixum Cadle, 2012
Dendrophidion rufiterminorum Cadle & Savage, 2012
Dendrophidion vinitor H.M. Smith, 1941

DENISONIA Krefft, 1869 (Elapidae)

Denisonia devisi Waite & Longman, 1920
Denisonia maculata (Steindachner, 1867)

†DIABLOPHIS Caldwell, Nydam, Palci and Apesteguía, 2015. Nature Communic. 6(5996): 2–3. (Ophidia: incerta sedis)

Type species: *Parviraptor gilmorei*, Evans 1996 by monotypy.

Distribution: Upper Jurassic of Colorado, USA.

†*Diablophis gilmorei* Evans, 1996. Northern Arizona Mus. Bull 60. (*Parviraptor gilmorei*)

Holotype: LACM 4684/140572, a right maxilla, right mandible, and axis vertebra, as restricted by Caldwell et al. (2015).

Type locality: Fruita locality, Mesa Co., Colorado, USA.

Distribution: Upper Jurassic (Kimmeridgian) of Colorado, USA.

Comments: Originally described as an anguimorph lizard.

DIADOPHIS Baird & Girard, 1853 (Dipsadidae: subfamily unnamed)

†*Diadophis elinorae* Auffenberg, 1963
Diadophis punctatus (Linnaeus, 1766)

DIAPHOROLEPIS Jan, 1863 (Dipsadidae: Dipsadinae)

Comments: Pyron et al. (2015) place *Diaphorolepis* in a new tribe (Diaphorolepidini) with *Emmochliophis* and *Synophis*.

Diaphorolepis laevis F. Werner, 1923

Comments: Pyron et al. (2015) provide a revised description of the holotype, from which this taxon remains known.

Diaphorolepis wagneri Jan, 1863

Type locality: restricted to "Milpé, Pichincha province, Ecuador (0.035, −78.87; 1076 m)," by Pyron et al. (2015).

Distribution: Upper elevation of 1600 m, Vera-Pérez et al. (2018).

Comments: Pyron et al. (2015) provide a species account (description and map of localities).

DIEUROSTUS Berg, 1901 (Homalopsidae)

Comments: J.C. Murphy & Voris (2014) provide a generic diagnosis.

Dieurostus dussumierii (A.M.C. Duméril, Bibron & Duméril, 1854)

Comments: J.C. Murphy & Voris (2014) provide a description, and suggest that the synonym *Hypsirhina malabarica* may be a valid species.

†*DINILYSIA* Woodward, 1901 (Dinilysiidae)

†*Dinilysia patagonica* Woodward, 1901

Fossil Record: Scanferla & Canale (2007) describe a specimen from the early Campanian, being the youngest record for the species.

Comments: Palci & Caldwell (2014) conclude that the species lacks a crista circumfenestralis, the ambiguous presence thereof having produced marginal conflicts in its phylogenetic position.

DIPSADOBOA Günther, 1858 (Colubridae: Colubrinae)

Comments: Branch et al. (2019c) use DNA sequence data to produce a phylogeny of nine of the species.

Dipsadoboa aulica (Günther, 1864)

Dipsadoboa duchesnii (Boulenger, 1901)

Synonyms: Add *Dipsadomorphus brevirostris* Sternfeld, 1908.

Distribution: Add Gabon (Ogooué-Ivindo), Carlino & Pauwels (2015).

Comments: Trape & Baldé (2014) note that two of the syntypes of *D. brevirostris* are specimens of *D. duchesnii*, and the third has characteristics similar to those of both *D. duchesnii* and *D. guineensis*. In addition, specimens assignable to *D. guineensis* are not known from the geographic range of *D. duchesnii*. They recommend synonymizing *D. brevirostris* with *D. duchesnii*, and recognizing *D. guineensis* for West African populations.

Dipsadoboa flavida (Broadley & Stevens, 1971)

Dipsadoboa guineensis Chabanaud, 1920. Bull. Comité Études Hist. Scient. Afrique Occid. Franç. 4: 491–492. (*Leptodira guineensis*)

Syntypes: MNHN, lost according to Trape & Baldé (2014).

Type locality: "Diéké (cercle de N'Zèrékoré)," Guinea.

Distribution: West Africa: Guinea, Sierra Leone, Liberia, Ivory Coast and Ghana.

Dipsadoboa kageleri Uthmöller, 1939. Zool. Anz. 125(5/6): 108–112. (*Crotaphopeltis hotamboiea kageleri*)

Holotype: ZSM 254/1937, a 615 mm male (W. Uthmöller, 1937).

Type locality: "'Sanya', Tanzania."

Distribution: E Tanzania (Arusha, Kilimanjaro, Lindi), 1526–1840 m.

Comments: Branch et al. (2019c) discover that populations from N Tanzania are sister taxon to a group of species that includes *D. shrevei*, and elevate the former to species level.

Dipsadoboa montisilva Branch, Conradie & Tolley *in* Branch et al., 2019c. Zootaxa 4646(3): 551–554, fig. 3.

Holotype: PEM R21122, a 1084 mm male (K.A. Tolley, S.P. Loader, W. Conradie, G.B. Bittencourt-Silva, M. Menegon, H.M. Engelbrecht & C. Nanvonamuquitxo, 21 November 2014).

Type locality: "Mt Mabu Forest Base Camp (16°17′10.4″S, 36°24′00.2″E, 919 m above sea level ~a.s.l.), Zambezia Province, Mozambique."

Distribution: N Mozambique (Zambezia), 919–1644 m.

Dipsadoboa shrevei (Loveridge, 1932)

Synonyms: Remove *Crotaphopeltis hotamboiea kageleri* Üthmöller, 1939.

Distribution: C Africa. S Democratic Republic of the Congo (Bandundu, Bas-Congo, Katanga, Kinshasa, Sud-Kivu), Zambia (Central, Copperbelt, Luapula, Northern, North-Western, Western), E Angola (Bie, Huambo, Moxico).

Dipsadoboa underwoodi J.B. Rasmussen, 1993

Distribution: Add Liberia, Trape & Baldé (2014); Gabon (Estuaire), Pauwels et al. (2016b), Gabon (Haut-Ogooue), Tarn et al. (2018a); Gabon (Ogooué-Ivindo), Carlino & Pauwels (2015); Congo (Niari), Branch et al. (2019c).

Dipsadoboa unicolor Günther, 1858

Dipsadoboa viridis (W.C.H. Peters, 1869)

Distribution: Add Gabon (Haut-Ogooue), Tarn et al. (2018b).

Dipsadoboa weileri (Lindholm, 1905)

Distribution: Add Senegal (Kedougou, as *D.* aff. *weileri*), Monasterio et al. (2016); Nigeria (Cross River), Branch et al. (2019c). Upper elevation of 2044 m, Ineich et al. (2015).

Comments: Trape & Baldé (2014) note that Guinea specimens are a species distinct from *D. weileri* based on habitat and coloration, but there are no clear differences in scalation.

Dipsadoboa werneri (Boulenger, 1897)

Distribution: Add Tanzania (Morogoro), Menegon et al. (2008), Lyakurwa (2017).

DIPSAS Laurent, 1768 (Dipsadidae: Dipsadinae)

Synonyms: Add *Sibynomorphus* Fitzinger, 1843, *Anholodon* A.M.C. Duméril, Bibron & Duméril, 1854, *Cochliophagus* A.M.C. Duméril, Bibron & Duméril, 1854, *Pseudopareas* Boulenger, 1896.

Comments: Arteaga et al. (2018) use DNA sequence-data to produce a phylogeny using 58% of known Dipsadini species. Their results place *Sibynomorphus* as a new synonym of *Dipsas*.

Dipsas albifrons (Sauvage, 1884)

Distribution: Add Brazil (S Bahia), I.R. Dias et al. (2018); Brazil (SE Minas Gerais), and upper elevation of 1020 m, Silveira et al. (2018); Brazil (Rio de Janeiro: Ilha Grande), C.F.D. Rocha et al. (2018). Silveira et al. (2018) provide a map of known localities.

Dipsas alternans (Fischer, 1885)

Distribution: Add Brazil (Rio de Janeiro: Ilha Grande), C.F.D. Rocha et al. (2018).

Dipsas andiana (Boulenger, 1896)

Distribution: Add Ecuador (Cañar, Cotopaxi, El Oro), and low elevation of 7 m, Arteaga et al. (2018); Ecuador (Manabi), Almendáriz et al. (2012).

Dipsas articulata (Cope, 1868)

Distribution: Add Nicaragua (Rio San Juan), and low elevation of 10 m, Sunyer et al. (2014); Panama (Coclé, Panama), Vecchiet et al. (2014).

Dipsas baliomela Harvey, 2009

Dipsas bicolor (Günther, 1895 *in* 1885–1902)

Dipsas bobridgleyi Arteaga, Salazar-Valenzuela, Mebert, Peñafiel, Aguiar, Sánchez-Nivicela, Pyron, Colston, Cineros-Heredia, Yánez-Muñoz, Venegas, Guayasamin & Torres-Carvajal, 2018. ZooKeys 766:108–109, figs. 1, 9, 10.

Holotype: MZUTI 5417, a 533 mm male (M. Hollanders, 1 August 2017).

Type locality: "Reserva Buenaventura, province of El Oro, Ecuador (S3.65467, W79.76794; 524 m)."

Distribution: Ecuador (Azuay, El Oro), Peru (Tumbes), 39–572 m.

Dipsas brevifacies (Cope, 1866)

Dipsas bucephala (Shaw, 1802)

Synonyms: Remove *Leptognathus cisticeps* Boettger, 1885.

Distribution: Limited to SE Brazil (Bahia, Espirito Santo, Rio Grande do Sul, Sao Paulo), N Argentina (Misiones), and E Paraguay (Alto Parana, Canindeyu, Itapua, Misiones). Add Brazil (Mato Grosso), M.C. Silva et al. (2015); Brazil (Parana), Dainesi et al. (2019).

Comments: See under *D. cisticeps*.

Dipsas catesbeji (Seetzen, 1796)

Distribution: Add Ecuador (Orellana, Sucumbios), and upper elevation of 1476 m, Arteaga et al. (2018), Brazil (Amapa), Pedroso-Santos et al. (2019); Brazil (Roraima), Moraes et al. (2017).

Dipsas chaparensis R.P. Reynolds & Foster, 1992

Dipsas cisticeps Boettger, 1885. Z. Naturwiss. 58: 237–238. (*Leptognathus* [*Dipsadomorus*] *cisticeps*)

Holotype: unlocated according to Uetz et al. (2019), a 605 mm specimen.

Type locality: "Paraguay, Amer. Merid."

Distribution: E Bolivia (Cochabamba, Santa Cruz), S Paraguay (Central, Paraguari). Add Brazil (Mato Grosso do Sul), Ferreira et al. (2017); Paraguay (San Pedro), Atkinson et al. (2017).

Comments: Cacciali et al. (2016b) conclude that *D. cisticeps* and *D. bucephalus* are not subspecies, based on the very different color patterns and the wide hiatus in their ranges, and they recognize both as species.

Dipsas copei (Günther, 1872)

Dipsas elegans (Boulenger, 1896)

Distribution: Add Ecuador (Santo Domingo), and upper elevation of 2579 m, Arteaga et al. (2018).

Dipsas ellipsifera (Boulenger, 1898)

Distribiton: Add Ecuador (Carchi), Arteaga et al. (2018).

Dipsas gaigeae (Oliver, 1937)

Distribution: Add Mexico (Guerrero), Palacios-Aguilar & Flores-Villela (2018).

Dipsas georgejetti Arteaga, Salazar-Valenzuela, Mebert, Peñafiel, Aguiar, Sánchez-Nivicela, Pyron, Colston, Cineros-Heredia, Yánez-Muñoz, Venegas, Guayasamin & Torres-Carvajal, 2018. ZooKeys 766:110–111, figs. 1, 11, 12.

Holotype: MZUTI 5411, a 402 mm male (M. Costales, 31 August 2017).

Type locality: "Cabuyal, province of Manabí, Ecuador (S0.19698, W80.29059; 15 m)."

Distribution: Ecuador (Guayas, Manabi), 5–317 m.

Dipsas gracilis (Boulenger, 1902)

Distribution: Add Ecuador (Cañar, Santo Domingo), and elevation range of 14–1638 m, Arteaga et al. (2018); Ecuador (Manabi), Almendáriz et al. (2012).

Dipsas inaequifasciata (A.M.C. Duméril, Bibron & Duméril, 1854)

Dipsas incerta (Jan, 1863)

Dipsas indica Laurenti, 1768)

Distribution: Add Ecuador (Orellana), and upper elevation of 1355 m, Arteaga et al. (2018), Brazil (Rio de Janeiro: Ilha Grande), C.F.D. Rocha et al. (2018); Brazil (Roraima), Moraes et al. (2017).

Dipsas jamespetersi Arteaga, Salazar-Valenzuela, Mebert, Peñafiel, Aguiar, Sánchez-Nivicela, Pyron, Colston, Cineros-Heredia, Yánez-Muñoz, Venegas, Guayasamin & Torres-Carvajal, 2018. ZooKeys 766: 96. Substitute name for *Sibynomorphus petersi* Orcés-V. & Almendáriz, 1989.

Synonyms: *Sibynomorphus petersi* Orcés-V. & Almendáriz, 1989.

Distribution: Upper elevation of 3148 m, Arteaga et al. (2018).

Dipsas klebbai Arteaga, Salazar-Valenzuela, Mebert, Peñafiel, Aguiar, Sánchez-Nivicela, Pyron, Colston, Cineros-Heredia, Yánez-Muñoz, Venegas, Guayasamin, & Torres-Carvajal, 2018. ZooKeys 766: 120–124, figs. 1, 15, 16.

Holotype: MZUTI 5412, an 870 mm male (P. Torres, 28 April 2016).

Type locality: "Pacto Sumaco, province of Napo, Ecuador (S0.66377, W77.59895; 1556 m)."

Distribution: Ecuador (Napo, Sucumbios), 1182–2110 m.

Dipsas latifrontalis Boulenger, 1905. Ann. Mag. Nat. Hist. Series 7, 15(90): 561. (*Leptognathus latifrontalis*)

Holotype: BMNH 1946.1.20.98, an 800 mm female (Briceño).

Type locality: "Aricagua, 1000 m. altitude," Mérida, Venezuela.

Distribution: Colombia, W Venezuela (Barinas, Merida), 1000–1400 m.

Comments: Resurrected from the synonymy of *D. peruana* by Arteaga et al. (2018) based on distinctive color pattern and DNA-based phylogenetic placement.

Dipsas lavillai Scrocchi, Porto & Rey, 1993

Dipsas maxillaris (F. Werner, 1909)

Dipsas mikanii (Schlegel, 1843)

Distribution: Add Brazil (Ceara), Guedes et al. (2014); Brazil (Mato Grosso do Sul), Ferreira et al. (2017); Brazil (Paraiba), R. França et al. (2012); Brazil (Piauí), Dal Vechio et al. (2013); Brazil (Tocantins), Dal Vechio et al. (2016).

Comments: M.A. Freitas et al. (2014b) report the rediscovery of *S. m. septentrionalis* and add Pará to its range.

Dipsas neivai Amaral, 1926

Dipsas neuwiedi (Ihering, 1011)

Distribution: Add Brazil (Alagoas), Parnazio & Vrcibradic (2018); Brazil (Paraiba), R. França et al. (2012, as *S.* cf. *neuwiedi*); Brazil (Sergipe), Andrade et al. (2019); Brazil (Paraíba), and upper elevation of 1030 m, Guedes et al. (2014). Andrade et al. (2019) list and map known localities.

Dipsas nicholsi (Dunn, 1933)

Dipsas oligozonata (Orcés-V. & Almendáriz, 1989)

Distribution: Upper elevation of 2891 m, Arteaga et al. (2018).

Dipsas oneilli (Rossman & J.P.R. Thomas, 1979)

Dipsas oreas (Cope, 1868)

Distribution: Add Ecuador (Pichincha), Amendáriz & Orcés (2004), Ecuador (Azuay), Almendáriz (2007); Ecuador (El Oro), and upper elevation of 524 m, Arteaga et al. (2018).

Dipsas oswaldobaezi Arteaga, Salazar-Valenzuela, Mebert, Peñafiel, Aguiar, Sánchez Nivicela, Pyron, Colston, Cincros-Heredia, Yánez-Muñoz, Venegas, Guayasamin & Torres-Carvajal, 2018. ZooKeys 766: 117–119, figs. 1, 13, 14.

Holotype: QCAZ 10369, a 362 mm female (S. Aldás & G. Zapata, 3 March 2010).

Type locality: "Quebrada El Faique, province of Loja, Ecuador (lS4.17889, W80.04226; 1004 m)."

Distribution: Ecuador (El Oro, Loja), Peru (Piura), 32–1289 m.

Dipsas pakaraima MacCulloch & Lathrop, 2004

Dipsas palmeri (Boulenger, 1912). Ann. Mag. Nat. Hist. Series 8, 10(58): 422. (*Leptognathis palmeri*)

Synonyms: *Leptognathus latifasciatus* Boulenger, 1913.

Holotype: BMNH 1946.1.20.77, a 950 mm male (M.G. Palmer)

Type locality: "El Topo, Rio Pastaza, E. Ecuador, 4200 feet."

Distribution: Ecuador (Morona Santiago, Pastaza, Tungurahua, Zamoro Chinchipe), Peru (Cajamarca), 1211–2282 m.

Comments: Ressurrected from the synonymy of *D. peruana* by Arteaga et al. (2018) based on distinctive color pattern and DNA-based phylogenetic placement.

Dipsas pavonina Schlegel, 1837

Distribution: Add Ecuador (Zamora-Chinchipe), Almendáriz et al. (2014); increase elevation to 981 m, Arteaga et al. (2018), Brazil (Roraima), Moraes et al. (2017).

Dipsas peruana (Boettger, 1898)

Synonyms: Remove *Leptognathus latifrontalis* Boulenger, 1905, *Leptognathis palmeri* Boulenger, 1912, *Leptognathus latifasciatus* Boulenger, 1913.

Distribution: Limited to Peru (Amazonas, Cuzco, Huanuco, Pasco, Puno, San Martin), 1502–2670 m; Add Peru (Cajamarca), Koch et al. (2018).

Dipsas petersi Hoge & Romano-Hoge, 1975

Dipsas praeornata (F. Werner, 1909)

Dipsas pratti (Boulenger, 1897)

Dipsas sanctijoannis (Boulenger, 1911)

Dipsas sazimai Fernandes, Marques & Argólo, 2010

Distribution: Add Brazil (Alagoas, Pernambuco), Roberto et al. (2014).

Dipsas schunkii (Boulenger, 1908)

Dipsas temporalis (F. Werner, 1909)

Distribution: Add Panama (Panama Oeste), Ray & Ruback (2015); Colombia (Cauca), and upper elevation of 1141 m, Vera-Pérez et al. (2018); Ecuador (Imbabura), Arteaga et al. (2018).

Dipsas tenuissima Taylor, 1954

Distribution: Upper elevation of 1127 m, Ryan et al. (2015).

Dipsas trinitatis H.W. Parker, 1926

Distribution: Upper elevation of 300 m, J.C. Murphy & Rutherford (2014).

Dipsas turgida (Cope, 1868)

Distribution: Add Brazil (Mato Grosso do Sul), Ferreira et al. (2017); Paraguay (Misiones, Neembucu), Cabral & Weiler (2014).

Dipsas vaga (Jan, 1863)

Dipsas vagrans (Dunn, 1923)

Distribution: Add Peru (Amazonas), and elevation range of 487–1326 m, Koch et al. (2018); Peru (San Martin), low elevation of 316 m, Arteaga et al. (2018).

Dipsas variegata (A.M.C. Duméril, Bibron & Duméril, 1854a)

Distribution: Add Trinidad & Tabago (Trinidad), J.C. Murphy & Rutherford (2014); Ecuador (Napo), Arteaga et al. (2018); Brazil (Pernambuco), E.G. Dias et al. (2019); Brazil (Rondônia), Bernarde et al. (2012b) and Ferrão et al. (2012). Remove Panama, Cadle & C.W. Myers, 2003.

Comments: Murphy & Rutherford describe differences between this species and *D. trinitatis* on Trinidad.

Dipsas ventrimaculata (Boulenger, 1885)

Distribution: Add Brazil (Mato Grosso do Sul), Ferreira et al. (2017); Brazil (Marinhciros I., Rio Grande do Sul), Quintela et al. (2011); Paraguay (Canindeyu), Cacciali et al. (2015b); Paraguay (Paraguari), Cabral & Weiler (2014).

Dipsas vermiculata J.A. Peters, 1960

Distribution: Add Ecuador (Zamora Chinchipe), and upper elevation of 1476 m, Arteaga et al. (2018).

Dipsas viguieri (Bocourt, 1884)

Dipsas williamsi (Carrillo de Espinoza, 1974)

DIPSINA Jan, 1863 (Psammophiidae)

Dipsina multimaculata (A. Smith, 1847 *in* 1838–1849)

DISPHOLIDUS Durvernoy, 1832 (Colubridae: Colubrinae)

Dispholidus typus (A. Smith, 1828)

Distribution: Add Angola (Benguela, Bie, Cuanza Norte, Cuanza Sul, Cunene, Huambo, Huila, Luanda, Lunda Sul, Malanje, Zaire), M.P. Marques et al. (2018).

DITAXODON Hoge, 1958 (Dipsadidae: Xenodontinae)

Ditaxodon taeniatus (Hensel, 1868)

DITYPOPHIS Günther, 1881 (Pseudoxyrhophiidae)

Ditypophis vivax Günther, 1881

Comments: Using DNA barcoding, Vasconcelos et al. (2016) find some intraspecific pairwise distance between specimens (mean 1.54).

DJOKOISKANDARUS J.C. Murphy, 2011. Raffles Bull. Zool. 59(2): 233. (Homalopsidae)

Type species: *Cantoria annulata* Jong, 1926 by monotypy.

Distribution: S coast of New Guinea.

Comments: Molecular data remain unavailable to test the relationship of *Djokoiskandarus* apart from *Cantoria*, but the genus is generally recognized, i.e., Zaher et al. (2019).

Djokoiskandarus annulatus (Jong, 1926)

Comments: J.C. Murphy & Voris (2014) provide a description and illustrations.

DOLICHOPHIS Gistel, 1868 (Colubridae: Colubrinae)

Fossil records: Add Late Miocene (Messinian), Georgialis et al. (2017, as cf. *Dolichophis* sp.).

Dolichophis caspius (Gmelin, 1789)

Distribution: Add Romania (Dolj), Lazăr et al. (2005); Romania (Giurgiu, Olt), Covaciu-Marcov & David (2010), Iftime & Iftime (2008); Romania (Teleorman), Iftime & Iftime (2016); Bulgaria (Sofia Region), Popgeorgiev et al. (2014); Greece (Agathonisi, Fourni, Kalymnos, Kos, Nisyros and Patmos Islands), Cattaneo (2018); Greece (Kefallinia), M.J. Wilson (2006); Turkey (Afyon), Eser & Erismis (2014); Turkey (Bartin), Çakmak et al. (2017); Turkey (Burdur), Ege et al. (2015); Turkey (Canakkale), Tok et al. (2006); Turkey (Karabük), Kumlutaş et al. (2017); Turkey (Gökçeada I.), Yakin et al. (2018); Turkey (Tenedos I.), Tosunoğlu et al. (2009); Kazakhstan (Atyrau), Ostrovskikh et al. (2010); Iran (Ardabil, East Azerbaijan, Gilan, Golestan, Hamedan, Mazandaran, North Khorasan, Qazvin, Tehran, West Azerbaijan, Zanjan), Safaei-Mahroo et al. (2015). Babocsay (2013), based on misidentification, eliminate the Zselic region of SW Hungary from the range. Sahlean et al. (2014) map known localities.

Dolichophis cypriensis (Schätti, 1985)

Comments: In the gene-based phylogeny of Zaher et al. (2019) *D. cypriensis* is within the *Hierophis* subclade. Of course, the sample on which the stem is based could be misidentified.

Dolichophis gyarosensis (Mertens, 1968)

Dolichophis jugularis (Linnaeus, 1758)

Synonyms: Add *Dolichophis jugularis cypriacus* Zinner, 2018 (*nomen nudum*).

Distribution: Add Greece (Halki I.), Cattaneo (2018); Greece (Kastellorizo I.), Paysant (1999) and Kakaentzis et al. (2018); Turkey (Burdur), Ege et al. (2015); Turkey (Aydin, Bitlis, Erzincan, Hatay, Kahramanmaraş, Karaman, Mersin, Nigde, Osmaniye, Sanliurfa, Siirt), Göçmen et al. (2013); Iraq (Diyalah, Najaf), Abbas-Rhadi et al. (2017); Iran (Alborz, Ardabil, East Azerbaijan, Gilan, Golestan, Hamedan, Ilam, Isfahan, Khorasan Razavi, Kermanshah, Kordestan, Lorestan, Markazi, Mazandaran, North Khorasan, Qazvin, Semnan, Tehran, West Azerbaijan, Zanjan), Safaei-Mahroo et al. (2015). Iran (Kermanshah), Sadeghi et al. (2014b). Göçmen et al. (2013) map known localities in Turkey.

Dolichophis mesopotamicus Afrasiab, Mohammad & Hussein, 2016. J. Biodivers. Environ. Sci. 8(4): 16–17, fig. 1.

Holotype: IMN 435, a 117 cm male.

Type locality: "Tarmyah north of Baghdad," Iraq.

Distribution: NC Iraq (Al-Sulaimaniyah, Baghdad, Diyala), 45–900 m.

Comments: A weakly defined species, probably a synonym of *D. jugularis*.

Dolichophis schmidti (Nikolsky, 1909)

Distribution: Add Turkey (Aksaray, Akşehir, Diyarbakir, Igdir, Isparta, Kahramanmaraş, Kayseri, Mardin, Muş, Sanliurfa, Siirt), Göçmen et al. (2013); Iraq (Najaf), and low elevation of 22 m, Abbas-Rhadi et al. (2017); Iran (Alborz, Ardabil, East Azerbaijan, Gilan, Golestan, Khorasan Razavi, Khuzestan, Mazandaran, North Khorasan, Qazvin, Semnan, Tehran, West Azerbaijan, Zanjan), Safaei-Mahroo et al. (2015). Göçmen et al. (2013) map known localities in Turkey.

DREPANOIDES Dunn, 1928 (Dipsadidae: Xenodontinae)

Drepanoides anomalus (Jan, 1863)

Distribution: Add Peru (Huanuco), Schlüter et al. (2004); Brazil (Acre), Bernarde et al. (2011b).

DROMICODRYAS Boulenger, 1893 (Pseudoxyrhophiidae)

Dromicodryas bernierii (A.M.C. Duméril, Bibron & Duméril, 1854)

Dromicodryas quadrilineata (A.M.C. Duméril, Bibron & Duméril, 1854)

Distribution: Add Madagascar (Nosy Komba), Roberts & Daly (2014).

†*DRYINOIDES* Auffenberg, 1958 (Dipsadidae: subfamily unnamed)

†*Dryinoides oxyrhachis* Auffenberg, 1958

DRYMARCHON Fitzinger, 1843 (Colubridae: Colubrinae)

Fossil records: Pleistocene (Rancholbrean) of USA (South Carolina), Knight & Cicimurri (2019, as *Drymarchon* sp.).

Drymarchon caudomaculatus Wüster, Yrausquin & Mijares-Urrutia, 2001

Drymarchon corais (F. Boie, 1827)

Distribution: Add Peru (Amazonas, Cajamarca), Koch et al. (2018); Peru (Huanuco), Schlüter et al. (2004); Brazil (Acre), Bernarde et al. (2011b); Brazil (Goias), Vaz-Silva et al. (2007); Brazil (Mato Grosso do Sul), Ferreira et al. (2017); Brazil (Minas Gerais, Sergipe), Guedes et al. (2014); Brazil (Paraíba), R. França et al. (2012); Brazil (Rio Grande do Norte), Sales et al. (2009); Brazil (Roraima), Prudente et al. (2014).

Drymarchon couperi (Holbrook, 1842)

Synonyms: *Drymarchon kolpobasileus* Krysko, Granatosky, Nuñez & Smith, 2016. Zootaxa 4138(3): 564–566, figs. 10, 11.

Holotype: UF 52751, a 2321 mm male (D.M. Sargent, August 1981).

Type locality: "on Mill Terrace and Riverwood Avenue, Sarasota, Sarasota County, Florida, USA (27.29291 N, 82.52453 W…)."

Comments: Krysko et al. (2016a-b) use molecular and morphological data to recognize Atlantic and Gulf (*D. kolpobasileus*) coastal plain populations as distinct species. Folt et al. (2019) were unable to resolve two species using microsatellite data, and recommend synonymy of *D. kolpobasileus* with *D. couperi*, as do Guyer et al. (2019) based on a lack of supposed morphological differences.

Drymarchon margaritae Roze, 1959

Drymarchon melanurus (A.M.C. Duméril, Bibron & Duméril, 1854)

Distribution: Add Honduras (Isla del Tigre), McCranie & Gutsche (2016); Panama (Panama, Panama Oeste), Ray & Ruback (2015); Colombia (Huila), Moreno-Arias & Quintero-Corzo (2015); Ecuador (Azuay, El Oro, Esmeraldas, Guayas, Imbabura, Pichincha) and elevation to 2200 m, Almendáriz & Brito (2012); Ecuador (Loja, Manabí), Cisneros-Heredia (2006a); Villa et al. (2015) extend the known range in Sonora, Mexico.

DRYMOBIUS Fitzinger, 1843 (Colubridae: Colubrinae)

Drymobius chloroticus (Cope, 1886)

Distribution: Add Mexico (Hidalgo), Badillo-Saldaña et al. (2014).

Drymobius margaritiferus (Schlegel, 1837)

Distribution: Add Panama (Panama Oeste), Ray & Ruback (2015). Martínez-Fuentes et al. (2017) confirm the species in Mexico, Mexico.

Drymobius melanotropis (Cope, 1875)

Drymobius rhombifer (Günther, 1860)

Distribution: Add Colombia (Antioquia), Padilla-Pérez et al. (2015); Ecuador (Manabi), Almendáriz et al. (2012); Peru (Huanuco), Schlüter et al. (2004); Brazil (Acre), Bernarde et al. (2011b); Brazil (Pará), Mendes-Pinto & Souza (2011).

DRYMOLUBER Amaral, 1930 (Colubridae: Colubrinae)

Comments: H. Costa et al. (2013) present a taxonomic summary of the genus.

Drymoluber apurimacensis Lehr, Carrillo & Hocking, 2004

Drymoluber brazili (Gomés, 1918)

Type locality: H. Costa et al. (2013) believe that the holotype came from the vicinity of the Tronco-Catalão Railway, rather than the point associated with the holotype.

Distribution: Add Paraguay (San Pedro), P. Smith et al. (2016). Upper elevation of 1100 m, Guedes et al. (2014).

Drymoluber dichrous (W.C.H. Peters, 1863)

Lectotype: ZMB 1661, designated by H. Costa et al. (2013).

Type locality: suggested as SE Brazil by H. Costa et al. (2013).

Distribution: Add Ecuador (Zamora-Chinchipe), Torres-Carvajal et al. (2019); Peru (Cajamarca), Koch et al. (2018); Bolivia (La Paz, Pando), Colombia (Amazonas), Guyana (East Berbice), H. Costa et al. (2013).

Comments: H. Costa et al. (2013) provide a description, illustrations, and distribution map.

DRYOCALAMUS Günther, 1858 (Colubridae: Colubrinae)

Comments: Figueroa et al. (2016), based on DNAsequence-data, synonymize *Dryocalamus* with *Lycodon* due to paraphyly. Also using sequence data, Zaher et al. (2019) recognize *Dryocalamus* as part of a polyphyletic *Lycodon-Stegonotus* group.

Dryocalamus davisoni (Blanford, 1878)

Distribution: Add Cambodia (Siem Reap), Geissler et al. (2019); Vietnam (Ba Ria-Vung Tau, Gia Lai, Kon Tum, Ninh Binh), Luu et al. (2019); Vietnam (Hoa Binh), Nguyen et al. (2018).

Dryocalamus gracilis (Günther, 1864)

Dryocalamus nympha (Daudin, 1803)

Dryocalamus philippinus Griffin, 1909

Dryocalamus subannulatus (A.M.C. Duméril, Bibron & Duméril, 1854)

Distribution: Add Myanmar (Tanintharyi: Linn Lune Kyun I.), J. Lee et al. (2018a); West Malaysia (Pahang), Zakaria et al. (2014); Indonesia (Pulau Bangkuru off Sumatra), Tapley & Muurmans (2016).

Dryocalamus tristrigatus Günther, 1858

DRYOPHIOPS Boulenger, 1896 (Colubridae: Ahaetullinae)

Dryophiops philippina Boulenger, 1896

Distribution: Add Philippines (Marinduque, Panay, Siquijor, possibly Samar Is.), Leviton et al. (2018).

Dryophiops rubescens Gray 1835 *in* Gray & Hardwicke, 1830–1835

DRYSDALIA Worrell, 1961 (Elapidae)

Drysdalia coronoides (Günther, 1858)

Drysdalia mastersii (Krefft, 1866)

Drysdalia rhodogaster (Jan, 1873 *in* Jan & Sordelli, 1870–1881)

DUBERRIA Fitzinger, 1826 (Pseudoxyrhophiidae)

Duberria lutrix (Linnaeus, 1758)

Comments: Kulenkampff et al. (2019) produce a DNA-sequence based phylogeny for the nominate race, which reveals five allopatric clades in South Africa. The authors suggest that *D. l. lutrix* is a species complex that is in need of further study.

Duberria rhodesiana Broadley, 1958

Duberria shirana (Boulenger, 1894)

Distribution: Add Tanzania (Morogoro), and low elevation of 1429 m, Lyakurwa (2017); Mozambique (Nampula/Nlassa), Conradie et al. (2016a).

Duberria variegata (W.C.H. Peters, 1854)

†***DUNNOPHIS*** Hecht *in* McGrew, 1959 (Charinidae: Ungaliophiinae)

†***Dunnophis cadurcensis*** Rage, 1974
†***Dunnophis matronensis*** Rage, 1973
†***Dunnophis microechinis*** Hecht *in* McGrew, 1959

ECHINANTHERA Cope, 1894 (Dipsadidae: Xenodontinae)

Echinanthera amoena (Jan, 1863)

Distribution: Add Brazil (Espirito Santo, Santa Catarina), Azevedo et al. (2018).

Echinanthera cephalomaculata Di-Bernardo, 1994

Distribution: Add Brazil (Pernambuco), and upper elevation of 850 m, Freitas, M.A. de et al. (2019b).

Echinanthera cephalostriata Di-Bernardo, 1996

Distribution: Add Brazil (Rio de Janeiro: Ilha Grande), C.F.D. Rocha et al. (2018).

Echinanthera cyanopleurus (Cope, 1885)
Echinanthera melanostigma (Wagler *in* Spix, 1824)
Echinanthera undulata (Wied-Neuwied, 1824)

ECHIOPSIS Fitzinger, 1843 (Elapidae)
Echiopsis curta (Schlegel, 1837)

ECHIS Merrem, 1820 (Viperidae: Viperinae)
Comments: Trape (2018) reviews the West African species.

Echis borkini Cherlin, 1990

Echis carinatus (***Schneider***, 1801)

Distribution: Add Iran (Chaharmahal and Bakhtiari, Golestan, North Khorasan, Semnan, South Khorasan, Yazd), Safaei-Mahroo et al. (2015); Iran (Razavi Khorasan), Yousefkhani et al. (2014) and Nasrabadi et al. (2016); Afghanistan (Balkh, Farah, Herat, Kandahar, Nangahar, Nimroz), Wagner et al. (2016b). Increase upper elevation limit to 2700 m, Moradi et al. (2013).

Comments: Abbas-Rhadi et al. (2015a) are unable to find morphological differences between two samples in SE Iraq. Abbas-Rhadi et al. (2016) compare mtDNA sequence data between populations in United Arab Emirates, Iraq, Pakistan and India, revealing India as the sister group to the others.

Echis coloratus Günther, 1878

Distribution: Add Saudi Arabia (Ha'il), Alshammari et al. (2017).

Comments: Alshammari (2011) evaluated variation in RNA sequence data between Arabian populations, which indicated two population clades.

Echis hughesi Cherlin, 1990

Echis jogeri Cherlin, 1990

Distribution: Add Senegal (Kedougou), Monasterio et al. (2016); Senegal (Thies) and Mali (Kayes, Koulikoro, Segou, Sikasso), Trape (2018).

Comments: Trape (2018) provides a description, color photographs, and locality map.

Echis khosatzkii Cherlin, 1990

Distribution: Add Oman (Dhofar), Ball & Borrell (2016).

Echis leucogaster Roman, 1972

Distribution: Add Morocco (Ouarzazate, Tata), Marmol-Marin & Fernandez (2012); Mauritania (Nouakchott), Padial (2006); Mali (Kidal), Trape & Mané (2017); Niger (Tillaberi), Trape & Mané (2015); Chad (Guera), Trape (2018). Low elevation of 41 m, Trape & Mané (2017).

Echis megalocephalus Cherlin, 1990

Echis ocellatus Stemmler, 1970

Distribution: Add Burkina Faso (Boucle du Mouhon, Cascades, Centre-Sud), Trape (2018); Mali (Mopti), Trape (2018); Niger (Tillaberi), Trape & Mané (2015). Remove region east of western Nigeria (Kwara, Sokoto) to *E. romani*.

Comments: Trape (2018) provides a description, color photographs, and locality map, after transferring the eastern portion of the range to his new species *Echis romani*.

Echis omanensis Babocsay, 2004

Distribution: Add Oman (Al-Batinah South) and upper elevation of 1843 m, Grossmann et al. (2013).

Echis pyramidum Geoffroy St.-Hilaire, 1827 *in* Savigny, 1809–1829

Distribution: Add Saudi Arabia (Farasan Al-Kebir I.), Cunningham (2010); Saudi Arabia (Sarso I.), Masseti (2014).

Echis romani Trape, 2018. Bull. Soc. Herpétol. France 167: 24–29, figs. 9–14.

Holotype: MNHN 2018.0006, a 517 mm male (villager, 20 May 2015).

Type locality: "Kumao au Tchad (region du Logone Oriental, département de Baibokoum, 07°36'N / 15°36'E)."

Distribution: C Africa. Nigeria (Anambra, Benue, Kaduna, Plateau), S Chad (Chari-Baguirmi, Mayo-Kebbi Est, Tandjile), N Cameroon (Extreme Nord, Nord), NW Central African Republic (Ombella-Mpoko, Ouham-Pende). Add SC Niger (Zinder), Nigeria (Gombe), and Chad (Logone Occidental, Logone Oriental, Mayo-Kebbi Ouest), Trape (2018).

Echis varius A. Reuss, 1834

EIRENIS Jan, 1863 (Colubridae: Colubrinae)
Comments: Rajabizadeh et al. (2015b) revise *E. persicus* to contain six species. Wagner et al. (2016b) mention two

specimens from Kandahar, Afghanistan, that are of the *E. persicus* complex, but are not referred to a particular species.

Eirenis africanus (Boulenger, 1914)

Eirenis angusticeps Boulenger, 1894. Cat. Snakes Brit. Mus. 2: 262. (*Contia angusticeps*)

Holotype: ZSI, a 340 mm specimen, lost according to M.A. Smith (1943)

Type locality: "Cherat, Baluchistan."

Distribution: N Pakistan (Federally Administered Tribal Areas, Khyber-Pakhtunkhwa, Punjab).

Comments: Rajabizadeh et al. (2015b) revive this name for certain northern Pakistan populations of *E. persicus*. They provide a description, locality map and color photograph.

Eirenis aurolineatus (Venzmer, 1919)

Eirenis barani J.F. Schmidtler, 1988

Eirenis collaris (Ménétriés, 1932)

Distribution: Add Turkey (Adana), Winden et al. (1997); Iran (Fars), Gholamifard et al. (2012); Iran (Alborz, Ilam, Isfahan, Hamedan, Kermanshah, Kurdistan, Markazi), Sadeghi et al. (2014a); Iran (Ardabil, Bushehr, Qazvin, Tehran), Safaei-Mahroo et al. (2015). Sadeghi et al. (2014a) mapped known localities in Iran.

Eirenis coronella (Schlegel, 1837)

Distribution: Add Iran (Gilan), Safaei-Mahroo et al. (2015); Syria (Homs), Sindaco et al. (2014).

Eirenis coronelloides (Jan, 1862)

Distribution: Add: Iraq (Al-Sulaimaniyah), Afrasiab & Mohamad (2014); Iraq (Anbar), Mohamad & Afrasiab (2015).

Eirenis decemlineatus (A.M.C. Duméril, Bibron & Duméril, 1854)

Eirenis eiselti J.J. Schlmidtler & J.F. Schlmidtler, 1978

Distribution: Add Turkey (Kahramanmaraş, Kilis, Muş, Şanliurfa, Van), Göçmen et al. (2013); Turkey (Tunceli), İğci et al. (2015). Göçmen et al. (2013) map known localities in Turkey.

Eirenis hakkariensis J.F. Schmidtler & Eiselt, 1991. Salamandra 27(4): 230–232, figs. 5–7.

Holotype: ZSM 3/1991, a 400 mm female (J.F. & H. Schmidtler, 11 June 1976).

Type locality: "oberhalb Hakkâri-Stadt, 1 900 m ü. M.," Turkey.

Distribution: E Turkey (Hakkari), 1900 m.

Comments: Mahlow et al. (2013) regard *hakkariensis* to be a distinct species from *E. thospitis* based on morphological differences.

Eirenis kermanensis Rajabizadeh, J.F. Schmidtler, Orlov & Soleimani, 2012

Comments: See under *E. medus*.

Eirenis levantinus J.F. Schmidtler, 1993

Eirenis lineomaculatus K.P. Schmidt, 1939

Distribution: Add Iraq, Al-Barazengy et al. (2015); Iraq (Erbil), Mohamad & Afrasiab (2015); Turkey (Kahramanmaraş,

Malatya, Mardin, Mersin), and upper elevation of 1200 m, Göçmen et al. (2014c).

Eirenis mcmahoni Wall, 1911. J. Bombay Nat. Hist. Soc. 20(4): 1037–1038. (*Contia mcmahoni*)

Syntypes: ZSI 16624, three others in the Quetta Museum destroyed. I. Das et a. (1998) refer to the ZSI specimen as holotype, and do not state which locality it is from.

Type locality: "Quetta, Loralai, Mach and Spingtangi," Pakistan.

Distribution: C Pakistan (Baluchistan).

Comments: Rajabizadeh et al. (2015b) revive this name for central Pakistan populations of *E. persicus*. They provide a description and a locality map.

Eirenis medus (Chernov *in* Terentjev & Chernov, 1940)

Distribution: Add Iran (Ardabil, East Ajerbaizan, Markazi, North Khorasan, Tehran), Safaei-Mahroo et al. (2015).

Comments: Mahlow et al. (2013) believe that *E. kermanensis* may be a synonym of *E. medus* based on locality and similar morphology.

Eirenis modestus (Martin, 1838)

Distribution: Add Greece (Kastellorizo I.), Paysant (1999) and Kakaentzis et al. (2018); Turkey (Bartin), Çakmak et al. (2017); Turkey (Karabük), Kumlutaş et al. (2017); Turkey (Kutahya, Usak), Özdemir & Baran (2002); Turkey (Mersin), Winden et al. (1997); Iran (Golestan, Markazi), Safaei-Mahroo et al. (2015).

Eirenis nigrofasciatus Nikolsky 1907. Ann. Mus. Zool. Acad. Imp. Sci. St. Petersborg 10(3/4): 298–299. (*Contia persica nigrofasciata*)

Holotype: ZISP 10323, (16 March 1904).

Type locality: "Urbis Dizful (Arabistan)," Iran.

Distribution: E Iraq (Diyala, Halabjah, Kirkuk), W Iran (Fars, Hormozgan, Ilam, Kerman, Khuzestan, Yazd).

Comments: Rajabizadeh et al. (2015b) revive this name for certain southwestern populations of *E. persicus*. They provide a description, locality map and color photograph.

Eirenis occidentalis Rajabizadeh, Nagy, Adriaens, Avci, Masroor, Schmidtler, Nazarov, Esmaeili & Christiaens, 2015b. Zool. J. Linn. Soc. 176(4): 900–901, fig. 14. (*Eirenis [Pseudocyclophis] occidentalis*)

Holotype: ZDEU 136/2005, a 362 mm male (İ. Baran, Y. Kumlutaş, Ç. Ilgaz & A. Avci, 1 May 2005).

Type locality: "Between Diyarbakır and Sivarek, 48 km west of Diyarbakır (37°49'12.3″N, 39°37'58.5″E), Şanlıurfa Province, Turkey.... Altitude 1100 m a.s.l."

Distribution: S Armenia, SE Turkey (Adiyaman, Diyarbakir, Hakkari, Malatya, Mardin, Sanliurfa, Siirt, Sirnak, Urfa), W Iran (Kermanshah, and possibly Markazi and Tehran), 1100 m. Add Turkey (Erzincan), İğci et al. (2015); Iraq (Erbil, Sulaimaniyah), Mohamad & Afrasiab (2015, this species based on geography, though possibly *E. persicus*).

Comments: Formerly the westernmost populations of *E. persicus*, Rajabizadeh et al. (2015b).

Eirenis persicus (J. Anderson, 1872)

Synonyms: Remove *Pseudocyclophis walteri* Boettger, 1888, *Contia angusticeps* Boulenger, 1894, *Contia persica nigrofasciata* Nikolsky, 1907, *Contia mcmahoni* Wall, 1911.

Distribution: SW Iran (Bushehr, Chaharmahal and Bakhtiari, Fars, Isfahan, Khuzestan, Lorestan). Add Iran (Markazi), Sabbaghzadeh & Mashayekhi (2015, also based on geography) and Iran (Qom), S.M. Kazemi et al. (2015), though assignment to species is tentative.

Comments: Rajabizadeh et al. (2015b) revise this taxon using morphological and genetic data. They recover six species, distributed within primary eastern and western clades, including recognition of the former synonyms *P. walteri*, *C. angusticeps*, *C. p. nigrofasciata*, and *C. mcmahoni*, plus a new species *E. occidentalis*. They provide a revised description and color photograph.

Eirenis punctatolineatus (Boettger, 1893)

Distribution: Add Turkey (Tunceli), İğci et al. (2015); Iraq (Erbil, Sulaimaniyah), Mohamad & Afrasiab (2015); Iran (Kerman), Moradi et al. (2013); Iran (Ardabil, Bushehr, Fars, Gilan, Hormozgan, Ilam, Isfahan, Kermanshah, Khuzestan, Kohgiluyeh and Boyer Ahmad, Markazi, North Khorasan, Qazvin, Semnan, Tehran), Safaei-Mahroo et al. (2015); Iran (Qom), S.M. Kazemi et al. (2015).

Eirenis rechingeri Eiselt, 1971

Distribution: Add Iran (Esfahan), Zadhoush et al. (2016).

Comments: Gholamhosseini et al. (2009) report a second specimen, also from Fars Province.

Eirenis rothii Jan, 1863

Distribution: Add Iraq (Ninevah), Mohamad & Afrasiab (2015).

Eirenis thospitis J.J. Schmidtler & Lanza, 1990

Synonyms: Remove *Eirenis hakkariensis* J.F. Schmidtler & Eiselt, 1991.

Distribution: Add Iraq (Erbil), and upper elevation limit of 2106 m, Mohamad & Afrasiab (2015).

Comments: The Iraq specimen described by Mohamad & Afrasiab (2015) is intermediate in morphological characters between *E. hakkariensis* and *E. thospitis*. See under *E. hakkariensis*.

Eirenis walteri Boettger, 1888. Zool. Anz. 11(279): 262–263. (*Pseudocyclophis walteri*)

Holotype: Cauc. Mus. Tibilisi, a 394 mm specimen (G. von Radde, 1886).

Type locality: "Neu-Serachs [=Serakhs] an der Nordostspitze Persiens," Iran.

Distribution: SC Asia. S Turkmenistan (Ahal, Ashkabad, Krasnovodsk, Mary), E Iran (Golestan, Kerman, Northern Khorasan, Sistan and Baluchestan, Southern Khorasan), W Pakistan (Baluchistan).

Comments: Rajabizadeh et al. (2015b) revive this name for certain eastern populations of *E. persicus*. They provide a description, locality map and color photograph.

Eirenis yassujicus Fathinia, Rastegar-Pouyani & Shafaeipour, 2019. Zool. Middle East 65(4): 3–5, figs. 1–3.

Holotype: YUZM-CE.1, a 505 mm female (3 May 2016).

Type locality: "Damkoreh Region (30.63057N and 51.41985E), Dasht-e-Room, Kohgilouyeh and Boyer Ahmad Province, Iran."

Distribution: Iran (Kohgilouyeh and Boyer Ahmad). Known only from the holotype.

ELAPHE Fitzinger *in* Wagler, 1833 (Colubridae: Colubrinae)

Distribution: Eurasia.

Fossil records: Add middle Pleistocene (Likhvinian) of Ukraine, Ratnikov (2005).

Comments: X. Chen et al. (2017) determine that the placement of *Elaphe zoigeensis* and *Orthriophis taeniurus* in a novel phylogeny leave the respective genera paraphyletic, which the authors resolve by placing *Orthriophis* within an expanded *Elaphe*. The phylogeny of Zaher et al. (2019) also recognizes *Orthriophis* within a separate clade that is in a hierarchical pattern as sister clade to *Zamenis* > *Coronella* > *Oocatochus* > *Elaphe*. For this reason we continue to recognize *Orthriophis*.

Elaphe anomalus (Boulenger, 1916)

Elaphe bimaculata K.P. Schmidt, 1925

Comments: Simonov et al. (2018) find several published mtDNA sequences to be nested within samples of *E. dione*. They suggest that the *E. bimaculata* samples are misidentified.

Elaphe carinata (Günther, 1864)

Comments: Guo et al. (2012b) synonymize *E. c. deqenensis* with the nominate race after finding no molecular or morphological differences to support recognition as two subspecies.

Elaphe climacophora (H. Boie, 1826)

Distribution: Add Japan (Kinkasan I.), Mori & Nagata (2016).

Elaphe davidi (Sauvage, 1884)

Elaphe dione (Pallas, 1773)

Distribution: Add Iran (Gilan, Golestan, West Azerbaijan), Safaei-Mahroo et al. (2015). Upper elevation record of 3520 m, S. Hofmann et al. (2016). Tupikov & Zinenko (2015a) detail the distribution in Ukraine.

Comments: S. Hofmann et al. (2016) report significant variation in DNA sequence data from various populations. See under *E. bimaculata*.

Elaphe quadrivirgata (H.Boie, 1826)

Distribution: Add Japan (Kinkasan I.), Mori & Nagata (2016).

Comments: Kuriyama et al. (2011) use mtDNA sequence data to demonstrate that the Izu Archipelago was colonized multiple times from mainland populations, and that morphological variation between island populations evolved from post-colonization selection.

Elaphe quatuorlineata (Lacépède, 1789)

Distribution: Add Croatia (Dugi Otok I.), Madl (2017); Greece (Folegandros Island), Itescu et al. (2017), Greece (Ithakos I.),

Strachinis & Artavanis (2017); Greece (Kefallinia, Zakinthos), M.J. Wilson (2006). Cattaneo & Cattaneo (2016) document an eastward range extension into East Macedonia & Thrace, Greece.

Comments: Kornilios et al. (2013b) present information on mtDNA phylogeography, intraspecific diversity and phenotypic convergence. The mtDNA data resolve three clades that roughly approximate the three subspecies: *muenteri*, *quatuorlineata*, and *scyrensis*.

Elaphe rechingeri F. Werner, 1932

Elaphe sauromates (Pallas, 1814)

Distribution: Remove portions of the range inhabited by *E. urartica* (i.e., E Turkey and the Caucasus region to NW Iran). Add Turkey (Adana), Sarikaya et al. (2017); Turkey (Canakkale), Turkey (Hatay), Jablonski et al. (2019); Tok et al. (2014); Turkey (Karabük), Kumlutaş et al. (2017).

Comments: See under *E. urartica*.

Elaphe schrenckii (Strauch, 1873)

Elaphe urartica Jablonski, Kukushkin, Avcı, Bunyatova, Igaz, Tuniyev & Jandzik *in* Jablonski et al., 2019. PeerJ 7(e6944): 22–31, figs. 4, 6–8.

Synonyms: Jablonski et al. (2019, suppl.) note that a number of old names under *E. sauromates* could have priority over *E. urartica*, but they are declared *nomina dubia* due to imperfect descriptions and/or geographic origins.

Holotype: ZDEU 26/2012, a 1013 mm male (S.B. Tuniyev, 16 July 2012).

Type locality: "Bitlis Province, Turkey (Kısıkh Village, Süphan Mts.; 38.93°N, 42.91°IE, 2,394 m a. s. l.)."

Distribution: SW Asia. E Turkey (Agri, Bitlis, Diyarbakir, Erzurum, Igdir, Kars, Van), Armenia, Georgia, Azerbaijan, Moldavia, SW Russia (Dagestan), Nagorno-Karabakh, NW Iran (Bakhtaran, East Azarbaijan, Hamdan, West Azarbaijan, Zanjhan). −25–2600 m. Add Iran (Ardabil, Golestan, Kermanshah, Mazandaran, Semnan, Tehran), Safaei-Mahroo et al. (2015).

Comments: This species represents a SE portion of the range formerly included within *E. sauromates*, separated morphologically and genetically from that species, Jablonski et al. (2019).

Elaphe zoigeensis S. Huang, Ding, Burbrink, Yang, Huang, Ling, Chen & Zhang, 2012

ELAPOGNATHUS Boulenger, 1896 (Elapidae)

Elapognathus coronatus (Schlegel, 1837)

Elapognathus minor (Günther, 1863)

ELAPOIDIS H Boie *in* F. Boie, 1827 (Natricidae)

Comments: Vogel et al. (2018) review the genus. They note that the three specimens that they examined from Borneo differ from other populations, as well as one other specimen from Sumatra. But they do not describe them as new with so little material.

Elapoidis fusca H. Boie *in* F. Boie, 1827

Synonyms: Remove *Elaphis sumatranus* Bleeker, 1860.

Distribution: Limited to Java, and possibly populations on Borneo, Vogel et al. (2018).

Elapoidis sumatrana Bleeker, 1860. Nat. Tijd. Nederland. Indie. Series 5, 21(1): 297–298.

Holotype: BMNH, not numbered, a female (Ludeking), now lost according to Vogel et al. (2018).

Neotype: BMNH 1928.2.18.17, a 434 mm male (E. Jacobson).

Type locality: stated as "Fort de Kock [=Bukittingi]," Sumatra. Neotype from "Fort de Kock, 920 m, West coast of Sumatra."

Distribution: Sumatra.

Comments: Vogel et al. (2018) revive this species based on morphological differences from Java specimens.

ELAPOMORPHUS Weigmann *in* Fitzinger, 1843 (Dipsadidae: Xenodontinae)

Elapomorphus quinquelineatus (Raddi, 1820)

Elapomorphus wuchereri Wucherer *in* Günther, 1861

Distribution: Add Brazil (Minas Gerais), Entiauspe-Neto et al. (2017c).

Comments: Entiauspe-Neto et al. (2017c) provide a species account, map, and color photographs.

ELAPOTINUS Jan, 1862 (Pseudoxyrhophiidae)

Synonyms: Add *Exallodontophis* Cadle, 1999.

Distribution: Madagascar.

Comments: Kucharzewski et al. (2014) find *Exallodontophis* to be a junior synonym of *Elapotinus* based on morphological comparison of specimens of the former with the old, poorly studied specimens of the latter.

Elapotinus picteti Jan, 1862

Synonyms: Add *Parahadinaea albignaci* Domergue, 1984.

Distribution: Madagascar (Antsiranana, Mahajanga, Toamasina), elevation 350–1009 m, Kucharzewski et al. (2014).

Comments: Kucharzewski et al. (2014) compare the two specimens of *Elapotinus picteti* to specimens of *Exallodontophis albignaci* and could find no morphological characters with which to recognize two species.

ELAPSOIDEA Bocage, 1866 (Elapidae)

Elapsoidea boulengeri Boettger, 1895

Elapsoidea broadleyi Jakobsen, 1997

Elapsoidea chelazziorum Lanza, 1979

Elapsoidea guentherii Bocage, 1866

Elapsoidea laticincta (F. Werner, 1919)

Elapsoidea loveridgei Parker, 1949

Elapsoidea nigra Günther, 1888

Elapsoidea semiannulata Bocage, 1882

Distribution: Add Senegal (Kedougou, Sedhiou, Ziguinchor), Mané & Trape (2017); Angola (Bengo, Cabinda, Luanda), M.P. Marques et al. (2018).

Elapsoidea sundevalli (A. Smith 1848 *in* 1838–1849)

Elapsoidea trapei Mané, 1999

Distribution: Add Senegal (Kolda), Mané & Trape (2017).

EMMOCHLIOPHIS Fritts & Smith, 1969 (Dipsadidae: Dipsadinae)

Comments: Pyron et al. (2015) place *Emmochliophis* in a new tribe (Diaphorolepidini) with *Diaphoralepis* and *Synophis*.

Emmochliophis fugleri Fritts & Smith, 1969

Pyron et al. (2015) provide a species account (description and locality map).

Emmochliophis miops (Boulenger, 1898)

Type locality: Pyron et al. (2015) state that the type locality is in the province of Imbabura rather than in Carchi, Ecuador.

Distribution: Add Colombia (Cauca), and upper elevation of 1188 m, Vera-Pérez et al. (2018, as *E.* cf. *miops*).

Comments: Pyron et al. (2015) provide a species account (description and locality map).

EMYDOCEPHALUS Krefft, 1869 (Elapidae)

Emydocephalus annulatus Krefft, 1869

Emydocephalus ijimae Stejneger, 1898

Emydocephalus szczerbaki Dotsenko, 2011. Zbirnik Prats' Zool. Mus. 41: 128–138.

Holotype: NMNHU 27.

Type locality: Ba Ria-Vung Tau, Vietnam.

Distribution: Vicinity of the type locality.

Comments: We do not have a copy of this paper; data are from Uetz et al. (2019, The Reptile Database).

ENHYDRIS Latreille *in* Sonnini & Latreille, 1801 (Homalopsidae)

Comments: J.C. Murphy & Voris (2014) provide a generic diagnosis and key to the species.

Enhydris chanardi J.C. Murphy & Voris, 2005

Distribution: Add Thailand (Nakhon Nayak), J.C. Murphy & Voris (2014).

Comments: J.C. Murphy & Voris (2014) provide a diagnosis and photograph.

Enhydris enhydris (Schneider, 1799)

Distribution: Add Nepal (Nawalparasi), Pandey et al. (2018); India (Arunachal Pradesh), Purkayastha (2018).

Comments: J.C. Murphy & Voris (2014) provide a diagnosis and photograph.

Enhydris innominata (Morice, 1875)

Comments: J.C. Murphy & Voris (2014) provide a diagnosis and photograph.

Enhydris jagorii (W.C.H. Peters, 1863)

Synonyms: Add *Hypsirhina smithii* (Boulenger, 1914).

Distribution: Add Thailand (Prachuap Khiri Khan), J.C. Murphy & Voris (2014); Laos (Champasak), Teynié et al. (2004).

Comments: J.C. Murphy & Voris (2014) provide a diagnosis and photograph. They consider *E. smithii* to be a synonym of *E. jagorii*.

Enhydris longicauda (Bourret, 1934)

Comments: J.C. Murphy & Voris (2014) provide a diagnosis and photograph.

Enhydris subtaeniata (Bourret, 1934)

Distribution: Add Thailand (Nakhon Ratchasima, Nakhon Sawan), Voris et al. (2012); Thailand (Surin), Chuaynkern et al. (2019); Cambodia (Battambang), Voris et al. (2012); Vietnam (Quang Ngai), Nemes et al. (2013).

Comments: Voris et al. (2012) evaluate geographic variation. J.C. Murphy & Voris (2014) provide a diagnosis and photograph.

ENULIOPHIS McCranie and Villa, 1993 (Dipsadidae: subfamily unnamed)

Comments: C.W. Myers & McDowell (2014) recognize *E. sclateri* as the sister taxon to all *Enulius* species, and recommend that it be returned to *Enulius* in order to conservatively eliminate monotypic genera.

Enuliophis sclateri (Boulenger, 1894)

Distribution: Add Nicaragua (Atlantico Norte, Rio San Juan), Sunyer et al. (2014); Panama (Panama, Panama Oeste), Ray & Ruback (2015).

ENULIUS Cope, 1871 (Dipsadidae: subfamily unnamed)

Enulius bifoveatus McCranie & G. Köhler, 1999

Enulius flavitorques (Cope, 1869)

Distribution: Add Mexico (Tabasco), Hernández-Valadez et al. (2016); Honduras (Olancho), Solís et al. (2014b); Honduras (Valle: Isla Exposición), McCranie et al. (2013b); Honduras (Isla Inglesera), McCranie & Gutsche (2016); Panama (add Veraguas: Isla Canales de Afuera), Flores et al. (2016a); Panama (Panama, Panama Oeste), Ray & Ruback (2015); Colombia (Córdoba), Carvajal-Cogollo et al. (2007). Abarca-Alvarado & Bolaños (2017) extend the distribution southeastward within Puntarenas Province, Costa Rica.

Enulius oligostichus H.M. Smith, Arndt & Sherbrooke, 1967

Distribution: Low elevation of 81 m, Reynosa et al. (2014).

Enulius roatanensis McCranie & G. Köhler, 1999

†*EOANILIUS* Rage, 1974 (Aniliidae)

Distribution: Add middle Miocene (Orleanian) of Germany, Čerňanský et al. (2017).

†*Eoanilius europae* Rage, 1974

Comments: K.T. Smith (2013) cites a personal communication from J.-C. Rage that this taxon probably belongs to the Booidea.

†*Eoanilius oligocenicus* Szyndlar, 1994

Distribution: Add Lower Miocene of Italy, Venczel & Sanchíz (2006); Lower Oligocene (Rupelian, MP 22) of France, Sigé et al. (1998, as *E.* aff. *oligocenicus*).

†**EOMADTSOIA** R.O. Gómez, Garberoglio & Rougier, 2019. Comptes Rendus Palevol 18: 774. (†Madtsoiidae)

Type species: †*Eomadtsoia ragei* R.O. Gómez, Garberoglio & Rougier, 2019 by monotypy.

Distribution: Late Cretaceous of C Argentina.

†**Eomadtsoia ragei** R.O. Gómez, Garberoglio & Rougier, 2019. Comptes Rendus Palevol 18: 774–776, figs. 2, 3.

Holotype: MPEF-PV 2378, a mid-trunk vertebra.

Type locality: "El Uruguayo fossil site…, southeastern slopes of the Somún Curá Massif, Chubut Province, Patagonia, Argentina."

Distribution: Late Cretaceous (Maastrichtian-Danian), central Argentina (Chubut).

†**EOPHIS** Caldwell, Nydam, Palci and Apesteguía, 2015. Nature Communic. 6(5996): 3–5. (Ophidia: incerta sedis)

Type species: *Eophis underwoodi* Caldwell, Nydam, Palci and Apesteguía, 2015 by monotypy.

Distribution: Middle Jurassic of England.

†**Eophis underwoodi** Caldwell, Nydam, Palci and Apesteguía, 2015. Nature Communic. 6(5996): 3–5, figs. 1, 2.

Syntypes: NHMUK R12354, 12355, 12370, right dentaries.

Type locality: "Kirtlington Cement Works Quarry, Oxfordshire, England."

Distribution: middle Jurassic (Bathonian) of England.

EPACROPHIS Hedges, Adalsteinsson & Branch *in* Adalsteinsson, Branch, Trape, Vitt & Hedges, 2009 (Leptotyphlopidae: Leptotyphlopinae)

Epacrophis boulengeri (Boettger in Voeltzkow, 1913)
Epacrophis drewesi (Wallach, 1996)
Epacrophis reticulatus (Boulenger, 1906)

EPHALOPHIS M.A. Smith, 1931 (Elapidae)

Ephalophis greyae M.A. Smith, 1931

EPICRATES Wagler, 1830 (Boidae)

Synonyms: Remove *Chilabothrus* A.M.C. Duméril & Bibron, 1844, *Homalochilus* J.G. Fischer, 1856, *Piesigaster* Seoane, 1880, *Boella* H.M. Smith & Chiszar, 1992.

Distribution: Remove West Indies.

Fossil records: Add late Miocene of Brazil, Albino & Brizuela (2014).

Comments: See under *Chilabothrus*.

Epicrates alvarezi Abalos, Báez & Nader, 1964

Epicrates assisi Machado, 1944

Distribution: Add Brazil (Amapa, Minas Gerais), and upper elevation of 985 m, Guedes et al. (2014).

Epicrates cenchria (Linnaeus, 1758)

Distribution: Add Peru (Puno), Llanqui et al. (2019); Brazil (Acre), D.P.F. França et al. (2017).

Comments: Michels & Bauer (2004) correct the spelling of the subspecies *E. c. gaigei* to *E. c. gaigeae*.

Epicrates crassus Cope, 1862

Distribution: Add Paraguay (Canindeyu), Cacciali et al. (2015b); Paraguay (San Pedro), P. Smith et al. (2016). Upper elevation of 541 m, Guedes et al. (2014).

Epicrates maurus Gray, 1849

Distribution: Add Colombia (La Guajira), Aya-Cuero et al. (2019).

EPICTIA Gray, 1845 (Leptotyphlopidae: Epictinae)

Comments: Esqueda-González et al. (2015) present a morphology-based phylogram from which they define three species groups (*albifrons, goudotii, undecimstriata*). McCranie & Hedges (2016) provide a DNA sequence-based phylogeny using six species, as well as a revised taxonomy of the *E. goudotii* complex that confirms the recognition of *E. ater*, *E. bakewelli*, *E. magnamaculata* and *E. phenops* as species. Wallach (2016) further revises the *E. phenops* species group, within the *E. goudotii* complex, describing six additional species. The etymology of *Epictia* is not known, and Gray lists two species in the original description, one with a masculine and the other with a feminine ending. We follow its common usage as feminine.

Epictia albifrons (Wagler *in* Spix, 1824)

Synonyms: *Leptotyphlops tenella* Klauber, 1939.

Proposed Neotypes: MCZ R2885, a 152 mm specimen, from the state of "Pará, Brasil," designated by Natera-Mumaw et al. (2015).

BYU 11490, a 154 mm female (Mormon missionaries between 1946 and 1953), "vicinity of Belém, state of Pará, Brazil," designated by Wallach (2016).

Distribution: Cis-Andean South America. E Colombia (Cesar, Meta, Vichada), E Venezuela (Amazonas, Bolivar, Distrito Federal, Monagas, Sucre), Guyana (Barima-Waini, Cuyuni-Mazaruni, Essequibo Islands-West Demerara, Mahica-Berbice, Potaro-Siparuni, Upper Demerara-Berbice, Upper Takutu-Upper Essequibo), Suriname (Brokopondo, Coronie, Marowijne, Nickerie, Paramaribo), French Guiana (Cayenne, Saint Laurent du Maroni), Brazil (Amapá, Amazonas, Mato Grosso, Pará, Rondônia, Roraima), S Ecuador (Azuay), NE Peru (Cajamarca, Cuzco, Huánuco, Loreto, San Martin), N Bolivia (La Paz), Trinidad and Tabago (Trinidad); NSL-1750 m. Add Brazil (Tocantins), Esqueda-González et al. (2015).

Comments: Wallach (2016) gives a redescription and color photographs, based on the assumption that the BYU Belém series represents true, topotypic material of the taxon described by Wagler. He rejects the holotype of Natera-Mumaw et al. (and refers readers to Recommendation 75B of the Code, also see his remarks). Pinto et al. (2018) counter that Wallach's neotype designation is not valid under ICZN article 75.4, and that the specimen is of an unnamed species, and its stated locality is in error. They uphold the neotype designation of Natera-Mumaw et al. (2015), and provide morphological evidence that *E. albifrons* and *E. tenella* are synonymous. J.C. Murphy et al. (2016) find little divergence in DNA sequence data between samples from Trinidad and Guyana.

Epictia albipuncta (Burmeister, 1861)

Distribution: Add Argentina (San Juan), Acosta et al. (2017).

Epictia alfredschmidti (Lehr, Wallach & Aguilar, 2002)

Epictia amazonica Orejas-Miranda, 1969. Comun. Zool. Mus. Hist. Nat. Montevideo 10(124): 1–4, fig. 1. (*Leptotyphlops amazonicus*)

Holotype: AMNH 36664, a 120 mm specimen (10 March 1929).

Type locality: "Esmeralda, Territorio Federal Amazonas, Venezuela."

Distribution: S Colombia, S Venezuela (Amazonas, Bolivar), perhaps the Guianan region, Natera-Mumaw et al. (2015).

Comments: Natera-Mumaw et al. (2015) list a number of morphological differences between *E. amazonica* specimens and the holotype of *E. signata*. They refer the latter to populations in Colombia and Ecuador.

Epictia antoniogarciai C. Koch, Venegas & Böhme, 2015. Zootaxa 3964(2): 236–239, figs. 2, 6.

Holotype: CORBIDI 7678, a 144 mm adult (M. Enciso, S. Duran & C. Koch, 4 May 2009).

Type locality: "vicinities of Santa Rosa de la Yunga Village, Jaén Province, Cajamarca Region, Peru (S 05°25′53.3″, W 078°33′47.0″, 1268 m.a.s.l.)."

Distribution: NW Peru (Amazonas, Cajamarca), 934–1268 m. See also Koch et al. (2018).

Epictia ater (Taylor, 1940)

Distribution: Add Guatemala (Izabal), Wallach (2016); Honduras (El Paraíso, Ocotepeque), McCranie (2014b), McCranie et al. (2014b); Honduras (Valle: Islas Conejo, Inglesera, Zacate Grande), McCranie & Gutsche (2016); Nicaragua (Isla Ometepe, Rivas), Stark et al. (2014); Costa Rica (San Jose), Wallach (2016); elevation to 1350 m, Wallach (2016).

Comment: McCranie & Hedges (2016) provide a species account. Wallach (2016) gives a species account, map and color photographs.

Epictia australis (Freiberg & Orejas-Miranda, 1968)

Distribution: Add Paraguay (Asuncion), Esqueda-González et al. (2015); Paraguay (Chaco), Koch et al. (2019).

Epictia bakewelli (Oliver, 1937)

Distribution: Mexico (add México, and remove Jalisco and Oaxaca), elevation to 1950 m, Wallach (2016).

Comments: McCranie & Hedges (2016) provide a species account. Wallach (2016) gives a species account, map and color photographs.

Epictia borapeliotes (Vanzolini, 1996)

Distribution: Upper elevation of 938 m, Guedes et al. (2014).

Epictia clinorostris Arredondo & Zaher, 2010

Distribution: Add Brazil (Mato Grosso do Sul), Ferreira et al. (2017).

Epictia columbi (Klauber, 1939)

Comment: Wallach (2016) gives a species account, map and color photographs.

Epictia diaplocia (Orejas-Miranda, 1969)

Distribution: Add Peru (Cusco, Huanuco), Koch et al. (2019); Peru (Puno), Llanqui et al. (2019).

Epictia fallax (W.C.H. Peters, 1857). Montsb. Preuss. Akad. Wiss. Berlin 1857: 402. (*Stenostoma fallax*)

Holotype: ZMB 9550, a 140 mm specimen (C.F.E. Otto).

Type locality: "Laguayra," Venezuela.

Distribution: Netherlands Antilles (Bonaire I.), N Venezuela (Anzoátegui, Aragua, Carabobo, Distrito Federal, Falcón, Lara, Miranda, Monagas, Nueva Esparta [Isla Margarita], Portuguesa, Sucre, Vargas, Yaracuy).

Comments: Natera-Mumaw et al. (2015) detail differences between Venezuela and Colombia specimens formerly assigned to *E. goudotii*, and revalidate *fallax* for the former.

Epictia goudotii (A.M.C. Duméril & Bibron, 1844)

Synonyms: Remove *Stenostoma fallax* W.C.H. Peters, 1857.

Distribution: Restricted mostly to Colombia with the recognition of *E. fallax*. Add Colombia (Huila), Moreno-Arias & Quintero-Corzo (2015, as *E.* cf. *goudotii*); Colombia (La Guajira), Wallach (2016); elevation modified to NSL-1280 m, Wallach (2016). Add Venezuela (Aragua), Koch et al. (2019).

Comments: McCranie & Hedges (2016) provide a species account. Wallach (2016) provides a synomymy, description, photographs (including the holotype), and distribution map. He concurs with differences between Colombian and Venezuelan populations noted by Natera-Mumaw et al., but does not address the validity of *E. fallax*, which see.

Epictia hobartsmithi Esqueda-González, Schlüter, Machado, Castelaín-Fernández & Natera-Mumaw *in* Natera-Mumaw et al., 2015. Atlas Serp. Venezuela: 415–420, figs. 1–3.

Holotype: EBRG 2523, 140 mm specimen of unknown sex (H. Mägdefrau & A. Schlüter, February 1990).

Type locality: "Cima del Tepui Guaiquinima, a 1180 m de altitude, coordenadas geográficas 5°54′N y 63°33′O, Campamento 2", Bolivar, Venezuela.

Distribution: C Venezuela (Bolivar), 1180 m. Known only from the type locality.

Epictia magnamaculata (Taylor, 1940)

Distribution: Upper elevation of 200 m, Wallach (2016).

Comments: McCranie & Hedges (2016) and Wallach (2016) give species accounts, maps and color photographs.

Epictia martinezi Wallach, 2016. Mesoamerican Herpetol. 3(3): 256–259, fig. 6.

Holotype: FMNH 283740 (orig. JRMC 20118), a 153 mm male (J.R. McCranie, 18 June 2012).

Type locality: "between Río Lempa and Antigua, Departamento de Ocotepeque, Honduras, ca. 14°24.3″N, 89°12.2″W, elev. 730 m asl."

Distribution: Western Honduras (Ocotepeque), 730 m. Known only from the holotype.

Epictia melanoterma (Cope, 1862)

Epictia melanura (K.P. Schmidt & Walker, 1943)

Epictia munoai (Orejas-Miranda, 1961)

Epictia pauldwyeri Wallach, 2016. Mesoamerican Herpetol. 3(3): 263–268, fig. 7.

Holotype: FMNH 130672, a 112 mm female (N. Gale, 20 March 1961).

Type locality: "Curundú, Ciudad de Panama, Canal Zone, Provincia de Panamá, Panama, ca. 8°59′08″N, 79°32′26″W, elev. 35 m asl."

Distribution: Pacific coast of central Panama (Colon, Panamá, Panamá Oeste), NSL-250 m.

Epictia peruviana (Orejas-Miranda, 1969)

Distribution: Add Peru (Ayacucho), Koch et al. (2019); Peru (Ucayali), Wallach (2016). Upper elevation of 1200 m, Esqueda-González et al. (2015).

Epictia phenops (Cope, 1875)

Distribution: Add Mexico (Guerrero), Palacios-Aguilar & Flores-Villela (2018); Belize (Stann Creek), Guatemala (Baja Verapaz, Escuintla, Guatemala, Izabal, Zacapa), El Salvador (Cuscatlán, La Libertad, Morazán, San Miguel, San Salvador, San Vicente, Santa Ana, Sonsonate), Honduras (Copán, Lempira), Wallach (2016); in Mexico only in Oaxaca and Chiapas according to Wallach (2016); elevation to 1670 m, Wallach (2016).

Comments: McCranie & Hedges (2016) and Wallach (2016) give species accounts, maps and color photographs.

Epictia resetari Wallach, 2016. Mesoamerican Herpetol. 3(3): 271–273, fig. 8.

Holotype: FMNH 178600 (orig. E.H. Taylor 2194), a 116 mm specimen (E.H. Taylor, 17 July 1932).

Type locality: "Paso de Ovejas, Veracruz, Mexico, 19°17′00″N, 96°26′26″W, elev. 55 m asl."

Distribution: NE Mexico (S Tamaulipas, NE Hidalgo, and Veracruz), NSL-900 m.

Epictia rioignis Koch, Martins & Schweiger, 2019. PeerJ 7(7411): 5–24, figs. 1–9.

Holotype: NMW 15446.6, a 211 mm specimen (collector uncertain, 1907).

Type locality: "Corinto, presumably Nicaragua (12°29′N, 87°11′W…)."

Distribution: Supposedly NW Nicaragua. The locality is deduced by elimination of alternatives.

Epictia rubrolineata (F. Werner, 1901)

Epictia rufidorsa (Taylor, 1940)

Distribution: Add Peru (Ancash), and low elevation of 1320 m, Koch et al. (2019); Peru (Callao), Wallach (2016).

Epictia schneideri Wallach, 2016. Mesoamerican Herpetol. 3(3): 279–281, 283, fig. 9.

Holotype: TCWC 9450 (orig. RWA 543), a 141 mm specimen (R.W. Axtell, 23 June 1953).

Type locality: "1.7 km SW Colotlipa, Guerrero, Mexico, 17°23′46″N, 99°10′48″W, elev. 825 m asl."

Distribution: S Mexico (Guerrero, Oaxaca), 825–1675 m.

Epictia septemlineata C. Koch, Venegas & Böhme, 2015. Zootaxa 3964(2): 230–232, figs. 1, 2.

Holotype: CORBIDI 14683, a 181 mm adult (A. Garcia-Bravo & C. Koch, 28 April 2009).

Type locality: "Limon Village, Celendín Province, Cajamarca Region, Peru (S 06°52′34.2″, W 078°05′10.5″, 2053 m.a.s.l.)."

Distribution: NW Peru (Cajamarca), 2053 m. Known only from the holotype.

Epictia signata (Jan, 1861)

Synonyms: Remove *Leptotyphlops amazonicus* Orejas-Miranda, 1969.

Distribution: E Ecuador and parts of Colombia, Natera-Mumaw et al. (2015).

Comments: See under *E. amazonica.*

Epictia striatula (H.M. Smith & Laufe, 1945)

Distribution: Add Peru (Cuzco), Wallach (2016); Peru (Puno), Koch et al. (2019); Bolivia (Chaco, Santa Cruz), Koch et al. (2019); Brazil (Amazonas), Francisco et al. (2018).

Epictia subcrotilla (Kaluber, 1939)

Distribution: Add Peru (Apurimac), Wallach (2016); Peru (Cajamarca), and increase elevation range to 447 m, Koch et al. (2018); Peru (Lambayeque), and low elevation of sea level, Venegas (2005).

Comments: See under *Trilepida guayaquilensis.*

Epictia teaguei (Orejas-Miranda, 1964)

Epictia tesselata (Tschudi, 1845)

Distribution: Add Peru (Ica), Koch et al. (2019).

Epictia tricolor (Orejas-Miranda & Zug, 1974)

Distribution: Add Peru (La Libertad), Koch et al. (2019).

Epictia undecimstriata (Schlegel, 1839 *in* 1837–1844)

Distribution: Add Bolivia (Cochabamba), and upper levation of 2200 m, Esqueda-González et al. (2015).

Epictia unicolor (F. Werner, 1913)

Epictia vanwallachi C. Koch, Venegas & Böhme, 2015. Zootaxa 3964(2): 233–235, figs. 2, 4.

Holotype: CORBIDI 14682, a 107 mm adult (E. Hoyas-Granda, A. Beraún & C. Koch, 10 January 2010).

Type locality: "Vijus Village, Pataz Province, La Libertad Region, Peru (S 07°43′11.6″, W 077°39′51.1″, 1290 m.a.s.l.)."

Distribution: NW Peru (La Libertad). Known only from the holotype.

Epictia vellardi (Laurent, 1984)

Distribution: Add Paraguay (Boqueron), and upper elevation of 190 m, Cabral & Sisa (2016); Brazil (Mato Grosso), Francisco et al. (2018).

Epictia venegasi C. Koch, Santa-Cruz & Cárdenas, 2016. Zootaxa 4150(2): 103–113, figs. 1–9.

Holotype: MUSA 4252, a 192 mm adult (R. Santa-Cruz & H. Cárdenas, 21 March 2015).

Type locality: "Cachachi-Moyan, Province Cajabamba, Region Cajamarca, Peru (07°37′2.348″ S, 078°10′47.565″ W, 2551 m a.s.l.)."

Distribution: NW Peru (Cajamarca), 2551–2736 m.

Epictia vindumi Wallach, 2016. Mesoamerican Herpetol. 3(3): 289–291, 293, fig. 10.

Holotype: FMNH 153536, a 153 mm specimen, (E.W. Andrews, February 1959).

Type locality: "Chichén Itza, Yucatán, Mexico, 20°41′03″N, 88°34′06″W, elev. 35 m asl."

Distribution: SE Mexico (N part of the Yucatan Peninsula in Quintana Roo, Yucatán), NSL-35 m.

Epictia vonmayi C. Koch, Santa-Cruz & Cárdenas, 2016. Zootaxa 4150(2):115–117, figs. 1, 11, 12.

Holotype: MUSA 4342, a 129 mm adult (R. Santa-Cruz & H. Cárdenas, 15 January 2013).

Type locality: "La Granja-Río Tinto, District Querocoto, Province Chota, Region Cajamarca, Peru (06°20′30.592″ S, 079°06′41.058″ W, 2069 m a.s.l.)."

Distribution: NW Peru (Cajamarca), 1985–2069 m.

Epictia weyrauchi (Orejas-Miranda, 1964)

Epictia wynni Wallach, 2016. Mesoamerican Herpetol. 3(3): 298–300, 302, fig. 11.

Holotype: TCWC 32899 (orig. RLB 137), a 163 mm specimen (R.L. Beck, 7 June 1970).

Type locality: "Río Ayutla, 30 km north of Jalpan de Serra, Querétaro, Mexico, 21°25′01″N, 99°36′15″W, elev. 575 m asl."

Distribution: NE Mexico (Sierra Madre Oriental of N Querétaro, NW Hidalgo), 315–2500 m.

ERISTICOPHIS Alcock & Finn, 1897 (Viperidae: Viperinae)

Eristicophis macmahonii Alcock & Finn, 1897

Distribution: Add Afghanistan (Kandahar), Wagner et al. (2016b).

ERPETON Lacépède, 1801 (Homalopsidae)

Erpeton tentaculatum Lacépède, 1801

Comments: J.C. Murphy & Voris (2014) provide a diagnosis and photograph.

ERYTHROLAMPRUS F. Boie, 1826 (Dipsadidae: Xenodontinae)

Synonyms: Add *Iaculatrix* Boddaert, 1783, *Dromicus* Bibron, 1843 *in* Ramon de la Sagra, 1838–1857, *Calophis* Fitzinger, 1843, *Leimadophis* Fitzinger, 1843, *Opheomorphus* Fitzinger, 1843, *Pariopeltis* Fitzinger, 1843, *Pseudophis* Fitzinger, 1843, *Umbrivaga* Roze, 1964.

Distribution: Central and South America, and Lesser Antilles.

Comments: In combining *Liophis* and *Umbrivaga* with *Erythrolamprus*, we follow Grazziotin et al. (2012) and Zaher et al. (2019). Ascenso et al. (2019) revise the *E. reginae* group (*E. dorsocorallinus*, *E. macrosomus*, *E. oligolepis*. *E. reginae*, *E. rochai*, *E. zweifeli*), and provide a key to the species.

Erythrolamprus aesculapii (Linnaeus, 1758)

Distribution: Add Colombia (Amazonas), Venezuela (Santa Elena), Guyana (Mazaruni-Potaro), Surinam (Marowijne, Nickerie), Brazil (Acre, Alagoas, Distrito Federal, Espirito Santo, Maranhao, Mato Grosso do Sul, Roraima, Tocantins), Paraguay (Guaira), Peru (Ayacucho, Cajamarca, Huanuco), Ecuador (Morona-Santiago, Pastaza), all Curcio et al. (2015); Colombia (Antioquia, Arauca, Caqueta, Cesar, Choco, Cundinamarca, La Guajira), Serrano & Díaz-Ricaurte (2018); Brazil (Pernambuco), Roberto et al. (2017); Brazil (Rio de Janeiro: Ilha Grande), C.F.D. Rocha et al. (2018); Paraguay (San Pedro), P. Smith et al. (2016).

Erythrolamprus albertguentheri Grazziotin, Zaher, R.W. Murphy, Scrocchi, Benavides, Zhang & Bonatto, 2012. Substitute name.

Synonyms: Add *Liophis guentheri* Peracca, 1897.

Comments: The name substitution is necessitated by the inclusion of *Liophis* within *Erythrolamprus*, in which *E. guentheri* Garman, 1884 has priority.

Erythrolamprus almadensis (Wagler *in* Spix, 1824)

Distribution: Add Brazil (Piauí, Tocantins), Dal Vechio et al. (2016); Brazil (Rondônia), França et al. (2006) and Bernarde et al. (2012b).

Erythrolamprus andinus (Dixon, 1983)

Erythrolamprus atraventer (Dixon & R.A. Thomas, 1985)

Erythrolamprus bizonus Jan, 1863

Types: Lectotype: MSNM Re 2871, from Popayan, Cauca, Colombia, designated by Curcio et al. (2015).

Distribution: Add Costa Rica (Limón), Panama (Chiriqui, Cocle, Colón, Herrera, La Joya, Los Santos, Panama, Panama Oeste), Colombia (Boyacá, Casanare, Risaralda, Tolima, Vichada), Curcio et al. (2015), Ray & Ruback (2015); Herrera & Ray (2016).

Comments: Curcio et al. (2015) present a redescription, details of geographic variation, and designate and illustrate a lectotype.

Erythrolamprus breviceps (Cope, 1860)

Synonyms: *Liophis longiventris* Amaral, 1925. See under *E. longiventris*.

Erythrolamprus carajasensis (Cunha, Nascimento & Ávila-Pires, 1985)

Erythrolamprus ceii (Dixon, 1991)

Erythrolamprus cobellus (Linnaeus, 1758)

Erythrolamprus cursor (Bonnaterre, 1790)

Comment: Believed to be extinct. Jower et al. (2013) were able to sequence DNA from preserved specimens, and found that it was a sister taxon to *E. juliae*.

Erythrolamprus dorsocorallinus Esqueda, Natera, La Marca & Ilija-Fistar, 2005. Herpetotropicos 2(2): 96–99, figs. 1, 3, 4. (*Liophis dorsocorallinus*)

Holotype: ULABG 6691, a 639 mm male (L.F. Esqueda, 3 November 2004).

Type locality: "aproximadamente 40 Km de la población de El Cantón, 100 m.s.n.m., 07°27′85″ N y 71°02′98″ W, Municipio Andrés Eloy Blanco, Estado Barinas, Venezuela."

Distribution: WC South America. Venezuela (Barinas), Colombia (Arauca), Peru (Huanuco, Loreto, Madre de Dios, San Martin, Ucayali), Brazil (Acre, Amazonas, Mato Grosso, Rondonia ?), Bolivia (Beni, Cochabamba, La Paz, Pando, Santa Cruz).

Comments: Based on morphology, Ascenso et al. (2019) remove *dorsocorallinus* from the synonymy of *E. reginae*. Bernarde et al. (2011b) and J. da S. Araújo et al. (2012) document new localities in Brazil, and Eversole et al. (2016) report it from Bolivia. Souto et al. (2017) describe the skull and morphological variation, document known localities, and present evidence for the distinctiveness of this species. Ascenso et al. (2019) provide a description, color photographs and a map of known localities.

Erythrolamprus epinephelus (Cope, 1862)

Distribution: Add Panama (Panama Oeste), Ray & Ruback (2015); Colombia (Quindio), Quintero-Ángel et al. (2012); Venezuela (Barinas), Escalona (2017); Ecuador (Chimborazo, Napo, Sucumbio), Almendáriz & Orcés (2004).

Erythrolamprus festae (Peracca, 1897)

Erythrolamprus frenatus (F. Werner, 1909)

Distribution: Add Brazil (Goias), Vaz-Silva et al. (2007); Brazil (Tocantins), Dal Vechio et al. (2016, as *L. cf. frenatus*).

Erythrolamprus guentheri Garman, 1884

Distribution: Add Ecuador (Chimborazo), Curcio et al. (2015); Peru (Cajamarca), and upper elevation of 1239 m, Koch et al. (2018).

Erythrolamprus ingeri (Roze, 1958)

Erythrolamprus jaegeri (Günther, 1858)

Distribution: Add Brazil (Marinheiros I., Rio Grande do Sul), Quintela et al. (2011); Paraguay (Presidente Hayes), Cabral & Weiler (2014).

Erythrolamprus janaleeae (Dixon, 2000)
Erythrolamprus juliae (Cope, 1879)

Erythrolamprus longiventris Amaral, 1925. Commissão de Linhas Telegraphicas Estrategicus de Matto Grosso ao Amazonas (84): 16, pl. 1–3. (*Liophis longiventris*)

Holotype: IB 43, now likely destroyed by fire.

Type locality: Not stated, but Mato Grosso according to the IB catalog.

Distribution: C Brazil (Amazonas, Mato Grosso). Add Brazil (Rondônia), França et al. (2006) and Bernarde et al. (2012b).

Comments: Grazziotin et al. (2012) resume recognition of *longiventris* as distinct from *L. breviceps*.

Erythrolamprus macrosomus Amaral, 1935. Mem. Inst. Butantan 9: 238. (*Leimadophis reginae macrosoma*)

Synonyms: *Leimadophis reginae maculicauda* Hoge, 1953.

Syntypes: IB 9130-9133, 2 male and 2 female adults, all destroyed in the 2010 fire.

Type locality: "Canna Brava, Goiás," Brazil.

Distribution: SE Brazil (S Goias, S Mato Grosso, Mato Grosso do Sul, S Minas Gerais, Parana, Sao Paulo), Paraguay (Amambay, Central, Presidente Hayes, San Pedro), N Argentina (Formosa, Misiones, Salta). Add Paraguay (Canindeyu), Cacciali et al. (2015b); Paraguay (Alto Parana, Itapua, Paraguari), Arzamendia (2016); Argentina (Entre Rios, Jujuy), Arzamendia (2016).

Comments: Based on morphology, Ascenso et al. (2019) recognize populations formerly assigned to the southeastern distribution of *E. reginae* as a distinct species. They provide a description, color photographs and a map of known localities.

Erythrolamprus maryellenae (Dixon, 1985)

Erythrolamprus melanotus (Shaw, 1802)

Distribution: Add Colombia (Santander), E. Ramos and Meza-Joya (2018).

Erythrolamprus mertensi (Roze, 1964)

Erythrolamprus miliaris (Linnaeus, 1758)

Synonyms: Remove *Liophis mossoroensis* Hoge & Lima-Verde, 1973.

Distribution: Remove the Caatinga portions of the range to *E. mossoroensis*.

Erythrolamprus mimus (Cope, 1869)

Distribution: Add Nicaragua (Atlantico Norte, Altlantico Sur), Panama (Chiriqui, Panamá), Ecuador (Carchi), Curcio et al. (2015); Ecuador (Morona-Santiago), Almendáriz et al. (2014).

Erythrolamprus mossoroensis Hoge & Lima-Verde, 1973. Mem. Inst. Butantan 36: 215–217, fig. 2. (*Liophis mossoroensis*)

Holotype: IB 32078, a 606 mm male, likely destroyed in the 2010 fire.

Type locality: "Mossoró, R[io Grande do]N[orte]," Brazil.

Distribution: Brazil (Alagoas), Dixon (1983); Brazil (Ceara), Roberto & Loebmann (2016); Brazil (Bahia, N Minas Gerais, Pernambuco, Rio Grande do Norte), Guedes et al. (2014).

Comments: Guedes et al. (2014) consider *Erythrolamprus mossoroenis* Hoge & Lima-Verde, 1973 as a valid species.

Erythrolamprus ocellatus W.C.H. Peters, 1868

Erythrolamprus oligolepis (Boulenger, 1905)

Distribution: Add Brazil (Amapa, Amazonas, Mato Grosso), Ascenso et al. (2019). D. França et al. (2013) map known localities.

Comments: Ascenso et al. (2019) provide a description, color photographs and a map of known localities.

Erythrolamprus ornatus (Garman, 1887)

Comments: Williams et al. (2016) provide a revised description.

Erythrolamprus perfuscus (Cope, 1862)

Erythrolamprus poecilogyrus (Wied-Neuwied, 1824)

Distribution: Add Brazil (Marinheiros I., Rio Grande do Sul), Quintela et al. (2011). Upper elevation of 1100 m, Guedes et al. (2014).

Erythrolamprus pseudocorallus Roze, 1959

Distribution: Add Colombia (Antioquia), Restrepo et al. (2017); Colombia (Norte Santander, Santander, Tolima), Curcio et al. (2015).

Erythrolamprus pseudoreginae J.C. Murphy, Braswell, Charles, Auguste, Rivas, Borzée, Lehtinen & Jowers, 2019a. ZooKeys 817: 138–141, figs. 2, 3, 6.

Holotype: UWIZM 2016.22.45, a 525 mm male (A.L. Braswell & R.J. Auguste, 13 June 2016).

Type locality: "Gilpin Trace Trail, 8.5 km NNW Roxborough, St. John, Tobago (~ 11°16′55″N; 60°37′12″W, about 493 m ASL)."

Distribution: Trinidad and Tobago (Tobago), 430–500 m.

Comments: The phylogeny produced by J.C. Murphy et al. (2019a) used two mtDNA genes, and has *E. pseudoreginae*, with Costa Rican *E. epinephelus*, as sister taxa to *E. melanotus* populations from Trinidad and Tobago.

Erythrolamprus pyburni (Markezich & Dixon, 1979)

Erythrolamprus pygmaeus (Cope, 1868)

Distribution: Add Brazil (Acre), Bernarde et al. (2011b); Brazil (Rondonia) and upper elevation of 110 m, Dal Vechio et al. (2015a); Ecuador (Zamora-Chinchipe), Dal Vechio et al. (2015a); Peru (Amazonas, Ucayali), Dal Vechio et al. (2015a).

Erythrolamprus reginae (Linnaeus, 1758)

Synonyms: *Coluber graphicus* Shaw, 1802, *Natrix semilineata* Wagler in Spix, 1824, *Liophis reginae maculata* Steindachner, 1867, *Liophis miliaris intermedius* Henle & Ehrl, 1991. Ascenso et al. (2019) consider *Coluber violaceus* Lacépède, 1789 to be a *nomen dubium*, not applicable to *E. reginae*.

Distribution: Add Peru (Amazonas), Ascenso et al. (2019); Suriname (Paramaribo), Ascenso et al. (2019); Brazil (Amapa, Roraima), Ascenso et al. (2019); Brazil (Piauí), Dal Vechio et al. (2013); Brazil (Sergipe), Guedes et al. (2014); Brazil (Tocantins), Dal Vechio et al. (2016). Remove Trinidad and Tobago (Tobago) to *E. pseudoreginae*, J.C. Murphy et al. (2019a), and Argentina, Paraguay and SE Brazil (S Goias, S Mato Grosso, Mato Grosso do Sul, S Minas Gerais, Parana, most of Sao Paulo) to *E. macrosomus*, Ascenso et al. (2019).

Comments: Ascenso et al. (2019) redefine *E. reginae* by removing southeast populations to *E. macrosomus*, and recognizing *E. dorsocorallinus*. They provide a description, color photographs and a map of known localities.

Erythrolamprus rochai Ascenso, Costa & Prudente, 2019. Zootaxa 4586(1): 78–79, 87, figs. 5, 7.

Holotype: MPEG 25680, a 281 mm male (D. Silvano, B. Pimenta & U. Galatti, Sept.-Nov. 2000).

Type locality: "Urucum region (00°50′ N, 51°53′ W; ca. 171 m asl), municipality of Serra do Navio, state of Amapá, Brazil."

Distribution: Brazil (Amapa), 171 m. Known from two specimens from the type locality.

Erythrolamprus sagittifer (Jan, 1863)

Distribution: Add Paraguay (Alto Paraguay, Boqueron, Presidente Hayes), Cabral & Weiler (2014).

Erythrolamprus semiaureus (Cope, 1862)

Distribution: Add Paraguay (Misiones, Paraguari, Presidente Hayes), Cabral & Weiler (2014).

Erythrolamprus subocularis Boulenger, 1902. Ann. Mag. Nat. Hist. Series 7, 9(49): 56. (*Rhadinaea subocularis*)

Syntypes: BMNH 1946.1.5.81-82, the largest 370 mm.

Type locality: "Paramba, 3500 feet," Ecuador.

Distribution: W Ecuador.

Comments: Dixon (1980) refers *subocularis* to *Rhadinaea decorata*, although C.W. Myers (1974) does not mention it, nor was it mentioned in Wallach et al. (2014). Grazziotin et al. (2012) refer *subocularis* to *Erythrolamprus* s.l.

Erythrolamprus taeniogaster (Jan, 1863)

Distribution: Add Brazil (Acre), Bernarde et al. (2011b); Brazil (Alagoas), Guedes et al. (2014); Brazil (Maranhão), J.P. Miranda et al. (2012); Brazil (Mato Grosso do Sul), Ferreira et al. (2017); Brazil (Paraiba), R. França et al. (2012); Brazil (Piauí), Dal Vechio et al. (2013); Brazil (Rondônia), Bernarde et al. (2012b).

Erythrolamprus taeniurus (Tschudi, 1845)

Distribution: Low elevation of 1239 m, Koch et al. (2018).

Erythrolamprus torrenicolus (Donnelly & C.W. Myers, 1991)

Erythrolamprus trebbaui (Roze, 1958)

Erythrolamprus triscalis (Linnaeus, 1758)

Erythrolamprus typhlus (Linnaeus, 1758)

Distribution: Add Brazil (Marajo I., Para), G.M. Rodrigues et al. (2015); Peru (Puno), Llanqui et al. (2019).

Erythrolamprus viridis (Günther, 1862)

Erythrolamprus vitti (Dixon, 2000)

Erythrolamprus williamsi (Roze, 1958)

Erythrolamprus zweifeli (Roze, 1959)

Distribution: Add Venezuela (Anzoategui, Caracas, Lara, Tachira, Vargas), Ascenso et al. (2019); Venezuela (Monagas, Sucre, Yaracuy), J.C. Murphy et al. (2019a); Trinidad and Tobago (Trinidad), J.C. Murphy et al. (2019a).

Comments: Ascenso et al. (2019) provide a description, color photographs and a map of known localities.

ERYX Daudin, 1803 (Erycidae)

Fossil records: Add lower Pleistocene (Villanyian) of Bulgaria, Boev (2017, as *Eryx* sp.).

Comments: E. Rastegar-Pouyani et al. (2014) evaluate relationships of taxa in northeast Iran using mtDNA sequence data. They conclude that samples of *E. elegans* are synonymous with *E. jaculus*, and those of *E. tataricus* with *E. miliaris*. However, Safaei-Mahroo et al. (2015) continue to recognize those four species as distinct. Reynolds et al. (2014b) produce a DNA sequence-based phylogeny that fails to support recognition of *Gongylophis* for rough-tailed species. Their phylogeny has *E. jayakari* and *E. muelleri* forming a sister clade to the remaining species. *Eryx miliaris* and *E. tataricus* are polyphyletic, but one of the specimens was from the pet trade, and they admitted the possibility of misidentification. Zarrintab et al. (2017) evaluate morphological characters by multivariate analysis, and conclude that four species occur in Iran. In a renewed effort, Eskandarzadeh et al. (2020) evaluate two mtDNA genes for populations of *miliaris* and *tataricus* from Iraq, Iran and Kazakhstan. Their phylogeny indicates a sister clade from NE Iran, followed by clades from Iraq + W Iran, and a clade consisting of several subclades that offer a mix of specimens, assigned to *miliaris* or *tataricus*, from N and E Iran, and Kazakhstan. From these results the authors suggest the NE clade to represent a new, unnamed species, and that *miliaris* and *tataricus* are conspecific. They note that populations within the complex are morphologically non-diagnosible. See below under *E. tataricus*.

Eryx borrii Lanza & Nistri, 2005

Eryx colubrinus (Linnaeus, 1758)

Distribution: Add Niger (Zinder), Trape & Mané.

Eryx conicus (Schneider, 1801)

Distribution: Add Nepal (Bara, Nawalparasi), Pandey et al. (2018); Bangladesh (Rajshahi Division, Naogaon District), 23 m, Ahmad et al. (2015b).

Eryx elegans (Gray, 1849)

Distribution: Add Iran (Ardabil, Khuzestan, North Khorasan, South Khorasan, West Azarbaijan), Shafaei-Mahroo et al. (2015); Afghanistan (Bamyan, Ghazni), and upper elevation of 3000 m, Wagner et al. (2016b).

Comments: E. Rastegar-Pouyani et al. (2014) consider *E. elegans* to be conspecific with *E. jaculus* in northeast Iran. DNA-sequence data evaluated by Reynolds et al. (2014b) has *E. elegans* as sister taxon to *E. miliaris*, *E. tataricus* and *E. vittatus*. For that reason we continue to recognize it as a species.

Eryx jaculus (Linnaeus, 1758)

Distribution: Add Libya (Darnah, Jafara, Marj, Sirte), Bauer et al. (2017); Greece (Euboea), Christopoulos et al. (2019); Greece (Folegandros Island), Itescu et al. (2017); Greece (Patmos Island), Roussos (2016); Greece (Samos Island), Speybroek et al. (2014); Italy (Sicily), Insacco et al. (2015, likely an ancient introduction); Romania (Olt), Covaciu-Marcov et al. (2012); Turkey (Canakkale), Tok et al. (2014); Turkey (Gökçeada I.), Yakin et al. (2018); Turkey (Tenedos I.), Tosunoğlu et al. (2009); Iran (Alborz, Ardabil, Gilan, Golestan, Hamedan, Isfahan, Kerman, North Khorasan, Semnan, South Khorasan, Tehran), Shafaei-Mahroo et al. (2015); Iran (Qazvin, Zanjan), Eskandarzadeh et al. (2018a); Iran (Ilam, Kurdistan, Lorestan, Markazi, Qom), S.M. Kazemi et al. (2015); Syria (Homs), Sindaco et al. (2014). Sahlean et al. (2015) map localities in Romania.

Fossil records: upper Pliocene (Ruscinian, MN 15) of Italy (Sardinia), Delfino et al. (2011, as cf. *E. jaculus*).

Comments: See comment under *E. elegans*. Abbas-Rhadi et al. (2015b) examine morphological variation in a series from central Iraq, concluding that some specimens are referable to *E. miliaris* due to a relatively short second supralabial. Faraone et al. (2017) report additional specimens from Sicily, but intermediate morphological characters prohibit assignment to subspecies. Reynolds & Henderson (2018) do not recognize subspecies.

Eryx jayakari Boulenger, 1888

Distribution: Add Iraq (Al-Basrah), Habeeb & Rastegar-Pouyani (2016a); Saudi Arabia (Ha'il), Alshammari et al. (2017). Upper elevation of 2359 m, Eskandarzadeh et al. (2018b).

Comments: Al-Sadoon & Al-Otaibi (2014) present a morphological description.

Eryx johnii (Russell, 1802 *in* 1801–1810)

Distribution: Add India (Madhya Pradesh), Manhas et al. (2018a); Nepal (Chitwan), Pandey (2012).

Eryx miliaris (Pallas, 1773)

Distribution: Add Egypt (Tabuk), Aloufi & Amr (2015); Iran (Ardabil, East Azerbaijan, Golestan, Kermanshah, Kurdistan, Markazi, North Khorasan, Qazvin, South Khorasan), Shafaei-Mahroo et al. (2015). Remove Afghanistan, Wagner et al. (2016b).

Comments: See under *E. tataricus*. Rhadi F. et al. tentatively consider this species occurring in the Bahr Al-Najaf depression, Al-Najaf Prov., Iraq along with *E. jaculus*.

†***Eryx mongoliensis*** (Gilmore, 1943)

Eryx muelleri (Boulenger, 1892)

Distribution: Add Mauritania (Inchiri), Padial (2006); Burkina Faso (Boucle du Mouhoun, Nord, Plateau-Central), Vignoli et al. (2015); Niger (Tahoua, Tillaberi), Trape & Mané (2015); Benin (Donga), Vignoli et al. (2015); Nigeria (Kebbi, Zamfara), Vignoli et al. (2015). Delete Sierra Leone, Trape & Baldé (2014). Low elevation of 41 m, Trape & Mané (2017); upper elevation of 510 m, Trape & Mané (2015).

†***Eryx primitivus*** Szyndlar & Schleich, 1994

Eryx somalicus Scortecci, 1939

Eryx tataricus (Lichtenstein in Eversmann, 1823)

Distribution: Add Iran (Razavi Khorasan), Nasrabadi et al. (2016); Iran (Ardabil, Kurdistan, Gilan, Golestan, Hamedan, Kohgiluyeh and Boyer Ahmad, Markazi, North Khorasan, Qazvin, Sistan and Baluchistan), Shafaei-Mahroo et al. (2015); Afghanistan (Faryab, Ghor, Helmand, Herat, Kabul, Kunduz, Nangahar), Wagner et al. (2016b).

Comments: Eskandarzadeh et al. (2013) consider this taxon conspecific with *E. miliaris* in northeast Iran at least, with *E. miliaris* being the species present. They indicate the need for further study to determine the validity of the species *tataricus* in other areas. E. Rastegar-Pouyani et al. (2014) also consider *E. tataricus* to be conspecific with *E. miliaris* in northeast Iran based on mtDNA sequence-data. Zarrintab et al. (2017) use multivariate analysis of morphology to conclude that Iranian *E. tataricus* is distinct from *E. miliaris*, the latter which they do not consider to be present in Iran. Eskandarzadeh et al.

(2020) present an expanded phylogeny that has several clades with mixed samples from regions of Iran and Kazakhstan. From these results they recommend that *tataricus* be synonymized with *miliaris*. However, topotypic samples of both species occupy different, shallow clades. Due to the lack of recent consensus from this group of researchers as to the taxonomy of Iranian *Eryx* of the *jaculus-miliaris-tataricus* complex, we refrain from rearranging taxa at this time.

Eryx vittatus Chernov, 1959

Distribution: Safaei-Mahroo et al. (2015) list the provinces for Iran.

Eryx whitakeri I. Das, 1991

Distribution: Upper elevation of 57 m, Chowdhury & Chaudhuri (2017).

ETHERIDGEUM Wallach, 1988 (Calamariidae)

Etheridgeum pulchrum (F. Werner, 1924)

EUNECTES Wagler, 1830 (Boidae)

Eunectes beniensis Dirksen, 2002

Eunectes deschauenseei Dunn & Conant, 1936

Eunectes murinus (Linnaeus, 1758)

Distribution: Add Peru (Huanuco), Schlüter et al. (2004); Brazil (Acre), Bernarde et al. (2011b); Brazil (Parana), Dainesi et al. (2019); Brazil (Tocantins), Dal Vechio et al. (2016); Paraguay (San Pedro), P. Smith et al. (2016).

Eunectes notaeus Cope, 1862

Distribution: Add Paraguay (Concepcion, Itapua, San Pedro), Cabral & Weiler (2014).

Comments: McCartney-Melstad et al. (2012) describe intra- and interpopulation structure in N Argentina, using mtDNA sequence data.

†***Eunectes stirtoni*** Hoffstetter & Rage, 1977

†***EUPODOPHIS*** Rage & Escuillié, 2002 (†Simoliophiidae)

Comments: Palci et al. (2013b) reevaluate anatomy and discuss relationships.

†***Eupodophis descouensi*** (Rage & Escuillié, 2000)

Comments: Palci et al. (2013b) provide an emended diagnosis based on new interpretation of fossil specimens.

EUPREPIOPHIS Fitzinger, 1843 (Colubridae: Colubrinae)

Euprepiophis conspicillatus (H. Boie, 1826)

Distribution: Add Japan (Kinkasan I.), Mori & Nagata (2016).

Euprepiophis mandarinus (Cantor, 1842)

Distribution: Add India (Nagaland), Lele et al. (2018); Vietnam (Ha Giang), Ziegler et al. (2014); Vietnam (Hoa Binh), Nguyen et al. (2018).

Euprepiophis perlaceus (Stejneger, 1929)

EUTRACHELOPHIS C.W. Myers & McDowell, 2014. Bull. Amer. Mus. Nat. Hist. (385): 6. (Dipsadidae: Xenodontinae)

Type species: *Eutrachelophis bassleri* C.W. Myers & McDowell, 2014 by original designation.

Distribution: NW South America.

Comments: *Eutrachelophis* is assigned to its own tribe (Eutrachelophini) to include two named species and one under description, C.W. Myers & McDowell (2014).

Eutrachelophis bassleri C.W. Myers & McDowell, 2014. Bull. American Mus. Nat. Hist. (385): 8–14, figs. 1, 2.

Holotype: AMNH 52926, a 345 mm male (H. Bassler, 15 January 1927).

Type locality: "Pisqui Hills, [upper] Rio Pisqui, Province of Loreto, Peru."

Distribution: Ecuador (Pastaza); Peru (Huanuco, Loreto), 80–510 m.

Eutrachelophis steinbachi Boulenger, 1905. Ann. Mag. Nat. Hist. Series 7, 15(89): 454–455. (*Rhadinaea steinbachi*)

Lectotype: BMNH 1946.1.21.62, designated by C.W. Myers & McDowell (2014), a 558 mm female (J. Steinbach).

Type locality: "the Province Sara, Department Santa Cruz de la Sierra," Bolivia.

Distribution: Bolivia (Santa Cruz), 50–500 m.

Comments: Listed as Incerta Sedis, *Rhadinaea steinbachi* by Wallach et al. (2014). C.W. Myers & McDowell (2014) present a revised description and photographs of the syntypes.

EXILIBOA Bogert, 1968 (Charinidae: Ungaliophiinae)

Exiliboa placata Bogert, 1968

†***FALSERYX*** Szyndlar & Rage, 2003 (Tropidophiidae)

Distribution: Add middle Miocene (Orleanian) of Germany, Čerňanský et al. (2017, as cf. *Falseryx*).

†***Falseryx neervelpensis*** Szyndlar, R. Smith & Rage, 2008

†***Falseryx petersbuchi*** Szyndlar & Rage, 2003

FARANCIA Gray, 1842 (Dipsadidae: subfamily unnamed)

Farancia abacura (Holbrook, 1836)

Farancia erytrogramma (Palisot de Beauvois *in* Sonnini & Latreille, 1801)

FERANIA Gray, 1842 (Homalopsidae)

Comments: J.C. Murphy & Voris (2014) provide a generic diagnosis.

Ferania sieboldii (Schlegel, 1837)

Distribution: Add Nepal (Nawalparasi), and upper elevation range of 215 m, Pandey (2012).

Comments: J.C. Murphy & Voris (2014) provide a diagnosis and photograph.

FICIMIA Gray, 1849 (Colubridae: Colubrinae)

Ficimia hardyi Mendoza-Quijano & H.M. Smith, 1993

Distribution: Add Mexico (Tamaulipas), Farr et al. (2013).

Ficimia olivacea Gray, 1849

Ficimia publia Cope, 1866

Distribution: Add Mexico (Michoacan), Torres-Pérez-Coeto et al. (2016).

Ficimia ramirezi H.M. Smith & Langebartel, 1950

Ficimia ruspator H.M. Smith & Taylor, 1941

Distribution: Upper elevation to 1300 m, Palacios-Aguilar & Flores-Villela (2018).

Ficimia streckeri Taylor, 1931

Ficimia variegata (Günther, 1858)

FIMBRIOS M.A. Smith, 1921 (Xenodermidae)

Fimbrios klossi M.A. Smith, 1921

Distribution: Add Vietnam (Quang Ngai), Nemes et al. (2013).

Fimbrios smithi Ziegler, David, Miralles, Kien & Truong, 2008.

†*FLORIDAOPHIS* Holman, 2000 (Colubroidea: incerta sedis)

†*Floridaophis auffenbergi* Holman, 2000

FORDONIA Gray, 1842 (Homalopsidae)

Comments: J.C. Murphy & Voris (2014) provide a generic diagnosis.

Fordonia leucobalia (Schlegel, 1837)

Distribution: Add Bangledesh (Khulna Division, Bagerhat District), Freed et al. (2015); Myanmar (Tanintharyi), Mulcahy et al. (2018). Leviton et al. (2018) are unaware of any confirmed records from Luzon, Philippines.

Comments: J.C. Murphy & Voris (2014) provide a diagnosis and photograph.

FOWLEA Theobald, 1868. J. Asiatic Soc. Bengal 37: 57. (Natricidae)

Synonyms: *Diplophallus* Cope, 1893.

Type species: *Fowlea peguensis* Theobald, 1868 by monotypy.

Distribution: Southern Asia.

Comments: Formerly the *Xenochrophis piscator* species group; see under *Xenochrophis*.

Fowlea asperrima (Boulenger, 1891)

Fowlea flavipunctata (Hallowell, 1861)

Distribution: Add Thailand (Koh Chang I., Rayong), Chanard and Makchai (2011); China (Nan Ao Island), Qing et al. (2015); Vietnam (Hai Phong: Cat Ba Island), T.Q. Nguyen et al. (2011).

Fowlea melanzostus (Gravenhorst, 1807)

Fowlea piscator (Schneider, 1799)

Distribution: Add Afghanistan (Nangarhar), Mebert et al. (2013); Nepal (Nawalparasi), Pandey et al. (2018); India (Goa, Telangana), Purkayastha et al. (2018); Myanmar (Tanintharyi), Mulcahy et al. (2018).

Fowlea punctulata (Günther, 1858)

Distribution: Add Myanmar (Ayeyarwady), Platt et al. (2018).

Fowlea sanctijohannis (Boulenger, 1890)

Fowlea schnurrenbergeri Kramer, 1977

Distribution: Add India (Arunachal Pradesh), Purkayastha (2018); India (Bihar), Purkayastha et al. (2010).

Fowlea tytleri (Blyth, 1863)

FURINA A.M.C. Duméril, 1853 (Elapidae)

Furina bernardi (Kinghorn, 1939)

Furina diadema (Schlegel, 1837)

Furnima dunmalli (Worrell, 1955)

Furina ornata (Gray, 1842)

Furina tristis (Günther, 1858)

†*GAIMANOPHIS* Albino, 1996 (Booidea: incerta sedis)

†*Gaimanophis tenuis* Albino, 1996

†*GANSOPHIS* Head, 2005 (Colubridae: incerta sedis)

†*Gansophis potwarensis* Head, 2005

GARTHIUS Malhotra & Thorpe, 2004 (Viperidae: Crotalinae)

Garthius chaseni (M.A. Smith, 1941)

GEAGRAS Cope, 1875 (Colubridae: Colubrinae)

Comments: L.D. Wilson & Mata-Silva (2015) summarize data on the genus and single species (*G. redimitus* Cope). They, and Ramírez-Bautista et al. (2014), discuss unpublished data that place *Geagras* within *Tantilla*.

Geagras redimitus Cope, 1875

Distribution: L.D. Wilson & Mata-Silva (2015) limit the elevational range to 0–400 m.

GEOPHIS Wagler, 1830 (Dipsadidae: Dipsadinae)

Comments: Canseco-Márquez et al. (2016) provide a key to the *G. dubius* group. The gene-based phylogeny of Zaher et al. (2019) has one clade of *Geophis* as sister clade to *Atractus*, but other species of *Geophis* are sister taxa to the *Atractus* species.

Geophis anocularis Dunn, 1920

Geophis bellus C.W. Myers, 2003

Distribution: Add Panama (Cocle, Colon, Veraguas), and upper elevation of 1100 m, Lara et al. (2015); Panama (Panama), Ray & Ruback (2015).

Geophis betaniensis Restrepo & Wright, 1987

Geophis bicolor Günther, 1868

Geophis blanchardi Taylor & H.M. Smith, 1939

Geophis brachycephalus (Cope, 1871)

Distribution: Add Panama (Panama Oeste), Ray & Ruback (2015); upper elevation of 2230 m, Köhler et al. (2013).

Geophis cancellatus H.M. Smith, 1941

Geophis carinosus L.C. Stuart, 1941

Geophis chalybeus (Wagler, 1830)

Geophis championi Boulenger, 1894

Geophis damiani L.D. Wilson, McCranie & Williams, 1998
Distribution: Add Honduras (Atlantida), and low elevation of 1075 m, Townsend et al. (2010).

Geophis downsi Savage, 1981
Geophis dubius (W.C.H. Peters, 1861)
Geophis duellmani H.M. Smith & Holland, 1969

Geophis dugesii Bocourt, 1883 *in* A.H.A. Duméril, Bocourt & Mocquard, 1870–1909
Distribution: Add Mexico (Nayarit), Luja & Grünwald (2015).

Geophis dunni K.P. Schmidt, 1932
Geophis fulvoguttatus Mertens, 1952
Geophis godmani Boulenger, 1894

Geophis hoffmanni (W.C.H. Peters, 1859)
Distribution: Add Panama (Panama Oeste), Ray & Ruback (2015).

Geophis immaculatus Downs, 1967
Geophis incomptus Duellman, 1959
Geophis isthmicus (Boulenger, 1894)
Geophis juarezi Nieto-Montes de Oca, 2003
Geophis juliai Pérez-Higareda, H.M. Smith & López-Luna, 2001

Geophis laticinctus H.M. Smith & Williams, 1963
Distribution: Add Mexico (Tabasco), Barragán-Vázquez et al. (2017); Canseco-Márquez & Ramírez-Gonzalez (2015) discuss a third specimen from Oaxaca.

Geophis laticollaris H.M. Smith, Lynch & Altig, 1965

Geophis latifrontalis Garman, 1884
Distribution: Add Mexico (Guanajuato), Hernández-Arciga et al. (2013).

Geophis lorancai Canseco-Márquez, Pavón-Vázquez, López-Luna & Nieto-Montes de Oca, 2016. Zookeys 610: 134–139, fig. 2, 3.
Holotype, MZFC 28401, a 321 mm male (M. A. de la Torre-Laranca, 6 April 2008).
Type locality: "Instituto Tecnológico Superior de Zongolica, vicinity of Atlanca, Municipality of Los Reyes, Sierra de Zongolica, Veracruz, Mexico (18°41'48″ N, 97°03'21″W), 1700 m elevation."
Distribution: EC Mexico (WC Veracruz and adjacent Puebla), 1210–1700 m.
Comment: A member of the *G. dubius* group.

Geophis maculiferus Taylor, 1942
Geophis mutitorques (Cope, 1885)
Geophis nasalis (Cope, 1868)
Geophis nephodrymus Townsend & L.D. Wilson, 2006

Geophis nigroalbus Boulenger, 1908
Distribution: Add Colombia (Cauca), and elevation range of 900–1916 m, Vera-Pérez et al. (2018).

Geophis nigrocinctus Duellman, 1959

Geophis occabus Pavón-Vázquez, García-Vázquez, Blancas-Hernández & Nieto-Montes de Oca, 2011

Geophis omiltemanus Günther, 1893 *in* 1885–1902

Distribution: Low elevation of 2300 m, Palacios-Aguilar & Flores-Villela (2018).

Geophis petersii Boulenger, 1894
Distribution: Add Mexico (Guerrero), 1705 m, Carmona-Torres & González-Hernández (2014); Mexico (Jalisco), Muñoz-Nolasco et al. (2015).

Geophis pyburni Campbell & J.B. Murphy, 1977
Geophis rhodogaster (Cope, 1868)
Geophis rostralis (Jan, 1865 *in* Jan & Sordelli, 1860–1866)
Geophis russatus H.M. Smith & Williams, 1966
Geophis ruthveni F. Werner, 1925
Geophis sallaei Boulenger, 1894

Geophis semidoliatus (A.M.C. Duméril, Bibron & Duméril, 1854)
Distribution: Add Mexico (Oaxaca), Vázquez-Vega et al. (2016).

Geophis sieboldi (Jan, 1862)
Distribution: Add Mexico (Colima, Jalisco), 1320–1635 m, Ahumada-Carrillo et al. (2014); Mexico (Guerrero), and upper elevation of 2370 m, Palacios-Aguilar & Flores-Villela (2018).

Geophis talamancae Lips & Savage, 1994
Geophis tarascae Hartweg, 1959
Geophis tectus Savage & Watling, 2008

Geophis turbidus Pavón-Vázquez, Canseco-Márquez & Nieto-Montes de Oca, 2013. Herpetologica 69(3): 361–367, figs. 2, 3.
Holotype, MZFC 27254, a 285 mm female (I.F. Patiño, 24 March 1998).
Type locality: "3.5 km W of Xocoyolo, municipality of Cuetzalan, Puebla, Mexico (19°59.432'N, 97°33.325'W…), 1225 m elevation."
Distribution: Mexico (Sierra Norte of Puebla), 1215–1345 m. Add Hidalgo, upper elevation of 2255 m, Cruz-Elizalde et al. (2015)
Comments: In the *G. dubius* group.

Geophis zeledoni Taylor, 1954
Distribution: Upper elevation of 2230 m, Köhler et al. (2013).

GERARDA Gray, 1849 (Homalopsidae)
Comments: J.C. Murphy & Voris (2014) provide a generic diagnosis.

Gerarda prevostiana (Eydous & Gervais, 1837)
Distribution: Add Sri Lanka (Northern, North Western), Ukuwela et al. (2017); Sri Lanka (Southern), S. Karunarathna et al. (2018). S. Karunarathna et al. (2018) document records for Sri Lanka.
Comments: I. Das et al. (2013) confirm the presence in Sarawak, Malaysia. Ukuwela et al. (2017) note moderate genetic divergence between Sri Lanka and Singapore specimens. Vyas et al. (2013) comment on the distribution at the western edge of range. J.C. Murphy & Voris (2014) provide a diagnosis and photograph.

†*GERINGOPHIS* Holman, 1976 (Booidea: incerta sedis)
†*Geringophis depressus* Holman, 1976
†*Geringophis robustus* Holman & Harrison, 2001
†*Geringophis vetus* Holman, 1982
†*Geringophis yatkolai* Holman, 1977

GERRHOPILUS Fitzinger, 1843 (Gerrhopilidae)

Comments: Pyron & Wallach (2014) provide a revised diagnosis for the genus.

Gerrhopilus addisoni Kraus, 2017a. Zootaxa 4299(1): 80–81, fig. 1.

Holotype: USNM 195953, a 304 mm female (H. Heatwole, 24 March 1969).

Type locality: "Panaete Island, Deboyne Group (10.68° S, 152.35° E; 0–50 m a.s.l.), Milne Bay Province, Papua New Guinea."

Distribution: SE Papua New Guinea (Milne Bay: Panaete I.), 0–50 m. Known only from the holotype.

Gerrhopilus andamanensis (Stoliczka, 1871)
Gerrhopilus ater (Schlegel, 1839 in 1837–1844)
Gerrhopilus beddomii (Boulenger, 1890)
Gerrhopilus bisubocularis (Boettger, 1893)
Gerrhopilus ceylonicus (M.A. Smith, 1943)
Gerrhopilus depressiceps (Sternfeld, 1913)

Distribution: Restricted to NE Papua New Guinea (Madang, Morobe), Kraus (2017a).

Comments: Kraus (2017a) provides a revised diagnosis and detailed description of the holotype.

Gerrhopilus eurydice Kraus, 2017a. Zootaxa 4299(1): 81–84, fig. 1.

Holotype: MCZ 145954, a 319 mm female (F. Parker, 1971).

Type locality: "Trobriand Islands, Milne Bay Province, Papua New Guinea."

Distribution: SE Papua New Guinea (Milne Bay: Trobriand Is.). Known from two specimens.

Gerrhopilus floweri (Boulenger in Flower, 1899)
Gerrhopilus fredparkeri (Wallach in O'Shea, 1996)
Gerrhopilus hades (Kraus, 2005)
Gerrhopilus hedraeus (Savage, 1950)

Distribution: Add Philippines (Cebu I.), Supsup et al. (2016); Philippines (Pacijan), Leviton et al. (2018); Philippines (Tablas I.), Siler et al. (2012).

Gerrhopilus inornatus (Boulenger, 1888)

Distribution: Add Papua New Guinea (Oro), O'Shea et al. (2018b, suppl.).

Gerrhopilus lestes Kraus, 2017a. Zootaxa 4299(1): 89–90, fig. 3.

Holotype: BPBM, a 298 mm specimen (A. Allison, 18 January 1994).

Type locality: "Weitin River Valley, 13 km N, 10.5 km W of river mouth (4.5035° S, 152.9374° E, 240 m a.s.l.), New Ireland, New Ireland Province, Papua New Guinea."

Distribution: Papua New Guinea (New Ireland), 240 m. Known only from the holotype.

Gerrhopilus manilae (Taylor, 1919)

Comments: Hedges et al. (2014) recommend placing *manilae* in *Malayotyphlops*, but Pyron & Wallach (2014) argue that it is morphologically closest to *Gerrhopilus*.

Gerrhopilus mcdowelli (Wallach in O'Shea, 1996)

Gerrhopilus mirus (Jan, 1860 in Jan & Sordelli, 1860–1866)

Distribution: Add Sri Lanka (Sabaragamuwa), Peabotuwage et al. (2012).

Gerrhopilus oligolepis (Wall, 1909)

Distribution: Add India (Sikkim), and low elevation to 550 m, Chettri et al. (2010).

Gerrhopilus persephone Kraus, 2017a. Zootaxa 4299(1): 84–88, figs. 3–5.

Holotype: UMMZ 242536, a 254 mm female (F. Francisco, 10 September 2013).

Type locality: "forest near Normanby Mining Camp, above Awaiara Bay (10.0592° S, 151.0722° E, 620 m a.s.l.), Normanby Island, D'Entrecasteux Archipelago, Milne Bay Province, Papua New Guinea."

Distribution: SE Papua New Guinea (Milne Bay: Normanby I.), 620 m. Known only from the holotype.

Gerrhopilus thurstoni (Boettger, 1890)

Comments: Hedges et al. (2014) recommend placing *thurstoni* in *Gerrhopilus* based on morphological similarity.

Gerrhopilus tindalli (M.A. Smith, 1943)

†*GIGANTOPHIS* Andrews, 1901 (†Madtsoiidae)

†*Gigantophis garstini* Andrews, 1901

Fossil records: Some African material emended to late Eocene (Priabonian) by McCartney & Seiffert (2016), with other retained as Bartonian by Rio & Mannion (2017).

Comments: McCartney & Seiffert (2016) describe additional fossil material. Rio & Mannion (2017) provide a detailed description of the type and ancillary material. They also provide evidence that the Paleocene of Pakistan specimens referred to *G. garstini* by Rage et al. (2014) are not *Gigantophis*.

GLOYDIUS Hoge & Romano-Hoge, 1981 (Viperidae: Crotalinae)

Fossil records: Add late Middle Miocene (Astaracian) of NE Kazakhstan, Ivanov et al. (2019).

Comments: Orlov et al. (2014) discuss distribution and taxonomy of the *G. blomhoffi* complex in eastern Russia. Shi et al. (2016, 2017) provide a DNA sequence-based phylogeny that is structured as a number of stepwise clades. The innermost clade contains *G. changdaoensis* (which they elevate to species) as sister taxon to *G. halys*, *G. intermedius*, *G. rickmersi*, and *G. shedaoensis*. Due to paraphyly within *G. halys* subspecies, they elevate *G. caraganus*, *G. cognatus* and *G. stejnegeri* to species. In particular, *G. intermedius* and *G. shedaoensis* lie within *G. halys*. Conversely, the entire group could be referred to *G. halys*. We recognize the subgroups as proposed by Shi et al. to be species, knowing that the geographic limits and population composition of each is imperfectly known. Shi et al. (2018) follow with an mtDNA-based phylogeny of 18 of the taxa within *Gloydius*.

Gloydius angusticeps Shi, Yang, Huang, Orlov & Li in Shi et al., 2018. Russian J. Herpetol. 25(2): 129–133, figs. 1–4.

Holotype: IVPP OV2634, a 439 mm male (J. Shi, 15 August 2015).

Type locality: "Xiaman Village, Zoige Country, Aba Tibetan and Qiang Autonomous Prefecture, Sichuan Province (33.74°N, 102.50°E, 3569 m)," China.

Distribution: C China (Qinghai, NW Sichuan).

Gloydius blomhoffi (H. Boie, 1826)

Distribution: Add Japan (Kinkasan I.), Mori & Nagata (2016). Orlov et al. (2014) document occurrence on Kunashir I., Kuril Is., Russia. Delete Russia (Sakhalin Island), as not based on specimens, Orlov et al. (2014).

Comments: Orlov et al. (2014) provide a synonymy.

Gloydius brevicaudus (Stejneger, 1907)

Distribution: Qing et al. (2015) record two specimens from China (Nan Ao Island, Guangdong), far from the nearest records, and note that it is popular in the restaurant trade, implying that the records may be based on escaped specimens.

Comments: Orlov et al. (2014) agree that *A. blomhoffii siniticus* Gloyd, 1977, is a synonym of *G. brevicaudus*, citing parapatry with *G. blomhoffi*.

Gloydius caraganus Eichwald, 1831. Zool. Specialis: 170. (*Trigonocephalus caraganus*)

Synonyms: *Ancistrodon halys paramonovi* Nikolsky, 1931.

Neotype: ZISP 2200, a 442 mm male (K. Baer, 1854), designated by Orlov & Barabanov (1999).

Type locality: "ora orientali caspii maris Tjuk-karaganensi" [=Cape Karagan, Mangyshlak Peninsula]; Neotype locality: Mangyshlak Peninsula, eastern edge of the Caspian Sea, Kazakhstan.

Distribution: C Asia. NE shore of the Caspian Sea through Kazakhstan, Uzbekistan, NW Tadzhikistan and Kirgiza, Wagner et al. (2016a).

Comments: Wagner et al. (2016a) note reports of sympatry between *G. caraganus* and *G. halys*, and recommend species status for the former. See under *G. halys*.

Gloydius changdaoensis Li, 1999. Acta Zootaxon. Sinica 4:454–460. (*Gloydius saxatilis changdaoensis*)

Holotype: perhaps in the Natural History Museum of Snake Island, Dalian, China.

Type locality: Daheishan Island, Shandong, China.

Distribution: China (Jiangsu, Shandong).

Comments: Shi et al. (2016, 2017) recognize this species as it is the sister taxon to the entire *G. halys-G. intermedius* complex based on DNA sequence data. We do not have this paper, so for details we have relied on Uetz et al. (2019, The Reptile Database).

Gloydius cognatus Gloyd, 1977. Proc. Biol. Soc. Washington 90(4): 1002–1004. (*Agkistrodon halys cognatus*)

Holotype: USNM 68586, a 375 mm male (F.R. Wulsin, 24–31 August 1923).

Type locality: "Choni (on Tao River), Kansu Province, China."

Distribution: China (Gansu, Nei Mongol, Ningxia, Qinghai, Xinjiang).

Comments: Re-elevated to species based on genetic and morphological characteristics, Shi et al. (2016, 2017), although David & Vogel (2015) consider it to be a synonym of *G. intermedius*.

Gloydius halys (Pallas, 1776)

Synonyms: Remove *Trigonocephalus caraganus* Eichwald, 1831, *Ancistrodon halys paramonovi* Nikolsky, 1931, *Ancistrodon halys stejnegeri* Rendahl, 1933, *Agkistrodon halys cognatus* Gloyd, 1977.

Add ***Gloydius halys ubsunurensis*** Kropachev & Orlov, 2017. Proc. Zool. Inst. St. Petersburg 321(2):133–153, figs. 2, 3, 13.

Holotype: ZMMU 6153.1, a 604 mm female (E.E. Syroechkovsky, July 1959).

Type locality: "Hoolu River, [Tes-Khemsky District], Tuva Republic, Russia."

Remove *Ancistrodon halys stejnegeri* Rendahl, 1933, *Agkistrodon halys cognatus* Gloyd, 1977, according to David & Vogel (2015), Shi et al. (2016, 2017).

Distribution: Add Iran (Alborz, North Khorasan, Qazvin, Semnan, Tehran), Safaei-Mahroo et al. (2015); Iran (Golestan), Khani et al. (2017); China (Heilongjiang, Liaoning), Shi et al. (2016). Remove China (Gansu, Qinghai), Shi et al. (2016, 2017). Remove portions of C Asia assigned to *G. caraganus* (Uzbekistan, Turkmenistan, and all but E Kazakhstan) though exact boundaries are imprecise, Wagner et al. (2016a).

Comments: Simonov & Wink (2012) evaluate intrapopulation genetics in SW Siberia using mtDNA. Wagner et al. (2016a) provide diagnoses for six subspecies that they recognize. Kropachev & Orlov (2017) describe a new subspecies from western Mongolia and SC Russia based on external morphology (scalation and color pattern). Using those characters, they recognize one undescribed plus 7 named subspecies. However, two of the subspecies, *cognatus* and *stejnegeri*, are recognized as species by Shi et al. (2016, 2017). They also recognize *G. h. caraganus* as a species, but based only on one sample in their haplotype network tree. Additional mtDNA data (Shi et al, 2018) have *G. caraganus* as the sister taxon to *G. rickmersi*. Khani et al. (2017) report morphological variation across the Elburz Mountains, N Iran.

Gloydius himalayanus (Günther, 1864)

Distribution: Add India (West Bengal), and low elevation of 1320 m, Chaudhuri et al. (2018); Bhutan (Gasa), Koirala et al. (2016).

Gloydius huangi K. Wang, Ren, Dong, Jiang, Siler & Che in K. Wang et al., 2019b. J. Herpetol. 53(3): 228–233, figs. 2–5.

Holotype: KIZ 027654, a 519 mm female (K. Wang & G. Nima, 10 June 2016).

Type locality: "Jinduo, Chagyab County, Chamdo, Tibet, China (30.2050°N, 97.2869°E,…3,046 m elevation)."

Distribution: China (E Tibet), 3046–3307 m.

Gloydius intermedius (Strauch, 1868)

Gloydius lijianlii F. Jiang & Zhao, 2009

Gloydius liupanensis Liu, Song & Luo, 1989. J. Lanzhou Univ. (Nat. Sci.) 25(1): 114–117. (*Agkistrodon halys liupanensis*)

Holotype: Unlocated according to McDiarmid et al. (1999).

Type locality: translated as "Liupan Mountain, Ningxia Hui Autonomous Region, China; 2,100 meters" by Zhao & Adler (1993).

Distribution: China (Ningxia).

Comments: Considered a valid species by Cai et al. (2015), Shi et al. (2016, 2017), K. Wang et al. (2019).

Gloydius monticola (F. Werner, 1922)

Gloydius qinlingensis Song & Chen, 1985. Animal World 2(2): 99–103. (*Agkistrodon halys qinlingensis*)

Holotype: Unlocated according to McDiarmid et al. (1999).

Type locality: translated as "Huangbaiyuan Xiang (33°45′ N, 107° E), Taibai Co., Shaanxi Prov., China; 1,500 meters," by Zhao & Adler (1993).

Distribution: China (Shaanxi).

Comments: Considered a valid species by Cai et al. (2015), K. Wang et al. (2019).

Gloydius rickmersi Wagner, Tiutenko, Borkin & Simonov *in* Wagner et al., 2016a. Amphibia-Reptilia 37(1): 21–27, figs. 2–4.

Holotype: ZMB 80360, a 479 mm female (locals, 4 August 2013).

Type locality: "Kul-Otek at the Sary-Buka Valley in Kyrgyzstan, about 25 km (by air) NE of Daroot-Korgon (Chong-Alai District) near the town Kyzyl-Eshme in direction to the Shuman-Bol pass and the Kichi-Alai River at an elevation of 3000 m a.s.l."

Distribution: SC Kyrgyzstan, 2798–3200 m.

Gloydius rubromaculatus Shi, Li & Liu in Shi et al., 2017. Amphibia-Reptilia 38(4): 522–528, figs. 1, 2.

Holotype: IOZ 032317, a 554 mm male (J. Shi & X. Chen, 9 July 2016).

Type locality: "mid-upper reaches of the Tongtianhe River, Qumarleb, Qinghai Province," China.

Distribution: China (Qinghai, NW Sichuan, NE Xixang), 3300–4770 m.

Gloydius saxatilis (Emelianov, 1937)

Synonyms: Remove *Gloydius saxatilis changdaoensis* Li, 1999.

Comments: David & Vogel (2015), Shi et al. (2016) consider *G. saxatilis* to be a synonym of *G. intermedius*, but without phylogenetic inference.

Gloydius shedaoensis (Zhao, 1979)

Holotype: Corrected to CIB 012816, Guo et al. (2016a).

Distribution: Add mainland of Lioning, China, opposite Shedao Island, Shi et al. (2016).

Comments: Wang et al. (2015) found low genetic diversity and moderate inbreeding in the Shedao Island population.

Gloydius stejnegeri Rendahl, 1933. Arkiv Zool. 25A(8): 18–21, fig. 4. (*Ancistrodon halys stejnegeri*)

Holotype: NHRM 1923.809.5780 (formerly 2780a), (D. Sjölander, 1923).

Type locality: "China."

Distribution: NE China (Beijing, Shanxi, Shaanxi), Shi et al. (2016, 2017).

Comments: David & Vogel (2015) consider this taxon to be a synonym of *G. brevicaudus*, but Shi et al. (2016, 2017) recognize it based on paraphyly within a DNA sequence-based phylogeny, in which it is related to *G. cognatus* and *G. halys*.

Gloydius strauchi (Bedriaga, 1912)

Distribution: Range restricted by the recognition of some populations as distinct species (i.e., *G. qinlingensis*, *G. liupanensis*). Confirmed for China (W Sichuan).

Gloydius tsushimaensis (Isogawa, Moriya & Mitsui, 1994)

Comments: Orlov et al. (2014) provide a brief taxonomic summary.

Gloydius ussuriensis (Emelianov, 1929)

Type locality: restricted to "Primorskiy kray, Tetjukhe River valley, Vladimiro-Monomakhovo village," Orlov et al. (2014).

Distribution: Add Mongolia (Dornod), Kropachev et al. (2016). Elevation to NSL-1300 m, Orlov et al. (2014).

Comments: Orlov et al. (2014) provide a synonymy and a map of localties.

†*GOINOPHIS* Holman, 1976 (Alethinophidia: incerta sedis)

†*Goinophis minusculus* Holman, 1976

GOMESOPHIS Hoge & Mertens, 1959 (Dipsadidac: Xenodontinae)

Gomesophis brasiliensis (Gomés, 1918)

Distribution: Add Brazil (Santa Catarina), Fortes et al. (2010); elevational range 427–1235 m, R.C. Gonzalez et al. (2014a).

Comments: R.C. Gonzalez et al. (2014a) describe morphological variation and provide a locality map.

GONGYLOSOMA Fitzinger, 1843 (Colubridae: Colubrinae)

Gongylosoma baliodeira (F. Boie, 1827)

Gongylosoma longicaudum (W.C.H. Peters, 1871)

Distribution: Add Thailand (Narathiwat), Pauwels & Grismer (2016).

Gongylosoma mukutense Grismer, I. Das & Leong, 2003

Gongylosoma nicobariense (Stoliczka, 1870)

Gongylosoma scriptum (Theobald, 1868)

Distribution: Add Cambodia (Preah Sihanouk, Pursat), and increase elevation range to 974 m, Neang et al. (2015).

GONIONOTOPHIS Boulenger, 1893 (Lamprophiidae)

Comments: Broadley et al. (2018) use three genes to produce a phylogeny of ten of the 15 species of the African "file snakes," genera *Gonionotophis* and *Mehelya*, which resolves four clades that they recognize as distinct genera: the two recognized genera, and two new genera, *Gracililima* and *Limaformosa*. They provide a revised genus description.

Gonionotophis brussauxi (Mocquard, 1889)

Distribution: Add Congo (Niari), Broadley et al. (2018); Democratic Republic of the Congo (Bas-Congo), Nagy et al. (2013).

Gonionotophis grantii (Günther, 1863)

Distribution: Add Mali (Koulikoro, Sikasso), Trape & Mané (2017); Benin, Trape & Baldé (2014).

Gonionotophis klingi Matschie, 1893

Distribution: Add Liberia (Grand Gedeh), Rödel & Glos (2019).

GONYOSOMA Wagler, 1828 (Colubridae: Colubrinae)

Synonyms: Add *Gonyophis* Boulenger, 1891, *Rhynchophis* Mocquard, 1897, *Rhadinophis* Vogt, 1922.

Comments: X. Chen et al. (2014) combine four species (*Gonyophis margaritatus*, *Rhadinophis frenatus*, *R. prasinus* and *Rhynchophis boulengeri*) with the genus *Gonyosoma* due to paraphyly of the two *Rhadinophis* species.

Gonyosoma boulengeri (Mocquard, 1897)

Distribution: Add Vietnam (Bac Giang), Hecht et al. (2013); Vietnam (Hai Phong: Cat Ba Island), T.Q. Nguyen et al. (2011); Vietnam (Yen Bai), Le et al. (2018).

Comments: Formerly one of three species in *Rhynchophis*.

Gonyosoma frenatum (Gray, 1853)

Distribution: Add Vietnam (Son La), Pham et al. (2014).

Comments: Formerly one of three species in *Rhynchophis*.

Gonyosoma jansenii Bleeker, 1858

Gonyosoma margaritatum W.C.H.Peters, 1871

Comment: Formerly in the monotypic genus *Gonyophis*.

Gonyosoma oxycephalum (Boie, 1827)

Distribution: Add Thailand (Chanthaburi), Chan-ard et al. (2011); Philippines (Batan, Calayan, Camiguin Norte, Leyte, Maringuque, Sabtang Is.), Leviton et al. (2018); Philippines (Sibuyan I.), Emerson & Tan (2013); Philippines (Mindanao), Sanguila et al. (2016).

Gonyosoma prasinum (Blyth, 1854)

Distribution: Add Vietnam (Bac Giang), Hecht et al. (2013); Vietnam (Dien Bien), Dung et al. (2014); Vietnam (Son La), Pham et al. (2014); Vietnam (Yen Bai), Le et al. (2018).

Comments: Formerly one of three species in *Rhynchophis*.

GRACILILIMA Broadley, Tolley, Conradie, Wishart, Trape, Burger, Kusamba, Zassi Boulou & Greenbaum, 2018 African J. Zool. 67(1): 47. (Lamprophiidae)

Type species: *Simocephalus nyassae* (Günther, 1888), by original designation.

Distribution: E Africa.

Comments: See under *Gonionotophis*.

Gracililima nyassae (Günther, 1888)

Distribution: Add Mozambique (Sofala), Broadley et al. (2018).

Comments: Lanza & Broadley (2014) provide a species account for NE Africa.

GRAYIA Günther, 1858 (Grayiidae)

Grayia caesar (Günther, 1863)

Distribution: Add Angola (Cabinda), M.P. Marques et al. (2018).

Grayia ornata (Bocage, 1866)

Distribution: Add Angola (Cabinda), M.P. Marques et al. (2018).

Grayia smythii (Leach *in* Tuckey, 1818)

Distribution: Add Mali (Sikasso), Trape & Mané (2017); Angola (Bengo, Cabinda, Cuanza Norte), M.P. Marques et al. (2018).

Grayia tholloni Mocquard, 1897

Distribution: Add Guinea, Trape & Baldé (2014).

GRYPOTYPHLOPS W.C.H. Peters, 1881 (Typhlopidae: Asiatyphlopinae)

Comments: Hedges et al. (2014), Pyron & Wallach (2014) provide generic diagnoses. Based on internal and external morphology that resemble some species of *Letheobia*, Pyron & Wallach (2014) place *Grypotyphlops* in the Afrotyphlopinae.

Grypotyphlops acutus (A.M.C. Duméril & Bibron, 1844)

Comments: Wallach et al. (2014) list *Onychocephalus unilineatus* A.M.C. Duméril & Bibron, 1844 as a synonym, which is correct, and the name should not have also been listed as a valid species of *Typhlops* in the same book.

GYALOPION Cope, 1860 (Colubridae: Colubrinae)

Gyalopion canum Cope, 1860
Gyalopion quadrangulare (Günther, 1893 *in* 1885–1902)

GYIOPHIS J.C. Murphy & Voris, 2014. Fieldiana Life Earth Sci. (8): 21. (Homalopsidae)

Type species: *Hypsirhina maculosa* Blanford, 1879 by original designation.

Distribution: S Myanmar.

Comments: J.C. Murphy & Voris (2014) provide a diagnosis and photograph. Quah et al. (2017), using mtDNA sequence data, place *Gyiophis salweenensis* as sister taxon to *Myrrophis chinensis*, which is contrary to the supposed placement of *Gyiophis* on the basis of morphological characters.

Gyiophis maculosus (Blandford, 1879)

Comments: J.C. Murphy & Voris (2014) provide a diagnosis and photograph.

Gyiophis salweenensis Quah, L. Grismer, Wood, Thura, Zin, Kyaw, Lwin, M. Grismer & Murdoch, 2017. Zootaxa 4238(4): 574–575, figs. 3, 4.

Holotype: LSUHC 12960, a 416 mm female (M. Thura, T. Zin, E. Quah, L. Grismer, P. Wood, M. Grismer, M. Murdoch and H. Kyaw, 8 October 2016).

Type locality: "close to Sanpel Cave, Mawlamyine, Mon State, Myanmar (N16°22.427, E97°46.388; 44 m in elevation)."

Distribution: Myanmar (Mon). Known only from the holotype.

Gyiophis vorisi (J.C. Murphy 2007)

Comments: J.C. Murphy & Voris (2014) provide a diagnosis and photograph.

†*HAASIOPHIS* Tchernov, Rieppel, Zaher, Polcyn & Jacobs, 2000 (†Simoliophiidae)

Comments: Palci et al. (2013b) reevaluate its anatomy and discuss relationships.

†*Haasiophis terrasanctus* Tchernov, Rieppel, Zaher, Polcyn & Jacobs, 2000

Comments: Palci et al. (2013b) provide an emended diagnosis based on a new interpretation of fossil specimens.

HABROPHALLOS A.Martins, Koch, Pinto, Folly & Passos in A. Martins et al., 2019. J. Zool. Syst. Evol. Res. 57(4): 842. (Leptotyphlopidae: Epictinae)

Type species: *Leptotyphlops collaris* Hoogmoed, 1977 by original designation.

Distribution: NC South America.

Comments: A. Martins et al. (2019) create a new genus for *Epictia collaris* based on anatomical characters and placement in a DNA sequence-based phylogeny as sister taxon to *Siagonodon*.

Habrophallos collaris (Hoogmoed, 1977)

Distribution: Add Brazil (Amapá) and French Guiana (Saint-Laurent du Maroni), Hoogmoed & Lima (2018). Upper elevation of 830 m, Hoogmoed & Lima (2018).

Comments: A. Martins et al. (2019) provide a revised description, color photographs and a locality map.

HAITIOPHIS Hedges & Vidal *in* Hedges et al., 2009. Zootaxa 2067: 17. (Dipsadidae: Xenodontinae)

Type species: *Zamenis anomalus* W.C.H. Peters, 1863 by original designation.

Distribution: Hispaniola.

Comments: See under *Caraiba*.

Haitiophis anomalus (W.C.H. Peters, 1863)

Distribution: Add Dominican Republic (Azua, Independencia, Pedernales, San Jose de Ocoa), Landestoy-T. (2017), who provides a map of known and new localities.

HALDEA Baird & Girard, 1853 (Natricidae)

Comments: *Haldea* is resurrected for *H. striatula* (formerly in *Virginia*) due to paraphyly, McVay & Carstens (2013).

Haldea striatula (Linnaus, 1766)

HAPSIDOPHRYS J.G. Fischer, 1856 (Colubridae: Colubrinae)

Hapsidophrys lineatus J.G. Fischer, 1856

Distribution: Add South Sudan (Central Equatoria), Ullenbruch & Böhme (2017).

Hapsidophrys principis (Boulenger, 1906)

Hapsidophrys smaragdinus (Schlegel, 1837)

Distribution: Add Angola (Bengo), Branch (2018); Angola (Cuanza Norte), M.P. Marques et al. (2018). Low elevation of 560 m, Portik et al. (2016).

†*HEADONOPHIS* Holman, 1993 (Colubroidea: incerta sedis)

†*Headonophis harrisoni* Holman, 1993

HEBIUS Thompson, 1913. Proc. Zool. Soc. London 1913(3): 424. (Natricidae)

Type species: *Tropidonotus vibakari* H. Boie, 1826 by monotypy.

Distribution: Eastern Asia.

Synonyms: *Parahelicops* Bourret, 1934, *Pararhabdophis* Bourret, 1934, *Paranatrix* Mahendra, 1984.

Comments: Guo et al. (2014) present a DNA sequence-based phylogeny that shows *Amphiesma stolatum* to be paraphyletic with all other species of *Amphiesma*, the latter of which they partition into the resurrected genera *Hebius* and *Herpetoreas*. Guo et al. present a generic diagnosis, and state that some species currently in *Hebius* may belong in *Herpetoreas* and/or *Amphiesma*. Kaito & Toda (2016) and Liu et al. (2019) present DNA sequence-based phylogenies using 21 and 18 species, respectively, that have paraphyly for several species, provided identifications of vouchers were correct. They recover a Taiwan-Ryukyu clade of four species, *H. concelarus, H. ishigakiensis, H. pryeri*, and an undescribed species from Taiwan. Kizirian et al. (2018), using DNA sequence data, find that *Parahelicops annamensis* and *Pararhabdophis chapaensis* are nested within *Hebius*, and synonymize *Parahelicops* and *Pararhabdophis* with *Hebius*. Ren et al. (2018) reach the same conclusion for their own study of *Pararhabdophis*. David et al. (2013) reviewed species allied with *H. khasiense*. David et al. (2015b) present a revised diagnosis for *H. annamensis*, after removing *Parahelicops boonsongi* to the genus *Isanophis*.

Hebius andreae (Ziegler & Le, 2006)

Distribution: Add Laos (Khammouane), 537 m, Ziegler et al. (2019a).

Comments: Ziegler et al. (2019a) describe a second specimen, and using combined DNA sequence data, find it to be the sister taxon to other *Hebius* species.

Hebius annamensis (Bourret, 1934)

Distribution: Add Vietnam (Ha Tinh, Quang Binh, Quang Tri, Thua Thien-Hue), and low elevation of 330 m, David et al. (2015b).

Comments: David et al. (2015b) present a revised description.

Hebius arquus (David & Vogel, 2010)

Hebius atemporalis (Bourret, 1934)

Distribution: Upper elevation of 2100 m, Ren et al. (2018).

Hebius beddomii (Günther, 1864)

Hebius bitaeniatus (Wall, 1925)

Distribution: Add Vietnam (Yen Bai), Le et al. (2018).

Hebius boulengeri (Gressitt, 1937)

Distribution: Add Vietnam (Bac Thai, Gia Lai, Lang Son, Quang Binh, Quang Nam), and elevation 80–1450 m, David et al. (2013); Vietnam (Quang Ngai), Nemes et al. (2013); Vietnam (Yen Bai), Le et al. (2018).

Comments: David et al. (2013) provide a revised diagnosis.

Hebius celebicus (W.C.H. Peters & Doria, 1878)

Hebius chapaensis (Bourret, 1934)

Distribution: Add China (Yunnan), Laos (Houaphan, Louangphabang) and Vietnam (Cao Bang, Son La), Ren et al. (2018); Vietnam (Yen Bai), and elevation range of 1134–2046 m, Le et al. (2018).

Comments: Ren et al. (2018) provide a revised description, color photographs and a locality map.

Hebius clerki Wall, 1925. J. Bombay Nat. Hist. Soc. 30(4): 809. (*Natrix clerki*)

Holotype: BMNH 1946.1.13.50, a 565 mm male (Clerk, 1924).

Type locality: "Sinlum Kaba, Kachin Hills," Kachin, Myanmar.

Distribution: Upper Southest Asia. Nepal (?), India (Arunachal Pradesh, Nagaland, Sikkim, West Bengal), Myanmar (Kachin), China (Yunnan).

Comments: David et al. (2015a) recognize *H. clerki* for northern and eastern populations previously referred to *H. parallelus*.

Hebius concelarus (Malnate, 1963)

Hebius craspedogaster (Boulenger, 1899)

Distribution: Add China (Guizhou), David et al. (2013).

Hebius deschauenseei (Taylor, 1934)
Hebius flavifrons (Boulenger, 1887)
Hebius frenata (Dunn, 1923)
Hebius groundwateri (M.A. Smith, 1922)

Hebius inas (Laidlaw, 1902)

Distribution: David et al. (2013) associate Thailand records from north of the Isthmus of Kra with other species.

Comments: David et al. (2013) provide a revised diagnosis.

Hebius ishigakiensis (Malnate & Munsterman, 1960)
Hebius johannis (Boulenger, 1908)

Hebius kerinciense (David & I. Das, 2003)

Distribution: Add Lampung, Sumatra

Comments: David et al. (2013) provide a revised diagnosis. Known from two specimens.

Hebius khasiensis (Boulenger, 1890)

Distribution: Add Myanmar (Kayah) and Thailand (Tak), and elevation 600–1400 meters, David et al. (2013). David et al. (2013) remove records for Cambodia and Vietnam to other species.

Comments: David et al. (2013) provide a revised diagnosis.

Hebius lacrima Purkayastha & David, 2019. Zootaxa 4555(1): 81–83, figs. 1, 2.

Holotype: VR/ERS/ZSI-610, a 487 mm male (villager, 7 August 2010).

Type locality: "Basar (27.980559°N, 94.688496°E), West Siang District, State of Arunachal Pradesh, India, at ca. 600 metres a.s.l."

Distribution: NE India (Arunachal Pradesh), approx. 600 m. Known only from the holotype.

Hebius leucomystax (David, Bain, Truong, Orlov, Vogel, Thanh & Ziegler, 2007)

Distribution: Add Thailand (Roi Et), Pauwels et al. (2015).

Hebius metusia (Inger, Zhao, Shaffer & Wu, 1990)
Hebius miyajimae (Maki, 1931)
Hebius modestus (Günther, 1875)

Distribution: Add Myanmar (Shan), Purkayastha & David (2019). Upper elevation of 1620 m, Ren et al. (2018).

Hebius monticolus (Jerdon, 1854)
Hebius nicobarensis (Sclater, 1891)
Hebius octolineatus (Boulenger, 1904)

Distribution: Elevation range of 1300–2500 m, Ren et al. (2018).

Hebius optatus (Hu & Zhao, 1966)

Holotype: Corrected to CIB 008397, Guo et al. (2012a).

Hebius parallelus (Boulenger, 1890)

Distribution: India (Meghalaya, Nagaland, possibly Sikkim), David et al. (2015a).

Comments: David et al. (2015a) provide a revised diagnosis, and remove *Natrix clerki*, along with northern and eastern populations, from the synonymy and hypodigm.

Hebius pealii (Sclater, 1891)
Hebius petersii (Boulenger, 1893)
Hebius popei (K.P. Schmidt, 1925)
Hebius pryeri (Boulenger, 1887)
Hebius sanguineus (Smedley, 1932)

Hebius sangzhiensis Zhou, Qi, Lu, Lyu & Li in Zhou et al., 2019. Zootaxa 4674(1): 71–76, figs. 3–5.

Holotype: SYNU 08070350, a 562 mm male (Z.-Y. Zhou & S.-Y. Mi, 28 July 2008).

Type locality: "Sangzhi County, Hunan Province, China, ca. 110°07′ E, 29°47′ lN, 1,430 m a.s.l."

Distribution: C China (Hunan) at the type locality, 1420–1430 m.

Hebius sarasinorum (Boulenger, 1896)
Hebius sauteri (Boulenger, 1909)
Hebius taronensis (M.A. Smith, 1940)

Hebius venningi (Wall, 1910)

Distribution: Add India (Mizoram), Lalbiakzuala & Lalremsanga (2019a); Myanmar (Sagaing), Purkayastha & David (2019).

Hebius vibakeri (H. Boie, 1826)
Hebius viperinus (Schenkel, 1901)
Hebius xenura (Wall, 1907)

Hebius yanbianensis Liu, Zhong, Wang, Liu & Guo, 2018. Zootaxa 4483(2): 388–391, figs. 3–5.

Holotype: YBU 15018, a 565 mm male (G. Zhong & P. Wang, March 2015).

Type locality: "Zemulong Town, Yanbian County (N 101°35'04.43", E 27°13'52.00"), South Sichuan Province, China, at an elevation of 1,974 m above sea level."

Distribution: C China (S Sichuan), 1974 m. Known only from the holotype.

†*HECHTOPHIS* Rage, 2001 (Booidea: incerta sedis)

†*Hechtophis austrinus* Rage, 2001

†*HELAGRAS* Cope, 1883 (Booidea: incerta sedis)

†*Helagras orellanensis* Holman, 1983
†*Helagras prisciformis* Cope, 1883

HELICOPS Wagler, 1828 (Dipsadidae: Xenodontinae)
Comments: Moraes-da-Silva et al. (2019) present a DNA sequence-based phylogeny for 11 of the species.

Helicops angulatus (Linnaeus, 1758)

Distribution: Add Peru (Huanuco), Schlüter et al. (2004); Brazil (Mato Grosso do Sul), Ferreira et al. (2017); Brazil (Paraíba), R. França et al. (2012); Brazil (Piauí), Dal Vechio et al. (2013); Colombia (Arauca, Caqueta, Cauca, Guainia, Vichada), Aponte-Gutiérrez et al. (2017); Ecuador (Morona-Santiago), Cisneros-Heredia (2006b).

Helicops apiaka Kawashita-Ribeiro, Ávila & Morais, 2013. Herpetologica 69(1): 83–86, figs. 1–4.

Holotype: UFMT-R 8512, a 551 mm male (C.L. Cavlac, 27 October 2009).

Type locality: "on the left margin (downstream) of the Teles Pires River (9°27'04.3"S; 56°30'51.9"W, datum SAD 69), municipality of Paranáita, state of Mato Grosso, Brazil."

Distribution: Brazil (NC Mato Grosso).

Comment: Perhaps closely related to *H. angulatus*.

Helicops boitata Moraes-da-Silva, Amaro, Nunes, Strüssmann, Junior, Andrade-Jr., Sudré, Recoder, Rodrigues & Curcio, 2019. Zootaxa 4651(3): 450–455, figs. 2–4.

Holotype: UFMT R11940, a 642 mm male (F.F. Curcio, M. Teixeira, R. Recoder & V. Sudré, 1 May 2016).

Type locality: "Transpantaneira Road (16°25'21.18"S, 56°40'12.64"W; 124 m above sea level), municipality of Poconé, Mato Grosso state, Brazil."

Distribution: C Brazil (S Mato Grosso), 124 m. Known only from the type locality.

Helicops carinicaudus (Wied-Neuwied, 1825)

Distribution: Add Brazil (Minas Gerais), Moraes-da-Silva et al. (2019).

Helicops danieli Amaral, 1938

Distribution: Add Colombia (Cesar), Medina-Rangel (2011); Colombia (Magdalena), Rueda-Solano & Castellanos-Barliza (2010).

Helicops gomesi Amaral, 1921

Helicops hagmanni Roux, 1910

Distribution: Add Brazil (Acre), Moraes-da-Silva et al. (2019); Brazil (Marajo I., Para), G.M. Rodrigues et al. (2015).

Helicops infrataeniatus Jan, 1865

Distribution: Add Brazil (Goias), Moraes-da-Silva et al. (2019); Brazil (Paraíba), Guedes et al. (2014, as *H*. aff. *infrataeniatus*); Brazil (Marinheiros I., Rio Grande do Sul), Quintela et al. (2011). Thaler et al. (2018) report a range extension into the Brazilian Cerrado of Mato Grosso do Sul.

Helicops leopardinus (Schlegel, 1837)

Distribution: Add Ecuador (Orellana, Napo, Sucumbíos), Cisneros-Heredia (2006b); Brazil (Alagoas, Maranhao, Minas Gerais, Paraiba, Sergipe), and upper elevation of 750 m, Guedes et al. (2014); Brazil (Amazonas), Moraes-da-Silva et al. (2019); Brazil (Mato Grosso do Sul), Ferreira et al. (2017); Brazil (Rondônia), Bernarde et al. (2012b); Paraguay (Presidente Hayes), Cabral & Weiler (2014).

Helicops modestus Günther, 1861

Distribution: Add Brazil (Mato Grosso do Sul), Ferreira et al. (2017); Brazil (Parana), Dainesi et al. (2019); Brazil (Santa Catarina), Moraes-da-Silva et al. (2019).

Helicops nentur H. Costa, Santana, Leal, Koroiva & Garcia, 2016. Herpetologica 72(2): 158–162, figs. 1–4.

Holotype: UFMG 2486, a 411 mm female (F. Leal, 8 November 2013).

Type locality: "a swamp in Fazenda Papa Capim (18.329398°S, 42.092017°W, 260 m above sea level [a.s.l.]…), São José da Safira, state of Minas Gerais, Brazil."

Distribution: SE Brazil (Minas Gerais), 260–480 m.

Helicops pastazae Shreve, 1934

Distribution: Add Colombia (Boyaca), García-Cobos & Gómez-Sánchez (2019).

Helicops petersi Rossman, 1976

Helicops polylepis Günther, 1861

Distribution: Add Brazil (Mato Grosso do Sul), Ferreira et al. (2017); Brazil (Rondônia), Bernarde et al. (2012b); Peru (Puno), Llanqui et al. (2019).

Helicops scalaris Jan, 1863
Helicops tapajonicus Frota, 2005

Helicops trivittatus (Gray, 1849)

Distribution: Add Brazil (Maranhao), Moraes-da-Silva et al. (2019). Confirmed record for Amapa, Brazil, by D.S. Oliveira et al. (2015).

Helicops yacu Rossman & Dixon, 1975

HELMINTHOPHIS W.C.H. Peters, 1860 (Anomalepididae)

Helminthophis flavoterminatus (W.C.H. Peters, 1857)

Helminthophis frontalis (W.C.H. Peters, 1860)

Distribution: Add Panama (Panama), Ray & Ruback (2015).

Helminthophis praeocularis Amaral, 1924

HELOPHIS Witte & Laurent, 1942 (Natricidae)

Comments: Nagy et al. (2014) thought *Helophis* could be a junior synonym of *Hydraethiops* based on variability of the division or lack of one of the internasal(s).

Helophis schoutedeni (de Witte, 1922)

Comments: Nagy et al. (2014) review the natural history.

HEMACHATUS Fleming, 1822 (Elapidae)

Hemachatus haemachates (Bonnaterre, 1790)

HEMEROPHIS Schätti & Utiger, 2001 (Colubridae: Colubrinae)

Hemerophis socotrae (Günther, 1881)

Comments: Using DNA barcoding, Vasconcelos et al. (2016) find slight intraspecific pairwise distance between specimens (mean 0.30).

HEMIASPIS Fitzinger, 1861 (Elapidae)

Hemiaspis damelii (Günther, 1876)
Hemiaspis signata (Jan, 1859)

HEMIBUNGARUS W.C.H. Peters, 1862 (Elapidae)

Comments: Leviton et al. (2014) and subsequent authors recognize three species within *Hemibungarus*.

Hemibungarus calligaster (Wiegmann *in* Meyen, 1834)

Synonyms: Remove *Hemibungarus gemianulis* W.C.H. Peters, 1872, *Hemibungarus mcclungi* Taylor, 1922.

Distribution: Philippines (Catanduanes, Luzon, Masbate, Mindoro, Samar Is.), 7–908 m.

Hemibungarus gemianulis W.C.H. Peters, 1872. Monatsb. Preuss. Akad. Wiss. Berlin 1872(7): 587.

Holotype: ZMB 7405, a 60 cm specimen (Wallis).

Type locality: "Philippinen."

Distribution: Philippine Islands (Cebu, Guimaras, Masbate, Negros, Panay Is.), 10–1137 m.

Hemibungarus mcclungi Taylor, 1922. Snakes Philippine Islands: 272–273, plates 33, 34.

Holotype: PBS 24, a 190 mm specimen (C. Canonizado, October 1909), destroyed.

Type locality: "Polillo Island," Philippines.

Distribution: Philippines (Catanduanes, Luzon, Polillo Island), 5–41 m.

HEMIRHAGERRHIS Boettger, 1893 (Psammophiidae)

Hemirhagerrhis hildebrandtii (W.C.H. Peters, 1878)
Hemirhagerrhis kelleri Boettger, 1893
Hemirhagerrhis nototaenia (Günther, 1864)
Hemirhagerrhis viperina (Bocage, 1873)

Distribution: Add Angola (Cunene), M.P. Marques et al. (2018).

HEMORRHOIS F. Boie, 1826 (Colubridae: Colubrinae)

Hemorrhois algirus (Jan, 1863)

Distribution: Add Algeria (Tindouf), Donaire et al. (2000); Libya (Al Wahat, Butnan, Jabal al Gharbi, Jura, Marj, Nalut, Tripoli, Zawiyah), Bauer et al. (2017).

Hemorrhois hippocrepis (Linnaeus, 1758)

Distribution: Add Spain (Baleares: Formentera, Ibiza, and Mallorca Is.), Mateo (2015); Spain (Cataluña), Meijide (1985); Spain (Teruel, Zaragoza), Moreno-Rodríguez (1995); Algeria (Chlef, Relizane, Tiaret), Ferrer et al. (2016), Algeria (El Tarf), Rouag & Benyacoub (2006). Increase upper elevation to 2700 m, Martínez-Freiría et al. (2017b).

Comments: Orriols (2014) attempts to identify the geographic source of Balearic specimens. He provides an mtDNA-based phylogeny that identifies 11 haplotypes. One haplotype is unique to the island of Mallorca, and the other Balearic haplotype matches that from specimens around the Strait of Gibralter. Silva-Rocha et al. (2015) use DNA sequence data in another effort to determine the source population for specimens introduced to the Balearic Islands, but were unable to restrict the source beyond the Iberian Peninsula.

Hemorrhois nummifera (A. Reuss, 1834)

Distribution: Add Greece (Chios and Samothrace Islands), Strachinis & Lymberakis (2013); Greece (Samos Island), Speybroek et al. (2014); Turkey (Afyon, Konya), Cihon & Tok (2014). Safaei-Mahroo et al. (2015) list the provinces for Iran.

Hemorrhois ravergieri (Ménétriés, 1832)

Distribution: Add Iraq, Al-Barazengy et al. (2015); Iran (Kerman), Moradi et al. (2013) and Safaei-Mahroo et al. (2015) list the provinces for Iran.

†*HERENSUGEA* Rage, 1996 (Ophidia: incerta sedis)

Comments: Gómez et al. (2019) exclude †*Herensugea* from the Madtsoiidae, and consider it to be in a pre-Serpentes scheme with †*Seismophis*, †*Dinylisia* and †*Najash*.

†*Herensugea caristiorum* Rage, 1996

HERPETOREAS Günther, 1860. Proc. Zool. Soc. London 1860(1): 156. (Natricidae)

Type species: *Herpetoreas sieboldii* Günther, 1860 by monotypy.

Distribution: S Asia.

Comments: Guo et al. (2014) present a DNA sequence-based phylogeny that shows *Amphiesma stolatum* to be paraphyletic with all other species of *Amphiesma*, the latter of which they partition into the resurrected genera *Hebius* and *Herpetoreas*. Guo et al. present a generic diagnosis, and state that some species currently in *Hebius* may belong in *Herpetoreas*.

Herpetoreas burbrinki Guo, Zhu, Liu, Zhang, Li, Huang & Pyron, 2014. Zootaxa 3873(4): 433–434, fig. 3.

Holotype: YBU 071128, a 625 mm male (Sept. 2007).

Type locality: "Zayu County, Xixang A.[utonomous] R.[egion], China, at an elevation of 1889 m above sea level."

Distribution: China (Xixang), 1889 m. Known only from the holotype.

Herpetoreas platyceps (Blyth, 1854)

Distribution: Add Pakistan (Punjab), Rais et al. (2012); India (Himachal Pradesh, Sikkim), David et al. (2015a); China (Yunnan), Ren et al. (2018); low elevation to 500 m, Chettri et al. (2010).

Herpetoreas sieboldii Günther, 1860

Distribution: Add India (Jammu & Kashmir), Manhas et al. (2018b); China (Xixang) and India (West Bengal), David et al. (2015a).

HETERODON Latreille *in* Sonnini & Latreille, 1801 (Dipsadidae: subfamily unnamed)

†*Heterodon brevis* Auffenberg, 1963
Heterodon kennerlyi Kennicott, 1860
Heterodon nasicus Baird & Girard, 1852
Heterodon platirhinos Latreille *in* Sonnini & Latreille, 1801
†*Heterodon plionasicus* J.A. Peters, 1953
Heterodon simus (Linnaeus, 1766)

HETEROLIODON Boettger, 1913 (Pseudoxyrhophiidae)

Heteroliodon fohy Glaw, Vences & Nussbaum, 2005
Heteroliodon lava Nussbaum & Raxworthy, 2000
Heteroliodon occipitalis (Boulenger, 1896)

HEURNIA Jong, 1926 (Homalopsidae)

Comments: J.C. Murphy & Voris (2014) provide a generic diagnosis.

Heurnia ventromaculata Jong, 1926

Comments: J.C. Murphy & Voris (2014) provide a diagnosis and photograph.

HIEROPHIS Fitzinger *in* Bonaparte, 1834 (Colubridae: Colubrinae)

Hierophis andreanus (F. Werner, 1917)

Distribution: Add Iran (Ilam, Khuzestan), Safaei-Mahroo et al. (2015); Iran (Markazi), Sabbaghzadeh & Mashayekhi (2015).

†*Hierophis arcuatus* (H. Meyer, 1845)
†*Hierophis cadurci* (Rage, 1974)

Hierophis carbonarius Bonaparte, 1833 *in* 1832–1841. Icon. Fauna Ital. (5): 21, plate. (*Coluber viridiflavus carbonarius*)

Synonyms: *Coluber sardus* Suckow, 1798, *Coluber xanthurus* Rafinesque-Schmaltz, 1810, *Coluber melanepis* Rafinesque-Schmaltz, 1814, *Coluber pustulatus* Rafinesque-Schmaltz, 1814, *Coluber uccellator* Rafinesque-Schmaltz, 1814, *Zamenis aristotelis* Gistel, 1868, *Zamenis viridiflavus ocellata* Betta, 1874, *Coluber viridiflavus antoniimanujeli* Capolongo, 1984.

Syntypes: ANSP 5427–5431.

Type locality: "Italica"; Restricted to Monti Euganei, Padua, Italy by Mertens & Wermuth (1928).

Distribution: S and E Italy (including Sicily, Favignano, Lipari, Vulcano Islands), Slovenia, NW Croatia. Add Italy (Friuli-Venezia Giulia, Veneto), Rassati (2012).

Comments: Zuffi (2008) describes geographic variation in Italian populations. Using DNA sequence- and color pattern data, Mezzasalma et al. (2015) find that *H. viridiflavus* populations separate into two clades eastward and westward from the middle of Italy, and apply this name to the eastern clade.

†*Hierophis caspioides* (Szyndlar & Schleich, 1993)
†*Hierophis dolnicensis* (Szyndlar, 1987)
†*Hierophis freybergi* (Brunner, 1954)

Hierophis gemonensis (Laurenti, 1768)

Distribution: Add Croatia (Plavnik I.), Tóth et al. (2017); Macedonia, Sterijovski et al. (2014); Greece (Elafonisos I.), Broggi (2016a); Greece (Aegina, Dhokos, Karpathos, Psili, Stavronissos, Trikkeri, Tsougriá Islands, and Crete), Grano et al. (2013); Greece (Diapontia Islets), Stille & Stille (2016).

Comments: Jablonski et al. (2017) describe color and pattern variation. Kyriazi et al. (2012) use mtDNA sequence data to determine that populations on Crete originated from southern continental Greece. See Mezzasalma et al. (2015) under *H. viridiflavus*.

†*Hierophis hungaricus* (Bolkay, 1913)

Distribution: Add late Middle Miocene (Astaracian) of NE Kazakhstan, Ivanov et al. (2019, as *Coluber* cf. *hungaricus*).

†*Hierophis pouchetii* (Rochebrune, 1880)
†*Hierophis robertmertensi* (Mlynarski, 1964)
†*Hierophis suevicus* (Fraas, 1870)

Hierophis viridiflavus (Lacépède, 1789)

Synonyms: *Coluber vulgaris* Bonnaterre, 1790, *Coluber communis* Donndorff, 1798, *Coluber franciae* Suckow, 1798, *Coluber luteostriatus* Gmelin, 1799, *Coluber petolarius* Georgi, 1800, *Coluber atroviriens* G. Shaw, 1802, *Coluber personatus* Daudin, 1803c, *Coluber plutonius* Daudin, 1803c, *Coluber glaucoides* Millet de la Turtandière, 1828, †*Coluber fossilis* Pomel, 1853, †*Coluber gervaisii* Pomel, 1853, *Zamenis atrovirens caudaelineata* F. Müller, 1878, †*Coluber etruriae* Portis, 1890, *Coluber viridiflavus kratzeri* Kramer, 1971.

Distribution: SW Europe. NE Spain, Andorra, France, Lichtenstein, S Switzerland, W Italy (N Calabria, Emilia-Romagna, Liguria, Tuscany; and Palmarola, Ponza, Ventotene Islands); elevation to 2265 m, Arribas (2014); add France (Hauts-de-Seine), Rivallin et al. (2017); Italy (Campania: Procida and Vivara Is.), Nappi et al. (2007) and Cipolla & Nappi (2008); Italy (Piemonte, Valle d'Aosta), Seglie & Sindaco (2013); Italy (Lazio), Piccoli et al. (2017); Italy (Lombardia), Schiavo & Ferri (1996); Italy (Toscana), Verducci & Zuffi (2015).

Fossil records: Add Upper Pleistocene (Tarantian) of Italy, Delfino (2004).

Comments: Zuffi (2008) describes geographic variation in Italian populations. Joger et al. (2010) construct a gene tree that indicates eastern and western lineages that diverge from Italy. Mezzasalma et al. (2015) compare the phylogeography and taxonomy between this species and *H. gemonensis*. They confirm that the two species are distinct, and that *H. viridiflavus* should be divided into an eastern and western species, with the name *H. viridiflavus* representing the western, and *H. carbonarius* the eastern species.

HIMALAYOPHIS Malhotra & Thorpe, 2004 (Vipera: Crotalinae)

Comment: See comment under *Trimeresurus*. Both Alencar et al. (2016) and Zaher et al. (2019) resolve *Himalayophis tibetanus* as the sister taxon to members of *Popeia*. As such, *Himalayophis* and *Popeia* could be considered congeneric. Using mtDNA only, Captain et al. (2019) resolve *Himalayophis* and their new species, *Trimersurus arunachalensis*, as sister taxa, but which are then sister taxa to *Parias*, *Popeia* and *Viridovipera*. Due to these conflicting results we leave *T. arunachalensis* in *Trimeresurus*.

Himalayophis tibetanus (Huang, 1982)

Distribution: Add Nepal (Bagmati), Z. Chen et al. (2019).

†HISPANOPHIS Szyndlar, 1985 (Colubridae: Colubrinae)

†Hispanophis coronelloideus Szyndlar, 1985

†HOFFSTETTERELLA Rage, 1998 (Aniliidae)

†Hoffstetterella brasiliensis Rage, 1998

HOLOGERRHUM Günther, 1858 (Cyclocoridae)

Hologerrhum dermali R.M. Brown, Leviton, Ferner & Sison, 2001

Distribution: Add Philippines (Sibuyan I.), Leviton et al. (2018).

Hologerrhum philippinum Günther, 1858

HOMALOPHIS W.C.H. Peters, 1871. Monatsb. Preuss. Akad. Wiss. Berlin 1871: 577. (Homalopsidae)

Type species: *Homalophis doriae* Peters, 1871 by monotypy.
Distribution: Borneo.
Comments: J.C. Murphy & Voris (2014) revive this genus based on morphology, and provide a generic diagnosis.

Homalophis doriae W.C.H. Peters, 1871

Comments: J.C. Murphy & Voris (2014) provide a diagnosis and photograph.

Homalophis gyii (Murphy, Voris & Auliya, 2005)

Comments: J.C. Murphy & Voris (2014) provide a diagnosis and photograph.

HOMALOPSIS Kuhl & Hasselt, 1822 (Homalopsidae)

Comments: J.C. Murphy & Voris (2014) provide a generic diagnosis.

Homalopsis buccata (Linnaeus, 1758)

Distribution: Add West Malaysia (Kedah), Shahriza et al. (2013); West Malaysia (Penang), Quah et al. (2013).
Comments: J.C. Murphy & Voris (2014) provide a diagnosis and photograph.

Homalopsis hardwickii Gray, 1842

Comments: J.C. Murphy & Voris (2014) provide a diagnosis.

Homalopsis mereljcoxi J.C. Murphy, Voris, Murthy, Traub & Cumberbatch, 2012

Distribution: Add Thailand (Chanthaburi), Chan-ard et al. (2011); Vietnam (Hau Giang, Kien Giang), J.C. Murphy & Voris (2014).
Comments: J.C. Murphy & Voris (2014) provide a diagnosis and photograph of the holotype.

Homalopsis nigroventralis Deuve, 1970

Comments: J.C. Murphy & Voris (2014) provide a diagnosis and photograph.

Homalopsis semizonata Blyth, 1855

Distribution: Add Myanmar (Kayin), J.C. Murphy & Voris (2014); Myanmar (Tanintharyi), Mulcahy et al. (2018); Thailand (Phuket, Ranong), Pauwels & Sumontha (2016).
Comments: J.C. Murphy & Voris (2014) provide a diagnosis and photograph.

HOMOROSELAPS Jan, 1858 (Atractaspididae: Atractaspidinae)

Comments: Using molecular data, Portillo et al. (2019a) confirm the monophyly of *Homoroselaps*.

Homoroselaps dorsalis (A. Smith, 1849 *in* 1838–1849)
Homoroselaps lacteus (Linnaeus, 1758)

HOPLOCEPHALUS Wagler, 1830 (Elapidae)

Hoplocephalus bitorquatus (Jan, 1859)
Hoplocephalus bungaroides (Schlegel, 1837)
Hoplocephalus stephensii Krefft, 1869

†HORDLEOPHIS Holman, 1996 (Booidea: incerta sedis)
†Hordleophis balconae Holman, 1996

HORMONOTUS Hallowell, 1857 (Lamprophiidae)

Hormonotus modestus (A.M.C. Duméril, Bibron & Duméril, 1854)

Distribution: Add Liberia, Trape & Baldé (2014); Gabon (Ogooue-Ivindo), Pauwels et al. (2016b); Gabon (Ogooué-Lolo), Carlino & Pauwels (2014).

†HUBEROPHIS Holman, 1977 (Booidea: incerta sedis)
†Huberophis georgiensis Holman, 1977

HYDRABLABES Boulenger, 1891 (Natricidae)

Hydrablabes periops (Günther, 1872)
Hydrablabes praefrontalis (Mocquard, 1890)

HYDRAETHIOPS Günther, 1872 (Natricidae)

Comments: See comment under *Helophis*.

Hydraethiops laevis Boulenger, 1904
Hydraethiops melanogaster Günther, 1872

HYDRELAPS Boulenger, 1896 (Elapidae)

Hydrelaps darwiniensis Boulenger, 1896

HYDRODYNASTES Fitzinger, 1843 (Dipsadidae: Xenodontinae)

Hydrodynastes bicinctus (Hermann, 1804)

Neotype: MPEG 24628, a 1635 mm male (M.S. Hoogmoed, M.A. Ribeiro-Jr., & C. Oliveira-Araújo, 24 Nov. 2005), designated by Murta-Fonseca et al. (2015).

Neotype locality: "the municipality of Novo Progresso (07°02′25″S, 55°24′55″W, about 240 m above sea level), state of Pará, Brazil."

Distribution: Add Brazil (Amapa, Maranhao, Mato Grosso, Tocantins), Murta-Fonseca et al. (2015); Brazil (Mato Grosso do Sul), Ferreira et al. (2017).

Comments: Murta-Fonseca et al. (2015) review geographic variation and conclude that two color morphs occupy the Cerrado and Amazon rainforest, but broad intergradation does not warrant recognition of subspecies. They provide a species account including description, locality map, and color photographs.

Hydrodynastes gigas (A.M.C. Duméril, Bibron & Duméril, 1854)

Hydrodynastes melanogigas Franco, Fernandes & Bentim, 2007

Distribution: Add Brazil (Maranhão) and low elevation of 175 m, Jorge da Silva et al. (2012); Brazil (Mato Grosso), Santos-Jr et al. (2017).

HYDROMORPHUS W.C.H. Peters, 1859 (Dipsadidae: Dipsadinae)

Hydromorphus concolor W.C.H. Peters, 1859

Distribution: Add Panama (Panama Oeste), Ray & Ruback (2015).

Hydromorphus dunni Slevin, 1942

HYDROPHIS Latreille *in* Sonnini & Latreille, 1801 (Elapidae)

Synonyms: Add *Pelamis* Daudin, 1803, *Disteira* Lacépède, 1804, *Leioselasma* Lacépède, 1804, *Ophinectes* Rafinesque-Schmaltz, 1817, *Nauticophis* Lesson, 1832 *in* Bélanger, 1831–1834, *Polyodontes* Lesson, 1832 *in* Bélanger, 1831–1834, *Lapemis* Gray, 1835 in Gray & Hardwicke, 1830–1835, *Aturia* Gray, 1842, *Elaphrodytes* Gistel, 1848, *Noterophis* Gistel, 1848, *Chitulia* Gray, 1849, *Enhydrina* Gray, 1849, *Kerilia* Gray, 1849, *Acalyptus* A.M.C. Duméril, 1853, *Colubrinus* A.M.C. Duméril, Bibron & Duméril, 1854, *Astrotia* Fischer, 1855, *Pelamydoidis* Fitzinger, 1861, *Acalyptophis* Boulenger, 1896, *Pelamydrus* Stejneger, 1910, *Dolichodira* Wall, 1921, *Melanomystax* Wall, 1921, *Micromastophis* Wall, 1921, *Polyodontognathus* Wall, 1921, *Polypholophis* Wall, 1921, *Porrecticollis* Wall, 1921, *Praescutata*, Wall, 1921, *Pseudodistira* Kinghorn, 1926, *Thalassophina* M.A. Smith, 1926, *Mediohydrophis* Kharin, 2004.

Distribution: Tropical and subtropical zones and shorelines of the Pacific and Indian Oceans.

Comments: The DNA-sequence based phylogeny produced by Sanders et al. (2013) renders most classically recognized

sea snake genera as polyphyletic with *Hydrophis*. They recommend combining those genera with *Hydrophis*, s.l., which involves twelve of the genera recognized by Wallach et al. (2014).

Hydrophis atriceps Günther, 1864

Distribution: Add Philippines (Mindanao, Samar, Sulu Archip.), Leviton et al. (2018).

Hydrophis belcheri (Gray, 1849)

Distribution: Add Australia (Arafura Sea), A.R. Rasmussen et al. (2014).

Comments: A.R. Rasmussen et al. (2011a) describe specimens from Vietnam.

Hydrophis bituberculatus W.C.H. Peters, 1873

Hydrophis brookii Günther, 1872

Distribution: Add West Malaysia (Selangor), Voris (2017).

Hydrophis caerulescens (Shaw, 1802)

Distribution: Add India (Gujarat), Patel & Vyas (2019).

Hydrophis coggeri (Kharin, 1984)

Comments: see Distribution under *H. melanocephalus.*

Hydrophis curtus (Shaw, 1802)

Distribution: Add Iran (Bushehr), Rezaie-Atagholipour et al. (2016); India (Gujarat), Patel & Vyas (2019); Philippines (Mindanao), Leviton et al. (2018).

Hydrophis cyanocinctus Daudin, 1803

Distribution: Add Iran (Sistan and Baluchestan), Rezaie-Atagholipour et al. (2016); India (Gujarat), Parmar (2019); Philippines (Mindanao), Leviton et al. (2018).

Hydrophis czeblukovi (Kharin, 1984)

Hydrophis donaldi Ukuwela, Sanders & Fry, 2012

Hydrophis elegans (Gray, 1842)

Hydrophis fasciatus (Schneider, 1799)

Distribution: Add West Malaysia (Johor), Voris (2015).

Hydrophis hendersoni Boulenger, 1903. J. Bombay Nat. Hist. Soc. 14(4): 719, plate. (*Distira hendersoni*)

Holotype: BMNH 1946.1.10.9, a 940 mm female (T. Beath Henderson).

Type locality: "Coast of Rangoon," Myanmar.

Distribution: Myanmar (Yangon).

Comments: A.R. Rasmussen et al. (2011b) revive this taxon due to discrete differences between it and *H. nigrocinctus*, of which it was formerly considered a synonym.

Hydrophis inornatus (Gray, 1849)

Distribution: Australian record is in error, likely based on *H. major*, A.R. Rasmussen et al. (2014).

Hydrophis jerdonii (Gray, 1849)

Hydrophis kingii Boulenger, 1896

Hydrophis klossi Boulenger, 1912

Hydrophis laboutei A.R. Rasmussen & Ineich, 2000

Hydrophis lamberti M.A. Smith, 1917

Distribution: Add Philippines (Luzon), Leviton et al. (2018).

Hydrophis lapemoides Gray, 1849

Hydrophis macdowelli Kharin, 1983

Hydrophis major (Shaw, 1802)

Hydrophis mamillaris (Daudin, 1803)

Distribution: Add India (Gujarat), Patel & Vyas (2019).

Hydrophis melanocephalus Gray, 1849

Distribution: A.R. Rasmussen et al. (2014) refer all Australian records of this species to *H. coggeri*.

Hydrophis melanosoma Günther, 1864

Hydrophis nigrocinctus Daudin, 1803

Synonyms: Add *Disteira walli* Kharin, 1989. Remove *Distira hendersoni* Boulenger, 1903.

Distribution: India (Bengal), West Malaysia, and possibly Bangladesh.

Comments: A.R. Rasmussen et al. (2011b) synonymize *D. walli* due to overlap in morphological characters.

Hydrophis obscurus Daudin, 1803

Hydrophis ocellatus Gray, 1849. Cat. Spec. Snakes Coll. Brit. Mus.: 53–54. (*Hydrophis ocellata*)

Synonyms: *Hydrophis godeffroyi* W.C.H. Peters, 1873, *Distira mjoebergi* Lönnberg & Andersson, 1913.

Holotype: BMNH 1946.1.3.91, a halfgrown female (Macgillivray).

Type locality: "Australian seas?"

Distribution: N Australia, Aru Islands, New Guinea, Gilbert Islands.

Comments: Sanders et al. (2013) produce a DNA sequence-based phylogeny in which *H. ornatus* and *H. ocellatus* are paraphyletic with respect to *H. lamberti*, and all three are considered separate species.

Hydrophis ornatus (Gray, 1842)

Synonyms: Remove *Hydrophis ocellata* Gray, 1849, *Hydrophis godeffroyi* W.C.H. Peters, 1873, *Distira mjoebergi* Lönnberg & Andersson, 1913.

Distribution: Excludes Australia, A.R. Rasmussen et al. (2014), and New Guinea, Aru and Gilbert Islands. Add Iran (Bushehr, Sistan and Baluchestan), Rezaie-Atagholipour et al. (2016).

Comments: See under *H. ocellatus*.

Hydrophis pachycercos J.G. Fischer, 1855

Hydrophis pacificus Boulenger, 1896

Distribution: Add northern Australia, A.R. Rasmussen et al. (2014).

Comments: Rasmussen et al. (2014) note that morphological and molecular data of Australian specimens are conspecific with populations of *H. cyanocinctus* from eastern Asia.

Hydrophis parviceps M.A. Smith, 1935

Comments: A.R. Rasmussen et al. (2011a) describe additional specimens from Vietnam.

Hydrophis peronii (A.M.C. Duméril, 1853)

Hydrophis platura (Linnaeus, 1766)

Synonyms: ***Hydrophis platurus xanthos*** Bessesen & Galbreath, 2017. ZooKeys 686: 115–117, fig. 2.

Holotype: MZUCR 20614, a 52 cm female (A. Solórzano, 13 Feb. 2009).

Type locality: "Costa Rica: Golfo Dulce: inner basin, 08°35.76′N; 083°13.25′W."

Distribution: Add Iran (Sistan and Baluchestan), Rezaie-Atagholipour et al. (2016); India (Goa), Palot and Radhakrishnan (2010); India (Gujarat), Parmar (2018); West Malaysia (Johore), Voris (2015); Philippines (Mindanao), Sanguila et al. (2016); Philippines (Jolo, Sibutu, Surigao), Leviton et al. (2018); Panama (Panama Oeste), Ray & Ruback (2015). Palot & Radhakrishnan (2010) confirm the presence in Kerala, India. Quiñones et al. (2014) report a second specimen from Peru.

Comments: Solórzano (2011) describes the color pattern variation in the Golfo Dulce, Costa Rica, which was later used in describing *H. p. xanthos*.

Hydrophis schistosus Daudin, 1803

Distribution: Exclude Australia, A.R. Rasmussen et al. (2014).

Comments: See under *H. zweifeli*.

Hydrophis semperi Garman, 1881

Hydrophis sibauensis A.R. Rasmussen, Auliya & Böhme, 2001

Hydrophis spiralis (Shaw, 1802)

Distribution: Add Iran (Hormozgan), Rezaie-Atagholipour et al. (2016).

Hydrophis stokesii (Gray *in* Stokes, 1846)

Hydrophis stricticollis Günther, 1864

Hydrophis torquatus Günther, 1864

Hydrophis viperinus (P. Schmidt, 1852)

Hydrophis vorisi Kharin, 1984

Hydrophis zweifeli (Kharin, 1985)

Distribution: Add Australia (Northern Territory, Queensland).

Comments: A.R. Rasmussen et al. (2014) find that all examined Australian specimens of *H. schistosus* are referable to *H. zweifeli* based on morphology.

HYDROPS Wagler, 1830 (Dipsadidae: Xenodontinae)

Hydrops caesurus Scrocchi, Ferreira, Giraudo, Avila & Motte, 2005

Distribution: Add Paraguay (San Pedro), Buongermini & Cacciali (2017).

Hydrops martii (Wagler *in* Spix, 1824)

Distribution: Add Colombia (Guainia), Entiauspe-Neto (2017b); Brazil (Amapá), Entiauspe-Neto (2017b); Brazil (Marajo I., Para), G.M. Rodrigues et al. (2015); Peru (Ucayali), Entiauspe-Neto (2017b).

Hydrops triangularis (Wagler *in* Spix, 1824)

Distribution: Add Brazil (Acre), D.P.F. França et al. (2017); Brazil (Mato Grosso), Albuquerque & Lema, 2008.

HYPNALE Fitzinger, 1843 (Viperidae: Crotalinae)

Hypnale hypnale (Merrem, 1820)

Distribution: Add India (Goa), Sawant et al. (2010); India (Maharashtra), Sayyed (2016).

Hypnale nepa (Laurenti, 1768)
Hypnale zara (Gray, 1849)

HYPOPTOPHIS Boulenger, 1908 (Atractaspididae: Aparallactinae)
Hypoptophis wilsonii Boulenger, 1908
Distribution: Add Angola (Lunda Norte), Branch (2018).

HYPSIGLENA Cope, 1860 (Dipsadidae: Dipsadinae)
Comment: Mulcahy et al. (2014) present a revised DNA sequence-based phylogeny, from which they recognize nine named and two undescribed species. J.L. Lee et al. (2016) discuss the utility of some morphological data in discerning species. E.A. Myers et al. (2017a) provide a DNA sequence-based phylogeny of USA populations, with three primary clades: partially overlapping eastern and western clades passing through central Arizona, USA, and a third clade in SE Arizona.

Hypsiglena affinis Boulenger, 1894

Hypsiglena catalinae Tanner, 1966. Trans. San Diego Soc. Nat. Hist. 14(15): 192–193, fig. 1. (*Hypsiglena torquata catalinae*)
Holotype: SDSNH 44680, a 371 mm male (G.E. Lindsay, 25 June 1964).
Type locality: "Santa Catalina Island, approximately 25° 38′ N, 110° 47′ W, Gulf of California, Baja California, Mexico."
Distribution: NW Mexico (Santa Catalina I., Baja California Sur).
Comments: Mulcahy et al. (2014) elevate *catalinae* to species based on DNA-based phylogenetic placement in a clade containing *H. unaocularis* and an undescribed species.

Hypsiglena chlorophaea Cope, 1860
Synonyms: Remove *Hypsiglena torquata catalinae* Tanner, 1966.
Comment: E.A. Myers et al. (2013a) present a phylogeographic analysis between populations of the subspecies *H. c. deserticola*. Mulcahy et al. (2014) recover an undescribed species from SE Arizona, USA populations based on DNA sequence data, and also elevate *catalinae* to species.

Hypsiglena jani (Dugès, 1865)

Hypsiglena ochrorhyncha Cope, 1860
Synonyms: Remove *Hypsiglena ochrorhynchus unaocularus* Tanner, 1946.
Distribution: Remove Mexico (Baja California Sur: Santa Catalina Island), see *H. catalinae*.

Hypsiglena slevini Tanner, 1943
Distribution: Northern-most record in Baja California, Mexico, S. Murray et al. (2015).

Hypsiglena tanzeri Dixon & Lieb, 1972
Duistribution: Add Mexico (Hidalgo), and increase elevation range to 1600 m, Morales-Capellán et al. (2016); Mexico (Tamaulipas), Rautsaw et al. (2018). Flores-Hernández et al. (2017) provide an additional record.

Hypsiglena torquata (Günther, 1860)
Disribution: Upper elevation of 1500 m, Palacios-Aguilar & Flores-Villela (2018).
Comments: Mulcahy et al. (2014) recover an undescribed species from S Sonora, N Sinaloa, Mexico populations, based on DNA sequence data.

Hypsiglena unaocularis Tanner, 1946. Great Basin Nat. 5(3–4): 74–76, pl. 3. (*Hysiglena ochrorhynchus unaocularus*)
Holotype: AMNH 62756, a 480 mm male.
Type locality: "Clarion Island, the most southwestern of the Revilla Gigedo Island Group, Mexico."
Distribution: Mexico (Clarion I., Colima).
Comments: Mulcahy et al. elevate *unaocularis* to species based on DNA-based phylogenetic placement in a clade containing *H. catalinae* and an undescribed species.

HYPSIRHYNCHUS Günther, 1858 (Dipsadidae: Xenodontinae)
Comments: See under *Schwartzophis*.
Hypsirhynchus ferox Günther, 1858
Hypsirhynchus parvifrons (Cope, 1862)
Hypsirhynchus scalaris Cope, 1862

HYPSISCOPUS Fitzinger, 1843. Syst. Rept.: 25. (Homalopsidae)
Type species: *Homalopsis plumbea* F. Boie *in* H. Boie, 1827, by monotypy.
Distribution: SE Asia and Indonesia.
Comments: J.C. Murphy & Voris (2014) revive this genus based on morphology, and provide a generic diagnosis.
Hypsiscopus matannensis (Boulenger, 1897)
Comments: J.C. Murphy & Voris (2014) provide a diagnosis and photograph of the holotype.
Hypsiscopus plumbeus (H. Boie *in* F. Boie, 1827)
Distribution: Add China (Nan Ao Island), Qing et al. (2015).
Comments: J.C. Murphy & Voris (2014) provide a diagnosis and photograph.

IALTRIS Cope, 1862 (Dipsadidae: Xenodontinae)
Synonyms: Add *Darlingtonia* Cochran, 1935.
Comments: Using DNA sequence-data, Krysko et al. (2015) and Zaher et al. (2019) subsume *Darlingtonia* within *Ialtris*, despite the comparison with only *I. dorsalis*.
Ialtris agyrtes Schwartz & Rossman, 1976
Ialtris dorsalis (Günther, 1858)
Distribution: Add Dominican Republic (Peravia), Disla et al. (2019).
Ialtris haetiana (Cochran, 1935)
Ialtris parishi Cochran, 1932

IGUANOGNATHUS Boulenger, 1898 (? Natricidae)
Iguanognathus werneri Boulenger, 1898

IMANTODES A.M.C. Duméril, 1853 (Dipsadidae: Dipsadinae)

Imantodes cenchoa (Linnaeus, 1758)

Distribution: Add Mexico (Guerrero), Palacios-Aguilar & Flores-Villela (2018); Honduras (Choluteca, Santa Bárbara), Espinal et al. (2014a), McCranie et al. (2013b); Honduras (Francisco Morazán), Solís et al. (2015); Panama (Panama Oeste), Ray & Ruback (2015); Colombia (Cesar), Medina-Rangel (2011); Colombia (Gorgona I.), Urbina-Cardona et al. (2008); Colombia (Valle del Cauca: Isla Palma), Giraldo et al. (2014); Ecuador (Manabi), Almendáriz et al. (2012); Ecuador (Zamora-Chinchipe), Almendáriz et al. (2014); Peru (Huanuco), Schlüter et al. (2004); Peru (Puno), Llanqui et al. (2019); Brazil (Maranhao, Roraima, Tocantins), Missassi & Prudente (2015); Brazil (Mato Grosso do Sul), Missassi & Prudente (2015) and Ferreira et al. (2017); Brazil (Paraiba), R. França et al. (2012); Brazil (Alagoas), R. Marques et al. (2013); Brazil (Pernambuco), Guedes et al. (2014).

Imantodes chocoensis Torres-Carvajal, Yánez-Muñoz, Quirola, E.N. Smith & Almendáriz, 2012

Distribution: Add Colombia (Antioquia, Choco), Missassi et al. (2015); Colombia (Valle del Cauca), and low elevation of 25 m, Jaramillo-Martínez et al. (2013).

Imantodes gemmistratus Cope, 1861

Distribution: Add Mexico (Tabasco), Charruau et al. (2015), Mexico (Zacatecas), Bañuelos-Alamillo et al. (2017); Nicaragua (Masaya), Sunyer et al. (2014); Panama (Panama Oeste), Ray & Ruback (2015); Colombia (Boyaca, Cesar, Guajira), Missassi & Prudente (2015).

Imantodes guane Missassi & Prudente, 2015. Zootaxa 3980(4): 563–570, figs. 2–6.

Holotype, ICN 5730, a 1096 mm male (R. Hernández, 1981).

Type locality: "headwater of the Luisito River, Virolín (6°18′18″N, 73°10′25″W; 1750 m asl), municipality of Charalá, Department of Santander, Colombia."

Distribution: Eastern Cordillera of Colombia (Santander), 1750–2324 m.

Imantodes inornatus Boulenger, 1896

Distribution: Add Nicaragua (Atlantico Norte, Atlantico Sur, Rio San Juan), Sunyer et al. (2014); Panama (Panama), Ray & Ruback (2015); Colombia (Antioquia), Missassi & Prudente (2015).

Imantodes lentiferus Cope, 1894

Distribution: Add Colombia (Caqueta, Meta), Missassi & Prudente (2015); SE-most record, Brazil (Pará), and low elevation of 85 m, Ascenso & Missassi (2015), Frota et al. (2015) map known localities and describe specimens from distribution gaps in Brazil.

Imantodes phantasma C.W. Myers, 1982

Distribution: Add Colombia (Choco), and upper elevation of 1175 m, Medina-Rangel et al. (2018).

Imantodes tenuissimus Cope, 1867

†*INCONGRUELAPS* Scanlon, Lee & Archer, 2003 (Elapidae)

†*Incongruelaps iteratus* Scanlon, Lee & Archer, 2003

†*INDOPHIS* Rage & Prasad, 1992 (†Nigerophiidae)

Comments: Pritchard et al. (2014) provide a revised generic diagnosis based on new material.

†*Indophis fanambinana* Pritchard, McCartney, Krause & Kley, 2014. J. Vert. Paleontol. 34(5): 1084–1087, fig. 2.

Holotype: UA 9942, a posterior trunk vertebra.

Type locality: "Locality MAD10-24, Lac Kinkony Study Area, Mahajanga Basin, northwestern Madagasgar."

Distribution: Upper Cretaceous (Maastrichtian) of NW Madagascar (Mahajanga).

†*Indophis sahnii* Rage & Prasad, 1992

INDOTYPHLOPS Hedges, Marion, Lipp, Marin & Vidal, 2014. Caribbean Herpetol. 49:37. (Typhlopidae: Asiatyphlopinae)

Type species: *Typhlops pammeces* Günther, 1864 by original designation.

Distribution: S Asia.

Comments: Hedges et al. (2014), Pyron & Wallach (2014) provide generic diagnoses.

Indotyphlops albiceps (Boulenger, 1898)

Distribution: Add Cambodia (Kampot), Neang et al. (2017).

Indotyphlops braminus (Daudin, 1803)

Distribution: Add Cuba, Díaz & Cádiz (2014); Montserrat, Snyder et al. (2019); French West Indies (La Désirade I.), Lorvelec et al. (2016); Mexico (Baja California), Valdez-Villavicencio et al. (2016); Mexico (Chihuahua), Carbajal-Márquez et al. (2015a); Mexico (Tamaulipas), Farr et al. (2013); Mexico (Yucatan), Paradiz-Dominguez (2016); Mexico (Zacatecas), Bañuelos-Alamillo & Carbajal-Márquez (2016); Mexico (Cozumel I., Quintana Roo), Pavón-Vázquez et al. (2016a); Honduras (Atlantida), McCranie & Valdés-Orellana (2015); Honduras (Cortés), McCranie et al. (2014a); Honduras (Ocotepeque), McCranie (2014b); Iran (Hormozgan), Afroosheh et al., (2010); Myanmar (Tanintharyi), Mulcahy et al. (2018); Thailand (Chanthaburi), Chan-ard et al. (2011); Thailand (Rayong: Koh Man Ni Island), Chan-ard and Makchai (2011); Indonesia (Yamdena Island), Lang (2013); Philippines (Camiguin Norte, Ivojos Is.), Leviton et al. (2018); Philippines (Tablas I.), Siler et al. (2012); Papua New Guinea (Woodlark Island), Kraus (2013); Papua New Guinea (Manam Island), Clegg & Jocque (2016); Papua New Guinea (Oro), O'Shea et al. (2018b, suppl.); Portugal (Madeira Island), Jesus et al. (2013); Spain (Baleares: Mallorca), Mateo (2015); Vanuatu (Toga Island), Ineich (2009); Canaries (Fuerteventura, La Gomera, Lanzarote and Tenerife Islands), Urioste & Mateo (2011); Libya (Gharyan), Joger et al. (2008).

Comments: Rato et al. (2015) were unable to determine the source of introduced European populations due to identical rRNA fragments with other samples.

Indotyphlops exiguus (Jan, 1864 *in* Jan & Sordelli, 1860–1866)

Indotyphlops filiformis (A.M.C. Duméril & Bibron, 1844)

Indotyphlops jerdoni (Boulenger, 1890)

Indotyphlops lankaensis (Taylor, 1947)

Indotyphlops lazelli (Wallach & Pauwels, 2002)

Indotyphlops leucomelas (Boulenger, 1890)

Comments: Hedges et al. (2014) assign *leucomelas* and *tenuicollis* to *Argyrophis*, but Pyron & Wallach (2014) note that they are morphologically closer to species of *Indotyphlops*.

Indotyphlops loveridgei (Constable, 1949)

Indotyphlops madgemintonae (Khan, 1999)

Indotyphlops malcolmi (Taylor, 1947)

Indotyphlops meszoelyi (Wallach, 1999)

Indotyphlops ozakiae (Wallach *in* Niyomwan, 1999)

Indotyphlops pammeces (Günther, 1864)

Indotyphlops porrectus (Stoliczka, 1871)

Indotyphlops schmutzi (Auffenberg, 1980)

Indotyphlops tenebrarum (Taylor, 1947)

Indotyphlops tenuicollis (W.C.H. Peters, 1864)

Comments: See under *I. leucomelas*.

Indotyphlops veddae (Taylor, 1947)

Indotyphlops violaceus (Taylor, 1947)

INYOKA Branch & Kelly *in* Kelly et al., 2011 (Lamprophiidae)

Inyoka swazicus (Schaefer, 1970)

Holotype: Updated to PEM R13502 by Conradie et al. (2019).

ISANOPHIS David, Pauwels, Nguyen & Vogel, 2015b. Zootaxa 3948(2): 205–206. (Natricidae)

Type species: *Parahelicops boonsongi* Taylor & Elbel, 1958, by original designation.

Distribution: Northeastern Thailand.

Comments: David et al. (2015b) present a diagnosis and revised description for the genus and only species.

Isanophis boonsongi (Taylor & Elbel, 1958)

Comments: Known from three specimens according to David et al. (2015b).

†*ITABORAIOPHIS* Rage, 2011 (Booidea: incerta sedis)

†*Itaboraiophis depressus* Rage, 2011

ITHYCYPHUS Günther, 1873 (Pseudoxyrhophiidae)

Ithycyphus blanci Domergue, 1988

Ithycyphus goudotii (Schlegel, 1837)

Ithycyphus miniatus (Schlegel, 1837)

Distribution: Add Madagascar (Nosy Komba), Roberts & Daly (2014).

Ithycyphus oursi Domergue, 1986

Ithycyphus perineti Domergue, 1986

KARNSOPHIS J.C. Murphy & Voris, 2013 (Homalopsidae)

Type species: *Karnsophis siantaris* J.C. Murphy & Voris, 2013.

Distribution: W Indonesia.

Comments: Related to *Brachyorrhos* and *Calamophis* according to J.C. Murphy & Voris (2013). J.C. Murphy & Voris (2014) provide a diagnosis.

Karnsophis siantaris J.C. Murphy & Voris, 2013. Asian Herpetol. Res. 4(2): 142–144, fig. 2.

Holotype: USNM 103578, a 465 mm male (George Vanderbilt Sumatran Expedition, 1937).

Type locality: "Siantar, Sumatera Utara, Sumatra, Indonesia (~ 1°58′ N, 99°47′ E)."

Distribution: Indonesia (Sumatra). Known only from type locality.

Comments: J.C. Murphy & Voris (2014) provide a description and illustrations of the holotype.

†*KATARIA* Scanferla, Zaher, Novas, Muizon & Céspedes, 2013. PLoS One 8(e57583): 2. (Booidea: incerta sedis)

Type species: †*Kataria anisodonta* Scanferla, Zaher, Novas, Muizon & Céspedes, 2013 by monotypy.

Comments: A phylogenetic analysis provided by Scanferia et al. (2013) places *Kataria* as sister taxon to the Caenophidia.

†*Kataria anisodonta* Scanferla, Zaher, Novas, Muizon & Céspedes, 2013. PLoS One 8(e57583): 2–5, figs. 1–3.

Holotype: MHNC 13323, an incomplete skull.

Type locality: "Tiupampa locality, Mizque province of the department of Cochabamba, Bolivia."

Distribution: Early Paleocene (Danian) of Cochabamba, Bolivia.

†*KELYOPHIS* LaDuke, Krause, Scanlon & Kley, 2010 (†Nigerophiidae)

†*Kelyophis hechti* LaDuke, Krause, Scanlon & Kley, 2010

Comments: Pritchard et al. (2014) describe additional vertebral material from near the type locality.

KLADIROSTRATUS Conradie, Keates & Edwards *in* Keates et al., 2019 (Psammophiidae)

Type species: *Psammophis acutus* Günther, 1888 by original designation.

Distribution: S Africa.

Comments: Keates et al. (2019) create a new genus for *Psammophis acutus* based on its unique morphology, and sister-relationship to all species of *Psammophylax*.

Kladirostratus acutus (Günther, 1888)

Distribution: Add Angola (Cauando-Cubango), Conradie et al. (2016b); Angola (Lunda Norte, Moxico), M.P. Marques et al. (2018).

Kladirostratus togoensis (Matschie, 1893)

Comments: Keates et al. (2019) provisionally transfer *Psammophylax togoensis* to *Kladirostratus* based on morphological similarity.

KOLPOPHIS M.A. Smith, 1926 (Elapidae)

Kolpophis annandalei (Laidlaw, 1902)

†***KREBSOPHIS*** Rage & C. Werner, 1999
(†Anomalophiidae)

†***Krebsophis thobanus*** Rage & C. Werner, 1999

KUALATAHAN J.C. Murphy & Voris, 2014. Fieldiana Life
Earth Sci. (8): 27. (Homalopsidae)

Type species: *Enhydris pahangensis* Tweedie, 1946 by original designation.

Distribution: West Malaysia.

Kualatahan pahangensis (Tweedie, 1946)

Comments: J.C. Murphy & Voris (2014) provide a diagnosis and photograph of the holotype.

LACHESIS Daudin, 1803 (Viperidae: Crotalinae)

Lachesis acrochorda (García, 1896)

Distribution: Add Panama (Panamá), Fuentes & Corrales (2016); Ecuador (Pichincha, Santo Domingo), and upper elevation of 1070, Valencia et al. (2016).

Lachesis melanocephala Solórzano & Cerdas, 1986

Lachesis muta (Linnaeus, 1766)

Distribution: Add Colombia (Caqueta, Putumayo), Díaz-Ricaurte et al. (2017); Ecuador (Pastaza, Sucumbios, Zamora-Chinchipe), and upper elevation of 1100 m, Valencia et al. (2016); Peru (Cusco), Carrasco et al. (2019); Peru (Huanuco), Schlüter et al. (2004); Brazil (Roraima), Moraes et al. (2017). R. Rodrigues et al. (2013) confirm the distribution in Paraiba, Brazil.

Comments: Díaz-Ricaurte et al. (2017) provide a species account for Colombia.

Lachesis stenophrys Cope, 1875a

Distribution: Add Nicaragua (Rio San Juan), Sunyer et al. (2014); Panama (Panama Oeste), Ray & Ruback (2015).

LAMPROPELTIS Fitzinger, 1843 (Colubridae: Colubrinae)

Comments: Systematic studies are by E.A. Myers et al. (2013, revision of *L. zonata*), Ruane et al. (2014, revision of the *L. triangulum* group); McKelvy & Burbrink (2017, revision of *L. calligaster*); Krysko et al. (2017, revision of the *L. getula* group), Hansen & Salmon (2017, revision of the *L. mexicana* group). Chen et al. (2017) present a phylogeny that has the species of *Lampropeltis* in two clades: species that are blotched or cross-banded, and species that are ringed and/or tricolor.

Lampropeltis abnorma (Bocourt, 1886 *in* A.H.A. Duméril, Bocourt & Mocquard, 1870–1909). Miss. Sci. Mex. Rept., livr. 10: 614, pl. 39, fig. 4. (*Coronella formosa* var. *abnorma*)

Synonyms: *Coronella formosa anomala* Bocourt, 1886, *Coronella formosa oligozona* Bocourt, 1886; *Lampropeltis polyzona blanchardi* Stuart, 1935; *Lampropeltis triangulum hondurensis* Williams, 1978; *Lampropeltis triangulum stuarti* Williams, 1978.

Holotype: MNHN 1888.129, a 1285 mm female (M.F. Bocourt).

Type locality: "Haute Vera Paz" (=Alta Verapaz, Guatemala); stated as "Coban" in MNHN register.

Distribution: Mexico (SE Guerrero, Oaxaca, Chiapas, S Veracruz, Tabasco, Campeche, Yucatan, Quintana Roo), Guatemala (Alta Verapaz, Escuentla, Jutiapa, Retalhuleu, Petén), Belize (Cayo, Orange Walk, Stann Creek), El Salvador (La Libertad, La Paz, Morazán, San Miguel, San Salvador, Santa Ana), Honduras (Atlántida, Choluteca, Cortés, El Paraiso, Francisco Morazán, Olancho, Vallé, Yoro), Nicaragua (Esteli, Granada, Managua, Matagalpa, Rio San Juan, Zelaya), Costa Rica (Alajuela, Guanacaste, Heredia, Puntarenas). Much of the distribution is speculative due to sparse molecular sampling by Ruane et al. (2014). See also Garcia-Padilla & Mata-Silva (2014a), Solis et al. (2014).

Comments: Removed from the synonymy of *L. triangulum* by Ruane et al. (2014).

Lampropeltis alterna (A.E. Brown, 1902)

Distribution: Delete Mexico (Chihuahua), Hansen & Salmon (2017), who give an elevational range of 384–2311 m. Nevárez-de los Reyes et al. (2016c) present new records from Nuevo Leon, Mexico. Hansen & Alamillo (2018) correct a misidentified record for Zacatecas by Campos-Rodríguez et al. (2017) to *L. greeri*. Hansen & Salmon (2017) map known localities.

Comments: Ruane et al. (2014) discuss phylogenetic relationship. Hansen & Salmon (2017) provide a species account, map, and color photographs. E.A. Myers et al. (2019) analyze genetic material that detects three lineages: Trans-Pecos, Mapimian, and Sierra Madre Occidental.

Lampropeltis annulata Kennicott *in* Cope, 1860e. Proc. Acad. Nat. Sci. Philadelphia 12: 257.

Synonyms: *Coronella doliata conjuncta* Jan 1863b, *Lampropeltis triangulum dixoni* Quinn, 1983.

Holotype: ANSP 3613, (Lt. Couch).

Type locality: stated as "Texas," corrected to "Matamoros, Tamaulipas," Mexico by Kennicott (1861).

Distribution: SC USA (S Texas), NE Mexico (Hidalgo, Nuevo Leon, Queretaro, E San Luis Potosi, Tamaulipas, perhaps also Coahuila). Ruane et al. (2014) give the range as primarily Tamaulipas and possible to the south. However the range undoubtedly extends northward into southern Texas. See Powell et al. (2016) for the range in southern Texas. Fernández-Badillo et al. (2017) confirm the distribution in Hidalgo at elevations of 961 to 1365 m.

Comments: Removed from the synonymy of *L. triangulum* by Ruane et al. (2014).

Lampropeltis californiae (Blainville, 1835)

Synonyms: Remove *L. g. nigritus* Zweifel & Norris, 1955.

Distribution: Add Mexico (Baja California Sur: Isla del Carmen), Frick et al. (2016b). Remove Sinaloa and all but NW Sonora (see Comments).

Comments: Krysko et al. (2017) remove *L. g. nigritus* from the synonymy. E.A. Myers et al. (2017a) provide a DNA-sequence based phylogeny with three primary clades.

Lampropeltis calligaster (Harlan, 1827)

Distribution: Central United States "... all populations west of the Mississippi River in the plains states and crosses the Mississippi River embayment east to Illinois, Indiana, Kentucky, Western Tennessee and northern Mississippi," McKelvy & Burbrink, (2017).

Comments: McKelvy & Burbrink (2017) subdivide *L. calligaster* into three species: nominal, plus *L. rhombomaculata* and *L. occipitolineata*.

Lampropeltis elapsoides (Holbrook, 1838)

Lampropeltis extenuata (A. E. Brown, 1890)

Lampropeltis floridana Blanchard, 1919. Occas. Pap. Univ. Michigan Mus. Zool. (70): 1–5, fig. 1. (*Lampropeltis getulus floridana*)

Synonym: *Lampropeltis getulus brooksi* Barbour, 1919.

Holotype: USNM 22368, an 1138 mm female (W. Palmer, March 1895).

Type locality: "Orange Hammock, De Soto County (northeast portion), Florida," USA.

Distribution: Southeastern USA (peninsular Florida).

Comments: Removed from the synonymy of *L. getula* by Krysko et al. (2017).

Lampropeltis gentilis (Baird & Girard, 1853). Cat. N. Amer. Rept., part I, Serpents: 90–91. (*Ophibolus gentilis*)

Synonyms: *Lampropeltis amaura* Cope, 1860e, *Lampropeltis multistriata* Kennicott, 1861, *Ophibolus doliatus occipitalis* Cope, 1889a, *Ophibolus doliatus parallelus* Cope, 1889a, *Lampropeltis pyrrhomelaena celaenops* Stejneger, 1902b, *Lampropeltis doliata taylori* Tanner & Loomis, 1957.

Lectotype, USNM 1853, a 682 + mm male (R.B. Marcy, 14 July 1852), designated by Blanchard (1921).

Type locality: "Red River, Ark.," emended to "north fork of the Red River, near Sweetwater Creek, Wheeler County, Texas" by Stejneger & Barbour (1917).

Distribution: USA (Louisiana, W Arkansas, extreme WC Missouri, Texas, Oklahoma, Kansas, Nebraska, W South Dakota, Montana, E Wyoming, Colorado, New Mexico, NE Nevada, Utah, E Arizona). Add N Mexico (Coahuila), Baeza-Tarin et al. (2018a).

Comments: Removed from the synonymy of *L. triangulum* by Ruane et al. (2014).

Lampropeltis getula (Linnaeus, 1766)

Distribution: Remove S and C Florida from range (see Comments).

Comments: Krysko et al. (2017) remove *L. floridana* and *L. meansi* from the synonymy.

Lampropeltis greeri Webb, 1961. Copeia 1961(3): 326–328, fig. 1.

Holotype: MSU 190, a 364 mm male (J.K. Greer, 18 July 1958).

Type locality: "Rancho Santa Barbara (Weicher Ranch), 29 miles west-southwest of Ciudad Durango, Durango, México,...*ca.* 7400 feet."

Distribution: Central Mexico (Aguascalientes, S Durango, N Jalisco, E Nayarit, Zacatecas), 2104–2603 m. Hansen & Salmon (2017) map known localities.

Comments: Hansen & Salmon (2017) provide a species account, map, and color photographs. They resurrect *L. greeri* from the synonymy of *L. mexicana* based on allopatric distribution, distinct color pattern, and lack of convergence in areas of parapatry.

Lampropeltis holbrooki Stejneger, 1902

Lampropeltis knoblochi Taylor, 1940

Lampropeltis leonis Günther, 1893 *in* 1885–1909. Biol. Centr. Amer. Rept. Batr.: 110, pl. 39. (*Coronella leonis*)

Holotype: BMNH 1946.1.4.10, a 597 mm specimen (W. Taylor).

Type locality: "Mexico, Nuevo Leon."

Distribution: Mexico (SE Coahuila, S Nuevo Leon, ext. W Tamaulipas), 1036–2268 m. Nevárez-de los Reyes et al. (2016c) present new records from Coahuila, Mexico (as *L. mexicana*). Hansen & Salmon (2017) map known localities.

Comments: Hansen & Salmon (2017) provide a species account, map, and color photographs. They resurrect *O. leonis* from the synonymy of *L. mexicana* based on allopatric distribution, distinct color pattern, and lack of convergence in areas of parapatry.

Lampropeltis meansi Krysko & Judd, 2006. Zootaxa 1193: 24, 26, fig. 12. (*Lampropeltis getula meansi*)

Holotype: UF 73433, an 1195 mm male (D. B. Means, 9 June 1970).

Type locality: "Apalachicola National Forest on FH-13 ca. 3.2 km W SR 67, Liberty County, Florida, United States."

Distribution: Southeastern USA (NC Florida).

Comments: Removed from the synonymy of *L. getula* by Krysko et al. (2017).

Lampropeltis mexicana (Garman, 1884)

Synonyms: Remove *Coronella leonis* Günther, 1893 in 1885–1909, *Lampropeltis greeri* Webb, 1961.

Distribution: C Mexico (Aguascalientes, Guanajuato, Hidalgo, Mexico, San Luis Potosi), 1194–2438 m. Records for states of Hidalgo and México are confirmed by Hansen et al. (2016). Hansen & Salmon (2017) map known localities. Quintero-Díaz & Carbajal-Márquez (2019) report new records.

Comments: Hansen & Salmon (2017) provide a species account, map, and color photographs.

Lampropeltis micropholis Cope, 1860. Proc. Acad. Nat. Sci. Philadelphia 12: 257–258.

Synonyms: *Coronella doliata formosa* Jan 1863b, *Lampropeltis triangulum gaigae* Dunn, 1937a; *Lampropeltis triangulum andesiana* Williams, 1978.

Type: Holotype, ANSP 3427, a 391 mm male (J.L. LeConte).

Type locality: "Panama."

Distribution: E Costa Rica (Alajuela, Cartago, Limon, Puntarenas, San José), Panama (Canal Zone, Chiriquí, Darién, Panamá, Panamá Oeste), W Colombia (Antioquia, Atlántico, Caldas, Caqueta, Cauca, Cesar, Chocó, Cundinamarca, Huila, Magdalena, Tolima, Valle), W Venezuela (Distrito Federal, Merida, Zulia) and W Ecuador (Bolivar, Guayas, Imbabura, Los Rios [Almendáriz & Carr, 2012], Pichincha). Add Colombia (Quindio), Quintero-Ángel et al. (2012)

Comments: Removed from the synonymy of *L. triangulum* by Ruane et al. (2014), who placed *Lampropeltis t. andesiana* as a synonym without supporting molecular data.

Lampropeltis multifasciata (Bocourt, 1886 *in* A.H.A. Duméril, Bocourt & Mocquard, 1870–1909). Miss. Sci. Mex. Rept., livr. 10: 616–617, pl. 40, fig. 2. (*Coronella multifasciata*)

Synonyms: *Lampropeltis agalma* Van Denburgh & Slevin, 1923, *Lampropeltis herrerae* Van Denburgh & Slevin, 1923, *Lampropeltis zonata parvirubra* Zweifel, 1952, *Lampropeltis zonata pulchra* Zweifel, 1952.

Holotype: MNHNP 1884.326, a 446 mm specimen (M. de Cessac, 31 July 1884).

Type locality: "Californie," USA; from San Luis Obispo according to MNHNP register.

Distribution: USA (SW California), Mexico (Baja California, Todos Santos I.).

Comments: Removed from the synonymy of *L. zonata* by E.A. Myers et al. (2013b).

Lampropeltis nigra (Yarrow, 1882)

Lampropeltis nigrita Zweifel and Norris, 1955. Amer. Midl. Nat. 54: 238–239, pl. 1, upper. (*Lampropeltis getulus nigritus*)

Holotype: MVZ 50814, a 1081 mm male (K.S. Norris & R.G. Zweifel, 3 August 1950).

Type locality: "30.6 miles (by road) south of Hermosillo, Sonora, on the main highway," Mexico.

Distribution: Northwestern Mexico (N Sinaloa, Sonora), extreme SW USA (SC Arizona).

Comments: Removed from the synonymy of *L. californiae* by Krysko et al. (2017).

Lampropeltis occipitolineata Price, 1987. Bull. Chicago Herpetol. Soc. 22: 148. (*Lampropeltis calligaster occipitolineata*)

Holotype: FMNH 48265, an adult male (R. Paulk, 31 May 1942).

Type locality: "Okeechobee, Okeechobee County, Florida," USA.

Distribution: Southeastern USA (SC Florida).

Comments: Removed from the synonymy of *L. calligaster*, of which it was formerly a subspecies, by McKelvy & Burbrink (2017).

Lampropeltis polyzona Cope, 1860. Proc. Acad. Nat. Sci. Philadelphia 12: 258.

Synonyms: *Lampropeltis micropholis arcifera* Werner, 1893a, *Lampropeltis triangulum nelsoni* Blanchard, 1920a, *Lampropeltis triangulum schmidti* Stuart, 1935a, *Lampropeltis triangulum conanti* Williams, 1978, *Lampropeltis triangulum sinaloae* Williams, 1978, *Lampropeltis triangulum*

smithi Williams, 1978, *Lampropeltis triangulum campbelli* Quinn, 1983.

Holotype: ANSP 9770, an 1115 mm male (Mr. Pease).

Type locality: "Quatupe, near Jalapa," Veracruz, Mexico.

Distribution: Mexico (SW Chihuahua, Colima, Guanajuato, Guerrero, Hidalgo, Jalisco, Michoacan, Morelos, Nayarit [including Tres Marias Is.], Oaxaca, Puebla, Sinaloa, S Sonora, NC Veracruz). Add Mexico (Aquascalientes), Quintero-Diaz et al. (2014a). Leyte-Manrique et al. (2017) confirm and add records for Guanajuato.

Comments: Removed from the synonymy of *L. triangulum* by Ruane et al. (2014).

Lampropeltis pyromelana (Cope, 1867)

Lampropeltis rhombomaculata (Holbrook, 1840). N. Amer. Herpetol., 1st ed., 4: 103–104, pl. 20. (*Coluber rhombo-maculatus*)

Type: none designated according to Blaney (1979b).

Type locality: "Georgia and Alabama"; restricted to Atlanta, Georgia, USA by Schmidt (1953a).

Distribution: Southeastern USA (SE Louisiana, S Mississippi, Alabama, panhandle of Florida, N Georgia, E Tennessee, South Carolina, North Carolina, Virginia, Washington, D.C., S Maryland).

Comments: Removed from the synonymy of *L. calligaster*, of which it was formerly a subspecies, by McKelvy & Burbrink (2017).

Lampropeltis ruthveni Blanchard, 1920

Distribution: Add Mexico (Guanajuato), Hansen et al. (2015), and range extension in Jalisco, Grünwald et al. (2016). Elevational range is 1925–2667 m, Hansen & Salmon (2017). Hansen & Salmon (2017) map known localities.

Comments: Hansen & Salmon (2017) provide a species account, map, and color photographs.

†***Lampropeltis similis*** Holman, 1964

Lampropeltis splendida (Baird & Girard, 1853)

Comments: E.A. Myers et al. (2017a) provide a DNA-sequence based phylogeny, mostly of USA populations, that supports the separation of this species from *L. californiae* in SE Arizona, USA.

Lampropeltis triangulum (Lacépède, 1789)

Synonyms: *Coluber eximius* Harlan, 1827, *Pseudoelaps Y* Berthold, 1842 (invalid name); *Ophibolus clericus* Baird & Girard, 1853; *Ophibolus doliatus syspila* Cope, 1889a; *Ophibolus doliatus collaris* Cope, 1889a; *Ophibolus doliatus temporalis* Cope, 1893b.

Distribution: SE Canada (SE Ontario and extreme SW Quebec), E USA (N Alabama, Arkansas, Connecticut, Delaware, District of Columbia, N Georgia, Illinois, Indiana, Iowa, Kentucky, NE Louisiana, SE Maine, Maryland, Massachusetts, Michigan, SE Minnesota, N Mississippi, Missouri, New Hampshire, New Jersey, New York, W North Carolina, Ohio, NE Oklahoma, Pennsylvania, Rhode Island, extreme NW South Carolina, Tennessee, Vermont, Virginia, West Virginia, S Wisconsin).

Comment: Ruane et al. (2014) distribute western and Mesoamerican subspecies/populations among five species: *L. abnorma, L. annulata, L. gentilis, L. micropholis, L. polyzona.*

†*Lampropeltis vetustum* (Auffenberg, 1963)

Lampropeltis webbi Bryson, Dixon & Lazcano-Villareal, 2005

Distribution: Elevation to 2394 m, Hansen & Salmen (2017). Hansen & Salmon (2017) map known localities.

Comments: Hansen & Salmon (2017) provide a species account, map, and color photographs.

Lampropeltis zonata (Blainville, 1835)

Distribution: Remove central and southern Coast Ranges of California and Baja California from range (see Comments).

Comments: E.A. Myers et al. (2013b) removed *L. multifasciata* from the synonymy.

LAMPROPHIS Fitzinger, 1843 (Lamprophiidae)

Lamprophis aurora (Linnaeus, 1758)
Lamprophis fiski Boulenger, 1887
Lamprophis fuscus Boulenger, 1893
Lamprophis guttatus (A. Smith, 1843 *in* 1838–1849)

Comment: Kelly et al. (2011) retained *L. guttatus* within *Lamprophis*, rather than transferring it to *Bouaedon* as indicated in error by Wallach et al. (2014).

LANGAHA Bonnaterre, 1790 (Pseudoxyrhophiidae)

Langaha alluaudi Mocquard, 1901

Langaha madagascariensis Bonnaterre, 1790

Distribution: Rosa et al. (2012) map known localities.

Langaha pseudoalluaudi Domergue, 1988

†*LAOPHIS* Owen, 1857. Quart. J. Geol. Soc. London 13: 199. (Viperidae: incerta sedis)

Type species: †*Laophis crotaloides* Owen, 1857 by monotypy.

Distribution: Early Pliocene of Greece.

†*Laophis crotaloides* Owen, 1857. Quart. J. Geol. Soc. London 13: 196–199, plate 4.

Holotype: 13 vertebrae (Capt. Spratt, 1846 or 1847), now lost according to Georgialis et al. (2013).

Type locality: "The Promontory of Karabournou," Macedonia, Greece.

Distribution: Early Pliocene (Ruscinian) of Greece.

Comments: Georgialis et al. (2016) describe additional material, confirming that the taxon is a viperid of imprecise relationships.

†*LAPPARENTOPHIS* Hoffstetter, 1960 (Lapparentophiidae)

Distribution: Add Middle Cretaceous of Morocco, Vullo (2019).

†*Lapparentophis defrennei* Hoffstetter, 1960

†*Lapparentophis ragei* Vullo, 2019. Comptes Rendus Palevol 18(7): 766–768, figs. 1, 2.

Holotype: MHNM KK387, a trunk vertebra.

Type locality: "El Begâa locality near Taouz, southeastern Morocco."

Distribution: Middle Cretaceous (late Albian/Cenomanian) of SE Morocco (Ksar es Souk).

LATICAUDA Laurenti, 1768 (Elapidae)

Laticauda colubrina (Schneider, 1799)

Distribution: Add Myanmar (Ayeyarwady), Platt et al. (2018); Thailand (Krabi: Phi Phi Don and Phi Phi Ley Is.), Milto (2014); Philippines (Barit, Calayan, Dalupiri, Mabag, Mindanao, Siquijor, Sitanki Is.), Leviton et al. (2018); Philippines (Maestre de Campo I.), Siler et al. (2012). Ghergel et al. (2016) provide a map of known localities.

Laticauda crockeri Slevin, 1934

Laticauda frontalis (De Vis, 1905)

Distribution: Ghergel et al. (2016) provide a map of known localities.

Laticauda guineai Heatwole, Busack & Cogger, 2005

Distribution: Ghergel et al. (2016) provide a map of known localities.

Laticauda laticaudata (Linnaeus, 1758)

Distribution: Add South Korea (Jeju Island), Park et al. (2017a); Japan (Kuchinoerabu, Kodakara, Amamioshima, Tsuken, Zamami, Miyako, Taketomi, Iriomote, Kuroshima Is.), Tandavanitj et al. (2013a); Japan (Honshu), Tandavanitj et al. (2013b); Taiwan (Lyudao and Lanyu Is.), Tandavanitj et al. (2013a); Philippines (Cebu I.), Supsup et al. (2016); Philippines (Calayan I.), Leviton et al. (2018). Ghergel et al. (2016) provide a map of known localities.

Comments: Tandavanitj et al. (2013a) find four cytochrome *b* haplotypes in specimens from the Ryukyus and Taiwan that have some correlation with deep water barriers.

Laticauda saintgironsi Cogger & Heatwole, 2006

Distribution: Ghergel et al. (2016) provide a map of known localities.

LEIOHETERODON A.M.C. Duméril, Bibron & Duméril, 1854 (Pseudoxyrhophiidae)

Leioheterodon geayi Mocquard, 1905

Leioheterodon madagascariensis (A.M.C. Duméril, Bibron & Duméril, 1854)

Distribution: Add Madagascar (Nosy Komba), Roberts & Daly (2014).

Leioheterodon modestus (Günther, 1863)

Distribution: Add Madagascar (Fianarantsoa), and upper elevation range of 870 m, Rosa et al. (2010).

LEIOPYTHON Hubrecht, 1879 (Pythonidae)

Comments: Reynolds et al. (2014b) recommend combining *Leiopython* within *Bothrochilus* due to close phylogenetic relationship of *B. boa* with *L. albertisii* and *L. hoserae* (other *Leiopython* species were unavailable). We continue to recognize *Leiopython* as distinct from *Bothrochilus* due to the position of *B. boa* as sister taxon to *Leiopython*, and based on the arguments for molecular and morphological distinctiveness provided by Barker et al. (2015).

Leiopython albertisii (W.C.H. Peters & Doria, 1878)

Distribution: Add Papua New Guinea (Duke of York, Emirau, Manam, New Britain and New Ireland Islands), Clegg & Jocque (2016).

Leiopython bennettorum Hoser, 2000

Synonym: *Leiopython montanus* Schleip, 2014 (substitute name).

Leiopython biakensis Schleip, 2008
Leiopython fredparkeri Schleip, 2008

Leiopython hoserae Hoser, 2000

Synonym: *Leiopython meridionalis* Schleip, 2014 (substitute name).

Distribution: Add Papua New Guinea (Northern: D'Entrecasteaux Is.), Kraus (2013, as *L. albertisii*).

Leiopython huonensis Schleip, 2008

LEPTODEIRA Fitzinger, 1843 (Dipsadidae: Dipsadinae)

Comments: Barrio-Amorós (2019) provides a summary of the genus, with some taxonomic revisions.

Leptodeira annulata (Linnaeus, 1758)

Synonyms: *Coronella taeniata* Laurenti, 1768, *Coluber epidaurius* Hermann, 1804, *Eteirodipsas annulata rhomboidalis* Jan, 1863, *Dipsas approximans* Günther, 1872, *Eteirodipsas wieneri* Sauvage, 1884, *Tarbophis dipsadomorphoides* Ahl, 1925 (?), *Leptodeira annulata pulchriceps* Duellman, 1958. Remove *Leptodira nycthemera* Werner, 1901, which was shown by J. Costa et al. (2015) to be a specimen of *Oxyrhopus petolarius*.

Distribution: Limited to the Amazon Basin of Colombia, Ecuador, Peru, Bolivia and Brazil, eastward to the Atlantic forest of Brazil, and southward into Paraguay and N Argentina. Add Brazil (Ceara), Mesquita et al. (2013); Brazil (Minas Gerais), Sousa et al. (2010); Brazil (Alagoas, Piaui, Rio Grande do Norte, Sergipe), Guedes et al. (2014); Peru (Puno), Llanqui et al. (2019).

Leptodeira ashmeadii Hallowell, 1845. Proc. Acad. Nat. Sci. Philadelphia 2: 244

Synonyms: *Urotheca aureorostris* Briceño-Rossi, 1934, *Leptodeira rhombifera kugleri* Shreve, 1947.

Lectotype: ANSP 10093, a 470 mm specimen (S. Ashmead), designated by Duellman (1958).

Type locality: "Republic of Columbia, within two hundred miles of Caraccas"; restricted to Caracas, Venezuela by Duellman (1958).

Distribution: N South America. N Colombia, N Venezuela, Trinidad and Tobago, and N Brazil, Barrio-Amorós (2019), also Guyana, Suriname and French Guiana?

Comments: Barrio-Amorós (2019) recognizes *ashmeadi* as a species distinct from *L. annulata* based on polyphyly of the DNA-based phylogeny of Daza et al. (2009), and different color patterns between the two taxa.

Leptodeira bakeri Ruthven, 1936

Leptodeira frenata (Cope *in* Ferrari-Pérez, 1886)

Leptodeira larcorum K.P. Schmidt & Walker, 1943. Zool. Series Field Mus. Nat. Hist. 24(27): 311–312.

Holotype: FMNH 34302, a 630 mm male (Magellanic Expedition, 1939).

Type locality: "Chiclin, Libertad, Peru."

Distribution: Arid regions of SW Ecuador and NW Peru, Barrio-Amorós (2019). Add Peru (Lambayeque), Venegas (2005).

Comments: Barrio-Amorós (2019) considers *larcorum* to be a species distinct from *L. ornata* based on the habitat hiatus and different color patterns.

Leptodeira maculata Hallowell, 1861. Proc. Acad. Nat. Sci. Philadelphia 12: 488. (*Megalops maculatus*)

Synonyms: *Leptodeira personata* Cope, 1869, *Leptodeira smithi* Taylor, 1938, *Leptodeira annulata cussiliris*, Duellman, 1958.

Holotype: USNM 7367, a 470 mm specimen (C.B. Adams).

Type locality: "Tahiti," in error; likely from Nicaragua according to Cochran (1961).

Distribution: W Mexico (Chiapas, Colima, Guerrero, Jalisco, Michoacan, Nayarit, Oaxaca, Sinaloa). Elevation NSL-2000 m, Palacios-Aguilar & Flores-Villela (2018). Add Mexico (Morelos), Aréchaga-Ocampo et al. (2008).

Comments: Barrio-Amorós (2019) recognizes *maculata* as a species distinct from *L. annulata* based on polyphyly of the DNA-based phylogeny of Daza et al. (2009).

Leptodeira nigrofasciata Günther, 1868

Leptodeira ornata (Bocourt, 1884)

Distribution: Add E Honduras and E Nicaragua, Barrio-Amorós (2019); Panama (Panama Oeste), Ray & Ruback (2015); Colombia (Cesar), Medina-Rangel (2011); Colombia (Huila), Moreno-Arias & Quintero-Corzo (2015); Ecuador (Manabi), Almendáriz et al. (2012); Peru (Amazonas), Koch et al. (2018).

Comments: Barrio-Amorós (2019) notes that populations in Colombia and N Ecuador are probably an undescribed species.

Leptodeira polysticta Günther, 1895 *in* 1885–1902
Leptodeira punctata (W.C.H. Peters, 1866)

Leptodeira rhombifera Günther, 1872. Ann. Mag. Nat. Hist., Series 4, 9(49): 32. (*Leptodira rhombifera*)

Synonyms: *Leptodira occelata* Günther, 1895 *in* 1885–1902.

Holotype: BMNH 1946.1.9.92, a 584 mm specimen (collector for O. Salvin).

Type locality: "Rio Chisoy, near the town of Cubulco," Depto. Baja Verapaz, Guatemala.

Distribution: Guatemala, Honduras, Nicaragua, Costa Rica, Panama. Add Honduras (Isla del Tigre), McCranie & Gutsche (2016); Panama (Panama Oeste), Ray & Ruback (2015).

Comments: Barrio-Amorós (2019) recognizes *rhombifera* as a species distinct from *L. annulata* based on polyphyly of the DNA-based phylogeny of Daza et al. (2009), and different color patterns between the two taxa.

Leptodeira rubricata (Cope, 1894)

Leptodeira septentrionalis (Kennicott *in* Baird, 1859)

Leptodeira splendida Günther, 1895 *in* 1885–1902

Distribution: Low elevation of NSL, Palacios-Aguilar & Flores-Villela (2018).

Leptodeira uribei (Ramírez-Bautista & H.M. Smith, 1992)

LEPTODRYMUS Amaral, 1927 (Colubridae: Colubrinae)

Leptodrymus pulcherrimus (Cope, 1874)

Distribution: Add Honduras (Santa Bárbara), Espinal et al. (2014b); Honduras (Valle: Isla Comandante), McCranie et al. (2013b); Honduras (Isla Zacate Grande), McCranie & Gutsche (2016); Nicaragua (León), Sunyer & Leonardi (2015); Nicaragua (Matagalpa), Sunyer et al. (2014).

LEPTOPHIS Bell, 1825 (Colubridae: Colubrinae)

Leptophis ahaetulla (Linnaeus, 1758)

Synonyms: Remove *Leptophis coeruleodorsus* Oliver, 1942.

Distribution: Add Mexico (Yucatan), Torres-Solís et al. (2017); Honduras (Santa Bárbara), Espinal et al. (2014b); Nicaragua (Rivas), Sunyer et al. (2014); Panama (Panama, Panama Oeste), Ray & Ruback (2015); Colombia (Cesar), Medina-Rangel (2011); Colombia (Huila), Moreno-Arias & Quintero-Corzo (2015); Colombia (Santander), E. Ramos and Meza-Joya (2018); Colombia (Valle del Cauca: Isla Palma), Giraldo et al. (2014); Ecuador (Manabi), Almendáriz et al. (2012); Peru (Huanuco), Schlüter et al. (2004); Brazil (Acre), Bernarde et al. (2011b); Brazil (Alagoas, Sergipe), Guedes et al. (2014); Brazil (Ceara), Mesquita et al. (2013); Brazil (Mato Grosso do Sul), Ferreira et al. (2017); Brazil (Paraiba), R. França et al. (2012); Brazil (Piauí), Dal Vechio et al. (2013); Brazil (Rondônia), Bernarde et al. (2012b). Remove from the distribution: Trinidad and Tobago, and NC Venezuela (Apure, Distrito Federal, Miranda, Monagas, Nueva Esparta, and Vargas), J.C. Murphy et al. (2013).

Leptophis coeruleodorsus Oliver, 1942. Occas. Pap. Mus. Zool. Univ. Michigan (462): 4–7.

Holotype: AMNH 209022, a 1375 mm female.

Type locality: "Trinidad, British West Indies"; restricted to "Mt. St. Benedict, Tunapuna, Trinidad (~10°39′N 61°23′W)" by J.C. Murphy et al. (2013).

Distribution: Trinidad and Tobago, and NC Venezuela (Apure, Distrito Federal, Miranda, Monagas, Nueva Esparta, and Vargas).

Comments: J.C. Murphy et al. (2013) revalidate *L. coeruleodorsus* (from *L. ahaetulla*) based on monophyly derived from mtDNA sequence-data, and morphological differences. They redescribe the holotype, and describe geographic variation between island and mainland samples.

Leptophis cupreus (Cope, 1868)

Distribution: Add Ecuador (Pichincha), Arteaga et al. (2013); Peru (Huanuco), Schlüter et al. (2004); Panama (Darien), Batista & L.W. Wilson (2017).

Leptophis depressirostris (Cope, 1861)

Distribution: Add Panama (Panama, Panama Oeste), Ray & Ruback (2015).

Leptophis diplotropis (Günther, 1872)

Distribution: Add Mexico (Hidalgo), Berriozabal-Islas et al. (2012).

Leptophis haileyi J.C. Murphy, Charles, Lehtinen & Koeller, 2013. Zootaxa 3718(6): 569–570, fig. 3.

Holotype: CAS 245313, a 1406 mm male (P.G. Frank, P.A. Frank, R. Lawson, 2006).

Type locality: "Tobago near Roxborough at 11° 15′ 05.8″N, 60° 34′ 04.7″W."

Distribution: Trinidad and Tobago (Tobago). Known only from the holotype.

Leptophis mexicanus A.M.C. Duméril, Bibron & Duméril. 1854

Distribution: Mexico (Nuevo Leon, northernmost locality), Nevárez-de Los Reyes et al. (2017a); add Mexico (Guerrero), and upper elevation of 1720 m, Palacios-Aguilar & Flores-Villela (2018); El Salvador (Santa Ana), Juárez-Peña et al. (2016); Honduras (Choluteca) and upper elevation of 1635 m, Espinal et al. (2014a); Honduras (La Paz), McCranie & Gutsche (2013b); Honduras (Santa Bárbara), Espinal et al. (2014b).

Comments: Schätti & Kucharzewski (2017) discuss the type material.

Leptophis modestus (Günther, 1872)

Leptophis nebulosus Oliver, 1942

Leptophis riveti Despax, 1910

Distribution: Add Panama (Panama Oeste), Ray & Ruback (2015).

Leptophis stimsoni Harding, 1995

LEPTOTYPHLOPS Fitzinger, 1843 (Leptotyphlopidae: Leptotyphlopinae)

Leptotyphlops aethiopicus Broadley & Wallach, 2007

Leptotyphlops albiventer Hallermann & Rödel, 1995

Distribution: Add Mali (Sikasso), and upper elevation of 386 m, Trape & Mané (2017); Burkina Fasso (Comoe), Böhme & Heath (2018).

Comments: Böhme & Heath (2018) remove *albiventer* from *Myriopholis* and return it to the genus *Leptotyphlops* based on cranial morphology.

Leptotyphlops conjunctus (Jan, 1861)

Leptotyphlops distanti (Boulenger, 1892)

Leptotyphlops emini (Boulenger, 1890)

Leptotyphlops howelli Broadley & Wallach, 2007

Leptotyphlops incognitus Broadley & Watson, 1976

Leptotyphlops jacobseni Broadley & Broadley, 1999

Leptotyphlops kafubi (Boulenger, 1919)

Leptotyphlops keniensis Broadley & Wallach, 2007

Leptotyphlops latirostris (Sternfeld, 1912)

Leptotyphlops macrops Broadley & Wallach, 1996

Leptotyphlops mbanjensis Broadley & Wallach, 2007

Leptotyphlops monticola (Chabanaud, 1917)

Leptotyphlops nigricans (Schlegel, 1839 *in* 1837–1844)
Leptotyphlops nigroterminus Broadley & Wallach, 2007
Leptotyphlops nursii (J. Anderson *in* Boulenger, 1896)

Leptotyphlops pembae Loveridge, 1941
Distribution: Add Tanzania (Mafia Island), Boundy (2014).

Leptotyphlops pungwensis Broadley & Wallach, 1997

Leptotyphlops scutifrons (W.C.H. Peters, 1854)
Distribution: Add Zambia (Western), Pietersen et al. (2017).

Leptotyphlops sylvicolus Broadley & Wallach, 1997
Leptotyphlops tanae Broadley & Wallach, 2007
Leptotyphlops telloi Broadley & Watson, 1976

LETHEOBIA Cope, 1869 (Typhlopidae: Afrotyphlopinae)
Distribution: Much of central and eastern sub-Saharan Africa, with one species reaching Turkey and another the Middle East.

Comments: Hedges et al. (2014), Pyron & Wallach (2014) provide generic diagnoses. Kornilios et al. (2013a) produce an nDNA phylogeny that confirms the sister-relationship of *L. espiscopa* and *L. simoni*, which share a most recent common ancestor with *L. feae/ L. newtonii.*

Letheobia acutirostrata (Andersson, 1916)

Letheobia akagerae Dehling, Hinkel, Ensikat, Babilon & Fischer, 2018. Zootaxa 4378(4): 482–487, figs. 1, 2, 4, 5.

Holotype: ZFMK 100862, a 458 mm specimen (unnamed construction workers, 8 July 2012).

Type locality: "Rwanda, Akagera National Park, Ruzizi Tented Lodge (1°54′24.90″S, 30°42′58.42″E; approx. 1290 m)."

Distribution: Rwanda, 1290–1600 m.

Letheobia angeli (Guibé, 1952)

Comments: Hedges et al. (2014) transfer *angeli* to *Letheobia* from *Afrotyphlops*, which is supported by Wallach & Gemel (2018).

Letheobia caecus (A.H.A. Duméril, 1856)
Distribution: Add Nigeria (Bayelsa), Wallach & Gemel (2018).

Letheobia crossii (Boulenger, 1893)
Distribution: Add Nigeria (Nasarawa, Niger), Wallach & Gemel (2018).

Letheobia debilis (Joger *in* G. Peters & Hutterer, 1990)

Letheobia episcopus (Franzen & Wallach, 2002)
Distribution: Upper elevation of 640 m, Göçmen et al. (2009a).

Letheobia erythraeus (Scortecci, 1928)
Letheobia feae (Boulenger, 1906)
Letheobia gracilis (Sternfeld, 1910)

Comments: Wallach *in* Wallach & Gemel (2018) states that the synonym *Typhlops katangensis* Witte 1933 appears to be a valid species. Dehling et al. (2018) redescribe the holotype of *T. gracilis.*

Letheobia graueri (Sternfeld, 1913)
Comments: Dehling et al. (2018) redescribe the holotype.

Letheobia jubana Broadley & Wallach, 2007
Comments: Hedges et al. (2014) transfer *jubana* to *Afrotyphlops* on the basis of subocular scale presence and

nasal suture characteristics. Pyron & Wallach (2014) counter that the vague eyespot and a number of visceral characters ally it with *Letheobia.*

Letheobia kibarae (Witte, 1953)
Letheobia largeni Broadley & Wallach, 2007
Letheobia leucosticta (Boulenger, 1898)
Distribution: Add Guinea (Conakry), Trape & Baldé (2014) and Wallach & Gemel (2018).

Letheobia logonensis Trape, 2019. Bull. Soc. Herpétol. France 169: 35–38, figs. 8–10.

Holotype: MNHN 2018.0015, a 390 mm specimen of unknown sex (villager, 29 May 2015).

Type locality: "Baibokoum (07°44′N / 15°40′E) au Tchad."

Distribution: S Chad (Logone-Oriental). Known only from the holotype.

Letheobia lumbriciformis (W.C.H. Peters, 1874)

Letheobia manni (Loveridge, 1941)
Distribution: Add Guinea, Trape & Baldé (2014).

Comments: Hedges et al. (2014) transfer *angeli* to *Letheobia* from *Afrotyphlops*, which is supported by Wallach & Gemel (2018).

Letheobia mbeerensis Malonza, Bauer & Ngwava, 2016. Zootaxa 4093(1): 145–146, figs 2–4.

Holotype: NMK S2927, a 280 mm adult (local farmer, 29 April 2014).

Type locality: "Kenya, Embu County, Siakago-Mbeere (00°35′S; 037°38′E, 1221 m)."

Distribution: C Kenya (Eastern Prov.). Known only from the type locality, 1220 m.

Letheobia newtoni (Bocage, 1890)
Letheobia pallida Cope, 1869
Letheobia pauwelsi Wallach, 2005
Letheobia pembana Broadley & Wallach, 2007

Letheobia praeocularis (Stejneger, 1894)
Distribution: Add Nigeria, Trape & Baldé (2014).

Letheobia rufescens (Chabanaud, 1917)
Letheobia simoni (Boettger, 1879)
Letheobia somalica (Boulenger, 1895)
Letheobia stejnegeri (Loveridge, 1931)
Letheobia sudanensis (K.P. Schmidt, 1923)
Letheobia swahilica Broadley & Wallach, 2007
Letheobia toritensis Broadley & Wallach, 2007
Letheobia uluguruensis (Barbour & Loveridge, 1928)

Letheobia weidholzi Wettstein *in* Wallach & Gemel, 2018. Herpetozoa 31(1/2): 31, 38–39, figs. 2, 13–17.

Holotype: NMW 23492, a 376 mm female (A. Weidholz, 1938 or 1939).

Type locality: "Poli, Département de Faro, Région du Nord, Cameroun (8° 27′ 16.4″ N, 13° 15′ 33.7″ E, elevation 525 m)."

Distribution: N Cameroon (Nord), 525 m. Known only from the holotype.

Letheobia wittei (Roux-Estève, 1974)

LIASIS Gray, 1842 (Pythonidae)

Synonyms: Add *Apodora* Kluge, 1993.

Comments: DNA sequence data analyzed by Reynolds et al. (2014b) resolve *L. papuanus* and *L. olivaceus* as sister taxa, which form a sister group to remaining *Liasis* species. Barker et al. (2015) argue that the genetic and morphological distinctiveness of *L. papuanus* warrant recognition of the genus *Apodora*. However, doing so would require inclusion of *L. olivaceus*, which has not yet been proposed. We await such a proposal before recognizing *Apodora*. The gender of *Liasis* is unknown – when Gray proposed the genus he listed one species as masculine, the other as feminine. We follow the common usage as masculine.

†*Liasis dubudingala* Scanlon & Mackness, 2002

Liasis dunni Stull 1932. Occas. Pap. Boston Soc. Nat. Hist. 8: 25–26, plate 1. (*Liasis mackloti dunni*)

Holotype: AMNH 32263, a female (E.R. Dunn).

Type locality: "Uhak, north coast of Wetar, Dutch East Indies," Maluku Prov., Indonesia.

Distribution: Indonesia (Wetar).

Comments: Barker et al. (2015) elevate this taxon to species on the basis of morphological, genetic and behavioral differences, combined with its allopatric, insular distribution.

Liasis fuscus W.C.H. Peters, 1873

Liasis mackloti A.M.C. Duméril & Bibron, 1844

Distribution: Remove Indonesia (Savu, Wetar) from the distribution, Barker et al. (2015)

Liasis olivaceus Gray, 1842

Comments: Ellis (2015) corrected the holotype number for *Liasis olivaceus barroni*: WAM 55383 rather than 55384.

Liasis papuanus W.C.H. Peters & Doria, 1878

Distribution: Add Papua New Guinea (Karkar Island), Clegg & Jocque (2016).

Comments: Barker et al. (2015) reject the synonymy of *Apodora* with *Liasis* based on some ambiguity in the sequence-based phylogenies of Reynolds et al. (2014), and morphological differences observed by the Barkers (see comments under *Liasis*).

Liasis savuensis Brongersma 1956. Proc. Koninkl. Nederl. Akad. Wetensch. C59(3): 296–297. (*Liasis mackloti savuensis*)

Holotype: BMNH 1896.6.21.33, (A. Everett).

Type locality: "Savu [=Sawu] I[slan]d.," Nusa Tenggara Timur, Indonesia.

Distribution: Indonesia (Savu).

Comments: Barker et al. (2015) elevate this taxon to species on the basis of morphological, genetic and behavioral differences, combined with its allopatric, insular distribution.

LICHANURA Cope, 1861 (Charinidae: Charininae)

Lichanura orcutti Stejneger, 1889

Lichanura trivirgata Cope, 1861

Distribution: Add Mexico (Baja California Sur: Isla Espiritu Santo, Isla San José), Frick et al. (2016a), Schoenig (2017); Mexico (Baja California Sur: Isla Coronados), Arnaud & Blázquez (2018).

LIMAFORMOSA Broadley, Tolley, Conradie, Wishart, Trape, Burger, Kusamba, Zassi-Boulou & Greenbaum, 2018. African J. Zool. 67(1): 46–47. (Lamprophiidae)

Synonyms: *Heterolepis* A. Smith, 1847, *Simocephalus* Gray *in* Günther, 1858.

Type species: *Heterolepis capensis* A. Smith, 1847, by original designation.

Distribution: Central and S Africa.

Comments: See under *Gonionotophis*.

Limaformosa capensis (A. Smith, 1847 *in* 1838–1849)

Distribution: Add Mozambique (Zambezia), Broadley et al. (2018); Angola (Huila), M.P. Marques et al. (2018).

Limaformosa chanleri (Stejneger, 1894)

Distribution: Add Congo (Kouilou), Broadley et al. (2018).

Limaformosa crossii (Boulenger, 1895)

Distribution: Lower elevation of 193 m, Trape & Mané (2015).

Limaformosa guirali (Mocquard, 1887)

Distribution: Add Guinea (Faranah), Broadley et al. (2018); Ghana, Trape & Baldé (2014).

Limaformosa savorgnani (Mocquard, 1877)

Distribution: Add Congo (Niari), Broadley et al. (2018).

Limaformosa vernayi (Bogert, 1940)

Distribution: Add Angola (Lunda Norte), Branch (2018).

Comments: Provisionally included in *Limaformosa* based on morphological similarity, Broadley et al. (2018).

LIMNOPHIS Günther, 1865 (Natricidae)

Limnophis bangweolicus (Mertens, 1936)

Distribution: Add Angola (Cauando-Cubango), Conradie et al. (2016b).

Limnophis bicolor Günther, 1865

Distribution: Add Angola (Lunda Sul), Branch (2018); Angola (Bie, Huambo), M.P. Marques et al. (2018).

LIODYTES Cope, 1885 (Natricidae)

Synonyms: Add *Seminatrix* Cope, 1895.

Comments: *Liodytes* is resurrected for *L. alleni* and *L. rigida* (formerly in *Regina*) due to paraphyly, and *Seminatrix pygaea* is included due to monophyly, McVay & Carstens (2013).

Liodytes alleni (Garman, 1874)

†*Liodytes intermedia* (Meylan, 1982)

Liodytes pygaea (Cope, 1871)

Liodytes rigidus (Say, 1825)

LIOHETEROPHIS Amaral, 1934 (Dipsadidae: subfamily unnamed)

Lioheterophis iheringi Amaral, 1934

Distribution: The type locality is stated as 560 m, Guedes et al. (2014).

LIOPELTIS Fitzinger, 1843 (Colubridae: Colubrinae)

Comments: Poyarkov et al. (2019) provide a key to the species.

Liopeltis calamaria (Günther, 1858)

Distribution: Add Nepal (Chitwan), Bhattarai et al. (2017a); Nepal (Makwanpur), Bhattarai et al. (2018a); northward range extension in Western Ghats, Maharashtra, India, Chunekar & Alekar (2015).

Liopeltis frenata (Günther, 1858)

Distribution: Add Thailand (Nan), Hauser (2018); Laos (Houaphan), Hauser (2018); Vietnam (Lai Chau), Hauser (2018); Vietnam (Son La), Pham et al. (2014). The Dong Nai and Gia Lai, Vietnam, records are doubtful according to Hauser (2018).

Liopeltis pallidonuchalis Poyarkov, Nguyen & Vogel, 2019. J. Nat. Hist. 53(27–28): 1655–1660, figs. 1, 2, 4, 5.

Holotype: ZMMU R15682, a 409 mm male (N.A. Poyarkov, 29 May 2016).

Type locality: "Kon Chu Rang N.R. (14.5034° N, 108.5383° E, at elevation of 1010 m asl), Gia Lai Province, central Vietnam."

Distribution: C Vietnam (Gia Lai, Quang Binh, Thua Thien Hue), 950–1010 m.

Liopeltis philippina (Boettger, 1897)

Distribution: Add Philippines (Calauit I.), Leviton et al. (2018).

Liopeltis rappii (Günther, 1860)

Distribution: Add Nepal (Chitwan, Khotang, Palpa, Sankhuwasabha, Tehrathum), and upper elevation of 2972 m, Bhattarai et al. (2017d).

Liopeltis stoliczkae (Sclater, 1891)

Distribution: Add India (Mizoram), Hauser (2019); Thailand (Chiang Mai, Loei, Mae Hong Son, Phetchabun, Phrae, Tak), and upper elevation of 1150 m, Hauser (2019); Vietnam (Lam Dong), Hauser (2018).

Liopeltis tricolor (Schlegel, 1837)

Distribution: Add West Malaysia (Pulau Tioman), K.K.P. Lim & Lim (1999). Remove Vietnam (Lam Dong), Hauser (2018).

LIOPHIDIUM Boulenger, 1896 (Pseudoxyrhophiidae)

Liophidium apperti Domergue, 1984
Liophidium chabaudi Domergue, 1984
Liophidium maintikibo Franzen, Jones, Raselimanana, Nagy, Cruze, Glaw & Vences, 2009
Liophidium mayottensis W.C.H. Peters, 1874. Monatsb. Preuss. Akad. Wiss. Berlin 1874: 793–794. (*Ablabes* [*Enicognathus*] *rhodogaster mayottensis*)

Syntypes: ZMB 8024 (n=2), young examples.

Type locality: "Mayotte," Comoro Islands.

Distribution: Comoro Islands.

Comments: This species was inadvertently neglected by Wallach et al. (2014).

Liophidium pattoni Vieites, Ratsoavina, Randrianianina, Nagy, Glaw & Vences, 2010

Comments: Miinala (2011) reports an additional specimen and locality.

Liophidium rhodogaster (Schlegel, 1837)
Liophidium therezieni Domergue, 1984

Liophidium torquatum (Boulenger, 1888)
Liophidium trilineatum Boulenger, 1896
Liophidium vaillanti (Mocquard, 1901)

LIOPHOLIDOPHIS Mocquard, 1904 (Pseudoxyrhophiidae)

Comments: Glaw et al. (2014) review systematics and produce a DNA sequence-based phylogeny of the genus, which partitions four species each in two clades.

Liopholidophis baderi Glaw, Kucharzewski, Nagy, Hawlitschek & Vences, 2014. Organisms Diversity & Evolution 14: 125–127, fig. 2.

Holotype: ZFMK 62235, a 283 mm male (F. Glaw, 31 January 1996).

Type locality: "on a trail near Hotel Feon'ny Ala (18° 56.845'S, 48° 25.078'E, ca. 940 m a. s. l.), at ca. 2.5 km distance from the village of Andasibe, central eastern Madagascar."

Distribution: EC Madagascar (Toamasina), 940 m. Known only from two specimens.

Liopholidophis dimorphus Glaw, Nagy, Franzen & Vences, 2007a

Liopholidophis dolicocercus (Peracca, 1892)
Liopholidophis grandidieri Mocquard, 1904

Liopholidophis oligolepis Glaw, Kucharzewski, Nagy, Hawlitschek & Vences, 2014. Organisms Diversity & Evolution 14: 127–129, figs. 3, 4.

Holotype: ZSM 153/2005 (field number FGZC 2796), a 234 mm female (F. Glaw, M. Vences & R.D. Randrianiaina, 15 February 2005).

Type locality: "Marojejy National Park, near a campsite locally known as 'Camp Mantella' (14°26.260'S, 49°46.533'E; 481 m a.s.l.), northeastern Madagascar."

Distribution: NE Madagascar (Antsiranana), 481 m. Known only from the holotype.

Liopholidophis rhadinaea Cadle, 1996
Liopholidophis sexlineatus (Günther, 1882)
Liopholidophis varius (J.G. Fischer, 1884)

LIOTYPHLOPS W.C.H. Peters, 1881 (Anomalepididae)

Liotyphlops albirostris (W.C.H. Peters, 1857)

Distribution: Add Panama (Panama Oeste), Ray & Ruback (2015).

Liotyphlops anops (Cope, 1899)
Liotyphlops argaleus Dixon & Kofron, 1984
Liotyphlops beui (Amaral, 1924)

Distribution: Add Brazil (Santa Catarina), F.J.M. Santos & Reis (2018).

Liotyphlops caissara Centeno, Sawaya & Germano, 2010
Liotyphlops haadi Silva-Haad, Franco & Maldonado, 2008
Liotyphlops schubarti Vanzolini, 1948
Liotyphlops sousai F.J.M. Santos & Reis, 2018. Copeia 106(3): 507–508, figs. 2, 3.

Holotype: UFRGS 6274, a 191 mm specimen (S. Leonardi, January 2012).

Type locality: "Brazil, Santa Catarina State, Municipality of Passos Maia, Passos Maia small hydroelectric power plant, 26°42′14″S, 51°55′05″W."

Distribution: SE Brazil (Santa Catarina). Known only from the holotype.

Liotyphlops taylori F.J.M. Santos & Reis, 2018. Copeia 106(3): 508–510, figs. 5, 6.

Holotype: MZUSP 14975, a 239 mm specimen (C. Nogueira, 22 October 2002).

Type locality: "Brazil, Mato Grosso state, Municipality of Porto Estrela, Serra das Araras Ecological Station, 15°38′31″S, 57°11′23″W."

Distribution: C Brazil (Mato Grosso). Known only from the holotype.

Liotyphlops ternetzii (Boulenger, 1896)

Distribution: Add Brazil (Mato Grosso do Sul), Ferreira et al. (2017); Brazil (Rio Grande do Sul), F.J.M. Santos & Reis (2018).

Liotyphlops trefauti Freire, Caramaschi & Argôlo, 2007

Distribution: Add Brazil (Pernambuco), Abegg et al. (2017a).

Liotyphlops wilderi (Garman, 1883)

Distribution: Add Brazil (Bahia), F.J.M. Santos & Reis (2018); Brazil (Mato Grosso do Sul), Ferreira et al. (2017).

†*LITHOPHIS* Marsh, 1871 (Booidea: incerta sedis)

†*Lithophis sargenti* Marsh, 1871

LOVERIDGELAPS McDowell, 1970 (Elapidae)

Loveridgelaps elapoides (Boulenger, 1890)

LOXOCEMUS Cope, 1861 (Loxocemidae)

Loxocemus bicolor Cope, 1861

Distribution: Add Nicaragua (Rivas: Isla Ometepe), Stark et al. (2014). Upper elevation of 1200 m, Palacios-Aguilar & Flores-Villela (2018).

†*LUNAOPHIS* Albino, Carrillo-Briceño & Neenan, 2016. PeerJ 2016(2027): 8. (Serpentes: incerta sedis)

Type species: *Lunaophis aquaticus* Albino, Carrillo-Briceño & Neenan, 2016 by monotypy.

Distribution: Late Cretaceous of Venezuela.

Comment: Oldest known snake from northern South America. Family status is not given, but some similarities to the simoliophiids exist.

†*Lunaophis aquaticus* Albino, Carrillo-Briceño & Neenan, 2016. PeerJ 2016(2027): 8–14, figs. 4–11.

Holotype: MCNC 1827, vertebral remains in a small block of black shale, which belongs to a single individual.

Type locality: "Cement quarry (Cementos Andinos company), located east of Lake Maracaibo, 10 km northeast of Monay, Trujillo State, Venezuela…The fossiliferous horizon is a black shale layer ∼28 m above the base of the La Aguada Member of the La Luna Formation (Cenomanian…)."

Distribution: Late Cretaceous (Cenomanian) of Venezuela. Known only from the type locality.

LYCODON H. Boie *in* Fitzinger, 1826 (Colubridae: Colubrinae)

Synonyms: Add *Dinodon* A.M.C. Duméril, Bibron & Duméril, 1854, *Eumesodon* Cope, 1860, *Lepidocephalus* Hallowell, 1861, *Lepturophis* Boulenger, 1900, *Adiastema* F. Werner, 1925.

Fossil records: Early Pleistocene of Japan (Okinawa), Ikeda et al. (2016, as *Dinodon* sp.).

Comments: Guo et al. (2013) present a DNA sequence-based phylogeny that shows *Dinodon* to be paraphyletic within a suite of *Lycodon* species. Also, there is no clear dichotomy in maxillary dentition that had been considered diagnostic between the two genera. For these reasons they synonymize *Dinodon* with *Lycodon*. The results of Guo et al. are supported by similar analyses by Siler et al. (2013), and by Lei et al. (2014) that include additional species, and show the existence of three or four species clades within *Lycodon* s.l. Wostl et al. (2018) produce a phylogeny of *Lycodon* and related genera based on one mtDNA gene. The phylogeny reveals two primary clades within *Lycodon*: 1) SE Asian *Lycodon* plus *Dinodon*, and 2) Philippine *Lycodon* plus *Lepturophis* and *Dryocalamus*. They recommend synonymizing *Dryocalamus* with *Lycodon*, but because their conclusions are based on results from only one mtDNA gene, we await further studies to make that taxonomic change. Zaher et al. (2019) produce a DNA sequence-based phylogeny that is similar to that of Wostl et al., but also includes *Stegonotus* in a position that emphasizes the paraphyly within *Lycodon* s.s.

Lycodon albofuscus (A.M.C. Duméril, Bibron & Duméril, 1854)

Distribution: Add West Malaysia (Penang), Wostl et al. (2018).

Comments: Figueroa et al. (2016), based on DNA sequence data, synonymize *Lepturophis* with *Lycodon* due to paraphyly.

Lycodon alcalai Ota & Ross, 1994

Distribution: Add Philippines (Babuyan Claro, Calayan, Camiguin Norte Is.), Leviton et al. (2018); Philippines (Sabtang I.), Siler et al. (2013).

Comments: Leviton et al. (2018) believe *L. alcalai* and *L. chrysoprateros* may be conspecific.

Lycodon anamallensis Günther, 1864. Rept. Brit. India: 318–319.

Synonyms: *Lycodon osmanhilli* Taylor, 1950.

Holotype: BMNH 1946.1.14.92, a 508 mm specimen (B.H. Beddome).

Type locality: "Anamallay Mountains," Tamil Nadu, India.

Distribution: India S of 21°N and Sri Lanka. India (Kerala, Tamil Nadu), Sri Lanka (Western).

Comments: Ganesh & Vogel (2018) revive *L. anamallensis* from *L. aulicus* on the basis of head shape and scale counts. They list some specimens that were examined, but do not detail the known distribution. Thus, some intermediate states in India (Goa, Orissa) may have one or both species, and the partition of distribution between the two species, if it exists, is unknown.

Lycodon aulicus (Linnaeus, 1758)

Synonyms: *Coluber scalaris* Gravenhorst, 1807, *Lycodon subfuscus* Cantor, 1839, *Lycodon atropurpureus* Cantor, 1839, *Lycodon aulicus oligozonatus* Wall, 1909.

Distribution: SC Asia. Pakistan (Punjab, Sindh), Nepal (Bardiya, Bhojpur, Chitwan, Jhapa, Kanchanpur, Kaski, Kathmandu, Latlipur, Myagdi, Palpa Parsa, Rupandehi, Sindhupalchok), India (Andhra Pradesh, Assam, Bihar, Gujarat, Jammu, Karnataka, Kashmir, Kerala, Madhya Pradesh, Maharashtra, Manipur, Mizoram, Nagaland, Rajasthan, Sikkim, Tamil Nadu, Uttar Pradesh, West Bengal), Sri Lanka (Eastern, Sabaragamuwa), Bangladesh, Myanmar (Kachin, Sagaing, Shan, Tanintharyi, Yangon), Mascarenes (Mauritius, Reunion), NSL-2130 m. Add India (Arunachal Pradesh), Purkayastha (2018); India (Jharkhand), Vogel & Harikrishnan (2013); Bhutan (Punakha), Koirala et al. (2016); Bhutan (Trongsa, Wangdue Phodrang), Tshewang & Letro (2018); Nepal (Bara, Nawalparasi), Pandey et al. (2018); Nepal (Dolakha), Bhattarai & Neupane (2017); Myanmar (Ayeyarwady), Vogel & Harikrishnan (2013); Myanmar (Rakhine), Siler et al. (2013).

Comments: See under *L. anamallensis*.

Lycodon banksi Luu, Bonkowski, Nguyen, Le, Calame & Ziegler, 2018. Revue Suisse Zool. 125(2): 266–271, figs. 2–5.

Holotype: VNUF R2015.20, a 465+ mm male (V.Q. Luu & T. Calame, 4 April 2015).

Type locality: "Phou Hin Poun NPA, Hinboun District, Khammouane Province, central Laos, at an elevation of 167 m a.s.l."

Distribution: C Laos (Khammouane), 167 m. Known only from the holotype.

Lycodon bibonius Ota & Ross, 1994

Distribution: Add Philippines (Babuyan Claro I.), Leviton et al. (2018).

Lycodon butleri Boulenger, 1900

Lycodon capucinus H. Boie *in* F. Boie, 1827

Synonyms: Remove *Lycodon atropurpureus* Cantor, 1839, see comments under *L. aulicus*; *Tytleria hypsirhinoides* Theobald, 1868, now *Lycodon hypsirhinoides*.

Distribution: Add Thailand (Chanthaburi), Chan-ard et al. (2011); Cambodia (Phnom Penh, Siem Reap), Neang et al. (2014); Vietnam (Thu Dau Mot), Luu et al. (2019); West Malaysia (Johor), Siler et al. (2013); Indonesia (Ambon, Banda Besar, Bawean, Buru, Karimunjawa, Kisar, Selayar, Serua Islands), Lang (2013); Indonesia (Lombok), Vogel & Harikrishnan (2013); Indonesia (Papua), introduced, O'Shea et al. (2018c); Timor-Leste (Ataúro Island), H. Kaiser et al. (2013b); Philippines (Bohol I., Scmirara I.), Siler et al. (2013); Philippines (Camiguin I., Dinagat I.), Sanguila et al. (2016); Philippines (Carabao I., Sibuyan I., Tablas I.), Siler et al. (2012); Indonesia (Papua). Remove India (Andaman and Nicobars), Vogel & Harikrishnan (2013).

Comments: O'Shea et al. (2013) discuss the taxonomic history and geographic distribution.

Lycodon cardamomensis Daltry & Wüster, 2002

Distribution: Add Vietnam (Gia Lai), Vogel & David (2019); Vietnam (Phu Yen), Do et al. (2017).

Lycodon carinatus (Kuhl, 1820)

Lycodon cavernicolus Grismer, Quah, Anuar M.S., Muin, Wood & Nor, 2014. Zootaxa 3815(1): 56–58, figs. 3–5.

Holotype: LSUHC 9985, a 508 mm female (E.S.H. Quah & S.Anuar M.S., 12 March 2011).

Type locality: "Gua Wang Burma, Perlis State Park, Perlis, Peninsular Malaysia (6°41.594N 100°11.400E at 175 m in elevation."

Distribution: Peninsular Malaysia (Perlis), 175 m. Known only from the vicinity of the type locality.

Lycodon chrysoprateros Ota & Ross, 1994

Comments: Leviton et al. (2018) believe *L. alcalai* and *L. chrysoprateros* may be conspecific.

Lycodon davidi Vogel, Nguyen, Kingsada & Ziegler, 2012

Lycodon dumerilii (Boulenger, 1893)

Distribution: Add Philippines (Leyte I.), Siler et al. (2013); Philippines (Dinagat), Sanguila et al. (2016); Philippines (Samar, Siargao Is.), Leviton et al. (2018).

Comments: Michels & Bauer (2004) corrected the matronym of the synonym *Dryocalamus mccroryi* to *mccroryae*.

Lycodon effraenis Cantor, 1847

Distribution: Add West Malaysia (Kedah), Siler et al. (2013); West Malaysia (Pulau Tioman), K.K.P. Lim & Lim (1999).

Lycodon fasciatus (J. Anderson, 1879)

Distribution: Add India (Mizoram), Lalbiakzuala & Lalremsanga (2017); Myanmar (Mandalay, Yangon), David & Vogel (2019); Thailand (Nan), David & Vogel (2019); Laos (Champasak), Luu et al. (2018).

Lycodon fausti Gaulke, 2002

Lycodon ferroni Lanza, 1999

Lycodon flavicollis Mukherjee & Bhupathy, 2007

Distribution: Add India (Andhra Pradesh, Karnataka, Telangana), and elevation range of 175–850 m, Narayana et al. (2018). Upper elevation of 929 m, Muliya et al. (2018). Narayana et al. (2018) map known localities.

Lycodon flavomaculatus Wall, 1907

Distribution: Add India (Chhattsghar), Deshmukh et al. (2018); India (Gujarat, Madhya Pradesh), Sharma et al. (2015); India (Tamil Nadu), and low elevation of 425 m, Melvinselvan et al. (2016). Narayanan et al. (2017b) map known localities.

Lycodon flavozonatus (Pope, 1928)

Distribution: Add China (Guangxi), Wostl et al. (2018); Vietnam (Lao Cai), Luu et al. (2019).

Lycodon futsingensis (Pope, 1928)

Distribution: Add China (Guangdong, Zhejiang), Laos (Xiangkheang); Wostl et al. (2018); Vietnam (Bac Giang, Cao Bang, Da Nang), Hecht et al. (2013); Vietnam (Dien Bien), and upper elevation of 1057 m, Dung et al. (2014); Vietnam (Ha Giang), Pham et al. (2017); Vietnam (Ha Tinh), Luu et al. (2018); Vietnam (Hoa Binh), and upper elevation of 901 m,

Nguyen et al. (2018); Vietnam (Ninh Binh, Quang Tri), Luu et al. (2019); Vietnam (Vinh Phuc), Janssen et al. (2019).

Lycodon gammiei (Blandford, 1878)

Distribution: Elevation 1070–2300 m, Chettri & Bhupathy (2009).

Lycodon gibsonae Vogel & David, 2019. Zootaxa 4577(3): 518–521, figs. 1–4.

Holotype: FMNH 180146, an 1100+ mm male (W.R. Heyer, 5 May 1969).

Type locality: "the Sakaerat Environmental Research Station, Amphoe Wang Nam Khiao, at the south-western edge of Khorat Plateau, Nakhon Ratchasima Province, Thailand."

Distribution: C Thailand (Nakhon Ratchasima).

Lycodon gongshan Vogel & Luo, 2011

Lycodon hypsirhinoides Theobald, 1868. J. Asiatic Soc. Bengal 37: 66. (*Tytleria hypsirhinoides*)

Holotype: ZSI 8145, a 550+ mm adult (R.C. Tytler).

Type locality: "Andamans," India.

Distribution: India (Andaman and Nicobars: Little Andaman, Neil, South Andaman and Tarmugli Islands).

Comments: Vogel & Harikrishnan (2013) consider *hypsirhinoides* to be distinct from *L. capucinus* based on morphology.

Lycodon jara (G. Shaw, 1802)

Synonyms: Add ***Lycodon odishii*** S. Mallik, Parida, Mohanty, Mallik, Purohit, Mohanty, Nanda, Sindura, Purohit, Mishra & Sahou, 2014. Russian J. Herpetol. 21(3): 206–210, figs. 1–9.

Holotype: ZSI 25992, a 329 mm male (volunteers of Snake Helpline).

Type locality: "Jail Training School, Lanjipalli, Industrial Estate, Berhampur, Odisha, India (19°18′06.1″ N 84°48′53.7″ E)."

Distribution: Add India (Jharkhand), and upper elevation of 617 m, A.A. Khan & Sharma (2018); India (Manipur, Meghalaya, Mizoram), Chaudhuri et al. (2015); India (Odisha), S. Mallik et al. (2014); Nepal (Bara), Pandey et al. (2018); Nepal (Parsa), Bhattarai et al. (2018b).

Comments: Chaudhuri et al. (2015) compare specimens of *L. jara* with the type series of *L. odishii* and, via multivariate analysis, determine that characteristics of the latter fall within variation of *L. jara*.

Lycodon kundui M.A. Smith, 1943

Lycodon laoensis Günther, 1864

Distribution: Add Thailand (Chanthaburi), Chan-ard et al. (2011); Laos (Bolikhamxai, Champasak), Luu et al. (2018); Laos (Salavan), Siler et al. (2013); Cambodia (Phnom Penh, Pursat), Neang et al. (2014).

Lycodon liuchengchaoi Zhang, Jiang, Vogel & Rao, 2011

Lycodon mackinnoni Wall, 1906

Distribution: Add India (Jammu and Kashmir), Manhas et al. (2018b).

Lycodon meridionalis Bourret, 1935

Distribution: Add Laos (Houaphan), Luu et al. (2018); Vietnam (Bac Giang), Hecht et al. (2013); Vietnam (Bac Kan), Luu et al. (2019); Vietnam (Hai Phong: Cat Ba Island), T.Q. Nguyen et al. (2011); Vietnam (Hoa Binh), Nguyen et al. (2018); Vietnam (Quang Ninh), Gawor et al. (2016); Vietnam (Thanh Hoa), Luu et al. (2018).

Lycodon muelleri A.M.C. Duméril, Bibron & Duméril, 1854

Distribution: Add Philippines (Marinduque I.), Leviton et al. (2018).

Lycodon multifasciatus (Maki, 1931)

Lycodon multizonatus (Zhao & Jiang, 1981)

Holotype: Corrected to CIB 009964, Guo et al. (2012a).

Comments: Transferred from *Oligodon* on the basis of a DNA sequence analysis by Lei et al. (2014).

Lycodon namdongensis Luu, Ziegler, Ha, Le & Hoang, 2019. Zootaxa 4586(2): 267–271, figs. 2–4.

Holotype: VNUF R2017.23, a 723 mm male (V.Q. Luu, N.V. Ha, O.V. Lo & N.V. Ha, 13 June 2017).

Type locality: "Nam Dong Nature Reserve, Quan Son District, Thanh Hoa Province, Vietnam (20°18.298′N; 104°54.776′E, at an elevation of 616 m a.s.l.)."

Distribution: N Vietnam (Thanh Hoa), 616 m. Known only from the holotype.

Lycodon ophiophagus Vogel, David, Pauwels, Sumontha, Norval, Hendrix, Vu & Ziegler, 2009

Lycodon orientalis (Hilgendorf, 1880)

Lycodon paucifasciatus Rendahl *in* M.A. Smith, 1943

Distribution: Add Vietnam (Hue), Luu et al. (2019).

Lycodon pictus Janssen, Pham, Ngo, Le, Nguyen & Ziegler, 2019. ZooKeys 875: 5–14, figs. 2–6.

Holotype: IEBR 4166, a 597+ mm male (T.Q. Nguyen et al., 18 April 2012).

Type locality: "(altitude 701 m a.s.l.), Trung Khanh District, Cao Bang Province," Vietnam.

Distribution: N Vietnam (Cao Bang), 588–701 m. Known from two specimens.

Lycodon rosozonatus (Hu & Zhao *in* Zhao, 1972)

Holotype: Corrected to CIB 009081, Guo et al. (2012a).

Distribution: Add Vietnam (Ha Tinh, Nghe An), Luu et al. (2019).

Lycodon rufozonatus Cantor, 1842

Distribution: Add China (Nan Ao Island), Qing et al. (2015); Vietnam (Ninh Binh), Luu et al. (2019).

Lycodon ruhstrati (J.G. Fischer, 1886)

Distribution: Add China (Hainan), Siler et al. (2013); Laos (Khammouane), Luu et al. (2019); Vietnam (Ninh Binh, Quang Ninh), Luu et al. (2019).

Lycodon sealei Leviton, 1955. Philippine J. Sci. 84(2): 195–198. (*Lycodon subcinctus sealei*)

Holotype: CAS 15819, a 789 mm male (A. Seale, 20 May 1908).

Type locality: "Puerto Princesa, Palawan Island, Philippines."

Distribution: Philippines (Palawan).

Comments: Leviton et al. (2018) elevate this taxon to species due to substantial genetic difference from Malaysia and

Thailand samples of *L. subcinctus*, as demonstrated by Siler et al. (2013).

Lycodon semicarinatus (Cope, 1860)

Lycodon septentrionalis (Günther, 1875)

Distribution: Add Bhutan (Trongsa), Tshewang & Letro, 2018); Laos (Phongsali), Luu et al. (2018); Cambodia (Ratanakiri), Neang et al. (2014); Vietnam (Hue), Luu et al. (2019).

Lycodon sidiki Wostl, Hamidy, Kurniawan & E.N. Smith, 2017. Zootaxa 4276(4): 547–551, figs. 2, 3.

Holotype: MZB 5980, a 715 mm male (E. Wostl, I. Fonna & M. Iksan, 5 August 2015).

Type locality: "between Takengon and Isaq, Aceh Province, Sumatra 04.50611°N, 96.86061 °E, 1614 m asl."

Distribution: Indonesia (N Sumatra), 1614 m. Known only from the holotype.

Lycodon solivagus Ota & Ross, 1984

Lycodon stormi Boettger, 1892

Lycodon striatus (Shaw, 1802)

Distribution: Add Iran (Hormozgan), Shafaei-Mahroo et al. (2015); India (Gujarat), Patel & Vyas (2019); Nepal (Chitwan, Nawalparasi), Pandey et al. (2018); Sri Lanka (Sabaragamuwa), Peabotuwage et al. (2012).

Lycodon subcinctus F. Boie, 1827

Distribution: Add India (Great Nicobar I., Andaman & Nicobars), Harikrishnan et al. (2010); Laos (Bolikhamxai, Champasak, Vientiane), Luu et al. (2018); Laos (Khammouane), Luu et al. (2019); Vietnam (Dong Nai), Janssen et al. (2019); Vietnam (Ninh Binh), Luu et al. (2019); West Malaysia (Terengganu), Sumarli et al. (2015); Indonesia (Bengkulu), Wostl et al. (2018). Remove Philippines (Palawan), elevated to species, *L. sealei*, by Leviton et al. (2018).

Lycodon synaptor Vogel & David, 2010

Lycodon tessellatus Jan, 1863

Comments: Leviton et al. (2018) hesitatingly refer *L. tessellatus* to the synonymy of *L. capucinus*, although acknowledge a color pattern difference.

Lycodon tiwarii Biswas & Sanyal, 1965

Distribution: Add India (Car Nicobar I., Andaman & Nicobars), Harikrishnan et al. (2010).

Lycodon travancoricus (Beddome1870)

Distribution: Add India (Gujarat), 129 m, Sharma & Jani (2015).

Lycodon zawi Slowinski, Pawar, Win, Thin, Gyi, Oo & Tun, 2001

Distribution: Add India (Arunachal Pradesh), Purkayastha (2018); India (West Bengal), Ghosh et al. (2017). D. Dutta et al. (2013) report range extensions and an altitude record of 750 m for NE India.

Lycodon zoosvictoriae Neang, Hartmann, Hun, Souter & Furey, 2014. Zootaxa 3814(1): 70–73, figs. 2–4, 6.

Holotype: CBC 02238, a 521 mm female (T. Neang & S. Hun, 15 June 2013).

Type locality: "N12°09′22.0″, E102°59′18.7″, 1,284 m above sea level (a.s.l.) in Phnom Samkos Wildlife Sanctuary, Cardamom Mountains, Pursat Province, southwest Cambodia."

Distribution: Cambodia (Pursat), 1284 m. Known only from the holotype.

LYCODONOMORPHUS Fitzinger, 1843 (Lamprophiidae)

Lycodonomorphus bicolor (Günther, 1894)

Lycodonomorphus inornatus (A.M.C. Duméril, Bibron & Duméril, 1854)

Lycodonomorphus laevissimus (Günther, 1862)

Lycodonomorphus leleupi (Laurent, 1950)

Lycodonomorphus mlanjensis Loveridge, 1953

Lycodonomorphus obscuriventris V.F.M. FitzSimons, 1964

Lycodonomorphus rufulus (Lichtenstein, 1853)

Lycodonomorphus subtaeniatus Laurent, 1954

Lycodonomorphus whytii (Boulenger, 1897)

LYCODRYAS Günther, 1879 (Pseudoxyrhophiidae)

Synonyms: Add *Stenophis* Boulenger, 1896.

Comments: Vences et al. (2004) review most of the species as the subgenus *Stenophis* of *Stenophis*. Nagy et al. (2010), using DNA sequence-data, determine that *Stenophis* (*Stenophis*) and *Lycodryas* are congeneric.

Lycodryas carleti (Domergue, 1995)

Comments: Vences et al. (2004) provide a description, and note that this taxon may represent *L. gaimardii*, differing only by lower ventral and subcaudal counts.

Lycodryas citrina (Domergue, 1995)

Comments: Vences et al. (2004) provide a description and color photographs.

Lycodryas cococola Hawlitschek, Nagy & Glaw, 2012

Lycodryas gaimardii (Schlegel, 1837)

Distribution: Add Madagascar (Taomasina), Vences et al. (2004).

Comments: Vences et al. (2004) provide a description and color photographs.

Lycodryas granuliceps (Boettger, 1877)

Distribution: Add Madagascar (Nosy Komba), Roberts & Daly (2014).

Comments: Vences et al. (2004) provide a description and color photographs.

Lycodryas guentheri (Boulenger, 1896)

Comments: Vences et al. (2004) provide a description.

Lycodryas inopinae (Domergue, 1995)

Comments: Vences et al. (2004) provide a description and color photographs.

Lycodryas inornata (Boulenger, 1896)

Comments: Vences et al. (2004) provide a description.

Lycodryas maculata (Günther, 1858)

Lycodryas pseudogranuliceps (Domergue, 1995)

Comments: Vences et al. (2004) provide a description and color photographs.

LYCOGNATHOPHIS Boulenger, 1893 (Natricidae)

Lycognathophis seychellensis (Schlegel, 1837)

Distribution: Add Seychelles (La Digue), S. Rocha et al. (2009).

LYCOPHIDION Fitzinger, 1843 (Lamprophiidae)

Lycophidion acutirostre Günther, 1868

Lycophidion albomaculatum Steindachner, 1870

Distribution: Add Sierra Leone, Trape & Baldé (2014). Upper elevation of 395 m, Trape & Mané (2017).

Lycophidion capense (A. Smith, 1831)

Distribution: South Africa (Western Cape, record from De Hoop Nature Preserve), Fantuzzi (2016).

Lycophidion depressirostre Laurent, 1968

Lycophidion hellmichi Laurent, 1964

Distribution: In Namibia it is confined to Kunene; other localities are referred to *L. namibianum*, Herrmann & Branch (2013).

Lycophidion irroratum (Leach *in* Bowdich, 1819)

Distribution: Low elevation of 386 m, Trape & Mané (2017).

Lycophidion laterale Hallowell, 1857

Distribution: Add Angola (Cuanza Norte), M.P. Marques et al. (2018).

Lycophidion meleagre Boulenger, 1893

Distribution: Add Angola (Bengo, Cuanza Sul, Luanda), M.P. Marques et al. (2018).

Lycophidion multimaculatum Boettger, 1888

Distribution: Add Nigeria, Trape & Baldé (2014); Cameroon (Nord-Ouest), Ineich et al. (2015); Angola (Luanda, Lunda Sul, Zaire), M.P. Marques et al. (2018).

Lycophidion namibianum Broadley, 1991

Distribution: Add Angola (Namibe), Herrmann & Branch (2013).

Lycophidion nanum (Broadley, 1968)

Lycophidion nigromaculatum (Schlegel *in* W.C.H. Peters, 1863)

Distribution: Add Nigeria, Trape & Baldé (2014).

Lycophidion ornatum Parker, 1936

Lycophidion pembanum Laurent, 1968

Lycophidion pygmaeum Broadley, 1996

Lycophidion semiannule W.C.H. Peters, 1854

Lycophidion semicinctum A.M.C. Duméril, Bibron & Duméril, 1854

Distribution: Lower elevation of 193 m, Trape & Mané (2015).

Lycophidion taylori Broadley & Hughes, 1993

Lycophidion uzungwense Loveridge, 1932

Lycophidion variegatum Broadley, 1969

LYGOPHIS Fitzinger, 1843 (Dipsadidae: Xenodontinae)

Lygophis anomala (Günther, 1858)

Distribution: Add Paraguay (unspecified province), Panzera et al. (2017); Uruguay (Colonia, Lavalleja), Panzera & Maneyro (2014). Panzera et al. (2017) map known localities.

Lygophis dilepis Cope, 1862

Distribution: Upper elevation of 850 m, Guedes et al. (2014).

Lygophis elegantissimus (Koslowsky, 1896)

Lygophis flavifrenatus Cope, 1862

Lygophis lineatus (Linnaeus, 1758)

Distribution: Add Colombia (Cesar), Medina-Rangel (2011).

Lygophis meridionalis (Schenkel, 1901)

Distribution: Add Brazil (Bahia, Tocantins), Dal Vechio et al. (2016) and T.M. Castro & Oliveira (2017); Brazil (Maranhão), J.P. Miranda et al. (2012); Brazil (Marajo I., Para), G.M. Rodrigues et al. (2015).

Lygophis paucidens Hoge, 1953

Distribution: Add Brazil (Ceara), Roberto & Loebmann (2016); Brazil (Maranhao), M.B. Silva et al. (2016); Brazil (Tocantins), Dal Vechio et al. (2016); Paraguay (San Pedro), Cacciali et al. (2013).

Lygophis vanzolinii (Dixon, 1985)

LYTORHYNCHUS W.C.H. Peters, 1862 (Colubridae: Colubrinae)

Comments: Based on the published description and locality, *Lytorhynchus levitoni* Torki, 2017a is identical to *Rhynchocalamus ilamensis* Fathinia, Rastegar-Pouyani, Rastegar-Pouyani & Darvishnia, 2017. Based on stated publication dates, *levitoni* has priority over *ilamensis* by three months. Fathinia et al. (2017) use DNA sequence data to show that the taxon belongs within *Rhynchocalamus*.

Lytorhynchus diadema (A.M.C. Duméril, Bibron & Duméril, 1854)

Distribution: Add Morocco (Figuig, Tiznit), Damas-Moreira et al. (2014), Jablonski et al. (2014); Mauritania (Nouakchott), Padial (2006); Algeria (El Oued), Mouane et al. (2014); Algeria (Tindouf), Donaire et al. (2000); Libya (Al Wahat, Butnan, Jabal al Gharbi, Jufra, Kufrah, Marj, Misratah, Murqub, Nalut, Sabha), Bauer et al. (2017); Jordan (Aqaba), Sindaco et al. (2014); Iran (Hormozgan, Ilam), Shafaei-Mahroo et al. (2015); Iran (Kerman), Moradi et al. (2013). Mediani et al. (2015a) report new records from C Western Sahara.

Lytorhynchus gasperetti Leviton, 1977

Lytorhynchus kennedyi K.P. Schmidt, 1939

Lytorhynchus maynardi Alcock & Finn, 1897

Distribution: Add Iran (Baluchistan and Sistan), Shafiei et al. (2015). Low elevation of 500 m, Salemi et al. (2018).

Lytorhynchus paradoxus (Günther, 1875)

Comments: Agarwal & Srikanthan (2013) discuss additional Indian specimens.

Lytorhynchus ridgewayi Boulenger, 1887

Distribution: Add Iran (Alborz, Bushehr, Golestan, Hormozgan, Isfahan, Markazi, North Khorasan, Qazvin, Semnan, South Khorasan, Tehran), Safaei-Mahroo et al. (2015); Iran (Qom), S.M. Kazemi et al. (2015); Iran (Razavi Khorasan), Yousefkhani et al. (2014) and Nasrabadi et al. (2016).

MACRELAPS Boulenger, 1896 (Atractaspididae: Aparallactinae)

Macrelaps microlepidotus (Günther, 1860)

MACROCALAMUS Günther, 1864 (Calamariidae)

Comments: Quah et al. (2019b) provide a review of the genus, including a key to the species. They produce a DNA sequence-based phylogeny using six of the species.

Macrocalamus chanardi David & Pauwels, 2004

Distribution: Add West Malaysia (Kedah), and low elevation of about 800 m, Quah et al. (2019b).

Comments: Quah et al. (2019b) provide a revised description, locality map, and color photographs. Based on DNA analysis, the species occupies three genetically distinct, allopatric sites, but morphology is conserved.

Macrocalamus emas Quah, Anuar, Grismer, Wood & Azizah, 2019b. Zool. J. Linnean Soc. 188: 7–13, figs. 4–6.

Holotype: USMHC 1866, a 241 mm male (E.S.H. Quah, 20 August 2015).

Type locality: "Gunung Brinchang, Cameron Highlands, Pahang, West Malaysia (04°31.105N 101°22.571E; at 1811 m a.s.l.)."

Distribution: West Malaysia (Pahang), about 1500 to over 1800 m.

Macrocalamus gentingensis Norsham & Lim, 2003

Comments: Quah et al. (2019b) provide a revised description and color photographs.

Macrocalamus jasoni Grandison, 1972

Comments: Quah et al. (2019b) provide a description and color photographs of the types.

Macrocalamus lateralis Günther, 1864

Distribution: Add West Malaysia (Langkawi Island, Kedah, Perak), and upper elevation of 800 m, Quah et al. (2019b).

Comments: Quah et al. (2019b) provide a revised description, locality map and color photographs.

Macrocalamus schulzi Vogel & David, 1999

Comments: Quah et al. (2019b) provide a revised description and color photographs.

Macrocalamus tweediei Lim, 1963

Distribution: Limited to West Malaysia (Pahang, only on Gunong Brinchang between 1500–1800 m), other records based on misidentifications, Quah et al. (2019b).

Comments: Quah et al. (2019b) provide a revised description and color photographs.

Macrocalamus vogeli David & Pauwels, 2004

Comments: Quah et al. (2019b) provide a description and color photographs of the holotype.

MACROPROTODON Guichenot, 1850 (Colubridae: Colubrinae)

Macroprotodon abubakeri Wade, 2001

Macroprotodon brevis (Günther, 1862)

Distribution: Add Portugal (Faro), Malkmus (2011); Portugal (Guarda), Malkmus & Loureiro (2010); Spain (Granada), Pleguezuelos (1989); Morocco (Tan Tan), Kane et al. (2019). Mediani et al. (2015b) map N Morocco records. See under *M. mauritanicus.*

Fossil Records: Add Upper Pleistocene (Tarantian) of Spain, Barroso-Ruiz and Bailon (2003).

Macroprotodon cucullatus (Geoffroy-Saint-Hilaire, 1827 *in* Savigny, 1809–1829)

Distribution: Add Algeria (Tiaret), Ferrer et al. (2016); Morocco (Taza), Mediani et al. (2015b); Libya (Al Wahat, Jabal al Gharbi, Marj, Nalut, Tripoli), Bauer et al. (2017).

Macroprotodon mauritanicus Guichenot, 1850

Distribution: Add Algeria (El Tarf), Rouag & Benyacoub (2006).

Comments: Mateo (2015) states that populations on the Baleares, Spain, are this species, not *M. brevis*. Silva-Rocha et al. (2015) use DNA sequence-data to determine that the source population for specimens introduced to the Balearic Islands was probably North Africa.

MACROVIPERA A.F.T. Reuss, 1927 (Viperidae: Viperinae)

Fossil records: Add middle Miocene (Orleanian) of Germany, Čerňanský et al. (2017, as *Vipera* sp. "Oriental viper"); Add Miocene/Pliocene transition (late Turolian) of Greece, Georgialis et al. (2019, a "Oriental Vipers").

†**Macrovipera burgenlandica** (Bachmayer & Szyndlar, 1987)

†**Macrovipera gedulyi** (Bolkay, 1913)

†**Macrovipera kuchurganica** (Zerova *in* Zerova, Lungu & Chkhikvadze, 1987)

Macrovipera lebetina (Linnaeus, 1758)

Distribution: Add Iran (Alborz, Ardabil, Kohgiluyeh and Boyer Ahmad, North Khorasan, Qazvin), Safaei-Mahroo et al. (2015); Iran (Bushehr, Golestan, Ilam, Kermanshah, Markazi, Sistan and Baluchistan, Tehran), Moradi et al. (2014); Iran (Qom), S.M. Kazemi et al. (2015). Some of the S Iran provinces likely represent *M. razii*. Coşkun et al. (2012) document a westward range extension to Sivas, Turkey. Mebert et al. (2015) document localities on the northern edge of its distribution in NE Turkey. Moradi et al. (2013) state an upper elevation range of 3000 m in Iran.

Comments: Moradi et al. (2014) find a general, north-south cline in Iran of external morphological characters, and they suggest that populations in S Iran are a provisional distinct species, which they do not describe. Oraie et al. (2018) use mtDNA sequence-data to confirm the southern populations as a distinct species, *M. razii*, and find two clades in northern populations referable to two subspecies of *M. lebetina*.

Macrovipera razii Oraie, E. Rastegar-Pouyani, Khosravani, Moradi, Akbari, Sehhatisabet, Shafiei, Stümpel & Joger, 2018. Salamandra 54(4): 241–243, figs. 5–9.

Holotype: SUHC 143, a 1270 mm male (E. Rastegar-Pouyani, 3–4 June 2004).

Type locality: "at 105 km on the road from Jiroft to Bam near Bab-Gorgi village and Valley, Kerman Province, 29°05′054″ N, 57°34′120″ E; altitude 3150 m."

Distribution: S Iran (Fars, Kerman, Yazd), 1500–3150 m.

Macrovipera schweizeri (F. Werner, 1935)

†*Macrovipera ukrainica* (Zerova, 1992)

MADAGASCAROPHIS Mertens, 1952
(Pseudoxyrhophiidae)

Comments: Ruane et al. (2016) describe a new species with comparison to other species in the genus, and produce a phylogeny and key to members of the genus.

Madagascarophis colubrinus (Schlegel, 1837)

Comments: Glaw et al. (2013) evaluate morphological data of various populations, and conclude that three subspecies are recognizable (*colubrinus, citrinus, septentrionalis*).

Madagascarophis fuchsi Glaw, Kucharzewski, Köhler, Vences & Nagy, 2013. Zootaxa 3630(2): 324–327, figs. 2, 4, 5.

Holotype: ZSM 2130/2007 (field no. FGZC 1152), a 514 mm female (F. Glaw, J. Köhler, H. Enting, P. Bora & A. Knoll, 27 February 2007).

Type locality: "close to the remains of the French Fort, Montagne des Français (12°19′34″S, 49°20′09″E, ca. 300 m above sea level), Antsiranana province, northern Madagascar."

Distribution: Extreme N Madagascar (Antsiranana), 300 m.

Madagascarophis lolo Ruane, Burbrink, Randriamahatantsoa & Raxworthy, 2016. Copeia 2016(3): 716–718, figs.1, 3a, 4, 5a.

Holotype, AMNH 176422, a 491 mm adult male (B. Randriamahatantsoa, C. Raxworthy & S. Ruane, 9 February 2014).

Type locality: "Madagascar, Antsiranana Province, Diana Region, Ankarana National Park, ~4 km northwest of the village of Mahamasina, tsingy karst trail, 102 m elevation, 49.11507°E, 12.94210°S."

Distribution: N Madagascar (extreme NW Antsiranana Province).

Madagascarophis meridionalis Domergue, 1987

Madagascarophis ocellatus Domergue, 1987

Comments: Glaw et al. (2013) illustrate and redescribe the holotype.

MADATYPHLOPS Hedges, Marion, Lipp, Marin & Vidal, 2014. Caribbean Herpetol. 49: 42. (Typhlopidae: Madatyphlopinae)

Synonyms: *Lemuriatyphlops* Pyron & Wallach, 2014.

Type species: *Onychocephalus arenarius* Grandidier, 1872 by original designation.

Distribution: NE Africa, Comoros, and Madagascar.

Comments: Hedges et al. (2014), Pyron & Wallach (2014) provide generic diagnoses. Nagy et al. (2014) were unable to recover a monophyletic *Lemuriatyphlops* in their phylogeny of Madagascar typhlopids, and recommend that it be synonymized with *Madatyphlops*.

Madatyphlops albanalis (Rendahl, 1918)

Comments: Pyron & Wallach (2014) provide a revised diagnosis and description.

Madatyphlops andasibensis (Wallach & Glaw, 2009)

Madatyphlops arenarius (Grandidier, 1872)

Comments: Nagy et al. (2015) note that, based on analysis of DNA sequence-data for 16 specimens, there are multiple cryptic species within *M. arenarius*.

Madatyphlops boettgeri (Boulenger, 1893)

Madatyphlops calabresii (Gans & Laurent, 1965)

Comments: Hedges et al. (2014) recommend this and several other species to *Afrotyphlops*, but Pyron & Wallach (2014) argue, on the basis of internal and external morphology, that they are assignable to *Madatyphlops*. Nagy et al. (2015) agree with Hedges et al., recommending a reversal of Pyron & Wallach on the basis of snout shape and potential complexity of biogeographic history.

Madatyphlops comorensis (Boulenger, 1889)

Comments: See under *M. calabresii*.

Madatyphlops cuneirostris (W.C.H. Peters, 1879)

Comments: See under *M. calabresii*.

Madatyphlops decorsei (Mocquard, 1901)

Madatyphlops domerguei (Roux-Estève, 1980)

Comments: See under *M. reuteri*.

Madatyphlops leucocephalus (H.W. Parker, 1930)

Comments: Based on morphological characters, Hedges et al. (2014) assign *leucocephalus* to *Rhinotyphlops*, but Pyron & Wallach (2014) argue that it most closely resembles members of *Madatyphlops*.

Madatyphlops madagascariensis (Boettger, 1877)

Madatyphlops microcephalus (F. Werner, 1909)

Comments: See under *M. reuteri*.

Madatyphlops mucronatus (Boettger, 1880)

Madatyphlops ocularis (H.W. Parker, 1927)

Madatyphlops platyrhynchus (Sternfeld, 1910)

Comments: See under *M. calabresii*.

Madatyphlops rajeryi (Renoult & Raselimanana, 2009)

Madatyphlops reuteri (Boettger, 1881)

Comments: Hedges et al. (2014) assign *domerguei, microcephalus* and *reuteri* to *Madatyphlops* on the basis of geography and morphology.

†*MADTSOIA* Simpson, 1933 (†Madtsoiidae)

†*Madtsoia bai* Simpson, 1933

†*Madtsoia camposi* Rage, 1998

†*Madtsoia madagascariensis* Hoffstetter, 1961

†*Madtsoia pisdurensis* Mohabey, Head & J.A. Wilson, 2011

MAGLIOPHIS Zaher, Grazziotin, Cadle, R.W. Murphy, Moura-Leite & Bonatto, 2009 (Dipsadidae: Xenodontinae)

Magliophis exiguus (Cope, 1862)

Magliophis stahli (Stejneger, 1904)

MALAYOPYTHON Reynolds, Niemiller & Revell, 2014b. Molec. Phylogen. Evol. 71: 211. (Pythonidae)

Synonyms: *Broghammerus* Hoser, 2004, a *nomen nudum* according to Reynolds et al. (2014b).

Type species: *Constrictor schneideri* Wagler, 1833 (=*Boa reticulata* Schneider, 1801), designated by Reynolds et al. (2014b).

Distribution: SE Asia and East Indies.

Comments: Recognition of the generic distinctiveness of *reticulatus* and *timoriensis* is supported by genetic data (Reynolds et al., 2014b). Barker et al. (2015) provide a morphological diagnosis.

Malayopython reticulatus (Schneider, 1801)

Distribution: Add Myanmar (Rakhine, Lampi I.), Platt et al. (2018); Thailand (Chanthaburi), Chan-ard et al. (2011); Thailand (Koh Man Nok I., Rayong), Chan-ard and Makchai (2011); Cambodia (Siem Reap), Geissler et al. (2019); West Malaysia (Jerejak Island), Quah et al. (2011); West Malaysia (Pahang), Zakaria et al. (2014); West Malaysia (Pulau Singa Besar), B.L. Lim et al. (2010); Indonesia (Bisa, Gorong, Tidore Islands), Lang (2013); Indonesia (Salibabu Island), A. Koch et al. (2009); Philippines (Bongao, Cagayan, Cagauit, Dalupiri, Marinduque, Siargao, Siasi, Siquijor Is.), Leviton et al. (2018); Philippines (Romblon, Tablas I.), Sy & Tan (2015b). Kalki et al. (2018) discuss the validity of mainland India records, concluding that one from West Bengal may be legitimate.

Comments: Murray-Dickson et al. (2017) use mtDNA sequence-data to evaluate haplotypes from throughout the geographic range. Populations east of Wallace's line, plus the Philippines, are distinct from western populations in possessing unique haplotypes. The sample from Ambon suggests that the island was recently populated with snakes from Singapore. Auliya (2006) discusses geographic distribution and variation.

Malayopython timoriensis (W.C.H. Peters, 1876)

MALAYOTYPHLOPS Hedges, Marion, Lipp, Marin & Vidal, 2014. Caribbean Herpetol. 49: 38. (Typhlopidae: Asiatyphlopinae)

Type species: *Typhlops luzonensis* Taylor, 1919 by original designation.

Distribution: Malay Archipelago southeast of Asia.

Comments: Hedges et al. (2014), Pyron & Wallach, 2014, provide generic diagnoses. Wynn et al. (2016) provide a key to Philippine species.

Malayotyphlops andyi Wynn, Diesmos & Brown, 2016. J. Herpetol. 50(1): 164–166, fig. 5.

Holotype: PNM 9779 (field no. ACD 3231, formerly KU 328597), a 243 mm specimen (A.C. Diesmos, K. Hesed, J. Fernandez & party, 30 June 2006).

Type locality: "Luzon island, Cagayan Province, Municipality of Gattaran, Sierra Madre Mountain Range, Barangay Nassiping, Nassiping Reforestation Project area (18.054N, 121.641E"; Philippines.

Distribution: Philippines (N Luzon Island). Known only from the holotype.

Malayotyphlops canlaonensis (Taylor, 1917)
Malayotyphlops castanotus (Wynn & Leviton, 1993)
Malayotyphlops collaris (Wynn & Leviton, 1993)
Malayotyphlops denrorum Wynn, Diesmos & Brown, 2016. J. Herpetol. 50(1): 163–164, fig. 4.

Holotype: PNM 9813 (field no. ACD 2084, formerly KU 328594), a 185 mm specimen (A.C. Diesmos & R.V. Sison, 25 February 2005).

Type locality: "Luzon island, Isabela Province, Municipality of San Mariano, Sierra Madre Mountain Range, Apaya Creek area, Barangay Dibuluan, Sitio Apaya (17.029N, 122.1928E"; Philippines.

Distribution: Phillipines (N Luzon Island). Known only from the holotype.

Malayotyphlops hypogia (Savage, 1950)
Malayotyphlops koekkoeki (Brongersma, 1934)
Malayotyphlops kraalii (Doria, 1875)

Comments: Lang (2013) suggests that, based on morphological differences between Seram and Kai specimens, that *M. kraalii* may be a species complex.

Malayotyphlops luzonensis (Taylor, 1919)

Distribution: Add Philippines (Babuyan Claro, Camiguin Norte, Cebu, Masbate, Pacijan, Poro, Semirara, Siquijor), Leviton et al. (2018), but see comment below.

Comment: Wynn et al. (2016) re-describe and illustrate the holotype. They caution about the assignment of specimens from islands other than the type locality of Luzon Island to this species.

Malayotyphlops ruber (Boettger, 1897)

Comment: Wynn et al. (2016) re-describe and illustrate the holotype. They caution about the assignment of specimens from islands other than the type locality of Samar Island to this species.

Malayotyphlops ruficaudus (Gray, 1845)

Distribution: Add Philippines (Camiguin Norte), Leviton et al. (2018).

MALPOLON Fitzinger, 1826 (Psammophiidae)

Distribution: Bakhouche & Escoriza (2018) describe a specimen from southern Algeria (Tamanrasset), but are unable to assign it to a species.

Comments: Mangiacotti et al. (2014) describe morphological differences in head shape between three of the species of *Malpolon*. See under *Rhagerrhis*.

Malpolon fuscus (Fleischmann, 1831)

Distribution: Add Greece (Diapontia Islets), Stille & Stille (2016); Greece (Elafonisos I.), Broggi (2016a); Greece (Kastellorizo I.), Paysant (1999) and Kakaentzis et al. (2018); Greece (Limnos Island), Strachinis & Roussos (2016); Turkey (Antalya), Kucharzeski (2015); Turkey (Burdur), Ege et al. (2015); Turkey (Bursa), Uğurtaş et al. (2000a); Turkey

(Canakkale), Tok et al. (2006); Turkey (Mersin, Mugla), Winden et al. (1997); Turkey (Gökçeada I.), Yakin et al. (2018); Turkey (Tenedos I.), Tosunoğlu et al. (2009); Iran (Alborz, Ardabil, Ilam, Isfahan, Kermanshah, Lorestan, Markazi, North Khorasan, Qazvin, Semnan, Tehran, Yazd), Safaei-Mahroo et al. (2015); Iran (Kurdistan), and upper elevation of 2432 m, Safai-Mahroo et al. (2017); Iran (Qom), S.M. Kazemi et al. (2015). Domozetski (2013) discuss several high elevation records in SW Bulgaria. Pulev et al. (2018b) detail the distribution in Bulgaria.

Malpolon insignitus (Geoffroy Saint-Hilaire, 1827 *in* Savigny, 1809–1829)

Distribution: Add Libya (Al Wahat, Butnan, Darnah, Jabal al Akhdar, Jabal al Gharbi, Jufra, Kufrah, Marj, Zawiyah), Bauer et al. (2017); Egypt (Ismailia), Ibrahim (2013); Tunisia (Small Kuriat Islet), Lo Cascio & Rivière (2014); Iraq (Babil, Karbala, Najaf, Qadisiyah), Abbas-Rhadi et al. (2017). Mediani et al. (2015b) map localities in Morocco.

†***Malpolon laurenti*** (Depéret, 1897)

†***Malpolon mlynarskii*** Szyndlar, 1988

Malpolon monspessulanus (Hermann, 1804)

Neotype: MNHN 2016.104, a 1264 mm male (1 Jan. 2000).

Type locality: restricted via neotype designation to "Les Cresses, commune de Vic-la-Gardiole, département de l'Hérault; environ 43°28′36″N, 3°47′49″E," France.

Distribution: Add Spain (Alava), Marcos et al. (2018); Spain (Leida), Martínez-Silvestre & Soler (2018); Spain (Navarra), Gosá & Bergerandi (1994); Spain (Soria), Amo (1994). Increase documented elevation to 2250 m, Fernández-Cardenete et al. (2000).

Fossil Records: Add upper Pliocene (Villanyian) of Spain, Blain (2009); lower Pleistocene (Calabrian) of Spain, Blain et al. (2007); upper Pleistocene (Tarantian) of Italy, Delfino (2004), and Spain, Barroso-Ruiz and Bailon (2003).

Comments: Silva-Rocha et al. (2015) use DNA sequence-data to determine the source population for specimens introduced to the Balearic Islands, but were unable to restrict the source beyond the Iberian Peninsula. Bour et al. (2017) were unable to locate the holotype of *Coluber monspessulanus* in the Strasbourg collection, so they designate a neotype and propose a type locality restriction on that basis.

MANOLEPIS Cope, 1885 (Dipsadidae: Xenodontinae)

Manolepis putnami (Jan, 1863)

MASTICOPHIS Baird & Girard, 1853 (Colubridae: Colubrinae)

Fossil record: Add upper Miocene (Hemphillian) of USA (Tennessee), Jasinski & Moscato (2017, as *Masticophis* sp.).

Comments: E.A. Myers et al. (2017b) use DNA sequence-data to conclude that *Coluber* is sister taxon to *Masticophis*. *Masticophis* resolves as two primary clades: those species with transverse markings and those with lineate markings. O'Connell & E.N. Smith (2018) use mtDNA and nDNA to evaluate relationships within *M. flagellum* and northern *M. mentovarius*, along with southern *M. bilineatus*. The resulting phylogeny has western *M. flagellum* as sister taxa to a clade containing eastern and central *M. flagellum*, and another containing northern *M. mentovarius* and southern *M. bilineatus*.

Masticophis anthonyi (Stejneger, 1901)

Masticophis aurigulus (Cope, 1861)

Masticophis barbouri (Van Denburgh & Slevin, 1921)

Masticophis bilineatus Jan, 1867 *in* Jan & Sordelli, 1866–1870

Masticophis flagellum (Shaw, 1802)

Synonyms: *Coluber flagelliformis* Daudin, 1803, *Coluber americanus* Gravenhorst, 1807, *Coluber testaceus* Say *in* James, 1823, *Herpetodryas psammophis* Schlegel, 1837, *Psammophis flavigularis* Hallowell, 1852, *Bascanium flagelliforme bicinctum* Yarrow, 1882, *Masticophis flagellum lineatulus* H.M. Smith, 1941.

Distribution: Limited to eastward from SE Arizona and Mexico E of the Sierra Madre Occidental. Add Mexico (Hidalgo), Reaño-Hernández et al. (2015).

Comments: E.A. Myers et al. (2017a) provide a DNA sequence-based phylogeny of W North American populations, with three primary clades: two in the Chihuahuan Desert and Great Plains, and a third in the American Southwest (referable to *M. piceus*). O'Connell & E.N. Smith (2018) also found eastern and central *M. flagellum* to be monophyletic with respect to western populations, which they elevate to species, *M. piceus*.

Masticophis fuliginosus (Cope, 1895)

Masticophis lateralis (Hallowell, 1853)

Distribution: Add Mexico (N Baja California Sur), Goodman et al. (2019).

Masticophis lineatus (Bocourt, 1890 *in* A.H.A. Duméril, Bocourt & Mocquard, 1870–1909). Miss. Scient. Mex. Amer. Centr. 3(1): 700–701, plate 48. (*Bascanion lineatus*)

Synonyms: *Coluber striolatus* Mertens, 1934, *Masticophis flagellum variolosus* H.M. Smith, 1943.

Lectotype: MNHN 1648, a 940 mm female (A.A. Dugès, 1868), designated by J.D. Johnson (1977).

Lectotype locality: "Colima," Mexico.

Distribution: W Mexico (Aguascalientes, Colima, Durango, Jalisco, Michoacan, Nayarit, Sinaloa, Sonora, Zacatecas, Maria Madre and Maria Magdalena Islands).

Comments: O'Connell & E.N. Smith (2018) elevate northern populations of *M. mentovarius* to this species based on paraphyly with geographically proximal populations of *M. bilineatus*. They provide a revised diagnosis.

Masticophis mentovarius (A.M.C. Duméril, Bibron & Duméril, 1854)

Synonyms: Remove *Bascanion lineatus* Bocourt, 1890 *in* A.H.A. Duméril, Bocourt & Mocquard, 1870–1909, *Coluber striolatus* Mertens, 1934, *Masticophis flagellum variolosus* H.M. Smith, 1943.

Distribution: Add Mexico (S Quintana Roo), Cedeño-Vázquez & Beutelspacher-García (2018); Honduras (Santa Bárbara), Espinal et al. (2014b); Panama (Herrera), Knight et al. (2016); Panama (Panama Oeste), Ray & Ruback (2015). Remove W Mexico N of Michoacan to *M. lineatus*.

Masticophis piceus Cope, 1875. Bull. U. S. Natl. Mus. (1): 40. (*Bascanion flagelliforme piceum*)

Synonyms: *Bascanion flagellum frenatum* Stejneger, 1893, *Masticophis flagellum ruddocki* Brattstrom & Warren, 1953, *Masticophis flagellum cingulum* Lowe & Woodin, 1954.

Holotype: USNM 7891, adult female? (E. Palmer).

Type locality: "Camp Grant, Arizona," USA.

Distribution: NW Mexico (NE Baja California, SW Chihuahua, N Sinaloa, Sonora), SW USA (Arizona, California, Nevada, SW New Mexico, SW Utah).

Comments: See under *M. flagellum*. O'Connell & E.N. Smith (2018) provide a revised diagnosis.

Masticophis schotti Baird & Girard, 1853

Masticophis slevini Lowe & Norris, 1955

Masticophis taeniatus (Hallowell, 1852)

MASTIGODRYAS Amaral, 1934 (Colubridae: Colubrinae)

Comments: Montingelli et al. (2019) produce a DNA sequence-based phylogeny of eight of the species. Seven species form a monophyletic group whereas *M. bifossatus* is paraphyletic as sister taxon to a clade containing *Simophis* and *Drymoluber*. For that reason the authors erect a new genus, *Palusophis*, for *M. bifossatus*.

Mastigodryas alternatus (Bocourt, 1884)

Distribution: Add Costa Rica (Alajuela, Heredia), Montingelli et al. (2019); Panama (Colon, Darien, Herrera, Veraguas), Montingelli et al. (2019); Panama (Panama, Panama Oeste), Ray & Ruback (2015). Eliminate Ecuador from range, Montingelli et al. (2011).

Mastigodryas amarali (Stuart, 1938)

Distribution: Add Venezuela (Bolivar, Guarico), Montingelli et al. (2019).

Mastigodryas boddaerti (Seetzen, 1796)

Synonyms: Remove *Herpetodryas reticulata* W.C.H. Peters, 1863.

Distribution: Add Guyana (Cuyuni-Mazaruni, East Demerara-West Berbice, West Demerara-Essequibo Islands), Montingelli et al. (2019); Suriname (Marawijne, Paramaribo), Montingelli et al. (2019); Brazil (Amapá), F.G.R. França et al. (2006); Brazil (Goias), Montingelli et al. (2019); Brazil (Maranhao, Tocantins), Siqueira et al. (2013); Brazil (Mato Grosso do Sul), Montingelli et al. (2011) and Ferreira et al. (2017); Brazil (Piauí), Dal Vechio et al. (2013); Brazil (Roraima), Montingelli et al. (2011); Brazil (Tocantins), Montingelli et al. (2011) and Dal Vechio et al. (2016); Bolivia (Cochabamba), Montingelli et al. (2011); Peru (Amazonas, Cajamarca, Ucayali), Cadle (2012c); Peru (La Libertad), Koch et al. (2018). Remove range of *M. reticulatus*: SW Ecuador (Guayas, Loja, Manabi), Montingelli et al. (2011). Upper elevation of 2560 m, Vera-Pérez et al. (2018).

Comments: Montingelli et al. (2011) discuss an undescribed species from SW Colombia and N Ecuador.

Mastigodryas bruesi (Barbour, 1914)

Mastigodryas cliftoni (L. Hardy, 1964)

Distribution: Mexico (range extensions within Jalisco & Nayarit), and upper elevation of 2000 m, Ahumada-Carrillo et al. (2014)

Mastigodryas danieli Amaral, 1934

Mastigodryas dorsalis (Bocourt, 1890 *in* A.H.A. Duméril, Bocourt & Mocquard, 1870–1909)

Distribution: Add Mexico (Chiapas), Montingelli et al. (2019); Guatemala (Chimaltenango, Escuintla, Huehuetenango, Sacatapequez, Solola), Montingelli et al. (2019); Honduras (Choluteca), Espinal et al. (2014a).

Mastigodryas heuthii (Cope, 1875)

Distribution: Add Ecuador (El Oro, Loja), Montingelli et al. (2011); Peru (Amazonas), Koch et al. (2018); Peru (Lambayeque), and low elevation of sea level, Venegas (2005); Peru (Piura), Cadle (2012c).

Mastigodryas melanolomus (Cope, 1868)

Distribution: Add Mexico (Quintana Roo: Cozumel I.), Escalante-Pasos & García-Padilla (2015); Honduras (Isla Cochino Menor), McCranie & Valdés-Orellana (2014). Elevation range 1–1900 m, Palacios-Aguilar & Flores-Villela (2018).

Mastigodryas moratoi Montingelli & Zaher, 2011

Distribution: Fraga et al. (2018) report a new locality, and provide a map of known localities.

Mastigodryas pleii (A.M.C. Duméril, Bibron & Duméril, 1854)

Distribution: Add Panama (Herrera), Montingelli et al. (2019); Panama (Panama), Ray & Ruback (2015); Colombia (Cesar), Medina-Rangel (2011); Colombia (Córdoba), Carvajal-Cogollo et al. (2007); Colombia (Huila, Valle del Cauca), Montingelli et al. (2019); Brazil (Pará), F.G.R. França et al. (2006).

Mastigodryas pulchriceps (Cope, 1868)

Distribution: Add Colombia (Cauca), and elevation range of 0–2800 m, Vera-Pérez et al. (2018); Ecuador (Chimborazo, Imbabura), Montingelli et al. (2019).

Mastigodryas reticulata (W.C.H. Peters, 1863). Monatsb. Preuss. Akad. Wiss. Berlin 1863: 285. (*Herpetodryas reticulata*)

Holotype: ZMB 4504, a 972 mm male (C. Reiss).

Type locality: "Umgebung von Guayaquil," Ecuador.

Distribution: SW Ecuador (Guayas, Loja, Manabi, Santa Elena, and La Plata and Puna Is.), NW Peru (Cajamarca, La Libertad), to 1239 m.

Comments: Removed from the synonymy of *M. boddaerti* based on morphological distinctiveness by Montingelli et al. (2011). See also Koch et al. (2018).

MEHELYA Csiki, 1903 (Lamprophiidae)

Synonyms: *Grobbenia* Poche, 1903, *Siebenrockia* Poche, 1904.

Distribution: C Africa.

Comments: See under *Gonionotophis*. Broadley et al. (2018) provide a revised genus description. *Mehelya egbensis* and

M. gabouensis are provisionally included here despite a lack of morphological or genetic corroboration, Broadley et al. (2018).

Mehelya egbensis Dunger, 1966

Mehelya gabouensis Trape & Mané, 2005

Mehelya laurenti Witte, 1959

Mehelya poensis (A. Smith, 1847 *in* 1838–1849)

Distribution: Add Guinea (Faranah), Broadley et al. (2018); Benin (Zou), Broadley et al. (2018); Gabon (Ogooue-Lolo), Pauwels et al. (2016a); Democratic Republic of the Congo (North Kivu), Broadley et al. (2018); Angola (Bengo), M.P. Marques et al. (2018); Angola (Cuanza Sur, Lunda Norte), Branch (2018).

Mehelya stenophthalma (Mocquard, 1887)

Distribution: Add Congo (Kouilou), Broadley et al. (2018).

MEIZODON J.G. Fischer, 1856 (Colubridae: Colubrinae)

Meizodon coronatus (Schlegel, 1837)

Distribution: Add Niger (Maradi), Trape & Mané (2015).

Meizodon krameri Schätti, 1985

Meizodon plumbiceps (Boettger, 1893)

Meizodon regularis J.G. Fischer, 1856

Distribution: Add Sierra Leone and Nigeria, Trape & Baldé (2014).

Meizodon semiornatus (W.C.H. Peters, 1854)

MELANOPHIDIUM Günther, 1864 (Uropeltidae)

Comments: Gower et al. (2016a) review the genus and provide a key to the species. Pyron et al. (2016c) confirm the monophyly of the genus through mtDNA and nDNA sequence-data, and provide a morphological diagnosis.

Melanophidium bilineatum Beddome, 1870

Lectotype: BMNH 1946.1.15.75, from "Peria peak, Wynad," designated by Gower et al. (2016a).

Comments: Pyron et al. (2016c) provide a description.

Melanophidium khairei Gower, Giri, Captain & Wilkinson, 2016a. Zootaxa 4085(4): 482–490, figs. 1–5.

Holotype: BNHS 3452, a 550 mm female (M. Bhise, H. Ogale & R. Koregaonkar, 15 July 2006).

Type locality: "Amboli, Sindudurg district, Maharashtra, India (15° 57′ N, 73° 59′ E, 715 m elevation)."

Distribution: W India (Goa, Karnataka, Maharashtra), 510–780 m.

Comments: Pyron et al. (2016c) provide a description.

Melanophidium punctatum Beddome, 1871

Lectotype: BMNH 1946.1.4.37, from "Agasthyamalai hills ('Mutikuli Vayal' valet at '4500 feet')," designated by Gower et al. (2016a).

Distribution: Restricted to the Agasthyamalai Hills and vicinity, Kerala and Karnataka, India, at elevations below 1000 m. Former northern populations belong to *M. khairei*.

Comments: Pyron et al. (2016c) provide a description and color photograph.

Melanophidium wynaudense (Beddome, 1863)

Lectotype: BMNH 1946.1.15.46, from "near Manantoddy (Mananthavady), Wyanad," designated by Gower et al. (2016a).

Comments: Pyron et al. (2016c) provide a description and color photograph, and correct the type locality to Tamil Nadu state, rather than Kerala.

†*MENARANA* LaDuke, Krause, Scanlon & Kley, 2010 (†Madtsoiidae)

†*Menarana laurasiae* Rage, 1996

†*Menarana nosymena* LaDuke, Krause, Scanlon & Kley, 2010

†*MESOPHIS* Bolkay, 1925. Glasn. Zemalj. Muz. Bosni Hercegovini 37:125. (†Simoliophiidae)

Type species: *Mesophis nopcsai* Bolkay, 1925 by monotypy.

Distribution: Middle/upper Cretaceous of Bosnia-Hercegovina.

Comments: Bardet et al. (2008) recognize *Mesophis* as a taxon distinct from *Pachyophis*, but noted that the former may be an ontogene of the latter.

†*Mesophis nopcsai* Bolkay, 1925. Glasn. Zemalj. Muz. Bosni Hercegovini 37: 125–130, figs. 1–4, plates 1, 2.

Holotype: a skeleton on a slab, unlocated according to Rage (1984:62),

Type locality: "östlich von Bilek, in dem Vorort Selišta," Bosnia-Hercegovina.

Distribution: Cretaceous (middle/upper) of Bosnia Hercegovina.

†*MESSELOPHIS* Baszio, 2004 (Tropidophiidae)

†*Messelophis ermannorum* Schaal & Baszio, 2004

†*Messelophis variatus* Baszio, 2004

†*MICHAUXOPHIS* Bailon, 1988 (Cylindrophiidae)

†*Michauxophis occitanus* Bailon, 1988

MICRELAPS Boettger, 1880 (Colubroidea: incerta sedis)

Comments: Molecular data studied by Portillo et al. (2018) affirm *Micrelaps* as not having close affinities to the Aparallactinae or Atractaspidinae, and they consider it *incerta sedis*. Zaher et al. (2019) find *Micrelaps* to be sister taxon to a clade containing the Prosymnidae, Psammophiidae and Pseudaspididae.

Micrelaps bicoloratus Sternfeld, 1908

Micrelaps muelleri Boettger, 1880

Micrelaps tchernovi Y.L. Werner, Babocsay, Carmely & Thuna, 2006

Micrelaps vaillanti (Mocquard, 1888)

MICROCEPHALOPHIS Lesson, 1832 *in* Bélanger, 1831–1834 (Elapidae)

Microcephalophis cantoris (Günther, 1864)

Distribution: Add Iran (Hormozgan), Rezaie-Atagholipour et al. (2016).

Microcephalophis gracilis (Shaw, 1802)

Distribution: Add United Arab Emirates, Buzás et al. (2018); West Malaysia (Johor), Voris (2017); Philippines (Luzon), Leviton et al. (2018). Australia excluded from the known geographic range, A.R. Rasmussen et al. (2014).

†*MICRONATRIX* Parmley & Hunter, 2010 (Natricidae)

†*Micronatrix juliescottae* Parmley & Hunter, 2010

MICROPECHIS Boulenger, 1896 (Elapidae)

Micropechis ikaheca (Lesson, 1829 *in* Duperrey, 1826–1832)

Distribution: Add Indonesia (Papua: Biak Island), O'Shea et al. (2015); Papua New Guinea (Boisa Island), Clegg & Jocque (2016); Papua New Guinea (Madang, Oro, Sandaun), O'Shea et al. (2015).

MICROPISTHODON Mocquard, 1894 (Pseudoxyrhophiidae)

Micropisthodon ochraceus Mocquard, 1894

MICRUROIDES K.P. Schmidt, 1928 (Elapidae)

Micruroides euryxanthus (Kennicott, 1860)

Distribution: Add Mexico (N Jalisco), Ahumada-Carrillo et al. (2018) and Dávalos-Martínez et al. (2019).

MICRURUS Wagler *in* Spix, 1824 (Elapidae)

Comments: Dashevsky & Fry (2018) use sequence data from three-finger toxin proteins to evaluate their variation within *Micrurus*. Although the data partition eight clades of the protein, there is imperfect sorting among the species of *Micrurus* species. Jowers et al. (2019) use mtDNA sequence-data to evaluate relationships among 33 *Micrurus* taxa, which separate into clades for monadal and triadal body color patterns.

Fossil records: Add Miocene/Pliocene transition (late Turolian) of Greece, Georgialis et al. (2019, as cf. *Micrurus*).

Micrurus albicinctus Amaral, 1925

Distribution: Add Peru (Huanuco), Feitosa et al. (2015).

Micrurus alleni K.P. Schmidt, 1936

Micrurus altirostris (Cope, 1860)

Distribution: Add Paraguay (Canindeyu), Cacciali et al. (2015b).

Micrurus ancoralis (Jan, 1872 *in* Jan & Sordelli, 1870–1881)

Distribution: Add Ecuador (Bolivar, Carchi, Cotopaxi, Imbabura, Manabi, Pastaza, Santo Domingo), and upper elevation of 1800 m, Valencia et al. (2016).

Micrurus annellatus (W.C.H. Peters, 1871)

Distribution: Add Brazil (Acre), Bernarde et al. (2012a, first Brazilian record for subspecies *annellatus*); Colombia (Meta), Feitosa et al. (2015); Peru (Pasco, San Martin, Ucayali), Feitosa et al. (2015).

Micrurus averyi K.P. Schmidt, 1939

Distribution: Add Brazil (Roraima), Feitosa et al. (2015).

Micrurus baliocoryphus (Cope, 1860)

Micrurus bernadi (Cope, 1887)

Distribution: Low elevation of 1249 m, Valencia-Herverth & Fernández-Badillo (2013).

Micrurus bocourti (Jan, 1872 *in* Jan & Sordelli, 1870–1881)

Distribution: Add Ecuador (El Oro), and upper elevation of 1800 m, Valencia et al. (2016); Ecuador (Manabi), Almendáriz et al. (2012).

Micrurus bogerti Roze, 1967

Micrurus boicora Bernarde, Turci, Abegg & Franco, 2018. Salamandra 54(4): 252–255, figs. 2–5.

Holotype: IBSP 77773, a 306 mm male (M.F. da Silva & A.M. da Silva, 23 December 2007).

Type locality: "Rondon II Hydroelectric Power Plant, Municipality of Pimenta Bueno, State of Rondônia, Brazil (11°57′S, 60°41′W; 325 m above sea level)."

Distribution: Brazil (N Mato Grosso, E Rondonia), 200–325 m.

Micrurus brasiliensis Roze, 1967

Distribution: Add Brazil (Maranhão), Pires et al. (2014). Upper elevation of 630 m, Guedes et al. (2014). Silveira (2014b) reports the third record of this species from Minas Gerais and the southern-most record.

Micrurus browni K.P. Schmidt & H.M. Smith, 1943

Distribution: Add Mexico (Jalisco), Grünwald et al. (2016).

Micrurus camilae Renjifo & Lundberg, 2003

Distribution: Add Colombia (Antioquia), Alzate (2014); Colombia (Santander), and elevation range of 60–176 m, Meneses-Pelayo & Caballero (2019). Meneses-Pelayo & Caballero (2019) list and map known localities.

Micrurus carvalhoi Roze, 1967

Distribution: Add Brazil (Alagoas, Paraiba, Tocantins), Pires et al. (2014); Brazil (Espirito Santo, Rio Grande do Sul), T.M. Castro et al. (2017); Paraguay (Amambay, Caaguazu), Pires et al. (2014).

Micrurus catamayensis Roze, 1989

Micrurus circinalis (A.M.C. Duméril, Bibron & Duméril, 1854)

Micrurus clarki K.P. Schmidt, 1936

Distribution: Add Panama (Panama), Ray & Ruback (2015).

Micrurus collaris (Schlegel, 1837)

Micrurus corallinus (Merrem, 1820)

Lectotype: The lectotype is AMNH R3911, not MNHN 3911, Vanzolini & Myers (2015).

Micrurus decoratus (Jan, 1858)

Distribution: Add Brazil (Espirito Santo), and modify elevational range to 400–1600 m, R.C. Gonzalez et al. (2014b), who also provide a locality map.

Micrurus diana Roze, 1983

Distribution: Add Brazil (Mato Grosso), and low elevation of 345 m, Pires et al. (2013); Bolivia (La Paz), and high elevation of 749 m, Pires et al. (2013).

Comments: Pires et al. (2013) provide an expanded description based on additional specimens.

Micrurus diastema (A.M.C. Duméril, Bibron & Duméril, 1854)

Micrurus dissoleucus (Cope, 1860)

Distribution: Add Panama (Panama Oeste), Ray & Ruback (2015).

Micrurus distans (Kennicott, 1860)

Distribution: Mexico (range extensions within Jalisco & Nayarit), and upper elevation of 2370 m, Ahumada-Carrillo et al. (2014)

Micrurus dumerilii (Jan, 1858)

Distribution: Add Colombia (Cesar), Rojas-Murcia et al. (2016); Colombia (Córdoba), Carvajal-Cogollo et al. (2007), Colombia (Gorgona I.), Urbina-Cardona et al. (2008); Ecuador (Bolivar, Carchi, El Oro, Guayas, Imbabura, Los Rios, Santo Domingo), Valencia et al. (2016). Prairie et al. (2015) document known records from Panama, including a new record for Panamá Province.

Micrurus elegans (Jan, 1858)

Micrurus ephippifer (Cope, 1886)

Micrurus filiformis (Günther, 1859)

Micrurus frontalis (Duméril, Bibron & Duméril, 1854)

Distribution: Add Brazil (SW Bahia), Guedes et al. (2014).

Micrurus frontifasciatus (F. Werner, 1927)

Micrurus fulvius (Linnaus, 1766)

†*Micrurus gallicus* Rage & Holman, 1984)

Distribution: Add late Middle Miocene (Astaracian or Vallesian) of Hungary, Venczel (2011).

Micrurus hemprichii (Jan, 1858)

Synonyms: Remove *Micrurus hemprichi ortoni* K.P. Schmidt, 1953, *Micrurus rondonianus* Roze & Silva, 1990.

Distribution: Limited to SE Colombia, S Venezuela, Guyana, Suriname, French Guiana, and N Brazil. Add Brazil (Mato Grosso, Roraima), Bernarde et al. (2018).

Micrurus hippocrepis (W.C.H. Peters, 1861)

Micrurus ibiboboca (Merrem, 1820)

Micrurus isozonus (Cope, 1860)

Synonyms: Add *Micrurus isozonus sandneri* Arenas-Vargas, 2015. Serp. Coral Una Nueva Subesp.: 13–20, figs. 2–9.

Holotype: Example 1 in the author's collection, a 522 mm female (I.S. Arenas-Vargas and I. Heredia de Arenas, 17 August 1998).

Type locality: "Región Central de Venezuela, Estado Aragua, Municipio Zamora, lugar poblado llamado La Pavona, donde se puede ver la Fila El Charal, cerca de la Población de San Francisco de Asis."

Distribution: Add Colombia (Bolívar), Feitosa et al. (2013).

Micrurus langsdorffi (Wagler *in* Spix, 1824)

Distribution: Add Colombia (Boyacá, Guainía), Feitosa et al. (2015); Ecuador (Sucumbíos), Feitosa et al. (2015); Peru (Ucuyali), Feitosa et al. (2015).

Micrurus laticollaris (W.C.H. Peters, 1870)

Micrurus latifasciatus K.P. Schmidt, 1933

Micrurus lemniscatus (Linnaeus, 1758)

Distribution: Add Ecuador (Orellana, Sucumbios), and upper elevation of 1300 m, Valencia et al. (2016); Brazil (Bahia), and upper elevation of 760 m; Brazil (Minas Gerais), Moura et al. (2012); Brazil (Paraiba), R. França et al. (2012); Brazil (Pernambuco), Roberto et al. (2018); Brazil (Mato Grosso do Sul), Ferreira et al. (2017).

Comments: Jowers et al. (2019) produce a phylogeny using one mtDNA gene that indicates paraphyly among some populations, along with *M. carvalhoi*, in which the subspecies *diutius* is sister taxon to *carvalhoi*, *helleri* and *lemniscatus*. The authors elevate *diutius* and *helleri* to species level, in part following recommendations in an unpublished dissertation by M.G. Pires. Due to the use of a single gene, and limited geographic scope of samples, we await further supportive data before redefining *helleri* and *diutius* as species.

Micrurus limbatus Fraser, 1964

Micrurus margaritiferus Roze, 1967

Micrurus medemi Roze, 1967

Distribution: Add Colombia (Vaupes), Feitosa et al. (2015).

Micrurus meridensis Roze, 1989

Micrurus mertensi K.P. Schmidt, 1936

Distribution: Add Ecuador (Azuay), and upper elevation of 2400 m, Valencia et al. (2016); Ecuador (Pichincha), Feitosa et al. (2015); Peru (Cajamarca, Lima), Feitosa et al. (2015).

Micrurus mipartitus (A.M.C. Duméril, Bibron & Duméril, 1854)

Synonyms: Remove *Elaps multifasciatus* Jan, 1858, *Elaps hertwigi* F. Werner, 1897.

Distribution: Add Panama (Panama Oeste), Ray & Ruback (2015); Colombia (Cesar, Cordoba), Rios-Soto et al. (2018); Colombia (Valle del Cauca: Isla Palma), Giraldo et al. (2014); Ecuador (Bolivar, Carchi, Chimborazo, El Oro, Imbabura, Santo Domingo), Valencia et al. (2016). Remove most of Central America, aside from E Panama.

Micrurus mosquitensis K.P. Schmidt, 1933. Zool. Series Field Mus. Nat. Hist. 20, 33. (*Micrurus nigrocinctus mosquitensis*)

Holotype: MCZ 19741, an adult male (S. Kress, 1924).

Type locality: "Limon, Costa Rica."

Distribution: Atlantic coast slope of E Nicaragua, E Costa Rica and NE Panama.

Comments: J. Fernández et al. (2015) recognize *M. mosquitensis* as distinct from *M. nigrocinctus* based on color pattern, scale counts, and unpublished "molecular divergences."

Micrurus multifasciatus Jan, 1858. Rev. Mag. Zool. 10(12): 516. (*Elaps multifasciatus*)

Synonyms: *Elaps hertwigi* F. Werner, 1897.

Holotype: MSNM, likely destroyed during WWII.

Type locality: "Amérique centrale."

Distribution: Lower Central America. Nicaragua, Costa Rica, Panama. NSL-1200 m.

Comments: We recognize this species as distinct from *M. mipartitus* due to morphological differences and allopatric distribution, together with genetic findings of Zaher et al. (2019).

Micrurus multiscutatus Rendahl & Vestergren 1940

Distribution: Add Ecuador (Esmeraldas), Valencia et al. (2016). Elevation range 50–1800 m, Vera-Pérez et al. (2018).

Micrurus narduccii (Jan, 1863)

Distribution: Add Ecuador (Chimborazo, Orellana, Sucumbios, Zamora-Chinchipe), Valencia et al. (2016).

Micrurus nattereri K.P. Schmidt, 1952

Micrurus nebularis Roze, 1989

Distribution: Add Mexico (Puebla), Pavón-Vázquez et al. (2015).

Micrurus nigrocinctus (Girard, 1854)

Synonyms: Remove *Micrurus nigrocinctus mosquitensis,* K.P. Schmidt, 1933.

Distribution: Add Honduras (Isla Zacate Grande), McCranie & Gutsche (2016); Panama (Panama Oeste), Ray & Ruback (2015). Remove range of *M. mosquitensis.*

Micrurus obscurus (Jan, 1872 *in* Jan & Sordelli, 1870–1881)

Distribution: Add Ecuador (Chimborazo, Orellana, Sucumbios), and upper elevation of 750 m, Valencia et al. (2016); Peru (Puno), Llanqui et al. (2019).

Micrurus oligoanellatus Ayerbe-González & López-López, 2005

Micrurus ornatissimus (Jan, 1858)

Distribution: Add Brazil (Rondônia), Bernarde et al. (2012b, *Micrurus* cf. *ornatissimus*); Colombia (Putumayo), Feitosa et al. (2015); Ecuador (Sucumbios, Tungurahua), and elevation range of 200–2100 m, Valencia et al. (2016); Ecuador (El Oro, Los Ríos, Orellana), Feitosa et al. (2015).

Micrurus ortoni K.P. Schmidt, 1953. Fieldiana Zool. 34(13): 166–168. (*Micrurus hemprichi ortoni*)

Synonyms: *Micrurus rondonianus* Roze & Silva, 1990.

Holotype: MCZ 12423, a 735 mm male (J. Orton, December 1867).

Type locality: "Pebas, Peru."

Distribution: S Colombia, E Ecuador, E Peru, W. Brazil and N Bolivia. Add Brazil (Acre), Bernarde et al. (2011b); Colombia (Meta), Bernarde et al. (2018); Ecuador (Esmeraldas, Orellana, Sucumbios), Valencia et al. (2016).

Comments: Valencia et al. (2016) recognize *ortoni* as a species distinct from *M. hemprichii* based on morphological differences.

Micrurus pacaraimae Carvalho, 2002

Micrurus pachecogili Campbell, 2000

Micrurus paraensis Cunha & Nascimento, 1973

Distribution: Add Brazil (Amazonas), S.M. Souza et al. (2011).

Micrurus peruvianus K.P. Schmidt, 1936

Distribution: Add Peru (La Libertad), Koch et al. (2018).

Micrurus petersi Roze, 1967

Micrurus potyguara Pires, Silva, Feitosa, Prudente, Filho & Zaher, 2014. Zootaxa 3811(4): 571–577, figs. 1, 2, 5, 6.

Holotype: UFPB 4359, a 935 mm male (G.A.P. Filho, 22 August 2006).

Type locality: "Mata do Buraquinho (7°8′ 42.08″S, 34° 51′42.40″W), municipality of João Pessoa, state of Paraíba," Brazil.

Distribution: Lowland coastal rainforests of NE Brazil (Paraiba, Pernambuco, Rio Grande do Norte).

Micrurus proximans H.M. Smith & Chrapliwy, 1958

Micrurus psyches (Daudin, 1803)

Micrurus putumayensis Lancini, 1962

Micrurus pyrrhocryptus (Cope, 1862)

Distribution: Add Brazil (Mato Grosso do Sul), Ferreira et al. (2017).

Micrurus remotus Roze, 1987

Distribution: Add Brazil (Acre), Bernarde et al. (2011b); Brazil (Rondonia), Feitosa et al. (2015); Colombia (Meta), Feitosa et al. (2015), Brazil (Roraima), Moraes et al. (2017).

Micrurus renjifoi (Lamar, 2003)

Micrurus ruatanus (Günther, 1895 *in* 1885–1902)

Micrurus sangilensis Nicéforo-María, 1942

Micrurus scutiventris (Cope, 1870)

Distribution: Add Ecuador (Sucumbios), Valencia et al. (2016).

Micrurus serranus Harvey, Aparicio & Gonzales, 2003

Micrurus silviae Di-Bernardo, Borges-Martins & Silva, 2007

Distribution: Add Argentina (Corrientes, Misiones); species confirmed for Paraguay by Cabral and Caballero (2013).

Comments: Giraudo et al. (2015) provide details for additional specimens and provide a revised morphological description.

Micrurus spixii Wagler *in* Spix, 1924

Micrurus spurrelli (Boulenger, 1914)

Micrurus steindachneri (F. Werner, 1901)

Distribution: Add Ecuador (Pichincha, Sucumbios, Zamora-Chinchipe), and low elevation of 500 m, Valencia et al. (2016).

Micrurus stewarti Barbour & Amaral, 1928

Distribution: Add Panama (Panama Oeste), Ray & Ruback (2015) and remove Panamá Province.

Micrurus stuarti Roze, 1967

Micrurus surinamensis (Cuvier, 1817)

Distribution: Add Ecuador (Orellana, Sucumbios), and upper elevation of 750 m, Valencia et al. (2016); Peru (Huanuco), Schlüter et al. (2004); Brazil (Acre), Fonseca et al. (2019); Brazil (Amapa, Maranhao, Tocantins), Morais et al. (2011). Upper elevation of 718 m, Díaz-Ricaurte & Fiorillo (2019).

Micrurus tamaulipensis Lavin-Murcio & Dixon, 2004

Micrurus tener (Baird & Girard, 1853)

Distribution: Upper elevation of 2015 m, Valdez-Rentería & Fernández-Badillo (2016).

Micrurus tikuna Feitosa, Silva, Pires, Zaher & Prudente, 2015. Zootaxa 3974(4): 539–547, figs. 1–2, 8.

Holotype: MPEG 18199, a 699.5 mm female (J.S. Haad, 1991).

Type locality: "INCRA neighborhood, municipality of Tabatinga (04°14'36"S, 69°54'15"W; ca. 80 m above sea level...), state of Amazonas, Brazil."

Distribution: Brazil (Amazonas) and Colombia (Amazonas), 80–110 m.

Micrurus tschudii (Jan, 1858)

Distribution: Add Ecuador (El Oro), Valencia et al. (2016).

MIMOPHIS Günther, 1868 (Psammophiidae)

Comments: Ruane et al. (2018a) produce a DNA sequence-based phylogeny that resolves two species-level clades. Ruane et al. also note that the several color pattern morphs occur in both sexes and both species.

Mimophis mahfalensis (Grandidier, 1867)

Distribution: S Madagascar (Antananarivo, Fianarantsoa, Toamasina, Toliara).

Mimophis occultus Ruane, E. Myers, Lo, Yuen, Welt, Juman, Futterman, Nussbaum, Schneider, Burbrink & Raxworthy, 2017. Syst. Biodiv. 16(3): 10–11, figs. 1, 4.

Holotype: UMMZ 237408, a 615 mm male (J. Spannring, J. Rafanomazantsoa & M. Rakotoarivelo, 20 April 2002).

Type locality: "Madagascar, Mahajanga Province, Tsiambara Forest (-15.920716 S., 45.9886 E), 41 m elevation."

Distribution: N Madagascar (Antsiranana, Mahajanga), to 700 m. Add Madagascar (Nosy Be I.), McLellan (2013).

MINTONOPHIS J.C. Murphy & Voris, 2014. Fieldiana Life Earth Sci. (8): 27–28. (Homolopsidae)

Type species: *Enhydris pakistanica* Mertens, 1959 by original designation.

Distribution: S Pakistan.

Mintonophis pakistanicus (Mertens, 1959)

Comments: J.C. Murphy & Voris (2014) provide a diagnosis and photograph.

†*MIOCOLUBER* Parmley, 1988 (Colubridae: Colubrinae)

†*Miocoluber dalquesti* Parmley, 1988

†*MIONATRIX* Sun, 1961 (Natricidae)

†*Mionatrix diatomus* Sun, 1961

MIRALIA Gray, 1842. Zool. Misc.: 68. (Homalopsidae)

Type species: *Brachyorrhos alternans* A. Reuss, 1834 by monotypy.

Distribution: Greater Sundas.

Comments: J.C. Murphy & Voris (2014) revive this genus based on morphology, and provide a generic diagnosis.

Miralia alternans (A. Reuss, 1834)

Comments: J.C. Murphy & Voris (2014) provide a diagnosis and photograph.

MITOPHIS Hedges, Adalsteinsson & Branch *in* Adalsteinsson, Branch, Trape, Vitt & Hedges, 2009 (Leptotyphlopidae: Epictinae)

Mitophis asbolepis (J.P.R. Thomas, McDiarmid & Thompson, 1985)

Mitophis calypso (J.P.R. Thomas, McDiarmid & Thompson, 1985)

Mitophis leptepileptus (J.P.R. Thomas, McDiarmid & Thompson, 1985)

Mitophis pyrites (J.P.R. Thomas, 1965)

MIXCOATLUS Jadin, E.N. Smith & Campbell, 2011 (Viperidae: Crotalinae)

Comments: Grünwald et al. (2015) discuss the biogeography of the species of *Mixcoatlus*.

Mixcoatlus barbouri (Dunn, 1919)

Distribution: Elevation range 1955–2608 m, Grünwald et al. (2015, 2016a).

Mixcoatlus browni (Shreve, 1938)

Distribution: Elevation range 1826–3296 m, Grünwald et al. (2015).

Mixcoatlus melanurus (L. Müller, 1924)

Distribution: Localities are mapped and elevation range is discussed by Grünwald et al. (2015).

MONTASPIS Bourquin, 1991 (Colubroidea incerta sedis)

Montaspis gilvomaculata Bourquin, 1991

MONTATHERIS Broadley, 1996 (Viperidae: Viperinae)

Montatheris hindii (Boulenger, 1910)

MONTIVIPERA Nilson, Tuniyev, Andrén, Orlov, Joger & Herrmann, 1999 (Viperidae: Viperinae)

Comments: Stümpel et al. (2016) obtain three clades from mtDNA sequence-data: a *bornmuelleri* clade, a *xanthina* complex, and a *raddei* complex. They choose to recognize isolated members of the *raddei* complex as distinct species due to isolation and genetic uniqueness, as opposed to some prior conclusions to "lump" all under *M. raddei*, such as recommended by E. Rastegar-Pouyani et al. (2014b).

Montivipera albizona (Nilson, Andrén & Flärdh, 1990)

Distribution: Range extends SW to Kahramanmaraş Province, Turkey, 1300 m, Göçmen et al. (2009c). Göçmen et al. (2014a) provide additional localities, between 1790 and 2200 m, outside of the previously known geographic range.

Comments: Göçmen et al. (2009c) use serum electrophoresis coupled with geographic barriers to argue that *M. albizona* is a species distinct from *M. xanthina*. Göçmen et al. (2014a) argue to retain *M. albizona* as a distinct species despite it being phylogenetically nested within *M. bulgardaghica*.

Montivipera bornmuelleri (F. Werner, 1898)

Comments: Stümpel et al. (2016) support the recognition as a full species from their mtDNA-based phylogeny.

Montivipera bulgardaghica (Nilson & Andrén, 1985)

Comments: Stümpel et al. (2016) support the recognition as a full species from their mtDNA-based phylogeny.

Montivipera kuhrangica Rajabizadeh, Nilson & Kami, 2011
Comments: See under *M. raddei*.

Montivipera latifi (Mertens, Darevsky & Klemmer, 1967)
Distribution: Add Iran (Alborz, Tehran), Safaei-Mahroo et al. (2015).
Comments: Rajabizadeh et al. (2012) discuss the morphology. See under *M. raddei*.

†*Montivipera platyspondyla* (Szyndlar, 1987)

Montivipera raddei (Boettger, 1890)
Synonyms: Add *Vipera albicornuta* Nilson & Andrén, 1985.
Distribution: Add Iran (Gilan, Hamedan, Kurdestan, Qazvin, Zanjan), Safaei-Mahroo et al. (2015), Iran (East Azerbaijan), Safdarian et al. (2016). Mebert et al. (2015, 2016) document localities near the contact zone with *M. wagneri*.
Comments: E. Rastegar-Pouyani et al. (2014b) find little variation in mtDNA sequences between Iranian *M. raddei* and related species (*M. albicornuta*, *M. kuhrangica*, and *M. latifi*). They conclude that the latter three are isolated color morphs, and all should be synonymized with *M. raddei*. Rajabizadeh et al. (2015a) recognize *M. kuhrangica* and *M. latifi* as species (they do not reference E. Rastegar-Pouyani et al., 2014a), but consider *M. albicornuta* to be a subspecies of *M. raddei*. Also, based on mtDNA sequence data, Stümpel et al. (2016) agree in recognizing *M. kuhrangica* and *M. latifi* as species, but conclude that *M. albicornuta* should not be recognized because its morphotype appears in multiple localities within the *M. raddei* phylogeny.

Montivipera wagneri (Nilson & Andrén, 1984)
Distribution: Add Turkey (Erzurum), Kumlutaş et al. (2015a); Turkey (Muş), and increase elevational range to 2146 m, Göçmen et al. (2014a). Mebert et al. (2015, 2016) document localities near the contact zone with *M. raddei*.
Comments: Stümpel et al. (2016) support the recognition as a full species from their mtDNA-based phylogeny.

Montivipera xanthina (Gray, 1849)
Synonyms: Add the following

Montivipera xanthina nilsoni Cattaneo, 2014. Naturalista Sicil. Ser. 4, 38: 69–72, figs. 4–8.
Holotype: MZCR VR-02358, a 1089 mm male (A. Vesci, 11 May 2001).
Type locality: "isola greca di Chios, località Dafnonas, c. 220 m s.l.m (Mar Egeo orientale)."

Montivipera xanthina dianae Cattaneo, 2014. Naturalista Sicil. Ser. 4, 38: 73–75, figs. 9–11.
Holotype: MZCR VR-02359, a 78 cm male (A. Cattaneo, 22 May 2007).
Type locality: "isola greca di Leros, località Ag. Petros, c. 110 m s.l.m. (Dodecaneso)."

Montivipera xanthina occidentalis Cattaneo, 2016. Naturalista Sicil. Ser. 4, 41(1): 69–72, figs. 8, 9.
Holotype: A. Cattaneo Collection MX/TR20/5, a 671 mm male (A. Cattaneo, 6 May 2015).
Type locality: "Tracia greca centrale, Regione dei Rodopi, località Proskinites, 121 m s.l.m."

Distribution: Add Greece (Samos Island), Speybroek et al. (2014); Turkey (Bursa), Uğurtaş et al. (2000a); Turkey (Canakkale), Tok et al. (2014).
Comments: Cattaneo (2014, 2016) evaluates morphological variation in populations, and describes three new subspecies. Stümpel et al. (2016) present an mtDNA-based phylogeny from numerous Middle Eastern populations.

MOPANVELDOPHIS Figueroa, McKelvy, Grismer, Bell & Lailvaux, 2016. PLoS ONE 11(9.0161070): 23. (Colubridae: Colubrinae)
Type species: *Coluber zebrinus* Broadley & Schätti, 2000 by original designation.
Distribution: SW Africa.

Mopanveldophis zebrinus (Broadley & Schätti, 2000)
Distribution: Add Angola (Namibe), Ceríaco et al. (2016).

MORELIA Gray, 1842 (Pythonidae)
Synonyms: Remove *Simalia* Gray, 1849, *Aspidopython* Meyer, 1875, *Hypaspistes* Ogilby, 1891, *Australiasis* Wells & Wellington, 1984, *Nyctophilopython* Wells & Wellington, 1985, *Lenhoserus* Hoser, 2000.
Distribution: Australia and New Guinea.
Comments: Reynolds et al. (2014b) analyze DNA sequence data, which sorts species of *Morelia* between two clades. *Simalia* is resurrected for the clade that contains *M. amethistina* and related species.

Morelia azurea (A.B. Meyer, 1874)
Neotype: UTA R61633, 121 cm female (1990), designated by Barker et al. (2015).
Neotype locality: "Biak Island," Irian Jaya, Indonesia.
Distribution: Add Papua New Guinea (Bagabag Island), Clegg & Jocque (2016).
Comments: Natusch & Lyons (2014, as *M. viridis*) describe some morphological variation between populations, but did not discuss the taxonomic relevance of *M. azurea*.

Morelia bredli (Gow, 1981)

Morelia carinata (L.A. Smith, 1981)

Morelia imbricata L.A. Smith 1981. Rec. West. Austr. Mus. 9: 222–223, fig. 7. (*Python spilotus imbricatus*)
Holotype: WAM R54340, (N. Lang, 22 February 1976).
Type locality: "Jurien Bay, Western Australia in 30°18′S, 115°02′E," Australia.
Distribution: Australia (SW Western Australia, including Garden, Mondrain, N Twin Peak and West Wallabi Is.).
Comments: Barker et al. (2015) recognize *M. spilota imbricata* as a species without comment. We follow by recognizing it due to morphological uniqueness and allopatry.

†*Morelia riversleighensis* (L.A. Smith & Plane, 1985)

Morelia spilota (Lacépède, 1804)
Synonyms: Remove *Python spilotus imbricatus* L.A. Smith, 1981.
Distribution: Remove SW Western Australia, Barker et al. (2015).

Morelia viridis (Schlegel, 1872)

Distribution: Add Papua New Guinea (Oro), O'Shea et al. (2018b, suppl.).

Comments: Natusch & Lyons (2014) describe some morphological variation between populations.

MUHTAROPHIS Avci, Ilgaz, Rajabizadeh, Yilmaz, Üzüm, Adriaens, Kumlutas & Olgun, 2015. Russ. J. Herpetol. 22(3): 165. (Colubridae: Colubrinae)

Type species: *Rhynchocalamus barani* Olgun, Avci, Ilgaz, Üzüm & Yilmaz, 2007 by original designation.

Comments: Avci et al. (2015), using mtDNA sequence-data, find *Rhynchocalamus barani* to be paraphyletic with respect to *Lytorhynchus* and *R. melanocephalus,* and erect a new genus for *R. barani.*

Muhtarophis barani (Olgun, Avci, Ilgaz, Üzüm & Yilmaz, 2007)

Distribution: Add Turkey (Osmaniye), and elevation range 550–1300 m, Kariş & Göçmen (2018).

Comments: Avci et al. (2009) report a second specimen. Avci et al. (2015) provide a revised diagnosis and decription.

MUSSURANA Zaher, Grazziotin, Cadle, R.W. Murphy, Moura-Leite & Bonatto, 2009 (Dipsadidae: Xenodontinae)

Mussurana bicolor (Peracca1904)

Distribution: Add Paraguay (Boqueron, Concepcion, Paraguari, San Pedro), Cabral & Weiler (2014).

Mussurana montana (Franco, Marques & Puorto, 1997)

Distribution: Add Brazil (Minas Gerais, Rio de Janeiro), and elevation range 750–1610 m, Costa et al. (2015a).

Mussurana quimi (Franco, Marques & Puorto, 1997)

Distribution: Add Brazil (Mato Grosso do Sul, Rio Grande do Sul), Entiauspe-Neto et al. (2017a).

MYERSOPHIS Taylor, 1963 (Cyclocoridae)

Myersophis alpestris Taylor, 1963

MYRIOPHOLIS Hedges, Adalsteinsson & Branch *in* Adalsteinsson, Branch, Trape, Vitt & Hedges, 2009 (Leptotyphlopidae: Leptotyphlopinae)

Myriopholis adleri (Hahn & Wallach, 1998)

Distribution: Add Mali (Koulikoro), Trape & Mané (2017); Ghana (Northern), Boundy (2014).

Myriopholis algeriensis (Jacquet, 1898)

Distribution: Add Morocco (Figuig), Barata et al. (2011); Morocco (Tan Tan), Broadley et al. (2014); Morocco (Tiznit), Bouazza et al. (2018). Upper elevation of 1407 m, García-Cardenete et al. (2015). García-Cardenete et al. (2015), Bouazza et al. (2018) map all known localities.

Myriopholis blanfordii (Boulenger, 1890)

Distribution: Add Afghanistan (Laghman), Wagner et al. (2016b).

Myriopholis boueti (Chabanaud, 1917)

Distribution: Add Mali (Kayes), Trape & Mané (2017); Niger (Zinder), Trape & Mané (2015); Cameroon (Northern), Ineich & Prudent (2014). Upper elevation of 456 m, Trape & Mané (2015).

Myriopholis braccianii (Scortecci, 1928)

Myriopholis burii (Boulenger, 1905)

Myriopholis cairi (A.M.C. Duméril & Bibron, 1844)

Comments: Reference as *"Myriolepis" cairi* by Wallach et al. (2014) is a *lapsus.*

Myriopholis erythraeus (Scortecci, 1928)

Myriopholis filiformis (Boulenger, 1899)

Comments: Using DNA barcoding, Vasconcelos et al. (2016) find no intraspecific pairwise distance between specimens.

Myriopholis hamulirostris (Nikolsky, 1907)

Distribution: Add Iraq (Al-Anbar, Al-Sulaimaniyah, Arbil, Diyala, Kirkuk, Maysan, Wasit), Afrasiab & Ali (1996); Iraq (Babil, Muthanna), and low elevation of 4 m, Abbas-Rhadi et al. (2017); Iran (Alborz, East Azerbaijan, Golestan, Kermanshah, Lorestan, Mazandaran, Qazvin, Tehran, West Azerbaijan), Safaei-Mahroo et al. (2015).

Myriopholis ionidesi (Broadley & Wallach, 2007)

Myriopholis lanzai Broadley, Wade & Wallach, 2014. Arnoldia Zimbabwe 10: 353–356, fig. 1.

Holotype, MZUF 36519, a 190 mm adult (G. Garganese, summer, 1934).

Type Locality: "Ghat Oasis, south-west Fezzan, Libya (24°59′N: 10°11E) at 700 m a.s.l."

Distribution: SW Libya (Awban), N Chad (Borkou-Ennedi-Tibesti), 700 m.

Myriopholis longicaudus (W.C.H. Peters, 1854)

Myriopholis macrorhynchus (Jan, 1860 *in* Jan & Sordelli, 1860–1866)

Myriopholis macrurus (Boulenger, 1903)

Comments: Using DNA barcoding, Vasconcelos et al. (2016) find some intraspecific pairwise distance between specimens (mean 1.23).

Myriopholis narirostris (W.C.H. Peters, 1867)

Myriopholis occipitalis Trape & Chirio, 2019. Bull. Soc. Herpétol. France 169: 46–50, figs. 1–4.

Holotype: MNHN 1997.3350, a 172 mm specimen of unknown sex (villager, between October 1995 and June 1996).

Type locality: "Kouki (07°09′N/17°18′E) en République centrafricaine."

Distribution: S Chad (Mandoul), NW Central African Republic (Ouham).

Myriopholis parkeri (Broadley, 1999)

Distribution: Add Kenya (Coast), Boundy (2013).

Comments: Boundy (2013) describes a second known specimen.

Myriopholis perreti (Roux-Estève, 1979)

Myriopholis phillipsi (Barbour, 1914)

Distribution: Add Turkey (Batman, Sirnak), Uğurtaş et al. (2006); Turkey (Kilis), and low elevation of 270 m, Göçmen et al. (2009a).

Myriopholis rouxestevae (J.-F. Trape & Mané, 2004)

Distribution: Add Guinea, Trape & Baldé (2014). Remove Mali (Koulikoro), Trape & Mané (2017), which is *M. boueti*.

Myriopholis wilsoni (Hahn, 1978)

Comments: Using DNA barcoding, Vasconcelos et al. (2016) find moderate intraspecific pairwise distance between specimens (mean 4.20).

Myriopholis yemenicus (Scortecci, 1933)

MYRON Gray, 1849 (Homalopsidae)

Comments: J.C. Murphy & Voris (2014) provide a generic diagnosis.

Myron karnsi J.C. Murphy, 2011

Comments: J.C. Murphy & Voris (2014) provide a diagnosis and photograph of the holotype.

Myron resetari J.C. Murphy, 2011

Comments: J.C. Murphy & Voris (2014) provide a diagnosis and photograph.

Myron richardsonii Gray, 1849

Comments: J.C. Murphy & Voris (2014) provide a diagnosis and photograph.

MYRROPHIS Kumar, Sanders, George & J.C. Murphy, 2012 (Homalopsidae)

Comments: J.C. Murphy & Voris (2014) provide a generic diagnosis.

Myrrophis bennettii (Gray, 1842)

Comments: J.C. Murphy & Voris (2014) provide a diagnosis and photograph.

Myrrophis chinensis (Gray, 1842)

Distribution: Add China (Nan Ao Island), Qing et al. (2015); Vietnam (Hai Phong: Cat Ba Island), T.Q. Nguyen et al. (2011); Vietnam (Tam Dao), J.C. Murphy & Voris (2014).

Comments: J.C. Murphy & Voris (2014) provide a diagnosis and photograph.

NAJA Laurenti, 1768 (Elapidae)

Synonyms: Add *Palaeonaja* Hoffstetter, 1939.

Comments: Kurniawan et al. (2017) present a DNA sequence-based phylogeny that supports the recognition of *N. sputatrix* and *N. sumatrana* as species distinct from *N. atra* and *N. kaouthia*. The gene-based phylogeny from Zaher et al. (2019) supports the partition of *Naja* into four genera, including *Afronaja*, *Boulengerina*, and *Uraeus*.

Naja atra Cantor, 1842

Distribution: Add China (Nan Ao Island), Qing et al. (2015); Vietnam (Hai Phong: Cat Ba Island), T.Q. Nguyen et al. (2011).

Naja kaouthia Lesson, 1831

Distribution: Add Nepal (Chitwan), Pandey (2012); Myanmar (Tanintharyi), Mulcahy et al. (2018); Thailand (Chumphon),

Kurniawan et al. (2017); Thailand (Ayutthaya, Bangkok, Prachuap Khiri Khan, Ranong, Sukhothai, Surat Thani, Pha Ngan Island), Ratnarathorn et al. (2019), West Malaysia (Pulau Singa Besar), B.L. Lim et al. (2010).

Comments: Ratnarathorn et al. (2019) examine geographic variation in Thailand populations using mtDNA sequence data. There is a hierarchical system of clades between regions, and the northeastern Thailand populations are likely a cryptic species, being paraphyletic with respect to *N. atra* and *N. naja*.

Naja mandalayensis Slowinski & Wüster, 2000

Naja naja (Linnaeus, 1758)

Distribution: Add Nepal (Bara), Pandey et al. (2018); Nepal (Nawalparasi), Pandey (2012).

Naja oxiana (Eichwald, 1831)

Distribution: Add Iran (Golestan, North Khorasan, Sistan and Baluchistan, South Khorasan), Shafaei-Mahroo et al. (2015); Afghanistan (Badghis, Ghazni, Herat, Kunduz, Logar), Wagner et al. (2016b). E. Kazemi et al. (2019) document localities in Iran.

Naja philippinensis Taylor, 1922

Distribution: Add Philippines (Benguet), and upper elevation of 1040 m, Sy & Gerard (2014); Philippines (Cantanduanes Island), Sy & Vargas (2017); Philippines (Lubang Island), Sy & Balete (2017).

†*Naja romani* (Hoffstetter, 1939)

Distribution: Add late Miocene (Vallesian) of Greece, Georgialis et al. (2018, as *Naja* cf. *romani*).

Comments: V. Wallach called my attention to the paper by Szyndlar & Rage (1990) in which they associate *Palaeonaja romani* with the Asiatic, rather than African, lineage of *Naja* s.l., requiring its transfer from *Afronaja* of Wallach et al. (2014), along with the generic synonym *Palaeonaja*.

Naja sagittifera Wall, 1913
Naja samarensis W.C.H. Peters, 1861
Naja siamensis Laurenti, 1768
Naja sputatrix F. Boie, 1827

Naja sumatrana F. Müller, 1887

Distribution: Add West Malaysia (Pahang), Zakaria et al. (2014); Philippines (Busuanga I.), Sy et al. (2016b); Philippines (Calauit I.), Leviton et al. (2018).

†*NAJASH* Apesteguía & Zaher, 2006 (Ophidia: incerta sedis)

Comments: Palci et al. (2013a) discuss relationships based on reinterpretation of fossil material.

†*Najash rionegrina* Apesteguía & Zaher, 2006

Comments: Zaher et al. (2009) provide additional descriptive data and an emended diagnosis. Palci et al. (2013a) provide a further emended diagnosis based on reinterpretation of fossil specimens. Garberoglio et al. (2019) discuss much additional material from the region of the type locality.

NAMIBIANA Hedges, Adalsteinsson & Branch *in* Adalsteinsson, Branch, Trape, Vitt & Hedges, 2009 (Leptotyphlopidae: Leptotyphlopinae)

Namibiana gracillor (Boulenger, 1910)

Namibiana labialis (Sternfeld, 1908)

Namibiana latifrons (Sternfeld, 1908)

Synonyms: Add *Stenostoma scutatum* Peters 1865.

Comment: Boundy (2014) supports retaining the name *N. latifrons* over the senior synonym, *S. scutatum*, which is based on the same types.

Namibiana occidentalis (V.F.M. FitzSimons, 1962)

Namibiana rostrata (Bocage, 1886)

Distribution: Add Angola (Huila), Butler et al. (2019).

†*NANOWANA* Scanlon, 1987 (†Madtsoiidae)

†*Nanowana godthelpi* Scanlon, 1987

†*Nanowana schrenki* Scanlon, 1987

NATRICITERES Loveridge, 1953 (Natricidae)

Natriciteres bipostocularis Broadley, 1962

Distribution: Add Angola (Huambo), M.P. Marques et al. (2018); Angola (Lunda Norte), Branch (2018).

Natriciteres fuliginoides (Günther, 1858)

Distribution: Add Liberia, Trape & Baldé (2014).

Natriciteres olivacea (W.C.H. Peters, 1854)

Distribution: Add Niger, Trape & Mané (2015); Angola (Cuando-Cubango), Conradie et al. (2016b).

Natriciteres pembana (Loveridge, 1935)

Natriciteres sylvatica Broadley, 1966

Natriciteres variegata (W.C.H. Peters, 1861)

Distribution: Add Tanzania (Morogoro), Lyakurwa (2017).

NATRIX Laurenti, 1768 (Natricidae)

Synonyms: Fossil records: Add lower Miocene (Agenian) of Germany, Čerňanský et al. (2015, as *Natrix* sp.); middle Miocene (Orleanian) of Germany, Čerňanský et al. (2017, as *Natrix* sp. 1); upper Pliocene (Villanyian) of Germany, Čerňanský et al. (2017, as *Natrix* sp. 2); lower Pleistocene (Villanyian) of Bulgaria, Boev (2017); middle Pleistocene of China, Li et al. (2019, as cf. *Natrix* sp.).

Comments: Systematics of *Natrix natrix* is revised by Kindler & Fritz (2014, et seq.).

Natrix astreptophora Seoane, 1884. Ident. Lacerta schreiberi Lacerta viridis Investig. Herpetol. Galicia: 15–16. (*Tropidonotus natrix astreptophorus*)

Synonyms: *Tropidonotus natrix algericus* G. Hecht, 1930.

Syntypes: MNHN 1889.0580–0583, 2012.0459.

Type locality: "Galicia"; restricted to Coruña, Galicia, Spain by Mertens and Müller (1928).

Distribution: SC France, Spain, Portugal, N Morocco, N Algeria, N Tunisia. Add Morocco (Al Hoceima), Mediani et al. (2015b); Algeria (El Tarf), Rouag & Benyacoub (2006). Increase elevation to 3060 m, Fernández-Cardenete et al. (2000).

Fossil Records: Upper Pliocene (Villanyan) of Spain, Blain (2009); lower Pleistocene (Calabrian) of Spain, Blain et al. (2007); middle Pleistocene (Ionian) of Spain, Blain (2009) and Blain et al. (2014a); upper Pleistocene (Tarantian) of Spain, Blain et al. (2013b).

Comments: Pokrant et al. (2016) elevate this taxon to species based on morphology, mtDNA sequence, and parapatry with *Natrix helvetica*. Kindler et al. (2018a) use mtDNA sequence-data to examine phylogeography in *N. atreptophora*. They recover three clades: Iberian, N Morocco, and NE Algeria/NW Tunisia.

†*Natrix borealis* (Szyndlar, 1981)

Natrix helveticus Lacépède, 1789. Hist. Nat. Serpens 2: 100, 326–327. (*Coluber helveticus*)

Synonyms: *Coluber helvetus* Donndorff, 1798, *Vipera vissena* Rafinesque, 1814, *Natrix hybridus* Merrem, 1820, *Coluber siculus* Cuvier, 1829, *Natrix torquata minax* Bonaparte, 1834 *in* 1832–1841, *Natrix torquata murorum* Bonaparte, 1834 *in* 1832–1841, *Natrix cetti* Gené, 1839, *Coluber jenisonii* Gistel *in* Gistel & Bromme, 1850, *Natrix torquata nigrescens* De Betta, 1853, *Tropidonotus fallax* Fatio, 1872, *Tropidonotus natrix concolor* Ninni, 1880, *Tropidonotus natrix lineata* Ninni, 1880, *Tropidonotus natrix nigritorquata* Ninni, 1880, *Tropidonotus natrix albo-torquata* Camerano, 1891, *Tropidonotus natrix britannicus* G. Hecht, 1930, *Tropidonotus natrix corsus* G. Hecht, 1930, *Natrix natrix lanzai* Kramer, 1971, *Natrix natrix calabra* Vanni & Lanza, 1983.

Holotype: Unlocated.

Type locality: "Mont-Jorat," Switzerland.

Distribution: W Europe (France, Great Britain, Lichtenstein, Switzerland, W Germany at and W of the Rhine River, Belgium, Netherlands, Italy, and including Corsica, Sardinia, and Sicily.

Comments: See comments under *N. natrix* for Kindler & Fritz (2014) and Kindler et al. (2017, 2018b). Kindler & Fritz (2018) use mtDNA and 12 microsatellite loci to evaluate phylogeography within *N. helvetica*, concluding that there are five lineages that they recognize as subspecies, their geographic ranges defined by genes rather than strict morphological correspondence.

Fossil Records: Upper Pliocene (Ruscinian, MN 15) of Italy (Sardinia), Delfino et al. (2011, as *Natrix* sp.); middle Pleistocene (Ionian) of Italy, Delfino (2004); upper Pleistocene (Tarantian) of Belgium, Blain et al. (2014b), Italy, Delfino (2004); Late Pleistocene (post-Tarantian) of Belgium, Blain et al. (2019).

†*Natrix longivertebrata* Szyndlar, 1984

Natrix maurus (Linnaus, 1758)

Distribution: Add Andorra, Orriols & Fernàndez (2003); Spain (Pontevedra: Faro, Monteagudo, Ons, Onza, San Martiño Islands), Galán (2012); Morocco (Al Hoceima, Chefchaouen, Taounate), Mediani et al. (2015b); Algeria (Chlef, Tiaret), Ferrer et al. (2016); Algeria (El Tarf), Rouag & Benyacoub (2006); Libya (Murqub, Tripoli), Bauer et al. (2017). Upper elevation of 2650 m, Martínez-Freiría et al. (2017b).

Fossil Records: Add upper Pliocene (Villanyian) of Spain, Blain (2009); lower Pleistocene (Calabrian) of Spain, Blain et al. (2008); upper Pleistocene (Tarantian) of Spain, Barroso-Ruiz and Bailon (2003).

Comments: Joger et al. (2010) construct a phylogeny that indicates a stepwise hierarchical pattern of lineages.

†*Natrix merkurensis* Ivanov, 2002

†*Natrix mlynarskii* Rage, 1988

†*Natrix natricoides* (Augé & Rage *in* Ginsberg, 2000)

***Natrix natrix* (Linnaeus, 1758)**

Synonyms: *Natrix gronoviana* Laurenti, 1768, *Natrix vulgaris* Laurenti, 1768, *Coluber scutatus* Pallas, 1773, *Coluber bipes* Gmelin, 1789, *Coluber tyrolensis* Gmelin, 1789, *Coluber gronovius* Bechstein, 1802, *Coluber scopolianus* Daudin, 1803e, *Coluber persa* Pallas, 1811, *Coluber minutus* Pallas, 1814, *Tropidonotus ater* Eichwald, 1831, *Tropidonotus persicus* Eichwald, 1831, *Coluber bilineatus* Bibron & Bory de Saint-Vincent, 1832, *Coluber niger* Dwigubsky, 1832, *Coluber ponticus* Ménétriés, 1832, *Coluber bilineatus* Bibron & Bory, 1833, *Coluber natrix dalmatina* Schinz, 1833, *Tropidonotus natrix colchica* Nordmann *in* Demidoff, 1840, *Tropidonotus natrix nigra* Nordmann *in* Demidoff, 1840, *Tropidonotus natrix bilineatus* Jan, 1864a, *Tropidonotus natrix nigra* Jan, 1864a, *Tropidonotus natrix picturata* Jan, 1864a, *Tropidonotus natrix subbilineata* Jan, 1864a, *Tropidonotus natrix moreoticus* Bedriaga, 1881–1882, *Tropidonotus natrix bulsanensis* Gredler, 1882, *Tropidonotus natrix bithynius* G. Hecht, 1930a, *Tropidonotus natrix bucharensis* G. Hecht, 1930a, *Tropidonotus natrix cephallonicus* G. Hecht, 1930a, *Tropidonotus natrix cypriacus* G. Hecht, 1930a, *Tropidonotus natrix dystiensis* G. Hecht, 1930a, *Tropidonotus natrix schirvanae* G. Hecht, 1930a, *Tropidonotus natrix syriacus* G. Hecht, 1930a, *Tropidonotus natrix syrae* G. Hecht, 1930a, *Natrix natrix schweizeri* L. Müller, 1932b, *Natrix natrix gotlandica* Nilson & Andrén, 1981a, *Natrix megalocephala* Orlov & Tuniyev, 1986, *Natrix natrix fusca* Cattaneo, 1990.

The following names are not associated specifically with *N. astreptophora*, *N. helvetica*, or *N. natrix* s.s.: *Coluber bipedalis* Scopoli, 1788, *Coluber arabicus* Gmelin,1789, *Coluber torquatus* Lacépède, 1789, *Coluber azureus* Donndorff, 1798, *Coluber aesculapii* Sturm, 1799, *Coluber decorus* Gravenhorst, 1807, *Coluber discinctus* Gravenhorst, 1807, *Coluber irroratus* Gravenhorst, 1807, *Coluber bicephalus* Hufeland & Osann, 1825, *Coluber natrix minax* Fitzinger, 1826a (*nomen nudum*), *Coluber natrix murorum* Fitzinger, 1826a (*nomen nudum*), *Coluber scopolii* Risso, 1826, *Tropidonotus sparsus* Schreiber, 1875, *Tropidonotus natrix albiventris* Dürigen, 1897, *Tropidonotus natrix fasciatus* Dürigen, 1897, *Natrix vibakari continentalis* Nikolsky, 1925.

Distribution: E and N Europe through C Asia. Western boundary of the range includes the Balkan Peninsula, E Italy, Austria, E Germany roughly at and E of the Rhine River, Denmark, S Norway and S Sweden. Add Denmark (Nordfriesland Islands: Fanø, Rømø), Grosse et al. (2015); Germany (Nordfriesland Islands: Amrum, Helgoland, St. Peter-Ording, Sylt), Grosse et al. (2015); Liechtenstein, Kühnis (2006); Montenegro, Polović & Čađenović (2013); Greece (Kithira I.), Broggi (2016b); Greece (Tigani I.), Mossman

et al. (2016); Slovakia (Prešov), Pančišin & Klembara (2003) and Jablonski (2011); Turkey (Bartin), Çakmak et al. (2017); Turkey (Canakkale), Tok et al. (2006); Turkey (Igdir), Kaya & Özuluğ (2017); Turkey (Karabük), Kumlutaş et al. (2017); Turkey (Kutahya), Özdemřr & Baran (2002); Turkey (Mersin), Winden et al. (1997); Turkey (Gökçeada I.), Yakin et al. (2018); Israel, Martens (1996); Syria (Hamah), Martens (1996); Iraq (Baghdad), Afrasiab & Ali (2012); Iran (Ardabil, Golestan, Markazi, Semnan, Tehran, West Azerbaijan, Zanjan), Safaei-Mahroo et al. (2015); Iran (Qom), S.M. Kazemi et al. (2015). Göçmen et al. (2011), referring to populations formerly known as *Natrix megalocephala*, report records from Ardahan, Artvin, Giresun, Kars, Rize and Zonguldak provinces, Turkcy, and increase the elevational range to 1950 m. Litvinchuk et al. (2013) detail records from Buryatia Republic, Russia. Milto (2003) plots and discusses the distribution in northwestern Russia.

Fossil records: Add lower Pliocene (Ruscinian) of Slovakia, Čerňanský (2011); upper Pliocene (Villanyan) of Slovakia, Čerňanský (2011); middle Pleistocene (Ionian) of Russia, Ratnikov, (2002), Ukraine, Ratnikov (2002, 2005); upper Pleistocene (Tarantian) of Romania, Venczel (2000).

Comments: Kindler et al. (2013) and Kindler & Fritz (2014) resolve three geographic clades from mtDNA sequence data: one on the Iberian Peninsula and NW Africa, and a western and eastern clade roughly separated from Germany, southward to the Balkan Peninsula. Their analysis places specimens of *N. megalocepala* within a subclade with individuals of the eastern clade of *N. natrix*, and they formally synonymize the former with the latter. They also found limited correspondence between 16 subclades and currently recognized subspecies. Kindler et al. (2014) use mtDNA sequence data to demonstrate multiple lineage invasions of Fennoscandia during the Holocene. Kindler et al. (2017) use DNA sequence data to delineate contact zones between *Natrix* populations in central Europe. Based on molecular, morphological, and divergence-time data, they recognize western European populations as *N. helvetica*. In addition, they support the recognition of *N. astreptophora* as a species, as well as subspecies with correspondence between morphology and genes. Kindler et al. (2018b) reconstruct the glacial refugia and postglacial re-dispersal routes in *N. natrix* and *N. helvetica*. Dubey et al. (2017) document the presence of *N. n. persa* genes and phenotypes in native *N. helvetica* populations in W Switzerland, supposedly originating from escaped pets.

†*Natrix parva* Szyndlar, 1984

†*Natrix rudabanyaensis* Szyndlar, 2005

Fossil records: Add Miocene/Pliocene transition (late Turolian) of Greece, Georgialis et al. (2019, as *Natrix* aff. *rudabanyaensis*).

†*Natrix sansaniensis* (Lartet, 1851)

***Natrix tessellata* (Laurenti, 1768)**

Distribution: Add Germany (Nordfriesland Islands: Sylt), Grosse et al. (2015); Poland (Silesia), Vlček et al. (2010); Czech Republic (Kralovehradecky, Pardubicky), Lemberk

(2013); Czech Republic (Severomoravsky), Vlček et al. (2010); Romania (Alba, Arad, Bistrita-Nasaud, Covasna, Harghita, Maramures, Salaj, Satu-Mare, Sibiu), Ghira et al. (2002); Romania (Botosani), Strugariu et al. (2016); Romania (Giurgiu), Iftime & Iftime (2008); Romania (Tulcea), Strugariu et al. (2008); Turkey (Afyon), Eser & Erismis (2014); Turkey (Bartin), Çakmak et al. (2017); Turkey (Hatay), Uğurtaş et al. (2000b); Turkey (Kutahya), Özdemïr & Baran (2002); Turkey (Mersin), Winden et al. (1997); Iran (Alborz, Ardabil, Chaharmahal and Bakhtiari, Gilan, Golestan, Ilam, Kermanshah, Kohgiluyeh and Boyer Ahmad, Mazandaran, North Khorasan, Qazvin, Sistan and Baluchistan, South Khorasan, Tehran), Safaei-Mahroo et al. (2015); Iran (Isfahan, Markazi, Semnan), E. Rastaegar-Pouyani et al. (2017); Iran (Qom), S.M. Kazemi et al. (2015); Afghanistan (Bamyan, Faryab, Kabul, Wardak), Mebert el al. (2013); Afghanistan (Badghis, Herat, Jowzjan, Kunduz), Wagner et al. (2016b); Egypt (Ismailia, Port Said), Ibrahim (2012). Nekrasova et al. (2013) discuss the range limits in Ukraine. Buric & Baskiera (2014) and Vlcek et al. (2015) report on insular populations off Croatia. Strugariu et al. (2016) map known localities for Romania.

Fossil records: Add middle Pleistocene (Likhvinian) of Ukraine, Ratnikov, 2005).

Comments: Joger et al. (2010) construct a phylogeny that indicates a hierarchical pattern of lineages. Werner & Shapira (2011) describe morphological variation within and between populations in Israel. Franzen et al. (2016) use DNA sequence-data to determine the origin of introduced populations in Bavaria, Germany, and conclude that they arose from multiple sources in the native range. Kyriazi et al. (2012) use mtDNA sequence-data to determine that populations on Crete originated from southwestern Turkey. Marosi et al. (2012) discover three clades in eastern European populations, using mtDNA sequence-data, which correspond to those found in earlier studies. E. Rastegar-Pouyani et al. (2017) use mtDNA sequence-data to evaluate relationships of Iranian populations, compared with previously studied samples from throughout the range. Samples from western Iran were sister to all other populations, which formed two major clades: those of northeast Iran and central Asia, and those of Europe.

†NEBRASKOPHIS Holman, 1973 (Colubroidea: incerta sedis)

†Nebraskophis oligocenicus Holman, 1999
†Nebraskophis skinneri Holman, 1973

†NEONATRIX Holman, 1973 (Natricidae)

Distribution: Add lower Miocene (Agenian) of Germany, Čerňanský et al. (2015, as *Neonatrix* sp.); upper Miocene (Hemphillian) of USA (Tennessee), Jasinski & Moscato (2017, as *Neonatrix* sp.).

†Neonatrix crassa Rage & Holman, 1984
†Neonatrix elongata Holman, 1973
†Neonatrix europaea Rage & Holman, 1984
†Neonatrix infera Holman, 1996

†Neonatrix magna Holman, 1982
†Neonatrix nova Szyndlar, 1987

NERODIA Baird & Girard, 1853 (Natricidae)

Fossil record: Add upper Miocene (Hemphillian) of USA (Tennessee), Jasinski & Moscato (2017, as *Nerodia* sp.).

Nerodia clarkii (Baird & Girard, 1853)
Nerodia cyclopion (Duméril, Bibron & Duméril, 1854)
Nerodia erythrogaster (Forster *in* Bossu, 1771)

Distribution: Add Mexico (Chihuahua), Uriarte-Garzón & García-Vázquez (2014).

Comments: Boundy & David (2015) discuss the status and origin of the synonym *Tropidonotus roulei* Chabanaud, 1917.

Nerodia fasciata (Linnaeus, 1766)
Nerodia floridana (Goff, 1936)
Nerodia harteri (Trapido, 1941)
†Nerodia hibbardi (Holman, 1968)
†Nerodia hillmani (R.L. Wilson, 1968)
Nerodia paucimaculata (Tinkle & Conant, 1961)
Nerodia rhombifer (Hallowell, 1852)
Nerodia sipedon (Linnaeus, 1758)
Nerodia taxispilota (Holbrook, 1838)

†NIDOPHIS Vasile, Csiki-Sava, & Venczel, 2013. J. Vert. Paleontol. 33(5): 1102. (Madtsoiidae)

Type species: †*Nidophis insularis* Vasile, Csiki-Sava & Venczel, 2013 by original designation.

Distribution: Upper Cretaceous of Romania.

†Nidophis insularis Vasile, Csiki-Sava & Venczel, 2013, J. Vert. Paleontol. 33(5): 1102–1108, figs. 2–4.

Holotype: LPB (FGGUB) v.547/1, a mid-trunk vertebra.

Type locality: "Tuştea nesting site, Oltoane Hill, Tuştea Village, Haţeg Basin, Hunedoara County, Romania."

Distribution: Upper Cretaceous (Maastrichtian) of C Romania (Hunedoara).

†NIGEROPHIS Rage, 1975 (†Nigerophiidae)

†Nigerophis mirus Rage, 1975

NINIA Baird & Girard, 1853 (Dipsadidae: Dipsadinae)

Comments: Angarita-Sierra (2014) provides a review of the hemipenial morphology of the genus, plus a key to the genus and distribution maps.

Ninia atrata (Hallowell, 1845)

Distribution: Add Colombia (Bolivar, Casanare), Angarita-Sierra (2017); Colombia (confirmed for Guaviare), Medina-Rangel (2015); Colombia (Guajira), Meza-Joya (2015); Ecuador (Manabi), Angarita-Sierra & Lynch (2017).

Comments: Angarita-Sierra (2017) provides a species account.

Ninia celata McCranie & L.D. Wilson, 1995

Ninia diademata Baird & Girard, 1953

Distribution: Add Belize (Stann Creek), Gray & Hofmann (2017); Honduras (Lempira), McCranie (2014b); Honduras (Francisco Morazán), Solís et al. (2015). Remove Pacific region of Guatemala, and possibly coastal Mexico (Chiapas and Oaxaca), now considered *N. labiosa* according to Angarita-Sierra (2014).

Ninia espinali McCranie & L.D. Wilson, 1995

Ninia franciscoi Angarita-Sierra, 2014. South American J. Herpetol. 9(2): 121–122, figs. 3, 4a.

Holotype: UTA-R 22316, a 319 mm male (W.B. Montgomery & D. Resnick, 6 March 1988).

Type locality: "ca. 7.2 km Arima, Simla Research Station, Province of St. George, Trinidad (10°41′1″N, 61°17′W)."

Distribution: Trinidad & Tobago (northern range of Trinidad, 240 m).

Ninia hudsoni Parker, 1940

Distribution: Add Colombia (Caqueta), Rojas-Morales et al. (2018a).

Ninia labiosa (Bocourt, 1883 *in* A.H.A. Duméril, Bocourt & Mocquard, 1870–1909). Miss. Sci. Mex. Amér. Cent., Rept. 3(9): 550–551, pl. 32, fig. 6. (*Streptophorus labiosus*)

Holoype: MNHN 5944, a 313 mm specimen (from the "Société économique du Guatemala").

Type locality: "Guatemala"; restricted to Yepocapa, Depto. Chimaltenango by Burger and Werler (1954).

Distribution: Pacific region of Guatemala, and likely also adjacent portions of Mexico (Chiapas, Oaxaca). Angarita-Sierra (2014) includes a specimen from Veracruz, Mexico, which may require adding synonyms (e.g., *N. d. nietoi*) from *N. diademata*.

Comment: Angarita-Sierra (2014) revives this taxon based on hemipeneal differences with *N. diademeta*, with which he compares.

Ninia maculata (W.C.H. Peters, 1861)

Distribution: Add Panama (Panama, Panama Oeste), Ray & Ruback (2015).

Ninia pavimentata (Bocourt, 1883 *in* A.H.A. Duméril, Bocourt & Mocquard, 1870–1909)

Distribution: Add Honduras (Atlantida), Townsend et al. (2012). Low elevation of 310 m, Antúnez-Fonseca (2019).

Ninia psephota (Cope, 1875)

Ninia sebae (A.M.C. Duméril, Bibron & Duméril, 1854)

Distribution: Confirmation for Panama (Chiriqui), Geiger et al. (2014), listed in Wallach et al. (2014) with a question mark. Add El Salvador (Santa Ana), Juárez-Peña et al. (2016).

Ninia teresitae Angarita-Sierra & Lynch, 2017. Zootaxa 4244(4): 481–487, figs. 2–5.

Holotype: ICN 12527, a 429 mm male (L. Barrientos, 9 April 2010).

Type locality: "Santa Helena oil palm plantation (01°37′30″N, 78°44′20″W; 20 m asl), municipality of Tumaco, km 28 of the Tumaco-Llorente road, 1 km S of Tumaco, department of Nariño, Colombia."

Distribution: Western Colombia (Antioquia, Boyaca, Nariño, Risaralda), 20–1404 m; Angarita-Sierra (2018) adds Ecuador (Cotopaxi, Esmeraldas, Imbabura, Manabi, Pichincha).

†***NORISOPHIS*** Klein, Longrich, Ibrahim, Zouhri & Martill, 2017. Cret. Res. 72:135. (Ophidia: incerta sedis)

Type species: *Norisophis begaa* Klein, Longrich, Ibrahim, Zouhri & Martill, 2017 by original designation.

Distribution: Middle Cretaceous of Morocco.

†***Norisophis begaa*** Klein, Longrich, Ibrahim, Zouhri & Martill, 2017. Cret. Res. 72:135–136, figs. 2, 3.

Holotype: FSAC-KK 7001, a posterior trunk veretebra.

Type locality: "Kem Kem beds of Begaa, east of Taouz, [Ksar Es Souk,] southeastern Morocco…, ? Cenomanian. Locality Aferdou N'Chaft (30°53′57″ N, 3°50′46 W), or Aferdou n'Bou Tarif (30°53′12″ W, 3°52′29″ W), in the Ifezouane Formation."

Distribution: Middle Cretaceous (? Cenomanian) of Morocco.

NOTECHIS Boulenger, 1896 (Elapidae)

Notechis ater (Krefft, 1866)

Notechis scutatus (W.C.H. Peters, 1861)

NOTHOPSIS Cope, 1871 (Dipsadidae: Dipsadinae)

Comments: Pyron et al. (2015) consider this genus as the only member of the tribe Nothopsini.

Nothopsis rugosus Cope, 1871

Distribution: Add Nicaragua (Atlantico Norte), Sunyer et al. (2014); Panama (Panama Oeste), Ray & Ruback (2015).

Comments: Pyron et al. (2015) provide a description.

†***NUBIANOPHIS*** Rage & C. Werner, 1999 (†Nigerophiidae)

†***Nubianophis afaahus*** Rage & C. Werner, 1999

OCYOPHIS Cope, 1866 (Dipsadidae: Xenodontinae)

Ocyophis ater (Gosse, 1851)

Ocyophis melanichnus (Cope, 1862)

OGMODON W.C.H. Peters, 1864 (Elapidae)

Ogmodon vitianus W.C.H. Peters, 1864

†***OGMOPHIS*** Cope, 1884 (Booidea: incerta sedis)

Comments: K.T. Smith (2013) assigns *O. compactus* to the Loxocemidae.

†***Ogmophis arenarum*** Douglass, 1903

†***Ogmophis compactus*** Lambe, 1908

Distribution: Add late Eocene (Chadronian) of USA (North Dakota), K.T. Smith (2013).

†***Ogmophis europaeus*** Szyndlar *in* Mlynarski, Szyndlar, Estes & Sanches, 1982

†***Ogmophis miocompactus*** Holman, 1976

†***Ogmophis oregonensis*** Cope, 1884

†***Ogmophis pliocompactus*** Holman, 1975

†***Ogmophis voorhiesi*** Holman, 1977

OLIGODON H. Boie *in* Fitzinger, 1826 (Colubridae: Colubrinae)

Comments: H.N. Nguyen et al. (2020) present a phylogeny of 22 of the species based on genetic sequence data.

Oligodon affinis Günther, 1862

Oligodon albocinctus (Cantor, 1839)

Distribution: Hasan et al. (2013) confirm the presence in Chittagong, Bangladesh.

Oligodon ancorus (Girard, 1858)

Distribution: Add Philippines (Luzon Island: Aurora Prov., 110 m), Supsup (2016).

Oligodon annamensis Leviton, 1953

Distribution: Add Cambodia (Pursat), and upper elevation of 916 m, Neang & Hun (2013), Vietnam (Dak Lak), H.N. Nguyen et al. (2020).

Comments: H.N. Nguyen et al. (2020) provide a revised description based on new specimens, and a locality map and color photographs.

Oligodon annulifer (Boulenger, 1893)

Oligodon arenarius Vassilieva, 2015. Zootaxa 4058(2): 211–220, figs. 2–6.

Holotype: ZMMU R14503 (field no. ZMMU ABV 813), a 351 mm male (A.B. Vassilieva, 14 November 2014).

Type locality: "Binh Chau-Phuoc Buu Nature Reserve, Xuen Moc District, Ba Ria-Vung Tau Province, southern Vietnam…, coordinates 10°29′46″N, 107°27′54″E, elevation 5 m a.s.l."

Distribution: S Vietnam (Ba Ria-Vung Tau), 5–43 m. Known only from the type locality.

Oligodon arnensis (Shaw, 1802)

Distribution: Add Pakistan (Punjab), Rais et al. (2012); India (Gujarat), Parmar & Tank (2019); India (Madhya Pradesh), Manhas et al. (2018a); Nepal (Nawalparasi), Pandey (2012).

Oligodon barroni (M.A. Smith, 1916)

Distribution: Add Thailand (Krabi: Phi Phi Don I.), Milto (2014).

Oligodon bitorquatus (F. Boie, 1827)

Distribution: Add Indonesia (Bali), 70 m, Lilly (2013).

Oligodon booliati Leong & Grismer, 2004

Oligodon brevicaudus Günther, 1862

Oligodon calamarius (Linnaeus, 1758)

Oligodon catenatus (Blyth, 1854)

Distribution: Add Vietnam (Son La), Pham et al. (2014).

Oligodon cattienensis Vassilieva, Geissler, Galoyan, Poyarkov, Van Devender & Böhme 2013. Zootaxa 3702(3): 235–240, figs. 1–5.

Holotype: ZMMU R13865, a 415 mm male (A.B. Vassilieva, 20 January 2011).

Type locality: "the environs of Ben Cu forest station, Nam Cat Tien sector, Cat Tien National Park, Dong Nai Province, southern Vietnam (11°26′ 03″ N, 107° 25′ 42″ E, 130 m a.s.l.)."

Distribution: S Vietnam (Dong Nai), 130–167 m. Known only from the type locality.

Comments: In the *cyclurus* species group.

Oligodon chinensis (Günther, 1888)

Distribution: Vietnam (Bac Giang), Hecht et al. (2013); Vietnam (Ha Giang), Ziegler et al. (2014); Vietnam (Hai Phong: Cat Ba Island), T.Q. Nguyen et al. (2011); Vietnam (Phu Tho), Vassilieva (2015); Vietnam (Quang Ngai), Nemes et al. (2013); Vietnam (Son La), Pham et al. (2014).

Oligodon cinereus (Günther, 1864)

Distribution: Add Vietnam (Ba Ria-Vung Tau [Con Dao Islands], Binh Phuoc, Ho Chi Minh, Thua Thien Hue), T.Q. Nguyen et al. (2014); Vietnam (Dong Nai), Vassilieva (2015); Vietnam (Vung Tau), S.N. Nguyen et al. (2017b).

Oligodon condaoensis S.N. Nguyen, Nguyen, Le & Murphy, 2016. Zootaxa 4139(2): 263–266, figs. 2, 3.

Holotype: ITBCZ 2595, a 531 mm male (S.N. Nguyen, 24 October 2015).

Type locality: Hon Ba Island, Con Dao District, Ba Ria-Vung Tau Prov., Vietnam; coordinates 8°39′03″N, 106°33′29″E; elevation 15 m a.s.l."

Distribution: Vietnam (Ba Ria-Vung Tau), 15 m. Known only from the type locality.

Oligodon cruentatus (Günther, 1868)

Oligodon culaochamensis S.N. Nguyen, Nguyen, Nguyen, Phan, Jiang & Murphy, 2017b. Zootaxa 4286(3): 335–340, figs. 2, 4, 5.

Holotype: ITBCZ 5646, a 543 mm male (L.T. Nguyen, V.D.H. Nguyen & S.N. Nguyen, 27 July 2016).

Type locality: "Hon Lao, Cu Lao Cham Islands, Quang Nam province, Vietnam;… 15°56′38″N, 108°30′44″E; elevation 41 m a.s.l."

Distribution: Vietnam (Hon Lo Island, Quang Nam), 41–106 m.

Oligodon cyclurus (Cantor, 1839)

Oligodon deuvei David, Vogel & Rooijen, 2008

Distribution: Add Thailand (Loei), and upper elevation of 270 m, Pauwels et al. (2017e); Vietnam (Ba Ria-Vung Tau), Vassilieva (2015).

Oligodon dorsalis (Gray, 1835 *in* Gray & Hardwicke, 1830–1835

Oligodon eberhardti Pellegrin, 1910

Oligodon erythrogaster Boulenger, 1907

Oligodon erythrorhachis Wall, 1910

Oligodon everetti Boulenger, 1893

Oligodon fasciolatus (Günther, 1864)

Distribution: Add Laos (Vientiane Capital), Teynié et al. (2004); Vietnam (Ba Ria-Vung Tau [Con Dao Islands], S.N. Nguyen et al. (2016); Vietnam (Dien Bien), Dung et al. (2014); Vietnam (Vung Tau), S.N. Nguyen et al. (2017b); Vietnam (Yen Bai), Le et al. (2018, as *O. cyclurus*).

Oligodon forbesi (Boulenger, 1883)

Distribution: Add Indonesia (Barat Daya, Selaru Islands), Lang (2013).

Oligodon formosanus (Günther, 1872)

Distribution: Add Vietnam (Hai Duong), S.N. Nguyen et al. (2016).

Oligodon hamptoni Boulenger, 1918

Oligodon huahin Pauwels, Larsen, Suthanthangjai, David & Sumontha, 2017e. Zootaxa 4291(3): 532–540, figs. 1–9.

Holotype: QSMI 1501, a 526 mm male (H. Larsen, 20 December 2016).

Type locality: "road 3218 to Pala-U waterfall (12.528768 N, 99.527812 E), about 2 km east of the entrance gate to Kaeng Krachan National Park, Hua Hin District, Prachuap Khiri Khan Province, peninsular Thailand."

Distribution: Thailand (Prachuap Khiri Khan), at the type locality.

Oligodon inornatus (Boulenger, 1914)

Distribution: Add Thailand (Chanthaburi), Chan-ard et al. (2011).

Oligoson jintakunei Pauwels, Wallach, David & Chanhome, 2002

Oligodon joynsoni (M.A. Smith, 1917)

Oligodon juglandifer (Wall, 1909)

Distribution: Add Bhutan (Punakha), Sapkota & Sharma (2017). Elevation 800–1700 m, Chettri et al. (2010).

Oligodon kampucheaensis Neang, Grismer & Daltry, 2012

Oligodon kheriensis Acharji & Ray, 1936

Distribution: Add Nepal (Chitwan), Pandey (2012); India (Assam), and low elevation of 54 m, Sutradhar & Nath (2013).

Comments: Sutradhar & Nath (2013) describe what may be a fourth specimen of *O. kheriensis*.

Oligodon lacroixi Angel & Bourret, 1933

Distribution: Add Vietnam (Phu Tho), Vassilieva (2015).

Oligodon lungshenensis Zheng & Huang *in* Huang, Zheng & Fang, 1978

Oligodon macrurus (Angel, 1927)
Oligodon maculatus (Taylor, 1918)
Oligodon mcdougalli Wall, 1905
Oligodon melaneus Wall, 1909
Oligodon melanozonatus Wall, 1922
Oligodon meyerinkii (Steindachner, 1891)
Oligodon modestus Günther, 1864
Oligodon moricei David, Vogel & Rooijen, 2008

Oligodon mouhoti (Boulenger, 1914)

Distribution: Add Thailand (Chanthaburi), Chan-ard et al. (2011).

Oligodon nagao David, Nguyen, Nguyen, Jiang, Chen, Teynié & Ziegler, 2012

Oligodon nikhili Whitaker & Dattatri, 1982

Oligodon notospilus Günther, 1873

Distribution: Add Philippines (Calauit I.), Leviton et al. (2018). Leviton et al. (2018) consider the type locality of Mindanao to be in error in light of the known distribution of the species.

Oligodon ocellatus (Morice, 1875)
Oligodon octolineatus (Schneider, 1801)
Oligodon ornatus Van Denburgh, 1909
Oligodon perkinsi (Taylor, 1925)
Oligodon petronellae Roux *in* Rooij, 1917
Oligodon planiceps (Boulenger, 1888)
Oligodon propinquus Jan, 1862

Oligodon pseudotaeniatus David, Vogel & Rooijen, 2008

Oligodon pulcherrimus F. Werner, 1909

Synonyms: Add *Oligodon durheimi* Baumann, 1913, fide Tillack & Günther (2010).

Oligodon purpurascens (Schlegel, 1837)

Distribution: Add West Malaysia (Pulau Tioman), K.K.P. Lim & Lim (1999).

Oligodon rostralis H.N. Nguyen, Tran, Nguyen, Neang, Yushchenko & Poyarkov, 2020. PeerJ 8(e8332): 12–20, figs. 3, 4, 6, 7.

Holotype: SIEZC 20201, a 582 mm male (B.V. Tran & L.H. Nguyen, 13 June 2017).

Type locality: "Bidoup-Nui Ba National Park, ca. 6 km northwards from Da Nhim village, Da Chais Commune, Lac Duong District, Lam Dong Province, southern Vietnam (12.1518°N, 108.5279°E; elevation 1,622 m a.s.l.)."

Distribution: S Vietnam (Lam Dong), 1622 m. Known only from the holotype.

Oligodon saintgironsi David, Vogel & Pauwels, 2008

Oligodon saiyok Sumontha, Kunya, Dangsri & Pauwels, 2017. Zootaxa 4294(3): 317–319, figs. 1–7.

Holotype: QSMI 1506, a 718 mm male (K. Kunya, M. Sumontha & S. Dangsri, 7 October 2009).

Type locality: "Wat Tham Benjarat Nakhon (= Benjarat Nakhon Cave Temple), Sai Yok District, Kanchanaburi Province, western Thailand."

Distribution: WC Thailand (Kanchanaburi). Known from two specimens from the type locality.

Oligodon signatus (Günther, 1864)

Distribution: Add West Malaysia (Pulau Tioman), K.K.P. Lim & Lim (1999).

Oligodon splendidus (Günther, 1875)

Oligodon sublineatus A.M.C. Duméril, Bibron & Duméril, 1854

Lectotype: MNHN 3238, a 289 mm female, collector unknown, designated by Amarasinghe et al. (2015a).

Type locality: Restricted to "Sri Lanka" by Amarasinghe et al. (2015a).

Distribution: Sri Lanka (Central, Eastern, North Central, North Western, Sabaragamuwa, Southern, Uva, Western), 10–1600 m.

Comment: Amarasinghe et al. (2015a) designate a lectotype, give detailed descriptions of the lectotype and paralectotype, and provide a diagnosis and description for the species. They believe it may represent a cryptic species complex due to the range of morphological variation.

Oligodon taeniatus (Günther, 1861)

Distribution: Add Vietnam (Binh Thuan), S.N. Nguyen et al. (2017b); Vietnam (Hoa Binh), Nguyen et al. (2018).

Comments: David et al. (2011) report upon and describe the holotype (MNHN 598) of the synonym *Simotes quadrilineatus* Jan, 1865 *in* Jan & Sordelli, 1860–1866.

Oligodon taeniolatus (Jerdon, 1854)

Distribution: Add Iran (Golestan, North Khorasan, West Azerbaijan, Zanjan), Shafaei-Mahroo et al. (2015); Afghanistan (Kandahar), Wagner et al. (2016b); India (Gujarat, Madhya Pradesh, Uttarakhand) and elevation range of 250–1068 m, Seetharamaraju et al. (2011); Bhutan (Trongsa, Wangdue Phodrang), Tshewang & Letro, 2018).

Oligodon theobaldi (Günther, 1868)

Distribution: Add Myanmar (Sagaing), Pauwels et al. (2017e).

Oligodon torquatus (Boulenger, 1888)

Oligodon travancoricus Beddome, 1877

Oligodon trilineatus (A.M.C. Duméril, Bibron & Duméril, 1854)

Oligodon unicolor (Kopstein, 1926)

Distribution: Add Indonesia (Selaru Island), Lang (2013).

Oligodon venustus (Jerdon, 1854)

Oligodon vertebralis (Günther, 1865)

Oligodon waandersi (Bleeker, 1860)

Oligodon wagneri David & Vogel, 2012

Oligodon woodmasoni (Sclater, 1891)

OMOADIPHAS Köhler, L.D. Wilson & McCranie, 2001 (Dipsadidae: Dipsadinae)

Omoadiphas aurula Köhler, L.D. Wilson & McCranie, 2001

Omoadiphas cannula McCranie & Cruz-Díaz, 2010

Omoadiphas texiguatensis McCranie & Castañeda, 2004

OOCATOCHUS Helfenberger, 2001 (Colubridae: Colubrinae)

Oocatochus rufodorsatus (Cantor, 1842)

Distribution: Add China (Nan Ao Island), Qing et al. (2015). Stein & Kalinina (2016) report a westward range extension in E Russia.

OPHEODRYS Fitzinger, 1843 (Colubridae: Colubrinae)

Opheodrys aestivus (Linnaeus, 1766)

Opheodrys vernalis (Harlan, 1827)

†*OPHIDIONISCUS* Kuhn, 1963. Fossil. Catal. (103): 35. Substitute name for †*Ophidion* Pomel 1853. (Booidea: incerta sedis)

Synonyms: †*Ophidion* Pomel, 1853.

Type species: †*Ophidion antiquus* Pomel, 1853 by monotypy.

Distribution: Lower Miocene of France.

†*Ophidioniscus antiquus* Pomel, 1853. Cat. Méth. Vert. Foss.: 128. (†*Ophidion antiquus*)

Holotype: lost according to Rage (1984).

Type locality: "Langy," Aquitane, France.

Distribution: Lower Miocene of France.

OPHIOPHAGUS Günther, 1864 (Elapidae)

Ophiophagus hannah (Cantor, 1836)

Distribution: Add Pakistan (Azad Kashmir), Faiz et al. (2017); India (Goa), Yadav & Yankanchi (2015); Nepal (Bara), Pandey et al. (2018); Nepal (Nawalparasi), Pandey (2012); West Malaysia (Pulau Singa Besar), B.L. Lim et al. (2010); Philippines (Bohol I. [Sy, 2017], Cebu I. [Sy, 2016a], Romblon I. [Sy & Wallbank, 2013], Leyte I. [Sy & Boos, 2015a], Luzon Island: Ilocos Norte [Urriza et al., 2017], Luzon Island: Neuva Ecija [Sy et al., 2016a], Luzon Island: Pangasinan [Sy et al., 2015], Luzon Island: Nueva Vizcaya [Sy & Labatos, 2017], Luzon Island: Sorsogon [Sy & Letana, 2017], Mindanao Island: Zamboanga del Norte [Sy, 2016b], Pangulasian I. [Sy & Dichaves, 2017]); Philippines (Catanduanes, Panay, Polillo Is.), Leviton et al. (2018); Philippines (Samar), Fernandez et al. (2019). Remove India (Gujarat), Patel & Vyas (2019).

Comments: Suntrarachun et al. (2014) use mtDNA sequence-data from a dozen Thai specimens that indicate the presence of five haplotypes distributed among two primary clades. They suggest the possibility that the southern and northern clades are specifically distinct.

OPHRYACUS Cope, 1887 (Viperidae: Crotalinae)

Comment: Grünwald et al. (2015) revise the genus with description of a new species and revival of another.

Ophryacus smaragdinus Grünwald, Jones, Franz-Chávez & Ahumada-Carrillo, 2015. Mesoamerican Herpetol. 2(4): 391–394, figs. 1–3, 4a, c, 9a, b, c.

Holotype, MZFC 29290, a 552 mm female (J.M. Jones & I. T. Ahumada-Carrillo, 2 November 2013).

Type locality: "Los Ocotes (20.601542, -98.469708…; elev. 2,150 m asl), Zilacatipan, Municipio de Huayacocotola, Veracruz, Sierra Madre Oriental, Mexico."

Distribution: EC Mexico (EC Hidalgo, WC Veracruz, NE Puebla, NC Oaxaca), 1400–2340 m. Martínez-Vaca et al. (2016) map known and new localities.

Ophryacus sphenophrys H.M. Smith, 1960. Trans. Kansas Acad. Sci. 62(4): 267–271, plates 1, 2. (*Bothrops sphenophrys*)

Holotype: UIMNH 6262, a 461 mm male (W.L. Burger, 22 July 1949).

Type locality: "La Soledad, Oaxaca, Mexico, about 6000 ft."

Distribution: Mexico (SC Oaxaca), 1340–1460 m.

Comment: Revived from the synonymy as a valid species, based on morphology of three known individuals, by Grünwald et al. (2015).

Ophryacus undulatus (Jan, 1859)

Synonyms: Remove *Bothrops sphenophrys* H.M. Smith, 1960.

Distribution: S Mexico (Guerrero, Oaxaca, SE Puebla, WC Veracruz), 1800–2800 m.

Comment: Grünwald et al. (2015) revise the hypodigm of *O. undulatus*.

OPISTHOTROPIS Günther, 1872 (Natricidae)

Synonyms: Remove *Trimerodytes* Cope, 1895, *Liparophis* Peracca, 1904.

Comments: Teynié et al. (2014) and Ren et al. (2017) each provide a key to the species. Y. Wang et al. (2017) use mtDNA sequence-data to produce a phylogeny for ten of the species.

Ren et al. (2019) do the same using 15 species, one of which, *O. balteata*, is removed to *Trimerodytes*.

Opisthotropis alcalai W.C. Brown & Leviton, 1961

Opisthotropis andersonii (Boulenger, 1888)

Distribution: Add China (Guangdong), and upper elevation of 740 m, Y. Wang et al. (2017).

Comments: Y. Wang et al. (2017) provide a revised diagnosis and color photographs.

Opisthotropis ater Günther, 1872

Opisthotropis cheni Zhao, 1999

Distribution: Add China (Guangxi), Ren et al. (2019).

Opisthotropis cucae David, Pham, Nguyen & Ziegler, 2011

Opisthotropis daoventieni Orlov, Darevsky & Murphy, 1998

Distribution: Upper elevation of 1170 m, Teynié et al. (2014).

Opisthotropis durandi Teynié, Lottier, David, T.Q. Nguyen & Vogel, 2014. Zootaxa 3774(2): 167–174, figs. 1–4.

Holotype: MNHN 2013l.1001, a 538 mm female (A. Teynié & A. Lottier, 18 Sept. 2012).

Type locality: "vicinity of Muang Ngoi (20°42′10″N, 102°41′21″E), Ngoi District, Louangphabang Province, Lao People's Democratic Republic (Laos), at an elevation of about 370 m a.s.l."

Distribution: Laos (Louangphabang), 370–700 m. Known only from the vicinity of the type locality.

Opisthotropis guangxiensis Zhao, Jiang & Huang, 1978

Holotype: Corrected to CIB 009972, Guo et al. (2012a).

Distribution: Add China (Guangdong), Y. Wang et al. (2017).

Opisthotropis haihaensis Ziegler, Pham, Nguyen, Nguyen, Wang, Wang, Stuart & Le, 2019c. Zootaxa 4613(3): 580–583, figs. 2, 3.

Holotype: IEBR A2016.34, a 509 mm female (C.T. Pham & T.V. Nguyen, 9 May 2016).

Type locality: "forest near Tai Chi Village, Quang Son Commune, Hai Ha District, Quang Ninh Province, 950 m asl., Vietnam."

Distribution: N Vietnam (Quang Ninh), 950 m. Known only from the holotype.

Opisthotropis jacobi Angel & Bourret, 1933

Comments: Ziegler et al. (2018b) provide a revised description using additional specimens.

Opisthotropis kikuzatoi (Okada & Takara, 1958)

Opisthotropis kuatunensis Pope, 1928

Distribution: Add China (Guangdong), Y. Wang et al. (2017).

Opisthotropis lateralis Boulenger, 1903

Distribution: Add China (Guangdong, Hainan), Y. Wang et al. (2017); China (Yunnan), Ren et al. (2019); Vietnam (Quang Nam), Teynié et al. (2014);

Vietnam (Hai Duong), Hecht et al. (2013).

Opisthotropis latouchii (Boulenger, 1899)

Opisthotropis laui Yang, Sung & Chan, 2013. Zootaxa 3646(3), 290–294, figs. 1–3.

Holotype: KIZ 060100, a 299 mm female (B.P.L. Chan & M.W.N. Lau, 26 July 2002).

Type locality: "Beifengshan Forest Park, Mt. Gudou, Jiangmen City, Guangdong Province, China (22°14′20″ N, 112°55′05″ E, ca. 300 m above sea level)."

Distribution: China (Guangdong, including Shangchuan Island), 108–300 m.

Comments: Based on additional specimens, J. Wang et al. (2017) provide a revised description and color photographs.

Opisthotropis maculosus B.L. Stuart & Chuaynkern, 2007

Distribution: Add Vietnam (Quang Ninh), Ren et al. (2019).

Opisthotropis maxwelli Boulenger, 1914

Distribution: Add China (Guangdong: Nan Ao I.), and elevation range of 425–995 m, Y. Wang et al. (2017).

Comments: Y. Wang et al. (2017) provide a revised diagnosis and color photographs.

Opisthotropis rugosus Lidth de Jeude, 1890)

Opisthotrops shenzhenensis Y. Wang, Guo, Liu, Lyu, Wang, Luo, Sun & Zhang, 2017. Zootaxa 4247(4): 402–405, figs. 5, 6.

Holotype: SYS r001018, a 407 mm male (J. Zhao & R.-L. Li, 15 September 2014).

Type locality: "Mt. Wutong (22°34′54.8″N, 114°12′2.7″E; 260 m a.s.l.), Shenzhen City, Guangdong Province, China."

Distribution: China (C Guangdong), 155–327 m.

Opisthotropis spenceri M.A. Smith, 1918

Opisthotropis tamdaoensis Ziegler, David & Vu, 2008

Comments: Ziegler et al. (2017) provide a revised description based on new specimens. Based on mtDNA sequence-data, *tamdaoensis* is the sister taxon to *O. lateralis*.

Opisthotropis typicus (Mocquard, 1890)

Opisthotropis voquyi Ziegler, David, Ziegler, Pham, Nguyen & Le, 2018b. Zootaxa 4374(4), 481–487, figs. 6, 8–10.

Holotype: IEBR 4326, a 443 mm male (C.T. Pham et al., 14 June 2013).

Type locality: "Tay Yen Tu Nature Reserve, Bac Giang Province, 437 m asl., Vietnam."

Distribution: Vietnam (Bac Giang), 437 m. Known only from the type locality.

Opisthotropis zhaoermii Ren, Wang, Jiang, Guo & Li, 2017. Zool. Res. 38(5): 257–259, figs. 1–4.

Holotype: CIB 109999, a 587 mm female (J.-L. Ren, S.-B. Su, 24 August 2017).

Type locality: "Zuolong Gorges, Guzhag, Tujia-Miao of western Hunan, China (N28°42′17.88″, E109°55′26.26″, 561 m a.s.l."

Distribution: C China (W Hunan), 561 m. Known only from the type locality.

OREOCALAMUS Boulenger, 1899 (Colubridae: Colubrinae)

Comments: Within the gene-based phylogeny of Zaher et al. (2019), *Oreocalamus* is sister taxon to the remaining Colubridae: Colubrinae, rather than the Calamariidae where it had long been assigned.

Oreocalamus hanitschi Boulenger, 1899

OREOCRYPTOPHIS Utiger, Schätti & Helfenberger, 2005 (Colubridae: Colubrinae)

Oreocryptophis porphyraceus (Cantor, 1839)

Distribution: Add Vietnam (Dien Bien), Dung et al. (2014); Vietnam (Ha Giang), Ziegler et al. (2014); Vietnam (Dien Bien, Lam Dong, Thanh Hoa, Tuyen Quang), Hoefer (2019); Vietnam (Yen Bai), Le et al. (2018); West Malaysia (Pulau Tioman), K.K.P. Lim & Lim (1999).

Comments: P. Wang et al. (2019) recognize four subspecies in mainland China based on morphology, in addition to five extralimital subspecies.

ORIENTOCOLUBER Kharin, 2011 (Colubridae: Colubrinae)

Orientocoluber spinalis (W.C.H. Peters, 1866)

ORTHRIOPHIS Utiger, Helfenberger, Schätti, Schmidt, Ruf & Ziswiler, 2002 (Colubridae: Colubrinae)

Comments: See under *Elaphe*.

Orthriophis cantoris (Boulenger, 1894)

Orthriophis hodgsonii (Günther, 1860)

Distribution: Low elevation of 831 m, Singh & Joshi (2018).

Orthriophis moellendorffi (Boettger, 1886)

Orthriophis taeniurus (Cope, 1861)

Synonyms: *Orthriophis taeniurus helfenbergeri* Schulz, 2010. Sauria 32(2): 5–7, substitute name for *Coluber taeniurus pallidus* Rendahl, 1937.

Orthriophis taeniurus callicyanous Schulz, 2010. Sauria 32(2): 10–16, figs. 8–12, 14–17.

Holotype: ZFMK 81450, a 1658 mm male (T. Ziegler, 29 July 1997).

Type locality: "Ky Anh-Ke Go (bei Chin-Xai), Ha Tinh, Vietnam."

Distribution: Add Myanmar (Kayin, Shan), Schulz et al. (2015); Thailand (Chaiyaphum, Kanchanaburi, Khno Kaen, Lampang, Lamphun, Mae Hong Son, Phitsanulok, Tak), Schulz (2010); Cambodia (Koh Kong, Pursat), Schulz (2010); Vietnam (Ha Giang), Ziegler et al. (2014); Vietnam (Nghe Anh, Thanh Ho), Schulz (2010).

Comments: Schulz (2010) discusses geographic variation, and recognizes nine subspecies. Schulz et al. (2015) report on their study of the type material of the synonyms *Coluber nuthalli* Theobald, 1868 and *Elaphis yunnanensis* Anderson, 1879. They conclude that the two names represent the same subspecies, and propose nomen protectum status for *E. yunnanensis*. They provide a key to the nine subspecies of *E. taeniurus*.

OVOPHIS Burger *in* Hoge & Romano-Hoge, 1981 (Viperidae: Crotalinae)

Comments: Fong et al. (2017) produce a phylogeny based on mtDNA sequence-data for five species (excluding *O. gracilis* and *O. okinavensis*). *Ovophis monticolus* is paraphyletic with respect to its former subspecies *O. convictus* and *O. zayuensis*.

Ovophis convictus (Stoliczka, 1870)

Distribution: Add Cambodia (Pursat), and low elevation range of 545 m, Neang et al. (2011)

Ovophis gracilis (Oshima, 1920)

Comments: This species and *O. okinavensis* are sister taxa to *Gloydius*, and are not members of *Ovophis* (Alencar et al., 2016).

Ovophis makazayazaya (Takahashi, 1922)

Distribution: Remove Hong Kong, Fong et al. (2017); see under *O. tonkinensis*.

Ovophis monticola (Günther, 1864)

Distribution: Add Vietnam (Dien Bien), Dung et al. (2014).

Ovophis okinavensis (Boulenger, 1892)

Comments: This species and *O. gracilis* are sister taxa to *Gloydius*, and are not members of *Ovophis* (Alencar et al., 2016).

Ovophis tonkinensis (Bourret, 1934)

Distribution: Add Hong Kong, Fong et al. (2017).

Ovophis zayuensis (Jiang *in* Zhao & Jiang, 1977)

Holotype: Corrected to CIB 013375, Guo et al. (2016a).

OXYBELIS Wagler, 1830 (Colubridae: Colubrinae)

Comments: Jadin et al. (2019) produce a DNA sequence-based phylogeny containing *O. aeneus*, *O. fulgidus* and *O. wilsoni*.

Oxybelis aeneus (Wagler *in* Spix, 1824)

Distribution: Add Mexico (Yucatan: Isla Contoy), Noguez & Ramírez-Bautista (2008); Guatemala (Zacapa), Jadin et al. (2019); Honduras (Isla del Tigre, Isla Zacate Grande), McCranie & Gutsche (2016); Honduras (Isla Cochino Mejor, Isla Cochino Menor), McCranie & Valdés-Orellana (2014); Nicaragua (Jinotega), Jadin et al. (2019); Panama (Coiba I.), Jadin et al. (2019); Panama (Panama Oeste), Ray & Ruback (2015); Colombia (Huila), Moreno-Arias & Quintero-Corzo (2015); Colombia (Norte de Santander), Armesto et al. (2011); Colombia (Valle del Cauca: Isla Palma), Giraldo et al. (2014); Peru (Amazonas, Cajamarca, La Libertad), Koch et al. (2018); Peru (Huanuco), Schlüter et al. (2004); Peru (Lambayeque), Venegas (2005); Brazil (Mato Grosso do Sul), Ferreira et al. (2017); Brazil (Paraiba), R. França et al. (2012); Brazil (Piauí), Dal Vechio et al. (2016); Brazil (Sergipe), Guedes et al. (2014). Upper elevation of 2381 m, Quintero-Díaz & Carbajal-Márquez (2017a).

Comments: Jadin et al. (2019), using multilocus sequence data, recover four lineages that correspond to Mexico, N Central America, Panama, and Trinidad and Tobago specimens.

Oxybelis brevirostris (Cope, 1861)

Distribution: Add Panama (Panama, Panama Oeste), Ray & Ruback (2015).

Oxybelis fulgidus (Daudin, 1803)

Distribution: Add Mexico (Chiapas, Veracruz), Jadin et al. (2019); Mexico (Guerrero), Palacios-Aguilar & Flores-Villela (2018); Guatemala (Izabal), Jadin et al. (2019); Honduras (Comayagua), Jadin et al. (2019); Honduras (Santa Bárbara),

Espinal et al. (2014b); Panama (Panama Oeste), Ray & Ruback (2015); Colombia (Magdalena), Rueda-Solano & Castellanos-Barliza (2010); Peru (Huanuco), Schlüter et al. (2004); Brazil (Acre), Bernarde et al. (2011b); Brazil (Tocantins), Scartozzoni et al. (2009).

Oxybelis wilsoni Villa & McCranie, 1995

OXYRHABDIUM Boulenger, 1893 (Cyclocoridae)

Oxyrhabdium leporinum (Günther, 1858)
Distribution: Add Philippines (Calayan, Marinduque, Mindoro Is.), Leviton et al. (2018).

Oxyrhabdium modestum (A.M.C. Duméril, Bibron & Duméril, 1854)
Distribution: Add Philippines (Biliran, Catanduanes, Maripipi Is.), Leviton et al. (2018); Philippines (Camiguin), Sanguila et al. (2016).

OXYRHOPUS Wagler, 1830 (Dipsadidae: Xenodontinae)

Oxyrhopus clathratus A.M.C. Duméril, Bibron & Duméril, 1854
Distribution: Add Brazil (Parana), Souza-Filho et al. (2015).

Oxyrhopus doliatus A.M.C. Duméril, Bibron & Duméril, 1854

Oxyrhopus erdisii (Barbour, 1913)
Distribution: Add Peru (Puno), Sheehy et al. (2014).

Oxyrhopus fitzingeri (Tschudi, 1845)
Distribution: Low elevation of sea level, Venegas (2005).

Oxyrhopus formosus (Wied-Neuwied, 1820)
Distribution: Add Peru (Arequipa), Luque-Fernández & Paredes (2017); Peru (Puno), Llanqui et al. (2019).

Oxyrhopus guibei Hoge & Romano-Hoge, 1978
Distribution: Add Brazil (Ceara), D.P. Castro et al. (2019); Brazil (Paraiba), R. França et al. (2012); Paraguay (Presidente Hayes), N. Martínez et al. (2019); Paraguay (San Pedro), P. Smith et al. (2016).

Oxyrhopus leucomelas (F. Werner, 1916)
Distribution: Add Ecuador (Morona-Santiago, Zamora-Chinchipe), Almendáriz et al. (2014).

Oxyrhopus marcapatae (Boulenger, 1902)

Oxyrhopus melanogenys (Tschudi, 1845)
Distribution: Upper elevation of 1239 m, Koch et al. (2018).

Oxyrhopus occipitalis (Wagler *in* Spix, 1824)
Distribution: Add Colombia (Caldas), Rojas-Morales et al. (2018b); Brazil (Acre), Bernarde et al. (2011b).

Oxyrhopus petolarius (Linnaeus, 1758)
Synonyms: Add *Leptodira nycthemera* Werner, 1901, which was shown by J. Costa et al. (2015) to be a specimen of *Oxyrhopus petolarius*, rather than *Leptodeira annulata*.
Distribution: Add Guatemala (Izabal, Quiche), Sheehy et al. (2014); Honduras (Copán), McCranie et al. (2013b); Panama (Panama Oeste), Ray & Ruback (2015); Colombia (Valle del Cauca: Isla Palma), Giraldo et al. (2014); Ecuador (Esmeraldas,

Tungurahua), Sheehy et al. (2014); Peru (Huanuco), Schlüter et al. (2004); Suriname (Marowijne), Sheehy et al. (2014); Brazil (Espirito Santo), Silva-Soares et al. (2011); Brazil (Minas Gerais), Moura et al. (2012); Brazil (Acre), Bernarde et al. (2011b); Brazil (Paraiba), R. França et al. (2012); Brazil (Parana), Dainesi et al. (2019); Brazil (Pernambuco), Roberto et al. (2018); Paraguay (Alto Paraná, Itapúa), Beconi & Scott (2014).

Oxyrhopus rhombifer A.M.C. Duméril, Bibron & Duméril, 1854
Distribution: Add Brazil (Amazonas), F.G.R. França et al. (2006); Brazil (Piauí), Dal Vechio et al. (2013); Brazil (Tocantins), Dal Vechio et al. (2016); Paraguay (San Pedro), Atkinson et al. (2017).

Oxyrhopus trigeminus A.M.C. Duméril, Bibron & Duméril, 1854
Distribution: Add Peru (Junin), Sheehy et al. (2014); Brazil (Maranhão), J.P. Miranda et al. (2012); Brazil (Mato Grosso), M.M. dos Santos et al. (2011); Brazil (Piauí), Dal Vechio et al. (2013); Brazil (Sergipe), and upper elevation of 1100 m, Guedes et al. (2014); Brazil (Tocantins), Dal Vechio et al. (2016).

Oxyrhopus vanidicus Lynch, 2009
Distribution: Add Brazil (Rondônia), Bernarde et al. (2012b).

OXYURANUS Kinghorn, 1923 (Elapidae)

Oxyuranus microlepidotus (F. McCoy, 1879)
Oxyuranus scutellatus (W.C.H. Peters, 1867)
Oxyuranus temporalis Doughty, Maryan, Donnellan & Hutchinson, 2007

†PACHYOPHIS Nopcsa, 1923 (†Simoliophiidae)
Comments: Đurić et al. (2017) report and describe a new fossil specimen that most closely resembles †*Pachyophis*. It is from near the type locality, but of Turonian age. See under †*Mesophis*.

†Pachyophis woodwardi Nopcsa, 1923
Synonyms: Remove †*Mesophis nopcsai* Bolkay, 1925.

†PACHYRHACHIS Haas, 1979 (†Simoliophiidae)
Comments: Palci et al. (2013b) reevaluate its anatomy and discuss relationships.

†Pachyrhachis problematicus Haas, 1979
Comments: Palci et al. (2013b) provide an emended diagnosis based on new interpretation of fossil specimens.

†PALAEONATRIX Szyndlar *in* Mlynarski, Szyndlar, Estes & Sanchiz, 1982 (Natricidae)

†Palaeonatrix lehmani (Rage & Rocek, 1983)
†Palaeonatrix silesiaca Szyndlar *in* Mlynarski, Szyndlar, Estes & Sanchiz, 1982

†PALAELAPHIS Rochebrune, 1884. Mém. Soc. Sci. Nat. Saône-et-Loire 5: 156. (Booidea: incerta sedis)
Type species: †*Palaelaphis antiquus* Rochebrune, 1884, designated by Rage (1984).

Distribution: Middle Tertiary of France.

Comments: Rage (1984) determined that material assigned to *Palaelaphis* represents Booidea.

†*Palaelaphis antiquus* Rochebrune, 1884. Mém. Soc. Sci. Nat. Saône-et-Loire 5: 156, plates 1, 2.

Lectotype: MNHN QU16339, a trunk vertebra, designated by Rage (1984).

Type locality: "Phosphorites du Quercy," France.

Distribution: Within the upper Eocene to Oligocene of France.

†*Palaelaphis robustus* Rochebrune, 1884. Mém. Soc. Sci. Nat. Saône-et-Loire 5: 157, plate 2.

Holotype: MNHN QU16341, a maxilla.

Type locality: "Phosphorites du Quercy," France.

Distribution: Within the upper Eocene to Oligocene of France.

†*PALAEOPHIS* Owen, 1840 (†Palaeophiidae: Palaeophiinae)

†*Palaeophis africanus* Andrews, 1924
†*Palaeophis casei* Holman, 1982
†*Palaeophis colossaeus* Rage, 1983

Comments: McCartney et al. (2018) and O'Leary et al. (2019) describe newly acquired veretebrae and provide a revised species diagnosis and description.

†*Palaeophis ferganicus* Averianov, 1997
†*Palaeophis grandis* (Marsh, 1869)
†*Palaeophis littoralis* Cope, 1869
†*Palaeophis maghrebianus* Arambourg, 1952

Comments: Houssaye et al. (2013) describe newly acquired veretebrae and provide a revised species diagnosis and description.

†*Palaeophis nessovi* Averianov, 1997
†*Palaeophis tamdy* (Averianov, 1997)
†*Palaeophis toliapicus* Owen, 1840
†*Palaeophis typhaeus* Owen, 1850
†*Palaeophis udovichenkoi* Averianov, 1997
†*Palaeophis vastaniensis* Bajpai & Head, 2007

Comments: T. Smith et al. (2016) describe additional material from near the type locality.

†*Palaeophis virginianus* Lynn, 1934
†*Palaeophis zhylan* (Nessov & Udovitschenko, 1984)
†*Palaeophis zhylga* (Averianov, 1997)

†*PALAEOPYTHON* Rochebrune, 1880 (Pythonidae)

†*Palaeopython cadurcensis* (Filhol, 1877)
†*Palaeopython ceciliensis* Barnes, 1927
†*Palaeopython filholii* Rochebrune, 1880
†*Palaeopython fischeri* Schaal, 2004

Distribution: Add middle to late Eocene ((Bartonian to Priabonian) of Switzerland, Georgalis & Sheyer (2019, as *P.* cf. *fischeri*).

† *Palaeopython helveticus* Georgalis & Scheyer, 2019. Swiss J. Geosciences 112: 8–17, figs. 5–11.

Holotype: PIMUZ A/III 634, a trunk vertebra.

Type locality: "Fissure A, Dielsdorf, Zurich Canton, Switzerland."

Distribution: Middle to late Eocene (Bartonian to Priabonian) of Switzerland.

†*Palaeopython neglectus* Rochebrune, 1884

†*PALEOFARANCIA* Auffenberg, 1963 (Dipsadidae: subfamily unnamed)

†*Paleofarancia brevispinosa* Auffenberg, 1963

†*PALEOHETERODON* Holman, 1964 (Dipsadidae: subfamily unnamed)

†*Paleoheterodon arcuatus* Rage & Holman, 1984
†*Paleoheterodon tiheni* Holman, 1964

†*PALERYX* Owen, 1850 (Booidea: incerta sedis)

†*Paleryx rhombifer* Owen, 1850
†*Paleryx spinifer* Barnes, 1927

PALUSOPHIS Montingelli, Grazziotin, Battilana, R.W. Murphy, Zhang & Zaher, 2019. J. Zool. Syst. Evol. Res. 57(2): 13. (Colubridae: Colubrinae)

Type species: *Coluber bifossatus* Raddi, 1820 by original designation.

Distribution: South America.

Comments: See under *Mastigodryas*.

Palusophis bifossatus (Raddi, 1820)

Distribution: Add Colombia (Arauca), Montingelli et al. (2019); Brazil (Alagoas, Amapa, Amazonas, Distrito Federal, Mato Grosso do Sul, Rio Grande do Norte, Roraima), Montingelli et al. (2019); Brazil (Ceara), Guedes et al. (2014); Brazil (Espirito Santo), Silva-Soares et al. (2011); Brazil (Mato Grosso do Sul), Ferreira et al. (2017); Brazil (Paraiba), Sampaio et al. (2018); Brazil (Parana), Dainesi et al. (2019); Brazil (Piauí), Dal Vechio et al. (2013); Brazil (Rio de Janeiro: Ilha Grande), C.F.D. Rocha et al. (2018); Brazil (Tocantins), Dal Vechio et al. (2016); Bolivia (Beni), Padial et al. (2003); Paraguay (Boqueron, Presidente Hayes), Cabral & Weiler (2014). Upper elevation of 900 m, Guedes et al. (2014).

PANTHEROPHIS Fitzinger, 1843 (Colubridae: Colubrinae)

Fossil records: Add upper Miocene (Hemphillian) of USA (Tennessee), Jasinski & Moscato (2017, as *Pantherophis* sp.).

Pantherophis alleghaniensis (Holbrook, 1836)
Pantherophis bairdi (Yarrow *in* Cope, 1880)
†*Pantherophis buisi* (Holman, 1973)

Pantherophis emoryi (Baird & Girard, 1853)

Distribution: Add Mexico (Aguascalientes), Quintero-Díaz et al. (2016). Quintero-Díaz et al. (2016) map known localities.

Pantherophis guttatus (Linnaeus, 1766)

Distribution: Add to introduced populations Anguilla, Antigua, Bahamas (Abaco I., Grand Bahama, New Providence), Bonaire, British Virgin Islands (Peter, Tortola), Curaçao, Saint-Barthelemy, Saint Martin (Guana Cay, Saint

Martin), US Virgin Islands (Little Saint James, Saint Croix), Giery (2013). Note that these are not necessarily established populations.

†Pantherophis kansensis (Gilmore, 1938)
Pantherophis obsoletus (Say *in* James, 1823)
†Pantherophis pliocenicus (Holman, 1968)
Pantherophis ramspotti Crother, White, Savage, Eckstut, Graham & Gardner, 2011

Comment: See under *P. vulpinus.*

Pantherophis slowinskii (Burbrink, 2002)
Pantherophis spiloides (A.M.C. Duméril, Bibron & Duméril, 1854)

Pantherophis vulpinus (Baird & Girard, 1853)

Comments: Row et al. (2011) evaluate microsatellites between various populations, which were primarily divided by the geographic hiatus of the former taxon *P. gloydi* and. *P vulpinus.*

†PARACOLUBER Holman, 1970 (Colubridae: Colubrinae)

†Paracoluber storeri Holman, 1970

†PARAEPICRATES Hecht *in* McGrew, 1959 (Charinidae: Charininae)

†Paraepicrates brevispondylus Hecht *in* McGrew, 1959

PARAFIMBRIOS Teynié, David, Lottier, Le, Vidal & T.Q. Nguyen, 2015. Zootaxa 3926(4): 527. (Xenodermidae)

Type species: *Parafimbrios lao* Teynié, David, Lottier, Le, Vidal & T.Q. Nguyen, 2015, by original designation.

Distribution: Indochina.

Parafimbrios lao Teynié, David, Lottier, Le, Vidal & T.Q. Nguyen, 2015. Zootaxa 3926(4): 527–535, figs. 2–6.

Holotype: MNHN 2013.1002, a 285 mm male (A. Teynié & A. Lottier, 25 September 2012).

Type locality: "vicinity of Muang Ngoi (20°42.005′N, 102°41.730′E;…), Ngoi District, Louangphabang Province, Laos, at an elevation of ca. 360 m.a.s.l."

Distribution: Laos (Houaphan, Louangphabang), 360–890 m. Nguyen et al. (2015) add Vietnam (Son La), 1470 m, and Teynié & Hauser (2017) add Thailand (Nan), 1387–1490 m.

Parafimbrios vietnamensis Ziegler, Ngo, Pham, Nguyen, Le & Nguyen, 2018a. Zootaxa 4527(2): 271–273, figs. 2, 3.

Holotype: IEBR A2018.7, a 266 mm male (H.V. Tu & H.N. Ngo, 2 June 2016).

Type locality: "near Hoang Ho Village, Phang So Lin Commune, Sin Ho District, Lai Chau Province (22°22.180′N, 103°14.485′E), at an elevation of 1,317 m above sea level."

Distribution: N Vietnam (Lai Chau), 1370 m. Known only from the holotype.

PARAHYDROPHIS Burger & Natsuno, 1974 (Elapidae)

Parahydrophis mertoni (Roux, 1910)

†PARAOXYBELIS Auffenberg, 1963 (Colubridae: Colubrinae)

†Paraoxybelis floridanus Auffenberg, 1963

PARAPHIMOPHIS Grazziotin, Zaher, R.W. Murphy, Scrocchi, Benavides, Zhang & Bonatto, 2012 (Dipsadidae: Xenodontinae)

Paraphimophis rusticus (Cope, 1878)

Distribution: Add Brazil (Mato Grosso do Sul), Ferreira et al. (2017); Brazil (Marinheiros I., Rio Grande do Sul), Quintela et al. (2011).

Fossil records: Late Pleistocene of Argentina (Albino & Brizuela, 2014).

PARAPISTOCALAMUS Roux, 1934 (Elapidae)

Parapistocalamus hedigeri Roux, 1934

†PARAPLATYSPONDYLIA Holman & Harrison, 1998 (Charinidae: Ungaliophiinae)

Comments: K.T. Smith (2013) believes that *P. batesi* is not differentiated from species of *Platyspondylia.*

†Paraplatyspondylia batesi Holman & Harrison, 1998

PARAPOSTOLEPIS Amaral, 1930 (Dipsadidae: Xenodontinae)

Parapostolepis polylepis (Amaral, 1922)

Distribution: Add Brazil (Bahia), D.P.F. França et al. (2018); Brazil (Maranhao), F.M. Santos et al. (2018); Brazil (Tocantins), Dal Vechio et al. (2016).

PARARHADINAEA Boettger, 1898 (Pseudoxyrhophiidae)

Pararhadinaea melanogaster Boettger, 1898

Comments: Labanowski and Lowin (2011) describe three new specimens of this poorly known species.

PARASTENOPHIS Domergue, 1995. Arch. Inst. Pasteur Madagascar 61(2): 121. (Pseudoxyrhophiidae)

Type species: Dipsas betsileanus Günther, 1880.

Distribution: E Madagascar.

Comments: Vences et al. (2004) review the genus as the subgenus *Parastenophis* of *Stenophis.* See under *Lycodryas.*

Parastenophis betsileanus (Günther, 1880)

Comments: Vences et al. (2004) provide a description and color photographs. They note that there is a small form and giant form that may represent different species.

PARASUTA Worrell, 1961 (Elapidae)

Comments: Zaher et al. (2019) resolve *P. monachus* and *P. spectabilis* paraphyletically mingled with *Suta suta* and *S. fasciata* in a single clade.

Parasuta dwyeri (Worrell, 1956)

Distribution: Range extended to southernmost New South Wales, Australia, Michael & Lindenmayer (2011).

Parasuta flagellum (F. McCoy, 1878)
Parasuta gouldii (Gray, 1841)
Parasuta monachus (Storr, 1964)
Parasuta nigriceps (Günther, 1863)
Parasuta spectabilis (Krefft, 1869)

PARATAPINOPHIS Angel, 1929 (Natricidae)

Comments: Ren et al. (2019) believe this taxon may belong to *Trimerodytes*.

Paratapinophis praemaxillaris Angel, 1929

†*PARAUNGALIOPHIS* Rage, 2011 (Charinidae: Ungaliophiinae)

†*Paraungaliophis pricei* Rage, 2011

†*PARAXENOPHIS* Georgialis, Villa, Ivanov, Vasilyan & Delfino, 2019a. Palaeontol. Electronica 22.3.68: 47. (Colubridae: Colubrinae)

Type species: **†*Paraxenophis spanios*** Georgialis, Villa, Ivanov, Vasilyan & Delfino, 2019a by monotypy.

Distribution: Miocene/Pliocene transition of Greece.

†*Paraxenophis spanios* Georgialis, Villa, Ivanov, Vasilyan & Delfino, 2019a. Palaeontol. Electronica 22.3.68: 47–54, figs. 27–31.

Holotype: UU MAA 7645, a mid-trunk vertebra.

Type locality: "Maramena 1 site, Maramena locality…, Central Macedonia, Greece."

Distribution: Latest Miocene or earliest Pliocene (late Turolian) of Greece

PAREAS Wagler, 1830 (Pareidae)

Comment: Savage (2015) discusses the correct family name for this genus. You et al. (2015) produce a DNA sequence-based phylogeny using ten species, which results in the recognition of two additional species on Taiwan.

Pareas atayal You, Poyarkov & Lin, 2015. Zool. Scripta 44(4): 358–359, fig. 4.

Holotype: NMNS 05594, a 560 mm adult, stated to be both male and female in the description (C.-W. You, 27 August 2009).

Type locality: "TAIWAN, Taoyuan County, Fuxing Township, Sileng (24.653570 N, 121.409266 E) with elevation *ca.* 925 m."

Distribution: NW Taiwan, 100–2000 m.

Pareas boulengeri (Angel, 1920)

Pareas carinatus Wagler, 1830

Distribution: Add Myanmar (Mon), Vogel (2015); Thailand (Chanthaburi), Chan-ard et al. (2011); Thailand (Mae Hong Son), Hauser (2017); Cambodia (Siem Reap), Geissler et al. (2019); Vietnam (Dien Bien), Dung et al. (2014); West Malaysia (Penang), Quah et al. (2013).

Pareas chinensis (Barbour, 1912)

Pareas formosensis (Van Denburgh, 1909)

Synonyms: Remove *Amblycephalus komaii* Maki, 1931.

Comments: You et al. confirm the synonymy of *Psammodynastes compressus*, and redefine *formosensis* after the removal of some populations to other species.

Pareas hamptoni (Boulenger, 1905)

Distribution: Add Thailand (Chiang Mai, Mae Hong Son, Nan), Hauser (2017); Vietnam (Bac Giang), Hecht et al.

(2013); Vietnam (Dien Bien), Dung et al. (2014); Vietnam (Hai Phong: Cat Ba Island, as *P.* cf. *hamptoni*), T.Q. Nguyen et al. (2011); Vietnam (Hoa Binh), Nguyen et al. (2018); Vietnam (Nghe An), Vogel (2015); Vietnam (Quang Ngai), Nemes et al. (2013); Vietnam (Yen Bai), and upper elevation of 2046 m, Le et al. (2018).

Comments: You et al. (2015) discuss the complex, range-wide variation.

Pareas iwasakii (Maki, 1937)

Pareas komaii Maki, 1931. Monograph Snakes Japan: 149, plate. (*Amblycephalus komaii*)

Holotype: NSMT H00529, a 520 mm male (M. Maki, August 1923).

Type locality: translated to "Mt. Arisan (=Alishan), Taiwan" by Zhao & Adler (1993).

Distribution: Taiwan.

Comments: You et al. (2015) re-resurrect this species from *P. formosensis* based on DNA sequence-data and morphological characteristics. See Zhao & Adler (1993).

Pareas macularius Theobald, 1868

Distribution: Add Myanmar (Bago), Hauser (2017); Myanmar (Chin, Sagaing), Vogel (2015); Thailand (Chiang Rai, Lampang, Mae Hong Son, Nakhon Si Thammarat, Phitsanulok, Ranong, Tak), Hauser (2017); Thailand (Loei), Vogel (2015); Vietnam (Ngan Son), Vogel (2015); China (Hainan I.), Vogel (2015). Hauser (2017) mapped known localities.

Comments: Hauser (2017) confirms the morphological distinctiveness of *P. macularius*, compared with *P. margaritophorus*, using material from northern Thailand. He provides details of the external morphology and hemipenes.

Pareas margaritophorus (Jan *in* Bocourt, 1866)

Distribution: Add India (Mizoram), Lalbiakzuala & Lalremsanga (2019b); Laos (Champasak), Teynié et al. (2004); Thailand (Chiang Rai, Kanchanaburi, Lampang, Mae Hong Son, Nakhon Si Thammarat, Nan, Phitsanulok, Tak), Hauser (2017); Cambodia (Siem Reap), Geissler et al. (2019); West Malaysia (Selangor), Vogel (2015); West Malaysia (Terengganu), Sumarli et al. (2015); Vietnam (Bac Giang), Hecht et al. (2013); Vietnam (Ha Tinh, Nghe An, Quang Binh), Vogel (2015); Vietnam (Quang Ngai), Nemes et al. (2013); Vietnam (Quang Ninh), Gawor et al. (2016). Hauser (2017) mapped known localities.

Comments: Hauser (2017) confirms the morphological distinctiveness of *P. margaritophorus*, compared with *P. macularius*, using material from northern Thailand. He provides details of the external morphology and hemipenes.

Pareas monticola (Cantor, 1839)

Distribution: Add Myanmar (Kachin), Vogel (2015).

Pareas nigriceps K. Guo & Deng, 2009

Pareas nuchalis (Boulenger, 1900)

Pareas stanleyi (Boulenger, 1914)

Pareas vindumi Vogel, 2015. Taprobanica 7(1): 2–5, figs. 1–2.

Holotype: CAS 248147 (field no. CAS-MHS 28983), a 657 mm female (M. Hlaing, S.L. Oo, Z.H. Aung, K.S. Lwin & Y.M. Win, 28 July 2009).

Type locality: "Chipwi Township, Lukpwi village (25°42′41.7996″N, 98°19′ 22.7994″E, 1890 m asl), Kachin State, Myanmar."

Distribution: Myanmar (Kachin), 1890 m. Known only from the holotype.

PARIAS Gray, 1849 (Vipera: Crotalinae)

Parias flavomaculatus (Gray, 1842b)

Distribution: Add Philippines (Biliran Island), Sy & Boos (2015b); Philippines (Siquijor Island), Beukema (2011); Philippines (Babuyan Claro, Calayan, Dalipiri, Samar Is.), Leviton et al. (2018).

Parias gunaleni Vogel, David & Sidik, 2014b. Amphib. Rept. Conserv. 8(2): 21–25, figs. 3–7, 13. (*Trimeresurus gunaleni*) **new combination**

Holotype: MZB 5452, an 1190 mm female (D. Gunalen, H. Miyake, C. Sangyeon & M. Suk Cha).

Type locality: "Mt. Sibayak, ca. 1500–2200 m a.s.l., west of Brastagi (Berastagi), Karo Regency (Kabupaten Karo), Sumatera Utara Province, Sumatra, Indonesia."

Distribution: C Sumatra (Sumatera Barat, Sumatera Utara), 1500–2200 m.

Parias hageni (Lidth de Jeude, 1886)

Distribution: Add Thailand (Yala), Vogel et al. (2014b); West Malaysia (Kedah), Vogel et al. (2014b); West Malaysia (Terengganu), Sumarli et al. (2015); Indonesia (Belitung), Vogel et al. (2014b).

Parias malcolmi (Loveridge, 1938)

Parias mcgregori (Taylor, 1919)

Distribution: Add Philippines (Calayan, Camiguin Norte Is.), Leviton et al. (2018).

Parias schultzei (Griffin, 1909)

Parias sumatranus (Raffles, 1822)

Neotype: ZFMK 76340, a 1050 mm female, designated by Vogel et al. (2014b).

Type locality: "Southwestern Sumatra" via neotype designation.

Distribution: Vogel et al. (2014b) modify the distribution as follows: Sumatra portion of range limited to S half of island, add Thailand (Yala) and West Malaysia (Pahang, Terengganu), and remove the islands of Bangka, Belitung, Mentawai Archip., Nias and Simeulue, which are assigned to *P. hageni*.

Comments: Vogel et al. (2014b) give justification for neotype designation, and provide a revised description and color photographs.

PAROPLOCEPHALUS Keogh, Scott & Scanlon, 2000 (Elapidae)

Paroplocephalus atriceps (Storr, 1980)

†*PARVIRAPTOR* Evans, 1994. Palaeontology 37: 35. (Ophidia: incerta sedis)

Type species: *Parviraptor estesi* Evans, 1994 by original designation.

Comments: *Parviraptor* was originally described as an anguimorph lizard based on the mixed associatation of snake and lizard bones on the holotype and associated fossil-bearing blocks. Caldwell et al. (2015:2) restrict the hypodigm of the taxon to a maxilla, frontal, and vertebra on two of the blocks.

†*Parviraptor estesi* Evans, 1994. Palaeontology 37: 35–46, figs. 1, 2.

Holotype: NHMUK R48388, restricted to the left maxilla by Caldwell et al. (2015).

Type locality: "Durlston Bay, Dorset," England.

Distribution: upper Jurassic (Tithonian) or early Cretaceous (Berriasian) of Dorset, England.

†*PATAGONIOPHIS* Albino, 1986 (†Matdsoiidae)

†*Patagoniophis australiensis* Scanlon, 2005

†*Patagoniophis parvus* Albino, 1986

†*PAULACOUTOPHIS* Rage, 2011 (Booidea: incerta sedis)

†*Paulacoutophis perplexus* Rage, 2011

PELTOPELOR Günther, 1864 (Vipera: Crotalinae)

Comments: See comment under *Trimeresurus*.

Peltopelor macrolepis (Beddome, 1862)

†*PERIERGOPHIS* Georgialis, Villa, Ivanov, Vasilyan & Delfino, 2019a. Palaeontol. Electronica 22.3.68: 39. (Colubridae: Colubrinae)

Type species: †*Periergophis micros* Georgialis, Villa, Ivanov, Vasilyan & Delfino, 2019a by monotypy.

Distribution: Miocene/Pliocene transition of Greece.

†*Periergophis micros* Georgialis, Villa, Ivanov, Vasilyan & Delfino, 2019a. Palaeontol. Electronica 22.3.68: 39–47, figs. 22–26.

Holotype: UU MAA 7615, a posterior trunk vertebra.

Type locality: "Maramena 1 site, Maramena locality…, Central Macedonia, Greece."

Distribution: Latest Miocene or earliest Pliocene (late Turolian) of Greece

PHALOTRIS Cope, 1862 (Dipsadidae: Xenodontinae)

Phalotris bilineatus (A.M.C. Duméril, Bibron & Duméril, 1854)

Phalotris concolor Ferrarezzi, 1994

Distribution: Add Brazil (Bahia), M.A. Freitas et al. (2016); Brazil (N Minas Gerais), and elevation of 519–536 m, Moura et al. (2013a).

Comments: M.A. Freitas et al. (2016), Moura et al. (2013a): three additional specimens are described for a species previously known from the holotype. Moura et al. provide a revised description, color photographs and distribution map.

Phalotris cuyanus (Cei, 1984)

Holotype: Re-numbered MHNSR 317, L.A. Martins & Lema (2017).

Distribution: Add Argentina (Cordoba, San Luis), L.A. Martins & Lema (2017); Argentina (San Luis), and low elevation of 540 m, Quiroga & Ferrer (2016).

Comments: L.A. Martins & Lema (2017) provide a species account (description, color photograph and locality map).

Phalotris labiomaculatus Lema, 2002

Distribution: Hamdan et al. (2013b) map all known localities, and report a low elevation of 100 m.

Comments: Hamdan et al. (2013b) provide a revised description based on numerous additional specimens.

Phalotris lativittatus Ferrarezzi, 1994

Distribution: Add Paraguay (San Pedro), P. Smith et al. (2016).

Phalotris lemniscatus (A.M.C. Duméril, Bibron & Duméril, 1854)

Distribution: Add Uruguay (Treinta y Tres), Prigioni et al. (2013).

Phalotris mattogrossensis Lema, Agostini & Cappellari, 2005

Distribution: Add Paraguay (Central, Cordillera), Cabral & Weiler (2014).

Phalotris mertensi (Hoge, 1955)

Distribution: Add Brazil (Mato Grosso do Sul), Ferreira et al. (2017).

Phalotris multipunctatus Puorto & Ferrarezzi, 1994

Distribution: Add Brazil (Mato Grosso do Sul), Ferreira et al. (2017); Paraguay (San Pedro), Atkinson et al. (2018).

Phalotris nasutus (Gomés, 1915)

Add Brazil (Rondônia), Bernarde et al. (2012b).

Phalotris nigrilatus Ferrarezzi, 1994

Phalotris normanscotti Cabral & Cacciali, 2015. Herpetologica 71(1): 72–76, figs. 1–2.

Holotype: MNHNP 5160, a 291 mm male (P. Freed & J. Furman, 21 February 1995).

Type locality: "3 km S of Filadelfia, 22°22′45″S, 60°01′48″W…, 138 m above sea level (asl), Boquerón Department, Paraguay."

Distribution: Western Paraguay (Boquerón, Presidente Hayes), 138 m.

Comments: In the *Phalotris bilineatus* group.

Phalotris punctatus (Lema, 1979)

Holotype: Re-numbered MLP 579, L.A. Martins & Lema (2017).

Distribution: Add Paraguay (Chaco), L.A. Martins & Lema (2017); Argentina (Jujuy, Mendoza, San Juan), L.A. Martins & Lema (2017); Argentina (Tucuman), Scrocchi et al. (2019).

Comments: L.A. Martins & Lema (2017) provide a species account (description, color photograph and locality map).

Phalotris reticulatus (W.C.H. Peters, 1860)

Phalotris sansebastiani Jansen & Köhler, 2008

Distribution: Add Argentina (Jujuy, Salta), and elevation range of 434–1445 m, Scrocchi & Giraudo (2012).

Phalotris tricolor (A.M.C. Duméril, Bibron & Duméril, 1854)

Synonyms: Remove *Elapomorphus punctatus* Lema, 1979, which was inadvertently left under *P. tricolor*, though also recognized as a species by Wallach et al. (2014).

Distribution: Add Brazil (Mato Grosso do Sul), Ferreira et al. (2017); Bolivia (Tarija), L.A. Martins & Lema (2017); Paraguay (Aregua, Chaco), L.A. Martins & Lema (2017).

Comments: L.A. Martins & Lema (2017) provide a species account (description, color photograph and locality map).

PHILODRYAS Wagler, 1830 (Dipsadidae: Xenodontinae)

Fossil records: Upper Miocene of La Pampa, Argentina, Scanferla & Agnolín (2015, as cf. *Philodryas*).

Philodryas aestiva (A.M.C. Duméril, Bibron & Duméril, 1854)

Distribution: Add Brazil (Marinheiros I., Rio Grande do Sul), Quintela et al. (2011); Paraguay (Canindeyu), Cacciali et al. (2015b).

Philodryas agassizii (Jan, 1863)

Distribution: Add Brazil (Bahia), Guedes et al. (2014); Argentina (Corrientes), Etchepare & Ingaramo (2008) and Di Pietro et al. (2013).

Comments: G.A.S. Filho & Plombon (2014) define the geographic range in Paraná.

Philodryas amaru Zaher, Arredondo, Valencia, Arbeláez, Rodrigues & Altamirano-Benavides, 2014. Zootaxa 3785(3): 470–475, figs. 1, 2.

Holotype: FHGO 4749, a 622 mm male (E. Arbeláez, 6 June 2006).

Type locality: "private land owned by Manuel Merchan, Termas de Aguas Calientes-Soldados (2° 55′ 55″ S, 79°12′ 37″ W, ca. 3196 m), Parroquia San Joaquín, Cantón Cuenca, Province of Azuay, Ecuador."

Distribution: Ecuador (Azuay), 3196 m. Known only from the type locality.

Philodryas argentea (Daudin, 1803)

Distribution: Add Peru (Huanuco), Schlüter et al. (2004); Brazil (Marajo I., Para), G.M. Rodrigues et al. (2015).

Philodryas arnaldoi (Amaral, 1933)

Phildryas baroni Berg, 1895

Philodryas boliviana Boulenger, 1896

Distribution: Add Bolivia (Potosi), Rivas et al. (2018). Upper elevation of 2700 m, O. Martínez (2017).

Philodryas chamissonis (Wiegmann *in* Meyen, 1834)

Distribution: Add Chile (Bio-Bio), Zaher et al. (2014).

Comments: Sallaberry-Pincheira et al. (2011) use mtDNA sequence-data to detect three geographic lineages.

Philodryas cordata Donnely & Myers, 1991

Philodryas erlandi Lönnberg, 1902. Ann. Mag. Nat. Hist. (7)10(60): 460–461.

Lectotype: NHRM 5097, a 692 mm male (Baron Erland Nordensköld & company), designated by Cacciali et al. (2016a).

Type locality: "Crevaux, Bolivian Chaco" [=Fortín Crevaux Nuero, Tarija Prov., 21.9°S, 63.51°W].

Distribution: SE Bolivia, W Paraguay, N Argentina.

Comments: Elevated to species status from the synonymy of *P. mattogrossensis* by Cacciali et al. (2016a).

Philodryas georgeboulengeri Grazziotin, Zaher, R.W. Murphy, Scrocchi, Benavides, Zhang & Bonatto, 2012

Distribution: Add Brazil (Rondônia), Bernarde et al. (2012b).

Philodryas laticeps F. Werner, 1900

Distribution: Upper elevation of 1344 m, Navarro-Cornejo & Gonzales (2013).

Philodryas livida (Amaral, 1923)

Distribution: Add Brazil (Mato Grosso), P. Smith et al. (2014); Paraguay (San Pedro), P. Smith et al. (2014).

Philodryas mattogrossensis Koslowsky, 1898

Synonyms: Remove *Philodryas erlandi* Lönnberg, 1902.

Neotype: SMF 49990, a 1088 mm male, designated by Cacciali et al. (2016a).

Neotype locality: "Taunay, Mato Grosso do Sul (−20.28608° S, −56.0847° W), Brazil."

Distribution: S Brazil and E Paraguay; add to department list for Paraguay (Amambay, San Pedro), Cabral & Weiler (2014). Add Brazil (Goias), Vaz-Silva et al. (2007).

Comments: Cacciali et al. (2016a) remove W and S populations, as *P. erlandi*, and present a revised description.

Philodryas nattereri Steindachner, 1870

Distribution: Add Brazil (Alagoas), Guedes et al. (2014); Brazil (Maranhão), J.P. Miranda et al. (2012); Brazil (Piauí), Dal Vechio et al. (2013); Paraguay (San Pedro), P. Smith et al. (2016). Upper elevation of 1100 m, Guedes et al. (2014).

Philodryas olfersii (Lichtenstein, 1823)

Distribution: Add Brazil (Acre), Bernarde et al. (2013); Brazil (Amapá), F.G.R. França et al. (2006); Brazil (Mato Grosso do Sul), Ferreira et al. (2017); Brazil (Marinheiros I., Rio Grande do Sul), Quintela et al. (2011); Brazil (Rio de Janeiro: Ilha Grande), C.F.D. Rocha et al. (2018). Upper elevation of 1100 m, Guedes et al. (2014).

Philodryas patagoniensis (Girard, 1858)

Distribution: Add Brazil (Pará), F.G.R. França et al. (2006); Brazil (Paraiba), R. França et al. (2012); Brazil (Marinheiros I., Rio Grande do Sul), Quintela et al. (2011); Brazil (Sergipe), Guedes et al. (2014); Brazil (Tocantins), Dal Vechio et al. (2016).

Fossil record: Late Pleistocene of Argentina (Albino & Brizuela, 2014).

Philodryas psammophidea Günther, 1872

Distribution: Add Brazil (Goias), Ramalho et al. (2018a); Brazil (Mato Grosso do Sul), Ferreira et al. (2017); Bolivia (Potosi), Rivas et al. (2018); Paraguay (Alto Paraguay), Cabral & Weiler (2014); Argentina (Chubut), Minoli et al. (2015).

Philodryas simonsii Boulenger, 1900

Distribution: Zaher et al. (2014) map known localities.

Philodryas tachymenoides (K.P. Schmidt & Walker, 1943)

Philodryas trilineata (Burmeister, 1861)

Distribution: Add Argentina (Buenos Aires Prov.), De Pietro et al. (2016).

Philodryas varia (Jan, 1863)

Philodryas viridissima (Linnaeus, 1758)

Distribution: Add Peru (Huanuco), Schlüter et al. (2004); Brazil (Amapá, Roraima), Entiauspe-Neto et al. (2018).

PHILOTHAMNUS A. Smith, 1847 *in* 1838–1849 (Colubridae: Colubrinae)

Comments: Engelbrecht et al. (2018) use DNA sequence-data to produce a phylogeny using 15 of the species. They state that there is no evidence for a short-tailed group formerly referred to as *Chlorophis*, but there are two clades, one of which does contain most of the *Chlorophis*-group taxa.

Philothamnus angolensis Bocage, 1882

Distribution: Add Angola (Bengo, Bie, Luanda, Uige), M.P. Marques et al. (2018).

Philothamnus battersbyi Loveridge, 1951

Philothamnus bellii Günther, 1866. Ann. Mag. Nat. Hist. Series 3, 18(103): 27–28, plate 7. (*Herpetaethiops bellii*)

Holotype: BMNH 1946.1.10.27, an 838 mm female (Lt. Bell).

Type locality: "Sierra Leone," specifically from "Victoria, Sherborough Island," Boulenger (1894).

Distribution: W Africa: Guinea, Sierra Leone and Liberia, and perhaps Guinea-Bissau.

Comments: Trape & Baldé (2014) find their 20 Guinea specimens to be identical to Günther's description of *H. bellii*, which they resurrect from the synonymy of *P. heterodermus*, with which it is known to be sympatric (Wallach et al., 2014, have *H. bellii* under *P. carinatus* in error). They provide a comparative description between *P. bellii* and *P. heterodermus*, and note that preliminary molecular data support recognition of *P. bellii*.

Philothamnus bequaerti (K.P. Schmidt, 1923)

Philothamnus carinatus (Andersson, 1901)

Synonyms: Remove *Herpetaethiops bellii* Günther, 1866.

Distribution: Add Liberia, Trape & Baldé (2014); Angola (Lunda Norte), M.P. Marques et al. (2018).

Comments: Trape & Baldé (2014) find that Guinea specimens indicated that *P. carinatus* and *P. heterodermus* can only be separated based on dorsal rows. Engelbrecht et al. (2018) conclude that *P. carinatus* is comprised of two species, but cannot assign these species until molecular data from the type locality of the nominate race can be evaluated.

Philothamnus dorsalis (Bocage, 1866)

Distribution: Add Angola (Bengo, Cuanza Norte, Cuanza Sul, Luanda, Malanje), M.P. Marques et al. (2018).

Philothamnus girardi Bocage, 1893

Comments: According to Engelbrecht et al. (2018), DNA sequence-data do not recognize *P. girardi* as a species distinct from *P. dorsalis*.

Philothamnus heterodermus (Hallowell, 1857)

Distribution: Add Gabon (Ogooue-Lolo), Pauwels et al. (2016b); Angola (Bengo, Lunda Norte), Branch (2018); Angola

(Cuanza Norte), M.P. Marques et al. (2018); Angola (Malanje), Ceríaco et. al. (2014). Remove far W Africa: Guinea-Bissau, Guinea, Sierra Leone, Liberia).

Comments: See above under *P. bellii*.

Philothamnus heterolepidotus (Günther, 1863)

Distribution: Add Angola (Cuando Cubango, Cuanza Norte, Luanda), M.P. Marques et al. (2018).

Philothamnus hoplogaster (Günther, 1863)

Distribution: Add Angola (Cabinda, Huambo, Malanje, Zaire), M.P. Marques et al. (2018); Angola (Cuando-Cubango), Conradie et al. (2016b); Angola (Lunda Sul), Branch (2018).

Philothamnus hughesi Trape & Roux-Estève, 1990

Philothamnus irregularis (Leach *in* Bowdich, 1819)

Philothamnus macrops (Boulenger, 1895)

Distribution: Add Kenya (Coast), Branch et al. (2019a); Tanzania (Morogoro), Menegon et al. (2008) and Lyakurwa (2017); Mozambique (Cabo Delgado, Sofala, Zambezia), Branch et al. (2019a).

Comments: Branch et al. (2019a) present an updated description, color photographs, and a map of known localities.

Philothamnus natalensis (A. Smith, 1847 *in* 1838–1849)

Synonyms: Remove *Philothamnus natalensis occidentalis* Broadley, 1966.

Distribution: Remove South Africa.

Comments: See under *P. occidentalis*.

Philothamnus nitidus (Günther, 1863)

Distribution: Add Liberia, Trape & Baldé (2014).

Philothamnus occidentalis Broadley, 1966. Ann. Natal Mus. 18(2): 419–422. (*Philothamnus natalensis occidentalis*)

Holotype: UM 4554, a 705 mm female (B.G. Donnelly, 4 October 1962).

Type locality: "Camperdown, Natal," South Africa.

Distribution: E South Africa, Lesotho.

Comments: Engelbrecht et al. (2018) conclude that *P. natalensis occidentalis* is a valid species based on molecular and morphological data.

Philothamnus ornatus Bocage, 1872

Distribution: Add Angola (Cauando-Cubango), Conradie et al. (2016b); Angola (Namibe), Ceríaco et al. (2016).

Philothamnus pobeguini (Chabanaud, 1917)

Distribution: Add Benin and Ivory Coast, Trape & Baldé (2014).

Comments: Examination of additional specimens by Trape & Baldé (2014) supports recognition as a distinct species.

Philothamnus punctatus W.C.H. Peters, 1867

Distribution: Add Tanzania (Morogoro), Menegon et al. (2008).

Philothamnus ruandae Loveridge, 1951

Philothamnus semivariegatus (A. Smith, 1847 *in* 1838–1848).

Distribution: Add Angola (Cuando Cubango), M.P. Marques et al. (2018).

Comments: Engelbrecht et al. (2018) conclude that *P. semivariegatus* is comprised of four species, but cannot assign these species until molecular data from the type locality of the nominate race can be evaluated.

Philothamnus thomensis Bocage, 1882

PHIMOPHIS Cope, 1860 (Dipsadidae: Xenodontinae)

Phimophis guerini (A.M.C. Duméril, Bibron & Duméril, 1854)

Distribution: Add Brazil (Amazonas), F.G.R. França et al. (2006); Brazil (Distrito Federal, Mato Grosso, Parana, Rio Grande do Norte), and low elevation of 5 m, R. Marques et al. (2012); Brazil (Paraíba), G.A.P. Filho et al. (2012); Brazil (Tocantins), Dal Vechio et al. (2016); Paraguay (San Pedro), Atkinson et al. (2017). Upper elevation of 830 m, Guedes et al. (2014).

Phimophis guianensis (Troschel *in* Schomburgk, 1848)

Distribution: Add Panama (Panama Oeste), Ray & Ruback (2015); Colombia (Cesar), Medina-Rangel (2011); Brazil (Amapá, Pará), F.G.R. França et al. (2006).

Phimophis vittatus (Boulenger, 1896)

PHISALIXELLA Domergue, 1995. Arch. Inst. Pasteur Madagascar 61(2): 121. (Pseudoxyrhophiidae)

Type species: *Heterurus arctifasciatus* A.M.C. Duméril, Bibron & Duméril, 1854 by original designation.

Distribution: Madagascar.

Comments: Vences et al. (2004) review the genus as the subgenus *Phisalixella* of *Stenophis*. See under *Lycodryas*.

Phisalixella arctifasciata (A.M.C. Duméril, Bibron & Duméril, 1854)

Comments: Vences et al. (2004) provide a description and color photographs.

Phisalixella iarakaensis Domergue, 1995

Comments: Vences et al. (2004) provide a description.

Phisalixella tulearensis Domergue, 1995. Arch. Inst. Pasteur Madagascar 61(2): 122. (*Stenophis (Phisalixella) tulearensis*)

Holotype: MNHN 1988.379, (C.A. Domergue, 2 January 1981).

Type locality: "Ankarafantsika à Toliara," Madagascar.

Distribution: SW Madagascar (S Mahajanga, Toliar).

Comments: Nagy et al. (2010) revalidate *tulearensis* on the basis of genetic and morphological distinctiveness from *P. variabilis*.

Phisalixella variabilis Boulenger, 1896

Distribution: N Madagascar (Antsiranan, Mahajanga). Add Madagascar (Nosy Be I.), Vences et al. (2004).

Comments: Vences et al. (2004) provide a description.

PHRYNONAX Cope, 1862. Proc. Acad. Nat. Sci. Philadelphia 14(5): 348. (Colubridae: Colubrinae)

Type species: *Tropididipsas lunulata* Cope, 1861 by monotypy.

Synonyms: *Synchalinus* Cope, 1894.

Distribution: Central and N South America.

Comments: Jadin et al. (2014) use DNA sequence-data to resolve relationships of *Pseustes* and *Spilotes*, which proved to be sister taxa within Colubrinae. *Pseustes sulphureus* resolves within a clade with *Spilotes*, and was transferred to that genus. *Phrynonax* is the next available genus name for remaining *Pseustes* species.

Phrynonax poecilonotus (Günther, 1858)

Distribution: Remove South American populations to *P. polylepis*. Add Panama (Panama Oeste), Ray & Ruback (2015); Colombia (Antioquia), Restrepo et al. (2017); Colombia (Valle del Cauca: Isla Palma), Giraldo et al. (2014). Upper elevation of 1610 m, Köhler et al. (2013).

Phrynonaxs polylepis W.C.H. Peters 1867. Monatsb. Preuss. Akad. Wiss. Berlin 1867: 709. (*Ahaetulla polylypis*)

Holotype: ZMB 5899, a 1 m specimen (A. Kappler).

Type locality: "Surinam."

Distribution: N South America. S Colombia, Venezuela, Guyana, Surinam, French Guiana, N Brazil, Ecuador, Peru, N Bolivia. Add Peru (Huanuco), Schlüter et al. (2004); Peru (Puno), Llanqui et al. (2019); Brazil (Acre), Bernarde et al. (2011b); Brazil (Maranhão), M.A. Freitas et al. (2017); Brazil (Rondônia), Bernarde et al. (2012b).

Comments: Jadin et al. (2014), using DNA sequence data, find Central and South American populations of *P. poecilonotus* to form distinct clades, and resurrect *A. polylepis* for the disjunct South American populations.

Pseustes sexcarinatus (Wagler *in* Spix, 1824)

Pseustes shropshirei Barbour & Amaral, 1924

Distribution: Add Colombia (Valle del Cauca: Isla Palma), Giraldo et al. (2014); Ecuador (Manabi), Almendáriz et al. (2012).

PHYTOLOPSIS Gray, 1849. Cat. Spec. Snakes Coll. British Mus.: 67–68. (Homalopsidae)

Type species: *Phytolopsis punctata* Gray, 1849 by monotypy.

Distribution: Malaysia and the Greater Sundas.

Comments: J.C. Murphy & Voris (2014) revive this name based on morphology, and provide a generic diagnosis.

Phytolopsis punctata Gray 1849

Comments: J.C. Murphy & Voris (2014) provide a diagnosis and photograph.

PHYLLORHYNCHUS Stejneger, 1890 (Colubridae: Colubrinae)

Phyllorhynchus browni Stejneger, 1890
Phyllorhynchus decurtatus (Cope, 1869)

PITUOPHIS Holbrook, 1842 (Colubridae: Colubrinae)

Fossil record: Add upper Miocene (Hemphillian) of USA (Tennessee), Jasinski & Moscato (2017, as *Pituophis* sp.).

Pituophis catenifer (Blainville, 1835)

Comments: E.A. Myers et al. (2017a) provide a DNA sequence-based phylogeny with four primary clades: 1) S Great Plains, 2) Chihuahuan Desert and Rocky Mountains, 3) S Arizona, USA, and 4) Pacific Coast and Great Basin.

Pituophis deppei (A.M.C. Duméril, Bibron & Duméril, 1854)

Distribution: Add Mexico (Nayarit), Loc-Barragán & Ahumada-Carrillo (2016).

Pituophis insulanus Klauber, 1946
Pituophis lineaticollis (Cope, 1861)
Pituophis melanoleucus (Daudin, 1803)
Pituophis ruthveni Stull, 1929
Pituophis vertebralis (Blainville, 1835)

PLAGIOPHOLIS Boulenger, 1893 (Pseudoxenodontidae)

Plagiopholis blakewayi Boulenger, 1893

Synonyms: Add *Plagiopholis unipostocularis* Zhao, Jiang & Huang, 1978, *Oligodon kunmingensis* Kou & Wu, 1993.

Comments: G.-H. Zhong et al. (2016) use molecular and morphological data to conclude that *P. unipostocularis* is conspecific with *P. blakewayi*. The holotype number for *P. unipostocularis* is corrected to CIB 010195, Guo et al. (2012a).

Plagiopholis delacouri Angel, 1929
Plagiopholis nuchalis (Boulenger, 1893)
Plagiopholis styani (Boulenger, 1899)

PLATYCEPS Blyth, 1860 (Colubridae: Colubrinae)

Comments: Schätti et al. (2014) revise the *P. rhodorachis* complex east of the Tigris River. They also provide a synopsis of the genus, assigning the species among three species groups.

Platyceps afarensis Schätti & Ineich, 2004

Platyceps atayevi (Tuniyev & Shammakov, 1993)

Comments: Schätti et al. (2014) consider *atayevi* to be a subspecies of *P. najadum*.

Platyceps bholanathi (Sharma, 1976)

Distribution: Add India (Tamil Nadu), and change elevation range to 115–945 m, Ganesh et al. (2013b), Smart et al. (2014) and Samson et al. (2017); India (Karnataka), Sharma et al. (2013); India (Telangana), Samson et al. (2017).

Comments: Deshwal & Becker (2017), Ganesh et al. (2013b), Guptha et al. (2012), Seetharamaraju & Srinivasulu (2013) and Sharma et al. (2013) report additional specimens.

Platyceps brevis (Boulenger, 1895)

Platyceps collaris (F. Müller, 1878)

Distribution: Add Turkey (Adana), Winden et al. (1997); Turkey (Erzincan), İğci et al. (2015); Turkey (Mersin, Mugla), Winden et al. (1997).

Platyceps elegantissimus (Günther, 1879)

Platyceps florulentus (Geoffroy St.-Hilaire, 1827 *in* Savigny, 1809–1829)

Synonyms: Remove *Platyceps florulentus perreti* Schätti, 1988.

Distribution: Remove Cameroon and Nigeria to *P. perreti*.

Platyceps gracilis (Günther, 1862)

Distribution: Add India (Gujarat), Patel & Vyas (2019); India (Rajasthan), Schätti et al. (2014).

Platyceps insulanus (Mertens, 1965)

Platyceps karelinii (Brandt, 1838)

Distribution: Add Libya (Benghazi, Darnah, Marj), Bauer et al. (2017); Iran (Chaharmahal and Bakhtiari, Isfahan, Lorestan, Mazandaran, North Khorasan, Tehran), Shafaei-Mahroo et al. (2015); Iran (Qom), S.M. Kazemi et al. (2015); Afghanistan (Badghis, Farah, Jowzjan), Wagner et al. (2016b); Kazakhstan (range extension into Betpak-Dala Desert, South Kazakhstan), Guillemin & Martin (2017).

Comments: See Bauer et al. (2017) for discussion of potential species status for African and Middle Eastern populations as *P. rogersi*.

Platyceps largeni (Schätti, 2001)

Distribution: Add Eritrea (Museri I.), Maza et al. (2015).

Platyceps manseri (Leviton, 1986)

Platyceps messanai (Schätti & Lanza, 1989)

Platyceps mintonorum (Mertens, 1969)

Distribution: Add Afghanistan (Helmand, Kandahar), Wagner et al. (2016b).

Platyceps najadum (Eichwald, 1831)

Distribution: Add Serbia, Tomović et al. (2014); Greece (Elafonisos I.), Broggi (2016a); Greece (Kastellorizo I.), Kakaentzis et al. (2018); Greece (Paxos I.), M.J. Wilson & Stille (2014); Turkey (Afyon, Konya), Cihon & Tok (2014). Shafaei-Mahroo et al. (2015) list provinces for Iran.

Platyceps noeli Schätti, Tillack & Kucharzewski, 2014. Vertebrate Zoology 64(3): 358–361, fig. 19.

Holotype: SMF 50458, a 611 mm male (M.G. Konieczny).

Type locality: "Spin Karez (Spinkares, Spinkarez), Quetta District, Baluchistan, 30°13′N 67°09′E, ca. 2000 m above sea level," Pakistan.

Distribution: Pakistan (NE Baluchistan), 2000–2400 m.

Platyceps perreti Schätti, 1988. Trop. Zool. 1(1): 100–101, figs. 5, 9. (*Coluber florulentus perreti*)

Holotype: MHNG 1465.45, a 768 mm male (J.-L. Perret, 10 May 1961).

Type locality: "Soulédé, Mokolo, N Kamerun."

Distribution: N Cameroon (Extreme Nord, Nord), NE Nigeria (Borno, Plateau, Taraba).

Comments: Schätti et al. (2014) confer species status on this taxon, but without explanation, but perhaps due to allopatry.

†*Platyceps planicarinatus* (Bachmayer & Szyndlar, 1985)

Platyceps rhodorachis (Jan *in* Filippi, 1865)

Synonyms: Add *Platyceps semifasciatus* Blyth, 1860. Remove *Zamanis ladacensis subnigra* Boettger, 1893. *Coluber karelini mintonorum* Mertens, 1969, listed in Wallach et al. (2014) in error.

Distribution: Add Iran (Alborz, Bushehr, Chaharmahal and Bakhtiari, East Azerbaijan, Hamedan, Hormozgan, Ilam, Isfahan, Kermanshah, Kohgiluyeh and Boyer Ahmad, Lorestan, North Khorasan, Qazvin, South Khorasan, Tehran, West Azerbaijan, Yazd, Zanjan), Shafaei-Mahroo et al. (2015); Iran (Kerman), Moradi et al. (2013); Iran (Markazi), Sabbaghzadeh & Mashayekhi (2015); Iran (Qom), S.M.

Kazemi et al. (2015). Remove Israel, Jordan and W Syria, Sinaiko et al. (2018). Schätti et al. (2014) state that the western range limit is NW Iran and the E Caspian shore, thus eliminating N Iraq, S Turkey and southward. Elevation to at least 3700 m, Schätti et al. (2014). Newly published records for Saudi Arabia (Farasan Al-Kebir I.), Masseti (2014); Saudi Arabia (Farasan Islands), Cunningham (2010) must pertain to other species.

Comments: Perry (2012) provides morphological evidence of two species in Israel: a southern species is *P. saharicus* (which he renamed *P. tessellata*), and a northern species that he elevates to species, *P. ladacensis*. The latter is considered to incorporate the nonred-striped individuals ranging from northern Israel, to Yemen and India. He also elevates to species *Z. l. subnigra* as *P. subnigra* based on consensus, which occurs on the Horn of Africa. Sinaiko et al. (2018) counter that only one species, *P. saharicus*, occurs in Israel, the color pattern differences being based on habitat, and with no genetic basis. Schätti et al. (2014) provide a revised description, color photographs and locality maps. They transfer the name *P. semifasciatus* to this species, and recognize two subspecies: *P. r. rhodorachis* and *P. r. ladacensis*.

Platyceps saharicus Schätti & McCarthy, 2004

Comments: Perry (2012) concludes that this taxon was previously described as *Zamenis rhodorachis tessellata* Werner, 1909, and proposed that it be recognized as *Platyceps tessellata*. Sinaiko et al. (2018) counter that the type locality of *tessellata* is likely in error, and that the color pattern of the holotype does not permit assignment to either *P. rhodrachis* or *P. saharicus*. See under *P. rhodorachis*.

Platyceps schmidtleri (Schätti & McCarthy, 2001)

Comments: Schätti et al. (2014) refer to *schmidtleri* as a subspecies of *P. najadum*.

Platyceps scortecci (Lanza, 1963)

Platyceps sinai (K.P. Schmidt & Marx, 1956)

Platyceps sindhensis Schätti, Tillack & Kucharzewski, 2014. Vertebrate Zoology 64(3): 361–364, fig. 21.

Holotype: SMF 57306, a 528 mm male (M.G. Konieczny).

Type locality: "'Gaj River,' Kirthar Range, Sindh, arbitrarily placed in the vicinity of Chhota Kund, Dadu District..., ca. 26°53′N 67°14′E, ca. 180 m above sea level," Pakistan.

Distribution: Pakistan (Sindh, Baluchistan) NSL-200 m.

Platyceps smithi (Boulenger, 1895)

Comments: Schätti et al. (2014) consider *smithi* to be a subspecies of *P. brevis*.

Platyceps subnigra Boettger, 1893. Zool. Anz. 16(416): 118–119. (*Zamenis ladacensis subnigra*)

Holotype: SMF 62595, male.

Type locality: "Ogadeen, Somaliland," = Ogaden region of E Ethiopia.

Distribution: NE Africa, Eritrea, Djibouti, E Ethiopia, N Somalia.

Comments: See under *P. rhodorhachis*.

Platyceps somalicus (Boulenger, 1896)

Platyceps taylori (Parker, 1949)

Platyceps thomasi (Parker, 1931)

Platyceps variabilis (Boulenger, 1905)

Platyceps ventromaculatus (Gray, 1834 *in* Gray & Hardwicke, 1830–1835)

Synonyms: Remove *Platyceps semifasciatus* Blyth, 1860 to *P. rhodorachis*.

Distribution: Add Iraq (Babil, Basrah, Dhi-Qar, Najaf, Qadisiyah), Abbas-Rhadi et al. (2017); Iran (Bushehr, Fars, Kerman, Golestan, Isfahan, Khuzestan, Kordestan, Markazi, Semnan, Sistan and Baluchistan, South Khorasan, Tehran), Shafaei-Mahroo et al. (2015); Iran (Qom), S.M. Kazemi et al. (2015); Iran (Razavi Khorasan), Yousefkhani et al. (2014) and Nasrabadi et al. (2016); India (Jammu and Kashmir), Manhas et al. (2018b). Moradi et al. (2013) added Iran (Kerman), 1100–2500 m. A. Narayanan & Satyanarayan (2012) report additional specimens from NW India.

PLATYPLECTRURUS Günther, 1868 (Uropeltidae)

Comments: Pyron et al. (2016c) provide a morphological diagnosis.

Platyplectrurus madurensis Beddome, 1877

Distribution: Eliminate Sri Lanka, which is believed to be in error, based on a specimen likely to have come from India, Pyron et al. (2016c).

Comments: Pyron et al. (2016c) provide a description and color photograph.

Platyplectrurus trilineatus (Beddome, 1867)

Comments: Pyron et al. (2016c) provide a description.

†*PLATYSPONDYLIA* Rage, 1974 (Charinidae: Ungaliophiinae)

Distribution: Add Lower Oligocene (Rupelian, MP 22) of France, Sigé et al. (1998, as *Platyspondylia* sp.). Vianey-Liaud et al. (2014) describe and illustrate vertebrae of *Platyspondylia* sp. from the Oligocene (Chattian) of France.

Comments: Based on caudal vertebrae, Vianey-Liaud et al. (2014) confirm that *Platyspondylia* belongs to the Tropidophiidae s.s.

†*Platyspondylia germanica* Szyndlar & Rage, 2003

†*Platyspondylia lepta* Rage, 1974

†*Platyspondylia sudrei* Rage, 1988

†*PLATYSPONDYLOPHIS* T. Smith, Kumar, Rana, Folie, Solé, Noiret, Steeman, Sahni & Rose, 2016. Geoscience Frontiers 7(6): 977. (†Madtsoiidae)

Type species: *Platyspondylophis tadkeshwarensis* T. Smith, Kumar, Rana, Folie, Solé, Noiret, Steeman, Sahni & Rose, 2016 by monotypy.

Distribution: Lower Eocene of W India.

† *Platyspondylophis tadkeshwarensis* T. Smith, Kumar, Rana, Folie, Solé, Noiret, Steeman, Sahni & Rose, 2016. Geoscience Frontiers 7(6): 977–980, fig. 11.

Holotype: WIF/A 2272, a trunk vertebra.

Type locality: "Tadkeshwar Lignite Mine (TAD-1), Surat District, Gujarat, India."

Distribution: Lower Eocene (Ypresian) of W India (Gujarat).

PLECTRURUS A.M.C. Duméril & Duméril, 1851 (Uropeltidae)

Comments: Pyron et al. (2016c) provide a morphological diagnosis.

Plectrurus aureus Beddome, 1880

Comments: Pyron et al. (2016c) provide a description, and note that it is known only from three specimens collected in the late 1800s.

Plectrurus guentheri Beddome, 1863

Comments: Pyron et al. (2016c) provide a description.

Plectrurus perrotetii A.M.C. Duméril & Bibron *in* A.M.C. Duméril & Duméril, 1851

Distribution: Lower elevation limit corrected to 1800 m, Pyron et al. (2016c).

Comments: Pyron et al. (2016c) provide a description.

PLESIODIPSAS Harvey, Rivas-Fuenmayor, Portilla & Rueda-Almonacid, 2009 (Dipsadidae: Dipsadinae)

Plesiodipsas perijanensis (Aleman, 1953)

Distribution: Upper elevation of 2320 m, Sánchez-Martínez & Rojas-Runjaic (2018).

Comments: Sánchez-Martínez & Rojas-Runjaic (2018) provide a species account with description, color photographs and locality map.

†*PLESIOTORTRIX* Rochebrune, 1884 (Booidea: incerta sedis)

†*Plesiotortrix edwardsi* Rochebrune, 1884

PLIOCERCUS Cope, 1860 (Dipsadidae: Dipsadinae)

Pliocercus elapoides Cope, 1860

Distribution: Add Honduras (Atlantida), Townsend et al. (2012).

Pliocercus euryzonus Cope, 1862

Distribution: Add Panama (Panama, Panama Oeste), Ray & Ruback (2015).

POECILOPHOLIS Boulenger, 1903 (Atractaspididae: Aparallactinae)

Poecilopholis cameronensis Boulenger, 1903

POLEMON Jan, 1858 (Atractaspididae: Aparallactinae)

Polemon acanthias (Reinhardt, 1861)

Distribution: Add Liberia (Nimba), Portillo et al. (2018).

Polemon ater Portillo, Branch, Tilbury, Nagy, Hughes, Kusamba, Muninga, Aristote, Behangana & Greenbaum, 2019b. Copeia 107(1): 24–27, figs. 4–6.

Holotype: PEM R20734, a 271 mm male (C. Tilbury, 12 February 2014).

Type locality: "Democratic Republic of the Congo, Lualaba Province, Fungurume, 10.5338°S, 26.3375°E, 1189 m."

Distribution: SE Democratic Republic of the Congo (Katanga), and probably N Zambia and WC Tanzania, 1189–1472 m.

Polemon barthii Jan, 1858
Distribution: Add Guinea, Trape & Baldé (2014).

Polemon bocourti Mocquard, 1897

Polemon christyi (Boulenger, 1903)
Distribution: Limited to NE Democratic Republic of the Congo, Uganda, South Sudan and W Kenya. Specimens from Burundi and Rwanda are unassigned either here or to *P. ater*.
Comments: Portillo et al. (2019b) remove southern populations to their new species *P. ater*.

Polemon collaris (W.C.H. Peters, 1881)
Distribution: Add Gabon (Ogooue-Lolo), Portillo et al. (2018); Congo (Niari), Portillo et al. (2018); Angola (Lunda Norte), Portillo et al. (2019b).

Polemon fulvicollis (Mocquard, 1887)
Distribution: Add Gabon (Ogooue-Lolo), Pauwels et al. (2016a); Gabon (Ogooue-Maritime), Portillo et al. (2018).

Polemon gabonensis (A.H.A. Duméril, 1856)

Polemon gracilis (Boulenger, 1911)

Polemon graueri (Sternfeld, 1908)
Distribution: Upper elevation of 2075 m, Portillo et al. (2019b).
Comments: Molecular data evaluated by Portillo et al. (2018) affirm that *P. graueri* is a species distinct from *P. fulvicollis*.

Polemon griseiceps (Laurent, 1947)

Polemon neuwiedi (Jan, 1858)
Distribution: Upper elevation of 386 m, Trape & Mané (2017).

Polemon notatus (W.C.H. Peters, 1882)

Polemon robustus (Witte & Laurent, 1943)
Distribution: Low elevation of 309 m, Portillo et al. (2019b).

†*POLLACKOPHIS* Holman, 1998 (Booidea: incerta sedis)

†*Pollackophis depressus* Holman, 1998

POPEIA Malhotra & Thorpe, 2004 (Vipera: Crotalinae)
Comments: Mulcahy et al. (2017) use additional specimens to those used by Wostl et al. (2016) to produce a phylogeny of *Popeia* that supports a northern (*P. popeorum*) and southern (*P. sabahi*) species, and they identify a sister clade to *P. popeorum* on the Isthmus of Kra as an undescribed species. See comments under *Trimeresurus*.

Popeia nebularis (Vogel, David & Pauwels, 2004)

Popeia phuketensis Sumontha, Kunya, Pauwels, Nitikul & Punnadee, 2011

Popeia popeorum (M.A. Smith, 1937)
Distribution: Add Bhutan, A. Das et al. (2016); China (Yunnan), Guo et al. (2015).

Popeia sabahi (Regenass & Kramer, 1981)
Synonyms: *Trimeresurus popeorum barati* Regenass & Kramer, 1981, *Trimeresurus fucatus* Vogel, David & Pauwels, 2004, *Popeia buniana* Grismer, Grismer & McGuire, 2006, *Trimeresurus toba* David, Petri, Vogel & Doria, 2009.
Distribution: Myanmar (Tanintharyi, Kanmaw Island), S Thailand (Chumphon, Krabi, Nakhon Si Thammarat, Phang Nga, Prachuap Khiri Khan, Surat Thani, Trang), West Malaysia (Johor, Kedah, Pahang, Perak, Pinang, Selangor, Terengganu, Langkawi Island, Penang Island, Tioman Island), East Malaysia (Sabah, Sarawak), Brunei, Western Indonesia (Sumatra, Pagai Island), 200–1600 m; upper elevation limit is 1660 m, Hinckley et al. (2017).
Comments: Wostl et al. (2016) present morphological and DNA sequence data that suggest all species of *Popeia* of the Sunda Shelf region represent a single species. Mulcahy et al. (2017) recommend that the four synonyms should be recognized as subspecies.

Popeia yingjiangensis Z. Chen, Ding, Shi & Zhang in Z. Chen et al., 2019. Asian Herpetol. Res. 10(1): 15–20, figs. 1–3. (*Trimersurus yingjiangensis*) **new combination**
Holotype: CIB 2017070101, an 864 mm male (L. Ding, 19 July 2017).
Type locality: "Heihe Village, Kachang Town, Yingjiang County, Yunnan Province (24.782° N, 97.878° E, 1 112 m a.s.l.)," China.
Distribution: SE China (WC Yunnan), 1074–1200 m. Known only from the vicinity of the type locality.
Comments: A phylogeny of *Popeia* taxa produced by Z. Chen et al. (2019) has *P. yingjiangensis* as sister taxon to *P. sabahi* and its subspecies.

PORTHIDIUM Cope, 1871 (Viperidae: Crotalinae)

Porthidium arcosae Schätti & Kramer, 1993
Distribution: Elevation range 0–300 m, Valencia et al. (2016).

Porthidium dunni (Hartweg & Oliver, 1938)

Porthidium hespere (Campbell, 1976)
Distribution: Add Mexico (Guerrero), Palacios-Aguilar et al. (2016).

Porthidium lansbergii (Schlegel, 1841)
Distribution: Add Panama (Panama Oeste), Ray & Ruback (2015); Colombia (Cesar), Medina-Rangel (2011); Colombia (Huila), Medina-Rangel & López-Perilla (2015).

Porthidium nasutum (Bocourt, 1868)
Distribution: Add Nicaragua (Boaco, Chontales), Obando & Sunyer (2016a); Panama (Panama Oeste), Ray & Ruback (2015); Ecuador (Imbabura), Cisneros-Heredia & Yánez-Muñoz (2005); Ecuador (Santo Domingo), Valencia et al. (2016).

Porthidium ophryomegas (Bocourt, 1868)
Distribution: Add: Mexico (Chiapas), Grünwald et al. (2016b); Honduras (Cortes, Santa Barbara), Espinal et al. (2014b).

Porthidium porrasi Lamar & Sasa, 2003

Porthidium volcanicum Solórzano, 1995

Porthidium yucatanicum (H.M. Smith, 1941)

†*PORTUGALOPHIS* Caldwell, Nydam, Palci and Apesteguía, 2015. Nature Communic. 6(5996): 5–6. (Ophidia: incerta sedis)

Type species: *Portugalophis lignites* Caldwell, Nydam, Palci and Apesteguía, 2015 by monotypy.

Distribution: Late Jurassic of Portugal.

†*Portugalophis lignites* Caldwell, Nydam, Palci and Apesteguía, 2015. Nature Communic. 6(5996): 5–6, figs. 1, 2.

Holotype: MG-LNEG 28091, a left maxilla.

Type locality: "Guimarota mine, Leiria, Portugal."

Distribution: upper Jurassic (Kimmeridgian) of Leiria, Portugal.

†*POUITELLA* Rage, 1988 (†Lapparentophiidae)

Comments: Vullo (2019) confirms, based on vertebral similarity, that *Pouitella* belongs to the Lapparentophiidae.

†*Pouitella pervetus* Rage, 1988

PROAHAETULLA A.K. Mallik, Achyuthan, Ganesh, Pal, Vijayakumar & Shanker, 2019. PLoS One 14(7)e0218851: 8. (Colubridae: Ahaetullinae)

Type species: *Proahaetulla antiqua* A.K. Mallik, Achyuthan, Ganesh, Pal, Vijayakumar & Shanker, 2019 by original designation.

Distribution: S India.

Proahaetulla antiqua A.K. Mallik, Achyuthan, Ganesh, Pal, Vijayakumar & Shanker, 2019. PLoS One 14(7)e0218851: 8–12, figs. 2, 4.

Holotype: CESS 259, an 1113 mm male (S. Pal & S.P. Vijayakumar, 28 August 2011).

Type locality: "near Agasthiyar peak (8° 37′09″N, 77° 14′57″E), Agasthyamalai hills, Kalakad Mundanthurai tiger reserve, Tamil Nadu, India."

Distribution: S India (Kerala, Tamil Nadu), over 1200 m.

PROATHERIS Broadley, 1996 (Viperidae: Viperinae)

Proatheris superciliaris (W.C.H. Peters, 1854)

†*PROCEROPHIS* Rage, Folie, Rana, Singh, Rose & T. Smith, 2008 (Colubroidea: incerta sedis)

†*Procerophis sahnii* Rage, Folie, Rana, Singh, Rose & T. Smith, 2008

Comments: T. Smith et al. (2016) describe additional material from near the type locality.

†*PROPTYCHOPHIS* Whistler & Wright, 1989 (Colubridae: incerta sedis)

†*Proptychophis achoris* Whistler & Wright, 1989

PROSYMNA Gray, 1849 (Prosymnidae)

Prosymna ambigua Bocage, 1873

Distribution: Add Angola (Namibe), Branch (2018).

Prosymna angolensis Boulenger, 1915

Distribution: Add Angola (Cuando Cubango), Branch (2018).

Prosymna bivittata F. Werner, 1903

Prosymna collaris Sternfeld, 1908. Mitt. Zool. Mus. Berlin 4(1): 216. (*Prosymna meleagris collaris*)

Syntypes: ZMB 21970 (n = 2), (Thierry).

Type locality: "Mangu [=Sansane Mango]," Togo.

Distribution: Sahelian Africa. N Senegal (Matam, Saint-Louis), Burkina Faso (Boule du Mouhoun, Centre, Centre-Est, Centre-Nord, Centre-Ouest, Est, Nord), N Benin (Alibori), N Togo (Kara, Savanes), SW Niger (Dosso, Tahoua, Tillaberi), N Nigeria (Bauchi, Borno, Jigawa, Kebbi, Sokoto), Cameroon (Extreme Nord, Nord), Chad (Chari-Baguirmi), C Sudan. Add Mali (Kayes), Trape & Mané (2017).

Comments: Trape & Mané (2017) elevate this species from a subspecies of *P. greigerti* due to its distinctive color pattern, and due to sympatry of the two taxa in Chad and Cameroon.

Prosymna frontalis (W.C.H. Peters, 1867)

Distribution: Add Angola (Namibe), Branch (2018).

Prosymna greigerti Mocquard, 1906

Synonyms: Remove *Prosymna meleagris collaris* Sternfeld, 1908.

Distribution: Subsahelian Africa. Senegal (Kedougou, Kaolack, Kolda, Tambacounda, Thies), Gambia, Guinea (Labe, Siguiri, Telimele), SW Mali (Kayes, Koulikoro, Mopti, Segou, Sikasso), Burkina Faso (Boucle du Mouhoun, Cascades, Centre-Oest, Sud-Ouest), N Ghana (Northern), Nigeria (Plateau), Cameroon (Extreme Nord, Nord), Chad (Mayo Kebbe), Central African Republic (Bamingui-Bangoran, Ombella-Mpoko, Ouham, Vakaga), NW Democratic Republic of the Congo (Bandundu), E Sudan, South Sudan (Central Equatoria, Eastern Equatoria), W Ethiopia (Wollega). Add Niger (Tillaberi), Trape & Mané (2015).

Comments: See under *P. collaris*.

Prosymna janii Bianconi, 1862
Prosymna lineata (W.C.H. Peters, 1871)
Prosymna meleagris (J.T. Reinhardt, 1843)
Prosymna ornatissima Barbour & Loveridge, 1928
Prosymna pitmani Battersby, 1951
Prosymna ruspolii (Boulenger, 1896)
Prosymna semifasciata Broadley, 1995
Prosymna somalica Parker, 1930
Prosymna stuhlmannii (Pfeffer, 1893)

Synonyms: Remove *Stenorhabdium temporale* Werner 1909.

Lectotype: ZMH R07910, designated by Kirchhof et al. (2016).

Distribution: Add Tanzania (Morogoro), Menegon et al. (2008).

Comments: Kirchhof et al. (2016) re-examine the holotype of *Stenorhabdium temporale* and find it to be a specimen of *Pseudorabdion longiceps*. They provide a detailed description of the lectotype and paralectotype of *L. stuhlmannii*.

Prosymna sundevallii (A. Smith, 1849 *in* 1838–1849)
Prosymna visseri V.F.M. FitzSimons, 1959

PROTOBOTHROPS Hoge & Romano-Hoge, 1983 (Viperidae: Crotalinae)

Synonyms: Add *Triceratolepidophis* Ziegler, Herrmann, David, Orlov & Pauwels, 2001, according to Guo et al. (2007),

Zhaoermia Gumprecht & Tillack, 2004, according to Guo et al. (2016b).

Fossil records: Early Pleistocene of Japan (Okinawa), Ikeda et al. (2016, as *Protobothrops* sp.).

Comments: B.-W. Zhang et al. (2013), Shibata et al. (2016) and Guo et al. (2016b) present DNA sequence-based phylogenies of 12, 9 and all 14 species of *Protobothrops*, respectively, which resolve into four clades:

I – *himalayanus, kaulbacki, sieversorum*
II – *mangshanensis* (the former genus *Zhaoermia*)
III – *elegans, maolanensis, mucrosquamatus, trunkhanhensis*
IV – *cornutus, dabieshanensis, flavoviridis, jerdoni, tokarensis, xiangchengensis*.

Protobothrops cornutus (M.A. Smith, 1930b)

Distribution: Add China (Guangdong), Gong et al. (2011); China (Guangxi), Luu et al. (2015); Vietnam (Cao Bang, Lai Chau, Lang Son, Ninh Binh, Quang Binh), and low elevation of 59 m, Luu et al. (2015).

Comments: Gong et al. (2011) present an mtDNA-based phylogeny with specimens from Guangdong, China (near topotypes for *Ceratrimeresurus shenlii*) forming a clade with Vietnamese specimens of *P. cornutus*. They suggest that *shenlii* might be a subspecies of the latter.

Protobothrops dabieshanensis X. Huang, Pan, Han, L.-A. Zhang, Hu, Yu, Zheng & B.-W. Zhang, 2012

Protobothrops elegans (Gray, 1849)

Protobothrops flavoviridis (Hallowell, 1861)

Comments: Based on DNA sequence-data Shibata et al. (2016) find that *P. flavoviridis* forms two deep clades corresponding to the Okinawa and Amami Island groups. They also resolve *P. tokarensis* within the latter clade.

Protobothrops himalayanus Pan, Chettri, Yang, Jiang, Wang, L.-A. Zhang & Vogel, 2013. Asian Herpetol. Res. 4(2): 110–111, figs. 1, 2, 4, 6.

Holotype: KIZ 012736, a 1218 mm female (K. Wang & H. Pan, 14 June 2012).

Type locality: "the Jilong Valley, Jilong County, southern Tibet, China (85.35360° E, 28°37996° N; elevation 2708 m)."

Distribution: S China (S Xizang), NE India (N Sikkim), Bhutan (Paro), 1300–2708 m. Add Bhutan (Gasa), Koirala et al. (2016); Bhutan (Trongsa), Tshewang & Letro, 2018).

Protobothrops jerdonii (Günther, 1875)
Protobothrops kaulbacki (M.A. Smith, 1940)

Protobothrops mangshanensis (Zhao *in* Zhao & Chen, 1990)

Neotype: Guo et al. (2012c) note that the holotype is "considerably damaged and cannot presently be used" due to poor preservation. As neotype, they designate CIB 098485, a 370 mm male, from the same locality as the holotype.

Comments: Formerly in the monotypic genus *Zhaoermia*, which has been found to be paraphyletic within *Protobothrops* by Guo et al. (2016b).

Protobothrops maolanensis Yang, Orlov & Wang, 2011

Distribution: Add Vietnam (Ha Giang), Pham et al. (2017); Vietnam (Lang Son), 230–250 m, Kropachev et al. (2015).

Protobothrops mucrosquamatus (Cantor, 1839)

Distribution: Add India (Mizoram), Lalremsanga et al. (2017); Thailand (Nan), Vasaruchapong et al. (2017); Vietnam (Hai Phong: Cat Ba Island), T.Q. Nguyen et al. (2011); Vietnam (Hoa Binh), Nguyen et al. (2018); Vietnam (Quang Ngai), Nemes et al. (2013).

Comments: G. Zhong et al. (2017) conclude that there is no significant, morphological variation between populations in China. P. Guo et al. (2019a) likewise find no species-level variation using DNA sequence-data on specimens from throughout the geographic range.

Protobothrops sieversorum (Ziegler, Herrmann, David, Orlov & Pauwels, 2001)

Protobothrops tokarensis (Nagai, 1928)

Comments: Based on DNA sequence-data Shibata et al. (2016) find that *P. tokarensis* is paraphyletic with respect to *P. flavoviridis*.

Protobothrops trungkhanhensis Orlov, Ryabov & Nguyen, 2009

Distribution: Add China (Guangxi), T.-B. Chen et al. (2013, as *P. maolanensis*); Guo et al. (2016b) re-identified the specimen(s) to *P. trungkhanhensis*.

Protobothrops xiangchengensis (Zhao, Jiang & Huang, 1978)

Holotype: Corrected to CIB 013918, Guo et al. (2016a).

PSAMMODYNASTES Günther, 1858 (Pseudaspididae)

Psammodynastes pictus Günther, 1858

Distribution: Add West Malaysia (Penang), Quah et al. (2013); Indonesia (Pulau Bangkuru off Sumatra), Tapley & Muurmans (2016).

Psammodynastes pulverulentus (H. Boie *in* F. Boie, 1827)

Distribution: Add Nepal (Chitwan), Bhattarai et al. (2017b); Nepal (Parsa), Bhattarai et al. (2018b); Myanmar (Ayeyarwady, Chin, Mandalay, Rakhaing), Miller & Zug (2016); Thailand (Lamphun, Nakhon Phanthum, Prachuap Kiri Khan, Trang), Miller & Zug (2016); Cambodia (Siem Reap), Geissler et al. (2019); China (Nan Ao Island), Qing et al. (2015); Vietnam (Bac Giang), Hecht et al. (2013); Vietnam (Dien Bien), Dung et al. (2014); Vietnam (Hai Phong: Cat Ba Island), T.Q. Nguyen et al. (2011); Indonesia (Mentawai Archipelago), Lang (2013); Philippines (Cebu I.), Supsup et al. (2016); Philippines (Batan I.), Leviton et al. (2018).

Comments: Miller & Zug (2016) report on morphological variation in Myanmar and Thailand snakes, but find no geographic variation.

PSAMMOPHIS H. Boie *in* Fitzinger, 1826 (Psammophiidae)

Synonyms: Add *Dromophis* W.C.H. Peters, 1869.

Comments: Branch et al. (2019b) and Trape et al. (2019) produce phylogenies based on genetic material using 22 species. See under *Taphrometopon*.

Psammophis aegyptius Marx, 1958

Distribution: Add Libya (Jabal al Gharbi, Tripoli), Bauer et al. (2017).

Comments: Gonçalves et al. (2018) use DNA sequences to produce a phylogeny that shows little diversification among populations from Egypt, Sudan and Mali.

Psammophis afroccidentalis Trape, Böhme & Mediannikov *in* Trape et al., 2019. Bonn Zool. Bull. 68(1): 68–70, figs. 4, 5, 13–17.

Holotype: MNHN 2018.0013, a 1090 mm male (J.-F. Trape, 10 December 2005).

Type locality: "Dakar Hann, Senegal (14°43′N, 17°26′W)."

Distribution: W Africa (former western portion of *P. rukwae* range): SW Mauritania, Senegal, Gambia, Guinea Bissau, Guinea, Mali, Burkina Faso, Ivory Coast, Ghana, Togo, Benin, Niger, and possibly S Algeria, Trape et al. (2019).

Psammophis angolensis (Bocage, 1872)

Psammophis ansorgii Boulenger, 1905

Distribution: Add Angola (Huila), Branch (2018). Upper elevation of 2286 m, Branch et al. (2019b).

Comments: Branch et al. (2019b) provide a revised description, color photographs and a locality map.

Psammophis biseriatus W.C.H. Peters, 1881

Psammophis brevirostris W.C.H. Peters, 1881

Psammophis elegans (Shaw, 1802)

Distribution: Add Mauritania (Assaba), Padial (2006); Mali (Segou, Sikasso, Tombouctou), Trape & Mané (2017); Niger (Diffa, Tillaberi), Trape & Mané (2015); Chad (Guera), Trape et al. (2019).

Psammophis jallae Peracca, 1896

Psammophis leightoni Boulenger, 1902

Psammophis leopardinus Bocage, 1887

Distribution: Add Angola (Cuando Cubango, Cunene, Huambo, Luanda), M.P. Marques et al. (2018).

Comments: Trape et al. (2019) provide a brief description.

Psammophis lineatus (A.M.C. Duméril, Bibron & Duméril, 1854)

Distribution: Add Mali (Sikasso), Trape & Mané (2017); Niger, Trape & Mané (2015); Nigeria (Gongola), Nneji et al. (2019); Chad (Logone Oriental), Trape et al. (2019).

Psammophis mossambicus W.C.H. Peters, 1882

Distribution: The range is expanded by Trape et al. (2019) into WC Africa, including SE Nigeria, Cameroon, S Chad, Central African Republic, Gabo, Republic of Congo, Rwanda and Burundi. Add Angola (Bie, Cuando-Cubango), and upper elevation of 1763 m, Conradie et al. (2016b); Angola (Namibe), Ceríaco et al. (2016); Angola (Bengo, Benguela, Cabinda, Cuanza Norte, Cuanza Sul, Cunene, Huambo, Luanda, Lunda Sul, Malanje, Zaire), M.P. Marques et al. (2018).

Comments: Trape et al. (2019) provide a revised description and color photographs.

Psammophis namibensis Broadley, 1975

Psammophis notostictus W.C.H. Peters, 1867

Psammophis occidentalis F. Werner, 1919

Distribution: Add Niger (Tahoua, Tillaberi), Trape & Mané (2015).

Psammophis orientalis Broadley, 1977

Comments: Trape et al. (2019) provide a brief description.

Psammophis phillipsii (Hallowell, 1844)

Synonyms: Add *Psammophis irregularis* J.G. Fischer, 1856 from *P. sibilans*.

Distribution: Add Sierra Leone, Trape & Baldé (2014); Mali (Sikasso), Trape & Mané (2017). Exclude range E of Nigeria, Trape et al. (2019).

Comments: Trape et al. (2019) provide a description and color photographs.

Psammophis praeornatus (Schlegel, 1837)

Distribution: Add Niger (Tahoua, Tillaberi), Trape & Mané (2015).

Psammophis pulcher Boulenger, 1895

Psammophis punctulatus A.M.C. Duméril, Bibron & Duméril, 1854

Psammophis rukwae Broadley, 1966

Distribution: Remove W part of range, west of Chad and Cameroon populations (now *P. afroccidentalis*), and add S Ethiopia, Trape et al. (2019).

Comments: Trape et al. (2019) provide a revised description and color photographs.

Psammophis schokari (Forskål, 1775)

Synonyms: Remove *Coluber gemmatus* Shaw, 1802, *Coluber lacrymans* A. Reuss, 1834 to P. sibilans, Trape et al. (2019).

Distribution: Add Morocco (Al Hoceima), Mediani et al. (2015b); Mauritania (Dakhlet-Nouadhibou, Nouakchott, Tiris Zemmour), Padial (2006); Algeria (Tindouf), Donaire et al. (2000); Libya (Al Wahat, Jabal al Akhdar, Jabal al Gharbi, Jafara, Murqub, Nuqat al Khams, Tripoli, Wadi al Hayaa, Wadi al Shath), Bauer et al. (2017); Iraq (Wasit), Abbas-Rhadi et al. (2017); Iran (Chaharmahal and Bakhtiari, Hormozgan, North Khorasan, South Khorasan, Tehran), Safaei-Mahroo et al. (2015); Iran (Markazi), Sabbaghzadeh & Mashayekhi (2015); Iran (Qom), S.M. Kazemi et al. (2015); Afghanistan (Farah, Wardak), Wagner et al. (2016b). Moradi et al. (2013) report an upper elevation limit of 3500 m in central Iran.

Comments: Hussien & Hussein (2013) report genetic differences between coastal and montane populations in Egypt. Rato et al. (2007) produce a phylogeny from mtDNA sequence-data that contains three clades in North Africa plus one in Israel. D.V. Gonçalves et al. (2018) use DNA sequence-data to produce a phylogeny that contains six clades (two in the Middle East and four in North Africa). The clades are believed to have resulted from Pleistocene refugia.

Psammophis sibilans (Linnaeus, 1758)

Synonyms: Add *Coluber gemmatus* Shaw, 1802, *Coluber lacrymans* A. Reuss, 1834. Remove *Psammophis irregularis* J.G. Fischer, 1856, Trape et al. (2019).

Distribution: Trape et al. (2019) limit the range of *sibilans* to Egypt, Sudan and Ethiopia. Add Egypt (Suez), Ibrahim (2013).

Comments: Trape et al. (2019) provide a revised description and color photographs.

Psammophis subtaeniatus W.C.H. Peters, 1882

Distribution: Add Angola (Cauando-Cubango), Conradie et al. (2016b); Angola (Huila), Butler et al. (2019).

Psammophis sudanensis F. Werner, 1919

Distribution: Add Ghana & Nigeria, Trape & Baldé (2014); Niger (Zinder), Trape & Mané (2015); Chad (Logone Oriental), Trape et al. (2019).

Comments: Trape et al. (2019) provide a description and color photographs.

Psammophis tanganicus Loveridge, 1940

Distribution: Add Libya (Ghat, Wadi al Shath), Bauer et al. (2017).

Psammophis trigrammus Günther, 1865

Psammophis trinasalis F. Werner, 1902

Psammophis zambiensis Hughes & Wade, 2002

Distribution: Add Angola (Cuando Cubango), Branch (2018); Angola (Lunda Norte, Moxico), M.P. Marques et al. (2018).

Comments: Trape et al. (2019) provide a brief description.

PSAMMOPHYLAX Fitzinger, 1843 (Psammophiidae)

Comments: Branch et al. (2019b), Keates et al. (2019) each produce DNA sequence-based phylogenies using all species except *P. togoensis*. Keates et al. transfer *P. acutus* and *P. togoensis* to a new genus, *Kladirostratus*.

Psammophylax kellyi Conradie, Keates & Edwards *in* Keates et al., 2019. J. Zool. Syst. Evol. Res. 57(4): 1053–1054, figs. 1, 4.

Holotype: PEM R23926, an 846+ mm female (C.M.R. Kelly, 17 July 2003).

Type locality: "Arusha Region near Oldonyo Sambu, on the foothills of Mount Meru (3.17°S; 36.68°E, ~1,850 m a.s.l.), northern Tanzania."

Distribution: N Tanzania (Arusha, Dodoma, Kilimanjaro, Shinyanga), SE Kenya (Coast).

Psammophylax multisquamis (Loveridge, 1932)

Psammophylax ocellatus Bocage, 1873. J. Sci. Math. Phys. Nat. 4(15): 221–222.

Holotype: an 83 cm specimen (J. Anchieta), not located, perhaps lost according to Branch et al. (2019b).

Type locality: "l'intérieur de Mossamedes (Gambos)," Angola.

Distribution: SW Angola (Benguela, Cunene, Huila), 1108–2286 m.

Comments: Branch et al. (2019b) re-elevate *ocellatus* to species based on morphological and genetic distinctiveness. They provide a description, color photographs and locality map.

Psammophylax rhombeatus (Linnaeus, 1758)

Synonyms: Remove *Psammophylax ocellatus* Bocage, 1873.

Distribution: Add Namibia (Damaraland), Branch et al. (2019b). Remove SW Angola (see under *P. ocellatus*)

Psammophylax tritaeniatus (Günther, 1868)

Psammophylax variabilis Günther, 1892

PSEUDAGKISTRODON Van Denburgh, 1909 (Natricidae)

Pseudagkistrodon rudis (Boulenger, 1906)

PSEUDALSOPHIS Zaher, Grazziotin, Cadle, R.W. Murphy, Moura-Leste & Bonatto, 2009 (Dipsadidae: Xenodontinae)

Comments: Zaher et al. (2018) review and revise populations on the Galapagos Islands, as well as produce a phylogeny for the species of the genus. Their phylogeny depicts a continental and content-derived clade that evolved into a dwarf clade and giant clade.

Pseudalsophis biserialis (Günther, 1860)

Synonyms: Remove *Dromicus chamissonis habelii* Steindachner, 1876, which was duplicated here in error, with *P. dorsalis*.

Pseudalsophis darwini Zaher, Yánez-Muñoz, Rodrigues, Graboski, Machado, Altamirano-Benavides, Bonatto & Grazziotin, 2018. Syst. Biodiv. 16(7): 633–634, figs. 4, 8.

Holotype: DHMECN 12703, a 510 mm male (INABIO-MZUSP Expedition, 12 June 2008).

Type locality: "Tortuga Island," Galapagos Islands, Ecuador.

Distribution: Ecuador (Galapagos Is: Fernandina, Isabela, Toruga).

Pseudalsophis dorsalis (Steindachner, 1876)

Distribution: Remove Rabida and Santiago Is. (=*P. darwini*).

Pseudalsophis elegans (Tschudi, 1845)

Pseudalsophis hephaestus Zaher, Yánez-Muñoz, Rodrigues, Graboski, Machado, Altamirano-Benavides, Bonatto & Grazziotin, 2018. Syst. Biodiv. 16(7): 632–633, figs. 4, 7.

Holotype: DHMECN 12701, a 468 mm male (INABIO-MZUSP Expedition, 18 June 2008).

Type locality: "Santiago Island, collected in front of the islet known as Sombrero Chino (00°21'55"S, 90°35'09"W), Galapagos Islands, Ecuador."

Distribution: Ecuador (Galapagos Is.: Rábida, Santiago).

Pseudalsophis hoodensis (Van Denburgh, 1912)

Pseudalsophis occidentalis (Van Denburgh, 1912)

Pseudalsophis slevini (Van Denburgh, 1912)

Distribution: Remove Fernandina and Isabela Is. (=*P. darwini*).

Pseudalsophis steindachneri (Van Denburgh, 1912)

Distribution: Remove Rabida and Santiago Is. (=*P. thomasi*).

Pseudalsophis thomasi Zaher, Yánez-Muñoz, Rodrigues, Graboski, Machado, Altamirano-Benavides, Bonatto & Grazziotin, 2018. Syst. Biodiv. 16(7): 625–631, figs. 4, 5.

Holotype: DHMECN 12694, a 726 mm male (INABIO-MZUSP Expedition, 18 June 2008).

Type locality: "Rábida Island (00°24'04"S, 90°42'28"W)," Galapagos Islands, Ecuador.

Distribution: Ecuador (Galapagos Is.: Rábida, Santiago).

PSEUDASPIS Fitzinger, 1843 (Pseudaspididae)

Pseudaspis cana (Linnaus, 1758)

Distribution: Add Angola (Huambo), M.P. Marques et al. (2018).

PSEUDECHIS Wagler, 1830 (Elapidae)

Comments: Maddock et al. (2017) produce a phylogeny of all nine species, using DNA sequence data. A tenth, undescribed form from the Northern Territory is included in the analysis. *Pseudechis porphyriacus* is the sister taxon to the remaining species, and *P. butleri* is sister taxon to *P. australis* plus the dwarf "*Pailsus*" species.

Pseudechis australis (Gray, 1842)
Pseudechis butleri L.A. Smith, 1982
Pseudechis colletti Boulenger, 1902
Pseudechis guttatus De Vis, 1905
Pseudechis pailsi (Hoser, 1998)
Pseudechis papuanus W.C.H. Peters & Doria, 1878
Pseudechis porphyriacus (Shaw, 1794)
Pseudechis rossignolii (Hoser, 2000)

Pseudechis weigeli Wells & Wellington, 1987. Austral. Herpetol. (503): 1. (*Cannia weigeli*)

Holotype: AZM 1000, an 1140 mm male (J. Weigel, 12 January).

Type locality: "Mitchell River, about two kilometers upstream from Mitchell Falls, in far northern Western Australia," Australia.

Distribution: NW Australia (NW Western Australia).

Comments: The DNA sequence-based phylogeny that Maddock et al. (2017) produce confirms the validity of *weigeli* as a species.

PSEUDELAPHE Mertens & Rosenberg, 1943 (Colubridae: Colubrinae)

Pseudelaphe flavirufa (Cope, 1867)

Synonyms: Remove *Elaphe flavirufa phaescens* Dowling, 1982.

Distribution: Add Nicaragua (Esteli, Matagalpa), Sunyer et al. (2014). Remove Mexico (N Quintana Roo, Yucatan).

Comment: See under *P. phaescens*.

Pseudelaphe phaescens Dowling 1952. Occas. Pap. Mus. Zool. Univ. Michigan (540): 7–9, plate 1. (*Elaphe flavirufa phaescens*)

Holotype: UMMZ 73074, a 744 mm female (Edwin Creaser, 23 June 1932).

Type locality: "Chichén Itzá, Distrito Valladolid, Yucatán," Mexico.

Distribution: Yucatan Peninsula of Mexico.

Comment: Considered a species distinct from *P. flavirufa* by González-Sánchez et al. (2017), confirmed through molecular data by Dahn et al. (2018).

PSEUDOBOA Schneider, 1801 (Dipsadidae: Xenodontinae)

Pseudoboa coronata Schneider, 1801

Distribution: H. Costa et al. (2015b) summarize known localities and add the following: Brazil (Bahia/Pernambuco, Goias, Mato Grosso, Roraima, Tocantins), Colombia (Caqueta, Casanare, Guaina), Ecuador (Orellana), French Guiana (Camopi, Maripasoula), Guyana (Cuyuni-Mazaruni, Potaro-Siparuni), Surinam (Brokopondo, Sipaliwini). Add Brazil (Mato Grosso do Sul), Ferreira et al. (2017).

Pseudoboa haasi (Boettger, 1905)

Distribution: Pavan et al. (2018) report a southmost record in Rio Grande do Sul, Brazil.

Pseudoboa martinsi Zaher, Oliveira & Franco, 2008

Distribution: Frazão et al. (2017) provide a map of known localities.

Pseudoboa neuwiedii (A.M.C. Duméril, Bibron & Duméril, 1854)

Distribution: Add Panama (Panama Oeste), Ray & Ruback (2015); Colombia (Cesar), Medina-Rangel (2011); Colombia (Arauca, Casanare, Guainia, Huila), Lozano & Sierra (2018); Brazil (Mato Grosso do Sul), Ferreira et al. (2017).

Comments: Lozano & Sierra (2018) provide a species account for Colombia.

Pseudoboa nigra (A.M.C. Duméril, Bibron & Duméril, 1854)

Distribution: Add Brazil (Espirito Santo), Silva-Soares et al. (2011); Brazil (Sergipe), and upper elevation of 900 m, Guedes et al. (2014); Paraguay (San Pedro), P. Smith et al. (2016).

Pseudoboa serrana Morato, Moura-Leite, Prudente & Bérnils, 1995

PSEUDOBOODON Peracca, 1897 (Lamprophiidae)

Pseudoboodon boehmei J.B. Rasmussen & Largen, 1992

Pseudoboodon gascae Peracca, 1897

Distribution: Add Ethiopia (Bale), Tiutenko (2018).

Pseudoboodon lemniscatus (A.M.C. Duméril, Bibron & Duméril, 1854)

Pseudoboodon sandfordorum Spawls, 2004

†PSEUDOCEMOPHORA Auffenberg, 1963 (Colubridae: Colubrinae)

†Pseudocemophora antiqua Auffenberg, 1963

PSEUDOCERASTES Boulenger, 1896 (Viperidae: Viperinae)

Comments: Fathinia et al. (2014) use mtDNA sequence-data to produce a phylogeny of the three *Pseudocerastes* species, mostly based on Iranian specimens.

Pseudocerastes fieldi K.P. Schmidt, 1930

Distribution: Add Iraq (Muthanna, Najaf), Abbas-Rhadi et al. (2017); Iran (Fars), Gholamifard & Esmaeili (2010).

Pseudocerastes persicus (A.M.C. Duméril, Bibron & Duméril, 1854)

Distribution: Add Iran (Kermanshah, North Khorasan, South Khorasan), Safaei-Mahroo et al. (2015); Iran (Ilam), Bok et al. (2017) and Fathinia et al. (2014); Iran (Qom), S.M. Kazemi et al. (2015).

Comments: Pous et al. (2016) use mtDNA sequence-data to determine that the isolated N Oman population is derived from the most proximate Iranian population across the Strait of Hormuz.

Pseudocerastes urarachnoides Bostanchi, S. Anderson, Kami & Papenfuss, 2006

PSEUDOERYX Fitzinger, 1826 (Dipsadidae: Xenodontinae)

Pseudoeryx plicatilis (Linnaeus, 1758)

Distribution: Add Paraguay (San Pedro), P. Smith et al. (2016).

Pseudoeryx relictualis Schargel, Rivas-Fuenmayor, Barros-Blanco, Pefaur & Navarrete, 2007

PSEUDOFERANIA Ogilby, 1890 (Homalopsidae)

Comments: J.C. Murphy & Voris (2014) provide a generic diagnosis.

Pseudoferania macleayi Ogilby, 1890

Pseudoferania polylepis (J.G. Fischer, 1886)

Comments: J.C. Murphy & Voris (2014) provide a diagnosis and photograph. They consider *P. macleayi* as a synonym, which Wallach et al. (2014) consider valid, although the authors indicate the species is most likely a species complex.

PSEUDOFICIMIA Bocourt, 1883 *in* A.H.A. Duméril, Bocourt & Mocquard, 1870–1909 (Colubridae: Colubrinae)

Pseudoficimia frontalis (Cope, 1864)

PSEUDOHAJE Günther, 1858 (Elapidae)

Pseudohaje goldii (Boulenger, 1895)

Distribution: Add Angola (Bengo), Branch (2018); Angola (Cuanza Norte), M.P. Marques et al. (2018).

Pseudohaje nigra Günther, 1858

PSEUDOLATICAUDA Kharin, 1984 (Elapidae)

Pseudolaticauda schistorhyncha (Günther, 1874)

Distribution: Ghergel et al. (2016) provide a map of known localities.

Pseudolaticauda semifasciata (Schlegel, 1837)

Distribution: Add Japan (Kuchinoerabu, Kodakara, Yoron, Ikei, Ikema, Tarama, Ishigaki Is.), Tandavanitj et al. (2013a); Japan (Honshu), Tandavanitj et al. (2013b); South Korea (including Jeju Island), Park et al. (2017b); Philippines (Babuyan Claro), Leviton et al. (2018). Ghergel et al. (2016) provide a map of known localities.

Comments: Tandavanitj et al. (2013a) find 16 cytochrome *b* haplotypes in specimens from the Ryukyus and Taiwan that have some correlation with deep water barriers.

PSEUDOLEPTODEIRA Taylor, 1938 (Dipsadidae: Dipsadinae)

Pseudoleptodeira latifasciata (Günther, 1894 *in* 1885–1902)

Distribution: Low elevation of 10 m, Palacios-Aguilar & Flores-Villela (2018).

PSEUDONAJA Günther, 1858 (Elapidae)

Pseudonaja affinis Günther, 1872

Pseudonaja aspidorhyncha (McCoy, 1879)

Pseudonaja guttata (Parker, 1926)

Pseudonaja inframacula (Waite, 1925)

Pseudonaja ingrami (Boulenger, 1908)

Pseudonaja mengdeni Wells & Wellington, 1985

Pseudonaja modesta (Günther, 1872)

Pseudonaja nuchalis Günther, 1858

Pseudonaja textilis (A.M.C. Duméril, Bibron & Duméril, 1854)

PSEUDOPLECTRURUS Boulenger, 1890 (Uropeltidae)

Comments: Pyron et al. (2016c) provide a morphological diagnosis for the genus.

Pseudoplectrurus canarica (Beddome, 1870)

Comments: Pyron et al. (2016c) provide a description, and state that the species remains known only from the six syntypes collected in the late 1800s.

PSEUDORABDION Jan, 1862 (Calamariidae)

Pseudorabdion albonuchale (Günther, 1896)

Pseudorabdion ater (Taylor, 1922)

Pseudorabdion collare (Mocquard, 1892)

Pseudorabdion eiselti Inger & Leviton, 1961

Pseudorabdion longiceps (Cantor, 1847)

Synonyms: Add *Stenorhabdium temporale* Werner 1909; remove *Rabdion torquatum* A.M.C. Duméril, Bibron & Duméril, 1854.

Distribution: Add West Malaysia (Terengganu), Sumarli et al. (2015, as *P.* cf. *longiceps*).

Comments: Vogel et al. (2016) remove *P. torquatum* from the synonymy as a valid species. Kirchhof et al. (2016) re-examine the holotype of *Stenorhabdium temporale* and find it to be a specimen of *Pseudorabdion longiceps*.

Pseudorabdion mcnamarae Taylor, 1917

Distribution: Add Philippines (Biliran, Masbate Is.), Leviton et al. (2018); Philippines (Cebu I.), Supsup et al. (2016); Philippines (Sibuyan I., Tablas I.), Siler et al. (2012).

Comments: R.M. Brown et al. (2013) tentatively refer a Luzon specimen to this species, but suggest the population may represent a new species.

Pseudorabdion modiglianii G. Doria & Petri, 2010

Pseudorabdion montanum Leviton & Brown, 1959

Distribution: Add Philippines (Cebu I.), Leviton et al. (2018).

Pseudorabdion oxycephalum (Günther, 1858)

Distribution: Add Philippines (Cebu I.), Supsup et al. (2016); Philippines (Masbate I.), Leviton et al. (2018).

Pseudorabdion sarasinorum (F. Müller, 1895)

Pseudorabdion saravacense (Shelford, 1901)

Pseudorabdion sirambense G. Doria & Petri, 2010

Pseudorabdion talonuran R.M. Brown, Leviton & Sison, 1999

Comments: R.M. Brown et al. (2013) tentatively identify a Luzon I. specimen as this species, but suggest that it is likely a new species.

Pseudorabdion taylori Leviton & Brown, 1959

Pseudorabdion torquatum (A.M.C. Duméril, Bibron & Duméril, 1854). Erpétol. Gén. 7: 119–122. (*Rabdion torquatum*)

Lectotype: MNHN 2007.2456 (formerly MNHN 7212a), a 245 mm male (Dutch naturalists prior to 1845).

Type locality: "Macassar," Sulawesi.

Vogel et al. (2016) consider the locality to be dubious.

Distribution: Supposedly Indonesia (Sulawesi).

Comments: Vogel et al. (2016) revive *P. torquatum* from the synonymy of *P. longiceps* based on morphological differences.

PSEUDOTOMODON Koslowsky, 1896 (Dipsadidae: Xenodontinae)

Pseodotomodon trigonatus (Leybold, 1873)

PSEUDOXENODON Boulenger, 1890 (Pseudoxenodontidae)

Comments: B.-L. Zhang & Huang (2013) test the relationship between *Pseudoxenodon* and Dipsadidae using two mtDNA and one nDNA sequences. Their conclusion that *Pseudoxenodon* is nested within the Dipsadidae is not supported by subsequent studies using additional genes (e.g., Zheng & Wiens, 2016, and Figueroa et al., 2016).

Pseudoxenodon bambusicolus Vogt, 1922

Pseudoxenodon baramensis (M. Smith, 1921)

Pseudoxenodon inornatus (H. Boie *in* F. Boie, 1827)

Synonyms: Remove *Pseudoxenodon jacobsonii* Lidth de Jeude, 1922.

Distribution: Remove Sumatra.

Comments: Rahadian and I. Das (2013) consider *P. jacobsonii* to be a distinct species.

Pseudoxenodon jacobsonii Lidth de Jeude, 1922. Zool. Meded. 6: 240.

Holotype: RMNH 4693, (E.Jacobson, July 1915).

Type locality: "Serapai, Korintji," Sumatra, Indonesia.

Distribution: W Indonesia (Sumatra).

Comments: Rahadian and I. Das (2013) consider *P. jacobsonii* to be a distinct species based on diagnosibility and distant geographic distribution.

Pseudoxenodon karlschmidti Pope, 1928

Pseudoxenodon macrops (Blyth, 1854)

Distribution: Add Vietnam (Dien Bien), Dung et al. (2014); Vietnam (Ha Giang), Ziegler et al. (2014); Vietnam (Quang Ngai), Nemes et al. (2013); Vietnam (Son La), Pham et al. (2014); Vietnam (Yen Bai), Le et al. (2018).

Comments: DNA sequence-data analyzed by B.-L. Zhang & Huang (2013) show an eastern and western clade in *P. macrops*, as well as a very divergent sample from Tibet.

Pseudoxenodon stejnegeri Barbour, 1908

PSEUDOXYRHOPUS Günther, 1881 (Pseudoxyrhophiidae)

Pseudoxyrhopus ambreensis Mocquard, 1894

Pseudoxyrhopus analabe Nussbaum, Andreone & Raxworthy, 1998

Pseudoxyrhopus ankafinaensis Raxworthy & Nussbaum, 1994

Pseudoxyrhopus heterurus (Jan, 1863)

Pseudoxyrhopus imerinae (Günther, 1890)

Pseudoxyrhopus kely Raxworthy & Nussbaum, 1994

Pseudoxyrhopus microps Günther, 1881

Pseudoxyrhopus oblectator Cadle, 1999

Pseudoxyrhopus quinquelineatus (Günther, 1881)

Distribution: Add Madagascar (Nosy Be I.), McLellan (2013).

Pseudoxyrhopus sokosoko Raxworthy & Nussbaum, 1994

Pseudoxyrhopus tritaeniatus Mocquard, 1894

PSOMOPHIS C.W. Myers & Cadle, 1994 (Dipsadidae: Xenodontinae)

Psomophis genimaculatus (Boettger, 1885)

Psomophis joberti (Sauvage, 1884)

Distribution: Add Brazil (Maranhão), J.P. Miranda et al. (2012), Moura et al. (2013b); Brazil (Piauí), Dal Vechio et al. (2013); Brazil (Tocantins), Moura et al. (2013b), Dal Vechio et al. (2016). Upper elevation of 830 m, Guedes et al. (2014). Moura et al. (2013b) map known localities.

Psomophis obtusus (Cope, 1863)

†*PTEROSPHENUS* Lucas, 1898 (†Palaeophiidae: Palaeophiinae)

†*Pterosphenus biswasi* Rage, Bajpai, Thewissen & Tiwari, 2003

†*Pterosphenus kutchensis* Rage, Bajpai, Thewissen & Tiwari, 2003

†*Pterosphenus schucherti* Lucas, 1898

†*Pterosphenus schweinfurthi* Andrews, 1901

Comments: McCartney & Seiffert (2016) present a revised diagnosis based on new material.

†*Pterosphenus sheppardi* Hoffstetter, 1958

†*PTERYGOBOA* Holman, 1976 (Booidea: incerta sedis)

Distribution: Add late Oligocene (Arikareean) and early Miocene (Hemingfordian) of Florida, USA, Mead & Schubert (2013, closest to *P. miocenica*).

†*Pterygoboa delawarensis* Holman, 1998

†*Pterygoboa miocenica* Holman, 1976

PTYAS Fitzinger, 1843 (Colubridae: Colubrinae)

Ptyas carinata (Günther, 1858)

Ptyas dhumnades (Cantor, 1842)

Ptyas dipsas (Schlegel, 1837)

Ptyas fusca (Günther, 1858)

Ptyas korros (Schlegel, 1837)

Distribution: Add Myanmar (Mandalay, Sagaing), Platt et al. (2018); China (Nan Ao Island), Qing et al. (2015); Vietnam (Hai Phong: Cat Ba Island), T.Q. Nguyen et al. (2011); West Malaysia (Kedah), Shahriza et al. (2013); West Malaysia (Pahang), Zakaria et al. (2014).

Ptyas luzonensis (Günther, 1873)

Ptyas mucosa (Linnaeus, 1758)

Distribution: Add Afghanistan (Ghazni, Nuristan), Wagner et al. (2016b); India (Arunachal Pradesh), Purkayastha (2018);

China (Nan Ao Island), Qing et al. (2015); Taiwan (Kinmen Island), Saenz et al. (2009); Thailand (Chanthaburi), Chanard et al. (2011).

Ptyas nigromarginata (Blyth, 1854)

Distribution: Add Myanmar (Sagaing), Vogel & Hauser (2013); Thailand (Nan), Vogel & Hauser (2013); Vietnam (Ha Giang), Vogel & Hauser (2013). Upper elevation of 2400 m, Koirala et al. (2016). Vogel & Hauser (2013) map known localities and discuss elevational range.

PTYCHOPHIS Gomés, 1915 (Dipsadidae: Xenodontinae)

Ptychophis flavovirgatus Gomés, 1915

Distribution: Elevational range 316–1358 m, R.C. Gonzalez et al. (2014a).

Comments: R.C. Gonzalez et al. (2014a) describe morphological variation and provide a locality map.

PYTHON Daudin, 1803 (Pythonidae)

Fossil records: Add middle Pliocene of Tanzania, Rage & Bailon (2011, *Python natalensis* or *P. sebae*); middle Pleistocene of Thailand, Suraprasit et al. (2016, as *Python* sp.).

Comments: Reynolds et al. (2014b) present a DNA sequence-based phylogeny for seven of the living species.

Python anchietae Bocage, 1887

Distribution: Upper elevation of 1288 m, Branch (2018).

Python bivittatus Kuhl, 1820

Distribution: Add Nepal (Nawalparasi), Pandey et al. (2018); India (Uttaranchal), Joshi & Singh (2015); Myanmar (Sagaing), Platt et al. (2018); Thailand (Chanthaburi), Chanard ct al. (2011); Vietnam (Hai Phong: Cat Ba Island), T.Q. Nguyen et al. (2011). Cota (2010) lists known records for Thailand, and adds the following provinces: Chiang Mai, Kanchanaburi, Lampang, Phrae, Prachuap Khiri Khan, Udon Thani.

Python breitensteini Steindachner, 1880

Python brongersmai Stull, 1938

Comments: Auliya (2006) discusses geographic distribution and variation.

Python curtus Schlegel, 1872

†***Python europaeus*** Szyndlar & Ragc, 2003

Python kyaiktiyo Zug, Gotte & Jacobs, 2011

†***Python maurus*** Rage, 1976

Python molurus (Linnaeus, 1758)

Distribution: Add India (Nagaland), Yanthungbeni et al. (2018).

Python natalensis A. Smith, 1840 *in* 1838–1849

Distribution: Add Angola (Cuando-Cubango), Conradie et al. (2016b); Angola (Namibe), Branch (2018); Tanzania (Morogoro), Lyakurwa (2017).

Python regius (Shaw, 1802)

Distribution: Add Nigeria (Gongola), Nneji et al. (2019).

†***Python sardus*** (Portis, 1901)

Python sebae (Gmelin, 1789)

Distribution: Add Mauritania (Assaba, Tagant), Padial (2006); Mali (Kayes), Rosado et al. (2015); Mali (Mopti), Trape & Mané (2017); Nigeria (Gongola), Nneji et al. (2019); Gabon (Nyanga), Pauwels et al. (2017b); Angola (Bengo, Cuanza Norte, Lunda Sul), M.P. Marques et al. (2018); Angola (Zaire), Branch (2018).

Fossil records: Add Upper Miocene of Chad, Vignaud et al. (2002, as cf. *P. sebae*); Lower/middle Pleistocene of Eritrea, Delfino et al. (2004, as cf. *P. sebae*).

PYTHONODIPSAS Günther, 1868 (Pseudaspididae)

Pythonodipsas carinata Günther, 1868

RABDION A.M.C. Duméril, 1853 (Calamariidae)

Comments: Amarasinghe et al. (2015c) provide an overview of the genus.

Rabdion forsteni A.M.C. Duméril, Bibron & Duméril, 1854

Type: Syntypes are MNHN 7210, 7211, and possibly one of RMNH 66 or 66a, according to Amarasinghe et al. (2015c).

Comment: Amarasinghe et al. (2015c) redescribe MNHN 7210, and discuss the other type material.

Rabdion grovesi Amarasinghe, Vogel, McGuire, Sidik, Supriatna & Ineich, 2015c. Herpetologica 71(3): 235 237, fig. 1.

Type: Holotype, MZB 2679, a 522 mm male (J. McGuire, R.M. Brown & M. Williams, 10 August 2000).

Type locality: "Awan Village, Rindingalo, Tana Torja, Province of South Sulawesi, Indonesia (2°51'20 53"S, 119°48'30.04"E,…; 2150 m above sea level)."

Distribution: Indonesia (South Sulawesi). Known only from the holotype.

RACLITIA Gray, 1842. Zool. Misc.67. (Homalopsidae)

Type species: *Raclitia indica* Gray, 1842 by monotypy.

Distribution: Malay Peninsula.

Comments: J.C. Murphy & Voris (2014) revive this name based on morphology, and provide a generic diagnosis. Quah et al. (2018c) provide a phylogenetic placement of *Raclitia* for the first time.

Raclitia indica Gray, 1842

Distribution: Add Singapore and West Malaysia (Pahang), Quah et al. (2018c).

Comments: J.C. Murphy & Voris (2014) provide a diagnosis and photograph of a syntype. Quah et al. (2018c) provide a revised description, color photographs and a locality map of this rare snake.

†***RAGEOPHIS*** Wallach, 1986. J. Herpetology 20(3): 449. Substitute name for *Scytalophis* Rochebrune, 1880. (Booidea: incerta sedis)

Synonyms: †*Scytalophis* Rochebrune, 1880.

Type species: †*Coluber lafonti* Filhol, 1877 by monotypy.

Distribution: Middle Tertiary of France.

†***Rageophis lafonti*** Rochebrune 1880. Nouv. Arch. Mus. Hist. Nat., Series 23: 278, plate 12. (†*Scytalophis lafonti*)

Synonyms: †*Coluber lafonti* Filhol, 1877 (*nomen nudum*).

Holotype: MNHN QU 16342, a section of fossilized skin with vertebrae and ribs.

Type locality: "Phosphorites du Quercy," Lot, France.

Distribution: Within the late Eocene to early Miocene of France.

RAMPHOTYPHLOPS Fitzinger, 1843 (Typhlopidae: Asiatyphlopinae)

Comments: Hedges et al. (2014), Pyron & Wallach (2014) provide generic diagnoses. Pyron & Wallach (2014) note that assignment of the following species to *Ramphotyphlops* is tentative pending examination of the hemipenes, or evaluation of genetic material: *bipartitus, conradi, lorenzi, mansuetus, marxi, similis, suluensis, supranasalis.*

Ramphotyphlops acuticaudus (W.C.H. Peters, 1877)

Ramphotyphlops adocetus Wynn, R.P. Reynolds, Buden, Falanruw & B. Lynch, 2012

Ramphotyphlops angusticeps (W.C.H. Peters, 1877)

Ramphotyphlops becki (V. Tanner, 1948)

Ramphotyphlops bipartitus (Sauvage, 1879)

Comments: Lang (2013) suggests that *bipartitus* may be a synonym of *R. flaviventer.*

Ramphotyphlops conradi (W.C.H. Peters, 1874)

Ramphotyphlops cumingii (Gray, 1845)

Distribution: Add Philippines (Cebu I.), Supsup et al. (2016); Philippines (Masbate, Panay Is.), Leviton et al. (2018).

Ramphotyphlops depressus

Distribution: Add Papua New Guinea (New Hanover Island), Clegg & Jocque (2016).

Ramphotyphlops exocoeti (Boulenger, 1887)

Ramphotyphlops flaviventer (W.C.H. Peters, 1864)

Ramphotyphlops hatmaliyeb Wynn, R.P. Reynolds, Buden, Falanruw & B. Lynch, 2012

Ramphotyphlops lineatus (Schlegel, 1839 *in* 1837–1844)

Ramphotyphlops lorenzi (F. Werner, 1909)

Ramphotyphlops mansuetus (Barbour, 1921)

Ramphotyphlops marxi (Wallach, 1993)

Ramphotyphlops multilineatus (Schlegel, 1839 *in* 1837–1844)

Distribution: Add Indonesia (Dulah Island), Lang (2013).

Ramphotyphlops olivaceus (Gray, 1845)

Distribution: Add Philippines (Basilan, Sibutu Is.), Leviton et al. (2018).

Ramphotyphlops similis (Brongersma, 1934)

Ramphotyphlops suluensis (Taylor, 1918)

Ramphotyphlops supranasalis (Brongersma, 1934)

Ramphotyphlops willeyi (Boulenger in Willey, 1900)

REGINA Baird & Girard, 1853 (Natricidae)

Synonyms: Remove *Liodytes* Cope, 1885.

Regina grahamii Baird & Girard, 1853

Regina septemvittata (Say, 1825)

RENA Baird & Girard, 1853 (Leptotyphlopidae: Epictinae)

Rena boettgeri F. Werner, 1899. Zool. Anz. 22(581): 116. (*Glauconia boettgeri*)

Synonyms: *Leptotyphlops humilis slevini* Klauber, 1931.

Holotype: NMW 15455, a 225 mm specimen.

Type locality: "Habitat.--?"; restricted to La Paz, Baja California Sur, México by Smith and Larsen (1974).

Distribution: S Baja California Sur, Mexico, including Isla Cerralvo.

Comments: Adalsteinsson et al. (2009) recognize *boettgeri* as a species based on genetic divergence from *R. humilis.*

Rena bressoni (Taylor, 1939)

Distribution: Add Mexico (Jalisco), Ahumada-Carrillo et al. (2014)

Rena dissecta (Cope, 1892)

Rena dugesii (Bocourt, 1881)

Distribution: Add Mexico (Nayarit), Koch et al. (2019).

Rena dulcis Baird & Girard, 1853

Rena humilis Baird & Girard, 1853

Synonyms: Remove *Glauconia boettgeri* Werner, 1899, *Leptotyphlops humilis slevini* Klauber, 1931.

Distribution: Remove S Baja California Sur, Mexico. Upper elevation of 1725 m, Hunt (2017).

Rena iversoni (H.M. Smith, Breukelen, Auth & Chiszar, 1998)

Rena maxima (Loveridge, 1932)

Rena myopica (Garman, 1884)

Distribution: Maximum elevation 2462 m, Reyes-Vera et al. (2017).

Rena segrega (Klauber, 1939)

†***RENENUTET*** McCartney & Seiffert, 2016. J. Vert. Paleontol. 34(1): 13. (Colubroidea: incerta sedis)

Type species: †*Renenutet enmerwer* McCartney & Seiffert, 2016 by original designation.

Distribution: Late Eocene of N Egypt.

†***Renenutet enmerwer*** McCartney & Seiffert, 2016. J. Vert. Paleontol. 34(1): 13–14, fig. 7.

Holotype: CGM 83731, a mid-trunk vertebra.

Type locality: "BQ-2, Fayum Depression, Egypt."

Distribution: Late Eocene (Priabonian) of N Egypt.

RHABDOPHIS Fitzinger, 1843 (Natricidae)

Synonyms: Add *Macropisthodon* Boulenger, 1893, *Balanophis* M.A. Smith, 1938.

Comments: Based on paraphyly, Takeuchi et al. (2018) synonymize *Balanophis* and *Macropisthodon* with *Rhabdophis.*

Rhabdophis adleri Zhao, 1997

Holotype: Corrected to CIB 010494, Guo et al. (2012a).

Rhabdophis akraios Doria, Petri, Bellati, Tiso & Pistarino, 2013. Ann. Mus. Civ. Stor. Nat. "Giacomo Doria" 105.

Holotype: MSNG 55942a (O. Beccari, July 1878).

Type locality: Mount Singalang, Sumatera Barat, Sumatra, Indonesia.

Distribution: Indonesia (Sumatra).

Comments: We do not have this paper, so another kudos to Uetz et al. (2019, The Reptile Database).

Rhabdophis angelii (Bourret, 1934)

Rhabdophis auriculatus (Günther, 1858)

Rhabdophis barbouri (Taylor, 1922)

Rhabdophis callichromus (Bourret, 1934)

Rhabdophis callistus (Günther, 1873)

Rhabdophis ceylonensis (Günther, 1858)

Comments: Formerly in the monotypic *Balanophis*.

Rhabdophis chrysargoides (Günther, 1858)

Distribution: A specimen is listed for Philippines (Palawan) by Zhu et al. (2014).

Rhabdophis chrysargos (Schlegel, 1837)

Distribution: Add Thailand (Chanthaburi), Chan-ard et al. (2011); Thailand (Koh Chang I., Rayong), Chan-ard and Makchai (2011); West Malaysia (Kedah), Shahriza et al. (2013), West Malaysia (Terengganu), Sumarli et al. (2015). The record for Thailand (Roi Et) is suspect, based on a misidentification, Pauwels et al. (2015).

Rhabdophis conspicillatus (Günther, 1872)

Comments: Based on DNA sequence-data in Takeuchi et al. (2018), *R. conspicillatus* is paraphyletic with respect to some *Xenochrophis* species and *Ceratophallus*.

Rhabdophis fluviceps (A.M.C. Duméril, Bibron & Duméril, 1854)

Distribution: Add Thailand (Trang), Meewattana (2010).

Comments: Formerly in *Macropisthodon*.

Rhabdophis formosanus Maki, 1931. Monogr. Snakes Japan: 44, plate. (*Natrix tigrina formosana*)

Type: NSMT H02967, an 866+ mm male (M. Maki, July 1923).

Type locality: translated as "Hattsukan (Patungkuan: alt. 3000 m), Nantou Co., Taiwan" by Ota (1997).

Distribution: Taiwan.

Comments: See under *R. tigrinus*.

Rhabdophis guandongensis Zhu, Wang, Takeuchi & Zhao, 2014. Zootaxa 3765(5): 473–477, figs. 2, 3.

Holotype: SYS R000018, a 537 mm female (Y.-Y. Wang, 26 May 2008).

Type locality: "Aizhai Village (24°56′16.58″ N, 113°39′57.82″ E; 132 m a.s.l....), Renhua County, Guangdong Province, China."

Distribution: China (SC Guangdong), 56–138 m. Known only from the holotype and several non-collected individuals.

Rhabdophis himalayanus (Günther, 1864)

Rhabdophis lateralis (Berthold, 1859). Nachr. Ges. Wiss. Göttingen 17: 180. (*Tropidonotus lateralis*)

Synonyms: *Tropidonotus orientalis* Günther, 1861, *Rhabdophis tigrinus multiventralis* G. Stewart, 1970.

Type: Ostensibly located in the Göttingen Museum, but neither a specimen nor catalog entry was found when the collection was transferred to the ZFMK, Böhme & Bischoff (1984).

Type locality: "China."

Distribution: Far East. E China, SE Russia, and the Korean Peninsula including adjacent islands.

Comments: See under *R. tigrinus*.

Rhabdophis leonardi (Wall, 1923)

Rhabdophis lineatus (W.C.H. Peters, 1861)

Distribution: Add Philippines (Biliran I.), Zhu et al. (2014); Philippines (Leyte I.), Leviton et al. (2018).

Rhabdophis murudensis (M.A. Smith, 1925)

Rhabdophis nigrocinctus (Blyth, 1855)

Distribution: Add Thailand (Koh Chang I., Rayong), Chan-ard and Makchai (2011).

Rhabdophis nuchalis (Boulnger, 1891)

Distribution: Add China (Sichuan), Zhu et al. (2014); Vietnam (Dien Bien), Dung et al. (2014).

Rhabdophis pentasupralabialis Jiang & Zhao, 1983

Holotype: Corrected to CIB 010714, Guo et al. (2012a).

Rhabdophis plumbicolor (Cantor, 1839)

Comments: Formerly in *Macropisthodon*.

Rhabdophis rhodomelas (H. Boie *in* F. Boie, 1827)

Distribution: Add West Malaysia (Terengganu), Sumarli et al. (2015).

Comments: Formerly in *Macropisthodon*.

Rhabdophis sarawacensis (Günther, 1872)

Rhabdophis spilogaster (F. Boie, 1827)

Rhabdophis subminiatus (Schlegel, 1837)

Distribution: Add Nepal (Parsa), Bhattarai et al. (2018b); Bangladesh (Rajshahi Division, Naogaon District), Ahmad et al. (2015d); Myanmar (Sagaing), Platt et al. (2018); Thailand (Koh Chang I., Rayong), Chan-ard and Makchai (2011); Cambodia (Siem Reap), Geissler et al. (2019); China (Sichuan), Zhu et al. (2014); Vietnam (Dien Bien), Dung et al. (2014); Vietnam (Ha Giang), Ziegler et al. (2014); Vietnam (Hai Phong: Cat Ba Island), T.Q. Nguyen et al. (2011); Vietnam (Quang Ngai), Nemes et al. (2013).

Rhabdophis swinhonis (Günther, 1868)

Rhabdophis tigrinus (H. Boie, 1826)

Synonyms: Remove *Tropidonotus lateralis* Berthold, 1859, *Tropidonotus orientalis* Günther, 1861, *Natrix tigrina formosana* Maki, 1931, *Rhabdophis tigrinus multiventralis* G. Stewart, 1970.

Distribution: Limited to Japan. Add Japan (Kinkasan I.), Mori & Nagata (2016).

Comments: Takeuchi et al. (2014) use mtDNA sequence-data to produce a phylogeny of *R. tigrinus* from throughout its range, which forms three primary clades that they recognize as species: 1) mainland Asia (*R. lateralis*), 2 Japan (*R. tigrinus*), and 3) Taiwan (*R. formosanus*).

RHABDOPS Boulenger, 1893 (Natricidae)

Comments: See under *Smithophis*.

Rhabdops aquaticus Giri, Deepak, Captain & Gower *In* Giri et al. 2017. Zootaxa 4319(1): 31–40, figs. 3–7, 10.

Holotype: NCBS AU163, a 770 mm female (V.B. Giri, S. Pawar & A. Khandekar, 15 July 2015).

Type locality: "Amboli, Sindudurg district, Maharashtra, India (N 15.955801, E 73.997517; 745 m)."

Distribution: Northern Western Ghats of India (Goa, N Karnataka, Maharashtra), 745–1000 m. Bhosale & Joshi (2014) cite new locality records, and extend the known elevation to 1240 m.

Comments: This taxon represents populations formerly considered northern *R. olivaceus*.

Rhabdops olivaceus (Beddome, 1863)

Distribution: Limited to S India (Kerala, S Karnataka).

RHACHIDELUS Boulenger, 1908 (Dipsadidae: Xenodontinae)

Rhachidelus brazili Boulenger, 1908

Distribution: Add Brazil (Tocantins), P. Smith et al. (2013), who also map known localities.

RHADINAEA Cope, 1863 (Dipsadidae: Dipsadinae)

Rhadinaea bogertorum C.W. Myers, 1974

Rhadinaea calligaster (Cope, 1875)

Rhadinaea cuneata C.W. Myers, 1974

Distribution: Add Mexico (SE Puebla), Luría-Manzano et al. (2014).

Comments: Luría-Manzano et al. (2014) provide a photograph, and describe 4th and 5th known specimens.

Rhadinaea decorata (Günther, 1858)

Distribution: Range extension in Puntarenas Province, Costa Rica, Köhler et al. (2013); add Panama (Panama Oeste), Ray & Ruback (2015); Colombia (Antioquia), Restrepo et al. (2017); Colombia (Caldas, Choco, Cordoba), Vásquez-Restrepo & Toro-Cardona (2019).

Rhadinaea eduardoi Mata-Silva, Rocha, Ramírez-Bautista, Berriozabal-Islas & Wilson, 2019. ZooKeys 813: 58–61, figs. 2–4.

Holotype: CIB 5457, a 286 mm male (E. Mata-Silva, 6 June 2018).

Type locality: "Mexico, Oaxaca, municipality of Santa Catarina Juquila, El Obispo, 1320 m (.....16.183573, −97.305614.....)."

Distribution: S Mexico (extreme SC Oaxaca), 1320 m. Known only from the holotype.

Rhadinaea fluvilata (Cope, 1871)

Distribution: McKelvy et al. (2016) list and plot most known localities.

Rhadinaea forbesi H.M. Smith, 1942

Rhadinaea fulvivittis Cope, 1875

Rhadinaea gaigeae Bailey, 1937

Rhadinaea hesperia Bailey, 1940

Rhadinaea laureata (Günther, 1868)

Rhadinaea macdougalli H.M. Smith & Langebartel, 1950

Rhadinaea marcellae Taylor, 1949

Distribution: Add Mexico (Veracruz), Sánchez-García et al. (2019).

Rhadinaea montana H.M. Smith, 1944

Rhadinaea myersi Rossman, 1965

Distribution: Add Mexico (Guerrero), and upper elevation of 2331 m, Palacios-Aguilar & Flores-Villela (2018).

Rhadinaea nuchalis García-Vázquez, Pavó-Vázquez, Blancas-Hernández, Blancas-Calva & Centenero-Alcalá, 2018. ZooKeys 780: 140–146, figs. 1–4.

Holotype: MZFC-HE 22161, a 379 mm male (J.C. Blancas-Hernández, 19 July 2006).

Type locality: "0.36 km SE of El Molote, municipality of Atoyac de Álvarez, Guerrero, México (17.4167°N; 100.1672°W), ca. 1720 m elevation."

Distribution: SW Mexico (C Guerrero), 1680–1720 m.

Rhadinaea omiltemana (Günther, 1894 *in* 1885–1902)

Rhadinaea pulveriventris Boulenger, 1896

Rhadinaea quinquelineata Cope, 1886

Rhadinaea sargenti Dunn & Bailey, 1939

Rhadinaea taeniata (W.C.H. Peters, 1863)

Distribution: Add Mexico (Nayarit), 1500 m, Luja & Grünwald (2015).

Rhadinaea vermiculaticeps (Cope, 1860)

RHADINELLA H.M. Smith, 1941 (Dipsadidae: Dipsadinae)

Rhadinella anachoreta (E.N. Smith & Campbell, 1994)

Rhadinella donaji Campbell, 2015. Zootaxa 3918(3): 398–402, fig. 1.

Holotype: UTA-4223 (field no. JAC 277), a 468+ mm male (J.A. Campbell, 8 June 1974).

Type locality: "13.6 km SW Villa Sola de Vega, 16.454873 N, -97.002701 W, 2195 m above sea level…[sometimes called San Miguel Sola de Vega], Oaxaca, Mexico."

Distribution: Mexico (Oaxaca), 2195 m. Known only from the holotype.

Rhadinella dysmica Campillo, Dávila-Galavíz, Flores-Villela & Campbell, 2016. Zootaxa 4103(2): 166–169, figs. 1, 2.

Holotype: ENCB 18951, a 341 mm female (L.F. Dávila-Galavíz, 17 July 2014).

Type locality: "near Cueva de Tepozonales (17.2853 N, -99.3662 W; 432 m above sea level…), Cuajilotla, 20 km S Mochitlán, Guerrero, Mexico."

Distribution: Mexico (Guerrero). Known only from the holotype.

Rhadinella godmani (Guünther, 1865)

Distribution: Add Nicaragua (Nueva Segovia), elevation 1955 m, Loja et al. (2017). J.K. Clause (2016) reports the westernmost record, from Chiapas, Mexico.

Rhadinella hannsteini (Stuart, 1949)

Rhadinella hempsteadae (Stuart & Bailey, 1941)

Rhadinella kanalchutchan (Mendelson & Kizirian, 1995)

Rhadinella kinkelini (Boettger, 1898)
Distribution: Add Nicaragua (Esteli), Sunyer & G. Köhler (2007). Low elevation of 835 m, García-Padilla & Mata-Silva (2014b)

Rhadinella lachrymans (Cope, 1870)
Distribution: Remove Honduras records, which are *R. lisyae*, McCranie (2017). Add Guatemala (Jalapa, Sacatepequez), McCranie (2017).

Rhadinella lisyae McCranie, 2017. Mesoamerican Herpetol. 4(2): 246–249, figs. 1–3.
Holotype: USNM 535870, a 414 mm male (J.R. McCranie, 4 August 1997).
Type locality: "Cerro La Picucha (14°58′N, 85°55′W; WGS 84); elevation 2050 m asl, Montaña de Babilonia, Sierra de Agalta, Departamento de Olancho, Honduras."
Distribution: C Honduras (Olancho), 1300–2295 m.

Rhadinella montecristi (Mertens, 1952)
Distribution: Add Honduras (La Paz, Yoro), Espinal et al. (2017).

Rhadinella pegosalyta (McCranie, 2006)
Rhadinella pilonaorum (Stuart, 1954)
Rhadinella posadasi (Slevin, 1936)
Distribution: Add Mexico (ext SE Chiapas), Ariano-Sánchez & Campbell (2018).

Rhadinella rogerromani (Köhler & McCranie, 1999)
Rhadinella schistosa H.M. Smith 1941
Rhadinella serperaster (Cope, 1871)
Rhadinella stadelmani L.C. Stuart & Bailey, 1941
Comments: García-Vázquez et al. (2018) transfer *stadelmani* here from *Rhadinaea* based on morphological similarity.

Rhadinella tolpanorum (Holm & Cruz-Díaz, 1994)
Rhadinella xerophila Ariano-Sánchez & Campbell, 2018. Zootaxa 4442(2): 339–341, figs 1–2.
Holotype: UVG R7003, a 335 mm male (D. Ariano-Sánchez, 29 October 2016).
Type locality: "the northern limit of Heloderma Natural Reserve (HNR) at El Arenal (14.868878 N, 89.790526 W; 580 m above sea level…), Cabañas, Zacapa, Guatemala."
Distribution: Guatemala (Zacapa). Known only from the holotype.

RHADINOPHANES C.W. Myers & Campbell, 1981 (Dipsadidae: Dipsadinae)
Rhadinophanes monticola C.W. Myers & Campbell, 1981
Distribution: Low elevation of 2500, Palacios-Aguilar & Flores-Villela (2018).

RHAGERHIS W.C.H. Peters, 1862 (Psammophiidae)
Comments: Figueroa et al. (2016), based on DNA sequence data, synonymize *Rhagerhis* with *Malpolon* due to paraphyly. Zaher et al. (2019) find the two to be sister taxa.

Rhagerhis moilensis (A. Reuss, 1834)
Distribution: Add Morocco (Figuig, Kaar es Souk), Barata et al. (2011); Mauritania (Nouakchott), Padial (2006); Algeria (El Oued), Mouane et al. (2014); Algeria (Tindouf), Donaire et al. (2000); Libya (Al Wahat, Butnan, Darnah, Jabal al Akhdar, Jabal al Gharbi, Kufrah, Marj, Misratah, Murzuq, Sabha), Bauer et al. (2017); Niger (Dosso, Tahoua), Trape & Mané (2015); Saudi Arabia (Riyadh), Al-Sadoon (1989); Saudi Arabia (Ha'il), Alshammari et al. (2017); Saudi Arabia (Northern), Sindaco et al. (2014); Iraq (Najaf), Abbas-Rhadi et al. (2017); Iran (Bushehr, Ilam), Safaei-Mahroo et al. (2015).

RHAMNOPHIS Günther, 1862 (Colubridae: Colubrinae)
Rhamnophis aethiopissa Günther, 1862
Distribution: Add Gabon (Nyanga), Pauwels et al. (2017b); Angola (Bengo), Branch (2018); Angola (Cuanza Norte), M.P. Marques et al. (2018).

Rhamnophis batesii (Boulenger, 1908)

RHAMPHIOPHIS W.C.H. Peters, 1854 (Psammophiidae)
Fossil records: Add middle Pliocene of Tanzania, Rage & Bailon (2011, as cf. *Rhamphiophis*).

Rhamphiophis maradiensis Chirio & Ineich, 1991
Rhamphiophis oxyrhynchus (J.T. Reinhardt, 1843)
Distribution: Add Mauritania (Trarza), Padial (2006); Mali (Scgou), Trape & Mané (2017); Niger (Tillaberi), Trape & Mané (2015).

Rhamphiophis rostratus W.C.H. Peters, 1854
Rhamphiophis rubropunctatus (J.G. Fischer, 1884)

RHINOBOTHRYUM Wagler, 1830 (Colubridae: Colubrinae)
Rhinobothryum bovallii Andersson, 1916
Distribution: Add Honduras (Gracias a Dios), Turcios-Casco et al. (2018); Nicaragua (Rio San Juan), Martínez-Fonseca et al. (2019); Panama (Veraguas), Flores et al. (2016b); Panama (Bocas del Toro, Cocle, Guna Yala, Ngobe-Bugle, San Blas), Martínez-Fonseca et al. (2019); Panama (Panama Oeste), Ray & Ruback (2015); Colombia (Cauca, Magdalena, Sucre), Martínez-Fonseca et al. (2019); Colombia (Huila), Vera-Pérez et al. (2019); Ecuador (Manabi), Martínez-Fonseca et al. (2019). Martínez-Fonseca et al. (2019) list and map known localities.

Rhinobothryum lentiginosum (Scopoli, 1788)
Distribution: Add Ecuador (Orellana, Pastaza), Orcés and Almendáriz (1994); Brazil (Maranhão), M.A. Freitas et al. (2017); Brazil (Amapá), Gomes de Arruda et al. (2015), who also map known localities in Brazil.

RHINOCHEILUS Baird & Girard, 1853 (Colubridae: Colubrinae)
Rhinocheilus lecontei Baird & Girard, 1853
Distribution: Eliminate Jalisco, Mexico from distribution, Cruz-Sáenz et al. (2017).
Comments: E.A. Myers et al. (2017a) provide a DNA-sequence based phylogeny of USA populations, with five primary clades. Molecular data studied by Dahn et al. (2018) indicate the existence of a western and eastern clade.

RHINOGUINEA Trape, 2014. Bull. Soc. Herpétol. France 152: 47. (Leptotyphlopidae: Epictinae)

Type species: *Rhinoguinea magna* Trape, 2014 by original designation.

Distribution: West Africa.

Rhinoguinea magna Trape, 2014. Bull. Soc. Herpétol. France 152: 47–52, figs. 1, 4–5.

Holotype: MNHN 2014.0026 (orig. IRD TR.3478), a 398 mm male (a villager, between June 2009 and March 2010).

Type locality: "Mamoroubougou (11°14′10″N, 05°28′55″W, altitude: 386 m)," Mali.

Distribution: S Mali, 386 m. Known from the vicinity of type locality.

RHINOLEPTUS Orejas-Miranda, Roux-Estève & Guibé, 1970 (Leptotyphlopidae: Epictinae)

Comment: Boundy (2014) modifies the terminology for the head scales.

Rhinoleptus koniagui (Villiers, 1956)

Distribution: Add Mali (Koulikoro), Trape & Mané (2017).

Comment: Boundy (2014) suggests that this species probably consists of more than one.

RHINOPHIS Hemprich, 1820 (Uropeltidae)

Synonyms: Add *Pseudotyphlops* Schlegel, 1839, *Crealia* Gray, 1858.

Comments: The Cyriac & Kodandaramaiah (2017) phylogeny has *Pseudotyphlops* nested within *Rhinophis*. Pyron et al. (2016c) confirm the monophyly of the genus through mtDNA and nDNA sequence-data, and provide a morphological diagnosis.

Rhinophis blythii Kelaart, 1853

Neotype: BMNH 1946.1.1.45, designated by Pyron et al. (2016c), from "Ceylon."

Comments: Pyron et al. (2016c) provide a morphological description and color photograph.

Rhinophis dorsimaculatus Deraniyagala, 1941

Distribution: Add Sri Lanka (Northern), Gower & Wickramasinghe (2016).

Comments: Pyron et al. (2016c) provide a morphological description. Gower & Wickramasinghe (2016) expand the description through the acquisition of ten additional specimens and color photographs of others, which confirm its taxonomic validity.

Rhinophis drummondhayi Wall, 1921

Comments: Pyron et al. (2016c) provide a morphological description.

Rhinophis erangaviraji Wickramasinghe, Vidanapathirana, Wickramasinghe & Ranwella, 2009

Distribution: Add Sri Lanka (Sabaragamuwa), Pyron et al. (2016c).

Comments: Pyron et al. (2016c) provide a morphological description.

Rhinophis fergusonianus Boulenger, 1896

Comments: Pyron et al. (2016c) provide a morphological description.

Rhinophis goweri Aengals & Ganesh, 2013. Russ. J. Herpetol. 20(1): 63–65, fig. 1.

Holotype: ZSI/SRC/VRS 256, a 270 mm female (K. Ilango, R. Aengals & party, 2 October 2010).

Type locality: Noolathu Kombai, Bodamalai hills (11°28′ N 78°10′ E; ca. 980 m a.s.l.) situated between Namakkal and Salem districts in Tamil Nadu state, India."

Distribution: India (S Tamil Nadu), 980 m. Known only from the holotype.

Comments: Pyron et al. (2016c) provide a morphological description and color photograph.

Rhinophis grandis Kelaart, 1853. J. Ceylon Branch Royal Asiatic Soc. 2(3): 106–107. (*Uropeltis grandis*)

Synonyms: *Uropeltis philippinus* Cuvier, 1829, *Uropeltis philippinus* J.P. Müller, 1832, *Uropeltis pardalis* Kelaart, 1853, *Uropeltis saffragamus* Kelaart, 1853.

Holotype: BMNH 1946.1.8.1, a 508 mm specimen (Balkhuysen).

Type locality: "Southern Province...Kerinday near Matura," Sri Lanka.

Comments: Pyron et al. (2016c) transfer *Pseudotyphlops philippinus* (J.P. Müller, 1832), to *Rhinophis* based on its nested placement within *Rhinophis* in their DNA sequence-based phylogeny. Because *Typhlops philippinus* Cuvier, 1829 has priority in *Rhinophis*, Pyron et al. select the name *Uropeltis saffragamus* Kelaart, 1853 as the new name for *P. phlippinus*. However, Deraniyagala (1955) had already selected another synonym, *Uropeltis grandis* Kelaart, 1853 as the replacement name. Pyron & Somaweera (2019) provide additional data on the nomenclatural history, as well as photographs of the types of *U. philippinus* and *U. pardalis*. Pyron et al. (2016c) provide a morphological description and color photograph.

Rhinophis homolepis (Hemprich, 1820)

Comments: Pyron et al. (2016c) provide a morphological description and color photograph.

Rhinophis lineatus Gower & Maduwage, 2011

Comments: Pyron et al. (2016c) provide a morphological description.

Rhinophis melanogaster (Gray, 1858)

Comments: Pyron et al. (2016c) provide a morphological description and color photograph. They transfer *Mytilia* (*Crealia*) *melanogaster* Gray, 1858, from *Uropeltis* to *Rhinophis* based on DNA-sequence and morphological data.

Rhinophis oxyrhynchus (Schneider, 1801)

Distribution: Add Sri Lanka (North-Western), Pyron et al. (2016c).

Comments: Pyron et al. (2016c) provide a morphological description and color photograph.

Rhinophis philippinus (Cuvier, 1829)

Comments: Pyron et al. (2016c) provide a morphological description and color photograph.

Rhinophis phillipsi (Nicholls, 1929)

Pyron et al. (2016c) provide a morphological description and color photograph. They transfer *Silybura phillipsi* Nicholls, 1929, from *Uropeltis* to *Rhinophis* based on DNA-sequence and morphological data.

Rhinophis porrectus Wall, 1929

Comments: Pyron et al. (2016c) provide a morphological description.

Rhinophis punctatus J.P. Müller, 1832

Comments: Pyron et al. (2016c) provide a morphological description and color photograph.

Rhinophis roshanpererai Wickramasinghe, Vidanapathirana, Rajeev & Gower, 2017a. Zootaxa 4263(1): 155–161, figs. 2–5.

Holotype: NMSL 2016.08.01, a 215 mm male (L.J.M. Wickramasinghe, D.R. Vidanapathirana & M.D.G. Rajeev, 10 May 2010).

Type locality: "Galkanda, Beragala, Badulla District, Uva Province, Sri Lanka (6° 45′ 07.98″ N, 80° 57′ 20.23″ E, elevation 940 m)."

Distribution: Sri Lanka (Uva), 750–940 m.

Rhinophis sanguineus Beddome, 1863

Lectotype: BMNH 1946.1.16.54, designated by Pyron et al. (2016c).

Comments: Pyron et al. (2016c) provide a morphological description and color photograph.

Rhinophis travancoricus Boulenger, 1893

Comments: Pyron et al. (2016c) provide a morphological description.

Rhinophis tricoloratus Deraniyagala, 1975

Distribution: Add Sri Lanka (Southern), Pyron et al. (2016c).

Comments: Pyron et al. (2016c) provide a morphological description.

Rhinophis zigzag Gower & Maduwage, 2011

Comments: Pyron et al. (2016c) provide a morphological description and color photograph.

RHINOPLOCEPHALUS F. Müller, 1885 (Elapidae)

Synonyms: Remove *Unechis* Worrell, 1961, *Cryptophis* Worrell, 1961.

Distribution: SW Australia.

Comments: All species except *R. bicolor* are returned to *Cryptophis*, Cogger (2014).

Rhinoplocephalus bicolor F. Müller, 1885

RHINOTYPHLOPS Fitzinger, 1843 (Typhlopidae: Afrotyphlopinae)

Distribution: E and S Africa.

Comments: Hedges et al. (2014), Pyron & Wallach (2014) provide generic diagnoses, and transfer several species from *Letheobia*.

Rhinotyphlops ataeniatus (Boulenger, 1912)

Comments: See under *R. unitaeniatus*.

Rhinotyphlops boylei (V.F.M. FitzSimons, 1932)
Rhinotyphlops lalandei (Schlegel, 1839 *in* 1837–1844)
Rhinotyphlops schinzi (Boettger, 1887)
Rhinotyphlops scorteccii (Gans & Laurent, 1965)

Comments: Based on morphology, Hedges et al. (2014) and Pyron & Wallach (2014) transfer *scorteccii* to *Rhinotyphlops* from *Letheobia*.

Rhinotyphlops unitaeniatus (W.C.H. Peters, 1878)

Comments: The DNA sequence-based phylogenies that Hedges et al. (2014) and Pyron & Wallach (2014) produce associate *unitaeniatus* with *Rhinotyphlops*, and those authors transfer both *unitaeniaus* and *ataeniatus* to *Rhinotyphlops* from *Letheobia*.

RHYNCHOCALAMUS Günther, 1864 (Colubridae: Colubrinae)

Comments: Avci et al. (2015) remove *R. barani* to a new genus (*Muhtarophis*) and compare it to *R. melanocephalus* and *satunini*. Šmid et al. (2015) produce a DNA sequence-based phylogeny that includes *R. arabicus*. Tamar et al. (2016) expand on the work of Šmid et al., and describe a new species from Israel. Fathinia et al. (2017) present a DNA sequence-based phylogeny that confirms *Muhtarophis* as a sister taxon to *Lytorhynchus* plus *Rhynchocalamus*. They also provide a key to the species of *Rhynchocalamus*.

Rhynchocalamus arabicus K.P. Schmidt, 1933

Distribution: Add Oman (Dhofar), Šmid et al. (2015).

Comments: Šmid et al. (2015) describe a second specimen.

Rhynchocalamus dayanae Tamar, Šmid, Göçmen, Meiri & Carranza, 2016. PeerJ 2016(2769): 17–23, figs. 3–5.

Holotype: HUJ R21704, a 319 mm male (G. Vine, 21 June 2008).

Type locality: "road no. 40 near Nafha Prison, Negev Mountain, Israel, 30.7317N 34.7709E…, 700 m above sea level."

Distribution: The Negev Desert of S Israel.

Rhynchocalamus levitoni Torki, 2017a. Zool. Middle East 63(2): 110–113, figs. 1–4. (*Lytorhynchus levitoni*) **new combination**

Holotype: MTD 49319, a 420 mm male (F. Torki, 31 May 2016).

Type locality: "1200–1400 m a.s.l. on the western slope of the Kabi-Koh mountains, Abdanan region, Illam Province, southwestern Iran (33°02′N, 47°18′E)."

Synonyms: *Rhynchocalamus ilamenis* Fathinia, Rastegar-Pouyani, Rastegar-Pouyani & Darvishnia, 2017. Zootaxa 4282(3): 477–479, figs. 2–4.

Holotype: YUZM CRh.1, a 405 mm male (B. Fathinia & M. Mansouri, 6 June 2011).

Type locality: "Bina & Bijar No-hunting area (33.640668N, 46.038949E, 724 m asl), Ilam Province, Iran."

Distribution: W Iran (Ilam), 724–1383 m.

Comments: Based on the published descriptions and locality, *Lytorhynchus levitoni* is identical to *Rhynchocalamus ilamensis*. The former has priority, being published online on

5 March, whereas the latter was published online on 27 June. Fathinia et al. (2017) use DNA sequence data to show that the taxon belongs within *Rhynchocalamus*.

Rhynchocalamus melanocephalus (Jan, 1862)

Distribution: SE Turkey (Hatay), Lebanon, W Syria (Damascus, Homs, Latakia), Israel, W Jordan, NE Egypt (South Sinai).

Comments: See comments under *R. satunini*.

Rhynchocalamus satunini (Nikolsky, 1899). Ann. Mus. Zool. Acad. Imp. Sci. St. Pétersbourgh 4:449–450. (*Contia satunini*)

Synonym: *Oligodon melanocephalus septentrionalis* Werner, 1905.

Holotype: ZIL 9343, a 185 mm specimen (K.A. Satunin, 1893)

Type locality: "prov. Elisabetpol in Transcaucas, prope flum. Arax" (=vicinity of Megri, Axes River, Armenia).

Distribution SE Turkey (Adana, Adiyaman, Diyarbakir, Gaziantep, Malatya, Mardin, Sirnak), W Iran (Central, East Azarbaijan, Khuzestan), N. Iraq (Ta'min), Armenia and Azerbajan. Add Iran (Bushehr, Lorestan, Markazi), Shafaei-Mahroo et al. (2015).

Comments: Avci et al. (2007) define the distribution in Turkey. Avci et al. (2015) use morphological and mtDNA sequence-data to elevate this taxon to species, formerly a subspecies of *R. melanocephalus*.

†***RIONEGROPHIS*** Albino, 1986 (†Madtsoiidae)

†***Rionegrophis madtsoioides*** Albino, 1986

RODRIGUESOPHIS Grazziotin, Zaher, R.W. Murphy, Scrocchi, Benavides, Zhang & Bonatto, 2012 (Dipsadidae: Xenodontinae)

Rodriguesophis chui (Rodrigues, 1993)

Distribution: The type locality is stated to be at 475 m, Guedes et al. (2014).

Rodriguesophis iglesiasi (Gomés, 1915)

Distribution: Add Brazil (Pernambuco), Pedrosa et al. (2014). Increase elevation range to 865 m, Silveira (2014d).

Rodriguesophis scriptorcibatus (Rodrigues, 1993)

†***ROTTOPHIS*** Szyndlar & Böhme, 1996 (Tropidophiidae)

†***Rottophis atavus*** (H. Meyer, 1855)

†***RUKWANYOKA*** McCartney, Stevens & O'Connor, 2014. PLoS One 9(e90514): 3. (Boidae)

Type species: †*Rukwanyoka holmani* McCartney, Stevens & O'Connor, 2014 by original designation.

Distribution: Late Oligocene of Tanzania.

†***Rukwanyoka holmani*** McCartney, Stevens & O'Connor, 2014. PLoS One 9(e90514): 3–5, figs. 2, 4.

Holotype: RRBP 10041, a mid-trunk vertebra.

Type locality: "Nsungwe Formation, locality Nsungwe 2, Rukwa Rift Basin, southwestern Tanzania."

Distribution: Late Oligocene of SW Tanzania.

†***RUSSELLOPHIS*** Rage, 1975 (†Anomalophiidae)

†***Russellophis crassus*** Rage, Folie, Rana, Singh, Rose & T. Smith, 2008

†***Russellophis tenuis*** Rage, 1975

SALOMONELAPS McDowell, 1970 (Elapidae)

Salomonelaps par (Boulenger, 1884)

SALVADORA Baird & Girard, 1853 (Colubridae: Colubrinae)

Comments: Hernández-Jiménez et al. (2019) produce a DNA sequence-based phylogeny of seven of the eight species that comprises the *mexicana* group (*lemniscata*, *mexicana*) and the *grahamiae* group (all other species).

Salvadora bairdi Jan, 1860 *in* Jan & Sordelli, 1860–1866

Salvadora deserticola K.P. Schmidt, 1940

Comments: See under *S. hexalepis*.

Salvadora grahamiae Baird & Girard, 1853

Distribution: Add Mexico (Hidalgo, Michoacan), Hernández-Jiménez et al. (2019). Upper elevation of 2224 m, Campos-Rodríguez et al. (2017).

Salvadora gymnorhachis Hernández-Jiménez, Flores-Villela & Campbell, 2019. Zootaxa 4564(2): 589–593, figs. 1, 2.

Holotype: MZFC 28775, a 612 mm female (22 August 2014).

Type locality: "near San Pedro y San Pablo Ayutla, Distrito Mixe, Oaxaca, Mexico (17.00159° N, -96.08443° W…), 2100 m above sea level."

Distribution: S Mexico (C Oaxaca), 1760–2364 m.

Salvadora hexalepis (Cope, 1867)

Comments: E.A. Myers et al. (2017a) provide a DNA-sequence based phylogeny with three primary clades: one in the American Southwest, and two in SE Arizona, USA, and the Chihuahuan Desert that support the recognition of *S. deserticola*. The phylogeny produced by Hernández-Jiménez et al. (2019) also recognizes *S. deserticola* as a distinct species.

Salvadora intermedia Hartweg, 1940

Salvadora lemniscata (Cope, 1895)

Salvadora mexicana (A.M.C. Duméril, Bibron & Duméril, 1854)

Distribution: Add Mexico (Zacatecas), Bañuelos-Alamillo et al. (2019). Upper elevation of 1825 m, Mata-Silva et al. (2017).

†***Salvadora paleolineata*** Holman, 1973

†***SANAJEH*** J.A. Wilson, Mohabey, Peters & Head, 2010 (†Madtsoiidae)

†***Sanajeh indicus*** J.A. Wilson, Mohabey, Peters & Head, 2010

†***SANJUANOPHIS*** Sullivan & Lucas, 1988 (Booidea: incerta sedis)

†***Sanjuanophis froehlichorum*** Sullivan & Lucas, 1988

SANZINIA Gray, 1849 (Sanziniidae)

Comments: Orozco-Terwengel et al. (2008) suggest that *S m. volontany* is likely a distinct species on the basis of mtDNA sequence-data, but recommend confirmation from nDNA. The latter is analyzed by Reynolds et al. (2014b) and they recommend recognition of *volontany* as a species. Mezzasalma et al. (2019) describe the chromosomes of both species.

Sanzinia madagascariensis (A.M.C. Duméril & Bibron, 1844)

Synonyms: Remove *Sanzinia madagascariensis volontany* Vences & Glaw, 2003.

Distribution: Limited to E Madagascar (Antananarivo, SE Antsiranana, Toamasina, SE Toliara).

Sanzinia volontany Vences & Glaw, 2003. Salamandra 39(3/4): 194–195, figs. 4, 6. (*Sanzinia madagascariensis volontany*)

Holotype: ZSM 804/2001, a 143 cm male (M. Vences, D.R. Vieites, G. García, V.H. Raherisoa & A. Rasoamamonjinirina, 1 March 2001).

Type locality: "Ampijoroa (Ankarafantsika Reserve)," Madagascar.

Distribution: W Madagascar (Antsiranana, Mahajanga, Toliara, including Nosy Be). Add Madagascar (Nosy Komba), Roberts & Daly (2014).

SAPHENOPHIS C.W. Myers, 1973 (Dipsadidae: Xenodontinae)

Saphenophis antioquiensis (Dunn, 1943)
Saphenophis atahuallpae (Steindachner, 1901)
Saphenophis boursieri (Jan, 1867 *in* Jan & Sordelli, 1866–1870)
Saphenophis sneiderni C.W. Myers, 1973

Distribution: Elevation range 1800–2533 m, Vera-Pérez et al. (2018).

Saphenophis tristriatus (Rendahl & Vestergren, 1940)

Distribution: Elevation range 1800–3280 m, Vera-Pérez et al. (2018).

†***SARDOPHIS*** Georgialis & Delfino *in* Georgialis et al., 2019b. Boll. Soc. Paleontol. Ital. 58(3): 278. (Colubridae: incerta sedis)

Type species: †*Sardophis elaphoides* Georgialis & Delfino *in* Georgialis et al., 2019b, by original designation.

Distribution: Early Pleistocene of Italy (Sardinia).

†***Sardophis elaphoides*** Georgialis & Delfino *in* Georgialis et al., 2019b. Boll. Soc. Paleontol. Ital. 58(3): 278–283, figs. 1–8.

Holotype: MT-S-VI-01, a posterior trunk vertebra.

Type locality: "Monte Tuttavista VI, Orosei, Sardinia, Italy."

Distribution: Early Pleistocene of Italy (Sardinia).

SCAPHIODONTOPHIS Taylor & H.M. Smith, 1943 (Sibynophiidae)

Scaphiodontophis annulatus (A.M.C. Duméril, Bibron & Duméril, 1854)

Scaphiodontophis venustissimus (Günther, 1894 *in* 1885–1902)

Distribution: Add Colombia (Caldas), Rojas-Morales et al. (2018b).

SCAPHIOPHIS W.C.H. Peters, 1870 (Colubridae: Colubrinae)

Scaphiophis albopunctatus W.C.H. Peters, 1870

Distribution: Add Ivory Coast, Trape & Baldé (2014). Delete Sierra Leone, Trape & Baldé (2014).

Scaphiophis raffreyi Bocourt, 1875

SCAPTOPHIS Rochebrune, 1880. Nouv. Arch. Mus. Natl. Hist. Nat. Paris, Series 2, 3: 279. (Colubroidea: incerta sedis)

Type species: *Scaptophis miocenicus* Rochebrune, 1880 by monotypy.

Distribution: France.

Comments: Rage (1984:28, 53) considers this taxon to be a recent colubrid, though described as a fossil.

Scaptophis miocenicus Rochebrune, 1880. Nouv. Arch. Mus. Natl. Hist. Nat. Paris, Series 2, 3: 279–280, plate 12.

Holotype: MNHN SA9880, a trunk vertebra.

Type locality: "Colline de Sansan," France.

Distribution: France.

SCHWARTZOPHIS Zaher, Grazziotin, Cadle, R.W. Murphy, Moura-Leite & Bonatto, 2009 (Dipsadidae: Xenodontinae)

Comments: Using DNA sequence-data, Krysko et al. (2015) and Zaher et al. (2019) synonymize *Schwartzophis* with *Hypsirhynchus*. Although the two genera form a clade apart from other West Indian xenodontines, the several species of each partition into two separate clades. For that reason, we continue to recognize both genera.

Schwartzophis callilaemus (Gosse, 1851)
Schwartzophis funereus (Cope, 1862)
Schwartzophis polylepis (Buden, 1966)

SCOLECOPHIS Fitzinger, 1843 (Colubridae: Colubrinae)

Comment: L.D. Wilson & Mata-Silva (2015) review the genus and single species (*S. atrocinctus*).

Scolecophis atrocinctus (Schlegel, 1837)

Distribution: Add Nicaragua (Chontales), Alemán & Sunyer (2015). L.D. Wilson & Mata-Silva (2015) revise the elevational range to 100–1530 m, and Juárez-Peña et al. (2016) to 1800 m.

†***SEISMOPHIS*** Hsiou, Albino, Madeiros & Santos, 2013. Acta Palaeontol. Polonica 59(3): 637. (Ophidia: incerta sedis)

Type species: †*Seismophis septentrionalis* Hsiou, Albino, Madeiros & Santos, 2013 by original designation.

Distribution: Late Cretaceous of NE Brazil.

†***Seismophis septentrionalis*** Hsiou, Albino, Madeiros & Santos, 2013. Acta Palaeontol. Polonica 59(3): 637, figs. 2, 3.

Holotype: CPHNA-MA VT-1221, a posterior trunk vertebra (collector not reported).

Type locality: "Falésia do Sismito (2°28′43.0″S; 44°28′10.3″W) of Cajual Island, Maranhão, northeastern Brazil."

Distribution: Late Cretaceous (Cenomanian) of NE Brazil.

Comments: Relationships are discussed by Onary et al. (2017).

SENTICOLIS Dowling & Fries, 1987 (Colubridae: Colubrinae)

Senticolis triaspis (Cope, 1866)

Distribution: Add Honduras (Santa Bárbara), Espinal et al. (2014b); Nicaragua (Atlantico Sur), Gutiérrez-Rodríguez & Sunyer (2017a); Nicaragua (Jinotega), Ubeda-Olivas & Sunyer (2015a); Nicaragua (Masaya), Martínez-Fonseca et al. (2016a).

Comments: Molecular data studied by Dahn et al. (2018) indicate the existence of several phylogenetic clades.

SIAGONODON W.C.H. Peters, 1881 (Leptotyphlopidae: Epictinae)

Comments: Francisco et al. (2018) provide a revised diagnosis for *Siagonodon*.

Siagonodon acutirostris Pinto & Curcio, 2011

Siagonodon borrichianus (Degerbøl, 1923)

Distribution: Add Argentina (Neuquen), and elevation range of 193–943 m, Perez et al. (2010).

Siagonodon cupinensis (Bailey & Carvalho, 1946)

Distribution: Add Brazil (Para), Avila-Pires et al. (2010).

Comments: Francisco et al. (2018) provide a species account, distribiton map and color photographs.

Siagonodon septemstriatus (Schneider, 1801)
Siagonodon unguirostris (Boulenger, 1902)

SIBON Fitzinger, 1826 (Dipsadidae: Dipsadinae)

Sibon annulatus (Günther, 1872)

Distribution: Add Nicaragua (Atlantico Norte, Matagalpa, Rio San Juan), Sunyer et al. (2014); Panama (Panama Oeste), Ray & Ruback (2015); Colombia (Choco, Santander), Meneses-Pelayo et al. (2016); Colombia (Antioquia), Meneses-Pelayo et al. (2018); Ecuador (Esmeraldas), Arteaga et al. (2018). Elevation range 2–1620 m, Meneses-Pelayo et al. (2018).

Comments: Meneses-Pelayo et al. (2018) provide a species account for Colombia.

Sibon anthracops (Cope, 1868)

Distribution: Add El Salvador (Santa Ana), Juárez-Peña et al. (2016); Honduras (Choluteca), and upper elevation of 1398 m, Espinal & Solís (2015a); Nicaragua (Rivas: Isla Ometepe), Stark et al. (2014); Panamá (Boquete), Dwyer (2015); range is extended within Puntarenas Province, Costa Rica, Acosta-Chaves et al. (2014).

Sibon argus (Cope, 1875)

Distribution: Add Panama (Panama Oeste), Ray & Ruback (2015).

Sibon ayerbeorum Vera-Pérez, 2019. Zootaxa 4701(5): 444–449, figs. 1–5.

Holotype: MHNUC He-Se-000659, a 417 mm female (L.E. Vera-Pérez, 18 October 2017).

Type locality: "sector La Cueva, Parque Nacional Natural Munchique, municipality of El Tambo, department of Cauca, Colombia (2°46′15.6″ N, 76°58′48.6″ W, 1135 m)."

Distribution: SW Colombia (Cauca), 1135–1400 m.

Sibon bevridgleyi Arteaga, Salazar-Valenzuela, Mebert, Peñafiel, Aguiar, Sánchez-Nivicela, Pyron, Colston, Cineros-Heredia, Yánez-Muñoz, Venegas, Guayasamin & Torres-Carvajal, 2018. ZooKeys 766: 97–107, figs. 2, 6, 7.

Holotype: MZUTI 5416, a 788 mm male (M. Hollanders, 1 August 2017).

Type locality: "Reserva Buenaventura, province of El Oro, Ecuador (S3.65467, W79.76794; 524 m)."

Distribution: Ecuador (Azuay, Chimborazo, El Oro, Guayas, Los Rios, Manabi), Peru (Tumbes), 3–1206 m.

Sibon carri (Shreve, 1951)

Sibon dimidiatus (Günther, 1872)

Distribution: Add Honduras (Atlantida), Townsend et al. (2012); El Salvador (Santa Ana), Morán et al. (2015); Belize (Stann Creek), E.P. Hofmann (2016); Nicaragua (Masaya), Sunyer et al. (2014).

Sibon dunni J.A. Peters, 1957

Sibon lamari Solórzano, 2001

Distribution: Upper elevation of 700 m, Bartuano & La Cruz (2014).

Sibon linearis Pérez-Higareda, López-Luna & H.M. Smith, 2002

Sibon longifrenis (Stejneger, 1909)

Distribution: Add Panama (Panama Oeste), Ray & Ruback (2015). Upper elevation of 1030 m, Köhler et al. (2013).

Sibon manzanaresi McCranie, 2007

Sibon merendonensis Rovito, Papenfuss & Vásquez-Almazán, 2012

Sibon miskitus McCranie, 2006

Sibon nebulatus (Linnaeus, 1758)

Distribution: Add Panama (Panama Oeste), Ray & Ruback (2015); Colombia (Cesar), Rojas-Murcia et al. (2016); Colombia (Huila), Moreno-Arias & Quintero-Corzo (2015); Ecuador (Cotopaxi, Santo Domingo), Arteaga et al. (2018); Ecuador (Manabi), Almendáriz et al. (2012); Brazil (Paraiba), R. França et al. (2012); Brazil (Pernambuco), R. França et al. (2018).

Sibon noalamina Lotzkat, Hert & Köhler, 2012
Sibon perissostichon Köhler. Lotzkat & Hertz, 2010
Sibon sanniolus (Cope, 1867)

SIBYNOPHIS Fitzinger, 1843 (Sibynophiidae)

Sibynophis bistrigatus (Günther, 1868)
Sibynophis bivittatus (Boulenger, 1894)
Sibynophis chinensis (Günther, 1889)

Distribution: Add Vietnam (Bac Giang), Hecht et al. (2013).

Sibynophis collaris (Gray, 1853)

Distribution: Add Nepal (Nawalparasi), Pandey et al. (2018); Vietnam (Hoa Binh), Nguyen et al. (2018); Vietnam (Quang Ngai), Nemes et al. (2013).

Sibynophis geminatus (H. Boie, 1826)

Distribution: Add Philippines (Tawi-Tawi I.), Leviton et al. (2018).

Sibynophis melanocephalus (Gray, 1835 *in* Gray & Hardwicke, 1830–1835)

Distribution: Remove Philippines (Tawi-Tawi), which Leviton et al. reassign to *S. geminatus*.

Sibynophis sagittaria (Cantor, 1839)

Sibynophis subpunctatus (A.M.C. Duméril, Bibron & Duméril, 1854)

Distribution: Add India (Gujarat), Patel & Vyas (2019); India (Telangana), Kumar et al. (2017).

Sibynophis triangularis Taylor & Elbel, 1958

SIMALIA Gray, 1849 (Pythonidae)

Synonyms: *Aspidopython* Meyer, 1875, *Hypaspistes* Ogilby, 1891, *Australiasis* Wells & Wellington, 1984, *Nyctophilopython* Wells & Wellington, 1985, *Lenhoserus* Hoser, 2000.

Type species: *Boa amethistina* Schneider, 1801, designated by Williams & Wallach (1989).

Distribution: E Indonesia, New Guinea, and N Australia.

Comments: Reynolds et al. (2014b) analyze DNA sequence data that sorts species of *Morelia* between two clades. *Simalia* is resurrected for the clade that contains *M. amethistina* and related species. Barker et al. (2015) provide a morphological diagnosis. The gender of the name is not indicated, but we treat it in the common usage as feminine.

Simalia amethistina (Schneider, 1801)

Distribution: Add Indonesia (Dulah, Gebe Islands), Lang (2013); Indonesia (Maluku: Kei Besar, Kei Cecil, Kur, Tam Islands), Karin et al. (2018); Papua New Guinea (Blupblup, Manam, Mussau, New Britain and New Ireland Islands), Clegg & Jocque (2016); Papua New Guinea (Oro), O'Shea et al. (2018b, suppl.).

Simalia boeleni (Brongersma, 1953)

Distribution: Add Papua New Guinea (Milne Bay), Kraus (2013).

Simalia clastolepis (Harvey, Barker, Ammerman & Chippindale, 2000)

Distribution: Confirmed for Indonesia (Haruku, Saparua Islands), Lang (2013).

Simalia kinghorni (Stull, 1933)

Simalia nauta (Harvey, Barker, Ammerman & Chippindale, 2000)

Distribution: Add Indonesia (Laibobar Island), Lang (2013).

Simalia oenpelliensis (Gow, 1977)

Simalia tracyae (Harvey, Barker, Ammerman & Chippindale, 2000)

†*SIMOLIOPHIS* Sauvage, 1880 (†Simoliophiidae)
†*Simoliophis libycus* Nessov, Zhegallo & Averianov, 1998
†*Simoliophis rochebrunei* Sauvage, 1880

Comments: Rage et al. (2016) amplify the description based on a variety of newly acquired vertebrae.

SIMOPHIS W.C.H. Peters, 1860 (Colubridae: Colubrinae)

Simophis rhinostomus (Schlegel, 1837).

Distribution: Upper elevation of 1100 m, Guedes et al. (2014).

SIMOSELAPS Jan, 1859 (Elapidae)

Simoselaps anomalus (Sternfeld, 1919)
Simoselaps bertholdi (Jan, 1859)
Simoselaps bimaculatus (A.M.C. Duméril, Bibron & Duméril, 1854)
Simoselaps littoralis (Storr, 1968)
Simoselaps minima (Worrell, 1960)

SINOMICRURUS Slowinski, Boundy & Lawson, 2001 (Elapidae)

Sinomicrurus hatori (Takahashi, 1930)

Sinomicrurus houi Peng, Wang, Ding, Zhu, Luo, Yang, R. Huang, Lu & S. Huang, 2018. Asian Herpetol. Res. 9(2): 67–70, figs. 1–3, 5.

Holotype: HUM 20170001, a 629 mm male (L. Wang & M. Hou, 17 June 2010).

Type locality: "a path near a gutterway at the side of Tianchi Lake, Jianfengling NNR, Hainan island, Hainan, China (108°46'E, 18°39' N; 805 m a.s.l.)."

Distribution: China (Hainan), 726–805 m.

Sinomicrurus japonicus (Günther, 1868)

Fossil records: Early Pleistocene of Japan (Okinawa), Ikeda et al. (2016, as *Sinomicrurus* cf. *japonicus*).

Comments: Kaito et al. (2017) evaluate DNA sequence and morphological data among the three recognized subspecies of *S. japonicus*. The subspecies *takarai* is synonymized with the subspecies *boettgeri*, due to it having both two independent derivations and weak morphological differentiation from the latter.

Sinomicrurus kelloggi (Pope, 1928)

Distribution: Add China (Yunnan), J.-H. Wang et al. (2019). Remove China (Hainan), Peng et al. (2018).

Sinomicrurus macclellandi (J.T. Reinhardt, 1844)

Distribution: Add Vietnam (Ha Giang), Ziegler et al. (2014); Vietnam (Hai Phong: Cat Ba Island), T.Q. Nguyen et al. (2011); Cambodia (Kampot), Neang et al. (2017).

Comments: Lalremsanga & Zothansiama (2015) describe color pattern morphs from northeast India.

Sinomicrurus sauteri (Steindachner, 1913)

SIPHLOPHIS Fitzinger, 1843 (Dipsadidae: Xenodontinae)

Comments: Sheehy et al. (2014) provide a key to the species of *Siphlophis*.

Siphlophis ayauma Sheehy, Yánez-Muñoz, Valencia & E.N. Smith, 2014. South Amer. J. Herpetol. 9(1): 32–37, figs. 1–4.

Holotype, DHMECN 4599 (field number ENS 12841), a 714 mm female (L.C.Tobar-Suárez, M.H. Yánez-Muñoz & E.N. Smith, 26 March 2008).

Type locality: "El Topo, Cantón Baños, Provincia de Tungurahua, Ecuador, 1594 m (1.355715°S, 78.21052°W)."

Distribution: Ecuador (Azuay, Tungurahua, Zamora-Chinchipe), 1250–2200 m; also Morona-Santiago according to Almendáriz et al. (2014).

Siphlophis cervina (Laurenti, 1768)

Distribution: Add Panama (Panama Oeste), Ray & Ruback (2015); Panama (Bocas del Toro), Crumb et al. (2015); Colombia (Antioquia, Boyaca, Caqueta, Santander), Aponte-Gutiérrez & Vargas-Salinas (2018); Venezuela (Sucre), Padrón et al. (2016); Guyana (Barina-Waini), Sheehy et al. (2014); Ecuador (Morona-Santiago), Sheehy et al. (2014); Peru (Huanuco), Schlüter et al. (2004); Brazil (Marajo I., Para); G.M. Rodrigues et al. (2015).

Siphlophis compressus (Daudin, 1803)

Distribution: Add Colombia (Antioquia, Arauca, Caqueta, Guainia, Risaralda, Vichada), Aponte-Gutiérrez & Vargas-Salinas (2018); Ecuador (Morona-Santiago, Orellana, Sucumbios), Sheehy et al. (2014); Peru (Puno), Llanqui et al. (2019); Brazil (Ceara), Guedes et al. (2011); Brazil (Acre), Bernarde et al. (2011b); Brazil (Alagoas, Minas Gerais, Roraima), Abegg et al. (2017c); Brazil (Paraiba), Guedes et al. (2011), R. França et al. (2012); Venezuela (Barinas, Tachira), Barrio-Amorós et al. (2010). Upper elevation of 550 m, Vilela et al. (2011).

Siphlophis leucocephalus (Günther, 1863)

Distribution: Add Brazil (Tocantins), and upper elevation of 1300 m, Thomassen et al. (2015). Thomassen et al. (2015) map known localities.

Siphlophis longicaudatus (Andersson, 1901)

Distribution: Add Brazil (Santa Catarina), Thomassen et al. (2015). Upper elevation of 1250 m, Alencar et al. (2009). Thomassen et al. (2015) map known localities.

Siphlophis pulcher (Raddi, 1820)

Distribution: Add Brazil (Para), Prudente et al. (2017).

Siphlophis worontzowi (Prado, 1940)

Distribution: Add Brazil (Acre), Matos & Melo-Sampaio (2013); Brazil (Tocantins), Dal Vechio et al. (2015b) and Prudente et al. (2017); Bolivia (Beni), Prudente et al. (2017); Peru (Madre de Dios), Prudente et al. (2017). Upper elevation of 272 m, Dal Vechio et al. (2015b).

Comment: Both Dal Vechio et al. (2015b) and Prudente et al. (2017) provide a description, color photographs and a range map.

SISTRURUS Garman, 1884 (Viperidae: Crotalinae)

Sistrurus catenatus (Rafinesque-Schmaltz, 1818)

Comments: McCluskey & Bender (2015) found fairly uniform genetic structure in the subspecies *tergeminus* in confluent populations, but divergence in those that are geographically isolated.

Sistrurus miliarius (Linnaeus, 1766)

Sistrurus tergeminus (Say *in* James, 1823)

Neotype: Crother et al. (2012) request to change the proposed neotype of *Crotalus tergeminus* to DU 3917 as it is from near the type locality.

Proposed Neotype locality: "4.5 miles north of Hastings, Mills County, Iowa, U.S.A."

†*SIVAOPHIS* Head, 2005 (Colubridae: incerta sedis)

†*Sivaophis downsi* Head, 2005

SMITHOPHIS Giri, Gower, Das, Lalremsanga, Lalronunga, Captain & Deepak, 2019. Zootaxa 4603(2): 245. (Natricidae)

Type species: *Smithophis atemporalis* Giri, Gower, Das, Lalremsanga, Lalronunga, Captain & Deepak, 2019, by original designation.

Distribution: Interior Southeast Asia.

Comments: Giri et al. (2019a) use DNA sequence-data to demonstrate that *Rhabdops bicolor* is the sister taxon to their new species and genus, transferring the former to *Smithophis*.

Smithophis atemporalis Giri, Gower, Das, Lalremsanga, Lalronunga, Captain & Deepak, 2019. Zootaxa 4603(2): 246–251, figs. 2, 3, 5, 6.

Holotype: BNHS 3523, a 440 mm male (H.T. Lalremsanga, 10 July 2014).

Type locality: "Mizoram University Campus, Aizawl, India (23.76338°N, 93.09916°E, 833 m)."

Distribution: NE India (Mizoram), 833–1025 m.

Smithophis bicolor (Blyth, 1854)

Distribution: Add India (Mizoram), Giri et al. (2019a).

SONORA Baird & Girard, 1853 (Colubridae: Colubrinae)

Synonyms: Add *Chilomeniscus* Cope, 1860, *Chionactis* Cope, 1860, *Sonora* (*Eosonora*) Cox, Rabosky, Holmes, Reyes-Velasco, Roelke, Smith, Flores-Villela, McGuire & Campbell, 2018.

Comments: Wood et al. (2014) use mtDNA sequence-data to produce a phylogeny for the *occipitalis* complex, which they interpret as two species (*occipitalis* and *annulata*), the latter with two subspecies. Cox et al. (2018) use DNA sequence-data to produce a phylogeny of all species of *Chilomeniscus*, *Chionactis* and *Sonora*. *Chilomeniscus* and *Chionactis* were paraphyletic with respect to *Sonora*, so all species were transferred to the latter genus.

Sonora aemula (Cope, 1879)

Sources: North-most locality in Chihuahua, Mexico, Van Devender & Holm (2014); northwestern-most locality, Sonora, Mexico, Van Devender et al. (2014).

Comments: Cox et al. (2018) provide a species account, distribution map, and color photograph.

Sonora annulata Baird, 1859. Rep. U.S. Mex. Boundary Survey 2:22, plate 21. (*Lamprosoma annulatum*)

Synonyms: *Sonora occipitalis klauberi* Stickel, 1941, *Chionactis saxatilis* Funk, 1967.

Syntypes: USNM 2105, 2106, (A. Schott, 1855).

Type locality: "Colorado Desert." Restricted to near Holtville, Imperial Co., California, USA by Smith & Taylor (1950).

Distribution: NW Mexico (NE Baja California, NW Sonora), SW USA (SW Arizona, SE California).

Comments: Wood et al. (2014) re-elevate this taxon to species based on mtDNA sequence-data, and recognize *klauberi* as a subspecies. Cox et al. (2018) provide a species account, distribution map, and color photograph.

Sonora cinctus Cope, 1861. Proc. Acad. Nat. Sci. Philadelphia 13:303. (*Chilomeniscus cinctus*)

Synonyms: *Chilomeniscus ephippicus* Cope, 1867.

Holotype: MCZ 24, evidently lost.

Type locality: "Near Guaymas, east coast Gulf of California," Sonora, Mexico.

Distribution: NW Mexico (Baja California, Bajoa California Norte, W Sonora), SW USA (S Arizona).

Comments: Revalidated by Cox et al. (2018), in part based on molecular data, and on unpublished morphological data. Cox et al. (2018) provide a species account, distribution map, and color photograph.

Sonora episcopa Kennicott *in* Baird, 1859. Rep. U.S. Mex. Boundary Survey 2:22, plate 8. (*Lamprosoma episcopum*)

Synonyms: *Contia episcopa torquata* Cope, 1880, *Contia nuchalis* Schenkel, 1901.

Lectotype: USNM 2042, female (A. Schott), designated by Stickel (1938).

Lectotype locality: "Eagle Pass," Texas, USA.

Distribution: SC USA (SE Colorado, Kansas, SW Missouri, NW Arkansas, Oklahoma, Texas, E New Mexico), NE Mexico (Coahuila, N Nuevo Leon).

Comments: Revalidated by Cox et al. (2018) based on molecular data. Cox et al. (2018) provide a species account, distribution map, and color photograph.

Sonora fasciata Cope, 1892. Proc. U. S. Natl. Mus. 14(882): 595. (*Chilomeniscus stramineus fasciatus*)

Synonyms: *Chilomeniscus stramineus esterensis* Hoard, 1939.

Holotype: USNM 12630, a 235 mm specimen (L. Belding, February 1882).

Type locality: "La Paz, Cal.," Baja California Sur, Mexico.

Distribution: S Baja California Sur, Mexico.

Comments: Revalidated by Cox et al. (2018), in part based on molecular data, and on unpublished morphological data. Cox et al. (2018) provide a species account, distribution map, and color photograph.

Sonora michoacanensis (Dugès *in* Cope, 1885)

Comments: Cox et al. (2018) provide a species account, distribution map, and color photograph.

Sonora mosaueri Stickel, 1938. Copeia 1938(4): 189–190.

Synonyms: *Sonora bancroftae* Klauber, 1943.

Holotype: MVZ 13772, a 328 mm male (C.C. Lamb, 2 April 1931).

Type locality: "Comondu, Lower California," Baja California Sur Mexico.

Distribution: NW Mexico (Baja California, Baja California Sur).

Comments: Revalidated by Cox et al. (2018) based on molecular data. Cox et al. (2018) provide a species account, distribution map, and color photograph.

Sonora mutabilis Stickel, 1943

Comments: Cox et al. (2018) provide a species account, distribution map, and color photograph.

Sonora occipitalis (Hallowell, 1854)

Synonyms: *Chionactis occipitalis talpina* Klauber, 1951.

Distribution: SW USA (WC Arizona, SE California, S Nevada), NSL-1500 m.

Comments: See under *S. annulata*. Transferred from *Chionactis* by Cox et al. (2018). Cox et al. (2018) provide a species account, distribution map, and color photograph.

Sonora palarostris Klauber, 1937

Comments: Transferred from *Chionactis* by Cox et al. (2018). Cox et al. (2018) provide a species account, distribution map, and color photograph.

Sonora punctatissima Van Denburgh & Slevin, 1921. Proc. California Acad. Sci. Series 4, 11(6): 98. (*Chilomeniscus punctatissimus*)

Holotype: CAS 49156, a 156 mm female (May 30 1921).

Type locality: "Isla Partida, Espiritu Santo Island, Gulf of California, Mexico."

Distribution: S Baja California Sur (Espirito Santo and Partida Islands).

Comments: Revalidated by Cox et al. (2018) based on unpublished morphological data. Cox et al. (2018) provide a species account.

Sonora savagei (Cliff, 1954)

Comments: Transferred from *Chilomeniscus* by Cox et al. (2018). Cox et al. (2018) provide a species account.

Sonora semiannulata Baird & Girard, 1853

Synonyms: Remove *Contia episcopa torquata* Cope, 1880, *Contia taylori* Boulenger, 1894, *Contia nuchalis* Schenkel, 1901.

Distribution: NW Mexico (NW Coahuila, Chihuahua, NE Baja California Norte), W USA (W Texas, SW New Mexico, Arizona, SW Utah, E California, Nevada, SW Idaho, SE Oregon).

Comments: Transferred from *Chionactis* by Cox et al. (2018). Cox et al. (2018) provide a species account, distribution map, and color photograph.

Sonora straminea (Cope, 1860)

Comments: Transferred from *Chilomeniscus* by Cox et al. (2018). Cox et al. (2018) provide a species account, distribution map, and color photograph.

Sonora taylori Boulenger, 1894. Cat. Snakes Brit. Mus. 2: 265, pl. 12. (*Contia taylori*)

Syntypes: BMNH 1946.1.5.39 (N.L.), female, BMNH 1946.1.5.57–59 (Texas), males (all W. Taylor, 1880–1881).

Type locality: "Duval Co., Texas," USA and "Nuevo Leon," Mexico.

Distribution: S USA (S Texas) and NE Mexico (Nuevo Leon, N Tamaulipas).

Comments: Revalidated by Cox et al. (2018) based on molecular data. Cox et al. (2018) provide a species account, distribution map, and color photograph.

SORDELLINA Procter, 1923 (Dipsadidae: Xenodontinae)

Sordellina punctata (W.C.H. Peters, 1880)

SPALEROSOPHIS Jan *in* Filippi, 1865 (Colubridae: Colubrinae)

Spalerosophis arenarius (Boulenger, 1890)

Spalerosophis atriceps (J.G. Fischer, 1885)

Spalerosophis cliffordii (Schlegel, 1837)

Distribution: Add Algeria (El Oued), Mouane et al. (2014); Libya (Ghat, Jabal al Akhdar, Jabal al Gharbi, Marj, Murqub, Tripoli, Wadi al Hayaa, Wadi al Shath), Bauer et al. (2017); Niger (Dosso), Trape & Mané (2015); Turkey (Kilis), Göçmen et al. (2009a); Iraq (Najaf), Abbas-Rhadi et al. (2017). Mediani et al. (2013) document localities southward on the central coast of Western Sahara.

Spalerosophis diadema (Schlegel, 1837)

Distribution: Add Iran (Alborz, Hamedan, Golestan, Ilam, Isfahan, Lorestan, Markazi, North Khorasan, Qazvin, Tehran), Shafaei-Mahroo et al. (2015); Iran (Hormozgan, South Khorasan, Yazd), Moadab et al. (2018); Iran (Qom), S.M. Kazemi et al. (2015); Afghanistan (Faryab, Herat, Kabul, Kandahar, Parwan), Wagner et al. (2016b). Moadab et al. (2018) model the distribution in Iran.

Spalerosophis dolichospilus (F. Werner, 1923)

Distribution: Add Morocco (Figuig), Barata et al. (2011); Morocco (Nador, Oujda, Taza), Mediani et al. (2015b); Morocco (Tata, Tiznit), Damas-Moreira et al. (2014).

Spalerosophis josephscorteccii Lanza, 1964

Spalerosophis microlepis Jan *in* Filippi, 1865.

Distribution: Add Iraq (Al-Sulaimaniyah), low elevation of 1400 m, Afrasiab & Mohamad (2014); Iran (Chaharmahal and Bakhtiari, Kohgiluyeh and Boyer Ahmad, Markazi), Shafaei-Mahroo et al. (2015); Iran (Ilam), Hosseinzadeh et al. (2017); Iran (Qom), S.M. Kazemi et al. (2015); Iran (Kerman), Moradi et al. (2013).

SPILOTES Wagler, 1830 (Colubridae: Colubrinae)

Synonyms: Add *Pseustes* Fitzinger, 1843, *Thamnobius* Fitzinger, 1843, *Paraphrynonax* Lutz & Mello, 1922.

Comment: Jadin et al. (2014) use DNA sequence-data to resolve relationships of *Pseustes* and *Spilotes*, which proved to be sister taxa within Colubrinae.

Spilotes megalolepis Günther, 1865

Spilotes pullatus (Linnaeus, 1758)

Distribution: Add Panama (Panama Oeste), Ray & Ruback (2015); Colombia (Cesar), Medina-Rangel (2011); Colombia (Huila), Moreno-Arias & Quintero-Corzo (2015); Peru (Amazonas, Cajamarca), Koch et al. (2018); Peru (Huanuco), Schlüter et al. (2004); Brazil (Alagoas), Guedes et al. (2014); Brazil (Maranhão), J.P. Miranda et al. (2012); Brazil (Mato Grosso do Sul), Ferreira et al. (2017); Brazil (Paraiba), R. França et al. (2012); Brazil (Piauí), Dal Vechio et al. (2013); Brazil (Rio Grande do Norte), Sales et al. (2009); Brazil (Sao Paulo: Ilha Anchieta), Cicchi et al. (2009).

Spilotes sulphureus (Wagler *in* Spix, 1824)

Distribution: Add Colombia (Santander), E. Ramos and Meza-Joya (2018); Ecuador (Sucumbios), Valencia & Garzon-Tello (2018); Peru (Huanuco), Schlüter et al. (2004); Brazil (Acre, Alagoas, Amapa, Goias, Maranhao, Sergipe), Andrade et al. (2017); Brazil (Paraiba), R. França et al. (2012); Brazil (Rondônia), Bernarde et al. (2012b).

Comments: Jadin et al. (2014) propose including *Pseustes sulphureus* in *Spilotes* due to paraphyly with other members of the genus discovered via DNA sequence-data analysis.

STEGONOTUS A.M.C. Duméril, Bibron & Duméril, 1854 (Colubridae: Colubrinae)

Comments: C.M. Kaiser et al. (2018) present a taxonomic overview with a description of the named taxa and their respective type material. Ruane et al. (2018b) present a DNA sequence-based phylogeny using twelve species.

Stegonotus admiraltiensis Ruane, Richards, McVay, Tjaturadi, Krey & Austin, 2018b. J. Nat. Hist. 52(13–16): 21–24, figs. 2, 4, 8.

Holotype: LSUMZ 93598, an 843 mm male (C.C. Austin, 1 September 2001).

Type locality: "Penchal Village on Rambutyo Island, Manus Province, Papua New Guinea, elevation 58 m asl, (2.3283333°S, 147.7666667°E)."

Distribution: Papua New Guinea (Los Negros and Rambutyo Islands, Manus), 10–100 m.

Stegonotus aruensis Doria, 1875. Ann. Mus. Civ. Stor. Nat. Genova 6: 352, plate 12. (*Lycodon aruensis*)

Holotype: MSNG 30186, probably a male, 815 mm in length (O. Beccari, mid-1872).

Type locality: "Wokan (Isole Aru)," =Wokam, Tanahbesar Island, Maluku Province, Indonesia.

Distribution: E Indonesia (Aru Islands).

Comments: C.M. Kaiser et al. (2018) remove this species from *S. modestus*, and provide a revised description plus photographs of the holotype.

Stegonotus australis Günther, 1872. Ann. Mag. Nat. Hist., Series 4, 9(49): 21. (*Zamenophis australis*)

Synonyms: *Lycodon darnleyensis* Macleay, 1877, *Herbertophis plumbeus* Macleay, 1884.

Holotype: BMNH 1946.1.14.93, a 638 mm male (E.T. Higgins).

Type locality: "Cape York," Queensland, Australia.

Distribution: NE Australia (N Northern Territory, N Queensland; Darnley, Groote Eylandt, Melville, Morington, Murray Is.).

Comments: C.M. Kaiser et al. (2018) recognize this taxon as distinct from *S. cucullatus*, and provide a revised diagnosis, and description and photographs of the holotype, plus those of its synonyms.

Stegonotus ayamaru C.M Kaiser, O'Shea & H. Kaiser, 2019. Zootaxa 4590(2): 208–222, figs. 2, 4–7.

Holotype: RMNH 31199, a 710 mm male (H. Marcus, February 1953).

Type locality: "'Komara' [Kamro], Aitinyo District, Maybrat Regency, West Papua Province, Indonesia (1.5103°S, 132.3763°E; elevation ca. 140 m…)."

Distribution: W New Guinea: W West Papua, Indonesia, 140 m. Known only from the holotype.

Stegonotus batjanensis (Günther, 1865)

Distribution: Add Indonesia (Kasiruta, Obi, Tidore Islands), Lang (2013).

Comments: C.M. Kaiser et al. (2018) provide a revised diagnosis, and a description and photographs of the holotype.

Stegonotus borneensis Inger, 1967

Comments: C.M. Kaiser et al. (2018) provide a revised diagnosis, and a description and photographs of the holotype.

Stegonotus cucullatus (A.M.C. Duméril, Bibron & Duméril, 1854)

Synonyms: Remove *Zamenophis australis* Günther, 1872, *Lycodon keyensis* Doria, 1875, *Lycodon darnleyensis* Macleay, 1877, *Herbertophis plumbeus* Macleay, 1884, *Stegonotus reticulatus* Boulenger, 1895.

Type locality: Corrected to "Doreri Bay, West Papua Province, West New Guinea."

Distribution: At least E Indonesia (West Papua, Papua, including Supiori I.), but many related populations remain to be allocated according to C.M. Kaiser et al. (2018).

Comments: C.M. Kaiser et al. (2013) redescribe and illustrate the holotype and provide a revised diagnosis. They discuss the type material of the synonym *Lycodon magnus* Meyer, 1874, and provide a photograph of a putative syntype.

Stegonotus derooijae Ruane, Richards, McVay, Tjaturadi, Krey & Austin, 2018b. J. Nat. Hist. 52(13–16): 13–18, figs. 2, 4, 6.

Holotype: MZB 3288, a 648 mm male (S. Richards, B. Tjaturadi & K. Krey, 26 June 2005).

Type locality: "Waibya Camp, Salawati Island, Raja Ampat Regency, West Papua, Indonesia, elevation 75 m asl (0.956383°S, 130.784333°E)."

Distribution: E Indonesia (Batanta, Salawati and Waigeo Islands, West Papua), 45–75 m.

Comments: C.M. Kaiser et al. (2018) provide a species description and photographs of the holotype.

Stegonotus diehli Lindholm, 1905

Synonyms: Remove *Stegonotus dorsalis* Werner, 1924.

Distribution: Add Papua New Guinea (Manam Island), Clegg & Jocque (2016).

Comments: C.M. Kaiser et al. (2018) provide a revised diagnosis and photographs of the holotype.

Stegonotus dorsalis Werner, 1924. Sber. Akad. Wiss. Wien, Math.- Naturwiss. Klasse 133(1–3): 32.

Holotype: NMW 14861, a 1050 mm male (R. Pöch, mid-1904).

Type locality: no locality given; probably "Potsdamhafen… in a small bay across from Manam Island, Madang Province, Papua New Guinea" according to C.M. Kaiser at al. (2018).

Disttribution: At least Papua New Guinea (Madang).

Comments: C.M. Kaiser et al. (2018) recognize this taxon as distinct from *S. diehli*, and provide a revised diagnosis, and description and photographs of the holotype.

Stegonotus florensis (Rooij, 1917)

Synonyms: Remove *Stegonotus sutteri* Forcart, 1953.

Distribution: Remove Sumba (=*S. sutteri*).

Comments: C.M. Kaiser et al. (2018) provide a revised diagnosis and photographs of the holotype.

Stegonotus guentheri Boulenger, 1895

Distribution: Add Papua New Guinea (Milne Bay: Kuia I.), C.M. Kaiser et al. (2019).

Comments: C.M. Kaiser et al. (2018) provide a diagnosis and photographs of the syntypes.

Stegonotus heterurus Boulenger, 1893

Comments: C.M. Kaiser et al. (2018) provide a diagnosis and photographs of the syntypes.

Stegonotus iridis Ruane, Richards, McVay, Tjaturadi, Krey & Austin, 2018b. J. Nat. Hist. 52(13–16): 8–13, figs. 2, 4, 5.

Holotype: MZB 3306, a 951 mm male (S. Richards, B. Tjaturadi & K. Krey, 9 June 2005).

Type locality: "Warinkabom, Batanta Island, Raja Ampat Regency, West Papua Province, Indonesia, elevation 50 m above sea level (asl) (0.836942°S, 130.72162°E)."

Distribution: E Indonesia (Batanta, Salawati and Waigeo Islands, West Papua), 25–75 m.

Comments: C.M. Kaiser et al. (2018) provide a species description and photographs of the holotype.

Stegonotus keyensis Doria, 1875. Ann. Mus. Civ. Stor. Nat. Genova 6: 351, plate 12. (*Lycodon keyensis*)

Holotype: MSNG 7521, a probable male, 815 mm in length (O. Beccari, mid-1872).

Type locality: "Isole Kei," Maluku Prov., Indonesia.

Distribution: Indonesia (Dulah, Kei Besar), Lang (2013).

Comments: C.M. Kaiser et al. (2018) remove this species from *S. cucullatus*, and provide a revised description plus photographs of the holotype.

Stegonotus lividus A.M.C. Duméril, Bibron & Duméril, 1854. Erpétol. Gén. 7:381–382. (*Lycodon lividum*)

Syntypes: RMNH 325a, b, probably females, 607 and 600 mm in length (S. Müller, H. Macklot & P. van Oort, mid-1829).

Type locality: "l'île de Pulo-Samao" (=Semau Island, East Nusa Tenggara Province, Indonesia).

Distribution: E Indonesia (Samao Island).

Comments: C.M. Kaiser et al. (2018) re-elevate this taxon from *S. modestus* and provide a revised diagnosis and illustrations of the syntypes.

Stegonotus melanolabiatus Ruane, Richards, McVay, Tjaturadi, Krey & Austin, 2018b. J. Nat. Hist. 52(13–16): 18–21, figs. 2, 7.

Holotype: AMR115343, an 807 mm male (S. Donnellan & K. Aplin, 22 April-4 May 1984).

Type locality: "Doido, Chimbu Province, Papua New Guinea, elevation 1300 m asl (6.550000°S, 144.833333°E)."

Distribution: C Papua New Guinea (Chimbu, Southern Highlands), 550–1300 m. Add Papua New Guinea (Hela), C.M. Kaiser et al. (2019).

Comments: C.M. Kaiser et al. (2018) provide a species description and photographs of the holotype.

Stegonotus modestus (Schlegel, 1837)

Synonyms: Remove *Lycodon lividum* A.M.C. Duméril, Bibron & Duméril, 1854, *Lycodon aruensis* Doria, 1875.

Distribution: E Indonesia (Ambon, Buru, Ceram Is.).

Comments: C.M. Kaiser et al. (2018) geographically restrict this taxon and provide a revised description, including photographs of the holotype plus those of remaining synonyms.

Stegonotus muelleri A.M.C. Duméril, Bibron & Duméril, 1854

Distribution: Add Philippines (Bohol), Barnes & Knierim (2018); Philippines (Dinagat), Sanguila et al. (2016).

Comments: C.M. Kaiser et al. (2018) provide a revised description and diagnosis, and photographs of the holotype plus that of the synonym *Spilotes samarensis* W.C.H. Peters, 1861.

Stegonotus parvus (A.B. Meyer, 1874)

Neotype: RMNH 46844, a 292 mm female (D.L. Leiker, 22 February 1952), designated by C.M. Kaiser et al. (2019).

Neotype locality: "Serui, Yapen Island, Papua Province, Indonesia (ca. 1.8807°S, 136.2386°E)."

Distribution: Add Papua New Guinea (Karkar Island), Clegg & Jocque (2016).

Comments: C.M. Kaiser et al. (2018) provide a revised diagnosis, and a description and photographs of the neotype.

Stegonotus poechi Werner, 1924. Sber. Akad. Wiss. Wien, Math.- Naturwiss. Klasse 133(1–3): 32.

Holotype: NMW 23406, a 1024+ female (R. Pöch, mid-1904).

Type locality: no locality given; probably "Potsdamhafen… in a small bay across from Manam Island, Madang Province, Papua New Guinea" according to C.M. Kaiser at al. (2018).

Distribution: Papua New Guinea (Madang).

Comments: C.M. Kaiser et al. (2018) recognize this taxon, not mentioned in Wallach et al. (2014), and provide a revised diagnosis, and description and photographs of the holotype.

Stegonotus reticulatus Boulenger, 1895. Ann. Mag. Nat. Hist., Series 6, 16(91): 31.

Lectotype: BMNH 1946.1.14.87–88, an 1118 mm female (A.S. Meek, late 1894), designated by C.M. Kaiser et al. (2018).

Type locality: "Ferguson Island," Milne Bay Province, New Guinea.

Distribution: Papua New Guinea (Central, East Sepik, Eastern Highlands, Gulf, Madang, Milne Bay, Northern, Simbu, Southern Highlands, Western, Western Highlands), C.M. Kaiser et al. (2019).

Comments: C.M. Kaiser et al. (2018) recognize this taxon as distinct from *S. cucullatus*, and provide a revised diagnosis, and description and photographs of the holotype.

Stegonotus sutteri Forcart, 1953. Verhandl. Nat. Ges. Basel 64(2): 379–382, figs. 1–3.

Holotype: NMBA 14872, a 711 mm male (Basel Museum Sumba Expedition, 30 September 1949).

Type locality: "Zentralsumba, Lindiwatju 430 m," Nusa Tenggara Barat, Indonesia.

Distribution: E Indonesia (Sumba I.).

Comments: C.M. Kaiser et al. (2018) recognize this taxon as distinct from *S. florensis*, and provide a revised diagnosis, and description and photographs of the holotype.

STENORRHINA A.M.C. Duméril, 1853 (Colubridae: Colubrinae)

Stenorrhina degenhardtii (Berthold, 1845)

Distribution: Add Panama (Panama Oeste), Ray & Ruback (2015); Colombia (Caldas), Díaz-Ricaurte, J.C. (2019).

Stenorrhina freminvillei A.M.C. Duméril, Bibron & Duméril, 1854

Distribution: Add El Salvador (Santa Ana), Juárez-Peña et al. (2016); Honduras (Isla del Tigre), McCranie & Gutsche (2016); Nicaragua (Carazo), Gutiérrez-López et al. (2015).

STICHOPHANES X. Wang, Messenger, Zhao & Zhu, 2014. Asian Herpetol. Res. 5(3): 145. (Dipsadidae: subfamily unnamed)

Type species: *Oligodon ningshaanensis* Yuan, 1983 by original designation.

Distribution: C China.

Comments: Based on DNA sequence-data, X. Wang et al. (2014) find that *Stichophanes* is, with *Thermophis*, sister taxa to the New World dipsadids.

Stichophanes ningshaanensis Yuan, 1983

Holotype: The holotype and paratypes are lost, X. Wang et al. (2014).

Distribution: Add China (Hubei), and elevation range of 1550–2022 m, X. Wang et al. (2014).

Comments: X. Wang et al. (2014) transfer *ningshaanensis* from *Oligodon*, and present a revised diagnosis and color photographs. Messenger & Wang (2015) summarize the natural history.

STOLICZKIA Jerdon, 1870 (Xenodermidae)

Stoliczkia borneensis Boulenger, 1899
Stoliczkia khasiensis Jerdon, 1870

STORERIA Baird & Girard, 1853 (Natricidae)

Comments: Pyron et al. (2016b) review species delimitation within the genus using DNA sequence and morphological data. Although the molecular data suggest eight lineages, the incorporation of morphological characters concurs with a four-species scenario. Subspecies are no longer recognized.

Storeria dekayi (Holbrook, 1839)

Synonyms: Remove *Storeria victa* Hay, 1892.

Distribution: Remove Florida peninsula, USA, to *S. victa*.

Comments: Pyron et al. (2016b) produce a DNA-sequence based phylogeny in which samples from the eastern and western thirds of the range form separate clades. No samples from Central America or Mexico were evaluated, but they conclude that no subspecies should be recognized.

Storeria hidalgoensis Taylor, 1942

Comments: Pyron et al. (2016b) conclude that this species should be synonymized with *S. occipitomaculata* due to lack of morphological distinction, although no genetic samples were evaluated.

Storeria occipitomaculata (Storer, 1839)

Comments: Pyron et al. (2016b) produce a DNA-sequence based phylogeney that reveals four clades: 1) Florida, 2) eastern USA, 3) Ozarks, 4) northwest (Nebraska and South Dakota). They recommend that subspecies not be recognized.

Storeria storerioides (Cope, 1865)

Distribution: Add Mexico (Nayarit), Luja & Grünwald (2015). Quintero-Díaz et al. (2014b) report a second and third speci men from Zacatecas.

Storeria victa Hay, 1892. Science 19(479): 199.

Holotype: unlocated, a 356 mm specimen (H. T. Mann).

Type locality: "banks of the Oklawaha River, Florida," USA.

Distribution: Southeast USA (peninsular Florida and SE Georgia).

Comments: Pyron et al. (2016b) revalidate this species based on morphological and genetic distinctiveness.

SUBSESSOR J.C. Murphy & Voris, 2014. Fieldiana Life Earth Sci. (8): 33–34. (Homalopsidae)

Type species: *Hypsirhina bocourti* Jan, 1865 by original designation.

Distribution: SE Asia.

Subsessor bocourti (Jan, 1865)

Distribution: Add Thailand (Chanthaburi), Chan-ard et al. (2011); Cambodia (Prey Veng), J.C. Murphy & Voris (2014).

Comments: J.C. Murphy & Voris (2014) provide a diagnosis and photograph.

SUMATRANUS J.C. Murphy & Voris, 2014. Fieldiana Life Earth Sci. (8): 34–35. (Homalopsidae)

Type species: *Homalopsis albomaculatus* A.M.C. Duméril, Bibron & Duméril, 1854 by original designation.

Distribution: W Indonesia.

Sumatranus albomaculatus (A.M.C. Duméril, Bibron & Duméril, 1854)

Comments: J.C. Murphy & Voris (2014) provide a diagnosis and photograph.

SUNDATYPHLOPS Hedges, Marion, Lipp, Marin & Vidal, 2014. Caribbean Herpetol. 49: 39. (Typhlopidae: Asiatyphlopinae)

Type species: *Typhlops polygrammicus* Schlegel, 1839 *in* 1837–1844 by original designation.

Distribution: Lesser Sundas of the East Indies.

Comments: Pyron & Wallach (2014) argue that a monotypic *Sundatyphlops*, positioned as the sister taxon to *Anilios*, was phylogenetically uninformative. However, the validity of *Sundatyphlops* is affirmed by Nagy et al. (2015), and Hedges et al. (2014) and Pyron & Wallach (2014) note the morphological distinctiveness of Sunda populations of *polygrammicus*.

Sundatyphlops polygrammicus (Schlegel, 1839 *in* 1837–1844)

Synonyms: Remove *Typhlops torresianus* Boulenger, 1889.

Distribution: Limited to the Lesser Sundas (Flores, Komodo, Lombok, Moyo, Sumba, Sumbawa, Timor).

Comments: Eastern populations are now referred to *Anilios torresianus*, Hedges et al. (2014) and Pyron & Wallach (2014).

SUTA Worrell, 1961 (Elapidae)

Comments: See under *Parasuta*.

Suta fasciata (Rosén, 1905)
Suta ordensis (Storr, 1984)
Suta punctata (Boulenger, 1896)
Suta suta (W.C.H. Peters, 1863)

Distribution: Range extended to southernmost New South Wales, Australia, Michael & Lindenmayer (2011).

SYMPHIMUS Cope, 1870 (Colubridae: Colubrinae)

Symphimus leucostomus Cope, 1870
Symphimus mayae (Gaige, 1936)

SYMPHOLIS Cope, 1862 (Colubridae: Colubrinae)

Sympholis lippiens Cope, 1862

SYNOPHIS Peracca, 1896 (Dipsadidae: Dipsadinae)

Comments: Pyron et al. (2015) placed *Synophis* in a new tribe (Diaphorolepidini) with *Emmochliophis* and *Diaphorolepis*. They state that there are "numerous undescibed species," four of which were described within the next year. Pyron et al. (2016a) define three species groups, and provide a key to the species.

Synophis bicolor Peracca, 1896

Type locality: Restricted to "Tobar Donoso, Carchi Province, Ecuador (1.19, −78.50)" by Pyron et al. (2015).

Distribution: Restricted to NW Ecuador (Carchi, Esmeraldas), 229–318 m, Pyron et al. (2016a).

Fossil record: Late Pleistocene of Colombia (Albino & Brizuela, 2014).

Comments: Pyron et al. (2015) provide a species account (description, color photographs, locality map). They provide evidence of at least three species: 1) *S. bicolor* s.s. of NW Ecuador, 2) Ecuadorian highlands populations, and 3)

Colombian highland populations, now *S. niceforomariai*. Pyron et al. (2016a) provide a revised description of *S. bicolor* after removal of most populations to other species.

Synophis bogerti Torres-Carvajal, Echevarría, Venegas, Chávez & Camper, 2015. ZooKeys 546: 158–165, figs. 1, 2, 4, 5.

Holotype: QCAZ 12791, a 551 mm male (J.D. Camper, 18 July 2014).

Type locality: "Ecuador: Provincia Napo:…from Wildsumaco Wildlife Sanctuary, sendero Coatí (0°38′8.40″S, 77°31′19.20″W, 1000 m)."

Distribution: C Ecuador (Morona-Santiago, Napo, Pastaza), 1000–1747 m.

Comments: Pyron et al. (2016a) provide a description.

Synophis calamitus Hillis, 1998

Distribution: Ecuador (Carchi, Chimborazo, Cotopaxi, Pichincha, Santo Domingo), 763–2272 m, Pyron et al. (2016a).

Comments: Pyron et al. (2015) provide a species account (description, color photograph and locality map). Pyron et al. (2016a) provide a revised description which includes an expanded geographic range.

Synophis insulomontanus Torres-Carvajal, Echevarría, Venegas, Chávez & Camper, 2015. ZooKeys 546: 170–174, figs. 5, 10, 11.

Holotype: CORBIDI 13940, a 516 mm male (G. Chávez, 1 December 2013).

Type locality: "Peru: Departamento Huánuco: Provincia Puerto Inca: Distrito Llullapichis:…from Campamento Peligroso-Reserva Communal El Sira (9°25′34.22″S, 74°44′6.60″W, 1507 m)."

Distribution: Peru (Huánuco, San Martín), 1122–1798 m.

Comments: Pyron et al. (2016a) provide a description.

Synophis lasallei (Nicéforo-María, 1950)

Comments: Pyron et al. (2015) provide a species account (description, color photograph and locality map). Pyron et al. (2016a) provide a revised description.

Synophis niceforomariai Pyron, Arteaga, Echevarría & Torres-Carvajal, 2016. Zootaxa 4171(2): 295, 302, figs. 2, 3. (*Synophis niceforomariae*)

Holotype: MHUA 14577, an adult male, (24 November 2007).

Type locality: "Finca La Esperanza, Valle de La Manguita, Municipio Amalfi, Departamento Antioquia, Colombia (N6.978611, W-75.044444; 1394 m)."

Distribution: N Colombia (Antioquia), 865–1656 m.

Comments: The specific epithet is changed herein because it is a patronym.

Synophis plectovertebralis Sheil & Grant, 2001

Comments: Pyron et al. (2015) provide a species account (description and locality map). Pyron et al. (2016a) also provide a description.

Synophis zaheri Pyron, Guayasamin, Peñafiel, Bustamante & Arteaga, 2015. ZooKeys 541: 120–123, figs. 3, 8.

Holotype: MZUTI 3353, an adult male, (A. Arteaga, L. Bustamante, R. Hidalgo, D. Mideros & D. Troya, 30 December 2013).

Type locality: "vicinity of Buenaventura Reserve (Fundación Jocotoco), near Piñas, El Oro Province, SW Ecuador, 874 m above sea level (−3.65, −79.76…)."

Distribution: SW Ecuador (El Oro), 874 m. Known only from the type locality.

Comments: Pyron et al. (2016a) provide an updated description.

Synophis zamora Torres-Carvajal, Echevarría, Venegas, Chávez & Camper, 2015. ZooKeys 546: 166–170, figs. 5, 7, 8.

Holotype: QCAZ 9174, a 534 mm male (E.E. Tapia, J. Deichmann & A.F. Jiménez, 19 April 2009).

Type locality: "Ecuador: Provincia Zamora Chinchipe:… from Las Orquídeas, 4 km from río Nangaritza (4°15′47.52″S, 78°41′27.93″W, 1843 m)."

Distribution: S Ecuador (Morona-Santiago, Zamora-Chinchipe), 752–1843 m.

Comments: Pyron et al. (2016a) provide a description.

†*SZYNDLARIA* Rage & Augé, 2010 (Tropidophiidae)

†*Szyndlaria aureomontensis* Rage & Augé, 2010

TACHYMENIS Wiegmann, 1834 (Dipsadidae: Xenodontinae)

Tachymenis affinis Boulenger, 1896

Tachymenis attenuata Walker, 1945

Tachymenis chilensis (Schlegel, 1837)

Distribution: Giraudo et al. (2012) detail known localities in Argentina.

Tachymenis elongata Despax, 1910

Tachymenis peruviana Wiegmann, 1834

Distribution: Add Argentina (Catamarca, La Rioja, Salta, Tucuman), G.A. Gallardo et al. (2019).

Tachymenis tarmensis Walker, 1945

†*TACHYOPHIS* Rochebrune, 1884. Mém. Soc. Sci. Nat. Saône-et-Loire 5: 159. (Booidea: incerta sedis)

Type species: †*Tachyophis nitidus* Rochebrune, 1884 by monotypy.

Distribution: Middle Tertiary of France.

†*Tachyophis nitidus* Rochebrune, 1884. Mém. Soc. Sci. Nat. Saône-et-Loire 5: 159, plates 1, 2.

Lectotype: MNHN QU16331, four trunk vertebrae, designated by Rage (1984).

Type locality: "Phosphorites du Quercy," France.

Distribution: Within the late Eocene to early Miocene of France

TAENIOPHALLUS Cope, 1895 (Dipsadidae: Xenodontinae)

Taeniophallus affinis (Günther, 1858)

Distribution: Add Brazil (Paraiba), R. França et al. (2012); Brazil (Pernambuco), Pedrosa et al. (2014).

Taeniophallus bilineatus (J.G. Fischer, 1885)

Taeniophallus brevirostris (W.C.H. Peters, 1863)

Distribution: Add Brazil (Maranhao, Mato Grosso), Morais et al. (2010), who provide a map of localities for Brazil.

Taeniophallus nebularis Schargel, Rivas-Fuenmayor & Myers, 2005

Taeniophallus nicagus (Cope, 1868)

Distribution: Add Brazil (Para), Avila-Pires et al. (2010).

Comments: Jairam (2017) details the distribution and morphology of new specimens from Surinam.

Taeniophallus occipitalis (Jan, 1863)

Distribution: Add Brazil (Amazonas), F.G.R. França et al. (2006); Brazil (Maranhão), J.P. Miranda et al. (2012); Brazil (Mato Grosso), Tavares et al. (2012); Brazil (Mato Grosso do Sul), Ferreira et al. (2017); Brazil (Pernambuco), and upper elevation of 1200 m, Guedes et al. (2014); Brazil (Rio Grande do Norte), Sales et al. (2009); Brazil (Tocantins), Dal Vechio et al. (2016); Paraguay (San Pedro), P. Smith et al. (2016).

Taeniophallus persimilis (Cope, 1869)

Taeniophallus poecilopogon (Cope, 1863)

Taeniophallus quadriocellatus Santos, Di-Bernardo & Lema, 2008

†*TALLAHATTAOPHIS* Holman & Case, 1988 (Booidea: incerta sedis)

†*Tallahattaophis dunni* Holman & Case, 1988

TANTALOPHIS Duellman, 1958 (Dipsadidae: Dipsadinae)

Tantalophis discolor (Günther, 1860)

Distribution: Elevation to 2851 m, Aldápe-López and Santos-Motreno (2016).

TANTILLA Baird & Girard, 1853 (Colubridae: Colubrinae)

Comments: L.D. Wilson and Mata-Silva (2014) review the Mexican species and provide a key. L.D. Wilson & Mata-Silva (2015) review all species of this genus, giving synonymies, distribution, species group, references, and remarks for each.

Tantilla albiceps Barbour, 1925

Distribution: Panama (Panama Oeste), Ray & Ruback (2015).

Tantilla alticola (Boulenger, 1903)

Distribution: Add Panama (Cocle), E.A. Myers et al. (2013c); Colombia (Caldas), Rojas-Morales et al. (2018b); Colombia (Cauca), Vera-Pérez et al. (2015); Colombia (Tolima), Parra-Hernández et al. (2019); Colombia (Valle del Cauca), Vanegas-Guerrero et al. (2015). E.A. Myers et al. (2013c) and Vanegas-Guerrero et al. (2015) provide maps of known localities.

Comments: L.D. Wilson & Mata-Silva (2015) provide an account.

Tantilla andinista L.D. Wilson & Mena, 1980

Comments: L.D. Wilson & Mata-Silva (2015) provide an account.

Tantilla armillata Cope, 1875

Distribution: Add Honduras (Valle: Isla del Tigre), Firneno et al. (2016); Honduras (Isla El Pacar), McCranie & Gutsche (2016); Nicaragua (Masaya), Sunyer et al. (2014); Panama (Panama Oeste), Ray & Ruback (2015); elevation 0–1435 m, L.D. Wilson & Mata-Silva (2015).

Comments: L.D. Wilson & Mata-Silva (2015) provide an account.

Tantilla atriceps (Günther, 1895 *in* 1885–1902)

Distribution: Elevation 0–2134 m, L.D. Wilson & Mata-Silva (2015).

Comments: L.D. Wilson & Mata-Silva (2014, 2015) provide a species account, map and color photograph.

Tantilla bairdi Stuart, 1941

Distribution: Elevation 1524–1550 m, L.D. Wilson & Mata-Silva (2015).

Comments: L.D. Wilson & Mata-Silva (2015) provide an account, and note that only two specimens are known.

Tantilla berguidoi Batista, Mebert, Lotzkat & Williams, 2016. Mesoamerican Herpetol. 3: 951–953, figs. 2–4.

Holotype: SMF 97636 (original number AB 1029), a 408 mm male (A. Batista & K. Mebert, 2 December 2012).

Type locality: "Panama, Provincia de Darién, Serranía de Majé, Cerro Chucantí (8.79904°N, 78.46158°W; elev. 1,376 m asl."

Distribution: Panama (Darién), 1376 m.

Comments: Known only from the holotype.

Tantilla bocourti (Günther, 1895 *in* 1885–1902)

Distribution: Add Mexico (SE Coahuila, San Luis Potosí), and elevation NSL-2750 m, L.D. Wilson & Mata-Silva (2014). García-Padilla & Mata-Silva (2013) provide a second record from Oaxaca.

Comments: L.D. Wilson & Mata-Silva (2014, 2015) provide a species account, map and color photograph.

Tantilla boipiranga Sawaya & Sazima, 2003

Distribution: Elevation 648–1361 m, L.D. Wilson & Mata-Silva (2015).

Comments: L.D. Wilson & Mata-Silva (2015) provide an account.

Tantilla brevicauda Mertens, 1952

Distribution: Elevation 1200–1800 m, L.D. Wilson & Mata-Silva (2015).

Comments: L.D. Wilson & Mata-Silva (2015) provide an account.

Tantilla briggsi Savitzky & Smith, 1971

Distribution: Elevation ~200 m, L.D. Wilson & Mata-Silva (2014).

Comment: L.D. Wilson & Mata-Silva (2014, 2015) provide a species account, and discuss the type locality. Still known only from the holotype.

Tantilla calamarina Cope, 1867

Distribution: Elevation to 1677 m, L.D. Wilson & Mata-Silva (2014).

Comments: L.D. Wilson & Mata-Silva (2014, 2015) provide a species account, map and color photograph. Ramírez-Bautista et al. (2014) describe morphological variation within a large sample from a single locality.

Tantilla capistrata Cope, 1875

Distribution: Add Peru (Ancash), L.D. Wilson & Mata-Silva (2015); Peru (Lambayeque), Venegas (2005).

Comments: L.D. Wilson & Mata-Silva (2015) provide an account.

Tantilla cascadae L.D. Wilson & Meyer, 1981

Distribution: Add Mexico (SE Jalisco), and elevation to 1858 m, Cruz-Sáenz et al. (2015), L.D. Wilson & Mata-Silva (2014).

Comments: L.D. Wilson & Mata-Silva (2014, 2015) provide a species account, map and color photograph. Now known from three specimens.

Tantilla ceboruca Canseco-Márquez, E.N. Smith, Flores-Villela & Campbell, 2007

Distribution: Add Mexico (NC Jalisco), Cruz-Sáenz et al. (2015), L.D. Wilson & Mata-Silva (2014); delete Colima (in error, Wallach et al. 2014); minimum elevation 1233 m, L.D. Wilson & Mata-Silva (2014).

Comments: L.D. Wilson & Mata-Silva (2014, 2015) provide a species account, map and color photograph.

Tantilla coronadoi Hartweg, 1944

Comments: L.D. Wilson & Mata-Silva (2014, 2015) provide a species account and map.

Tantilla coronata Baird & Girard, 1853

Distribution: Elevation to 600 m, L.D. Wilson & Mata-Silva (2015).

Comments: L.D. Wilson & Mata-Silva (2015) provide an account.

Tantilla cucullata Minton, 1956

Distribution: Add Mexico (Chihuahua), and upper elevation of 1841 m, Herr et al. (2017); Mexico (Coahuila), Baeza-Tarin et al. (2018b). Low elevation of 1189 m, L.D. Wilson & Mata-Silva (2015).

Comments: L.D. Wilson & Mata-Silva (2015) provide an account.

Tantilla cuniculator H.M. Smith, 1939

Distribution: Delete Guatemala (in error in Wallach et al., 2014).

Comments: L.D. Wilson & Mata-Silva (2014, 2015) provide a species account, map and color photograph.

Tantilla deppii (Bocourt, 1883 *in* A.H.A. Duméril, Bocourt & Mocquard, 1870–1909)

Comments: L.D. Wilson & Mata-Silva (2014, 2015) provide a species account, map and color photograph.

Tantilla excelsa McCranie & E.N. Smith, 2017. Herpetologica 73(4): 342–344, figs. 6, 7.

Holotype: USNM 579682, a 400 mm male (L. Marineros, 26 February 2011).

Type locality: "Lancetilla (15°44′N, 87°27′W), 30 m elevation, department of Atlántida, Honduras."

Distribution: W Honduras (Atlantida, Cortes, Yoro), 30–700 m.

Tantilla flavilineata H.M. Smith & Burger, 1950

Distribution: Elevation 1890–2476 m, L.D. Wilson & Mata-Silva (2014).

Comments: L.D. Wilson & Mata-Silva (2014, 2015) provide a species account, map and color photograph.

Tantilla gottei McCranie & E.N. Smith, 2017. Herpetologica 73(4): 344–346, figs. 8, 9.

Holotype: ROM 19996, a 367 mm subadult female (J. Porras, 12 December 1986).

Type locality: "El Picacho (14°07′N, 87°11′W), a zoological park located near Tegucigalpa, 1280 m elevation, department of Francisco Morazán, Honduras."

Distribution: S Honduras (El Paraiso, Francisco Morazan), 500–1280 m.

Tantilla gracilis Baird & Girard, 1853

Comments: L.D. Wilson & Mata-Silva (2014, 2015) provide an account.

Tantilla hendersoni Stafford, 2004

Distribution: Add Belize (Stann Creek), 194 m, E.P. Hofmann et al. (2017).

Comments: L.D. Wilson & Mata-Silva (2014, 2015) provide an account, and suggest that *T. hendersoni* may be conspecific with *T. impensa*. E.P. Hofmann et al. (2017) report a second specimen that supports the reported differences between *T. hendersoni* and *T. impensa*.

Tantilla hobartsmithi Taylor, 1937

Distribution: Add Mexico (WC Nuevo León), and low elevation of NSL, L.D. Wilson & Mata-Silva (2014).

Comments: L.D. Wilson & Mata-Silva (2014, 2015) provide a species account, map and color photograph.

Tantilla impensa Campbell, 1998

Distribution: Add Honduras (Santa Bárbara, 1700 m), McCranie et al. (2013b). Upper elevation of 1830 m, L.D. Wilson & Mata Silva (2014).

Comments: L.D. Wilson & Mata-Silva (2014, 2015) provide a species account.

Tantilla insulamontana L.D. Wilson & Mena, 1980

Distribution: Add Ecuador (Pichincha), L.D. Wilson & Mata-Silva (2015).

Comments: L.D. Wilson & Mata-Silva (2015) provide an account.

Tantilla jani (Günther, 1895 *in* 1885–1902)

Distribution: Delete Mexico, in error in Wallach et al., 2014; elevation 1050 m, L.D. Wilson & Mata Silva (2015). Mexican records are referred to *T. vulcani*.

Comments: L.D. Wilson & Mata-Silva (2015) provide an account.

Tantilla johnsoni L.D. Wilson, Vaughan & Dixon, 1999

Type locality: Modified by Johnson et al. (2014: 306) to "Musté, located in the Municipality of Motozintla, about

14 km north of Huixtla, Chiapas, Mexico, on the Pacific slope of the southeastern Sierra Madre de Chiapas, at an approximate elevation of 518 m."

Comment: L.D. Wilson & Mata-Silva (2014, 2015) provide a species account. Known only from the type locality.

Tantilla lempira L.D. Wilson & Mena, 1980

Comments: L.D. Wilson & Mata-Silva (2015) provide an account, and Espinal & Solís (2015b) describe an additional specimen.

Tantilla melanocephala (Linnaeus, 1758)

Synonyms: Add *Tantilla marcovani* Lema, 2004.

Distribution: Add Colombia (Córdoba), Carvajal-Cogollo et al. (2007); Colombia (Huila), Moreno-Arias & Quintero-Corzo (2015); Colombia (Quindio), Quintero-Ángel et al. (2012); Peru (Cajamarca), Koch et al. (2018); Brazil (Acre), Bernarde et al. (2011b); Brazil (Alagoas, Minas Gerais, Sergipe), Guedes et al. (2014); Brazil (Maranhão), J.P. Miranda et al. (2012); Brazil (Mato Grosso do Sul), Ferreira et al. (2017); Brazil (Paraiba), R. França et al. (2012); Brazil (Pernambuco), Pedrosa et al. (2014); Brazil (Piauí), Dal Vechio et al. (2013); Brazil (Rondônia), Bernarde et al. (2012b); Argentina (Jujuy), B. Ramos et al. (2013).

Comments: L.D. Wilson & Mata-Silva (2015) provide an account. Mata-Silva & L.D. Wilson (2016) synonymize *T. marcovani* with *T. melanocephala* due to lack of diagnostic characters of the former.

Tantilla miyatai L.D. Wilson & Knight *in* L.D. Wilson, 1987

Distribution: L.D. Wilson & Mata-Silva (2015) correct the type locality coordinates to 0°77′17″N, 79°15′09W, and elevation of 189 m.

Comments: L.D. Wilson & Mata-Silva (2015) provide an account.

Tantilla moesta (Günther, 1863)

Distribution: Add Mexico (Campeche), Neri-Castro et al. (2017); elevation to 283 m, L.D. Wilson & Mata-Silva (2014).

Comments: L.D. Wilson & Mata-Silva (2014, 2015) provide a species account, map and color photograph.

Tantilla nigra (Boulenger, 1914)

Comments: L.D. Wilson & Mata-Silva (2015) provide an account.

Tantilla nigriceps Kennicott, 1860

Distribution: Mexico (central instead of NE Nuevo León), Névarez-de Los Reyes et al. (2017b).

Comments: L.D. Wilson & Mata-Silva (2014, 2015) provide a species account, map and color photograph.

Tantilla oaxacae Wilson & Meyer, 1971

Distribution: Elevation to 2286 m, L.D. Wilson & Mata-Silva (2014).

Comments: L.D. Wilson & Mata-Silva (2014, 2015) provide a species account, map and color photograph.

Tantilla olympia Townsend, L.D. Wilson, Medina-Flores & Herrera-B., 2013a. J. Herpetol. 47(1): 194–197, figs. 2–4.

Holotype: USNM 574000, a 338 mm male (H. Vega-R. & P.R. House, 31 July 2010). Type locality: "La Liberación (15.541°N, 87.294°W), 1,150 m a.s.l., Refugio de Vida Silvestre Texíguat, Departamento de Atlántida, Honduras."

Distribution: Honduras (Atlántida), 1150 m.

Comment: Known only from the holotype. L.D. Wilson & Mata-Silva (2015) provide an account.

Tantilla oolitica Telford, 1966

Comments: L.D. Wilson & Mata-Silva (2015) provide an account.

Tantilla petersi L.D. Wilson, 1979

Comments: L.D. Wilson & Mata-Silva (2015) provide an account.

Tantilla planiceps (Blainville, 1835)

Distribution: Elevation to 1220 m, L.D. Wilson & Mata-Silva (2014).

Comments: L.D. Wilson & Mata-Silva (2014, 2015) provide a species account, map and color photograph.

Tantilla psittaca McCranie, 2011

Comments: L.D. Wilson & Mata-Silva (2015) provide an account.

Tantilla relicta Telford, 1966

Distribution: Elevation to 90 m, L.D. Wilson & Mata-Silva (2015).

Comment: Schrey et al. (2015) evaluate genetic structure in populations in central Florida, USA, and discover a northern and southern unit.

Tantilla reticulata Cope, 1860

Distribution: Add Nicaragua (Zelaya), Berghe et al. (2014).

Comments: L.D. Wilson & Mata-Silva (2015) provide an account.

Tantilla robusta Canseco-Márquez, Mendelson & Gutiérrez-Mayén, 2002

Comments: L.D. Wilson & Mata-Silva (2014, 2015) provide a species account.

Tantilla rubra Cope, 1875

Distribution: Add Mexico (Guerrero); and upper elevation of 2618 m, L.D. Wilson & Mata-Silva (2014). Low elevation of NSL, Palacios-Aguilar & Flores-Villela (2018).

Comments: L.D. Wilson & Mata-Silva (2014, 2015) provide a species account, map and color photograph.

Tantilla ruficeps (Cope, 1894)

Distribution: Add Panama (Panama, Panama Oeste), Ray & Ruback (2015).

Comments: L.D. Wilson & Mata-Silva (2015) provide an account.

Tantilla schistosa (Bocourt, 1883 *in* A.H.A. Duméril, Bocourt & Mocquard, 1870–1909)

Distribution: Add Mexico (Chiapas), L.D. Wilson & Mata-Silva (2014); Honduras (Santa Bárbara), Solís et al. (2015); Panama (Panama Oeste), Ray & Ruback (2015).

Comments: L.D. Wilson & Mata-Silva (2014, 2015) provide a species account and map.

Tantilla semicincta (A.M.C. Duméril, Bibron & Duméril, 1854)

Comments: L.D. Wilson & Mata-Silva (2015) provide an account.

Tantilla sertula L.D. Wilson & Campbell, 2000.

Distribution: Add Mexico (Oaxaca), 487 m, A. Rocha et al. (2016); delete Mexico (Nayarit), in error in Wallach et al. (2014).

Comments: L.D. Wilson & Mata-Silva (2014, 2015) provide a species account and map.

Tantilla shawi Taylor, 1949

Comments: L.D. Wilson & Mata-Silva (2014, 2015) provide a species account and map.

Tantilla slavensi Pérez-Higereda, H.M. Smith & R.B. Smith, 1985

Comments: L.D. Wilson & Mata-Silva (2014, 2015) provide a species account, map and color photograph.

Tantilla stenigrammi McCranie & E.N. Smith, 2017. Herpetologica 73(4): 339–342, figs. 3, 4.

Holotype: UTA R52591, a 173 mm female (E.N. Smith, J.A. Ferrar-Castro, J. Murillo, C. Chavez, A. Sosa & J.H. Malone, 3 February 2005).

Type locality: "Cuaca (15°23′01″N, 86°12′41.22″W…), 895 m elevation, department of Olancho, Honduras."

Distribution: NC Honduras (Olancho), 895–1180 m.

Comments: Known from two specimens.

Tantilla striata Dunn, 1928

Distribution: Elevation 152–1143 m, L.D. Wilson & Mata-Silva (2014).

Comments: L.D. Wilson & Mata-Silva (2014, 2015) provide a species account and map.

Tantilla supracincta (W.C.H. Peters, 1863)

Distribution: Add Panama (Panama Oeste), Ray & Ruback (2015); Colombia (Antioquia), Hurtado-Gómez et al. (2015); Ecuador (Manabí), Cisneros-Heredia (2005) and Almendáriz et al. (2012); elevation to 1323 m, L.D. Wilson & Mata-Silva (2015).

Comments: L.D. Wilson & Mata-Silva (2015) provide an account.

Tantilla taeniata (Bocourt, 1883 *in* A.H.A. Duméril, Bocourt & Mocquard, 1870–1909)

Distribution: Add Nicaragua (Esteli, Jinotega), Sunyer & G. Köhler (2007). Remove Honduras (see comments).

Comments: McCranie & E.N. Smith (2017) assign Honduran specimens to other species of *Tantilla*, recognizing those from Guatemala as true *T. taeniata*, and leaving El Salvador and Nicaragua populations in taxonomic limbo. L.D. Wilson & Mata-Silva (2015) provide an account.

Tantilla tayrae L.D. Wilson, 1983

Comments: L.D. Wilson & Mata-Silva (2014, 2015) provide a species account and map.

Tantilla tecta Campbell & E.N. Smith, 1997

Comments: L.D. Wilson & Mata-Silva (2015) provide an account; remains known only from the holotype.

Tantilla tjiasmantoi C. Koch & Venegas, 2016. PeerJ 2016(2767): 10–13, figs. 1, 3–7.

Holotype: CORBIDI 7726, a 638 mm female (E. Hoyos-Granda, A. Beraún & C. Koch, 15 January 2010).

Type locality: "Pías, Province Pataz, Department of La Libertad, Peru (07°53′56.6″S, 77°34′43.8″W, 1,726 m a.s.l.)."

Distribution: NW Peru (Cajamarca, La Libertad), 1154–1726 m.

Tantilla trilineata (W.C.H. Peters, 1880)

Comments: L.D. Wilson & Mata-Silva (2015) provide an account.

Tantilla triseriata H.M. Smith & P.W. Smith, 1951

Distribution: Low elevation of 914 m, L.D. Wilson & Mata-Silva (2014).

Comments: L.D. Wilson & Mata-Silva (2014, 2015) provide a species account and map.

Tantilla tritaeniata H.M. Smith & Williams, 1966

Comments: L.D. Wilson & Mata-Silva (2015) provide an account.

Tantilla vermiformis (Hallowell, 1861)

Distribution: Add Guatemala (Zacapa), Ariano-Sánchez (2015), Honduras (Valle: Isla Exposición, 20 m), McCranie et al. (2013b); Nicaragua (Masaya), L.D. Wilson & Mata-Silva (2015).

Comments: L.D. Wilson & Mata-Silva (2015) provide an account.

Tantilla vulcani Campbell, 1998

Distribution: Mexico (Chiapas), not Oaxaca (stated in error by Wallach et al., 2014); Elevation 305–960 m, L.D. Wilson & Mata-Silva (2014), but 518–610 m according to L.D. Wilson & Mata-Silva (2015).

Comments: L.D. Wilson & Mata-Silva (2014, 2015) provide a species account and map.

Tantilla wilcoxi Stejneger, 1902

Distribution: Add Mexico (Jalisco), Carbajal-Márquez et al. (2015d); delete Mexico (Tamaulipas), L.D. Wilson & Mata-Silva (2014).

Comments: L.D. Wilson & Mata-Silva (2014, 2015) provide a species account, map and color photograph.

Tantilla yaquia H.M. Smith, 1942

Distribution: Elevation to 1680 m, L.D. Wilson & Mata-Silva (2014).

Comments: L.D. Wilson & Mata-Silva (2014, 2015) provide a species account and map.

TANTILLITA H.M. Smith, 1941 (Colubridae: Colubrinae)

Comments: L.D. Wilson & Mata-Silva (2015) provide a diagnosis and review of the species. The authors state that whether or not this genus should be synonymized with *Tantilla*, as has been suggested, must await a molecular analysis.

Tantillita brevissima (Taylor, 1937)

Comments: L.D. Wilson & Mata-Silva (2015) provide an account.

Tantllita canula (Cope, 1875)

Distribution: Upper elevation of 450 m, L.D. Wilson & Mata-Silva (2015).

Comments: L.D. Wilson & Mata-Silva (2015) provide an account.

Tantillita lintoni (H.M. Smith, 1940)

Distribution: Add Mexico (Chiapas); westernmost record in Veracuz, Mexico, García-Morales et al. (2017).

TAPHROMETOPON Brandt, 1838 (Psammophiidae)

Comments: Zaher et al. (2019) find *Psammophis crucifer* to be sister taxon to *Psammophis* + *Taphrometopon*. Branch et al. (2019b), using more samples and taxa, confirm that the South African *Psammophis crucifer* is allied with Eurasian *Psammophis* (*Taphrometopon*), but caution against the use of *Taphrometopon* pending fuller taxon sampling.

Taphrometopon condanarum (Merrem, 1820)

Distribution: Ganesh et al. (2017) provide records that extend the known range further south in Andhra Pradesh and Karnataka, India. Remove India (Gujarat), Patel & Vyas (2019).

Taphrometopon crucifer (Daudin, 1803)

Taphrometopon indochinensis (M.A. Smith, 1943)

Distribution: Add Thailand (Chanthaburi), Chan-ard et al. (2011); Thailand (Phetchaburi), Hartmann et al. (2011). Hartmann et al. (2011) report specific Cambodian records.

Taphrometopon leithii (Günther, 1869)

Taphrometopon lineolatum (Brandt, 1838)

Distribution: Add Iran (Golestan, Markazi, North Khorasan, South Khorasan), Safaei-Mahroo et al. (2015); Afghanistan (Badghis), Wagner et al. (2016b). Doronin (2016) summarizes records for Azerbaijan.

Taphrometopon longifrons (Boulenger, 1890)

Distribution: Add India (Karnataka), Premkumar & Sharma (2017); India (Andhra Pradesh, Telangana), and upper elevation of 498 m, Visvanathan et al. (2017). Vyas & Patel (2013) and Visvanathan et al. (2017) map known localities.

†*TAUNTONOPHIS* Parmley & D. Walker, 2003 (Colubridae: Colubrinae)

†*Tauntonophis morganorum* Parmley & D. Walker, 2003

TELESCOPUS Wagler, 1830 (Colubridae: Colubrinae)

Distribution: Add middle Miocene (Orleanian) of Germany, Čerňanský et al. (2017, as *Telescopus* sp.).

Comments: Šmid et al. (2019) analyze DNA sequence data for ten of the species, which resolve into two primary clades: African and Arabian species, and Asia Minor/Middle East species.

Telescopus beetzii (Barbour, 1922)

†*Telescopus bolkayi* Szyndlar, 2005

Telescopus dhara (Forskal *in* Niebuhr, 1775)

Distribution: Add Saudi Arabia (Ha'il), Alshammari et al. (2017).

Telescopus fallax (Fleischmann, 1831)

Distribution: Add Croatia (Plavnik I.), Tóth et al. (2017); Greece (Chios I.), Kirchner (2009); Greece (Elafonisos I.), Broggi (2016a); Greece (Ithakos I.), Strachinis & Artavanis (2017); Greece (Kimolos I.), Broggi (2014b); Greece (Kastellorizo I.), Kakaentzis et al. (2018); Greece (Samos I.), Speybroek et al. (2014); Turkey (Burdur), Ege et al. (2015); Turkey (Canakkale), Tok et al. (2014); Turkey (Mersin), Winden et al. (1997); Iran (Alborz, Ardabil, Gilan, Golestan, Mazandaran, Qazvin, Tehran), Shafaei-Mahroo et al. (2015); Iran (Kermanshah, Kordestan, Markazi), Nilson & Rastegar-Pouyani (2013).

Comments: Kyriazi et al. (2012) use mtDNA sequence data to determine that populations on Crete originated from the southern end of Greece. Poulakakis et al. (2016) use similar methods to test the origin of the Cyprus population. A Balkan and an Asia Minor/Middle East clade are recovered, with the Cyprus population as sister group to the eastern clade. See under *T. nigriceps*.

Telescopus finkeldeyi Haacke, 2013. Zootaxa 3737(3): 281–285, figs. 1–4.

Holotype: TMP 53542, a 562 mm female (J.A. van Rooyen, December 1979).

Type locality: "Rössing Uranium mine area, Swako[p]mund district (2214Db) Namibia."

Distribution: SW Africa. Angola (Namibe), Namibia (Damaraland, Kaokoland, Karibib, Swakopmund, Windhoek).

Telescopus gezirae Broadley, 1994

Telescopus hoogstraali K.P. Schmidt & Marx, 1956

Telescopus nigriceps (Ahl, 1924)

Distribution: Add Iran (Kermanshah, Kordestan), Nilson & Rastegar-Pouyani (2013); Iran (Markazi), 1748 m, Beyhaghi (2016).

Comments: Specimens from S Turkey were paraphyletic with respect to populations of *T. fallax*, according to a DNA sequence-based phylogeny produced by Šmid et al. (2019).

Telescopus obtusus (A. Reuss, 1834)

Telescopus pulcher (Scortecci, 1935)

Distribution: Add Somalia (Woqooyi Galbeed), Mazuch et al. (2018).

Comments: Mazuch et al. (2018) describe and illustrate two additional specimens.

Telescopus rhinopoma (Blanford, 1874)

Distribution: Add Iran (Bushehr, Hormozgan, Isfahan, Kohgiluyeh and Boyer Ahmad, North Khorasan, South Khorasan), Shafaei-Mahroo et al. (2015); Iran (Qom), S.M. Kazemi et al. (2015); Iran (Kerman), Moradi et al. (2013); Afghanistan (Helmand), Jablonski & Masroor (2019); Pakistan (Federally Administered Tribal Areas, Kyber Pakhtunkhwa), Jablonski & Masroor (2019). Upper elevation of 2269 m, Jablonski & Masroor (2019).

Telescopus semiannulatus A. Smith, 1849 *in* 1838–1849

Distribution: Add Angola (Cuanza Sul, Lunda Norte), M.P. Marques et al. (2018); Angola (Huila, Namibe), Branch (2018).

Telescopus somalicus (Parker, 1949)

Telescopus tessellatus (Wall, 1908)

Distribution: Add Iran (Bushehr, Tehran), Shafaei-Mahroo et al. (2015); Iran (Ilam, Kermanshah, Markazi), Nilson & Rastegar-Pouyani (2013).

Comments: Nilson & Rastegar-Pouyani (2013) morphologically and geographically define two subspecies in Iran.

Telescopus tripolitanus (F. Werner, 1909)

Distribution: Add Morocco (Figuig), Barata et al. (2011); Libya (Tripoli, Zawiyah), Bauer et al. (2017); Niger (Tillaberi), Trape & Mané (2015).

Telescopus variegatus (J.T. Reinhardt, 1843)

Distribution: Add Mali (Sikasso), and low elevation of 170 m, Trape & Mané (2017).

TERETRURUS Beddome, 1886 (Uropeltidae)

Synonyms: Add *Brachyophidium* Wall, 1921.

Comments: Cyriac & Kodandaramaiah (2017) find *Brachyophidium* to be nested within *Teretrurus* in their DNA sequence-based phylogeny. Pyron et al. (2016c) provide a morphological diagnosis of the genus.

Teretrurus hewstoni Beddome, 1876. Proc. Zool. Soc. London 1876(1): 701. (*Platyplectrurus hewstoni*)

Holotype: BMNH 1946.1.15.77, a female (Dr. Hewston)

Type locality: "Manantoddy [=Mananthavady], in the Wynad, elevation 2700 feet," Kerala, India.

Distribution: S India (N Kerala, N of the Coimbatore lowlands).

Comments: Recognized by Cyriac & Kodandaramaiah (2017) based on sister-group relation to *T. sanguineus* + *T. rhodogaster*.

Teretrurus rhodogaster (Wall, 1921)

Comments: Transferred with *Brachyophidium* by Cyriac & Kodandaramaiah (2017). Pyron et al. (2016c) provide a description and color photograph.

Teretrurus sanguineus (Beddome, 1867)

Synonyms: Remove *Platyplectrurus hewstoni* Beddome, 1876.

Distribution: Remove N Kerala N of the Coimbatore lowlands, Cyriac & Kodandaramaiah (2017).

Comments: Pyron et al. (2016c) note that only one specimen was used for the original description, and that the holotype may be one of the MNHN or NHMW specimens. They provide a description and color photograph.

TETRACHEILOSTOMA Jan, 1861 (Leptotyphlopidae: Epictinae)

Tetracheilostoma bilineatum (Schlegel, 1839 *in* 1837–1844)
Tetracheilostoma breuili (Hedges, 2008)
Tetracheilostoma carlae (Hedges, 2008)

TETRALEPIS Boettger, 1892 (Colubroidea: incerta sedis)

Tetralepis fruhstorferi Boettger, 1892

†*TETRAPODOPHIS* Martill, Tischlinger & Longrich, 2015. Science 349(6246): 416. (Ophidia: incerta sedis)

Type species: *Tetrapodophis amplectus* Martill, Tischlinger & Longrich, 2015, by monotypy.

Distribution: Early Cretaceous of Brazil.

Comments: Martill et al. (2015) place it as the sister taxon to †*Coniophis*. However, Onary et al. (2017) consider it a putative snake in the broadest sense (Ophidia).

†*Tetrapodophis amplectus* Martill, Tischlinger & Longrich, 2015. Science 349(6246): 416–418, figs. 1–4.

Holotype: BMMS BK 2–2. The specimen is preserved on laminated limestone.

Type locality: "Nova Olinda Member of the Early Cretaceous (Aptian) Crato Formation, Ceará, Brazil."

Distribution: Early Cretaceous of Brazil (Ceará). Known only from the holotype.

Comment: Onary et al. (2017) advise caution in acceptance of the stated type locality. M.S.Y. Lee et al. (2016b) provide a further description of the appendicular skeleton.

†*TEXASOPHIS* Holman, 1977 (Colubroidea: incerta sedis)

†*Texasophis bohemiacus* Szyndlar, 1987

Distribution: Add late Middle Miocene (Astaracian) of NE Kazakhstan, Ivanov et al. (2019).

†*Texasophis fossilis* Holman, 1977
†*Texasophis galbreathi* Holman, 1984
†*Texasophis hecki* M. Böhme, 2008
†*Texasophis meini* Rage & Holman, 1984
†*Texasophis wilsoni* Holman, 1984

THALASSOPHIS P. Schmidt, 1852 (Elapidae)

Thalassophis anomalus P. Schmidt, 1852

THAMNODYNASTES Wagler, 1830 (Dipsadidae: Xenodontinae)

Comments: Coelho et al. (2013) discuss species in NE Brazil.

Thamnodynastes almae Franco & Ferreira, 2002

Comments: Coelho et al. (2013) provide a morphological and distributional summary.

Thamnodynastes ceibae Bailey & Thomas, 2007

Thamnodynastes chaquensis Bergna & Alvarez, 1993

Distribution: Add Brazil (Mato Grosso do Sul), Ferreira et al. (2017); Paraguay (Alto Paraguay, Boqueron, Concepcion), Cabral & Weiler (2014).

Thamnodynastes chimanta Roze, 1958

Thamnodynastes corocoroensis Gorzula & Ayarzagüena, 1996

Thamnodynastes dixoni Bailey & Thomas, 2007

Thamnodynastes duida C.W. Myers & Donnelly, 1996

Thamnodynastes gambotensis Pérez-Santos & Moreno, 1989

Distribution: Add Colombia (Córdoba), Carvajal-Cogollo et al. (2007).

Thamnodynastes hypoconia (Cope, 1860)

Distribution: Add Brazil (Bahia, Piaui), Guedes et al. (2014); Brazil (Mato Grosso), Tavares et al. (2012); Brazil (Mato Grosso do Sul), Ferreira et al. (2017); Argentina (Chaco, Corrientes), Franco et al. (2017).

Thamnodynastes lanei Bailey, Thomas & Silva, 2005

Distribution: Add Brazil (Amapa, Rondonia), Franco et al. (2017); Brazil (Rondônia), Bernarde et al. (2012b).

Thamnodynastes longicaudus Franco, Ferreira, Marques & Sazima, 2003

Thamnodynastes marahuaquensis Gorzula & Ayarzagüena, 1996

Thamnodynastes nattereri Mikan, 1828. Delectae Florae Faunae Brasil.: fig. 1. (*Coluber nattereri*)

Synonyms: *Thamnodynastes nattereri laevis* Boulenger, 1885.

Holotype: Lost according to Nogueira et al. (2019).

Type locality: "Lectus prope Sebastianopolim" [=Rio de Janeiro], Brazil.

Distribution: E Brazil and Uruguay. Brazil (Bahia, Alagoas, Pernambuco, Sergipe), and upper elevation of 800 m, Guedes et al. (2014, as *T.* cf. *nattereri*); Brazil (Espirito Santo), Silva-Soares et al. (2011); Brazil (Minas Gerais), Moura et al. (2012, as *T.* cf *nattereri*); Brazil (Paraiba), Franco et al. (2017, as *T.* cf *nattereri*). Also Rio de Janeiro, Sao Paulo, Rio Grande do Norte, Parana and Santa Catarina.

Comments: This species is revived in the literature, usually as *T.* cf *nattereri*, for populations along the Atlantic coast of South America.

Thamnodynastes pallidus (Linnaeus, 1758)

Synonyms: Remove *Coluber nattereri* Mikan, 1828.

Distribution: Add Brazil (Amazonia, Rondonia), Franco et al. (2017); Brazil (Paraiba), R. França et al. (2012); Brazil (Tocantins), Dal Vechio et al. (2016, as *T.* cf. *pallidus*).

Thamnodynastes paraguanae Bailey & Thomas, 2007

Distribution: Add Colombia (Cesar), Rojas-Murcia et al. (2016).

Thamnodynastes phoenix Franco, Trevine, Montingelli & Zaher, 2017. Salamandra 53(3): 340–347, figs. 1–4.

Holotype: IBSP 87527, a 473 mm male (L. de B. Ribeiro, 4 November 2011).

Type locality: "Brazil, Pernambuco, municipality of Petrolina, 09°19′29.00″ S, 40°32′50.00″ W, 389 m above sea level, Campus Ciências Agrárias, Universidade Federal do Vale do São Francisco (UNIVASF)."

Distribution: NE Brazil (Alagoas, Bahia, Ceara, Goias, Minas Gerais, Paraiba, Pernambuco, Piaui, Rio Grande do Norte, Sergipe, Tocantins).

Comments: Data in Coelho et al. (2013) on an undescribed species pertain to *T. phoenix*.

Thamnodynastes ramonriveroi Manzanilla & Sánchez, 2005

Thamnodynastes rutilus (Prado, 1942)

Distribution: Add Brazil (Distrito Federal, Rio de Janeiro), J.J. Magalhães et al. (2017); Brazil (Mato Grosso do Sul), Ferreira et al. (2017); Brazil (Minas Gerais), Franco et al. (2017). J.J. Magalhães et al. (2017) map known localities.

Thamnodynastes sertanejo Bailey, Thomas, & Silva, 2005

Distribution: Add Brazil (Paraíba), and upper elevation of 657 m, Guedes et al. (2014).

Comments: Coelho et al. (2013) provide a morphological and distributional summary.

Thamnodynastes strigatus (Günther, 1858)

Distribution: Add Brazil (Rio Grande do Sul), Franco et al. (2017); Paraguay (Neembucu), Cabral & Weiler (2014). Upper elevation of 2450 m, Winkler et al. (2011).

Thamnodynastes yavi C.W. Myers & Donnelly, 1996

THAMNOPHIS Fitzinger, 1843 (Natricidae)

Fossil records: Add upper Miocene (Hemphillian) of USA (Tennessee), Jasinski & Moscato (2017, as *Thamnophis* sp.).

Thamnophis atratus (Kennicott *in* Cooper, 1860)

Thamnophis bogerti Rossman & Burbrink, 2005

Thamnophis brachystoma (Cope, 1892)

Thamnophis butleri (Cope, 1889)

Comment: Kapfer et al. (2013) report natural hybridization with *T. sirtalis* in Wisconsin, USA.

Thamnophis chrysocephalus (Cope, 1885)

Thamnophis conanti Rossman & Burbrink, 2005

Thamnophis couchii (Kennicott *in* Baird, 1859)

Thamnophis cyrtopsis (Kennicott, 1860)

Thamnophis elegans (Baird & Girard, 1853)

Thamnophis eques (A. Reuss, 1834)

Distribution: Add Mexico (Guerrero), Palacios-Aguilar & Flores-Villela (2018).

Thamnophis errans H.M.Smith, 1942

Distribution: Add Mexico (Jalisco), Ahumada-Carrillo et al. (2014)

Thamnophis exsul Rossman, 1969

Thamnophis fulvus (Bocourt, 1893 *in* A.H.A. Duméril, Bocourt & Mocquard, 1870–1909)

Distribution: Add Guatemala (Huehuetenango) and upper elevation of 3710 m, Eisermann et al. (2016).

Thamnophis gigas Fitch, 1940

Comments: Wood et al. (2015) use microsatellites to evaluate genetic structure within and between populations. They recover five clusters that correspond to drainage basins.

Thamnophis godmani Günther, 1894 *in* 1885–1902)

Distribution: Upper elevation of 3018 m, Palacios-Aguilar & Flores-Villela (2018).

Thamnophis hammondi (Kennicott, 1860)

Thamnophis lineri Rossman & Burbrink, 2005

Thamnophis marcianus (Baird & Girard, 1853)
Comments: E.A. Myers et al. (2017a) provide a DNA-sequence based phylogeny of USA populations, with two clades separated at the Continental Divide.

Thamnophis melanogaster (W.C.H. Peters, 1864)
Thamnophis mendax C.F. Walker, 1955
Thamnophis nigronuchalis Thompson, 1957
Thamnophis ordinoides (Baird & Girard, 1852)
Thamnophis postremus H.M. Smith, 1942
Thamnophis proximus (Say *in* James, 1823)
Distribution: Mexico (Oaxaca, disjunct record), Mata-Silva et al. (2015b).

Thamnophis pulchrilatus (Cope, 1885a)
Distribution: Upper elevation of 3054 m, González-Hernández et al. (2016). Fernández-Badillo et al. (2016c) report additional records from Hidalgo, Mexico, and Ahumada-Carrillo et al. (2014) within Jalisco, Mexico.

Thamnophis radix (Baird & Girard, 1853)
Thamnophis rossmani Conant, 2000
Distribution: Additional and recent record for Nayarit, Mexico, 922 m, Luja & Grünwald (2015).

Thamnophis rufipunctatus (Cope *in* Yarrow, 1875)
Thamnophis saurita (Linnaeus, 1766)
Comments: Kraus & Cameron (2016) note that the correct specific epithet is *saurita*, not *sauritus*.

Thamnophis scalaris Cope, 1861
Distribution: Mexico (range extension in Hildalgo), Olvera & Badillo (2016b).

Thamnophis scaliger (Jan, 1865)
Distribution: Mexico (Jalisco), Grünwald et al. (2016a, range extension).

Thamnophis sirtalis (Linnaeus, 1758)
Comment: Kapfer et al. (2013) report natural hybridization with *T. butleri* in Wisconsin, USA.

Thamnophis sumichrasti (Cope, 1867)
Thamnophis unilabialis W.W. Tanner, 1985
Thamnophis validus (Kennicott, 1860)

THAMNOSOPHIS Jan, 1863 (Pseudoxyrhophiidae)
Thamnosophis epistibes (Cadle, 1996)
Thamnosophis infrasignatus (Günther, 1882)
Thamnosophis lateralis (A.M.C. Duméril, Bibron & Duméril, 1854)
Thamnosophis martae (Glaw, Franzen & Vences, 2005)
Thamnosophis mavotenda Glaw, Nagy, Köhler, Franzen & Vences, 2009
Thamnosophis stumpffi (Boettger, 1881)

†*THAUMASTOPHIS* Rage, Folie, Rana, Singh, Rose & T. Smith, 2008 (Colubroidea: incerta sedis)

†*Thaumastophis missiaeni* Rage, Folie, Rana, Singh, Rose & T. Smith, 2008

Comments: T. Smith et al. (2016) describe additional material from near the type locality.

THELOTORNIS A. Smith, 1849 *in* 1838–1849 (Colubridae: Colubrinae)
Fossil records: Add middle Pliocene of Tanzania, Rage & Bailon (2011, as cf. *Thelotornis*).

Thelotornis capensis A. Smith, 1849 *in* 1838–1849
Distribution: Add Angola (Benguela, Cunene), Branch (2018); Angola (Cauando-Cubango), Conradie et al. (2016b); Angola (Huila), M.P. Marques et al. (2018); Angola (Namibe), Ceríaco et al. (2016).

Thelotornis kirtlandii (Hallowell, 1844)
Distribution: Add Angola (Bengo), Branch (2018); Angola (Luanda, Malanje), M.P. Marques et al. (2018).

Thelotornis mossambicanus (Bocage, 1895)
Thelotornis usambaricus Broadley, 2001
Distribution: Add Tanzania (Nguru Mtns., Morogoro), Menegon et al. (2008).

THERMOPHIS Malnate, 1953 (Dipsadidae: subfamily unnamed)
Comments: S. Hofmann et al. (2015) present new morphological data for the species of *Thermophis*, along with a phylogeny of each using DNA sequence data.

Thermophis baileyi (Wall, 1907)
Distribution: S. Hofman et al. (2015) map known localities.

Thermophis shangrila Peng, Lu, Huang, Guo & Zhang, 2014. Asian Herpetol. Res. 5(4): 230–235, fig. 4.
Holotype: HUM 20120001, a 951 mm female (L. Peng et al., 23 August 2011).
Type locality: "near a hot spring about 500 m in Shangri-La, Northern Yunnan, China."
Distribution: China (N Yunnan), 500 m. Known only from the type locality.
Comments: S. Hofman et al. (2015), using DNA sequence-data, find *T. shangrila* to be undifferentiated from *T. zhaoermii*, and note that morphological characters overlapped between the two taxa.

Thermophis zhaoermii Guo, Liu, Feng & He, 2008
Distribution: S. Hofman et al. (2015) map known localities.

THRASOPS Hallowell, 1857 (Colubridae: Colubrinae)
Thrasops flavigularis (Hallowell, 1852)
Distribution: Upper elevation of 2050 m, Ineich et al. (2015).

Thrasops jacksonii Günther, 1895
Distribution: Add Nigeria, Trape & Baldé (2014); Gabon (Ogooué-Ivindo), Carlino & Pauwels (2013); Gabon (Woleu-Ntem), Pauwels et al. (2017d); South Sudan (Eastern Equatoria), Ullenbruch & Böhme (2017); Angola (Lunda Norte, Zaire), Branch (2018).

Thrasops occidentalis H.W. Parker, 1840
Thrasops schmidti Loveridge, 1936

†*TITANOBOA* Head, Bloch, Hastings, Bourque, Cadena, Herrera, Polly & Jaramillo, 2009 (Booidea: incerta sedis)
Comments: Head et al. (2013) discuss phylogenetic relationships.

†*Titanoboa cerrejonensis* Head, Bloch, Hastings, Bourque, Cadena, Herrera, Polly & Jaramillo, 2009
Comments: Head et al. (2013) describe cranial and additional vertebral elements.

TOMODON A.M.C. Duméril, 1853 (Dipsadidae: Xenodontinae)

Tomodon dorsatus A.M.C. Duméril, Bibron & Duméril, 1854
Distrubution: Add Brazil (Mato Grosso do Sul), Ferreira et al. (2017).

Tomodon ocellatus A.M.C. Duméril, Bibron & Duméril, 1854

Tomodon orestes Harvey & Muñoz, 2004
Distribution: Add Argentina (Jujuy), F.B. Gallardo et al. (2014); Argentina (Tucuman), Scrocchi et al. (2019).

†*TOTLANDOPHIS* Holman & Harrison, 1998 (Booidea: incerta sedis)

†*Totlandophis americanus* Holman & Harrison, 2001
†*Totlandophis thomasae* Holman & Harrison, 1998

TOXICOCALAMUS Boulenger, 1896 (Elapidae)
Comments: J.L. Strickland et al. (2016) provide a phylogeny for eight *Toxicocalamus* species, which they confirm to be monophyletic. However, the phylogeny produced by Zaher et al. (2019) has *T. loriae* paraphyletic with respect to *T. preussi*, which resolves in a clade with *Aspidomorphus*, *Furina* and *Demansia*. O'Shea et al. (2018b) provide a taxonomic history of the genus, and a key to the species.

Toxicocalamus buergersi (Sternfeld, 1913)
Distribution: Add Papua New Guinea (Sandaun), and upper elevation of 634 m, O'Shea et al. (2018b).

Toxicocalamus cratermontanus Kraus, 2017b. J. Herpetol. 51(4): 578–579, figs. 1, 2.
Holotype: USNM 562941, a 727 mm female (D. Bickford, 1 March 1996).
Type locality: "9.6 km east of Haia, Crater Mountain Wildlife Management Area, 6.7239°S, 145.0931°E, 920 m a.s.l., Chimbu Province, Papua New Guinea."
Distribution: Papua New Guinea (Chimbu), 920 m. Known only from the holotype.

Toxicocalamus ernstmayri O'Shea, Parker & H. Kaiser, 2015. Bull. Mus. Comp. Zool. 161(6): 243–248, figs 3, 4.
Holotype: MCZ 145946, a 1200 mm female (F. Parker, 23 December 1969).
Type locality: "Wangbin village (5°14'26.72"S, 141°15'31.92"E), elevation 1468 m (4800 ft), near the Ok Tedi River, in the Star Mountains of the North Fly District, Western Province, P[apua]N[ew]G[uinea]."

Distribution: Papua New Guinea (Western), 1468–1700 m. Known only from two individuals.
Comments: O'Shea et al. (2018a) report and illustrate a second specimen.

Toxicocalamus grandis (Boulenger, 1914)
Distribution: Elevation 25 m, O'Shea et al. (2018b).

Toxicocalamus holopelturus McDowell, 1969
Distribution: Low elevation of 55 m, O'Shea et al. (2015), who list specimens in addition to the holotype.

Toxicocalamus longissimus Boulenger, 1896
Distribution: Elevation range 12–80 m, O'Shea et al. (2015). O'Shea & Kaiser (2018) show that the record for Fergusson Island is in error.

Toxicocalamus loriae (Boulenger, 1898)
Distribution: Add Papua New Guinea (Jiwaka, Sandaun), and upper elevation range of 2140 m, O'Shea et al. (2015).
Comments: J.L. Strickland et al. (2016) find five genetic species within this complex.

Toxicocalamus mintoni Kraus, 2010 (2009). Herpetologica 65(4): 461–462, figs. 1, 2.
Holotype: BPBM 20822, a 611+ mm male (J. Slapcinsky, 21 April 2004).
Type locality: "W slope of Mt. Rio, 11.49610° S, 153.42413° E, 410 m, Sudest Island, Milne Bay Province, Papua New Guinea."
Distribution: Papua New Guinea (Milne Bay: Sudest Island), 410 m. Known only from the holotype.

Toxicocalamus misimae McDowell, 1969
Distribution: Elevation range 128–350 m, O'Shea et al. (2018b). O'Shea & Kaiser (2018) show that the record for Central Province is in error.

Toxicocalamus nigrescens Kraus, 2017b. J. Herpetol. 51(4): 575–576, figs. 1, 2.
Holotype: BPBM 16545, a 720 mm female (P. Robert, 10 September 2002).
Type locality: "S slope Oya Waka, 9.4562°S, 150.5596°E…, 980 m a.s.l., Fergusson Island, Milne Bay Province, Papua New Guinea."
Distribution: Papua New Guinea (Fergusson Island), NSL-980 m.

Toxicocalamus pachysomus Kraus, 2010 (2009). Herpetologica 65(4): 461–462, figs. 1, 2.
Holotype: BPBM 15771, a 546+ mm male (F. Kraus, 22 April 2002).
Type locality: "along Upaelisafupi Stream, 10.4970833° S, 150.2329666° E, 715 m, Cloudy Mountains, Milne Bay Province, Papua New Guinea."
Distribution: Papua New Guinea (Milne Bay), 715 m. Known only from the holotype.

Toxicocalamus preussi (Sternfeld, 1913)
Distribution: Add Papua New Guinea (Sandaun, Simbu), and upper elevation of 2325 m, O'Shea et al. (2015); Papua New

Guinea (West New Britain), O'Shea et al. (2018b). O'Shea & Kaiser (2018) reason that the species is limited to the highlands, and the coastal record, NSL, in Western Province is in error.

Toxicocalamus pumehanae O'Shea, Allison & H. Kaiser, 2018b. Amphibia-Reptilia 39(4): 410–420, figs. 2, 3, 5, 6.

Holotype: BPBM 36185, a 241 mm female (A. Allison, March 2010).

Type locality: "Jarefa Camp village (09°12′19″S, 148°14′15″E…), elevation 820 m, near Itokama (=Itogama), on the Managalas Plateau, Managalas Conservation Area, Ijvitari District, Oro Province, Papua New Guinea."

Distribution: SE New Guinea (Oro), 820 m. Known only from the holotype.

Toxicocalamus spilolepidotus McDowell, 1969

Toxicocalamus stanleyanus Boulenger, 1903

Distribution: Add Papua New Guinea (East Sepik, Sandaun, Simbu), and low elevation limit of 37 m, O'Shea et al. (2015).

TOXICODRYAS Hallowell, 1857 (Colubridae: Colubrinae)

Toxicodryas blandingii (Hallowell, 1844)

Distribution: Add Angola (Bengo, Cuanza Sul), Branch (2018).

Toxicodryas pulverulenta (J.C. Fischer, 1856)

Type locality: Ceríaco et al. (2018) affirm that the type locality as stated in the original description (Edina, Liberia) is correct, and that the species does not occur on São Tomé.

Distribution: Add Gabon (Ogooue-Lolo), Pauwels et al. (2017b); South Sudan (Eastern Equatoria), Ullenbruch & Böhme (2017).

TRACHISCHIUM Günther, 1858 (Natricidae)

Trachischium apteii Bhosale, Gowande & Mirza, 2019. Comptes Rendus Biol. 342(9–10): 325–327, figs. 1, 2.

Holotype: BNHS3550, a 331 mm female (H.S. Bhosale, G. Gowande, M. Savant & P. Phansalkar, 19 June 2019).

Type locality: "Pange camp, Talle Valley Wildlife Sanctuary, Arunachal Pradesh (27.549322°N, 93.897138°E, elevation 1890 m)," India.

Distribution: NE India (Arunachal Pradesh), 1890 m. Known only from the type locality.

Trachischium fuscum (Blyth, 1954)

Distribution: Add India (Assam), Raha et al. (2018).

Comments: Raha et al. (2018) redescribe the lectotype of *C. fusca*, as well as type of synonyms, and additional specimens.

Trachischium guentheri Boulenger, 1890

Distribution: Add China (S Tibet), K. Wang et al. (2019a). Upper elevation of 2700 m, Chettri et al. (2010).

Comments: K. Wang et al. (2019a) describe a specimen from China.

Trachischium leave Peracca, 1904

Trachischium monticola (Cantor, 1839)

Comments: K. Wang et al. (2019a) describe additional specimens from China.

Trachischium sushantai Raha, S. Das, Bag, Debnath & Pramanick, 2018. Zootaxa 4370(5): 550–553, figs. 1, 2.

Holotype: ZSI 25651a, a 311 mm female (Rajtilok, 4 August 1993).

Type locality: "'Jammu' (Jammu & Kashmir state, India)."

Distribution: N India (Jammu & Kashmir). Known only from the holotype.

Trachischium tenuiceps (Blyth, 1854)

TRACHYBOA W.C.H. Peters, 1860 (Tropidophiidae)

Trachyboa boulengeri Peracca, 1910

Trachyboa gularis W.C.H. Peters, 1860

†***TREGOPHIS*** Holman, 1975 (Booidea: incerta sedis)

†***Tregophis brevirachis*** Holman, 1975

TRETANORHINUS A.M.C. Duméril, Bibron& Duméril, 1854 (Dipsadidae: Dipsadinae)

Tretanorhinus mocquardi Bocourt, 1891

Distribution: Add Panama (Panama Oeste), Ray & Ruback (2015).

Tretanorhinus nigroluteus Cope, 1861

Distribution: Add Mexico (Chiapas), Hernández-Ordóñez et al. (2015); Panama (Panama), Ray & Ruback (2015).

Tretanorhinus taeniatus Boulenger, 1903

Tretanorhinus variabilis A.M.C. Duméril, Bibron& Duméril, 1854

TRICHEILOSTOMA Jan & Sordelli, 1860 *in* 1860–1866 (Leptotyphlopidae: Epictinae)

Tricheilostoma bicolor (Jan & Sordelli, 1860 in 1860–1866)

Distribution: Add Niger (Tillaberi), Trape & Mané (2015).

Tricheilostoma broadleyi (Wallach & Hahn, 1997)

Tricheilostoma dissimile (Bocage, 1886)

Tricheilostoma greenwelli (Wallach & Boundy, 2005)

Comment: Boundy (2014) provides a corrected illustration of the ventral head scalation.

Tricheilostoma kongoensis Trape, 2019. Bull. Soc. Herpétol. France 169: 38–42, figs. 11–13.

Holotype: MNHN 2018.0014, a 100 mm specimen of unknown sex (F. Nsingi, 2017).

Type locality: "près du Stanley Pool à Kinshasa (04°19'S / 15°15'E) en République démocratique du Congo."

Distribution: W Democratic Republic of the Congo (Kinshasa). Known only from the holotype.

Tricheilostoma sundewalli (Jan, 1861)

Distribution: Add Nigeria, Trape & Baldé (2014).

TRILEPIDA Hedges, 2011 (Leptotyphlopidae: Epictinae)

Comments: Pinto & Fernandes (2017) provide a key to the species of *Trilepida*.

Trilepida affinis Boulenger, 1884. Ann. Mag. Nat. Hist., Series 5, 13(77): 396–397. (*Stenostoma affine*)

Holotype: BMNH 1946.1.11.16, a 218 mm male.

Type locality: "province of Tachira, Venezuela."

Distribution: Venezuela (Tachira). Known only from the holotype.

Comments: Pinto & Fernandes (2017) re-validate *S. affine*, and redescribe the holotype.

Trilepida anthracina (Bailey, 1946)

Trilepida brasiliensis (Laurent, 1949)

Distribution: Add Brazil (Paraíba), Moura et al. (2013b). Moura et al. map known localities.

Trilepida brevissima (Shreve, 1964)
Trilepida dimidiata (Jan, 1861)
Trilepida dugandi (Dunn, 1944)

Trilepida fuliginosa (Passos, Caramaschi & Pinto, 2006)

Distribution: Add Brazil (Mato Grosso), Francisco et al. (2018); Brazil (Piauí), Dal Vechio et al. (2016, as *T.* cf. *fuliginosa*).

Trilepida guayaquilensis (Orejas-Miranda & G. Peters, 1970)

Comment: Salazar-Valenzuela et al. (2015) note that numerous records of *T. quayaquilensis* are referable to *Epictia subcrotilla*, and that *T. guayaquilensis* is still only represented by the holotype.

Trilepida jani (Pinto & Fernandes, 2012)
Trilepida joshuai (Dunn, 1944)
Trilepida koppesi (Amaral, 1955)

Distribution: Add Brazil (Bahia), Guedes et al. (2014); Brazil (Minas Gerais), 181 and 1100 m, Filogonio & Canelas (2015a, b).

Trilepida macrolepis (W.C.H. Peters, 1857)

Distribution: Add Venezuela (Trujillo), Esqueda-González et al. (2015); Ecuador (Esmeraldas), Salazar-Valenzuela et al. (2015). Pinto & Fernandes (2017) provide a map of known localities.

Comments: Pinto & Fernandes (2017) provide a revised diagnosis.

Trilepida nicefori (Dunn, 1946)

Trilepida pastusa Salazar-Valenzuela, Martins, Amador-Oyola & Torres-Carvajal, 2015. Amphibian & Reptile Conservation 8(1): 108–113, figs. 1–5.

Holotype: QCAZ 8690, a 315 mm female (O. Torres-Carvajal, S. Aldás-Alarcón, E. Tapia, A. Pozo & local people, 23 February 2009).

Type locality: "surroundings of Chilmá Bajo on the way to Tres Marías waterfall (0°51′53.82″ N, 78°2′59.23″ W; 2071 m), Tulcán County, Carchi province, Ecuador."

Distribution: Ecuador (Carchi), 2071 m. Add Colombia (Cauca), and low elevation of 1472 m, Vera-Pérez et al. (2018).

Trilepida salgueiroi (Amaral, 1955)

TRIMERESURUS Lacépède, 1804 (Vipera: Crotalinae)

Comments: Mrinalini et al. (2015) use DNA sequence-data to evaluate the phylogeny of the *T. macrops* complex. Alencar et al. (2016) use DNA sequence-data to resolve six clades in *Trimeresurus* sensu lato, one of which lacks an available genus name. They did not evaluate the monotypic genera *Peltopelor*, *Triceratolepidophis* or *Zhaoermia*.

Trimeresurus albolabris Gray, 1842

Distribution: Add Nepal (Bara), Pandey et al. (2018); Nepal (Chitwan, Nawalparasi), Pandey (2012); Nepal (Parsa), Bhattarai et al. (2018b); Thailand (Chanthaburi), Chan-ard et al. (2011); Vietnam (Quang Ngai), Nemes et al. (2013); Vietnam (Yen Bai), Le et al. (2018).

Trimeresurus andersoni Theobald, 1868

Trimeresurus arunachalensis Captain, Deepak, Pandit, Bhatt & Athreya, 2019. Russ. J. Herpetol. 26(2): 113–116, fig. 3.

Holotype: APF/SFRI 1871, a 658 mm male (W. Phiang & R. Pandit).

Type locality: "near Ramda, West Kameng, Arunachal Pradesh, northeastern India (27°15′ N 92°46′ E; 1876 m elevation)."

Distribution: NE India (Arunachal Pradesh), 1876 m. Known only from the holotype.

Comments: See under *Himalayophis*.

Trimeresurus cantori Blyth, 1846

Trimeresurus cardamomensis (Malhotra, Thorpe, Mrinalini & B.S. Stuart, 2011)

Comments: Mrinalini et al. (2015) confirm the validity of this species using DNA sequence-data.

Trimeresurus erythrurus (Cantor, 1839)
Trimeresurus fasciatus (Boulenger, 1896)
Trimeresurus honsonensis (L.L. Grismer, Ngo & Grismer, 2008)

Trimeresurus insularis Kramer, 1977

Distribution: Add Timor-Leste (Ataúro Island), H. Kaiser et al. (2013b).

Trimeresurus kanburiensis M.A. Smith, 1943

Comments: Alencar et al. (2016), using four mtDNA genes, resolved *T. kanburiensis* in a clade that shared a most common recent ancestor with *Sinovipera+Viridovipera*, and was not associated witrh *Trimeresurus* sensu stricto.

Trimeresurus labialis Fitzinger *in* Steindachner, 1867

Synonymy: Remove *Trimeresurus mutabilis* Stoliczka 1870, as a species.

Type locality: Restricted to Car Nicobar Island by Vogel et al. (2014a).

Distribution: Limited to Cat Nicobar Island, Andaman and Nicobars, India.

Comments: Vogel et al. (2014a) find two morphotypes from a multivariate analysis, one of which is referable to *T. labialis*, the other they resurrect as *T. mutabilis*. They provide a revised description and color photograph of *T. labialis*.

Trimeresurus macrops Kramer, 1977

Distribution: Add Thailand (Chanthaburi), Chan-ard et al. (2011); Thailand (Nakhon Ratchasima), Barnes et al. (2017); Cambodia (Siem Reap), Geissler et al. (2019, as *T.* cf. *macrops*).

Comments: Mrinalini et al. (2015) provide a DNA sequence-based phylogeny that shows some clades within the species.

Alencar et al. (2016), using four mtDNA genes, resolved *T. macrops* in a clade that shared a most common recent ancestor with *Sinovipera+Viridovipera*, and was not associated with *Trimeresurus* sensu stricto.

Trimeresurus mutabilis Stoliczka, 1870. Proc. Asiatic Soc. Bengal 1870(April): 107.

Lectotype: NMW 14863.1, designated by Vogel et al. (2014a).

Type locality: "Andamans and Nicobars," India. Restricted to Camorta Island by lectotype designation.

Distribution: India (Andaman and Nicobar Is.: Bompoka, Camorta, Chowra, Katchal, Nancowry, Tarasa and Trinkat Islands).

Comments: Vogel et al. (2014a) revive this taxon as a valid species, and provide a description and color photographs.

Trimeresurus purpureomaculatus (Gray, 1832 *in* Gray & Hardwicke, 1830–1835)

Distribution: Add West Malaysia (Jerejak Island), Quah et al. (2011); West Malaysia (Kedah), Shahriza et al. (2013); West Malaysia (Pulau Singa Besar), B.L. Lim et al. (2010).

Trimeresurus rubeus (Malhotra, Thorpe, Mrinalini & B.S. Stuart, 2011)

Comments: Mrinalini et al. (2015) confirm the validity of this species using DNA sequence-data.

Trimeresurus septentrionalis Kramer, 1977

Distribution: Add India (Uttarakhand), Singh et al. (2017).

Trimeresurus venustus Vogel, 1991

Comments: Mrinalini et al. (2015), using DNA-sequence data, results in *T. venustus* and *T. rubeus* as forming a clade relative to other *T. macrops* complex members. Alencar et al. (2016), using four mtDNA genes, resolve *T. venustus* in a clade that shared a most common recent ancestor with *Sinovipera+Viridovipera*, and was not associated with *Trimeresurus* sensu stricto. On the other hand, the phylogeny produced by Zaher et al. (2019) has *venustus* in a clade with *kanburiensis* and *macrops*.

TRIMERODYTES Cope, 1895. Proc. Acad. Nat. Sci. Philadelphia 46: 426. (Natricidae)

Synonyms: *Liparophis* Persacca, 1904, *Sinonatrix* Rossman & Eberle, 1977.

Type species: *Trimerodytes balteatus* Cope, 1895 by monotypy.

Distribution: SE Asia.

Comments: In an mtDNA-based phylogeny of *Opisthotropis* and related genera, Ren et al. (2019) resolve *O. balteata* within *Sinonatrix*. The name *Trimerodytes* is resurrected for the clade, having priority over *Sinonatrix*. Ren et al. (2018) provide a revised generic diagnosis and a key to the species. See *Paratapinophis*.

Trimerodytes aequifasciatus (Barbour, 1908)

Distribution: Add Vietnam (Hoa Binh), Nguyen et al. (2018).

Trimerodytes annularis (Hallowell, 1856)

Trimerodytes balteatus Cope, 1895

Distribution: Add Vietnam (Bac Giang), Teynié et al. (2014).

Comments: Ren et al. (2019) provide a revised description, color photographs and a locality map.

Trimerodytes percarinatus (Boulenger, 1899)

Distribution: Add Vietnam (Dien Bien, Kon Tum, Than Hoa), Le et al. (2015); Vietnam (Quang Ngai), Nemes et al. (2013); Vietnam (Yen Bai), Le et al. (2018); China (Shandong), J.L. Lee et al. (2018b).

Trimerodytes yapingi Guo, Zhu & Liu, 2019b. Zootaxa 4623(3): 538–539, figs. 3–5. (*Sinonatrix yapingi*) **new combination**

Holotype: YBU 15296, a 795 mm female (F. Zhu & Z.Q. Zhang, August 2015).

Type locality: "Jingdong (24.53°N, 100.94°E), Yunnan Province, China, at an elevation of 1500 m a.s.l."

Distribution: SE China (S Yunnan), 1500 m. Known only from the holotype.

Comments: We follow Ren et al. (2019) in transferring members of *Sinonatrix* to *Trimerodytes*.

Trimerodytes yunnanensis (Rao & Yang, 1998)

Distribution: Add Vietnam (Son La), Le et al. (2015).

TRIMETOPON Cope, 1885 (Dipsadidae: Dipsadinae)

Trimetopon barbouri Dunn, 1930

Distribution: Add Panama (Cocle), and upper elevation of 710 m, Ray et al. (2013); Panama (Chiriqui, Darien, Panama, Panama Oeste), Derry et al. (2015).

Trimetopon gracile (Günther, 1872)

Trimetopon pliolepis Cope, 1894

Distribution: Add Nicaragua (Rio San Juan), Gutiérrez-Rodríguez & Sunyer (2016).

Trimetopon simile Dunn, 1930

Trimetopon slevini (Dunn, 1940)

Distribution: Herse & Ray (2014) map known localities, and suspect that the reported low elevation of 120 m is a typographical error for 1200 m.

Trimetopon viquezi Dunn, 1937

TRIMORPHODON Cope, 1861 (Colubridae: Colubrinae)

Trimorphodon biscutatus (A.M.C. Duméril, Bibron & Duméril, 1854)

Trimorphodon lambda Cope, 1886

Comments: E.A. Myers et al. (2017a) provide a DNA sequence-based phylogeny of USA populations, with two clades on either side of the Continental Divide, which support the recognition of *T. vilkinsonii*.

Trimorphodon lyrophanes (Cope, 1860)

Trimorphodon paucimaculatus Taylor, 1938

Distribution: Add Mexico (Zacatecas), 1100 m, Bañuelos-Alamillo et al. (2015).

Trimorphodon quadruplex H.M. Smith, 1941

Distribution: Add Honduras (El Paraíso), McCranie et al. (2014b); Honduras (Gracias á Dios), McCranie (2014a);

Honduras (Islas Exposición, Inglesera, El Pacar, de la Vaca, Violín), McCranie & Gutsche (2016).

Trimorphodon tau Cope, 1870

Distribution: Upper elevation of 2711 m, Quintero-Díaz & Carbajal-Márquez (2017b).

Trimorphodon vilkinsonii Cope, 1886

Distribution: Add N Mexico (Coahuila), Baeza-Tarin et al. (2018c).

Comments: See under *T. lambda*.

TROPIDECHIS Günther, 1863 (Elapidae)

Tropidechis carinatus (Krefft, 1863)

TROPIDOCLONION Cope, 1860 (Natricidae)

Tropidoclonion lineatum (Hallowell, 1856)

TROPIDODIPSAS Günther, 1858 (Dipsadidae: Dipsadinae)

Tropidodipsas annulifera Boulenger, 1894

Tropidodipsas fasciata Günther, 1858

Distribution: Add Guatemala (El Peten), Griffin & Powell (2014). Upper elevation of 2130 m, Palacios-Aguilar & Flores-Villela (2018).

Tropididipsas fischeri Boulenger, 1894

Tropidodipsas philippii (Jan, 1863)

Distribution: Upper elevation of 1812 m, Mata-Silva et al. (2017).

Tropidodipsas repleta H.M. Smith, Lemos-Espinal, Hartman & Chiszar, 2005

Tropidodipsas sartorii Cope, 1863

Distribution: Add Honduras (Olancho), Medina-Flores & Townsend (2016); Nicaragua (Masaya), Sunyer et al. (2014).

Tropidodipsas zweifeli Liner & L.D. Wilson, 1970

Distribution: Add Mexico (Puebla), Grünwald et al. (2016a).

TROPIDODRYAS Fitzinger, 1843 (Dipsadidae: Xenodontinae)

Tropidodryas serra (Schlegel, 1837)

Tropidodryas striaticeps (Cope, 1870)

TROPIDOLAEMUS Wagler, 1830 (Viperidae: Crotalinae)

Tropidolaemus huttoni (M.A, Smith, 1949)

Tropidolaemus laticinctus Koch, Gumprecht & Melaun, 2007

Tropidolaemus philippinensis (Gray, 1842)

Distribution: Add Philippines (Dinagat, Leyte, Samar Is.), Leviton et al. (2018).

Comments: Leviton et al. (2018) write that the Zamboanga Peninsula, Mindanao population is morphologically very distinct.

Tropidolaemus subannulatus (Gray, 1842)

Distribution: Add Philippines (Siasi, Tawi-Tawi Is.), Leviton et al. (2018).

Tropidolaemus wagleri (H. Boie *in* F. Boie, 1827)

Distribution: Add West Malaysia (Jerejak Island), Quah et al. (2011); West Malaysia (Kedah), Shahriza et al. (2013), West Malaysia (Terengganu), Sumarli et al. (2015).

TROPIDONOPHIS Jan, 1863 (Natricidae)

Tropidonophis aenigmaticus Malnate & Underwood, 1988

Tropidonophis dahlii (F. Werner, 1899)

Distribution: Add Papua New Guinea (Duke of York Island), Clegg & Jocque (2016).

Tropidonophis dendrophiops (Günther, 1883)

Distribution: Add Philippines (Biliran, Luzon Is.), Leviton et al. (2018); Philippines (Siquijor I.), Beukema (2011).

Tropidonophis dolasii Kraus & Allison, 2004

Tropidonophis doriae (Boulenger, 1898)

Tropidonophis elongatus (Jan, 1865)

Tropidonophis halmahericus (Boettger, 1895)

Distribution: Add Indonesia (Tidore Island), Lang et al. (2013).

Tropidonophis hypomelas (Günther, 1877)

Distribution: Add Papua New Guinea (Duke of York Island), Clegg & Jocque (2016).

Tropidonophis mairii (Gray *in* Grey, 1841)

Tropidonophis mcdowelli Malnate & Underwood, 1988

Tropidonophis montanus (Lidth de Jeude, 1911)

Tropidonophis multiscutellatus (Brongersma, 1948)

Distribution: Add Papua New Guinea (Oro), O'Shea et al. (2018b, suppl.).

Tropidonophis negrosensis (Taylor, 1917)

Distribution: Add Philippines (Siquijor I.), Leviton et al. (2018).

Tropidonophis novaeguineae (Lidth de Jeude, 1911)

Tropidonophis parkeri Malnate & Underwood, 1988

Tropidonophis picturatus (Schlegel, 1837)

Distribution: Add Papua New Guinea (Karkar Island), Clegg & Jocque (2016).

Tropidonophis punctiventris (Boettger, 1895)

Tropidonophis statisticus Malnate & Underwood, 1988

Tropidonophis truncatus (W.C.H. Peters, 1863)

Distribution: Add Indonesia (Morotai, Tidore Islands), Lang (2013).

TROPIDOPHIS Bibron, 1840 *in* Ramón de la Sagra, 1838–1843 (Tropidophiidae)

Tropidophis battersbyi Laurent, 1949

Tropidophis bucculentus (Cope, 1868)

Tropidophis canus (Cope, 1868)

Tropidophis caymanensis Battersby, 1938

Tropidophis celiae Hedges, Estrada & Diaz, 1999

Tropidophis curtus (Garman, 1887)

Distribution: Add Bahamas (Elbow Cay, East Double Headed Shot Cay), Reynolds et al. (2018b).

Comments: Using mtDNA sequence-data, Reynolds et al. (2018b) concluded that Cay Sal Bank specimens were conspecific with those of the Great Bahamas Bank.

Tropidophis feicki Schwartz, 1957

Tropidophis fuscus Hedges & Garrido, 1992

Tropidophis galacelidus Schwartz & Garrido, 1975

Distribution: Low elevation of 152 m, Torres et al. (2016).

Tropidophis grapiuna Curcio, Sales-Nuñes, Suzart-Argôlo, Skuk & Rodrigues, 2012

Tropidophis greenwayi Barbour & Shreve, 1936

Tropidophis haetianus (Cope, 1879)

Distribution: Add Dominican Republic (Valverde), Gabot-Rodríguez & Marte (2019).

Tropidophis hardyi Schwartz & Garrido, 1975

Tropidophis hendersoni Hedges & Garrido, 2002

Comments: Díaz et al. (2014) describe additional specimens (previously known only from the holotype).

Tropidophis jamaicensis Stull, 1928

Tropidophis maculatus Bibron, 1840 *in* Ramón de la Sagra, 1838–1843

Tropidophis melanurus (Schlegel, 1837)

Tropidophis morenoi Hedges, Garrido & Diaz, 2001

Tropidophis nigriventris Bailey, 1937

Tropidophis pardalis (Gundlach, 1840)

Tropidophis parkeri Grant, 1940

Tropidophis paucisquamis (F. Müller *in* Schenkel, 1901)

Disribution: Add Brazil (Parana), Capela et al. (2017). Low elevation of 32 m, Tanaka et al. (2018).

Tropidophis pilsbryi Bailey, 1937

Tropidophis preciosus Curcio, Sales-Nuñes, Suzart-Argôlo, Skuk & Rodrigues, 2012

Tropidophis schwartzi J.P.R. Thomas, 1963

Tropidophis semicinctus (Gundlach & Peters *in* W.C.H. Peters, 1864)

Tropidophis spiritus Hedges & Garrido, 1999

Distribution: Fong-G. & Armas (2011) report a specimen from E Sancti Spiritus Province, Cuba, 140 m elev., and add that it is reported from Villa Clara Prov.

Tropidophis stejnegeri Grant, 1940

Tropidophis stullae Grant, 1940

Tropidophis taczanowskyi (Steindachner, 1880)

Tropidophis wrighti Stull, 1928

Tropidophis xanthogaster Domínguez, Moreno & Hedges, 2006

†*TUSCAHOMAOPHIS* Holman & Case, 1992 (Serpentes: incerta sedis)

†*Tuscahomaophis leggetti* Holman & Case, 1992

TYPHLOPHIS Fitzinger, 1843 (Anomalepididae)

Typhlophis squamosus (Schlegel, 1839 *in* 1837–1844)

TYPHLOPS Oppel, 1811 (Typhlopidae: Typhlopinae)

Synonyms: *Meditoria* Gray, 1845, and possibly *Ophthalmidion* A.M.C. Duméril & Bibron, 1844.

Distribution: Restricted to Western Caribbean Islands, African Gold Coast, and Cameroon.

Comments: Hedges et al. (2014), Pyron & Wallach (2014) provide generic diagnoses.

Typhlops agoralionis J.P.R. Thomas & Hedges, 2007

Typhlops capitulatus Richmond, 1964

Typhlops coecatus Jan, 1863

Distribution: Add Guinea, Trape & Baldé (2014).

Comments: Hedges et al. (2014) assigned *coecatus* and *zenkeri* to *Letheobia*, but Pyron & Wallach (2014) countered that internal and external morphology allied them with *Typhlops* s.l., though they could equivocally belong to *Antillotyphlops*.

Typhlops eperopeus J.P.R. Thomas & Hedges, 2007

Typhlops gonavensis Richmond, 1964

Typhlops hectus J.P.R. Thomas, 1974

Typhlops jamaicensis (Shaw, 1802)

Typhlops leptolepis Domínguez, Fong-G. & Iturriaga 2013. Zootaxa 3681(2): 137–143, figs. 1, 2.

Holotype: CZACC 4.5395, a 183 mm specimen (A. Fong-G., N. Viña-Dávila & N. Viña-Bayés, 5 October 1996).

Type locality: "Altiplanicie del Toldo (20°27′33″N, 74°53′60″ W, 830 m elevation), Moa Municipality, Holguín Province, Cuba."

Distribution: SE Cuba (Holguín), 800–900 m. Add Granma Province, Iturriaga (2015), and Guantanamo, Torres et al. (2019).

Comments: See under *T. lumbricalis.*

Typhlops lumbricalis (Linnaeus, 1758)

Distribution: Add Bahamas (Elbow Cay, Man-O-War Cay), Krysko et al. (2013).

Comments: Hedges et al. (2019) consider *T. leptolepis, T. oxyrhinus* and *T. pachyrhinus* to be synonyms of *T. lumbricalis* based on small sample sizes for each, and on unpublished data.

Typhlops oxyrhinus Domínguez & Diaz, 2011

Comments: See under *T. lumbricalis.*

Typhlops pachyrhinus Domínguez & Diaz, 2011

Comments: See under *T. lumbricalis.*

Typhlops proancylops J.P.R. Thomas & Hedges, 2007

Typhlops pusillus Barbour, 1914

Typhlops rostellatus Stejneger, 1904

Typhlops schwartzi J.P.R. Thomas *in* Woods, 1989

Typhlops silus Legler, 1959

Comment: Domínguez & Diaz (2015) formally revalidate its status as a distinct species, and include a rediscription.

Typhlops sulcatus Cope, 1868

Typhlops sylleptor J.P.R. Thomas & Hedges, 2007

Typhlops syntherus J.P.R. Thomas, 1965

Typhlops tetrathyreus J.P.R. Thomas *in* Woods, 1989

Typhlops titanops J.P.R. Thomas *in* Woods, 1989

Typhlops zenkeri Sternfeld, 1908

Comments: See under *T. coecatus.*

Incertae sedis

Comments: The following Old World species are not assigned to the current scheme of genera.

†*Typhlops cariei* Hoffstetter, 1946

†*Typhlops grivensis* Hoffstetter, 1946

Typhlops longissimus (A.M.C. Duméril & Bibron, 1844)

Comments: Pyron & Wallach (2014) note a similarity between the holotype of *longissimus* and *Indotyphlops tenuicollis*, but continue to consider *longissimus* incerta sedis.

UNGALIOPHIS F. Müller, 1880 (Charinidae: Ungaliophiinae)

Ungaliophis continentalis F. Müller, 1880

Ungaliophis panamensis K.P. Schmidt, 1933

Distribution: Add Nicaragua (Rio San Juan), Sunyer et al. (2014); Panamá (Panamá Oeste), Ray (2015).

URAEUS Wagler, 1830 (Elapidae)

Comments: See under *Naja*.

Uraeus anchietae (Bocage, 1879)

Distribution: Add Angola (Malanje), Ceriaco et al. (2014); Angola (Benguela, Cunene), M.P. Marques et al. (2018).

Uraeus annuliferus (W.C.H. Peters, 1854)

Uraeus arabicus (Scortecci, 1932)

Uraeus haje (Linnaeus, 1758)

Distribution: Add Libya (Al Wahat, Jabal al Gharbi, Marj, Sirte), Bauer et al. (2017); Uganda (Kaabong District), D. Hughes et al. (2017a).

Uraeus niveus (Linnaus, 1758)

Uraeus senegalensis (Trape, Chirio & Wüster *in* Trape et al., 2009)

Distribution: Add Senegal (Kolda), Mané & Trape (2017); Gambia and Guinea Bissau, and upper elevation of 395 m, Trape & Mané (2017); Ivory Coast, Trape & Baldé (2014); Niger (Dosso), Trape & Mané (2015).

UROMACER A.M.C. Duméril, Bibron & Duméril, 1854 (Dipsadidae: Xenodontinae)

Uromacer catesbyi (Schlegel, 1837)

Distribution: Add Dominican Republic (Valverde), Gabot-Rodríguez & Marte (2019).

Uromacer frenata (Günther, 1865)

Uromacer oxyrhynchus A.M.C. Duméril, Bibron & Duméril, 1854

UROPELTIS Cuvier, 1829 (Uropeltidae)

Synonyms: Remove *Crealia* Gray, 1858.

Distribution: Eliminate Sri Lanka from the geographic range, Pyron et al. (2016c).

Comments: Pyron et al. (2016c) confirm the monophyly of the genus through mtDNA and nDNA sequence-data, and provide a morphological diagnosis. Both the Pyron et al. (2016c) and Cyriac & Kodandaramaiah (2017) phylogenies have *Uropeltis* species resolved into two clades.

Uropeltis arcticeps (Günther, 1875)

Synonyms: Remove *Silybura madurensis* Beddome, 1878.

Distribution: S India (SW Tamil Nadu), 1200–1800 m, Pyron et al. (2016c).

Comments: Ganesh et al. (2014b) remove *S. madurensis* from synonymy. They note that the synonym *Silybura nilgherriensis picta* Beddome, 1886 pertains to neither *U. arcticeps* or *U. madurensis*. Pyron et al. (2016c) provide a description.

Uropeltis beddomii (Günther, 1862)

Distribution: Known elevation ~2000 m, Pyron et al. (2016c).

Comments: Pyron et al. (2016c) provide a description, and note that it has not been found since the late 1800s.

Uropeltis bhupathyi Jins, Sampaio & Gower, 2018. Zootaxa 4415(3): 403–410, figs. 2, 5, 6, 8.

Holotype: NCBS AU173, a 366 mm female (V.J. Jins, 25 November 2015).

Type locality: "Environs of the campus of the Salím Ali Centre for Ornithology and Natural History, Anaikatty (sometimes spelled Anaikatti), Coimbatore District, Tamil Nadu, India (11.09° N, 76.79° E, elevation 645 m)."

Distribution: S India (WC Tamil Nadu), 645 m.

Uropeltis bicatenata (Günther, 1864)

Distribution: Upper elevation of 1800 m, Pyron et al. (2016c).

Comments: Pyron et al. (2016c) provide a description and color photograph.

Uropeltis broughami (Beddome, 1878)

Holotype: Pyron et al. (2016c) note that only one specimen, BMNH 1946.1.16.29, was used in the original description.

Comments: Pyron et al. (2016c) provide a description.

Uropeltis ceylanica Cuvier, 1829

Synonyms: Remove *Silybura shorttii* Beddome, 1863.

Distribution: Remove the Eastern Ghats (Shevaroys) of SW Tamil Nadu, Ganesh et al. (2014b).

Comments: Ganesh et al. (2014b) remove *S. shorttii* from its synonymy. Pyron et al. (2016c) provide a description and color photograph.

Uropeltis dindigalensis (Beddome, 1877)

Comments: Pyron et al. (2016c) provide a description and color photograph.

Uropeltis ellioti (Gray, 1858)

Distribution: Add India (Maharashtra), Pyron et al. (2016c) and Sayyed (2016).

Comments: Pyron et al. (2016c) provide a description and color photograph.

Uropeltis liura (Günther, 1875)

Comments: Pyron et al. (2016c) provide a description and color photograph.

Uropeltis macrolepis (W.C.H. Peters, 1861)

Type locality: Restricted to "Mahableshwar, Satara District, Maharashtra state, India" by Pyron et al. (2016c).

Comments: Pyron et al. (2016c) provide a description and color photograph. They note that the supposed diagnostic color pattern of *U. macrolepis mahableshwarensis* Chari, 1955, is the same as that of the holotype of the nominate race.

They synonymize *mahableshwarensis*, and revise the type locality of *U. macrolepis* to that of *mahableshwarensis*.

Uropeltis macrorhyncha (Beddome, 1877)

Comments: Pyron et al. (2016c) provide a description.

Uropeltis maculata (Beddome, 1878)

Comments: Pyron et al. (2016c) provide a description.

Uropeltis madurensis Beddome, 1878. Proc. Zool. Soc. London 46(1): 802. (*Silybura madurensis*)

Syntypes: BMNH 1946.1.16.38–39, (R.H. Beddome).

Type locality: "'High Wavy' mountains, Madura district, elevation 5500 feet," Tamilnadu, S India.

Distribution: Southern India (S Tamil Nadu), endemic to HighWavys-Varushanad-Periyar hill complex, Ganesh et al. (2014b), 1300–1600 m.

Comments: Ganesh et al. (2014b) retrieve *madurensis* from the synonymy of *U. arcticeps* based on morphological differences. Pyron et al. (2016c) provide a description and color photograph.

Uropeltis myhendrae (Beddome, 1886)

Distribution: Revised to S tip of India in S Kerala by Pyron et al. (2016a).

Comments: Pyron et al. (2016c) provide a description.

Uropeltis nitida (Beddome, 1878)

Comments: Pyron et al. (2016c) provide a description.

Uropeltis ocellata (Beddome, 1863)

Comments: Pyron et al. (2016c) provide a description.

Uropeltis petersi (Beddome, 1878)

Comments: Pyron et al. (2016c) provide a description, and note that it has not been collected since the late 1800s.

Uropeltis phipsonii (Mason 1888)

Comments: Pyron et al. (2016c) provide a description and color photograph.

Uropeltis pulneyensis (Beddome, 1863)

Distribution: Pyron et al. (2016c) doubt that the species occurs on Sri Lanka.

Comments: Pyron et al. (2016c) provide a description and color photograph.

Uropeltis rubrolineata (Günther, 1875)

Comments: Pyron et al. (2016c) provide a description.

Uropeltis rubromaculata (Beddome, 1867)

Comments: Pyron et al. (2016c) provide a description.

Uropeltis shorttii (Beddome, 1863). Proc. Zool. Soc. London 31(1): 225. (*Silybura shorttii*)

Syntypes: BMNH 1946.1.15.91–94 and MNHNP 95.100, (J. Shortt).

Type locality: "Shevaroy Hills (4500 feet elevation)," Eastern Ghats, Tamil Nadu, India.

Distribution: Southern India (Shevaroy Hills, Tamil Nadu), 350–1600 m.

Comments: Ganesh et al. (2014b) retrieve *shorttii* from the synonymy of *U. ceylonica* based on morphological differences. Pyron et al. (2016c) provide a description and color photograph.

Uropeltis smithi Gans, 1966

Comments: Pyron et al. (2016c) provide a description.

Uropeltis woodmasoni (Theobald, 1876)

Synonyms: Add *Uropeltis ruhunae* Deraniyagala, 1954.

Comments: Pyron et al. (2016c) designate *U. ruhunae* as a synonym of *U. woodmasoni*, and indicate that the Sri Lankan type locality is in error. Pyron & Somaweera (2019) provide further evidence for the synonymy, as well as photographs of the holotype of *ruhunae*. Pyron et al. (2016c) provide a description and color photograph.

UROTHECA Bibron, 1843 *in* Ramón de la Sagra, 1838–1843 (Dipsadidae: Dipsadinae)

Urotheca decipiens (Günther, 1893 *in* 1885–1902)

Distribution: Add Colombia (Quindio), Quintero-Ángel et al. (2012); Colombia (Risaralda), Arias-Monsalve & Rojas-Morales (2013).

Urotheca dumerilii (Bibron, 1843 *in* Ramón de la Sagra, 1838–1843)

Distribution: Add Colombia (Cauca), Vera-Pérez & Zúñiga-Baos (2015); Colombia (Valle del Cauca: Isla Palma), Giraldo et al. (2014, as *Tantilla* sp.). Elevation range 40–1600 m, Vera-Pérez et al. (2018).

Urotheca fulviceps (Cope, 1886)

Distribution: Add Panama (Coclé), Ray & Santana (2014); Panama (Panama, Panama Oeste), Ray & Ruback (2015); Panama (Veraguas), and upper elevation of 1100 m, Sosa-Bartuano (2017); Colombia (Antioquia), Restrepo et al. (2017); Colombia (Cauca), Vera-Pérez et al. (2018).

Urotheca guentheri (Dunn, 1938)

Distribution: Add Panama (Darien), Elizondo (2016).

Urotheca lateristriga (Berthold, 1859)

Distribution: Add Colombia (Caldas), Rojas-Morales et al. (2018b).

Urotheca multilineata (W.C.H. Peters, 1863)
Urotheca myersi Savage & Lahanas, 1989
Urotheca pachyura (Cope, 1875)

†**VECTOPHIS** Rage & Ford, 1980 (Serpentes: incerta sedis)

†**Vectophis wardi** Rage & Ford, 1980

VERMICELLA Gray *in* Günther, 1858 (Elapidae)

Comments: Derez et al. (2018) present an mtDNA sequence-based phylogeny in which *V. intermedia* and *V. multifasciata* samples are not cladistically differentiated.

Vermicella annulata (Gray *in* Grey, 1841)
Vermicella calonotus (A.M.C. Duméril, Bibron & Duméril, 1854)
Vermicella intermedia Keogh & S.A. Smith, 1996

Vermicella mutlifasciata (Longman, 1915)

Vermicella parscauda Derez, Arbuckle, Ruan, Xie, Huang, Dibben, Shi, Vonk & Fry, 2018. Zootaxa 4446(1): 10–11, fig. 1.

Holotype: QM J95678, a 388 mm male (B.G. Fry & F.J. Vonk, August 2014).

Type locality: "boat ramp Weipa, Cape York, Queensland 12°31′53″ S 141°50′51″E," Australia.

Distribution: NE Australia (N Queensland).

Vermicella snelli Storr, 1968

Vermicella vermiformis Keogh & S.A. Smith, 1996

VIPERA Laurenti, 1768 (Viperidae: Viperinae)

Synonyms: Add *Pelias* Merrem, 1820, *Berus* Swainson, 1839, *Polygerrus* Bonaparte, 1840, *Echidnoides* Mauduyt, 1844, *Acridophaga* A.F.T. Reuss, 1927, *Mesocoronis* A.F.T. Reuss, 1927, *Tzarevscya* A.F.T. Reuss, 1929.

Distribution: NW Africa, Europe and N Asia.

Fossil records: Add lower Miocene (Agenian) of Germany, Čerňanský et al. (2015, as *V. aspix* complex); Lower Miocene of Italy, Venczel & Sanchíz (2006, as *Vipera* sp. 'aspis' group); late Middle Miocene (Astaracian or Vallesian) of Hungary and Romania, Venczel (2011, as *V. aspis* and *V. berus* group); late Middle Miocene (Astaracian) of NE Kazakhstan, Ivanov et al. (2019, as *V. aspis* complex); upper Pliocene (Ruscinian, MN 15) of Italy (Sardinia), Delfino et al. (2011, as *Vipera* sp.); upper Pleistocene (Tarantian) of Bulgaria, Boev (2017).

Comments: Alencar et al. (2016) and Zaher et al. (2019) show that *Pelias* species are a monophyletic group, but within *Vipera*, not as a sister clade as previously determined. Based on these findings, *Pelias* is returned to *Vipera* s.l. S.B. Tuniyev et al. (2013) describe geographic variation in the *V. ursinii* complex in the Caucasus Mountains region. B.S. Tuniyev (2016) summarizes the species of the Caucasus Mountains region. Zinenko et al. (2015) present an mtDNA-based phylogeny, and Mizsei et al. (2017a) review taxonomy, of the *Vipera renardi-ursinii* complex. B.S. Tuniyev et al. (2018a) produce an mtDNA-based phylogeny for taxa in the Lesser Caucasus.

†***Vipera aegertica*** Augé & Rage *in* Ginsberg, 2000

Vipera altaica Tuniyev, Nilson & Andrén, 2010

Comments: Zinenko et al. (2015), using mtDNA sequence data, find *dinniki* to appear in their phylogeny within the Central Asian clade of *V. renardi*.

Vipera ammodytes (Linnaeus, 1758)

Distribution: Add Greece (Cyclades: Rhinia I.), Roussos (2013); Greece (East Macedonia & Thrace), Cattaneo & Cattaneo (2013); Greece (Elafonisos I.), Broggi (2016a); Greece (Kefallinia), M.J. Wilson (2006). Göçmen et al. (2014a, 2015a) describe specimens from new localities in Turkey. Mulder (2017) lists and plots known localities in Turkey from the following provinces: Ankara, Artvin, Bartin, Bolu, Bursa, Cankiri, Erzurum, Istanbul, Karabuk, Kastamonu, Kocaeli, Ordu, Sakarya, Samsun, Tokat, and Zonguldak.

Fossil records: Add upper Pleistocene (Tarantian) of Bulgaria, Boev (2017).

Comment: Akkaya & Uğurtaş (2012) report on morphology of a population from Bursa, northwestern Turkey.

Vipera anatolica Eiselt & Baran, 1970

Synonyms: Add *Vipera anatolica senliki* Göçmen, Mebert, Kariş, Oğuz & Ursenbacher, 2017. Amphibia-Reptilia 38(3): 294–296, figs. 2, 3, 5, 6.

Holotype: ZMADYU 2016/97–2, a 415 mm male (Göçmen, Kariş, Oğuz, Şenlik & Bulut, 23 May 2016).

Type locality: "Serinyaka Plateau, Mühür Dağ, Gündoğmuş District, Antalya Province, Turkey, 1755 m asl. (36°51′N, 32°02′E)."

Distribution: Low elevation of 1559 m, Göçmen et al. (2017). Mebert el al. (2017) report a second population of *P. a. senliki*.

Comments: Göçmen et al. (2014b, 2017) and Zinenko et al. (2016b) present some ecological and morphological data on this poorly known viper.

†***Vipera antiqua*** Szyndlar, 1987

Vipera aspis (Linnaeus, 1758)

Distribution: Add Spain (Soria), Amo (1994).

Fossil records: Add lower Pleistocene (Calabrian) of Spain, Blain (2009, as *Vipera* cf. *aspis*); middle Pleistocene (Ionian) of Italy, Delfino (2004); upper Pleistocene (Tarantian) of Italy, Delfino (2004).

Comments: Ferquel et al. (2007) report a significant proportion of *V. aspis* with neurotoxic venom components among non-neurotoxic individuals. Guiller et al. (2016) report natural hybrids from a contact zone with *Pelias berus*. Tarroso et al. (2014) discover frequent hybridization with *V. latastei* at another contact zone. Masseti & Zuffi (2011) and Masseti & Böhme (2014) conclude that the population on Montecristo Island is the result of 6th century human introduction from Sicily. Zuazo et al. (2019) evaluate morphology of individuals and hybrids at a contact zone with *V. latastei* in Spain.

Vipera barani Böhme & Joger, 1984

Synonyms: Add *Vipera pontica* Billing, Nilson & Sattler, 1990.

Distribution: Revise to northern Turkey (Bilecik, Bolu, Bursa, Giresun, Kütahya, Ordu, Rize, Samsun, Trabzon, Zonguldak), NSL-2000 m, Göçmen et al. (2015b), Gül (2015), Gül et al. (2016b), Kumlutaş et al. (2013). Mebert et al. (2015) detail the easternmost edge of the range.

Comments: Kumlutaş et al. (2013) affirm the synonymy of *Vipera pontica* with *V. barani*.

Vipera berus (Linnaeus, 1758)

Distribution: Add Denmark (Nordfriesland Islands: Fanø, Langli), Grosse et al. (2015); Germany (Nordfriesland Islands: Amrum, Sylt), Grosse et al. (2015); Liechtenstein, Kühnis (2006); Italy (Veneto), Rassati (2012); Slovakia (Prešov), Pančišin & Klembara (2003) and Jablonski (2011); Romania (Alba, Arad, Bihor, Bistrita-Nasaud, Brasov, Cluj, Covasna, Harghita, Hunedoara, Maramures, Mures, Salaj, Satu-Mare,

Sibiu); Romania (Arges, Gorj), Iftime & Iftime (2010) and Iftime et al. (2009); Romania (Buzau, Vrancea), Strugariu et al. (2009); Romania (Suceava), Strugariu et al. (2006); Bulgaria (Sofia Region), Popgeorgiev et al. (2014); Serbia, Tomović et al. (2014). Remove Italy (Piemonte), Ghielmi et al. (2016). Solovyov et al. (2016) describe the distribution in the Vladimir Region, Russia.

Fossil Records: Add middle Pleistocene of Russia, Ratnikov (2002), Ukraine, Ratnikov (2002); upper Pleistocene (Tarantian) of Slovakia, Ivanov & Čerňanský (2017); Late Pleistocene (post-Tarantian) of Belgium, Blain et al. (2019).

Comments: Westerström (2005) discusses difficulty in recognizing the subspecies V. b. bosniensis based on morphology. Using microsatellite markers, Ursenbacher et al. (2015) determine the post-glacial recolonization pattern of V. berus in western Europe. Cui et al. (2016) describe a population in the Altai Mountains, Xinjiang, China, which is genetically within the northern clade of V. berus. Guiller et al. (2016) report natural hybrids from a contact zone with Vipera aspis. See comments under V. nikolskii. Molecular data analyzed by Mizsei et al. (2017a) support the synonymy (as a subspecies) of V. nikolskii with V. berus.

Vipera darevskii Vedmederja, Orlov & Tuniyev, 1986

Synonyms: Add Pelias olguni Tuniyev, Avci, Tuniyev, Agasian & Agasian, 2012.

Pelias darevskii uzumorum Tuniyev, Avcı, Ilgaz, Olgun, Petrova, Bodrov, Geniez & Teynié, 2018a. Proc. Zool. Inst. Russian Acad. Sci. 322(1): 13–16, figs. 11–13.

Holotype: SNP 904, a 440 mm female (S.B. Tuniyev, 11 July 2012).

Type locality: "Turkey, Artvin Province, the Yalnizçam Dağlari Ridge, vicinity of Zekeriya Village, (2000 m above sea level)."

Pelias darevskii kumlutasi Tuniyev, Avcı, Ilgaz, Olgun, Petrova, Bodrov, Geniez & Teynié, 2018a. Proc. Zool. Inst. Russian Acad. Sci. 322(1): 17–20, figs. 16–17.

Holotype: SNP 910, a 465 mm female (B.S. Tuniyev, 12 July 2012).

Type locality: "Turkey, Ardahan Province, the Yalnizçam Dağlari Ridge, vicinity of Bağdaşan Village."

Type locality: S.B. Tuniyev et al. (2014) correct the type locality of V. darevskii to "Mt. Sevsar, Dzhavakhetskiy (Kechutskiy) Ridge, Ashotsk (former Gukasyan) district, northwestern Armenia."

Distribution: Add southern Georgia, and low elevation of 1970 m, Geniez & Teynié (2005). Ardahan, Turkey is added, in part, by the synonymization of Pelias olguni, Göçmen et al. (2014a). Mebert et al. (2015, 2016) discuss margins of the range in Turkey. S.B. Tuniyev et al. (2014) and B.S. Tuniyev (2016) describe the distribution.

Comments: Geniez & Teynié (2005) provide an expanded description. Göçmen et al. (2014a) state that the recognition of P. olguni is premature based on the small sample size used in its description, compounded by "ecophenotypic correlations" and "mixed genotypes" between viperids in the Caucasus

region. The B.S. Tuniyev et al. (2018a) mtDNA-based phylogeny affirms that specimens referred to olguni are undifferentiated from some darevskii populations.

Vipera dinniki Nikolsky, 1913

Distribution: B.S. Tuniyev (2016) summarizes the known distribution. Zinenko et al. (2015), using mtDNA sequence-data, find dinniki to appear in their phylogeny within V. renardi s.l.

Vipera ebneri Knoepffler & Sochurek, 1955

Distribution: Add Iran (Alborz, Ardabil, East Azarbaijan, Tehran, West Azarbaijan, Zanjan), Safaei-Mahroo et al. (2015); Iran (Gilan), Rajabizadeh et al. (2011).

Comments: Validity as a species is supported by mtDNA sequence-data analyzed by Zinenko et al. (2015).

Vipera eriwanensis (A.F.T. Reuss, 1933)

Distribution: Add extr. S Georgia and NE Azerbaijan, with low elevation about 1000 m, Kukushkin et al. (2012); Turkey (Ardahan, Hanak), Mebert et al. (2015); Turkey (Gümüşhane), upper elevation of 2600 m, Kutrup et al. (2005); Iran (Ardabil, East Azarbaijan), Rajabizadeh et al. (2011). B.S. Tuniyev (2016) summarizes the known distribution.

Comments: Validity as a species is supported by mtDNA sequence-data analyzed by Zinenko et al. (2015).

Vipera graeca Nilson & Andrén, 1988. Zool. Scripta 17(3): 311–314, figs. 1–3, 5. (Vipera ursinii graeca)

Holotype: GNM 4942, a 412 mm female (G. Nilson & C. Andrén).

Type locality: "Lakmos mountains in southern Pindos mountain range, Greece…at 1900 m altitude."

Distribution: N Greece. Add S Albania, Korsós et al. (2008) and Mizsei et al. (2016). Mizsei et al. (2016) detail the distribution in S Albania.

Comments: Based on the DNA sequence-data analyzed by Mizsei et al. (2017a), V. graeca is the sister taxon to the other "meadow vipers" (V. renardi-V.ursinii complex), and they elevate it to full species.

Vipera kaznakovi Nikolsky, 1909

Synonyms: Add Vipera magnifica Tuniyev & Ostrovskikh, 2001.

Distribution: Add Georgia (Adjara, Guria, Samegrelo-Zemo Svaneti), Russia (Kraznodar Kray) and Turkey (Rize), Gül et al. (2016a); Georgia (South Ossetia), B.S. Tuniyev (2016). Gül et al. (2016a) map known localities, and document a low elevation of 203 m. Mebert et al. (2015) detail localities at the SW margin of the range in Turkey. B.S. Tuniyev (2016) summarizes the known distribution.

Comments: S.B. Tuniyev et al. (2016) describe additional specimens and morphological variation of populations referred to V. magnifica. Zinenko et al. (2016a) conclude that populations of V. magnifica originated via hybridization between V. renardi and V. kaznakovi, and should be considered a marginal population of the latter.

Vipera latastei Boscá, 1878

Synonyms: Add Vipera latastei montana Saint-Girons, 1953, Vipera latastei monticola Saint-Girons, 1954

Distribution: Add C Morroco. Add Portugal (Guarda), Malkmus & Loureiro (2010); Portugal (Leiria), Malkmus (2005); Portugal (Setúbal), Malkmus (2011); Portugal (Vila real), Malkmus & Loureiro (2012); Morocco (Al Hoceima, Chefchaouen, Taounate, Taza), Mediani et al. (2015b); Algiers (Batna, Bejaia, Bouira, Jijel, Skikda, Tizi Ouzou), Bouam et al. (2019). Ferrer et al. (2018) add records in the eastern part of the range in Catalonia, Spain.

Fossil records: Add upper Pliocene (Villanyian) of Spain, Blain (2009, as *Vipera* cf. *latasti*); lower Pleistocene (Calabrian) of Spain, Blain et al. (2007).

Comments: See under *V. aspis*. X. Santos et al. (2014) discuss geographic variation in dorsal color pattern. I. Freitas et al. (2018) use mtDNA sequence data to detect four clades in North African specimens. Those from central Morocco (*V. monticola*) were nested within N Moroccan, Algerian and Iberian specimens, and the authors synonymized *V. monticola* with *V. latastei*. Zuazo et al. (2019) evaluate morphology of individuals and hybrids at a contact zone with *V. aspis* in Spain. Alonso-Zarazaga (2013) argues that the correct spelling of the specific epithet is *latasti*. Salvador et al. (2014) request that the ICZN conserve the original spelling. The ICZN (2017) has ruled that the correct spelling of the specific epithet is *latastei*.

Vipera lotievi Nilson, Tuniyev, Orlov, Höggren & Andrén, 1995

Distribution: Add Russia (Chechnya, Ingushetia, Kabardino-Balkaria, Karachay-Cherkessia, North Ossetia), B.S. Tuniyev (2016) and S.B. Tuniyev et al. (2011). Iskenderov et al. (2017) confirm the presence in Azerbaijan. B.S. Tuniyev (2016) summarizes the known distribution.

Comments: S.B. Tuniyev et al. (2011) describe geographic variation, noting that two distinct populations are separated by Bogossky Ridge, Dagestan. Zinenko et al. (2015), using mtDNA sequence-data, find *lotievi* populatations to appear in their phylogeny at various points within *V. renardi* s.l.

†*Vipera meotica* Zerova in Szyndlar & Zerova, 1992
†*Vipera natiensis* Bailon, García-Porta & Quintana-Cardona, 2002

Vipera nikolskii Vedmederja, Grubant & Rudayeva, 1986
Distribution: Add E Ukraine, Milto & Zinenko (2005).

Comments: Milto & Zinenko (2005) and Sokolov (2005) evaluate morphology, and conclude that *nikolskii* is a subspecies of *Velias berus*, with which it includes a broad zone of intergradation with *V. b. berus*. Frolova & Gapanov (2016) discuss morphological variation.

Vipera orlovi Tuniyev & Ostrovskikh, 2001

Distribution: B.S. Tuniyev (2016) summarizes the known distribution, and reports a low elevation of 450 m.

Comments: Zinenko et al. (2016a) conclude that this species originated from hybridization between *V. kaznakovi* and *V. renardi*, and that should provisionally remain a recognized taxon pending further studies.

Vipera renardi (Christoph, 1861)

Synonyms: *Vipera renardi puzanovi* Kukushkin, 2009. Current Herpetol. 9(1–2): 21–24, figs. 1, 2.

Holotype: NNRM NANU 24/1, a 595 mm female (O. Kukushkin, 26 July 1997).

Type locality: "Crimea Simferopol District. Perevalnoe, northern slopes of Chatyr Dag ridge, Eagle Gorge (=Tuyu Tract) (~44°48′ N, 34°18′ E)," Russia.

Distribution: Eliminate N Azerbaijan, which are now *V. shemakhensis*, S.B. Tuniyev et al. (2013). Elevation range of NSL-1500 m, B.S. Tuniyev (2016). Mizsei et al. (2018) map the distribution, clarifying the boundary between *V. renardi* and *V. ursinii*. Tupikov & Zinenko (2015a) detail the distribution in Kharkiv, Ukraine.

Fossil records: Add middle Pleistocene (Muchkapian) of Russia, Ratnikov (2002).

Comments: Zinenko et al. (2015) detect four, shallow, monophyletic lineages, using mtDNA data, which correspond to two populations of *V. lotievi*, Central Asian populations of *renardi*, and western populations of *renardi*.

Vipera sachalinensis Tzarevsky, 1917

Vipera sakoi Tuniyev, Avcı, Ilgaz, Olgun, Petrova, Bodrov, Geniez & Teynié, 2018a. Proc. Zool. Inst. Russian Acad. Sci. 322(1): 12–13, figs. 4–7. (*Pelias sakoi*)

Holotype: SNP 911, a 398 mm male (A. Avcı, 10 July 2012).

Type locality: "Turkey, Gumuşhane District, vicinity of Erzincan, Çilhoroz Village (2000 m above sea level)"

Distribution: NC Turkey (Gumushane), 1850–2200 m. Known only from the type locality.

Vipera seoanei Lataste, 1879

Comments: Martínez-Freiría et al. (2015) find very little (shallow) genetic variation, using mtDNA sequence-data. Populations form two clades that evolved less than one mya, which roughly partition into the current concept of two subspecies (*V. s. seoanei* and *V. s. cantabrica*). The recognition of the two subspecies is also supported by morphological analysis by Martínez-Freiría & Brito (2013).

Vipera shemakhensis S.B. Tuniyev, Orlov, B.S. Tuniyev & Kidov, 2013. Russian J. Herpetol. 20(2): 140–141, figs. 1, 2, 11. (*Pelias shemakhensis*)

Holotype: ZISP 21720, a 173 mm male (E.A. Erukh, 25 April 1972).

Type locality: "Vicinity of Shemakhan (= Shemakha, northeastern Azerbaijan)."

Synonyms: *Pelias shemakhensis kakhetiensis* B.S. Tuniyev, Iremashvili, Petrova & Kravchenko, 2018b. Proc. Zool. Inst. Russian Acad. Sci. 322(2): 99–102, figs. 7, 8, 10, 11, 13, 14, 17.

Holotype: SNP 1059, a 350 mm male (N. Seturidze, June 2016).

Type locality: "Dedopliszkaro Village, Fortress Khornabudji, Shirak Plateau (Kakhetia, East Georgia)."

Distribution: E Georgia and N Azerbaijan. B.S. Tuniyev et al. (2018b) provide a map of known localities.

Comments: The specific epithet is sometimes spelled "*sha-makhenensis*" in the original description, and date of collection and collector of the holotype is alternatively given as 14 April 1974, L.A. Erukh.

Vipera ursinii (Bonaparte, 1835 *in* 1832–1841)

Synonyms: Remove *Vipera ursinii graeca* Nilson & Andrén, 1988.

Distribution: Add Hungary, Dely and Janisch (1959), Serbia, Tomović et al. (2014). Ghira (2007a) and Krecsák & Zamfirescu (2008) plot known localities in Romania. Mizsei et al. (2017a) remove populations from Greece and S Albania to *V. graeca*. Mizsei et al. (2018) map the distribution, clarifying the boundary between *V. renardi* and *V. ursinii*.

Comments: Ghira (2007a) reports the rediscovery in Transylvania. Halpern et al. (2007) report low genetic differences (RAPD and microsatellite markers), and no morphological differences between two populations in the Danube Delta of Romania. Zinenko et al. (2015) detect five monophyletic lineages, using mtDNA data, which correspond to four named and one unnamed subspecies.

Vipera walser Ghielmi, Menegon, Marsden, Laddaga & Ursenbacher, 2016. Zool. Syst. Evol. Res. 54(3): 164–168, figs. 5–7.

Holotype: MSNG 34485, a 570 mm female (A. Rosazza, summer 1930).

Type locality: "San Giovanni d'Andorno, strada per Oropa at 1300 m a.s.l. in the Alps north of town of Biella, a subrange of the Pennine Alps, north-western Italy."

Distribution: NW Italy (Piemonte), 1300–2070 m.

Comments: Ghielmi et al. (2006) discussed the distribution (as *Vipera berus*). An mtDNA-based phylogram produced by Ghielmi et al. (2016) places *V. walser* in a clade with other species from the Caucasus region.

VIRGINIA Baird & Girard, 1853 (Natricidae)

Virginia valeriae Baird & Girard, 1853

VIRIDOVIPERA Malhotra & Thorpe, 2004 (Vipera: Crotalinae)

Synonyms: Add *Sinovipera* P. Guo & Wang, 2011.

Comments: See comment under *Trimeresurus*. Alencar et al. (2016), using mtDNA sequence-data, resolve *S. sichuanensis* as the sister taxon to *Viridovipera*, which is herein considered congeneric with the latter.

Viridovipera gumprechti (David, Vogel, Pauwels & Vidal, 2002)

Distribution: Add India (Arunachal Pradesh), Purkayastha (2018); Vietnam (Hoa Bình), Nguyen et al. (2018).

Viridovipera medoensis (Zhao *in* Zhao & Jiang, 1977)

Holotype: Corrected to CIB 013612, Guo et al. (2016a).

Comments: Griffin et al. (2012) provide additional, morphological data.

Viridovipera sichuanensis (P.Guo & Wang, 2011) **new combination**

Comment: See comments above.

Viridovipera stejnegeri (K.P. Schmidt, 1925)

Distribution: Add Myanmar (Tanintharyi), Mulcahy et al. (2018); Vietnam (Bac Giang, Quang Binh), Hecht et al. (2013); Vietnam (Hai Phong: Cat Ba Island), T.Q. Nguyen et al. (2011).

Viridovipera truongsonensis (Orlov, Ryabov, Bui & Ho, 2004)

Viridovipera vogeli (David, Vidal & Pauwels, 2001)

Distribution: Vietnam (Quang Ngai), Nemes et al. (2013).

Viridovipera yunnanensis (K.P. Schmidt, 1925)

†***WAINCOPHIS*** Albino, 1987 (Booidea: incerta sedis)

†***Waincophis australis*** Albino, 1987
†***Waincophis cameratus*** Rage, 2001
†***Waincophis pressulus*** Rage, 2001

WALLACEOPHIS Mirza, Vyas, Patel, Maheta & Sanap, 2016. PLoS One 11(0148380): 3–4. (Colubridae: Colubrinae)

Type species: *Wallaceophis gujaratensis* Mirza, Vyas, Patel, Maheta & Sanep, 2016 by original designation.

Distribution: W India.

Comments: Mirza et al. (2016) produce a DNA sequence-based phylogeny for numerous Asian colubrines, which places *Wallaceophis* and *Lytorhynchus* as a sister clade to the remaining taxa.

Wallaceophis gujaratensis Mirza, Vyas, Patel, Maheta & Sanep, 2016. PLoS One 11(0148380): 4–8, figs. 1–6.

Holotype, NCBS HA-105, a 501 mm male (J. Maheta, 24 July 2014).

Type locality: "Khengariya village, Viramgam taluka, Ahmedabad district, Gujarat state, India (23.0217946 N, 72.0217584 E, altitude 21 m)."

Distribution: W India (Gujarat), 21 m.

WALLOPHIS F. Werner, 1929. Zool. Jahrb. Abt. Syst. 57(1/2): 126. (Colubridae: Colubrinae)

Type species: *Zamenis brachyurus* Günther 1866, designated by Williams & Wallach, 1989).

Distribution: SW Asia (India).

Comments: Mirza & Patel (2018) resurrect the genus *Wallophis* for *Coronella brachyura* based on DNA sequence-data. It is the sister taxon to *Wallaceophis*. They provide a revised morphological diagnosis for the genus.

Wallophis brachyurus (Günther, 1866)

Distribution: Patel et al. (2015) map known localities, and give an elevation range of 13–715 m.

WALTERINNESIA Lataste, 1887 (Elapidae)

Walterinnesia aegyptia Lataste, 1887

Distribution: Add Saudi Arabia (Ha'il), Alshammari et al. (2017; the authors state that their specimen is *W. aegyptia* rather than the more proximal *W. morgani*).

Walterinnesia morgani (Mocquard, 1905)

Distribution: Add Syria (Homs), Sindaco et al. (2014); Turkey (Kilis), Göçmen et al. (2009a); Iran (Bushehr, Hormozgan, Ilam), Shafaei-Mahroo et al. (2015).

†*WONAMBI* M.J. Smith, 1976 (†Madtsoiidae)
Comments: Palci et al. (2014) provide a revised diagnosis for the genus.
†*Wonambi barriei* Scanlon *in* Scanlon & Lee, 2000
†*Wonambi naracoortensis* M.J. Smith, 1976
Comments: Palci et al. (2014) describe pelvic elements for the species and provide a revised diagnosis.

†*WOUTERSOPHIS* Rage, 1980 (†Nigerophiidae)
†*Woutersophis novus* Rage, 1980

XENELAPHIS Günther, 1864 (Colubridae: Colubrinae)
Xenelaphis ellipsifer Boulenger, 1900
Xenelaphis hexagonotus (Cantor, 1847)

XENOCALAMUS Günther, 1868 (Atractaspididae: Aparallactinae)
Comments: Figueroa et al. (2016), based on DNA sequence-data, synonymize *Xenocalamus* with *Amblyodipsas* due to paraphyly. Portillo et al. (2018) obtained similar results, but recommended against synonymizing the two genera until additional species could be studied.
Xenocalamus bicolor Günther, 1868
Distribution: Add Angola (Moxico), Portillo et al. (2018).
Xenocalamus lineatus Roux, 1907
Xenocalamus mechowii W.C.H. Peters, 1881
Distribution: Add Angola (Moxico), Portillo et al. (2018).
Xenocalamus michelli L. Müller, 1911
Xenocalamus sabiensis Broadley, 1971
Xenocalamus transvaalensis Methuen, 1919

XENOCHROPHIS Günther, 1864 (Natricidae)
Synonyms: Remove *Fowlea* Theobald, 1868, *Diplophallus* Cope, 1893.
Comments: Purkayastha et al. (2018) and Takeuchi et al. (2018) each produce a DNA sequence-based phylogeny using numerous Asian natricid species, with a focus on *Xenochrophis* and *Rhabdophis*, respectively. The Purkayastha et al. analysis reveals that *X. cerasogaster* is sister taxon to a group that includes *Atretium schistosum* plus *X. piscator* and related species. The Takeuchi et al. analysis did not use *X. cerasogaster*, but found *Atretium yunnanensis* to be sister taxon to *X. asperrimus*. Purkayastha et al. recommend resurrecting *Fowlea* Theobald, 1868 for the *piscator* group, although this arrangement leaves *Atretium* paraphyletic. This issue could be resolved by including *yunnanensis* within *Fowlea*. The remaining species (*X. bellulus* not studied) form a sister clade to *Rhabdophis* in the Takeuchi et al. analysis, along with *Ceratophallus vittatus* and *Rhabdophis conspicillatus*. They leave the status of this clade to be determined.
Xenochrophis bellulus (Stoliczka, 1871)

Neotype: USNM 587200, a 592 mm female (S.W. Kyi & T. Win, July 2003).
Neotype locality: "Moyingyi Wetland Wildlife Sanctuary, Bago Region, Myanmar (17°35′27.4″ N, 96°34′24.5″ N [sic]; 3 m. in elevation)."
Distribution: S Myanmar (Bago, Yangon), 3–29 m.
Comments: J.L. Lee et al. (2018b) present a revised description, color photographs and locality map for the species, previously known only from the lost holotype.
Xenochrophis cerasogaster (Cantor, 1839)
Distribution: Add India (Arunachal Pradesh), Purkayastha (2018); India (Bihar, Meghalaya), Purkayastha et al. (2013); Bangladesh (Dhaka), Ahmad et al. (2015).
Xenochrophis maculatus (Edeling, 1864)
Xenochrophis trianguligerus (F. Boie, 1827)
Distribution: West Malaysia (Pahang), Zakaria et al. (2014); West Malaysia (Penang), Quah et al. (2013); Indonesia (Pulau Bangkuru off Sumatra), Tapley & Muurmans (2016).

XENODERMUS J.C.H. Reinhardt, 1836 (Xenodermidae)
Xenodermus javanicus J.C.H. Reinhardt, 1836

XENODON H. Boie *in* Fitzinger, 1826 (Dipsadidae: Xenodontinae)
Synonyms: Add *Rhinostoma* Fitzinger, 1826, *Rhinosiphon* Fitzinger, 1843, *Lystrophis* Cope, 1885.
Comments: The gene-based phylogeny of Zaher et al. (2019) has the species formerly placed in *Lystrophis* as a monophyletic sister clade to the species of *Xenodon* s.s.
Xenodon angustirostris W.C.H. Peters, 1864. Monatsb. Preuss. Akad. Wiss. Berlin 1864: 390.
Synonyms: Add *Xenodon mexicanus* H.M. Smith, 1940.
Holotype: ZMB 2334 (Warszewicz).
Type locality: "Veragua," Panama.
Distribution: Central America and W Colombia. Add Nicaragua (Boaco), Sunyer et al. (2014); Nicaragua (Chontales), Gutiérrez-Rodríguez & Sunyer (2017b); Panama (Panama, Panama Oeste), Ray & Ruback (2015); Colombia (Antioquia), Restrepo et al. (2017).
Comments: C.W. Myers *in* C.W. Myers & McDowell (2014) tentatively resurrects *angustirostris* for populations in Central America and W Colombia based on hemipeneal morphology. He cautions against wholesale assignment of populations pending study of the hemipenes.
Xenodon dorbignyi (Bibron *in* A.M.C. Duméril, Bibron & Duméril, 1854)
Distribution: Add Brazil (Marinheiros I., Rio Grande do Sul), Quintela et al. (2011).
Xenodon guentheri Boulenger, 1894
Xenodon histricus (Jan, 1863)
Distribution: Add Argentina (San Luis), D'Angelo et al. (2015). Alves et al. (2015) document the rediscovery in Rio Grande do Sul, Brazil.

Xenodon matogrossensis (Scrocchi & Cruz, 1993)
Comments: See below under *X. pulcher*

Xenodon merremii (Wagler *in* Spix, 1824)
Distribution: Add Brazil (Maranhão), J.P. Miranda et al. (2012); Brazil (Mato Grosso do Sul), Ferreira et al. (2017); Brazil (Minas Gerais), Sousa et al. (2010) and Moura et al. (2012); Brazil (Piauí), Dal Vechio et al. (2013); Brazil (Alagoas, Rio Grande do Norte, Sergipe), and upper elevation of 1200 m, Guedes et al. (2014); Brazil (Rio de Janeiro: Ilha Grande), C.F.D. Rocha et al. (2018); Bolivia (Potosi), Rivas et al. (2018).

Xenodon nattereri (Steindachner, 1867)

Xenodon neuwiedii Günther, 1863
Distribution: Add Brazil (Minas Gerais), Moura et al. (2012). Upper elevation of 900 m, Vaz & Ortega-Chinchilla (2019).

Xenodon pulcher (Jan, 1863)
Distribution: Add Brazil (Mato Grosso do Sul), Cabral et al. (2015).

Xenodon rabdocephalus (Wied-Neuwied, 1824)
Synonyms: Remove *Xenodon angustirostris* W.C.H. Peters, 1864, *Xenodon suspectus* Cope, 1868, *Xenodon mexicanus* H.M. Smith, 1940.
Lectotype: The "holotype," AMNH 3609, listed in Wallach et al. (2014) is not destroyed, but remains extant, and is figured by C.W. Myers *in* C.W. Myers & McDowell (2014), who designates it as lectotype.
Distribution: Add Brazil (Acre), Bernarde et al. (2013); Brazil (Espirito Santo), Silva-Soares et al. (2011); Brazil (Mato Grosso do Sul), Ferreira et al. (2017), Brazil (Roraima), Moraes et al. (2017).
Comments: C.W. Myers *in* C.W. Myers & McDowell (2014) discusses cryptic species within the *X. rabdocephalus* complex, and recommends some taxonomic changes based on morphology.

Xenodon semicinctus (Bibron *in* A.M.C. Duméril, Bibron & Duméril, 1854)

Xenodon severus (Linnaeus, 1758)
Distribution: Add Peru (Huanuco), Schlüter et al. (2004); Peru (Puno), Llanqui et al. (2019); Brazil (Acre), Bernarde et al. (2011b); Brazil (Rondônia), Bernarde et al. (2012b).

Xenodon suspectus Cope, 1868. Proc. Acad. Nat. Sci. Philadelphia 20(2): 133–134.
Holotype: MCZ 3649, a 569 mm female (Thayer Expedition, 1865–1866).
Type locality: "Lake Jose Assu," stated as Lake José Issu, Amazonas, Brazil in the MCZ catalog. C.W. Myers *in* C.W. Myers & McDowell (2014) considers the type locality "to be unknown, but probably in eastern Peru."
Distribution: E Peru (Junin, Loreto, San Martin).
Comments: Resurrected from the synonymy of *X. rabdocephalus*, based on morphological differences, by C.W. Myers *in* C.W. Myers & McDowell (2014).

Xenodon werneri Eiselt, 1963
Distribution: Add Brazil (Pará), Avila-Pires et al. (2010).

XENOPELTIS F. Boie, 1827 (Xenopeltidae)
Xenopeltis hainanensis Hu & Zhao *in* Zhao, 1972
Holotype: Corrected to CIB 008073, Guo et al. (2012a).

Xenopeltis unicolor F. Boie, 1827
Distribution: Add Thailand (Chanthaburi), Chan-ard and Makchai (2011); Thailand (Krabi: Phi Phi Don I.), Milto (2014); West Malaysia (Kedah), Shahriza et al. (2013); West Malaysia (Penang), Quah et al. (2013); Cambodia (Siem Reap), Geissler et al. (2019).

XENOPHIDION R. Günther & Manthey, 1995 (Xenophidiidae)
Distribution and Comments: Quah et al. (2018b) provide a photograph of a specimen of an unknown species of *Xenophidion* from W Sumatra, Indonesia.

Xenophidion acanthognathus R. Günther & Manthey, 1995
Distribution: Add Malaysia (Sarawak), Rowntree et al. (2017).

Xenophidion schaeferi R. Günther & Manthey, 1995
Distribution: Add Malaysia (Negeri Sembilan), and elevation range of 87–800 m, Quah et al. (2018b).
Comments: Quah et al. (2018b) describe three additional specimens of this species, previously only known from the holotype.

XENOPHOLIS W.C.H. Peters, 1869 (Dipsadidae: Xenodontinae)
Xenopholis scalaris (Wucherer, 1861)
Distribution: Add Peru (Huanuco), Schlüter et al. (2004); Brazil (Acre), Bernarde et al. (2011b); Brazil (Alagoas), R.C. França et al. (2019); Brazil (Amazonas, Mato Grosso, Rio de Janeiro, Sao Paulo), Hamdan et al. (2015); Brazil (Pernambuco), Roberto et al. (2018). R.C. França et al. (2019) map known localities.

Xenopholis undulatus (Jensen, 1900)
Distribution: Add Brazil (Alagoas, Paraíba), and low elevation of 20 m, Guedes et al. (2014).

Xenopholis werdingorum Jansen, Álvarez & Köhler, 2009
Distribution: Add Brazil (Mato Grosso do Sul), Ferreira et al. (2017); Bolivia (Beni), R.L. Powell et al. (2016).

XENOTYPHLOPS Wallach & Ineich, 1996 (Xenotyphlopidae)
Comments: Pyron & Wallach (2014) provide a generic diagnosis.

Xenotyphlops grandidieri (Mocquard, 1905)
Comments: Wegener et al. (2013) confirm that, due to infraspecific variation, *X. mocquardi* Wallach, Mercurio & Andreone, 2007, is a synonym of *X. grandidieri*. They also report on intrapopulation genetic structure.

XEROTYPHLOPS Hedges, Marion, Lipp, Marin & Vidal, 2014. Caribbean Herpetol. 49: 41. (Typhlopidae: Asiatyphlopinae)
Type species: *Typhlops vermicularis* Merrem, 1820 by original designation.

Distribution: SE Europe and SW Asia through the Middle East, S Mauritania, and Socotra Island.

Comments: Hedges et al. (2014), Pyron & Wallach (2014) provide generic diagnoses. Kornilios et al. (2013a) produce an nDNA phylogeny that confirms the sister-relationship of *X. socotranus* and *X. vermicularis*.

Xerotyphlops etheridgei (Wallach, 2002)

Xerotyphlops luristanicus Torki, 2017b. Herpetol. Bull. 140: 1–4, fig. 2.

Holotype: MNHN R2016.0040, a 216 mm male (F. Torki, 29 March 2016).

Type locality: "1750–2100 m a.s.l., on the western slope of the central Zagros Mountains, Badavar region, Nourabad, Lorestan Province, western Iran, 34°07′ N, 47°53′ E."

Distribution: C Iran (Lorestan), 1750–2100 m.

Xerotyphlops socotranus (Boulenger, 1889)

Xerotyphlops vermicularis (Merrem, 1820)

Distribution: Add Greece (Kastellorizo I.), Paysant (1999) and Kakaentzis et al. (2018); Greece (Symi Island), Strachinis & M. Wilson (2014); Iran (Alborz, Chaharmahal and Bakhtiari, Isfahan, Kohgiluyeh and Boyer Ahmad, North Khorasan, Semnan, Yazd), Safaei-Mahroo et al. (2015); Iran (Gilan, Kurdistan), N. Rastegar-Pouyani et al. (2010); Iran (Qom), S.M. Kazemi et al. (2015); Iraq (Al-Sulaimaniyah, Arbil, Dahuk, Kirkuk, Nineveh), Afrasiab & Ali (1996); Iraq (Dhi Qar, Salah al-Din), Habeeb & Rastegar-Pouyani (2016a); Turkey (Afyon, Amasya, Ankara, Balikesir, Eskesihir, Istanbul, Kars, Konya, Samsun, Sinop, Tekirdag, Trabzon), Afsar et al. (2016); Turkey (Burdur), Ege et al. (2015); Turkey (Bursa), Uğurtaş et al. (2007); Turkey (Hatay), Uğurtaş et al. (2000b); Turkey (Kutahya), Özdemĭr & Baran (2002); Afghanistan (Badakhshan, Bamyan, Kunduz), Wagner et al. (2016b). Pulev et al. (2018a) detail the distribution in Bulgaria. N. Rastegar-Pouyani et al. (2010) detail the distribution in Iran.

Comments: Afroosheh et al. (2013) found no geographic variation in morphology between populations in Iran and vicinity. Gül et al. (2015) map the distribution, and discuss isolating factors, of evolutionary lineages in Turkey. Jablonski & Balej (2015) document the rediscovered population in the west Bulgarian Rhodope Mts. Kornilios (2017) re-analyzes mtDNA sequence data, and the resulting phylogeny supports an ancient split between northern and Palestinian populations. Northern populations are represented by three clades. From these results, Kornilios makes no taxonomic conclusions.

Xerotyphlops wilsoni (Wall, 1908)

†*XIAOPHIS* Xing, Caldwell, Chen, Nydam, Palci, Simões, McKellar, Lee, Liu, Shi, Wang & Bai, 2018. Science Advances 2018(4)(eaat5042): 1. (Serpentes: incerta sedis)

Type species: †*Xiaophis myanmarensis* Xing, Caldwell, Chen, Nydam, Palci, Simões, McKellar, Lee, Liu, Shi, Wang & Bai, 2018 by monotypy.

Distribution: Mid-Cretaceous of Myanmar.

†*Xiaophis myanmarensis* Xing, Caldwell, Chen, Nydam, Palci, Simões, McKellar, Lee, Liu, Shi, Wang & Bai, 2018. Science Advances 2018(4)(eaat5042): 1–3, figs. 1–3.

Holotype: DIP-S-0907, a neonate axial skeleton preserved in amber.

Type locality: "Angbamo site, Tanai Township, Myitkyina District, Kachin Province, Myanmar."

Distribution: Mid Cretaceous (early Cenomanian) of Myanmar.

XYELODONTOPHIS Broadley & Wallach, 2002 (Colubridae: Colubrinae)

Xyelodontophis uluguruensis Broadley & Wallach, 2002

Distribution: Low elevation of 1429 m, Lyakurwa (2017).

XYLOPHIS Beddome, 1878 (Pareidae: Xylophiinae)

Comments: Deepak et al. (2018) evaluate the phylogenetic relationship of *Xylophis*, recovering the three species as sister taxa to the Pareatidae, thus removing them to their own subfamily within the pareatids.

Xylophis captaini Gower & Winkler, 2007
Xylophis perrotetii (A.M.C. Duméril, 1853)
Xylophis stenorhynchus (Günther, 1875)

†*YURLUNGGUR* Scanlon, 1992 (†Madtsoiidae)

†*Yurlungger camfieldensis* Scanlon, 1992

ZAMENIS Wagler, 1830 (Colubridae: Colubrinae)

Comments: Figueroa et al. (2016), based on DNA sequence-data, synonymize *Rhinechis* with *Zamenis* due to paraphyly, but X. Chen et al. (2017) revive it from the synonymy with *Zamenis*, though it remains the sister taxon to the latter. Salvi et al. (2018) obtain a similar result, with *scalaris* as sister taxon to *Zamenis*, but recommend retaining it as *Zamenis* based on morphological similarity, and the interest of avoiding monotypic genera when an alternative, clade-based arrangement is available.

†*Zamenis algorensis* (Szyndlar, 1985)

Zamenis hohenackeri (Strauch, 1873)

Synonyms: *Zamenis hohenackeri lyciensis* Hofmann, Mebert, Schulz, Helfenberger, Göçmen & Böhme, 2018. Zootaxa 4471(1): 144–145, figs. 3–7.

Holotype: MHNG 2403.007, a 602 mm male (W. Zinniker, 1987).

Type locality: "Kohu Dağ, Elmalı, Antalya, Turkey (N36.502690, E29.819595; 1700 m a.s.l.)."

Distribution: Add Turkey (Afyon, Artvin, Aydin, Bitlis, N. Cyprus, Erzincan, Hakkari, Hatay, Igci, Igdir, Izmir, Kahramanmaras, Karaman, Kars, Kastomonu, Malatya, Mersin, Mugla, Osmaniye, Sanliurfa, Sivas, Tokat, Tunceli, Van), Hofmann et al. (2018); Iran (Chaharmahal and Bakhtiari), Shafaei-Mahroo et al. (2015).

Comments: Jandzik et al. (2013) use mtDNA sequence-data to detect three lineages: 1) the nominate race of Transcaucasia, 2)

Z. h. *tauricus* of southern Turkey, and 3) a possible unnamed subspecies distributed from extreme SE Turkey to Lebanon. Hofmann et al. (2018) use many more samples, including nDNA, which also detects three lineages. However, the Z. h. *tauricus* and Levant samples are congruent. Their third lineage, in SW Turkey, is named as a new subspecies that is also morphologically diagnosable.

†*Zamenis kohfidischi* (Bachmayer & Szyndlar, 1985)

†*Zamenis kormosi* (Bolkay, 1913)

Zamenis lineatus (Camerano, 1981)

Comments: Salvi et al. (2017) use DNA sequence-data to determine that Z. *lineatus* has a smaller geographic range than has been assumed based on morphology. They report extensive hybridization with Z. *longissimus*.

Zamenis longissimus (Laurenti, 1768)

Synonyms: Add *Coluber lobsingensis* Heller, 1960.

Distribution: Add Andorra, Orriols & Fernàndez (2003); Spain (Alava), Ana (2014); Spain (Asturias), Meijide (1985); Italy (Elba I.), Vaccaro & Turrisi (2007); Italy (Latium), Corsetti & Romano (2008); Italy (Rome), Esposito & Romano (2011); Czech Republic (Jihocesky), Jablonski et al. (2011, distribution map); Slovakia (Prešov), Jablonski (2011); Greece (Paxos I.), M.J. Wilson & Stille (2014); Turkey (Artvin), Eksilmez et al. (2017); Turkey (Bartin), Çakmak et al. (2017); Turkey (Karabük), Kumlutaş et al. (2017); Iran (Golestan, Mazandaran, Zanjan), Shafaei-Mahroo et al. (2015).

Fossil records: Add lower Pliocene (Ruscinian) of Slovakia, Čerňanský (2011); middle Pleistocene of Czech Republic, Musilova et al. (2007); upper Pleistocene (Tarantian) of Belgium, Blain et al. (2014b). Musilova et al. (2007) map and list known recent and fossil localities outside of the current distribution.

Comments: G. Böhme (1997) considers *Zamenis lobsingensis* to be a synonym of Z. *longissimus*. Joger et al. (2010) construct a gene tree that indicates several lineages. Dotsenko et al. (2013) add numerous records for SW Ukraine. Strödicke & Gerish (1999) document geographic variation in isolated, northern populations. See under Z. *lineatus*.

†*Zamenis paralongissimus* (Szyndlar, 1984)

Zamenis persicus (F. Werner, 1933)

Distribution: Add Iran (Ardabil, East Azerbaijan, Golestan, West Azerbaijan), Shafaei-Mahroo et al. (2015). Kidov et al. (2009) provide records from the Talish Mountains, Astara district, Azerbaijan.

†*Zamenis praelongissima* (Venczel, 1994)

Zamenis scalaris (Schinz *in* Cuvier, 1822)

Distribution: Add Spain (Baleares: Formentera, Ibiza, and Mallorca Is.), Mateo (2015).

Fossil Records: Add Upper Pliocene (Villanyian) of Spain, Blain (2009); Upper Pleistocene (Tarantian) of Spain, Barroso-Ruiz and Bailon (2003).

Comments: Nulchis et al. (2008) report no variation in mtDNA across the species' range. Silva-Rocha et al. (2015) use DNA sequence-data to determine the source population for specimens introduced to the Balearic Islands, but were unable to restrict to the source beyond the Iberian Peninsula.

Zamenis situla (Linnaeus, 1758)

Distribution: Add Greece (Chalki I.), Grano & Cattaneo (2015); Greece (Kimolos I.), Broggi (2014b); Greece (Kithira I.), Broggi (2016b); Greece (Kos I.), Hofmann (2018); Turkey (Burdur), Ege et al. (2015); Turkey (Bursa, Canakkale, Manisa, Usak, Yalova), Hofmann (2018); Turkey (Mugla), Winden et al. (1997); Turkey (Gökçeada I.), Yakin et al. (2018).

Comments: Kyriazi et al. (2012) use mtDNA sequence data to determine that populations on Crete originated from southern continental Greece.

†*Zamenis szyndlari* (Venczel, 1998)

†**ZELCEOPHIS** Szyndlar, 1984 (Colubridae: Colubrinae)

†*Zelceophis xenos* Szyndlar, 1984

†**ZILANTOPHIS** Jasinski & Moscato, 2017. J. Herpetol. 51(2), 251. (Colubridae: Colubrinae)

Type species: *Zilantophis schuberti* Jasinski & Moscato, 2017, by original designation.

Distribution: Upper Miocene of USA (Tennessee).

Comments: *Zilantophis* is morphologically similar to Lampropeltini according to Jasinski & Moscato (2017).

†*Zilantophis schuberti* Jasinski & Moscato, 2017. J. Herpetol. 51(2), 251–255, fig. 4.

Type: ETMNH-9557, posterior trunk vertebra.

Type Locality: "Gray Fossil Site at East Tennessee State University and General Shale Natural History Museum, 2.9 km west of Gray, Washington County, northeastern Tennessee," USA

Distribution: Upper Miocene (or earliest Pliocene; Hemphillian) of USA (Tennessee).

Geographical References

WORLD: Shupe (2013, venomous snakes).

AFRICA
North Africa: Sindaco et al. (2013).
North and West Africa: J. Brito et al. (2008, records).
ALGERIA: Mamou et al. (2014, Kabylie region), Mouane et al. (2014, Souf region).
ANGOLA: Branch (2018), Butler et al. (2019, Huila), Ceríaco et al. (2014a, Malanje records; 2016, Namibe), Conradie et al. (2016b, southeast); M.P. Marques et al. (2018).
BENIN: Trape & Baldé (2014, list).
CAMEROON: Gonwouo et al. (2007, Mt. Cameroon area); Ineich et al. (2015, Bamenda Highlands).
DEMOCRATIC REPUBLIC OF THE CONGO: Jackson & Blackburn (2007, Nouabale-Ndoki National Park).
EGYPT: Ibrahim (2013, Suez Canal region); Milto (2017, records).
ETHIOPIA: Largen & Spawls (2011, Bale Mountains).
GABON: Carlino & Pauwels (2015, Ivindo N. P.).
GHANA: Trape & Baldé (2014, list).
GUINEA: Trape & Baldé (2014).
INDIAN OCEAN ISLANDS: S. Rocha et al. (2009, Seychelles).
IVORY COAST: Trape & Baldé (2014, list).
LIBERIA: Trape & Baldé (2014, list).
LIBYA: Bauer et al. (2017), Essghaier et al. (2015, Fezzan district); Frynta et al. (2000, records).
MADAGASCAR: Roberts & Daly (2014, Nosy Komba).
MALI: Trape & Mané (2017).
MAURITANIA: Padial (2006), Sow et al. (2014, P.N. du Banc d'Arguin).
MOROCCO: Barata et al. (2011, records), Damas-Moreira et al. (2014, records), Funk et al. (2007, 2009, records), Mediani et al. (2015b, northern).
MOZAMBIQUE: Conradie et al. (2016a, montane isolates); Jacobsen et al. (2010, San Sebastian Peninsula); Portik et al. (2013, southern Afromontane Archipelago).
NAMIBIA: Herrmann & Branch (2013).
NIGER: Ineich et al. (2014, Termit Massif); Trape & Mané (2015).
NIGERIA: Nneji et al. (2019, Gashaka Gumti N.P.), Trape & Baldé (2014, list).
RWANDA: Roelke & E.N. Smith (2010, Parc National des Volcans).
SENEGAL: Monasterio et al. (2016, Dindefelo Reserve).
SIERRA LEON: Trape & Baldé (2014, list).
SOUTH AFRICA: Jacobsen & Randall (2013, Garden Route National Park).
TANZANIA: Lyakurwa et al. (2019: Uzungwa Scarp Nature Forest Reserve), Menegon et al. (2008, Nguru Mtns.).
TOGO: Trape & Baldé (2014, list).
ZAMBIA: Pietersen et al. (2017, Ngonye Falls region).

EUROPE: Sindaco et al. (2013).
ALBANIA: Mizsei et al. (2017b).
ANDORRA: F.A. Orriols & Fernàndez (2003).
BELGIUM: Graitson (2013, Wallonie).
BULGARIA: Boev (2017, fossils), Malakova et al. (2018, Oranovski Prolom-Leshko), Petrov (2000, eastern Rhodopes), Popgeorgiev et al. (2014, Ponor SPA); Pulev & Sakelarieva (2011, Blagoevgrad).
CROATIA: Glavaš et al. (2016), Jelić (2014), Tóth et al. (2017, Cres-Lošinj Arch.).
DENMARK: Grosse et al. (2015, Nordfriesland Islands).
FRANCE: Beltra (2013, Provence-Alpes-Côte d'Azur); De Massary et al. (2019), Poitou-Charentes Nature (2002, Poitou-Charentes).
GERMANY: Buschmann et al. (2006, Schaumburg Land), Grosse et al. (2015, Nordfriesland Islands).
GREECE: Broggi (2014a, Gavdos I.), Broggi (2014b, Kimolos I.), Broggi (2016a, Elafonisis I.), Broggi (2016b, Kythera I.), Kalaentzis et al. (2018, Kastellorizo I.), Speybroek et al. (2014, Samos I.), Stille & Stille (2016, Diapontia Islets), Strachinis & Roussos (2016, Limnos and Agios Efstratios Is.), M.J. Wilson (2006, Kefallinia and Zakinthos Is.); M.J. Wilson & Stille (2014, Paxos).
ITALY: Ambrogio & Mezzadri (2003, Piacenza), Bassu et al. (2013, Sardinia), Cantini et al. (2013, S Tuscany), Di Tizio et al. (2010, Prov. Chieti), Fattizzo & Marzano (2002, Salentina Peninsula), Liuzzi & Scillitani (2010, Apulia), Pous et al. (2012, Sardinia), Rassati (2012, northeast), Salvi & Bombi (2010, Sardinia), Scalera et al. (2006, Majella National Park), Schiavo & Ferri (1996, Cremona), Seglie & Sindaco (2013, Piemonte, Valle d'Aosta); Turrisi (1996, Monti Iblei region), Turrisi & Vaccaro (1998, Sicily), Turrisi et al. (2008, Sicily).
LIECHTENSTEIN: Kühnis (2006).
MACEDONIA: Iković et al. (2016, central); Sterijovski et al. (2014), Uhrin et al. (2016).
MONTENEGRO: Polović & Čađenović (2013).
PORTUGAL: Malkmus (2005, parts of Leiria, Santarém; 2011, coast), Malkmus & Loureiro (2010, Guarda; 2012, Vila Real).
ROMANIA: Bogdan et al. (2011, Poiana Ruscă Mountains), Cogălniceanu et al. (2013), Covaciu-Marcov et al. (2005, Tăşnad Hills), Covaciu-Marcov et al. (2006a, 2007a, Arad), Covaciu-Marcov et al. (2006b, Dobrogea), Covaciu-Marcov et al. (2007b, Maramureş), Covaciu-Marcov et al. (2009a, Mehedinti), Covaciu-Marcov et al. (2009b, Jiului N.P.), Ghergel et al. (2007, Neamţ), Ghira (2007b, Sighişoara area); Iftime & Iftime (2008, Giurgiu; 2010, Jiu-Lotru basins; 2013, Ţarcu Massif; 2014b, Nordul Gorjului de Est; 2014a, Leaota Mountains; 2016, Teleorman County), Iftime et al. (2009, Iezer-Păpuşa Massif), Lazăr et al. (2005, Dolj), Sos (2007, Brasov), Strugariu et al. (2006, Suceava; 2008, Tulcea), Venczel (2000, Quaternary fossils).
SERBIA: Sterijovski (2014, Bosilegrad region), Tomović et al. (2014, 2015).
SLOVAKIA: Lác & Lechovič (1964).
SPAIN: Fernández-Guiberteau et al. (2015, Extremadura), Fuentes & Lizana (2015, Salamanca), González-Fernández (2006, Madrid); Gosá & Bergerandi (1994, Navarra), Matco (2015, Baleares), Pérez de Ana (2014, Basque), Pleguezuelos (1989, Granada), Rodríguez-Rodríguez et al. (2018, Seville).
SWITZERLAND: Dušej & Müller (1997, Zurich).

SOUTHWEST ASIA: Sindaco et al. (2013).
ARMENIA: Arakelyan et al. (2011).
AZERBAIJAN: Bunyatova (2014, Caucasus Mountains).
CYPRUS: Göçmen et al. (2009b).

IRAN: Gholamifard et al. (2012, Fars); S.M. Kazemi et al. (2015, Qom); Moradi et al. (2013, Khabr N.P.), Nasrabadi et al. (2016, Sabzevar District); Pour et al. (2016, NW Yazd), N. Rastegar-Pouyani et al. (2008), Sabbaghzadeh & Mashayekhi (2015, Nazmabad of Arak), Safaei-Mahroo et al. (2015), Yousefkhani et al. (2014, Sabzevar District).

IRAQ: Abbas-Rhadi et al. (2017, southern), Al-Barazengy et al. (2015, checklist), Habeeb & Rastegar-Pouyani (2016a, 2016b), Habeeb et al. (2016), Mohammad et al. (2013, Bahr Al-Najaf Depression).

JORDAN: Amr & Disi (2011).

OMAN: Ball & Borrell (2016, Dhofar).

PALESTINE: Albaba (2016), Handal et al. (2016, West Bank).

PERSIAN GULF: Rezaie-Atagholipour et al. (2016, Hydrophiinae).

SAUDI ARABIA: Aloufi & Amr (2015, Tabuk), Al-Sadoon (1989, Ar-Riyad), Alshammari et al. (2017, Ha'il), Cunningham (2010, Farasan Is.), Masetti (2014, Farasan Is.).

TURKEY: Afsar & Tok (2011, Sultan Mountains), Afsar et al. (2012, Camili Biosphere Reserve), Afsar et al. (2013, records), Arslan et al. (2018, Izmir), Baran et al. (2005), Çakmak et al. (2017, Bartin), Cumhuriyet & Ayaz (2015, Güllük Gulf region), Ege et al. (2015, Burdur Province), Eksilmez et al. (2017, Karçal Mountains), Eser & Erismis (2014, Afyon), Kumlutaş et al. (2015b, Fethiye region), Kumlutaş et al. (2017, Karabük), Özdemǐr & Baran (2002, Murat Mountain), Sarikaya et al. (2017, Adana), Tok et al. (2014, Canakkale), Tosunoğlu et al. (2009, Tenedos I.), Uğurtaş et al. (2000b, Hatay), Uysal & Tosunoğlu (2011, Kavak Delta), Yakin et al. (2018, Imbros I.), Winden et al. (1997, Goksu Delta).

UNITED ARAB EMIRATES: Buzás et al. (2018, Hydrophiinae).

YEMEN: Sindaco et al. (2009, Socotra Island).

CENTRAL & EASTERN ASIA
AFGHANISTAN: Wagner et al. (2016b).

RUSSIA: Adnagulov & Oleinikov (2006, southern Far East); Pavlov et al. (2013, Great Kokshaga Natural Reserve).

CHINA: Cai et al. (2015), Qing et al. (2015, Nan Ao Is.), J.-H. Wang et al. (2019, Gaoligongshan Nature Reserve, Yunnan), Yao (2004, Gansu), C.-F. Zhong (2004, Jiangxi).

GEORGIA: Bekoshvili & Doronin (2015, Caucasus), B.S. Tuniyev et al. (2017, South Ossetia).

KAZAKHSTAN: Borkin & Litvinchuk (2015, western area).

UZBEKISTAN: Martin et al. (2017, central).

SOUTH AND SOUTHEAST ASIA
BHUTAN: Tshewang & Letro (2018, Jigme Singye Wangchuck N.P.).

CAMBODIA: Geissler et al. (2019, Phnom Kulen N.P.), Hartmann et al. (2013, Kulen Promtep Wildlife Sanctuary).

INDIA: Chandra & Gajbe (2005, Chhattisgarh and Madhya Pradesh); A. Das et al. (2009, Barail Wildlife Sanctuary, Assam); Farhad & Varsha (2013, Pune District); Ganesh & Arumugam (2016, S Eastern Ghats); Ganesh et al. (2013a, Western Ghats; 2014a, High Wavy Mts.; 2018, Eastern Ghats; 2019, marine species); Harikrishnan et al. (2012, Long I., Andaman & Nicobars), Manhas et al. (2018a, Bhopal, 2018b, Doda), Mirza & Pal (2008, Sanjay Gandhi N.P.), Palot (2015, Kerala), Patel & Vyas (2019, Gujarat), Purkayastha (2018, Gawahati, Arunachal Pradesh), Rao et al. (2005, Nallamalai Hills), Sayyed (2016, Satara District, Maharashtra), Sen & Nama (2013, Mukundara Hills N.P., Rajasthan), Sengupta et al. (2019, E Assam); Srinivasulu et al. (2014, Western Ghats), Yanthungbeni et al. (2018, Dimapur, Nagaland).

MYANMAR: Mulcahy et al. (2018, S Tanintharyi); Platt et al. (2018, records).

NEPAL: Bhattarai et al. (2017c, Chitwan National Park), Bhattarai et al. (2018b, Parsa National Park), Pandey (2012, Chitwan National Park and vicinity), Pandey et al. (2018, Chitwan National Park and vicinity).

PAKISTAN: Jamal et al. (2018, Dir region), M.S. Khan (2012, checklist), Rais et al. (2012, North Punjab).

SRI LANKA: D.M.S.S. Karunarathna & Amarasinghe (2012, Beraliya Mukalana Reserve).

THAILAND: Chan-ard and Makchai (2011, Prasae Estuary and Koh Man Islands), Chan-ard et al. (2011, Khao Soi Dao Wildlife Sanctary; 2015), Milto (2014, Phi Phi Archipelago).

VIETNAM: Pham et al. (2015, Son La City, Son La), Gawor et al. (2016, Bai Tu Long N.P.), Le et al. (2018, Yen Bai), Nemes et al. (2013, Quang Ngai Prov.), T.Q. Nguyen et al. (2014, S Vietnam), Rasmussen et al. (2011a, marine Elapidae), Vassilieva et al. (2016).

WEST MALAYSIA: B.L. Lim et al. (2010, Pulau Singa Besar), K.K.P. Lim & Lim (1999, Pulau Tioman), Onn et al. (2009, Perlis State Park), Onn et al. (2010, Gunong Panti Forest Reserve), Quah et al. (2013, Bukit Panchor State Park), Sumarli et al. (2015, Gunung Tebu region), Zakaria et al. (2014, Krau Wildlife Reserve).

EAST INDIES
BORNEO: Auliya (2006, West Kalimantan), Stuebing et al. (2014).

INDONESIA: Asad et al. (2012, Nusa Penida), Asad et al. (2015, Sarawak records), Karin et al. (2018, Kei Islands), A. Koch et al. (2009, Talaud Archipelago), Lang (2013, Moluccas), Tapley & Muurmans (2011, Pulau Bangkuru).

PHILIPPINE ISLANDS: R.M. Brown et al. (2012, Ilocos Norte Prov., Luzon); R.M. Brown et al. (2013, Cagayan & Isabela provinces, Luzon); Gojo-Cruz et al. (2018, Caraballo Range, Luzon), Leviton et al. (2014, Elapidae and Viperidae; 2018); Sanguila et al. (2016, Mindanao); Siler et al. (2012, Romblon); Supsup et al. (2016, Cebu; 2017, Hamiguitin Range), Weinell et al. (2019).

TIMOR-LESTE: H. Kaiser et al. (2013b, Ataúro Island).

AUSTRALIA & PACIFIC ISLANDS
AUSTRALIA: Bush et al. (2007, southwestern), Cogger (2014), Hutchinson et al. (2001, Tasmania), Rasmussen et al. (2014, marine Elapidae).

OCEANIA: Zug (2013).

PAPUA NEW GUINEA: Foufopoulos & Richards (2007, New Britain); Kraus (2013, records).

VANUATU: Ineich (2009, Torres and Banks groups).

SOUTH AMERICA
ARGENTINA: Cano et al. (2015, Formosa); Giambelluca (2015, Buenos Aires), Kass et al. (2018, Talampaya National Park), Minoli et al. (2015, Chubut), Zaracho et al. (2014, R.N.P. Isla Apipé Grande).

ECUADOR: Almendáriz et al. (2014, Cordillera del Condor); Arteaga et al. (2013, Mindo Township).

BOLIVIA: Cortez-Hernandez (2005, P. N. Madidi), Mano-Cuellar et al. (2015, Mutun region), Pinto-Viveros et al. (2017, Cerro Mutún), Rivas et al. (2018, P. N. Torotoro).

BRAZIL: Guedes et al. (2014, Caatinga), Nogueira et al. (2019).

Acre: Bernarde et al. (2013, Rio Moa), Fonseca et al. (2019, Porto Walter area).

Amapa: Pedroso-Santos et al. (2019, Reserva Extrativista Beija-Flor Brilho de Fogo).

Amazonas: Pantoja & Fraga (2012, Reserva Extrativista do Rio Gregorio), Waldez et al. (2013, lower Rio Purus).

Bahia: M.A. Freitas (2014, northern); M.A. Freitas et al. (2012, Chapada Diamantina; 2016, SW Bahia; 2018, Serra da Jibóia; 2019a, Serra do Timbó), Garda et al. (2013, Raso da Catarina E.S.), F. Magalhães et al. (2015, Chapada Diamantina N.P.), R. Marques et al. (2011, 2016, N coast).

Ceara: Borges-Leite et al. (2014, São Gonçalo do Amarante), T.B. Costa et al. (2018, Aiuaba Ecological Station); D.P. Castro et al. (2019, Ubajara N.P.), Mesquita et al. (2013, Fazenda Experimental Vale do Curu), Roberto & Loebmann (2016).

Goias: Campos & Lage (2013, Nova Roma Ecological Station), Ramalho et al. (2018b, Parque Estadual Altamiro de Moura Pacheco), D.L. Santos et al. (2014, southeastern).

Maranhão: M.A. Freitas et al. (2017, northern), J.P. Miranda et al. (2012, Lençóis Maranhenses N. P.).

Mato Grosso do Sul: Ferreira et al. (2017).

Minas Gerais: H. Costa et al. (2014, Nova Ponte region); Menezes et al. (2018, Serra da Mantiqueira), Moura et al. (2012, Serra do Brigadeiro).

Pará: Avila-Pires et al. (2010); Bernardo et al. (2012b, Reserva Biológica do Tapirapé), Frota et al. (2005, lower Amazon), G.M. Rodrigues et al. (2015, Marajo Island); Santos-Costa et al. (2015, Floresta Nacional de Caxiuanã).

Paraiba: Sampaio et al. (2018, SE coast).

Parana: Dainesi et al. (2019, Londrina).

Pernambuco: Pedrosa et al. (2014, Catimbau National Park); Roberto et al. (2017, Serra do Urubu).

Piauí: Cavalcanti et al. (2014, Serra da Capivara).

Rio de Janeiro: Almeida-Gomes et al. (2014, Reserva Ecológica de Guapiaçu); C.F.D. Rocha et al. (2018, Ilha Grande).

Rio Grande do Sul: Moser et al. (2018, Sinos River Basin), Outeiral et al. (2018, Serra do Sudeste), Quintela et al. (2006, Rio Grande), Quintela et al. (2011, Marinheiros I.).

Rondônia: Bernarde et al. (2012b).

Roraima: Moraes et al. (2017, Serra da Mocidade).

São Paulo: Barbo et al. (2011, municipality of São Paulo), Ortiz et al. (2017, São José do Barreiro), Sawaya et al. (2008, Estação Ecológica de Itirapina), Trevine et al. (2014, Paranapiacapa).

COLOMBIA: Armesto et al. (2011, Norte de Santander); Moreno-Arias & Quintero-Corza (2015, Magdalena Valley, Huila), Vera-Pérez et al. (2018, P.N. Natural Munchique).

GUYANA: MacCulloch & Reynolds (2013, Kurupukari region).

PARAGUAY: Cabral & Weiler (2014), Cacciali (2013, San Rafael N.P.), Cacciali et al. (2015a), Cacciali et al. (2015b, R.N. Bosque Mbaracayu), P. Smith et al. (2016, R. N. Laguna Blanca), Trutnau et al. (2016).

PERU: Catenazzi et al. (2013, Manu N.P.), Koch et al. (2018, Marañon Valley); Llanqui et al. (2019, Bahuaja-Sonene N.P.), Pérez-Z. & Lleellish (2015, Isla San Lorenzo).

TRINIDAD & TOBAGO: Auguste (2019, protected areas), Auguste et al. (2015, Aripo Savannas Scientific Reserve), Mohammed et al. (2014).

VENEZUELA: Natera-Mumaw et al. (2015).

MESOAMERICA

COSTA RICA: Abarca-Alvarado (2012, El Rodeo zone, San Jose), Acosta-Chaves et al. (2015, Río Macho Biological Station, Cartago), Arias & Bolaños (2014, Reserva Forestal Los Santos), Chacón & Johnston (2013), Köhler et al. (2013, records).

EL SALVADOR: Juárez-Peña et al. (2016, P.N. Montecristo); Townsend (2014).

GUATEMALA: Townsend (2014).

HONDURAS: McCranie (2014b, Intibucá, Lempira, Ocotepeque), McCranie & Gutsche (2016, Golfo de Fonseca Is.), McCranie & Valdés-Orellana (2014, Islas de la Bahia), McCranie et al. (2017, Swan Is.), Solís et al. (2014a), Townsend et al. (2012, Refugio de Vida Silvestre Texiguat), Townsend (2014, Chortis Block Biogeographic Province), L.D. Wilson & Townsend (2007).

MEXICO: Grünwald et al. (2016a, records), Heimes (2016), Johnson et al. (2017), Ramírez-Bautista et al. (2009, Valley of Mexico), L.D. Wilson et al. (2013).

Aguascalientes: Carbajal-Márquez & Quintero-Díaz (2016).

Campeche: González-Sánchez et al. (2017).

Chiapas: A.G. Clause et al. (2016, records); García-Padilla & Mata-Silva (2014b, records); Hernández-Ordóñez et al. (2015, eastern), Johnson et al. (2015b).

Chihuahua: Hernandez et al. (2019, records), Lemos-Espinal et al. (2013; 2017).

Coahuila: Lemos-Espinal & G.R. Smith (2016).

Colima: Ahumada-Carrillo et al. (2014, records).

Durango: Lares et al. (2013), Lemos-Espinal et al. (2018a).

Guanajuato: Leyte-Manrique et al. (2018, Area Natural Protegida Las Musas).

Guerrero: Palacios-Aguilar & Flores-Villela (2018), Palacios-Aguilar et al. (2016, records).

Hidalgo: Fernández-Badillo et al. (2016, records), Lemos-Espinal & G.R. Smith (2015), Obregón-Esparza et al. (2018, records), Roth-Monzón et al. (2018, Nopala de Villagrán Mun.), Vite-Silva et al. (2010, Res. Biosfera Barrance de Metztitlán).

Jalisco: Ahumada-Carrillo et al. (2014, records); Cruz-Sáenz et al. (2017).

Michoacan: Alvarado-Díaz et al. (2013).

Morelos: Castro-Franco & Zagal (2006, Sierra de Huautla).

Nayarit: Luja & Grünwald (2015, records), Woolrich-Piña et al. (2016).

Nuevo Leon: Lemos-Espinal et al. (2016; 2018c), Nevárez-de los Reyes et al. (2016b).

Oaxaca: García-Grajales et al. (2016, P.N. Lagunas de Chacahua), García-Padilla & Mata-Silva (2014a, records); Martín-Regalado et al. (2011, Cerro Guiengola), Mata Silva et al. (2015a), Mata-Silva et al. (2017, records), Schätti & Stutz (2016, southern).

Puebla: Woolrich-Piña et al. (2017).

Queretaro: Cruz-Elizalde et al. (2019).

Quinatana Roo: González-Sánchez et al. (2017), Köhler et al. (2016a-c, 2017a-b).

San Luis Potosí: Lemos-Espinal & Dixon (2013), Lemos-Espinal et al. (2018b).

Sinaloa: Bezy et al. (2017, records).

Sonora: Bezy et al. (2017, records); Enderson et al. (2014, Yécora region), Lemos-Espinal et al. (2019); Nevárez-de los Reyes et al. (2014), Rorabaugh & Lemos-Espinal (2016), Rorabaugh et al. (2019, Mesa Tres Ríos region).

Tamaulipas: Farr et al. (2013, Tamaulipas, records), Terán-Juárez et al. (2015, records), Terán-Juárez et al. (2016).

Veracruz: Luz et al. (2016, Cuautlapan Valley).

Yucatan: Colston et al. (2015, Calakmul Biosphere Reserve), González-Sánchez et al. (2017).

NICARAGUA: Sunyer (2014).

PANAMA: Ray & Ruback (2015, Panama and Panama Oeste provinces).

TRINIDAD & TOBAGO: Mohammed et al. (2014).

NORTH AMERICA

WEST INDIES: Hedges et al. (2019).

Bahamas: Reynolds et al. (2018b, Cay Sal Bank).

Dominica: Daniells et al. (2008).

Malhotra & Thorpe (1999, eastern Caribbean)

CANADA: Crother et al. (2017), Powell et al. (2016, eastern).

Ontario: Weller et al. (2017, records).

Ontario & Quebec: Desroches et al. (2010, James Bay area).

UNITED STATES: Crother et al. (2017), Hubbs (2013, western), Hubbs & O'Connor (2009, 2012, venomous), J.M. Parker & Brito (2013, Mojave Desert), Powell et al. (2016, eastern).

Alabama: Burchill & Diamond (2014, Pike County), Folt et al. (2015), Holt et al. (2017, records), Sutton et al. (2014, records).

Arizona: Bezy & Cole (2014, Madrean Archipelago), Schuett et al. (2016a, b, Viperids), Turner et al. (2003, Whetstone Mtns.).

Florida: Holbrook (2012), Mays et al. (2017, Florida Keys records), Thomas et al. (2015, Suwannee region records).

Georgia: Hudson et al. (2015, records), Pierson & Sollenberger (2014, records), Pierson et al. (2014, records), Stevenson et al. (2014, 2015, records), J.M. Strickland & Hartman (2015, west-central); Thesing & Clause (2018, records).

Illinois: Kessler et al. (2013, 2015, state managed areas).

Iowa: Hubbs (2016, records), LeClere (2013).

Kansas: Hubbs (2013b, 2016, records).

Kentucky: Moore & Slone (2002).

Louisiana: Battaglia et al. (2015, records), Boundy & Carr (2017).

Michigan: Casper et al. (2015).

Minnesota: Hoaglund & Smith (2013, records).

Mississippi: Folt et al. (2013, Jasper County), Keiser (2010, northern).

Nebraska: D. Davis & Dilliard (2016, records), D. Davis et al. (2014, records), Hubbs (2013b, 2016, records).

Nevada: Setser et al. (2002, Snake Range).

New Mexico: Bezy & Cole (2014, Madrean Archipelago).

North Carolina: Beane (2013, records).

Ohio: Behrendt (2016a, b, Perry County).

Oklahoma: Sievert & Sievert (2011).

South Dakota: Austin et al. (2017, records), D.R. Davis (2018, records), D.R. Davis & Farkas (2018, records), D.R. Davis et al. (2016, 2017a-b, records).

Tennessee: Foster et al. (2013, records).

Texas: Adams et al. (2016, Rio Grande Valley records), Davis & LaDuc (2018, Sierra Vieja), Price & Dimler (2015).

Virginia: Fulton et al. (2014, records), Lassiter et al. (2017, records).

Wisconsin: Casper (2015), Kapfer et al. (2015).

Literature Cited

Abarca-Alvarado, J.G. (2012) La herpetofauna de un bosque pre-montano: Diversidad de anfibios y reptiles de El Rodeo. *Brenesia* 77, 251–269, 27 figs.

Abarca-Alvarado, J.G. & Bolaños, F. (2017) Distribution notes: *Enulius flavitorques. Mesoamerican Herpetology* 4(2), 468–469, 1 fig.

Abbas-Rhadi, F., Mohammed, R.G., Rastegar-Pouyani, N., Rastegar-Pouyani, E. & Yousefkhani, S.S.H. (2017) On the snake fauna of central and southern Iraq and some zoogeographic remarks. *Russian Journal of Herpetology* 24(4), 251–266, 6 figs.

Abbas-Rhadi, F., Rastegar-Pouyani, N., Karamiani, R. & Mohammed, R.G. (2015b) Taxonomic status of sand boas of the genus *Eryx* (Daudin, 1803) (Serpentes: Boidae) in Bahr Al-Najaf depression, Al-Najaf Province, Iraq. *Iranian Journal of Animal Biosystematics* 11(2), 149–156, 3 figs.

Abbas-Rhadi, F., Rastegar-Pouyani, N., Mohammed, R.G., Al-Fartosi, K., Browne, R.K. & Karamiani, R. (2015a) A study of the nomino typic form of saw scaled viper, *Echis carinatus* (Schneider, 1801) (Squamata: Ophidia: Viperidae) in southern Iraq. *Scholars Academic Journal of Biosciences* 3(10), 845–851, 4 figs.

Abbas-Rhadi, F., Rastegar-Pouyani, E., Rastegar-Pouyani, N., Mohammed, R.G. & Yousefkhani, S.S.H. (2016) Phylogenetic affinities of the Iraqi populations of saw-scaled Vipers of the genus *Echis* (Serpentes: Viperidae), revealed by sequences of mtDNA genes. *Zoology in the Middle East* 62(4), 299–305, 2 figs.

Abegg, A.D., Azevedo, W. dos S. & Duarte, M.R. (2019) Noteworthy insular records of burrowing reptiles in southeastern Brazil. *Herpetology Notes* 12, 221–224, 2 figs.

Abegg, A.D., Borges, L.M., Rosa, C.M. da, Entiauspe-Neto, O.M., Arocha, N.M. & Santos-Jr, A.P. (2016) Included, excluded and re-included: *Chironius brazili* (Serpentes, Colubridae) in Rio Grande do Sul, southern Brazil. *Neotropical Biology and Conservation* 11(3), 198–203, 3 figs.

Abegg, A.D., Freitas, M.A. de & Moura, G.J.B. de (2017a) New records of *Liotyphlops trefauti* Freire (Caramaschi & Argôlo, 2007) (Squamata: Anomalepididae). *Herpetology Notes* 10, 345–347, 2 figs.

Abegg, A.D., Freitas, M.A. de & Moura, G.J.B. de (2017b) First confirmed record of *Atractus maculatus* (Serpentes, Dipsadidae) from the state of Pernambuco, northeastern Brazil. *Check List* 13(2080), 1–3, 2 figs.

Abegg, A.D., Santos, A. de S., Ortiz, F.R., Moraes, C.M. de & Freitas, M.A. de (2017c) First record of *Siphlophis compressus* (Daudin, 1803) (Serpentes, Dipsadidae) from the state of Roraima, Brazil. *Check List* 13(4), 251–254, 2 figs.

Abtin, E., Nilson, G., Mobaraki, A., Hosseini, A.A. & Dehgannejhad, M. (2014) A new species of Krait, *Bungarus* (Reptilia, Elapidae, Bungarinae) and the first record of that genus in Iran. *Russian Journal of Herpetology* 21(4), 243–250, 13 figs.

Abukashawa, S.M.A., Papenfuss, T.J. & Alkhedir, I.S. (2018) Geographic distribution: *Cerastes vipera. Herpetological Review* 49(1), 75.

Acosta, R., Blanco, G.M., Galdeano, A.P., Alés, R.G. & Acosta, J.C. (2017) Primer registro de *Epictia albipuncta* (Burmeister, 1861) (Serpentes: Leptotyphlopidae) en la provincial de San Juan, Argentina. *Cuadernos de Herpetología* 31(2), 127–128, 2 figs.

Acosta-Chaves, V.J., Chaves, G., Abarca, J.G., García-Rodríguez, A. & Bolaños, F. (2015) A checklist of the amphibians and reptiles of Río Macho Biological Station, Provincia de Cartago, Costa Rica. *Check List* 11(1784), 1–10, 3 figs.

Acosta-Chaves, V.J., Garita, C., Conejo-Barboza, K., Ramírez-Fernández, J.D., Naranjo, B. & Jiménez, A. (2014) Geographic distribution. *Sibon anthracops. Herpetological Review* 45(3), 467.

Adams, C.S., Hibbitts, T.J. & Campbell, T.A. (2016) New amphibian and reptile county records from the lower Rio Grande Valley in Texas, USA. *Herpetological Review* 47(3), 430–431.

Adnagulov, E.V. & Oleinikov, A.Y. (2006) On the distribution and ecology of amphibians and reptiles in the south of the Russian Far East. *Russian Journal of Herpetology* 13(2), 101–116, 7 figs.

Aengals, R. & Ganesh, S.R. (2013) *Rhinophis goweri* – A new species of shieldtail snake from the southern Eastern Ghats, India. *Russian Journal of Herpetology* 20(1), 61–65, 3 figs.

Afrasiah, S.R., Al-Ganabi, M.I. & Al-Fartosi, K. (2012) Snake species new or rare to the herpetofauna of Iraq. *Herpetozoa* 24(3/4), 179–181, 3 figs.

Afrasiab, S.R. & Ali, H.A. (1996) Notes on scolecophidians (blind snakes) Reptilia – Serpentes, of Iraq. *Bulletin of the Iraq Natural History Museum* 8(4), 31–39, 2 figs.

Afrasiab, S.R. & Mohamad, S.I. (2014) New records of snakes from Iraq (Reptilia: Colubridae). *Zoology in the Middle East* 60(1), 92–94, 1 fig.

Afrasiab, S.R., Mohammad, M.K. & Hussein, A.M. (2016) Description of a new species of genus *Dolichophis* Gitstel [sic] from the upper Mesopotamian Plain-Iraq (Reptilia-Serpentes-Colubridae). *Journal of Biodiversity and Environmental Sciences* 8(4), 15–19, 3 figs.

Afroosheh, M., Rajabizadeh, M., Rastegar-Pouyani, N. & Kami, H.G. (2010) The Brahminy blind snake, *Ramphotyplops braminus* (Daudin, 1803), a newcomer to Iran (Ophidia: Typhlopidae). *Zoology in the Middle East* 50(1), 135–137, 2 figs.

Afroosheh, M., Rastegar-Pouyani, N., Ghoreishi, S.K. & Kami, H.G. (2013) Comparison of geographic variations in *Typhlops vermicularis* (Merrem, 1820) (Ophidia: Typhlopidae) from the Iranian plateau with Turkey and Turkmenistan. *Turkish Journal of Zoology* 37, 685–692, 3 figs.

Afsar, M., Ayaz, D., Afsar, B., Çiçek, K. & Tok, C.V. (2012) Herpetofauna of the Camili Biosphere Rezerve Area (Borçka, Artvin, Turkey). *Anadolu University Journal of Science and Technology Series C* 2(1), 41–49, 2 figs.

Afsar, M., Çiçek, K., Tayhan, Y. & Tok, C.V. (2016) New records of the Eurasian blind snake, *Xerotyphlops vermicularis* (Merrem, 1820) from the Black Sea region of Turkey and its updated distribution. *Biharean Biologist* 10(2), 98–103, 3 figs.

Afsar, M., Dinçaslan, Y.E., Ayaz, D. & Tok, C.V. (2013) New record localities of five snake species in Turkey. *Herpetozoa* 25(3/4), 179–183, 1 fig.

Afsar, M. & Tok, C.V. (2011) The herpetofauna of the Sultan Mountains (Afyon-Konya-Isparta), Turkey. *Turkish Journal of Zoology* 35(4), 491–501, 2 figs.

Agarwal, I. & Srikanthan, A.N. (2013) Further records of the Sindh Awl-headed snake, *Lytorhynchus paradoxus* (Günther, 1875), from India with notes on its habitat and natural history. *Russian Journal of Herpetology* 20(3), 165–170, 4 figs.

Ahmad, F. & Alam, S.M.I. (2015) Geographic distribution. *Chrysopelea ornata. Herpetological Review* 46(4), 572.

Ahmad, F., Alam, S.M.I. & Khondakar, T. (2015a) Geographic distribution. *Ahaetulla pulverulenta. Herpetological Review* 46(4), 571.

Ahmad, F., Alam, S.M.I. & Khondakar, T. (2015b) Geographic distribution. *Eryx conicus. Herpetological Review* 46(4), 572–573.

Ahmad, F., Alam, S.M.I. & Khondakar, T. (2015c) Geographic distribution. *Xenochrophis cerasogaster. Herpetological Review* 46(4), 578.

Ahmad, F., Alam, S.M.I. & Sarker, M.A.R. (2015d) Geographic distribution. *Rhabdophis subminiatus. Herpetological Review* 46(4), 576.

Ahsan, M.F. & Rahman, M.M. (2017) Status, distribution and threats of kraits (Squamata: Elapidae: *Bungarus*) in Bangladesh. *Journal of Threatened Taxa* 9(3), 9903–9910, 6 figs.

Ahumada-Carrillo, I.T., Carbajal-Marquez, R.A., López-Cuellar, M.A. & Weatherman, G.N. (2018) The Sonoran Coralsnake, *Micruroides euryxanthus* (Kennicott, 1860), in the state of Jalisco, Mexico. *Mesoamerican Herpetology* 5(1), 185–188, 3 figs.

Ahumada-Carrillo, I.T., Pérez-Rivera, N., Reyes-Velasco, J., Grünwald, C.L. & Jones, J.M. (2014) Notable records of amphibians and reptiles from Colima, Nayarit, Jalisco, and Zacatecas, México. *Herpetological Review* 45(2), 287–291, 4 figs.

Akkaya, A. & Uğurtaş, I.H. (2012) Rediscovery of *Vipera ammodytes* (Linnaeus, 1758) at Uludağ-Bursa, Turkey, after 62 years. *Herpetozoa* 24(3/4), 181–185, 2 figs.

Albaba, I. (2016) The herpetofauna of Palestine: A preliminary checklist. *Journal of Entomology and Zoology Studies* 4(4), 123–128, 3 figs.

Al-Barazengy, A.N., Salman, A.O. & Hameed, F.T.A. (2015) Updated list of amphibians and reptiles in Iraq 2014. *Bulletin of the Iraq Natural History Museum* 13(4), 29–40.

Albino, A.M. & Brizuela, S. (2014) An overview of the South American fossil squamates. *Anatomical Record* 297, 349–368, 4 figs.

Albino, A.M., Carrillo-Briceño, J.D. & Neenan, J.M. (2016) An enigmatic aquatic snake from the Cenomanian of northern South America. *PeerJ* 2027, 1–26, 11 figs.

Albino, A.M., Montalvo, C.I. & Brizuela, S. (2013) New records of squamates from the upper Miocene of South America. *Journal of Herpetology* 47(4), 590–598, 5 figs.

Albuquerque, N.R. de & Lema, T. de (2008) Taxonomic revision of the Neotropical water snake *Hydrops triangularis* (Serpentes, Colubridae). *Zootaxa* (1685), 55–66, 6 figs.

Aldape-López, C.T. & Santos-Moreno, A. (2016) Ampliación de la distribución geográfica de *Abronia oaxacae* (Squamata: Anguidae) y *Tantalophis discolor* (Squamata: Colubridae) en el estado de Oaxaca, México. *Acta Zoológica Mexicana (new series)* 33(1), 116–119, 2 figs.

Alemán, A.R. & Sunyer, J. (2015) Distribution notes: *Scolecophis atrocinctus. Mesoamerican Herpetology* 2(3), 359.

Alencar, L.R.V., Quental, T.B., Grazziotin, F.G., Alfaro, M.L., Martins, M., Venzon, M. & Zaher, H. (2016) Diversification in vipers: Phylogenetic relationships, time of divergence and shifts in speciation rates. *Molecular Phylogenetics & Evolution* 105(1), 50–62, 3 figs.

Alencar, L.R.V. de, Righi, A.F., Nascimento, L.B. & Morato, S.A.A. (2009) *Siphlophis longicaudatus* (Brazilian Spotted Night Snake): Habitat. *Herpetological Bulletin* 108, 37–39, 1 fig.

Al-Fares, A. (2014) *Classification Guide for Horned Viper in Kuwaiti Environment.* Kuwait Institute for Scientific Research, Kuwait. 1–80, 64 figs.

Almeida, P.C., Feitosa, D.T., Passos, P. & Prudente, A.L.C. (2014) Morphological variation and taxonomy of *Atractus latifrons* (Günther, 1868) (Serpentes: Dipsadidae). *Zootaxa* 3860, 64–80, 11 figs.

Almeida-Gomes, M., Siqueira, C.C., Borges-Júnior, V.N.T., Vrcibradic, D., Fusinatto, L.A. & Rocha, C.F.D. (2014) Herpetofauna of the Reserva Ecológica de Guapiaçu (REGUA) and its surrounding areas, in the state of Rio de Janeiro, Brazil. *Biota Neotropica* 14(3), 1–15, 8 figs.

Almendáriz, A. (2007) Primer registro de *Dipsas oreas* en el Provincia del Azuay, Ecuador. *Politecnica (Biologia 7)* 27(4), 136–137, 1 fig.

Almendáriz, A. & Brito, J. (2012) Ampliación del rango distribucional de *Drymarchon melanurus* (Colubridae) y *Basiliscus galeritus* (Iguanidae-Corytophanidae), hacia los bosques secos interandinos de norte del Ecuador. *Revista Politécnica* 30(3), 179–183, 4 figs.

Almendáriz, A. & Carr, J.L. (2012) Lista actualizada de los anfibios y reptiles registrados en los remanentes de bosque de la Cordillera de La Costa y áreas adyacentes del suroeste del Ecuador. *Revista Politécnica* 30(3), 184–194, 1 fig.

Almendáriz, A., Hamilton, P., Mouette, C. & Robles, C. (2012) Análisis de la herpetofauna de los bosques secos y de trasición de la Reserva Biológica Tito Santos, Manabí-Ecuador. *Revista Politécnica* 30(3), 62–82, 30 figs.

Almendáriz, A. & Orcés, G. (2004) Distribución de algunas especies de la herpetofauna de los Pisos: Altoandino, temperado y subtropical. *Politecnica (Biologia 5)* 25(1), 97–149, figs.

Almendáriz, A., Simmons, J.E., Brito, J. & Vaca-Guerrero, J. (2014) Overview of the herpetofauna of the unexplored Cordillera del Cóndor of Ecuador. *Amphibian & Reptile Conservation* 8(1), 46–64, 20 figs.

Alonso-Zarazaga, M.A. (2013) *Vipera latasti* vs. *V. latastei*: A poisoned affair. *Graellsia* 69(1), 129–131.

Aloufi, A.A. & Amr, Z.S. (2015) On the herpetofauna of the Province of Tabuk, northwest Saudi Arabia (Amphibia, Reptilia). *Herpetozoa* 27(3/4), 147–158, 8 figs.

Al-Sadoon, M.K. (1989) Survey of the reptilian fauna of the Kingdom of Saudi Arabia. I – The snake fauna of the central region. *Journal of King Saud University, Science* 1(1/2), 53–69, 4 plates.

Al-Sadoon, M.K. & Al-Otaibi, F.S. (2014) Ecology of the sand boa, *Eryx jayakari* in Riyadh region of Saudi Arabia. *Saudi Journal of Biological Sciences* 21, 391–393, 3 figs.

Alshammari, A.M. (2011) Molecular phylogeny of Viperidae family from different provinces in Saudi Arabia. *International Journal of Biology* 3, 56–63, 4 figs.

Alshammari, A.M., Busais, S.M. & Ibrahim, A.A. (2017) Snakes in the province of Ha'il, Kingdom of Saudi Arabia, including two new records. *Herpetozoa* 30(1/2), 59–63, 1 fig.

Alvarado-Díaz, J., Suazo-Ortuño, I., Wilson, L.D. & Medina-Aguilar, O. (2013) Patterns of physiographic distribution and conservation status of the herpetofauna of Michoacán, Mexico. *Amphibian & Reptile Conservation* 7(1), 129–170, numerous figs.

Alves, S. da S., Bolzan, A.M.R., Santos, T.G. do, Gressler, D.T. & Cechin, S.Z. (2013) Rediscovery, distribution extension and defensive behaviour of *Xenodon histricus* (Squamata: Serpentes) in the state of Rio Grande do Sul, Brazil. *Salamandra* 49(4), 219–222, 3 figs.

Alzate, E. (2014) Geographic distribution. *Micrurus camilae. Herpetological Review* 45(2), 285–286.

Amarasinghe, A.A.T., Campbell, P.D., Hallermann, J., Sidik, I., Supriatna, J. & Ineich, I. (2015b) Two new species of the genus *Cylindrophis* (Wagler, 1828) (Squamata: Cylindrophiidae) from southeast Asia. *Amphibian & Reptile Conservation* 9(1), 34–51, 17 figs.

Amarasinghe, A.A.T., Karunarathna, D.M.S.S., Campbell, P.D. & Ineich, I. (2015a) Systematics and ecology of *Oligodon sublineatus* Duméril, Bibron & Duméril, 1854, an endemic snake of Sri Lanka, including the designation of a lectotype. *Zoosystmatics and Evolution* 91(1), 71–80, 3 figs.

Amarasinghe, A.A.T., Vogel, G., McGuire, J.A., Sidik, I., Supriatna, J. & Ineich, I. (2015c) Description of a second species of the genus *Rabdion* (Duméril, Bibron & Duméril, 1854) (Colubridae: Calamariinae) from Sulawesi, Indonesia. *Herpetologica* 71(3), 234–239, 4 figs.

Amaro-Valdés, S. & Morell-Savall, E. (2017) First notes on reproduction of Cuban snakes in the endemic genus *Arrhyton* (Günther, 1858) (Squamata: Dipsadidae). *IRCF Reptiles & Amphibians* 24(1), 47–50, 6 figs.

Ambrogio, A. & Mezzadri, S. (2003) *Anfibi e rettili*. Museo Civico di Storia Naturale Piacenza, Piacenza. 1–62, numerous figs.

Amo, O.A. (1994) Primeras notas herpetológicas de la provincia de Soria. *Doñana, Acta Vertebrata* 9, 385–388, 1 fig.

Amr, Z.S. & Disi, A.M. (2011) Systematics, distribution and ecology of the snakes of Jordan. *Vertebrate Zoology* 61(2), 179–266, 111 figs.

Ana, J.M.P. de (2014) Nuevos datos de anfibios y reptiles en el País Vasco. *Munibe* 62, 135–144, 1 fig.

Ananjeva, N.B., Golynsky, E.E., Lin, S.-M., Orlov, N.L. & Tseng, H.-Y. (2015) Modeling habitat suitability to predict the potential distribution of the Kelung cat snake *Boiga kraepelini* Stejneger, 1902. *Russian Journal of Herpetology* 22(1), 197–205, 7 figs.

Andrade, H., Almeida, R.P.S. & Dias, E.J. dos R. (2017) *Spilotes sulphureus* (Wagler, 1824) (Squamata: Colubridae): Review of distribution and first record in the state of Sergipe, northeastern Brazil. *Check List* 13(2055), 1–5, 3 figs.

Andrade, H., Lima, J.O., Silva, A.F.O., Fernandes, B.F. & Dias, E.J.R. (2019) *Dipsas neuwiedi* (Ihering, 1911) (Squamata: Dipsadidae): Review of distribution extension and first record in the state of Sergipe, northeastern Brazil. *Herpetology Notes* 12, 409–417, 2 figs.

Andreone, F. & Gavetti, E. (2010) Metropolitan natural history museums in the study and conservation of herpetological biodiversity: The case of the Museo Regionale di Scienze Naturali of Turin. *Museologia Scientifica Memorie* (5), 49–61, 6 figs.

Angarita-Sierra, T. (2014) Hemipenial morphology in the semifossorial snakes of the genus *Ninia* and a new species from Trinidad, West Indies (Serpentes: Dipsadidae). *South American Journal of Herpetology* 9(2), 114–130, 8 figs.

Angarita-Sierra, T. (2017) *Ninia atrata* (Hallowell 1845) Viejita. *Catálago de Anfibios y Reptiles de Colombia* 3(2), 30–37, 4 figs.

Angarita-Sierra, T. (2018) Range expansion in the geographic distribution of *Ninia teresitae* (Serpentes: Dipsadidae): New localities from northwestern Ecuador. *Herpetology Notes* 11, 357–360, 3 figs.

Angarita-Sierra, T. (2019) On the second specimen of *Atractus alytogrammus* (Serpentes: Dipsadidae), with description of its unusual hemipeneal morphology. *Zootaxa* 4551(3), 396–400, 4 figs.

Angarita-Sierra, T. & Lynch, J.D. (2017) A new species of *Ninia* (Serpentes: Dipsadidae) from Chocó-Magdalena biogeographical province, western Colombia. *Zootaxa* 4244(4), 478–492, 6 figs.

Antúnez-Fonseca, C.A. (2019) Lowest altitudinal and fourth known record for *Ninia pavimentata* (Serpentes: Colubridae) in Honduras. *Cuadernos de Herpetología* 33(2), 91–93, 2 figs.

Aponte-Gutiérrez, A. & Vargas-Salinas, F. (2018) Geographic distribution of the genus *Siphlophis* (Fitzinger, 1843) (Colubridae, Dipsadinae, Serpentes) in Colombia. *Check List* 14(1), 195–201, 2 figs.

Aponte-Gutiérriez, A.F., Parra-Torres, F. & Velásquez-Suarez, A.J. (2017) *Helicops angulatus* (Linnaeus 1758) Mapaná de agua. *Catálago de Anfibios y Reptiles de Colombia* 3(1), 62–66, 2 figs.

Arakelyan, M.S., Danielyan, F.D., Corti, C., Sindaco, R. & Leviton, A.E. (2011) *Herpetofauna of Armenia and Nagorno-Karabakh*. Society for the Study of Amphibians and Reptiles. iv+149 pp, 34+152 figs.

Araújo, J. da S., Souza, M.B. de, Farias, T. de A., Silva, D.P. da, Venâncio, N.M., Maciel, J.M.L. & Melo-Sampaio, P.R. (2012) *Liophis dorsocorallinus* (Esqueda, Natera, La Marca and Ilija-Fistar, 2007) (Squamata: Dipsadidae): Distribution extension in southwestern Amazonia, state of Acre, Brazil. *Check List* 8(3), 518–519, 2 figs.

Araújo, P., França, R.C. de, Nascimento, F.S. do, Laranjeiras, D.O. & França, F.G.R. (2019) New records and range expansion of *Chironius carinatus* (Linnaeus, 1758) (Serpentes, Colubridae) from the state of Paraíba, northeast Brazil. *Check List* 15(5), 927–932, 2 figs.

Aréchaga-Ocampo, S., Montalbán-Huidobro, C.A. & Castro-Franco, R. (2008) Nuevos registros y ampliacion de la distribution de anfibios y reptiles en el Estado de Morelos, México. *Acta Zoológica Mexicana (new series)* 24(2), 231–233.

Arenas-Monroy, J.C. & Ahumada-Carrillo, I.T. (2015) *Agkistrodon bilineatus* (Günther, 1863) (Squamata: Viperidae): Confirmation of an inland locality for central Jalisco, Mexico. *Mesoamerican Herpetology* 2(3), 371–374, 2 figs.

Arenas-Vargas, I.S. (2015) *Serpientes de coral y una nueva subspecies – Micrurus isozonus sandneri ssp. nov.* Privately published. 54 pp, 9 figs.

Ariano-Sánchez, D. (2015) Geographic distribution. *Tantilla vermiformis*. *Herpetological Review* 46(2), 221–222.

Ariano-Sánchez, D. & Campbell, J.A. (2018) A new species of *Rhadinella* (Serpentes: Dipsadidae) from the dry forest of Motagua Valley, Guatemala. *Zootaxa* 4442(2), 338–344, 3 figs.

Arias, E. & Bolaños, F. (2014) A checklist of the amphibians and reptiles of San Isidro de Dota, Reserva Forestal Los Santos, Costa Rica. *Check List* 10(4), 870–877, 4 figs.

Arias-Monsalve, H.F. & Rojas-Morales, J.A. (2013) *Urotheca decipiens* (Serpentes: Colubridae): First record for the Department of Risaralda, Colombia. *Boletín Científico Museo de Historia Natural* 17(1), 144–146, 2 figs.

Armesto, L.O., Gutiérrez, D.R., Pacheco, R.D. & Gallardo, A.O. (2011) Reptiles del municipio de Cúcuta (Norte de Santander, Colombia). *Boletín Científico Museo de Historia Natural* 15(2), 157–168, 4 figs.

Arnaud, G. & Blázquez, M.C. (2018) First record of *Lichanura trivirgata* (Cope, 1868) (Squamata: Boidae) from Coronados Island, Gulf of California, Mexico. *Herpetology Notes* 11, 1025–1026, 2 figs.

Arquilla, A.M. & Lehtinen, R.M. (2018) Geographic variation in head band shape in juveniles of *Clelia clelia* (Colubridae). *Mesoamerican Herpetology* 5(1), 111–120, 4 figs.

Arribas, O.J. (2014) Altitudes notables en *Lacerta bilineata*, *Hierophis viridiflavus* y *Timon lepidus*. *Butlletí de la Societat Catalana d'Herpetologia* 21, 110–112.

Arslan, D., Olivier, A., Yaşar, Ç., Ismail, I.B., Döndüren, Ö., Ernoul, L., Beck, N. & Çiçek, K. (2018) Distribution and current status of herpetofauna in the Gediz Delta (Western Anatolia, Turkey). *Herpetology Notes* 11, 1–15, 6 figs.

Arteaga, A., Bustamante, L. & Guayasamin, J.M. (2013) *The Amphibians and Reptiles of Mindo – Life in the Cloudforest.* Universidad Tecnológica Indoamérica, Quito. 257 pp., numerous figs.

Arteaga, A., Mebert, K., Valencia, J.H., Cisneros-Heredia, D.F., Peñafiel, N., Reyes-Puig, C., Vieira-Fernandes, J.L. & Guayasamin, J.M. (2017) Molecular phylogeny of *Atractus* (Serpentes, Dipsadidae), with emphasis on Ecuadorian species and the description of three new taxa. *ZooKeys* 661, 91–123, 7 figs.

Arteaga, A., Salazar-Valenzuela, D., Mebert, K., Peñafiel, N., Aguiar, G., Sánchez-Nivicela, J.C., Pyron, R.A. et al. (2018) Systematics of South American snail-eating snakes (Serpentes, Dipsadini), with the description of five new species from Ecuador and Peru. *ZooKeys* 766, 79–147, 16 figs.

Arzamendia, V. (2016) New southern record of *Erythrolamprus reginae* (Linnaeus, 1758) (Serpentes: Dipsadidae), a vulnerable species in Argentina. *Check List* 12(1976), 1–4, 4 figs.

Asad, S., Mathai, J., Laird, D., Ong, N. & Buckingham, L. (2015) Preliminary herpetofaunal inventory of a logging concession in the Upper Baram, Sarawak, Borneo. *Herpetological Review* 46(1): 64–68, 1 fig.

Asad, S., McKay, J.L. & Putra, A.P. (2012) The herpetofauna of Nusa Penida, Indonesia. *Herpetological Bulletin* (122), 8–15, 4 figs.

Ascenso, A.C., Costa, J.C.L. & Prudente, A.L.C. (2019) Taxonomic revision of the *Erythrolamprus reginae* species group, with description of a new species from Guiana Shield (Serpentes: Xenodontinae). *Zootaxa* 4586(1), 65–97, 8 figs.

Ascenso, A.C. & Missassi, A.F.R. (2015) Geographic distribution. *Imantodes lentiferus Herpetological Review* 46(3), 386.

Atkinson, K., Smith, P., Dickens, J.K. & Lee-Zuck, C. (2018) Rediscovery of the 'lost' snake *Phalotris multipunctatus* (Serpentes: Dipsadidae) in Paraguay with behavioral notes and reference to the importance of Rancho Laguna Blanca for its conservation. *Current Herpetology* 37(1), 75–80, 3 figs.

Atkinson, K., Smith, P. & Sarvary, J. (2017) New and noteworthy snake species records (Colubridae and Dipsadidae) for the Reserva Natural Laguna Blanca, eastern Paraguay. *Check List* 13(2027), 1–5, 6 figs.

Auguste, R.J. (2019) Herpetofaunal checklist for six pilot protected areas in Trinidad and Tobago. *Herpetology Notes* 12, 577–585, 2 figs.

Auguste, R.J., Charles, S.P. & Murphy, J.C. (2015) An updated checklist of the amphibians and reptiles of the Aripo Savannas. *Living World, Journal of the Trinidad and Tobago Field Naturalists' Club* 2015, 37–43, 1 fig.

Auliya, M.A. (2006) *Taxonomy, life history and conservation of giant reptiles in West Kalimantan (Indonesian Borneo). Natur und Tier – Verlag GmbH, Münster,* 1–432, 647 figs.

Austin, S.D., Kerby, J.L. & Davis, D.R. (2017) Distributional records of amphibians and reptiles from Lake Oahe, South Dakota, USA. *Herpetological Review* 48(4), 817–820.

Avci, A., Ilgaz, Ç., Kumlutas, Y., Olgun, K. & Baran, I. (2007) Morphology and distribution of *Rhynchocalamus melanocephalus satunini* (Nikolsky, 1899) in Turkey. *Herpetozoa* 20(1/2), 82–86, 2 figs.

Avci, A., Ilgaz, Ç., Rajabizadoh, M., Yilmaz, C., Üzüm, N., Adriaens, D., Kumlutaş, Y. & Olgun, K. (2015) Molecular phylogeny and micro ct-scanning revealed extreme cryptic biodiversity in Kukri snake, *Muhtarophis* gen. nov., a new genus for *Rhynchocalamus barani* (Serpentes: Colubridae). *Russian Journal of Herpetology* 22(3), 159–174, 13 figs.

Avci, A., Üzüm, N., Ilgaz, Ç. & Olgun, K. (2009) A new finding of *Rhynchocalamus barani*, Baran's black-headed dwarf snake (Reptilia, Colubridae), in the Mediterranean region of Turkey

widens its distribution range. *Acta Herpetologica* 4(2), 177–182, 2 figs.

Avella, I., Castiglia, R. & Senczuk, G. (2017) Who are you? The genetic identity of some insular populations of *Hierophis viridiflavus* s.l. from the Tyrrhenian Sea. *Acta Herpetologica* 12(2), 209–214, 2 figs.

Avila-Pires, T.C.S., Hoogmoed, M.S. & Rocha, W.A. da (2010) Notes on the vertebrates of northern Pará, Brazil: A forgotten part of the Guianan Region, 1. Herpetofauna. *Boletin do Museu Paraense Emilio Goeldi, Ciéncias Naturais* 5(1), 13–112, 83 figs.

Aya-Cuero, C.A., Cáceres-Martínez, C.H. & Esquivel, D.A. (2019) First record of predation on greater sac-winged bat, *Saccopteryx bilineata* (Chiroptera: Emballonuridae), by the Colombian rainbow boa, *Epicrates maurus* (Serpentes: Boidae). *Herpetology Notes* 12, 815–817, fig. 1.

Azevedo, W. dos S., Amorim, L.G. dos S., Menezes, F. de A. & Abbegg, A.D. (2018) Natural history notes and geographic distribution of the poorly known *Echinanthera amoena* (Serpentes: Dipsadidae). *Herpetology Notes* 11, 925–928, 2 figs.

Babocsay, G. (2013) Misidentification of a snake responsible for an erroneous locality for *Dolichophis caspius* (Ophidia: Colubridae) in Hungary – a case resolved. *Folia Historico Naturalia Musei Matraensis* 37, 123–125, 2 figs.

Badillo-Saldaña, L.M., Berriozabal-Islas, C. & Ramírez-Bautista, A. (2014) New record of the snake *Drymobius chloroticus* (Cope, 1886) (Squamata: Colubridae) from Hidalgo, Mexico. *Check List* 10(1), 199–201, 3 figs.

Baeza-Tarin, F., Hernandez, T., Giovanetto, L., Trevino, A.S., Lazcano, D. & Graham, S.P. (2018a) Geographic distribution: *Lampropeltis gentilis. Herpetological Review* 49(3), 505.

Baeza-Tarin, F., Hernandez, T., Giovanetto, L., Trevino, A.S., Lazcano, D. & Graham, S.P. (2018b) Geographic distribution: *Tantilla cucullata. Herpetological Review* 49(3), 508.

Baeza-Tarin, F., Hernandez, T., Giovanetto, L., Trevino, A.S., Lazcano, D. & Graham, S.P. (2018c) Geographic distribution: *Trimorphodon vilkinsonii. Herpetological Review* 49(3), 508.

Bakhouche, B. & Escoriza, D. (2017) Genus *Malpolon*: New distribution area in Algeria. *The Herpetological Bulletin* 140, 35–36, 2 figs.

Ball, L.D. & Borrell, J.S. (2016) An inventory of herpetofauna from Wadi Sayq, Dhofar, Oman. *Journal of Threatened Taxa* 8(12), 9454–9460, 2 figs.

Bañuelos-Alamillo, J.A., Ahumada-Carrillo, I.T., Quintero-Díaz, G.E. & Carbajal-Márquez, R.A. (2019) The Mexican Patch-nosed Snake, *Salvadora mexicana* (Duméril, Bibron & Duméril, 1854; Squamata: Colubridae): A new state record for Zacatecas, Mexico, and a new prey species. *Cuadernos de Herpetología* 33(1), 45–47, 2 figs.

Bañuelos-Alamillo, J.A. & Carbajal-Márquez, R.A. (2016) Distribution notes: *Indotyphlops braminus. Mesoamerican Herpetology* 3(1), 204, 1 fig.

Bañuelos-Alamillo, J.A., Carbajal-Márquez, R.A., Quintero-Díaz, G.E. & Moreno-Ochoa, G. (2015) Geographic distribution. *Trimorphodon paucimaculatus. Herpetological Review* 46(3), 387.

Bañuelos-Alamillo, J.A., Carbajal-Márquez, R.A. & Rojo-Carrillo, G. (2016) Distribution notes: *Boa imperator. Mesoamerican Herpetology* 3(1), 184, 1 fig.

Bañuelos-Alamillo, J.A., Trujillo-De la Torre, I.Y., Quientero-Díaz & Carbajal-Márquez, R.A. (2017) The Central American tree snake, *Imantodes gemmistratus* (Cope, 1861): A new record for Zacatecas, Mexico. *Check List* 13(2115), 1–9.

Baran, I., Ilgaz, Ç., Avci, A., Kumlutaş, Y. & Olgun, K. (2005) *Türkiye amfibi ve sürüngenleri.* TÜBİTAK Popüler Bilim Kitaplari, Ankara. 1–204, numerous ill.

Barata, M., Perera, A., Harris, D.J., Van der Meijden, A., Carranza, S., Ceacero, F., García-Muñoz, E. et al. (2011) New observations of amphibians and reptiles in Morocco, with a special emphasis on the eastern region. *The Herpetological Bulletin* (116), 4–14, 3 figs.

Barbo, F., Gasparini, J.L., Almeida, A.P., Zaher, H., Grazziotin, F.G., Gusmão, R.B., Ferrarini, J.M.G. & Sawaya, R.J. (2016) Another new and threatened species of lancehead genus *Bothrops* (Serpentes, Viperidae) from Ilha dos Franceses, southeastern Brazil. *Zootaxa* 4097(4), 511–529, 7 figs.

Barbo, F., Marques, O.A.V. & Sawaya, R.J. (2011) Diversity, natural history, and distribution of snakes in the municipality of São Paulo. *South American Journal of Herpetology* 6(3), 135–160, 7 figs.

Barbosa, V.N., Amaral, J.M.S., Lima, L.F.L., França, R.C., França, F.G.R. & Santos, E.M. (2019) A case of melanism in *Dendrophidion atlantica* (Freire, Caramaschi & Gonçalves, 2010) (Colubridae) from northeastern Brazil. Herpetology *Notes* 12, 109–111, 1 fig.

Barker, D.G., Auliya, M. & Barker, T.M. (2018). *Pythons of the World, Volume III – The Pythons of Asia and the Malay Archipelago*. VPI Library, Boerne, xiii+371, 388 figs.

Barker, D.G., Barker, T.M., Davis, M.A. & Schuett, G.W. (2015) A review of the systematics and taxonomy of Pythonidae: An ancient serpent lineage. *Zoological Journal of the Linnean Society* 175(1), 1–19, 8 figs.

Barlow, A., Baker, K., Hendry, C.R., Peppin, L., Phelps, T., Tolley, K.A., Wüster, C.E. & Wüster, W. (2013) Phylogeography of the widespread African puff adder (*Bitis arietans*) reveals multiple Pleistocene refugia in southern Africa. *Molecular Ecology* 22(4), 1134–1157, 5 figs.

Barnes, C.H. & Knierim, T.K. (2018) A novel cave habitat use and range extension for the cryptic snake *Stegonotus muelleri* (Serpentes: Colubridae). *Phyllomedusa* 17(2), 295–297, 1 fig.

Barnes, C.H., Strinc, C.T., Suwanwaree, P. & Hill, J.G. (2017) Movement and home range of green pit vipers (*Trimeresurus* spp.) in a rural landscape in north-east Thailand. *The Herpetological Bulletin* 142, 19–28, 3 figs.

Barnestein, J.A.M., García-Cardenete, L., Jiménez-Cazalla, F., Valdeón, A., Escoriza, E., Martínez, G., Benavides, J. et al. (2012) Nuevas localidades de *Myriopholis algeriensis* y *Lamprophis fuliginosus*, y otras citas herpetológicas, en Marruecos. *Boletin de la Asociasión Herpetológica Española* 23(2), 63–68, 3 figs.

Barragán-Vázquez, M. del R., López-Luna, M.A. & Torrez-Pérez, M.A. (2017) Geographic distribution. *Geophis laticinctus*. *Herpetological Review* 48(1), 128.

Barrio-Amorós, C.L. (2019) On the taxonomy of snakes in the genus *Leptodeira*, with an emphasis on Costa Rican species. *IRCF Reptiles & Amphibians* 26(1), 1–15, 53 figs.

Barrio-Amorós, C.L., Chacón-Ortiz, A., Diasparra, J.P., Orellana, A.M., Bautista, J. & Molina, C. (2010) Distribution of *Siphlophis compressus* (Daudin, 1803) in Venezuela with a remarkable geographic extension. *Herpetozoa* 23(1/2), 100–103, 2 figs.

Barroso-Ruiz, C. & Bailon, S. (2003) Los anfíbios y los reptiles del Pleistoceno Superior de la Cueva del Boquete de Zafarraya (Málaga, España). In: Barroso-Ruiz, C. (Ed.), *El Pleistoceno Superior de la Cueva del Boquete de Zafarraya*. Junta de Andalucía, Consejería de Cultura, Dirección General de Bienes Culturales, Seville, 267–278, 2 figs.

Bartuano, A.S. & La Cruz, J. de (2014) Geographic distribution. *Sibon lamari*. *Herpetological Review* 45(4), 665.

Bassu, L., Nulchis, V., Satta, M.G., Fresi, C. & Corti, C. (2013) Atlas of Amphibians and Reptiles of Sardinia part III, Reptilia. Anfibi e Rettili di Sardegna III, Reptilia. In: Scillitani, G., Liuzzi, C., Lorusso, L., Mastropasqua, F. & Ventrella, P. (Eds.), *Atti IX Congresso Nazionale della Societas Herpetologica Italica (Bari - Conversano, 26–30 Settembre 2012)*. Conversano, Pineta, 108–113, 3 figs.

Bates, M.F. & Broadley, D.G. (2018) A revision of the egg-eating snakes of the genus *Dasypeltis* Wagler (Squamata: Colubridae: Colubrinae) in north-eastern Africa and south-western Arabia, with descriptions of three new species. *Indago* 34(1), 1–95, 34 figs.

Batista, A., Mebert, K., Lotzkat, S. & Wilson, L.D. (2016) A new species of centipede snake of the genus *Tantilla* (Squamata: Colubridae) from an isolated premontane forest in eastern Panama. *Mesoamerican Herpetology* 3(4), 948–960, 5 figs.

Batista, A. & Wilson, L.W. (2017) A new record for *Leptophis cupreus* (Cope, 1868) (Squamata: Colubridae) for Panama and Mesoamerica. *Mesoamerican Herpetology* 4(3), 671–673, 1 fig.

Battaglia, C.D., Faidley, C.R., Hudson, A.N., Brown, M.D., Pardue, T.M., Reid, M.L., Bass, A.A., Townsend, C.L. & Carr, J.L. (2015) Distribution records for Louisiana amphibians and reptiles. *Herpetological Review* 46(4), 579–581.

Bauer, A.M., DeBoer, J.C. & Taylor, D.J. (2017) Atlas of the reptiles of Libya. *Proceedings of the California Academy of Sciences, Series 4*, 64(8), 155–318, 66 figs.

Bauer, A.M., Vogel, G. & Campbell, P.D. (2015) A preliminary consideration of the dry snake skin specimens of Patrick Russell. *Hamadryad* 37(1/2), 73–84, 5 figs.

Bauer, A.M. & Wahlgren, R. (2013) On the Linck collection and specimens of snakes figured by Johann Jakob Scheuchzer (1735) – the oldest fluid-preserved herpetological collection in the world? *Bonn Zoological Bulletin* 62(2), 220–252, 12 figs, 8 plates.

Beane, J.C. (2013) New geographic distribution records for reptiles from North Carolina, USA. *Herpetological Review* 44(3), 478–481.

Beconi, H.C. & Scott, N.J. (2014) Oxyrhopus petolarius (Linnaeus, 1758) (Serpentes, Dipsadidae): Distribution extension and new departmental record for Paraguay. *Check List* 10(5), 1207–1209, 3 figs.

Behrendt, M. (2016a) New herpetological county records from Perry County, Ohio, USA. *Herpetological Review* 47(1), 90–91.

Behrendt, M. (2016b) New amphibian and reptile township records from Perry County, Ohio, USA: A call to action. *Herpetological Review* 47(2), 270–272.

Bekoshvili, D. & Doronin, I.V. (2015) New data on the distribution of snakes in Georgia (Caucasus). *Herpetological Review* 46(3), 388–390, 12 figs.

Beltra, S. (2013) Actualisation de la liste des amphibiens et reptiles de la region Provence-Alpes-Côte d'Azur. *Revue de CEN PACA* 2, 55–62.

Berghe, E van den, Sunyer, J. & Salazar-Saavedra, M. (2014) Miscellaneous notes. *Tantilla reticulata* Cope, 1868. *Mesoamerican Herpetology* 1(2), 304–305, 1 fig.

Bernarde, P.S., Abuquerque, S. de, Barros, T.O. & Turci, L.C.B. (2012b) Serpentes do estado de Rondônia, Brasil. *Biota Neotropica* 12(3), 154–182, 15 figs.

Bernarde, P.S., Albuquerque, S. de, Miranda, D.B. de & Turci, L.C.B. (2013) Herpetofauna do Floresta do baixo Rio Moa em Cruzeiro do Sul, Acre – Brasil. *Biota Neotropica* 13(1), 220–244, 23 figs.

Bernarde, P.S., Costa, H.C., Machado, R.A. & São-Pedro, V. de A. (2011a) *Bothriopsis bilineata bilineata* (Wied, 1821) (Serpentes: Viperidae): New records in the states of Amazonas, Mato Grosso and Rondônia, northern Brazil. *Check List* 7(3), 343–347, 3 figs.

Bernarde, P.S., Machado, R.A. & Turci, L.C.B. (2011b) Herpetofauna do área do Igarapé Experança na Reserva Extrativista da Liberdade, Acre – Brasil. *Biota Neotropica* 11(3), 117–144, 19 figs.

Bernarde, P.S., Souza, M.B. de, França, D.P.F.de & Freitas, M.A.de (2012a) *Micrurus annellatus annellatus* (Peters, 1871) (Serpentes: Elapidae): Distribution extension in the state of Acre, northern Brazil. *Check List* 8(3), 516–517, 3 figs.

Bernarde, P.S., Turci, L.C.B., Abegg, A.D. & Franco, F.L. (2018) A remarkable new species of coralsnake of the *Micrurus hemprichii* species group from the Brazilian Amazon. *Salamandra* 54(4), 249–258, 6 figs.

Bernardo, P.H., Guerra-Fuentes, R.A., Matiazzi, W. & Zaher, H. (2012b) Checklist of amphibians and reptiles of Reserva Biológica do Tapirapé, Pará, Brazil. *Check List* 8(5), 839–846, numerous figs.

Berriozabal-Islas, C., Ramírez-Bautista, A., Badillo-Saldaña, L.M. & Cruz-Elizalde, R. (2012) New records of the snake *Leptophis diplotropis* (Günther, 1872) (Squamata: Colubridae) from Hidalgo State, México. *Check List* 8(6), 1370–1372, 3 figs.

Bessesen, B.L. & Galbreath, G.J. (2017) A new subspecies of sea snake, *Hydrophis platurus xanthos*, from Golfo Duce, Costa Rica. *ZooKeys* 686, 109–123, 2 figs.

Beukema, W. (2011) First record of the genus *Tropidonophis* (Serpentes: Colubridae) and rediscovery of *Parias flavomaculatus* (Serpentes: Viperidae) on Siquijor Island, Philippines. *Herpetology Notes* 4, 177–179, 4 figs.

Beyhaghi P. (2016) Geographic deistribution. *Telescopus nigriceps*. *Herpetological Review* 47(1), 84.

Bezy, R.L. & Cole, C.J. (2014) Amphibians and reptiles of the Madrean Archipelago of Arizona and New Mexico. *American Museum Novitates* (3810), 1–23, 4 figs.

Bezy, R.L., Rosen, P.C., Van Devender, T.R. & Enderson, E.F. (2017) Southern distributional limits of the Sonoran Desert herpetofauna along the mainland coast of northwestern Mexico. *Mesoamerican Herpetology* 4(1), 137–167, 16 figs.

Bhattarai, S., Chalise, L., Gurung, A., Pokheral, C.P., Subedi, N. & Sharma, V. (2017a) Geographic distribution. *Liopeltis calamaria. Herpetological Review* 48(1), 129.

Bhattarai, S., Gurung, A., Chalise, L. & Pokheral, C.P. (2017b) Geographic distribution. *Psammodynastes pulverulentus. Herpetological Review* 48(1), 131.

Bhattarai, S. & Neupane, U. (2017) Geographic distribution. *Lycodon aulicus. Herpetological Review* 48(2), 392.

Bhattarai, S., Pokheral, C.P., Lamichhane, B.R., Regmi, U.R., Ram, A.K. & Subedi, N. (2018b) Amphibians and reptiles of Parsa National Park, Nepal. *Amphibian & Reptile Conservation* 12(1), 35–48, 39 figs.

Bhattarai, S., Pokheral, C.P., Lamichhane, B. & Subedi, N. (2017c) Herpetofauna of a Ramsar site: Beeshazar and associated lakes, Chitwan National Park, Nepal. *IRCF Reptiles & Amphibians* 24(1), 17–29, 9 figs.

Bhattarai, S., Pokheral, C.P. & Subedi, N. (2018a) New locality record of the Lined Stripe-necked Snake, *Liopeltis calamaria* (Günther, 1856) (Squamata: Colubridae) from Nepal. *IRCF Reptiles & Amphibians* 25(2), 125–126, 2 figs.

Bhattarai, S., Thapa, K.B., Chalise, L., Gurung, A., Pokheral, C.P., Subedi, N., Thapa, T.B. & Shah, K.B. (2017d) On the distribution of the Himalayan Stripe-necked Snake *Liopeltis rappi* (Günther, 1860) (Serpentes: Colubridae) in Nepal. *Amphibian & Reptile Conservation* 11(1), 88–92, 3 figs.

Bhosale, H.S., Gowande, G.G. & Mirza, Z.A. (2019) A new species of fossorial natricid snake of the genus *Trachischium* (Günther, 1858) (Serpentes: Natricidae) from the Himalayas of northeastern India. *Comptes Rendus Biologies* 342(9–10), 323–329, 3 figs.

Bhosale, H.S. & Joshi, D. (2014) Notes on distribution, natural history and habitat use of a Colubridae snake, *Rhabdops olivaceus* (Beddome, 1863). *Russian Journal of Herpetology* 21(3), 166–168, 3 figs.

Binaday, J.W.B. & Lobos, A.H.T. (2016) Geographic distribution. *Boiga drapiezii. Herpetological Review* 47(3), 425.

Blain, H.-A. (2009) Contribution de la paléoherpétofaune (Amphibia & Squamata) à la connaissance de l'évolution du climat et du paysage du Pliocène supérieur au Pléistocène moyen d'Espagne. *Treballs du Museu de Geologia de Barcelona* 16, 39–170, 31 figs.

Blain, H.-A., Agustí, J., López-García, J.M., Haddoumi, H., Aouraghe, H., El Hammouti, K., Pérez-González, A., Chacón, M.G. & Sala, R. (2013a) Amphibians and squamate reptiles from the late Miocene (Vallesian) of eastern Morocco (Guefaït-1, Jerada Province). *Journal of Vertebrate Paleontology* 33(4), 804–816, 6 figs.

Blain, H.-A., Bailon, S. & Agustí, J. (2007) Anurans and squamate reptiles from the latest early Pleistocene of Almenara-Casablanca-3 (Castellón, east of Spain). Systematic, climatic and environmental considerations. *Geodiversitas* 29(2), 269–295, 11 figs.

Blain, H.-A., Bailon, S. & Agustí, J. (2008) Amphibians and squamate reptiles from the latest early Pleistocene of Cueva Victoria Murcia, southeastern Spain, SW Mediterranean): Paleobiogeographic and paleoclimatic implications. *Geologica Acta* 6(4), 345–361, 8 figs.

Blain, H.-A., Laplana, C., Sevilla, P., Arsuaga, J.L., Baquedano, E. & Pérez-González, A. (2013b) MIS 5/4 transition in a mountain environment: Herpetofaunal assemblages from Cueva del Camino, central Spain. *Boreas* 43, 107–120, 6 figs.

Blain, H.-A., López-García, J.M., Cordy, J.-M., Pirson, S., Abrams, G., Di Modica, K. & Bonjean, D. (2014b) Middle to late Pleistocene herpetofauna from Scladina and Sous-Saint-Paul (Namur, Belgium). *Comptes Rendus Palevol* 13(8), 681–690, 6 figs.

Blain, H.-A., Monzón, A.M., López-García, J.M., Lozano-Fernández, I. & Folie, A. (2019) Amphibians and squamate reptiles from the late Pleistocene of the "Caverne Marie-Jeanne" (Hastière-Lavaux, Namur, Belgium): Systematics, paleobiogeography, and paleoclimatic and paleoenvironmental reconstructions. *Comptes Rendus Palevol* 18(7), 849–875, 11 figs.

Blain, H.-A., Santonja, M., Pérez-González, A., Panera, J. & Rubio-Jara, S. (2014a) Climate and environments during Marine Isotope Stage 11 in the central Iberian Peninsula: The herpetofaunal assemblage from the Acheulean site of Áridos-1, Madrid. *Quaternary Science Reviews* 94(1), 7–21, 6 figs.

Blair, C. & Sánchez-Ramírez, S. (2016) Diversity-dependent cladogenesis throughout western Mexico: Evolutionary biogeography of rattlesnakes (Viperidae: Crotalinae: *Crotalus* and *Sistrurus*). *Molecular Phylogenetics and Evolution* 97(1), 145–154, 3 figs.

Bochaton, C. & Bailon, S. (2018) A new fossil species of *Boa* (Linnaeus, 1758) (Squamata, Boidae), from the Pleistocene of Marie Galante Island (French West Indies). *Journal of Vertebrate Paleontology* 3(1462829), 1–14, 8 figs.

Bochaton, C., Boistel, R., Grouard, S., Ineich, I., Tresset, A. & Bailon, S. (2019) Fossil dipsadid snakes from the Guadeloupe Islands (French West-Indies) and their interactions with past human populations. *Geodiversitas* 41(12), 501–523, 9 figs.

Boev, Z. (2017) Fossil and subfossil records of reptiles (Reptilia Laurenti, 1768) in Bulgaria. *Historia Naturalis Bulgarica* 24, 165–178, 1 fig.

Bogdan, H.V., Ilies, D., Covaciu-Marcov, S.-D., Cicort-Lucaciu, A.-S. & Sas, I. (2011) Contributions to the study of the western region of the Poiana Ruscă Mountains and its surrounding areas. *North-Western Journal of Zoology* 7(1), 125–131, 1 fig.

Böhme, G. (1997) Bemerkungen zu einigen herpetofaunen aus dem Pleistozän Mittel- und Süddeutschlands. *Quartär* 47/48, 139–147.

Böhme, W. & Heath, J. (2018) Amphibian and reptilian records from south-central Mali and western Burkina Faso. *Bonn Zoological Bulletin* 67(1), 59–69, 13 figs.

Bok, B., Berroneau, M., Yousefi, M., Nerz, J., Deschandol, F., Berroneau, M. & Tiemann, L. (2017) Sympatry of *Pseudocerastes persicus* and *P. urarachnoides* in the western Zagros Mountains, Iran. *Herpetology Notes* 10, 323–325, 2 figs.

Bonte, C. (2012) Affirmation of *Coronella austriaca* (Laurenti, 1768) on the island of Cres, Croatia. *Herpetology Notes* 5, 65–66, 2 figs.

Borges-Leite, M.J., Rodrigues, J.F.M. & Borges-Nojosa, D.M. (2014) Herpetofauna of a coastal region of northeastern Brazil. *Herpetology Notes* 7, 405–413, 6 figs.

Borges-Nojosa, D.M., Lima, D.C., Bezerra, C.H. & Harris, D.J. (2017) Two new species of *Apostolepis* (Cope, 1862) (Serpentes: Elapomorphini) from brejos de altitude in northeastern Brazil. *Revista Nordestina de Zoologia* 10(2), 74–94, 4 figs.

Borkin, L.J. & Litvinchuk, S.N. (2015) Herpetological field research in the western part of Kazakhstan: P.S. Pallas and present. In: Borkin, L.J. & Golubev, A.V. (Eds.), *The Nature of western Kazakhstan and Peter Simon Pallas (Field Research 2012)*. Europeisky Dom, St. Petersburg, pp. 53–79, 1 fig.

Bouam, I., Benmokhtar, E. & Guechi, R. (2019) A fortuitous encounter with the vulnerable *Vipera latastei*: A new locality record from Algeria and distributional range extension. *Herpetology Notes* 12, 809–812, 2 figs.

Bouazza, A., Lansari, A., Marmol-Marin, G.M. del, Barthe, L., Berroneau, M. & Donaire, D. (2018) New records in Morocco and predictive distribution modeling for the rare Algerian Thread-snake: *Myriopholis algeriensis* (Jacquet, 1895). *Bulletin de la Société Herpétologique de France* 166, 43–50, 4 figs.

Boundy, J. (2013) Description of a second specimen of *Leptotyphlops parkeri* (Squamata: Leptotyphlopidae), with comments on its generic placement. *Zootaxa* 3637(4), 493–497, 3 figs.

Boundy, J. (2014) Comments on some African taxa of Leptotyphlopid snakes. *Occasional Papers of the Museum of Natural Science, Louisiana State University* (84), 1–7, 3 figs.

Boundy, J. & Carr, J.L. (2017) *Amphibians & Reptiles of Louisiana: An Identification and Reference Guide*. Louisiana State University Press, Baton Rouge. xi+386 pp, numerous figs.

Boundy, J. & David, P. (2015) The taxonomic status of the snake name *Tropidonotus roulei* Chabannaud, 1917. *Herpetological Review* 46(2), 295–297, 2 figs.

Bour, R., Cheylan, M. & Wandhammer, M.-D. (2017) Jean Hermann, l'holotype et le néotype de la couleuvre de Montpellier, *Coluber monspessulanus* (Hermann, 1804) (Reptilia, Squamata). *Zoosystema* 39(2), 273–284, 3 figs.

Branch, W.R. (2018) Snakes of Angola: An annotated checklist. *Amphibian & Reptile Conservation* 12(2), 41–82, 34 figs.

Branch, W.R., Baptista, N., Keates, C. & Edwards, S. (2019b) Rediscovery, taxonomic status, and phylogenetic relationships of two rare and endemic snakes (Serpentes: Psammophiinae) from the southwestern Angolan plateau. *Zootaxa* 4590(3), 4342–366, 6 figs.

Branch, W.R., Bayliss, J., Bittencourt-Silva, G.B., Conradie, W., Engelbrecht, H.M., Loader, S.P., Menegon, M., Nanvonamuquitxo, C. & Tolley, K.A. (2019c) A new species of tree snake (*Dipsadoboa*, Serpentes: Colubridae) from 'sky island' forests in northern Mozambique, with notes on other members of the *Dipsadoboa werneri* group. *Zootaxa* 4646(3), 541–563, 8 figs.

Branch, W.R., Verburgt, L., Bayliss, J., Kucharzewski, C., Rödel, M.-A. & Conradie, W. (2019a) New records of the Large-eyed Green Snake, *Philothamnus macrops* (Boulenger 1895), from Mozambique. *Herpetology Notes* 12, 19–29, 5 figs.

Brito, J.C., Rebelo, H., Crochet, P.-A. & Geniez, P. (2008) Data on the distribution of amphibians and reptiles from north and west Africa, with emphasis on *Acanthodactylus* lizards and the Sahara Desert. *Herpetological Bulletin* 105, 19–27, 1 fig.

Brito, P.S. de & Freire, E.M.X. (2012) New records and geographic distribution map of *Typhlops amoipira* (Rodrigues and Juncá, 2002) (Typhlopidae) in the Brazilian rainforest. *Check List* 8(6), 1347–1349, 2 figs.

Brito, P.S. de & Gonçalves, U. (2012) Squamata, Dipsadidae, *Boiruna sertaneja* Zaher, 1996: New records and geographic distribution map. *Check List* 8(5), 968–969, 2 figs.

Broadley, D.G. (2014) A new species of *Causus* Lichtenstein from the Congo/Zambezi watershed in north-western Zambia (Reptilia: Squamata: Viperidae). *Arnoldia Zimbabwe* 10(29), 341–350, 9 figs.

Broadley, D.G., Tolley, K.A., Conradie, W., Wishart, S., Trape, J.-F., Burger, M., Kusamba, C., Zassi-Boulou, A.-G. & Greenbaum, E. (2018) A phylogeny and genus-level revision of the African file snakes *Gonionotophis* Boulenger (Squamata: Lamprophiidae). *African Journal of Herpetology* 67(1), 43–60, 3 figs.

Broadley, D.G., Wade, E.O. & Wallach, V. (2014) A new species of *Myriopholis* from Ghat Oasis, south-western Libya (Squamata: Leptotyphlopidae). *Arnoldia Zimbabwe* 10(30), 351–359, 2 figs.

Broggi, M.F. (2014a) The herpetofauna of the isolated island of Gavdos (Greece). *Herpetozoa* 27(1/2), 83–90, 6 figs.

Broggi, M.F. (2014b) The herpetofauna of Kimolos (Milos Archipelago, Greece). *Herpetozoa* 27(1/2), 102–103.

Broggi, M.F. (2016a) The reptile fauna of the island of Elafonisos (Peloponnese, Lakonia, Greece). *Herpetozoa* 28(3/4), 198–203, 7 figs.

Broggi, M.F. (2016b) The herpetofauna of the island of Kythera (Attica, Greece). *Herpetozoa* 29(1/2), 37–46, 7 figs.

Brown, R.M., Oliveros, C.H., Siler, C.D., Fernandez, J.B., Welton, L.J., Buenavente, P.A.C., Diesmos, M.L.L. & Diesmos, A.C. (2012) Amphibians and reptiles of Luzon Island (Philippines), VII: Herpetofauna of Ilocos Norte Province, northern Cordillera mountain range. *Check List* 8(3), 469–490, 48 figs.

Brown, R.M., Siler, C.D., Oliveros, C.H., Welton, L.J., Rock, A., Swab, J., Van Weerd, M. et al. (2013) The amphibians and reptiles of Luzon Island, Philippines, VIII: The herpetofauna of Cagayan and Isabela Provinces, northern Sierra Madre Mountain Range. *ZooKeys* 266, 1–120, 104 figs.

Brown, R.M., Smart, U., Leviton, A.E. & Smith, E.N. (2018) A new species of long-glanded coralsnake of the genus *Calliophis* (Squamata: Elapidae) from Dinagat Island, with notes on the biogeography and species diversity of Philippine *Calliophis* and *Hemibungarus*. *Herpetologica* 74(1), 89–104, 9 figs.

Bryson, Jr., R.W., Linkem, C.W., Dorcas, M.E., Lathrop, A., Jones, J.M., Alvarado-Díaz, J., Grünwald, C.I. & Murphy, R.W. (2014) Multilocus species delimitation in the *Crotalus triseriatus* species group (Serpentes: Viperidae: Crotalinae), with the description of two new species. *Zootaxa* 3826(3), 475–496, 9 figs.

Bunyatova, S. (2013). The reptiles of Caucasian major within Azerbaijan. In: *International Caucasian Forestry Symposium*, Artvin, 611–616, 3 figs.

Buongermini, E. & Cacciali, P. (2017) Notas sobre un muestreo herpetológico en un ambiente ripario en al chaco húmedo de Paraguay. *Kempffiana* 13(1), 119–126, 2 figs.

Burbrink, F.T. & Guiher, T.J. (2015) Considering gene flow when using coalescent methods to delimit lineages of North American pitvipers of the genus *Agkistrodon*. *Zoological Journal of the Linnean Society* 173(2), 505–526, 5 figs.

Burchill, J. & Diamond, A. (2014) A herpetofaunal survey of Pike County, Alabama, USA. *Herpetological Review* 45(1), 98–99.

Burger, W.L. & Werler, J.E. (1954) The subspecies of the ring-necked coffee snake, Ninia diademata, and a short biological and taxonomic account of the genus. *University of Kansas Science Bulletin* 36(10), 643–672, 2 figs.

Buschmann, H., Scheel, B. & Brandt, T. (2006) *Amphibien und reptilien im Schaumberger Land und am Steinhuder Meer*. Natur & Text in Brandenburg GmbH, *Rangsdorf* 1–184, numerous ill.

Bush, B., Maryan, B., Browne-Cooper, R. & Robinson, D. (2007) *Reptiles and Frogs in the Bush: Southwestern Australia*. University of Western Australia Press, Crawley, ix+302, numerous color photos.

Bushar, L.M., Aborde, C.C.B., Gao, S., Gonzalez, M.V., Hoffman, J.A., Massaro, I.K., Savitzky, A.H. & Reinert, H.K. (2014) Genetic structure of Timber Rattlesnake (*Crotalus horridus*) populations: Physiographic influences and conservation implications. *Copeia* 2014(4), 694–706, 7 figs.

Bushar, L.M., Reynolds, R.G., Tucker, S., Pace, L., Lutterschmidt, W.I., Odum, R.A. & Reinert, H.K. (2015) Genetic characterization of an invasive Boa constrictor population on the Caribbean island of Aruba. *Journal of Herpetology* 48(4), 602–610, 3 figs.

Butler, B.O., Ceríaco, L.M.P., Marques, M.P., Bandeira, S., Júlio, T., Heinicke, M.P. & Bauer, A.M. (2019) Herpetofaunal survey of Huíla Province, southwest Angola, including first records from Bicuar National Park. *Herpetological Review* 50(2), 225–240, 6 figs.

Buzás, B., Farkas, B., Gulyás, E. & Géczy, C. (2018) The sea snakes (Elapidae: Hydrophiinae) of Fujairah. *Tribulus* 26, 4–31, 26 figs.

Cabral, H. & Caballero, A. (2013) Confirmación de *Tomodon ocellatus* Duméril, Bibron & Duméril, 1854 y primer ejemplar de referencia de *Micrurus silviae* Di-Bernardo, Borges-Martins & Silva, 2007 para Paraguay. *Boletin de la Museo Nacional de Historia Natural del Paraguay* 17(1), 63–66, 2 figs.

Cabral, H. & Cacciali, P. (2015) A new species of *Phalotris* (Serpentes: Dipsadidae) from the Paraguayan Chaco. *Herpetologica* 71(1), 73–77, 3 figs.

Cabral, H. & Cacciali, P. (2018) On the type locality of *Atractus paraguayensis* (Werner, 1924) (Serpentes: Dipsadidae). *Cuadernos de Herpetología* 32(1), 59–60.

Cabral, H., Lema, T. de & Renner, M.F. (2017) Revalidation of *Apostolepis barrioi* (Sepentes: Dipsadidae). *Phyllomedusa* 16(2), 243–254, 4 figs.

Cabral, H., Piatti, L., Souza, F.L. de, Scrocchi, G. & Ferreaira, V.L. (2015) *Xenodon pulcher* (Jan, 1863) (Serpentes: Dipsadidae) first record for Brazil and a distribution extension. *Herpetology Notes* 8, 361–364, 2 figs.

Cabral, H. & Sisa, F.N. (2016) Geographic distribution. *Epictia vellardi*. *Herpetological Review* 47(1), 83.

Cabral, H. & Weiler, A. (2014) Lista comentada de los reptiles de la Colección Zoologica de la Facultad de Ciencias Exactas y Naturales de Asunción, Paraguay. *Cuadernos de Herpetología* 28(1), 19–28, 2 figs.

Cacciali, P. (2013) Diversidad y selección de hábitat de la fauna de serpientes en Kangüery (área para Parque San Rafael). *Boletín del Museo Nacional de Historia Natural del Paraguay* 17(1), 29–39, 4 figs.

Cacciali, P., Bauer, F. & Martínez, N. (2015b) Herpetofauna de La Reserva Natural del Bosque Mbaracayú, Paraguay. *Kempffiana* 11(1), 29–47, 5 figs.

Cacciali, P., Cabral, H. & Caballero, A. (2015a) Riqueza de anfibios y reptiles de Paraguay: Evidencias de vacíos en el conocimiento y recomendaciones para su mejora. *Paraquaria Natural* 3(1), 12–16, 4 figs.

Cacciali, P., Cabral, H., Ferreira, V.L. & Köhler, G. (2016a) Revision of *Philodryas mattogrossensis* with the revalidation of *P. erlandi* (Reptilia: Squamata: Dipsadidae). *Salamandra* 52(4), 293–306, 15 figs.

Cacciali, P., Scott, N.J., Ortíz, A.L.A., Fitzgerald, L.A. & Smith, P. (2016b) The reptiles of Paraguay: Literature, distribution, and an annotated taxonomic checklist. *Special Publication of the Museum of Southwestern Biology* (11), 1–373, 5 figs, numerous maps.

Cacciali, P., Smith, P., Källberg, A., Pheasey, H. & Atkinson, K. (2013) Reptilia, Squamata, Serpentes, *Lygophis paucidens* Hige, 1952: First records for Paraguay. *Check List* 9(1), 131–132, 3 figs.

Cadle, J.E. (2012c) Rediscovery of the holotype of *Mastigodryas heathii* (Cope) (Serpentes: Colubridae) and additional notes on the species. *South American Journal of Herpetology* 7(1), 16–24, 2 figs.

Cai, B., Wang, Y., Chen, Y. & Li, J. (2015) A revised taxonomy for Chinese reptiles. *Biodiversity Science* 23(3), 365–382.

Caicedo-Portilla, J.R. (2011) Dimorfismo sexual y variación geográfica de la serpiente ciega *Typhlops reticulatus* (Scolecophidia: Typhlopidae) y distribución de otras especies del género en Colombia. *Caldasia* 33(1), 221–234, 6 figs.

Caicedo-Portilla, J.R. (2014) Redescubrimiento de *Mabuya berengerae*, *Mabuya pergravis* (Squamata: Scincidae) y *Coniophanes andresensis* (Squamata: Colubridae) y evaluación de su estado de amenaza en Las Islas de San Andrés y Providencia, Colombia. *Caldasia* 36(1), 181–201, 4 figs.

Çakmak, M., Akman, B. & Yildiz, M.Z. (2017) Herpetofauna of Bartin Province (Northwest Black Sea, Turkey). *South Western Journal of Horticulture, Biology and Environment* 8(2), 89–102, 2 figs.

Calcaño, D. & Barrio-Amorós, C.L. (2017) New distributional record for Corallus ruschenbergerii in sympatry with C. hortulanus. In: Barrio-Amorós, C.L. (Ed.), *Field Observations on Neotropical Treeboas of the Genus Corallus (Squamata: Boidae)*. IRCF Reptiles & Amphibians 24(1), 15, 1 fig.

Caldwell, M.W., Nydam, R.L., Palci, A. & Apesteguía, S. (2015) The oldest known snakes from the middle Jurassic-lower Cretaceous provide insights on snake evolution. *Nature Communications* 6(5996), 1–11, 4 figs.

Calzada-Arciniega, R.A. & Palacios-Aguilar, R. (2015) Distribution notes: *Coniophanes michoacanensis*. *Mesoamerican Herpetology* 2(1), 126–127, 1 fig.

Campbell, J.A. (2015) A new species of *Rhadinella* (Serpentes: Colubridae) from the Pacific versant of Oaxaca, Mexico. *Zootaxa* 3918(3), 397–405, 4 figs.

Campbell, J.A., Smith, E.N. & Hall, A.S. (2018) Caudals and calyces: The curious case of a consumed Chiapan colubroid. *Journal of Herpetology* 52(4), 458–471, 7 figs.

Campillo, G., Dávila-Galavíz, L.F., Flores-Villela, O. & Campbell, J. (2016) A new species of *Rhadinella* (Serpentes: Colubridae) from the Sierra Madre del Sur of Guerrero, Mexico. *Zootaxa* 4103(2), 165–173, 5 figs.

Campos, F.S. & Lage, A.R.B. (2013) Checklist of amphibians and reptiles from the Nova Roma Ecological Station, in the

Cerrado of the state of Goiás, central Brazil. *Herpetology Notes* 6, 431–438, 4 figs.

Campos-Rodríguez, J.I., Flores-Leyva, X., Lorenzo-Márquez, M.G. & Toledo-Jiménez, L.M. (2017) New records and distribution extensions of reptiles (Reptilia: Squamata) for the state of Zacatecas, Mexico. *Acta Zoológica Mexicana, New Series*, 33(1), 151–153.

Cano, P., Ball, H.A., Carpinetto, M.F. & Peña, G.D. (2015) Reptile checklist of Río Pilcomayo National Park, Formosa, Argentina. *Check List* 11(1658), 1–13, 46 figs.

Canseco-Márquez, L., Pavón-Vázquez, C.J., López-Luna, M.A. & Nieto-Montes de Oca, A. (2016) A new species of earth snake (Dipsadidae, *Geophis*) from Mexico. *ZooKeys* 610, 131–145, 3 figs.

Canseco-Márquez, L. & Ramírez-Gonzalez, C.G. (2015) New herpetofaunal records for the state of Oaxaca, Mexico. *Mesoamerican Herpetology* 2(3), 363–367, 2 figs.

Canseco-Márquez, L., Ramírez-González, C.G. & Campbell, J.A. (2018) Taxonomic review of the rare Mexican snake genus *Chersodromus* (Serpentes: Dipsadidae), with the description of two new species. *Zootaxa* 4399(2), 151–169, 9 figs.

Cantini, M., Menchetti, M., Vannini, A., Bruni, G., Borri, B. & Mori, E. (2013) Checklist of amphibians and reptiles in a hilly area of southern Tuscany (Central Italy): An update. *Herpetology Notes* 6, 223–228, 2 figs.

Capela, D.J.V., Morato, S.A.A., Moura-Leite, J.C. de, Prado, F. do, Borges, G.O. & Camilo, L.H.A. (2017) *Tropidophis paucisquamis* (Müller *in* Schenkel, 1901) (Serpentes: Tropidophiidae): First record from Paraná state and southern Brazil. *Check List* 13(6), 917–920, 2 figs.

Captain, A., Deepak, V., Pandit, R., Bhatt, B. & Athreya, R. (2019) A new species of pitviper (Serpentes: Viperidae: *Trimeresurus* Lacepède, 1804) from West Kameng District, Arunachal Pradesh, India. *Russian Journal of Herpetology* 26(2), 111–122, 4 figs.

Carbajal-Márquez, R.A., Arenas-Monroy, J.C., Domínguez-De la Riva, M.A. & Rivas-Mercado, E.A. (2015e) *Loxocemus bicolor* (Serpentes: Loxocemidae): Elevational and geographic range extension in Michoacan, Mexico. *Revista Mexicana de Herpetología* 1(1), 15–17, 2 figs.

Carbajal-Márquez, R.A., Arenas-Monroy, J.C., Quintero-Díaz, G.E., González-Saucedo, Z.Y. & García-Balderas, C.M. (2015d) First records of the Chihuahuan Black-headed Snake, *Tantilla wilcoxi* (Stejneger, 1902) (Squamata: Colubridae), in the Mexican state of Jalisco. *Check List* 11(1537), 1–3, 3 figs.

Carbajal-Márquez, R.A., Bañuelos-Alamillo, J.A., Rivas-Mercado, E.A., Quintero-Díaz, G.E. & Domínguez-de la Riva, M.A. (2015f) Geographic distribution. *Crotalus basiliscus*. *Herpetological Review* 46(3), 385.

Carbajal-Márquez, R.A., Gonzáles-Quiñones, F. & Quintero-Díaz, G.E. (2015a) Geographic distribution. *Indotyphlops braminus*. *Herpetological Review* 46(4), 573.

Carbajal-Márquez, R.A., González-Saucedo, Z.Y. & Arenas-Monroy, J.C. (2015b) *Crotalus aquilus* (Klauber, 1952) (Squamata: Viperidae), a new state record for Zacatecas, Mexico. *Acta Zoológica Mexicana* 31(1), 131–133, 2 figs.

Carbajal-Márquez, R.A. & Quintero-Díaz, G.E. (2016) The Herpetofauna of Aguascalientes, México. *Revista Mexicana de Herpetología* 2(1), 1–30, 9 figs.

Carbajal-Márquez, R.A., Quintero-Díaz, G.E. & Rivas-Mercado, E.A. (2015c) Geographic distribution. *Crotalus basiliscus*. *Herpetological Review* 46(2), 219.

Card, D.C., Schield, D.R., Adams, R.H., Corbin, A.B., Perry, B.W., Andrew, A.L., Pasquesi, G.I.M. et al. (2016) Phylogeographic and population genetic analyses reveal multiple species of

Boa and independent origins of insular dwarfism. *Molecular Phylogenetics and Evolution* 102(1), 104–116, 6 figs.

Cardwell, M.D., Gotte, S.W., McDiarmid, R.W., Gilmore, N. & Poindexter II, J.A. (2013) Type specimens of Crotalus scutulatus (Chordata: Reptilia: Squamata: Viperidae) re-examined, with new evidence after more than a century of confusion. *Proceedings of the Biological Society of Washington* 126(1), 11–16, 3 figs.

Carlino, P. & Pauwels, O.S.G. (2013) First documented record of *Thrasops jacksonii* (Günther, 1895) (Squamata: Colubridae) in Gabon. *North-Western Journal of Zoology* 9(1), 195–197, 3 figs.

Carlino, P. & Pauwels, O.S.G. (2014) Geographic distribution. *Hormonotus modestus*. *Herpetological Review* 45(4), 664.

Carlino, P. & Pauwels, O.S.G. (2015) An updated reptile list of Ivindo National Park, the herpetological hotspot of Gabon. *Bulletin of the Chicago Herpetological Society* 50(3), 25–39, 10 figs.

Carmona-Torres, F.H. & González-Hernández, A.J. (2014) Geographic distribution. *Geophis petersii*. *Herpetological Review* 45(2), 285.

Carrasco, P.A., Grazziotin, F.G., Farfán, R.S.C., Koch, C., Ochoa, J.A., Scrocchi, G.J., Leynaud, G.C. & Chaparro, J.C. (2019) A new species of *Bothrops* (Serpentes: Viperidae: Crotalinae) from Pampas del Heath, southeastern Peru, with comments on the systematics of the *Bothrops neuwiedi* species group. *Zootaxa* 4565(3), 301–344, 8 figs.

Carrasco, P.A., Venegas, P.J. & Valencia, J.H. (2016) First confirmed records of the endangered Andean pitviper *Bothrops lojanus* (Parker, 1930) (Viperidae: Crotalinae) from Peru. *Herpetology Notes* 9, 297–301, 2 figs.

Carvajal-Cogollo, J.E., Castaño-Mora, O.V., Cárdenas-Arévalo, G. & Urbina-Cardona, J.N. (2007) Reptiles de áreas asociadas a humedales de la planicie del Departamento de Córdoba, Colombia. *Caldasia* 29(2), 427–438, 3 figs.

Carvalho, V.T. de, Fraga, R. de, Eler, E.S., Kawashita-Ribeiro, R.A., Feldberg, E., Vogt, R.C., Carvalho, M.A. de, Noronha, J. da C. de, Condrati, L.H. & Bittencourt, S. (2013) Toad-headed pitviper *Bothrocophias hyoprora* (Amaral, 1935) (Serpentes, Viperidae): New records of geographic range in Brazil, hemipenial morphology, and chromosomal characterization. *Herpetological Review* 44(3), 410–414, 4 figs.

Casper, G.S. (2015) New county distribution records for amphibians and reptiles in Wisconsin. *Herpetological Review* 46(4), 582–586.

Casper, G.S., Rutherford, R.D. & Anton, T.G. (2015) Baseline distribution records for amphibians and reptiles in the Upper Peninsula of Michigan. *Herpetological Review* 46(3), 391–406.

Castro, D.P. de, Mângia, S., Magalhães, F. de M., Röhr, D.L., Cumurugi, F., Silveira-Filho, R.R. da, Silva, M.M.X. da et al. (2019) Herpetofauna of protected areas in the Caatinga VI: The Ubajara National Park, Ceará, Brazil. *Herpetology Notes* 12, 727–742, 9 figs.

Castro, T.M. de & Oliveira, J.C.F. de (2017) Range extension of *Lygophis meridionalis* (Schenkel, 1901) (Reptilia: Squamata: Dipsadidae: Xenodontinae) to Espírito Santo state, southeastern Brazil. *Check List* 13(2077), 1–4, 2 figs.

Castro, T.M. de, Oliveira, J.C.F. de, Gonzalez, R.C., Curcio, F.F. & Feitosa, D.T. (2017) First record of *Micrurus lemniscatus carvalhoi* (Roze, 1967) (Serpentes: Elapidae) from Espirito Santo state, southeastern Brazil. *Herpetology Notes* 10, 391–393, 2 figs.

Castro-Franco, R. & Zagal, M.G.B. (2006) Herpetofauna de las Áreas Naturales Protegidas Corredor Biologico Chichinautzin y la Sierra de Huautla, Morelos, Mexico. *Publicación Especial de Centro de Investigaciones Biológicas, UAEM* (1), 1–111, 47 figs.

Catenazzi, A., Lehr, E. & May, R. von (2013) The amphibians and reptiles of Manu National Park and its buffer zone, Amazon basin and eastern slopes of the Andes, Peru. *Biota Neotropica* 13(4), 269–283, 3 figs.

Cattaneo, A. (2014) Variabilità e sottospecie di *Montivipera xanthina* (Gray, 1849) nelle Isole Egee Orientali (Reptilia Serpentes Viperidae). *Naturalista Siciliano*, Series 4, 38(1), 51–83, 14 figs.

Cattaneo, A. (2017) Note sull'erpetofauna dell'Evros Sud-Occidentale (Grecia NE) e nuovo contributo all conoscenza di *Montivipera xanthina* (Gray, 1849) della Tracia Greca centro-orientale, con descrizione di *Montivipera xanthina occidentalis* subsp. Nova (Reptilia Serpentes Viperidae). *Naturalista Siciliano*, Series 4, 41(1), 53–74, 9 figs.

Cattaneo, A. (2018) Morpho-ecology of the Black Whip Snake *Dolichophis jugularis* (Linnaeus, 1758) of the southeast Aegean area (Dodecanese, SW Turkey), including previously unpublished growth rate data (Reptilia Serpentes). *Naturalista Siciliano*, Series 4, 42(1), 3–24, 4 figs.

Cattaneo, A. & Cattaneo, C. (2013) Sulla presenza di *Vipera ammodytes montandoni* Boulenger, 1904 nella Tracia Greca orientale e note eco-morfologiche sull'erpetofauna locale (Reptilia Serpentes). *Naturalista Siciliano*, Series 4, 37(2), 543–565, 8 figs.

Cattaneo, A. & Cattaneo, C. (2016) Osservazioni sull'erpetofauna (Reptilia) dell'entroterra di Kavala (Macedonia Greca Orientale). *Naturalista Siciliano*, Series 4, 40(2), 253–274, 13 figs.

Cavalcanti, L.B. de Q., Costa, T.B., Colli, G.R., Costa, G.C., França, F.G.R., Mesquita, D.O., Palmeira, C.N.S. et al. (2014) Herpetofauna of protected areas in the Caatinga II: Serra da Capivara National Park, Piauí, Brazil. *Check List* 10(1), 18–27, 12 figs.

Cedeño-Vázquez, J.R. & Beutelspacher-García, P.M. (2018) New records of *Masticophis mentovarius* (Squamata: Colubridae) from the state of Quintana Roo, Mexico. *Mesoamerican Herpetology* 5(1), 182–183, 1 fig.

Ceríaco, L.M.P., Bauer, A.M., Blackburn, D.C. & Lavres, A.C.F.C. (2014a) The herpetofauna of the Capanda Dam Region, Malanje, Angola. *Herpetological Review* 45(4), 667–674, 2 figs.

Ceríaco, L.M.P., Blackburn, D.C., Marques, M.P. & Calado, F.M. (2014b) Catalogue of the amphibian and reptile type specimens of the Museu de História Natural da Universidade do Porto in Portugal, with some comments on problematic taxa. *Alytes* 31, 13–36, 11 figs.

Ceríaco, L.M.P., Marques, M.P. & Bauer, A.M. (2018) Miscellanea Herpetologica Sanctithomae, with a provisional checklist of the terrestrial herpetofauna of São Tomé, Príncipe and Annobon islands. *Zootaxa* 4387(1), 91–108, 5 figs.

Ceríaco, L.M.P., Marques, M.P., Schmitz, A. & Bauer, A.M. (2017) The "Cobra-preta" of São Tomé Island, Gulf of Guinea, is a new species of *Naja* (Laurenti, 1768) (Squamata: Elapidae). *Zootaxa* 4324(1), 121–141, 6 figs.

Ceríaco, L.M.P., Sá, S. dos A.C. de, Bandeira, S., Valério, H., Stanley, E.L., Kuhn, A.L., Marques, M.P., Vindum, J.V., Blackburn, D.C. & Bauer, A.M. (2016) Herpetological survey of Iona National Park and Namibe Regional Natural Park, with a synoptic list of the amphibians and reptiles of Namibe Province, southwestern Angola. *Proceedings of the California Academy of Sciences*, series 4, 63(2), 15–61, 19 figs.

Čerňanský, A. (2011) New finds of the Neogene lizard and snake fauna (Squamata: Lacertilla; Serpentes) from the Slovak Republic. *Biologia* 66(5), 899–911, 9 figs.

Čerňanský, A., Rage, J.-C. & Klembara, J. (2015) The early Miocene squamates of Amöneburg (Germany): The first stages of modern squamates in Europe. *Journal of Systematic Palaeontology* 13(2), 97–128, 16 figs.

Čerňanský, A., Szyndlar, Z. & Mörs, T. (2017) Fossil squamate faunas from the Neogene of Hambach (northwestern Germany). *Palaeobio Palaeoenv* 97, 329–354, 19 figs.

Céspedez, J. & Alvarez, B.B. (2005) Lista de ejemplaires tipos y de especies de la Coleccion Herpetologica de la Universidad Nacional del Nordeste, Corrientes, Argentina (UNNEC). *Facena* 21, 137–142.

Chacón, F.M. & Johnston, R.D. (2013) *Amphibians and Reptiles of Costa Rica – A Pocket Guide*. Comstock Publishing Associates, Ithaca, xiii+172, numerous color photos.

Chan-ard, T. & Makchai, S. (2011) An inventory of reptiles and amphibians on the Koh Man Islands and in the coastal Prasae Estuaries, Rayong Province, South-eastern Thailand. *The Thailand Natural History Museum Journal* 5(1), 25–37, 3 plates.

Chan-ard, T., Parr, J.W.K. & Nabhitabhata, J. (2015) *A Field Guide to the Reptiles of Thailand*. Oxford University Press, New York, xxix+ 314, 352 figs.

Chan-ard, T., Seangthianchai, T. & Makchai, S. (2011) Monitoring survey of the distribution of herpetofauna in Khao Soi Dao Wildlife Sanctuary, Chanthaburi Province, with a note on faunal comparison with the Cardamom Mt. ranges. *The Thailand Natural History Museum Journal* 5(1), 39–56, 2 plates.

Chandra, K. & Gajbe, P.U. (2005) An inventory of Herpetofauna of Madhya Pradesh and Chhattisgarh. *Zoos' Print Journal* 20(3), 1812–1819.

Charruau, P., Escobedo-Galván, A.H., González, J.M.K. & Morales-Garduza, M.A. (2014) Geographic distribution. *Agkistrodon russeolus. Herpetological Review* 45(4), 662.

Charruau, P., Morales-Garduza, M.A., Reyes-Trinidad, J.G. & Ramírez-Pérez, M.A. (2015) Geographic distribution. *Imantodes gemmistratus. Herpetological Review* 46(1), 62.

Chaudhuri, A., Mukherjee, S., Chowdhury, S. & Purkayastha, J. (2018) Geographic distribution: *Gloydius himalayanus. Herpetological Review* 49(3), 505.

Chaudhuri, A., Sen, M. & Chowdhury, S. (2017) Geographic distribution. *Boiga forsteni. Herpetological Review* 48(2), 391.

Chaudhuri, A., Sharma, V. & Purkayastha, J. (2015) *Lycodon odishii*, a junior synonym of *Lycodon jara*, with notes on morphological variation in this species (Squamata, Colubridae). *Hamadryad* 37(1–2), 95–103, 7 figs.

Chávez-Arribasplata, J.C., Vásquez, D., Torres, C., Echevarría, L.Y. & Venegas, P.J. (2016) Confirming the presence of *Clelia equatoriana* (Amaral, 1924) (Squamata: Dipsadidae) in Peru. *Amphibian & Reptile Conservation* 10(1), 1–4, 2 figs.

Chen, T.-B., Luo, J., Meng, Y.-J., Wen, B.-H. & Jiang, K. (2013) Discovery of *Protobothrops maolanensis* in Guangxi, with taxonomic discussion. *Sichuan Journal of Zoology* 32(1), 116–118.

Chen, X., Lemmon, A.R., Lemmon, E.M., Pyron, R.A. & Burbrink, F.T. (2017) Using phylogenomics to understand the link between biogeographic origins and regional diversification in ratsnakes. *Molecular Phylogenetics and Evolution* 111, 206–218, 7 figs.

Chen, X., McKelvy, A.D., Grismer, L.L., Matsui, M., Nishikawa, K. & Burbrink, F.T. (2014) The phylogenetic position and taxonomic status of the rainbow tree snake *Gonyophis margaritatus* (Peters, 1871) (Squamata: Colubridae). *Zootaxa* 3881(6), 532–548, 1 fig.

Chen, Z., Zhang, L., Shi, J.-S., Tang, Y., Guo, Y., Song, Z. & Ding, L. (2019) A new species of the genus *Trimeresurus* from southwest China (Squamata: Viperidae). *Asian Herpetological Research* 10(1), 13–23, 6 figs.

Chettri, B. & Bhupathy, S. (2009) Occurrence of *Dinodon gammiei* (Blanford, 1878) in Sikkim, eastern Himalaya, India. *Journal of Threatened Taxa* 1(1), 60–61, 4 figs.

Chettri, B., Bhupathy, S. & Acharya, B.K. (2010) Distribution patterns of reptiles along an eastern Himalayan elevation gradient, India. *Acta Oecologia* 36(1), 16–22, 6 figs.

Chowdhury, S. & Chaudhuri, A. (2017) Natural history notes. *Bungarus caeruleus*. *Herpetological Review* 48(4), 856–857, 1 fig.

Chowdhury, S., Dutta, A., Chaudhuri, A. & Ghosh, S. (2017) Notes on the occurrence of the Bamboo Pitviper, *Trimeresurus gramineus* (Reptilia: Squamata: Viperidae), from southwestern West Benga, India. *IRCF Reptiles & Amphibians* 24(3), 168–171, 3 figs.

Christopoulos, A., Verokokakis, A.-G., Detsis, V., Nikolaides, I., Tsiokos, L., Pafilis, P. & Kapsalas, G. (2019) First records of *Eryx jaculus* (Linnaus, 1758) from Euboea Island, Greece (Squamata: Boidae). *Herpetology Notes* 12, 663–666, 2 figs.

Chuaynkern, Y., Duengkae, P., Chuaynkern, C., Pinthong, K. & Tanamthong, A. (2019) *Enhydris subtaeniata* (Bourret, 1934) (Serpentes: Homalopsidae): New distribution record and map. *Herpetology Notes* 12, 561–564, 3 figs.

Chunekar, H. & Alekar, S. (2015) Range extension of the calamaria reed snake, *Liopeltis calamaria* (Günther 1858) from the Western Ghats, Maharashtra, India. *IRCF Reptiles & Amphibians* 22(2), 79–80, 1 fig.

Cicchi, P.J.P., Serafim, J., Sena, M.A. de, Centeno, F. da C. & Jim, J. (2009) Herpetofauna emu ma área de Floresta Atlântica na Ilha Anchieta, município de Ubatuba, sudeste do Brasil. *Biota Neotropica* 9(2), 201–212, 9 figs.

Cihon, D. & Tok, C.V. (2014) Herpetofauna of the vicinity of Akşehir and Eber (Konya, Afyon), Turkey. *Turkish Journal of Zoology* 38(2), 234–241, 1 fig.

Cipolla, R.M. & Nappi, A. (2008) Check-list preliminare degli anfibi e dei rettili delle Isole Campane. In: Corti, C., (Ed.), *Herpetologia Sardiniae*. Societas Herpetologica Italica/Edizioni Belvedere, Latina, "Le Scienze" (8), pp. 145–147.

Cisneros-Heredia, D.F. (2005) Reptilia, Serpentes, Colubridae, *Tantilla supracincta*: Filling gap, first provincial record, geographic distribution map, and natural history. *Check List* 1(1), 23–26, 3 figs.

Cisneros-Heredia, D.F. (2006a) Reptilia, Colubridae, *Drymarchon melanurus*: Filling distribution gaps. *Check List* 2(1), 20–21, 1 fig.

Cisneros-Heredia, D.F. (2006b) Reptilia, Colubridae, *Helicops angulatus* and *Helicops leopardinus*: Distribution extension, new country record. *Check List* 2(1), 36–37, 1 fig.

Cisneros-Heredia, D.F. & Romero, A. (2015) First record of *Atractus medusa* (Serpentes, Dipsadidae) in Ecuador. *Herpetology Notes* 8, 417–420, 3 figs.

Cisneros-Heredia, D.F. & Touzet, J.-M. (2004) Distribution and conservation status of *Bothrops asper* (Garman, 1884) in Ecuador (Squamata: Serpentes: Viperidae: Crotalinae). *Herpetozoa* 17(3/4), 135–141, 1 fig.

Cisneros-Heredia, D.F. & Yánez-Muñoz, M. (2005) Reptilia, Viperidae, Crotalinae, *Porthidium nasutum*: Distribution extension and remarks on its range and records. *Check List* 1(1), 16–17, 1 fig.

Clarke III, E.O., Pattanarangsan, R. & Stoskopf, M.K. (2013) Applied anatomy of the eyelash viper (*Bothriechis schlegelii*). *Herpetological Review* 44(1), 40–44.

Clause, A.G., Pavón-Vázquez, C.J., Scott, P.A., Murphy, C.M., Schaad, E.W. & Gray, L.N. (2016) Identification uncertainty and proposed best-practices for documenting herpetofaunal geographic distributions, with applied examples from southern Mexico. *Mesoamerican Herpetology* 3(4), 976–1000, 4 figs.

Clause, J.K. (2016) Distribution notes: *Rhadinaea* (*Rhadinella*) *godmani*. *Mesoamerican Herpetology* 3(1), 199, 1 fig.

Clegg, J.R. & Jocque, M. (2016) The collection of snakes made by Benoît Mys and Jan Swerts in northern Papua New Guinea in 1982–85. *Journal of Herpetology* 51(3), 476–485, 1 fig.

Cocca, W., Rosa, G.M., Andreone, F., Aprea, G., Bergò, P.E., Mattioli, F., Mercurio, V. et al. (2018) The herpetofauna (Amphibia, Crocodylia, Squamata, Testudines) of the Isalo Massif, southwest Madagascar: Combining morphological, molecular and museum data. *Salamandra* 54(3), 178–200, 8 figs.

Coelho, R.D.F., Souza, K. de, Weider, A.G., Pereira, L.C.M. & Ribeiro, L.B. (2013) Overview of the distribution of snakes of the genus *Thamnodynastes* (Dipsadidae) in northeastern Brazil, with new records and remarks on their morphometry and pholidosis. *Herpetology Notes* 6, 355–360, 5 figs.

Cogălniceanu, D., Rozylowicz, L., Székely, P., Samoilă, C., Stănescu, F., Tudor, M., Székely, D. & Iosif, R. (2013) Diversity and distribution of reptiles in Romania. *ZooKeys* 341, 49–76, 28 figs.

Cogger, H.G. (2014) *Reptiles & Amphibians of Australia*. 7th ed. CSIRO Publishing, Collingwood, xxx+1033, numerous color photos.

Colli, G.R., Barreto-Lima, A.F., Dantas, P.T., Morais, C.J.S., Pantoja, D.L., Sena, A. de & Sousa, H.C. de (2019) On the occurrence of *Apostolepis phillipsi* (Serpentes, Elapomorphini) in Brazil, with the description of a new specimen from Mato Grosso. *Zootaxa* 4619(3), 580–588, 3 figs.

Colston, T.J., Barão-Nóbrega, J.A.L., Manders, R., Lett, A., Willmott, J., Cameron, G., Hunter, S. et al. (2015) Amphibians and reptiles of the Calakmul Biosphere Reserve, México, with new records. *Check List* 11(1759), 1–7, 8 figs.

Conrad, J.L. (2008) Phylogeny and systematics of Squamata (Reptilia) based on phylogeny. *Bulletin of the American Museum of Natural History* (310), 1–82, 64 figs.

Conradie, W., Bills, R. & Branch, W.R. (2016b) The herpetofauna of the Cubango, Cuito, and lower Cuando river catchments of south-eastern Angola. *Amphibian & Reptile Conservation* 10(2), 6–36, 9 figs.

Conradie, W., Bittencourt-Silva, G.B., Engelbrecht, H.M., Loader, S.P., Menegon, M., Nanvonamuquitxo, C., Scott, M. & Tolley, K.A. (2016a) Exploration into the hidden world of Mozambique's sky island forests: New discoveries of reptiles and amphibians. *Zoosystematics and Evolution* 92(2), 163–180, 4 figs.

Conradie, W., Branch, W.R. & Watson, G. (2019) Type specimens in the Port Elizabeth Museum, South Africa, including the historically important Albany Museum collection. Part 2: Reptiles (Squamata). *Zootaxa* 4576(1), 1–45, 14 figs.

Contreras-Lozano, J.A., Lazcano, D. & Contreras-Balderas, AJ. (2011) Distribución ecológica de la herpetofauna en gradients altitudinales superiors del Cerro El Potosi, Galeana, Nuevo León, México. *Acta Zoológica Mexicana (n.s.)* 27(2), 231–243, 4 figs.

Corsetti, L. & Romano, A. (2008) On the occurrence of the Italian Aesculapian snake, *Zamenis lineatus* (Camerano, 1891), in

Latium (central Italy). *Acta Herpetologica* 3(2), 179–183, 2 figs.

Cortez-Fernandez, C. (2005) Herpetofauna de la zona norte del Parque Nacional y Area Natural de Manejo Integrado Madidi (NANMI-Madidi). *Ecología en Bolivia* 40(2), 10–26, 1 fig.

Coşkun, Y., Coşkun, M. & Schweiger, M. (2012) A new locality record of the blunt-nosed viper, *Macrovipera lebetina obtusa* in central Anatolia, Turkey (Serpentes: Viperidae). *Fen Bilimleri Dergisi, Cilt* 33(1), 22–28, 2 figs.

Costa, H.C. & Bérnils, R.S. (2015) *Apostolepis parassimilis* Lema & Renner, 2012 an objective synonym of *A. tertulianobeui* (Lema, 2004) (Dipsadidae: Elapomorphini). *Zootaxa* 3957(2), 243–245, 1 fig.

Costa, H.C., Cotta, G.A. & MacCulloch, R.D. (2015b) New easternmost and southernmost records of *Pseudoboa coronata* (Schneider, 1801) (Serpentes: Dipsadidae: Pseudoboini), with a distribution map. *Check List* 11(1624), 1–7, 2 figs.

Costa, H.C., Moura, M.R. & Feio, R.N. (2013) Taxonomic revision of *Drymoluber* (Amaral, 1930) (Serpentes: Colubridae). *Zootaxa* 3716(3), 349–394, 24 figs.

Costa, H.C., Resende, F.C., Gonzalez, R.C., Cotta, G.A. & Feio, R.N. (2014) Checklist of the snakes of Nova Ponte, Minas Gerais, Brazil. *Salamandra* 50(2), 110–116, 1 fig.

Costa, H.C., Santana, D.J., Leal, F., Koroiva, R. & Garcia, P.C.A. (2016) A new species of *Helicops* (Serpentes: Dipsadidae: Hydropsini) from southeastern Brazil. *Herpetologica* 72(2), 157–166, 6 figs.

Costa, H.C., Santos, P.S., Laia, W.P., Garcia, P.C.A. & Bérnils, R.S. (2015a) *Mussurana montana* (Franco, Marques & Puorto, 1997) (Serpentes: Dipsadidae): Noteworthy records and an updated distribution map. *Check List* 11(1657), 1–3, 2 figs.

Costa, J.C.L., Kucharzewski, C. & Prudente, A.L. da C. (2015) The real identity of *Leptodira nycthemera* Werner, 1901 from Ecuador: A junior synonym of *Oxyrhopus petolarius* (Linnaeus, 1758) (Serpentes, Dipsadidae) *ZooKeys* 506, 119–125, 1 fig.

Costa, T.B., Laranjeiras, D.O., Caldas, F.L.S., Santana, D.O., Silva, C.F. da, Alcântara, E.P. de, Brito, S.V. et al. (2018) Herpetofauna of protected areas in the Caatinga VII: Aiuaba Ecological Station (Ceará, Brazil). *Herpetology Notes* 11, 929–941, 6 figs.

Cota, M. (2010) Geographical distribution and natural history notes on *Python bivittatus* in Thailand. *The Thailand Natural History Museum Journal* 4(1), 19–28, 6 figs.

Couper, P.J., Peck, S.R., Emery, J.-P. & Scott, J.S. (2016) Two snakes from eastern Australia (Serpentes: Elapidae); a revised concept of Antaioserpens warro (De Vis) and a redescription of A. albiceps (Boulenger). *Zootaxa* 4097(3), 396–408, figs.

Covaciu-Marcov, S.-D., Cicort-Lucaciu, A., Dobre, F., Ferenţi, S., Birceanu, M., Mihuţ, R. & Strugariu, A. (2009b) The herpetofauna of the Jiului Gorge National Park, Romania. *North-Western Journal of Zoology* 5(suppl. 1), S01–S78, 39 figs, 26 maps.

Covaciu-Marcov, S.-D., Cicort-Lucaciu, A., Gaceu, O., Sas, I., Ferenti, S. & Bogdan, H.V. (2009a) The herpetofauna of the south-western part of Mehedinti County, Romania. *North-Western Journal of Zoology* 5(1), 142–164, 8 figs.

Covaciu-Marcov, S.-D., Cicort-Lucaciu, A., Ile, R.-D., Pascondea, A. & Vatamaniuc, R. (2007a) Contributions to the study of the geographical distribution of the herpetofauna in the north-east area of Arad County in Romania. *Herpetologica Romanica* 1, 62–69, 3 figs.

Covaciu-Marcov, S.-D., Cicort-Lucaciu, A., Sas, I., Groza, M.-I. & Bordaş, I. (2007b) Contributions to the knowledge regarding the herpetofauna from the Maramureş county areas of

"Măgura Codrului," Romania. *Biharean Biologist* 1, 50–56, 3 figs.

Covaciu-Marcov, S.-D. & David, A. (2010) *Dolichophis caspius* (Serpentes: Colubridae) in Romania: New distribution records from the northern limits of its range. *Turkish Journal of Zoology* 34, 119–121, 2 figs.

Covaciu-Marcov, S.-D., Ferenti, S., Cicort-Lucaciu, A. & Sas, I. (2012) *Eryx jaculus* (Reptilia, Boidae) north of Danube: A road-killed specimen from Romania. *Acta Herpetologica* 7(1), 41–47, 3 figs.

Covaciu-Marcov, S.-D., Ghira, I., Cicort-Lucaciu, A., Sas, I., Strugariu, A. & Bogdan, H. (2006b) Contributions to knowledge regarding the geographical distribution of the herpetofauna of Dobrudja, Romania. *North-Western Journal of Zoology* 2(2), 88–125, 23 figs.

Covaciu-Marcov, S.-D., Sas, I., Cicort-Lucaciu, A., Achim, A. & Andritcu, A. (2005) The herpetofauna of the Tăşnad Hills (Satu-Mare County, Romania). Analele Stiintifice ale Universitătii "Al.I. *Cuza" Iaşi, s. Biologie Animală* 51, 159–168.

Covaciu-Marcov, S.-D., Sas, I., Kiss, A., Bogdan, H. & Cicort-Lucaciu, A. (2006a) The herpetofauna from the Teuz River hydrographic basin (Arad County, Romania). *North-Western Journal of Zoology* 2(1), 27–38, 8 figs.

Cox, C.L., Rabosky, A.R.D., Holmes, I.A., Reyes-Velasco, J., Roelke, C.E., Smith, E.N., Flores-Villela, O., McGuire, J.A. & Campbell, J.A. (2018) Synopsis and taxonomic revision of three genera in the snake tribe Sonorini. *Journal of Natural History* 52(13–16), 945–988, 8 figs.

Crochet, P.-A., Leblois, R. & Renoult, J.P. (2015) New reptile records from Morocco and Western Sahara. *Herpetology Notes* 8, 583–588, fig. 2.

Crother, B.I., Bonett, R.M., Boundy, J., Burbrink, F., Queiroz, K. de, Frost, D.R., Highton, R. et al. (2017) *Scientific and Standard English Names of Amphibians and Reptiles of North America North of Mexico, with Comments Regarding Confidence in our Understanding.* Eighth edition. Society for the Study of Amphibians and Reptiles, Herpetological Circular (43), 102 pp.

Crother, B.I., Savage, J.M. & Holycross, A.T. (2012) Comment on the proposed conservation of *Crotalinus catenatus* (Rafinesque, 1818) (currently *Sistrurus catenatus*) and *Crotalus tergeminus* (currently *Sistrurus tergeminus*; Reptilia, Serpentes) by designation of neotypes for both species (Case 3571: See BZN 68: 271–274). *Bulletin of Zoological Nomenclature* 69(1), 1–2.

Crumb, B., Duley, A., Ruback, P. & Ray, J.M. (2015) Geographic range extension and comments on the snake *Siphlophis cervinus* (Laurenti, 1768) (Serpentes: Dipsadidae) in Panama. *Mesoamerican Herpetology* 2(4), 579–581, 2 figs.

Cruz-Elizalde, R., Ramírez-Bautista, A., Hernández-Salinas, U., Berriozabal-Islas, C. & Wilson, L.D. (2019) An updated checklist of the herpetofauna of Querétaro, Mexico: Species richness, diversity, and conservation status. *Zootaxa* 4638(2), 273–290, 2 figs.

Cruz-Elizalde, R., Ramírez-Bautista, A. & Lara-Tufiño, D. (2015) New record of the snake *Geophis turbidus* (Squamata: Dipsadidae) from Hidalgo, Mexico, with annotations of a juvenile specimen. *Check List* 11(1724), 1–6, 3 figs.

Cruz-Sáenz, D., Muñoz-Nolasco, F.J., Lazcano, D. & Flores-Covarrubias, E. (2015) Noteworthy records for *Tantilla cascadae* and *T. ceboruca* (Squamata: Colubridae) from Jalisco, Mexico. *Check List* 11(1708), 1–5, 3 figs.

Cruz-Sáenz, D., Muñoz-Nolasco, F.J., Mata-Silva, V., Johnson, Jerry D., García-Padilla, E. & Wilson, L.D. (2017) The herpetofauna of Jalisco, Mexico: Composition, distribution, and conservation status. *Mesoamerican Herpetology* 4(1), 22–118, numerous color illus.

Cui, S., Luo, X., Chen, D., Sun, J., Chu, H., Li, C. & Jiang, Z. (2016) The adder (*Vipera berus*) in southern Altay Mountains: Population characteristics, distribution, morphology and phylogenetic position. *PeerJ* 2016(2342), 1–18, 3 figs.

Cumhuriyet, O. & Ayaz, D. (2015) Güllük Körfezi (Milas, Muğla) ve civarının herpetofaunası. *Anadolu Doğa Bilimleri Dergisi* 6(2), 163–168, 1 fig.

Cunningham, P. (2010) A contribution to the reptiles of the Farasan Islands, Saudi Arabia. *African Herp News* (50), 21–25, 1 fig.

Curcio, F.F., Scali, S. & Rodrigues, M.T. (2015) Taxonomic status of *Erythrolamprus bizona* (Jan, 1863) (Serpentes, Xenodontinae): Assembling a puzzle with many missing pieces. *Herpetological Monographs* 29, 40–64, 11 figs.

Cyriac, V.P. & Kodandaramaiah, U. (2017) Paleoclimate determines diversification patterns in the fossorial snake family Uropeltidae Cuvier, 1829. *Molecular Phylogenetics and Evolution* 116, 97–107, 3 figs.

Dahn, H.A., Strickland, J.L., Osorio, A., Colston, T.J. & Parkinson, C.L. (2018) Hidden diversity within the depauperate genera of the snake tribe Lampropeltini (Serpentes, Colubridae). *Molecular Phylogenetics and Evolution* 129, 214–225, 4 figs.

Dainesi, R.L.S., Abegg, A.D., Bernarde, P.S., Correa, B.P., Machado, L.P.C., Meneses, A.S.O. & Sena, A. de. (2019) Integrative overview of snake species from Londrina, state of Paraná, Brazil (Reptilia; Squamata). *Herpetology Notes* 12, 419–430, 5 figs.

Dal Vechio, F., Junior, M.T., Neto, A.M. & Rodrigues, M.T. (2015b) On the snake *Siphlophis worontzowi* (Prado, 1940): Notes on its distribution, diet and morphological data. *Check List* 11(1534), 1–5, 3 figs.

Dal Vechio, F., Junior, M.T., Recoder, R., Sena, M.A. de Souza, S.M. & Rodrigues, M.T. (2015a) Distribution extension and revised map of *Erythrolamprus pygmaeus* (Cope, 1868) (Serpentes: Dipsadidae). *Check List* 11(1719), 1–4, 2 figs.

Dal Vechio, F., Recoder, R., Rodrigues, M.T. & Zaher, H. (2013) The herpetofauna of the Estação Ecológia de Uruçuí-Una, state of Piauí, Brazil. *Papéis Avulsos de Zoologia* 53(16), 225–243, 7 figs.

Dal Vechio, F., Teixeira Jr., M., Recoder, R., Rodrigues, M.T. & Zaher, H. (2016) The herpetofauna of Parque Nacional da Serra das Confusões, state of Piauí, Brazil, with a regional species list from an ecotonal area of Cerrado and Caatinga. *Biota Neotropica* 16(e20150105), 1–19, 8 figs.

Damas-Moreira, I., Tomé, B., Harris, D.J. & Maia, J.P. (2014) Moroccan herpetofauna: Distribution updates. *Herpetozoa* 27(1/2), 96–102, 2 figs.

Dandge, P.H. & Tiple, A.D. (2016) Notes on natural history, new distribution records and threats of Indian Egg Eater Snake *Elachistodon westermanni* (Reinhardt, 1863) (Serpentes: Colubridae): Implications for conservation. *Russian Journal of Herpetology* 23(1), 55–62, 10 figs.

D'Angelo, J.S., Agnolin, F.L. & Godoy, F.A. (2015) *Xenodon histricus* (Jan, 1863) (Squamata: Dipsadidae): Distribution extension and new province record in Argentina. *Check List* 11(1737), 1–3, 2 figs.

Daniells, E.A., Ackley, J.W., Carter, R.E., Muelleman, P.J., Rudman, S.M., Turk, P.A., Vélez-Espinet, N.J., White, L.A. & Wyszynski, N.N. (2008) An annotated checklist of the amphibians and reptiles of Dominica, West Indies. *Iguana* 15(3), 130–141, numerous figs.

Das, A., Saikia, U., Murthy, B.H.C.K., Dey, S. & Dutta, S.K. (2009) A herpetofaunal inventory of Barail Wildlife Sanctuary and adjacent regions, Assam, north-eastern India. *Hamadryad* 34(1), 117–134, 2 figs.

Das, A., Sharma, P., Surendran, H., Nath, A., Ghosh, S., Dutta, D., Mondol, J. & Wangdi, Y. (2016) Additions to the herpetofauna of Royal Manas National Park, with six new country records. *Herpetology Notes* 9, 261–278, 24 figs.

Das, I. (2015) Patrick Russell (1727–1805), surgeon and polymath naturalist. *Hamadryad* 37(1–2), 1–11, 11 figs.

Das, I., Breuer, H. & Shonleben, S. (2013) *Gerarda prevostiana* (Eydoux and Gervais, 1837) (Squamata: Serpentes: Homalopsidae), a new snake for Borneo. *Asian Herpetological Research* 4(1), 76–78, 4 figs.

Das, S., Campbell, P.D., Roy, S., Mukerjee, S., Pramanick, K., Biswas, A. & Raha, S. (2019) Cranial osteology and molecular phylogeny of Argyrogena fasciolata (Shaw, 1802) (Colubridae: Serpentes). *Vertebrate Zoology* 69, 311–325, 6 figs.

Dashevsky, D. & Fry, B.G. (2018) Ancient diversification of three-finger toxins in *Micrurus* coral snakes. *Journal of Molecular Evolution* 86(1), 58–67, 4 figs.

Dávalos-Martínez, A., Cruz-Sáenz, D., Najar-Sánchez, J.A. & Muñoz-Nolasco, F.J. (2019) A new southern-most record of the Sonoran Coralsnake, *Micruroides euryxanthus* (Kennicott, 1860) (Squamata: Elapidae) from the state of Jalisco, Mexico. *Herpetology Notes* 12, 681–683, 2 figs.

David, P., Agarwal, I., Athreya, R., Mathew, R., Vogel, G. & Mistry, V.K. (2015a) Revalidation of *Natrix clerki* Wall, 1925, an overlooked species in the genus *Amphiesma* (Duméril, Bibron & Duméril, 1854) (Squamata: Natricidae). *Zootaxa* 3919(2), 375–395, 10 figs.

David, P., Pauwels, O.S.G., Nguyen, T.Q. & Vogel, G. (2015b) On the taxonomic status of the Thai endemic freshwater snake *Parahelicops boonsongi*, with the erection of a new genus (Squamata: Natricidae). *Zootaxa* 3948(2), 203–219, 4 figs.

David, P. & Vogel, G. (2015) An updated list of Asian pitvipers and a selection of recent publications. In: Visser, J. (Ed.), *Asian Pitvipers. Breeding Experience & Wildlife.* Edition Chimaira, Frankfurt am Main, 545–565.

David, P., Vogel, G. & Rooijen, J. van (2013) On some taxonomically confused species of the genus *Amphiesma* Duméril, Bibron & Duméril, 1854 related to *Amphiesma khasiense* (Boulenger, 1890) (Squamata: Natricidae). *Zootaxa* 3694(4), 301–335, 7 figs.

David, P., Vogel, G., Rooijen, J. van & Pierre, L. (2011) The name-bearing type of *Oligodon quadrilineatus* (Jan & Sordelli, 1865) (Squamata: Colubridae). *Zootaxa* 3111, 67–68.

Davis, D.R. (2018) Distributional records of amphibians and reptiles from the lower James River Valley, South Dakota, USA. *Herpetological Review* 49(4), 720–721.

Davis, D.R. & Dilliard, K.A. (2016) Historic amphibian and reptile county records from the A. Jewell Schock Museum of Natural History at Wayne State College. *Herpetological Review* 47(4), 635–637.

Davis, D.R. & Farkas, J.K. (2018) New county records of amphibians and reptiles from South Dakota, USA from 2017. *Herpetological Review* 49(2), 289–295.

Davis, D.R., Farkas, J.K., Johannsen, R.E., Leonard, K.M. & Kerby, J.L. (2017a) Distributional records for amphibians and reptiles from South Dakota, USA. *Herpetological Review* 48(1), 133–137.

Davis, D.R., Farkas, J.K., Johannsen, R.E. & Maltaverne, G.A. (2017b) Historic amphibian and reptile county records from South Dakota, USA. *Herpetological Review* 48(2), 394–406.

Davis, D.R., Ferguson, K.J., Koch, A.D., Berg, E.A., Vlcek, J.R. & Kerby, J.L. (2016) New amphibian and reptile county records from eastern South Dakota, USA. *Herpetological Review* 47(2), 267–270.

Davis, D.R. & LaDuc, T.J. (2018) Amphibians and reptiles of C. E. Miller Ranch and the Sierra Vieja, Chihuahuan Desert, Texas, USA. *ZooKeys* 735, 97–130, 7 figs.

Davis, D.R., Siddons, S.R. & Kerby, J.L. (2014) New amphibian and reptile county records from eastern Nebraska, USA. *Herpetological Review* 45(1), 99–100.

Davis, M.A., Douglas, M.R., Collyer, M.L. & Douglas, M.E. (2016) Deconstructing a species-complex: Geometric morphometric and molecular analyses define species in the Western Rattlesnake (Crotalus viridis). *PLOS ONE* 11(0146166), 1–21, 6 figs.

Debata, S. (2017) Natural history notes. *Bungarus caeruleus. Herpetological Review* 48(4), 857, 1 fig.

Deepak, V., Narayanan, S., Sarkar, V., Dutta, S.K. & Mohapatra, P.P. (2019) A new species of *Ahaetulla* (Link, 1807) (Serpentes: Colubridae: Ahaetullinae) from India. *Journal of Natural History* 53(9–10), 497–516, 8 figs.

Deepak, V., Ruane, S. & Gower, D.J. (2018) A new subfamily of fossorial colubroid snakes from the Western Ghats of peninsular India. *Journal of Natural History* 52(45–46), 2919–2934, 4 figs.

Dehling, J.M., Hinkel, H.H., Ensikat, H.-J., Babilon, K. & Fischer, E. (2018) A new blind snake of the genus *Letheobia* (Serpentes: Typhlopidae) from Rwanda with redescriptions of *L. gracilis* (Sternfeld, 1910) and *L. graueri* (Sternfeld, 1912) and the introduction of a non-invasive preparation procedure for scanning electron microscopy in zoology. *Zootaxa* 4378(4), 480–490, 5 figs.

Delfino, M. (2004) The middle Pleistocene herpetofauna of Valdemino Cave (Liguria, north-western Italy). *Herpetological Journal* 14, 113–128, 12 figs.

Delfino, M., Bailon, S. & Pitruzzella, G. (2011) The late Pliocene amphibians and reptiles from "Capo Mannu D1 Local Fauna" (Mandriola, Sardinia, Italy). *Geodiversitas* 33(2), 357–382, 7 figs.

Delfino, M., Segid, A., Yosief, D., Shoshani, J., Rook, L. & Libsekal, Y. (2004) Fossil reptiles from the Pleistocene Homo-bearing locality of Buia (Eritrea, northern Danakil Depresssion). *Rivista Italiana di Paleontologia e Stratigrafia* 110(supplement), 51–60, 6 figs.

Dely, O.G. & Janisch, M. (1959) La répartition des vipères de champs (*Vipera ursinii rákosiensis* Méhely) dans le Bassin des Carpathes. *Vertebrata Hungarica* 1(1), 25–34, 1 fig.

De Massary, J.-C., Bour, R., Cheylan, M., Crochet, P.-A., Dewynter, M., Geniez, P., Ineich, I., Ohler, A., Vidal, N. & Lescure, J. (2019) Nouvelle liste taxonomique de l'herpétofaune de la France métropolitaine. *Bulletin de la Société Herpétologique de France* 171, 37–56, 4 figs.

Derez, C.M., Arbuckle, K., Ruan, Z., Xie, B., Huang, Y., Dibben, L., Shi, Q., Vonk, F.J. & Fry, B.G. (2018) A new species of bandy-bandy (*Vermicella*: Serpentes: Elapidae) from the Weipa region, Cape York, Australia. *Zootaxa* 4446(1), 1–12, 3 figs.

Derry, J., Ruback P. & Ray, J.M. (2015) Range extension and notes on the natural history of *Trimetopon barbouri* (Dunn, 1930) (Serpentes: Colubridae). *Mesoamerican Herpetology* 2(1), 136–140, 4 figs.

Deshmukh, R.V., Deshmukh, S.A. & Badhekar, S.A. (2018) New state records of the Slender Coralsnake, *Calliophis melanurus* (Shaw 1802), and the Yellow-spotted Wolfsnake, *Lycodon flavomaculatus* Wall 1907, in Chhattighar, India. *IRCF Reptiles & Amphibians* 25(3), 194–196, 3 figs.

Deshwal, A. & Becker, B. (2017) New locality record of Nagarjunasagar Racer (*Coluber bholanathi*) (Squamata: Serpentes: Colubridae) from near Rishi Valley School, Andhra Pradesh, India. *Russian Journal of Herpetology* 24(3), 245–247, 2 figs.

Desroches, J.-F., Schueler, F.W., Picard, I. & Gagnon, L.-F. (2010) A herpetological survey of the James Bay area of Québec and Ontario. *The Canadian Field-Naturalist* 124(4), 299–315, 12 figs.

Dias, E.G., Santos, I.Y.G.S. dos, Silva, F.J., Lima, E.S.M., Rocha, E.B.G., Santos, R.L. dos & Santos, E.M. dos (2019) Geographic distribution: *Dipsas variegata variegata. Herpetological Review* 50(1), 105.

Dias, I.R., Costa, C.A.S., Solé, M. & Argôlo, A.J.S. (2018) Two new records of *Dipsas albifrons* (Sauvage, 1884) from northeastern Brazil (Squamata: Dipsadidae). *Herpetology Notes* 11, 77–80, 2 figs.

Díaz, L.M. & Cádiz, A. (2014) First record of the Brahminy Blindsnake, *Indotyphlops braminus* (Squamata: Typhlopidae), in Cuba. *IRCF Reptiles & Amphibians* 21(4), 140–141, 1 fig.

Díaz, L.M., Cádiz, A., Villar, S. & Bermudez, F. (2014) Notes on the ecology and morphology of the Cuban Khaki Trope, *Tropidophis hendersoni* Hedges and Garrido (Squamata: Tropidophiidae), with a new locality record. *IRCF Reptiles & Amphibians* 21(4), 116–119, 5 figs.

Díaz-Ricaurte, J.C. (2019) Predation attempt on the Degenhardt's Scorpion-eating Snake *Stenorrhina degenhardtii* (Berthold, 1846) (Serpentes: Colubridae) by the False Coral Snake *Erythrolamprus bizonus* (Jan, 1863) (Serpentes: Dipsadidae) in Caldas, Colombia. *Herpetology Notes* 12, 331–333, 1 fig.

Díaz-Ricaurte, J.C., Cubides-Cubillo, S.D. & Fiorillo, B.F. (2018a) *Bothrops asper* (Garman, 1884). *Catálogo de Anfibios y Reptiles de Colombia* 4(2), 8–22, 5 figs.

Díaz-Ricaurte, J.C. & Fiorillo, B.F. (2019) *Micrurus surinamensis* (Cuvier, 1817). *Catálogo de Anfibios y Reptiles de Colombia* 5(1), 30–35, 4 figs.

Díaz-Ricaurte, J.C., Fiorillo, B.F. & Maciel, J.H. (2018c) *Crotalus durissus* Linnaeus, 1758. *Catálogo de Anfibios y Reptiles de Colombia* 4(2), 29–36, 3 figs.

Díaz-Ricaurte, J.C., Guevara-Molina, S.C. & Cubides-Cubillos, S.D. (2017) *Lachesis muta* (Linnaeus 1766). *Catálogo de Anfibios y Reptiles de Colombia* 3(2), 20–24, 3 figs.

Díaz-Ricaurte, J.C., Serrano, F. & Fiorillo, B.F. (2018b) *Clelia clelia* (Daudin, 1803). *Catálogo de Anfibios y Reptiles de Colombia* 4(2), 23–32, 3 figs.

Di Pietro, D.O., Christie, M.I. & Williams, J.D. (2013) Nuevos registros de *Philodryas agassizii* (Serpentes: Dipsadidae: Xenodontinae) en la Argentina. *Cuadernos de Herpetología* 27(1), 59–62, 2 figs.

Di Pietro, D.O., Nenda, S.J. & Gómez, S.E. (2016) Geographic distribution. *Philodryas trilineata. Herpetological Review* 47(3), 428.

Disla, A.D., Landestoy, M.A., Abreu, O.Á., Cabrera-Pichardo, J.M. & Frías, E. (2019) Nuevo registro de localidad para *Ialtris dorsalis* (Günther, 1858) (Squamata: Dipsadidae) en República Dominicana. *Novitates Caribaea* 13, 125–127, 1 fig.

Di Tizio, L., Pellegrini, M., Cameli, A. & Di Francesco, N. (2010) Altlante erpetologico della Provincia di Chieti: Dati preliminari. In: Di Tizio, L., Di Cerbo, A.R., Di Francesco, N. & Cameli, A. (Eds.), *Atti VIII Congresso Nazionale Societas Herpetologica Italica (Chieti, 22–26 Settembre 2010)*. Ianieri Edizioni, Pescara, pp. 51–55.

Do, D.T., Ngo, C.D., Ziegler, T. & Nguyen, T.Q. (2017) First record of *Lycodon cardamomensis* (Daltry et Wüster, 2002) (Squamata: Colubridae) from Vietnam. *Russian Journal of Herpetology* 24(2), 167–170, 3 figs.

Doan, T.M., Mason, A.J., Castoe, T.A., Sasa, M. & Parkinson, C.L. (2016) A cryptic palm-pitviper species (Squamata: Viperidae: *Bothriechis*) from the Costa Rican highlands, with notes on the variation within *B. nigroviridis. Zootaxa* 4138(2), 271–290, 7 figs.

Domínguez, M. & Díaz, Jr., R.E. (2015) Resurrection and redescription of the *Typhlops silus* Legler, 1959 from Cuba

(Scolecophidia, Typhlopidae). *Journal of Herpetology* 49(2), 325–331, 2 figs.

Domínguez, M., Fong -G.A. & Iturriaga, M. (2013) A new blind snake (Typhlopidae) from northeastern Cuba. *Zootaxa* 3681(2), 136–146, 3 figs.

Domozetski, L. (2013) New localities of Eastern Montpellier Snake – *Malpolon insignitus* (Geoffroy Saint-Hilaire, 1827) from southwestern Bulgaria. *ZooNotes* 41, 1–4, 6 figs.

Donaire, D., Mateo, J.A., Hasi, M. & Geniez, P. (2000) Nuevos datos sobre la fauna reptilian de La Hamada de Tinduf (Argelia). *Boletín de la Asociación Herpetológia Española* 11(1), 8–12, 3 figs.

Doria, G. (2010) La collezione erpetologica del Museo Civico di Storia Naturale "G. Doria" di Genova. *Museologia Scientifica Memorie* 2010(5), 62–68, 2 figs.

Doria, G., Petri, M., Bellati, M., Tiso, M. & Pistarino, E. (2013) *Rhabdophis* in the Museum of Genova with description and molecular analysis of a new species from Sumatra (Reptilia, Serpentes, Colubridae, Natricinae). *Annali del Museo Civico di Storia Naturale "Giacomo Doria"* 105, 139–153. [not seen].

Doronin, I.V. (2016) [A new record of the Steppe Ribbon Racer, *Psammophis lineolatus* (Brandt, 1836) (Serpentes: Lamprophiidae) in the Caucasus]. *Current Studies in Herpetology* 16(3/4), 161–163, 3 figs. [in Russian].

Dotsenko, I.B. (2003) *[Catalogue of the Collection of Zoological Museum, National Museum of Natural History, Ukrainian Academy of Sciences. The snakes].* Natural History Museum, Kiev, 1–85. [in Russian].

Dotsenko, I.B. (2010) [*Emydocephalus szczerbaki* sp. n. (Serpentes, Elapidae, Hydrophiinae) - a new species of the turtleheaded sea snake genus from Vietnam]. *Zbirnik Prats' Zoologichnogo Museyu* 41, 128–138. [in Russian] [not seen].

Dotsenko, I.B., Vikirchak, A.K. & Drebet, M.V. (2013) [New records of Aeskulapian (Forest) Snake, *Zamenis longissimus* (Serpentes, Colubridae), and the recommendations of the species conservation in the territory of Ukraine]. *Proceedings of the Zoological Museum, Kiev* 44, 123–133. [in Russian].

Douglas, M.R., Davis,, M.A., Amarello, M., Smith, J.J., Schuett, G.W., Herrmann, H.–W., Holycross, A.T. & Douglas, M.E. (2016) Anthropogenic impacts drive niche and conservation metrics of a cryptic rattlesnake on the Colorado Plateau of western North America. *Royal Society Open Science* 3(160047), 1–11, 4 figs.

Duarte, M. de O., Freitas, T.M. da S. & Prudente, A.L. da C. (2015) Polychromatism of populations of *Corallus hortulanus* (Squamata: Boidae) from the southern Amazon Basin, Brazil. *Acta Amazonica* 45(4), 373–381, 7 figs.

Dubey, S., Ursenbacher, S., Schuerch, J., Golay, J., Aubert, P. & Dufresnes, C. (2017) A glitch in the *Natrix*: Cryptic presence of alien grass snakes in Switzerland. *Herpetology Notes* 10, 205–208, 1 fig.

Dung, L.T., Nguyen, L.H.S., Pham, T.C. & Nguyen, T.Q. (2014) New records of snakes (Squamata: Serpentes) from Dien Bien Province. *Tap Chi Sinh Hoc* 36(4), 460–470, 3 figs.

Đurić, D., Radosavljević, D., Petrović, D. & Vojnović, P. (2017) A new evidence for pachyostotic snake from Turonian of Bosnia-Herzegovina. *Annales Géologiques de la Péninsule Balkanique* 78, 17–21, 3 figs.

Durso, A.M. & Norberg, P. (2016) Distribution notes: *Coniophanes piceivittis*. *Mesoamerican Herpetology* 3(1), 194–197, 1 fig.

Dušej, G. & Müller, P. (1997) Reptilieninventar des Kantons Zürich. *Naturforschende Gesellschaft in Zürich*, 1–47, 22 figs.

Dutta, D., Sengupta, S., Das, A.K. & Das, A. (2013) New distribution of records of *Lycodon zawi* (Serpentes: Colubridae) from northeast India. *Herpetology Notes* 6, 263–265, 2 figs.

Dwyer, Q. (2015) Geographic distribution. *Sibon anthracops*. *Herpetological Review* 46(1), 63.

Džukić, G., Tomović, L., Anđelković, M., Urošević, A., Nikolić, S. & Kalezić, M. (2017) The herpetological collection of the Institute for Biological Research "Siniša Stanković," University of Belgrade. *Bulletin of the Natural History Museum* 10, 57–104, 1 fig.

Echavarría-Rentería, J.D., Mosquera-Moya, V. & Rengifo-Mosquera, J.T. (2015) Ampliación del rango de distribución de *Atractus depressiocellus* (Myers, 2003) (Serpentes: Dipsadidae) para Colombia. *Cuadernos de Herpetología* 29(2), 157–159, 2 figs.

Ege, O., Yakin, B.Y. & Tok, C.V. (2015) Herpetofauna of the Lake District around Burdur. *Turkish Journal of Zoology* 39(6), 1164–1168, 2 figs.

Eisermann, K., Avendaño, C., Acevedo, M. & Matías, E. (2016) Elevational range extension and new habitat for *Thamnophis fulvus* (Bocourt, 1893) (Squamata: Natricidae). *Mesoamerican Herpetology* 3(4), 1094–1097, 2 figs.

Eksilmez, H., Altunşic, A. & Özdemir, N. (2017) The herpetofauna of Karçal Mountains (Artvin/Turkey). *Biological Diversity and Conservation* 10(1), 1–5, 3 figs.

Elizondo, L. (2016) *Urotheca guentheri* in Darién, Panama, at the northern extreme of the Chocó biogeographic region. *Mesoamerican Herpetology* 3(2), 536–539, 1 fig.

Ellis, R.J. (2015) Corrections of the type specimens of *Liasis olivaceus barroni* (Smith, 1981) (Serpentes: Pythonidae). *Records of the Western Australian Museum* 30(1), 61–63.

Ellis, R.J. (2016) A new species of blindsnake (Scolecophidia: Typhlopidae: *Anilios*) from the Kimberley region of Western Australia. *Herpetologica* 72(3), 271–278, 3 figs.

Ellis, R.J., Doughty, P., Donnellan, S.C., Marin, J. & Vidal, N. (2017) Worms in the sand: Systematic revision of the Australian blindsnake *Anilios leptosoma* (Robb, 1972) species complex (Squamata: Scolecophidia: Typhlopidae) from the Geraldton Sandplain, with description of two new species. *Zootaxa* 4323(1), 1–24, 8 figs.

Enderson, E.F., Van Devender, T.R. & Bezy, R.L. (2014) Amphibians and reptiles of Yécora, Sonora and the Madrean tropical zone of the Sierra Madre Occidental in northwestern Mexico. *Check List* 10(4), 913–926, 15 figs.

Engelbrecht, H.M., Branch, W.R., Greenbaum, E., Alexander, G.J., Jackson, K., Burger, M., Conradie, W., Kusamba, C., Zassi-Boulou, A.-G. & Tolley, K.A. (2018) Diversifying into the branches: Species boundaries in African green and bush snakes, *Philothamnus* (Serpentes: Colubridae). *Molecular Phylogenetics & Evolution* 130, 357–365, 3 figs.

Entiauspe-Neto, O.M. & Abegg, A.D. (2013) New record and distribution extension of *Atractus paraguayensis* (Werner, 1924) (Serpentes: Dipsadidae). *Check List* 9(1), 104–105, 2 figs.

Entiauspe-Neto, O.M., Abegg, A.D., Pinheiro, R.T., Borges, L.M. & Loebmann, D. (2017b) *Hydrops martii* (Wagler, 1824) (Serpentes: Dipsadidae): First record in Amapá state, northern Brazil. *Check List* 13(5), 659–661, 2 figs.

Entiauspe-Neto, O.M., Abegg, A.D., Quintela, F.M. & Loebmann, D. (2017a) *Mussurana quimi* (Franco, Marques & Puorto, 1997) (Serpentes: Dipsadidae): First records for Rio Grande do Sul, southern Brazil. *Check List* 13(2053), 1–4, 2 figs.

Entiauspe-Neto, O.M., Abegg, A.D., Rocha, A.M. & Loebmann, D. (2018) Connecting the dots: Filling distribution gaps of *Philodryas viridissima* (Serpentes: Dipsadidae) in Brazil, with a new state record to Roraima. *Herpetology Notes* 11, 697–702, 2 figs.

Entiauspe-Neto, O.M. & Lema, T. de (2015) *Apostolepis christineae* (Lema, 2002) (Serpentes: Xenodontinae: Elapomorphini): First record for Bolivia. *Check List* 11(1814), 1–4, 3 figs.

Entiauspe-Neto, O.M., Lema, T. de. & Beconi, H.E.C. (2014) *Apostolepis intermedia* (Koslowsky, 1898) (Serpentes: Xenodontinae: Elapomorphini): First records for Paraguay. *Check List* 10(3), 600–601, 2 figs.

Entiauspe-Neto, O.M., Renner, M.F., Mario-da-Rosa, C., Abegg, A.D., Loebmann, D. & Lema, T. de (2017c) Redescription, geographic distribution and ecological niche modeling of *Elapomorphus wuchereri* (Serpentes: Dipsadidae). *Phyllomedusa* 16(2), 225–242, 6 figs.

Entiauspe-Neto, O.M., Sena, A. de, Tiutenko, A. & Loebmann, D. (2019) Taxonomic status of *Apostolepis barrioi* Lema, 1978, with comments on the taxonomic instability of *Apostolepis* (Cope, 1862) (Serpentes, Dipsadidae). *ZooKeys* 841, 71–78, 2 figs.

Escalante-Pasos, J.A. & García-Padilla, E. (2015) Distribution notes: *Mastigodryas melanolomus*. *Mesoamerican Herpetology* 2(2), 206.

Escalona, M. (2017) Range extension for *Erythrolamprus epinephelus bimaculatus* (Cope, 1899) and *E. e. opisthotaenius* (Boulenger, 1908) in Venezuela (Serpentes: Colubridae). *Herpetology Notes* 10, 511–515, 2 figs.

Eser, Ö. & Erismis, U.C. (2014) Research on the herpetofauna of Başkomutan Historical National Park, Afyonkarahisar, Turkey. *Biharean Biologist* 8(2), 98–101, 3 figs.

Eskandarzadeh, N., Darvish, J., Rastegar-Pouyani, E. & Ghassemzadeh, F. (2013) Reevaluation of the taxonomic status of sand boas of the genus *Eryx* (Daudin, 1803) (Serpentes: Boidae) in northeastern Iran. *Turkish Journal of Zoology* 37(3), 348–356, 6 figs.

Eskandarzadeh, N., Rastegar-Pouyani, N., Rastegar-Pouyani, E. & Nasrabadi, R. (2018b) Modelling the habitat suitability of the Arabian Sand Boa *Eryx jayakari* (Serpentes: Erycidae). *Zoology and Ecology* 28(4), 337–342, 5 figs.

Eskandarzadeh, N., Rastegar-Pouyani, N., Rastegar-Pouyani, E., Todehdehghan, F. & Rajabizadeh, M. (2018a) Sexual dimorphism in the Javelin sand Boa, *Eryx jaculus* (Linnaeus, 1758) (Serpentes: Erycidae), from western Iran. *Current Herpetology* 37(1), 88–92, 1 fig.

Eskandarzadeh, N., Rastegar-Pouyani, N., Rastegar-Pouyani, E., Todehdehghan, F., Rajabizadeh, M., Zarrintab, M., Abbas-Raddi, F. & Kami, H.G. (2020) Revised classification of the genus *Eryx* (Daudinm, 1803) (Serpentes: Erycidae) in Iran and neighboring areas, based on mtDNA sequences and morphological data. *Herpetological Journal* 30(1), 1–11, 6 figs.

Espinal, M. & Solís, J.M. (2015a) New locality and elevational record for the snake *Sibon anthracops* (Cope, 1868) in Honduras. *Mesoamerican Herpetology* 2(2), 218–219, 2 figs.

Espinal, M. & Solís, J.M. (2015b) Rediscovery of the Honduran endemic snake *Tantilla lempira* (Wilson and Mena, 1980) (Squamata: Colubridae). *Mesoamerican Herpetology* 2(4), 576–579, 2 figs.

Espinal, M., Solís, J.M. & Mora, J.M. (2017) Distribution notes: *Rhadinella montecristi*. *Mesoamerican Herpetology* 4(3), 676–677, 2 figs.

Espinal, M., Solís, J.M., O'Reilly, C., Marineros, L. & Vega, H. (2014b) New distributional records for amphibians and reptiles from the department of Santa Bárbara, Honduras. *Mesoamerican Herpetology* 1(2), 300–303, 1 fig.

Espinal, M., Solís, J.M., O'Reilly, C. & Valle, R. (2014a) New distributional records for amphibians and reptiles from the department of Choluteca, *Honduras*. *Mesoamerican Herpetology* 1(2), 298–300.

Esposito, C. & Romano, A. (2011) Extension of known range area of *Zamenis lineatus* (Camerano, 1891). New northern and western limits. *Herpetozoa* 23(3/4), 86–87.

Esqueda-González, L.F. (2011) A new semifossorial snake species (Dipsadidae: *Atractus* Wagler, 1828) from the Lara-Falcón mountainous system, northwestern Venezuela. *Herpetotropicos* 6(1–2), 35–41, 3 figs.

Esqueda-González, L.F. & McDiarmid, R. (2015) Un Nuevo taxon del género *Atractus* (Wagler, 1828) (Colubroidea: Dipsadidae) procedente de la region noroccidental de Venezuela. In: Natera-Mumaw, M., Esqueda-González, L.F. & Castelaín-Fernández, M., (Eds.), *Atlas serpientes Venezuela – una vision actual de su diversidad*. Dimacofi Negocios Avanzados S.A., Santiago, 406–413, 3 figs.

Esqueda-González, L.F., Schlüter, A., Machado, C., Castelaín-Fernández, M. & Natera-Mumaw, M. (2015) Una nueva especie de cieguita o serpiente de gusano (Serpentes: Leptotyphlopidae: Epictia) native del Tepui Guaiquinima, provincial Pantepui en el Escudo de Guayana, Venezuela. In: Natera-Mumaw, M., Esqueda-González, L.F. & Castelaín-Fernández, M., (Eds.), *Atlas serpientes Venezuela – una vision actual de su diversidad*. Dimacofi Negocios Avanzados S.A., Santiago, 414–433, 6 figs.

Essghaier, M.F.A., Taboni, I.M. & Etayeb, K.S. (2015) The diversity of wild animals at Fezzan Province (Libya). *Biodiversity Journal* 6(1), 245–252, 5 figs.

Etchepare, E.G. & Ingaramo, M. del R. (2008) *Pseudablabes agassizii* (Jan, 1863) (Serpentes: Colubridae). Primer registro para la Provincia de Corrientes (República Argentina). *Cuadernos de Herpetología* 22(1), 51.

Evans, S.E. (1994) A new anguimorph lizard from the Jurassic and lower Cretaceous of England. *Palaeontology* 37(1), 33–49, 16 figs.

Evans, S.E. (1996) Parviraptor (Squamata: Anguimorpha) and other lizards from the Morrison Formation at Fruita, Colorado. In: Morales, M. (Ed.), *The Continental Jurassic*. Northern Arizona Museum Bulletin (60), 243–248, figs. [not seen].

Eversole, C.B., Powell, R., Lizarro, D. & Bravo, R.C. (2016) *Erythrolamprus dorsocorallinus* (Esqueda, Natera, La Marca & Ilija-Fistar, 2005) (Squamata: Dipsadidae): Range extension, new country record, and comments on color pattern. *Check List* 12(1987), 1-1-4, 3 figs.

Faiz, A. ul H., Abbas Fakher I., Bagaturov, M.F., Faiz, L.Z. & Akhtar, Tanveer (2017) First sighting and occurrence record of King Cobra (*Ophiophagus hannah*) in Pakistan. *Herpetology Notes* 10, 349–350, 1 fig.

Fantuzzi, J.A. (2016) Geographic distribution. *Lycophidion capense*. *Herpetological Review* 47(1), 84.

Faraone, F.P., Chiara, R., Barra, S.A., Giacalone, G. & Lo Valvo, M. (2017) Nuovi dati sulla presenza di Eryx jaculus (Linnaeus, 1758) in Sicilia. In: Menegon, M., Rodriguez-Prieto, A. & Deflorian, M.C. (Eds.), *Atti XI Congresso Nazionale della Societas Herpetologica Italica Trento, 22–25 Settembre 2016*. Ianieri Edizioni, Pescara. Pp. 75–79, 1 fig.

Farhad, A. & Varsha, W. (2013) The snake fauna of Mulshi Taluka, Pune District, India. *Russian Journal of Herpetology* 20(1), 16–18, 1 fig.

Farr, W.L. & Lazcano, D. (2017) Distribution of *Bothrops asper* in Tamaulipas, Mexico and a review of prey items. *The Southwestern Naturalist* 62(1), 77–84, 1 fig.

Farr, W.L., Lazcano, D. & Murcio, P.A.L. (2013) New distributional records for amphibians and reptiles from the state of Tamaulipas, México III. *Herpetological Review* 44(4), 631–645, 1 fig.

Farr, W.L., Reyes, M.N. de los, Banda-Leal, J. & Lazcano, D. (2015) The distribution of *Crotalus totonacus* in Nuevo León, Mexico. *Mesoamerican Herpetology* 2(3), 243–251, 4 figs.

Fathinia, B., Rastegar-Pouyani, E., Rastegar-Pouyani, N. & Darvishnia, H. (2017) A new species of the genus

Rhynchocalamus (Günther, 1864) (Reptilia: Squamata: Colubridae) from Ilam province in western Iran. *Zootaxa* 4282(3), 473–486, 6 figs.

Fathinia, B., Rastegar-Pouyani, E. & Shafaeipour, A. (2019) A new species of *Eirenis* (Ophidia: Colubridae) from highland habitats in southern Iran. *Zoology in the Middle East* 65(4), 1–11 (pre-publication), 6 figs.

Fathinia, B., Rastegar-Pouyani, N., Rastegar-Pouyani, E., Toodeh-Dehghan, F. & Rajabizadeh, M. (2014) Molecular systematics of the genus *Pseudocerastes* (Ophidia: Viperidae) based on the mitochondrial cytochrome *b* gene. *Turkish Journal of Zoology* 38(5), 575–581, 2 figs.

Fattizzo, T. & Marzano, G. (2002) Dati distributive sull'erpetofauna del Salento. *Thalassia Salentina* 26, 113–132, 26 figs.

Feio, R.N. & Caramaschi, U. (2002) Contribução ao conhecimento da herpetofauna do nordeste do estado de Minas Gerais, *Brasil. Phyllomedusa* 1(2), 105–111, 5 figs.

Feitosa, D.T., Pires, M.G., Prudente, A.L. da C. & Silva, N.J. da (2013) Distribution extension in Colombia and new records for Brazil of *Micrurus isozonus* (Cope, 1860) (Squamata, Serpentes, Elapidae). *Check List* 9(5), 1108–1112, 3 figs.

Feitosa, D.T., Silva, N.J. da, Pires, M.G., Zaher, H. & Pudente, A.L. da C. (2015) A new species of monadal coral snake of the genus *Micrurus* (Serpentes, Elapidae) from western Amazon. *Zootaxa* 3974(4), 538–554, 10 figs.

Fernandes, D.S. & Hamdan, B. (2014) A new species of *Chironius* Fitzinger, 1826 from the state of Bahia, northeastern Brazil (Serpentes: Colubridae). *Zootaxa* 3881(6), 563–575, 6 figs.

Fernándes, J., Vargas-Vargas, N., Pla, D., Sasa, M., Rey-Suárez, P., Sanz, L., Gutiérrez, J.M., Calvete, J.J. & Lomonte, B. (2015) Snake venomics of *Micrurus alleni* and *Micrurus mosquitensis* from the Caribbean region of Costa Rica reveals two divergent compositional patterns in New World elapids. *Toxicon* 107, 217–233, 11 figs.

Fernandez, D.C., Herr, M.W., Weinell, J. & Brown, R.M. (2019) Geographic distribution: *Ophiophagus hannah*. *Herpetological Review* 50(2), 332.

Fernández, M., Martínez-Fonseca, J.G., Salazar-Saavedra, M., Gutiérrez, L., Loza, J. & Sunyer, J. (2017) First verified record of *Cerrophidion wilsoni* (Reptilia: Squamata: Viperidae) from Nicaragua. *Mesoamerican Herpetology* 4(2), 481–484, 2 figs.

Fernández-Badillo, L., Aguillón-Gutiérrez, D.R., Valdez-Rentería, S.Y., Hernández-Melo, J.A., Olvera-Olvera, C.R., Callejas-Jiménez, F.J., Hernández-Ramos, M., Iturbe-Morgado, J.C., Torres-Angeles, F. & Reaño-Hernández, I. (2016b) First records for amphibians and reptiles from the municipality of Atotonico el Grandes, Hidalgo, México. *Herpetological Review* 47(1), 91–93.

Fernández-Badillo, L., Morales-Capellán, N., Contreras-Patiño, D.R. & Carreño-Cervantes, A. (2016a) Confirmación de la presencia de la serpiente de cascabel, *Crotalus scutulatus* para el estado de México, México. *Acta Zoological Mexicana* 32(2): 202–205, 2 figs.

Fernández-Badillo, L., Olvera-Olvera, C.R., Valdez-Rentaría, S.Y., Torres-Ángeles, F. & Goyenechea, I. (2016c) New records of *Thamnophis pulchrilatus* (Squamata: Natricidae) from the state of Hidalgo, Mexico. *Mesoamerican Herpetology* 3(2), 519–523, 3 figs.

Fernández-Badillo, L., Valdez-Rentería, S.Y., Olvera-Olvera, C.R., Sánchez-Martínez, G., Manríquez-Morán, N.L. & Goyenechea, I. (2017) The snake *Lampropeltis annulata*, Kennicott, 1861, in Hidalgo, Mexico. *Mesoamerican Herpetology* 4(2), 502–506, 3 figs.

Fernández-Cardenete, J.R., Luzón-Ortega, J.M., Pérez-Contreras, J., Pleguezuelos, J.M. & Tierno de Figueroa, J.M. (2000) Nuevos límites altitudinales para seis especies de herpetos de la Península Ibérica. *Boletín de la Asociación Herpetólogica Española* 11(1), 20–21.

Fernández-Guiberteau, D., Carrero-Casado, F. & Vázquez-Graña, R. (2015) Nuevas aportaciones a la distribución de la fauna herpetológica de Extremadura. *Butlletí de la Societat Catalana d'Herpetologia* 22, 113–120, 3 figs.

Ferquel, E., Haro, L. de, Jan, V., Guillemin, I., Jourdain, S., Teynié, A., d'Alayer, J. & Choumet, V. (2007) Reappraisal of *Vipera aspis* venom neurotoxicity. *PLOS ONE* 2007(1194), 1–18, 15 figs.

Ferrão, M., Filho, J.A.S.R. & Silva, M.O. da (2012) Checklist of reptiles (Testudines, Squamata) from Alto Alegre dos Parecis, southwestern Amazonia, Brazil. *Herpetology Notes* 5, 473–480, 3 figs.

Ferraro, D.P. & Williams, J.D. (2006) Material tipo de la Colección de Herpetología del Museo de La Plata, Buenos Aires, Argentina. *Cuadernos de Herpetología* 19(2), 19–35.

Ferreira, V.L., Terra, J. de S., Piatti, L., Delatorre, M., Strüssmann, C., Béda, A.F., Kawashita-Ribeiro, R.A. et al. (2017) Répteis do Mato Grosso do Sul, Brasil. *Iheringia, Série Zoologia* 107(supl. 2017153), 1–13, 3 figs.

Ferreira-Silva, C., Ribeiro, S.C., Alcantara, E.P. de & Ávila, W. (2019) Natural history of the rare and endangered snake *Atractus ronnie* (Serpentes: Colubridae) in northeastern Brazil. *Phyllomedusa* 18(1), 77–87, 3 figs.

Ferrer, J., Dahmani, W., Ait Hammou, M., Camarasa, S., Maatoug, M. & Sanuy, D. (2016) Contribució al coneixement de l'herpetofauna del nord d'Algèria (regions de Tiaret I Chlef). *Butlletí de la Societat Catalana d'Herpetologia* 23, 44–63, 31 figs.

Ferrer, J., Fontelles, F., Sort, F., Guixé, D. & Vidal-Coll, Y. (2018) Confirmació de la presència de l'escurçó ibèric *Vipera latastei* al Solsonès i al sud est de l'Alt Urgell i descripció d'una nova zona de simpatria amb escurçó pirinenc *Vipera aspis*. *Butlletí de la Societat Catalana d'Herpetologia* 25, 88–101, 13 figs.

Figueroa, A., McKelvy, A.D., Grismer, L.L., Bell, C.D. & Lailvaux, S.P. (2016) A species-level phylogeny of extant snakes with description of a new colubrid subfamily and genus. *PLOS ONE* 11(9:0161070), 1–31, 10 figs.

Filho, G.A. de S., Moura-Leite, J.C. de, Matias, E.G. & Morato, S.A.A. (2012) *Chironius fuscus* (Linnaeus, 1758) (Serpentes: Colubridae): Distribution extension, new state record and variation in southern Brazil. *Check List* 8(6), 1315–1318, 2 figs.

Filho, G.A. de S. & Plombon, L.L. (2014) Rediscovery and geographic distribution of *Philodryas agassizii* (Jan, 1863) (Squamata: Dipsadidae) in the state of Paraná, southern Brazil. *Biotemas* 27(1), 155–158, 2 figs.

Filho, G.A.P., Santana, G.G., Vieira, W.L. da S., Alves, R.R. da N., Montenegro, P.F.G.P. & Freitas, M.A. de (2012) *Phimophis guerini* (Duméril, Bibron and Duméril, 1854) (Serpentes: Dipsadidae): Distribution extension in Paraiba, Brazil. *Check List* 8(5), 966–967, 2 figs.

Filogonio, R. & Canelas, M.A.S. (2015a) Geographic distribution. *Trilepida koppesi. Herpetological Review* 46(2), 222.

Filogonio, R. & Canelas, M.A.S. (2015b) Geographic distribution *Trilepida koppesi. Herpetological Review* 46(3), 386–387.

Firneno, T.J., Itgen, M.W., Pereira-Pereira, F.M. & Townsend, J.H. (2016) Geographic distribution. *Tantilla armillata. Herpetological Review* 47(2), 266.

Flores, E.E., Gracia, V. de & Peña, B. (2016b) Geographic distribution. *Rhinobothryum bovallii. Herpetological Review* 47(3), 429.

Flores, E.E., Peña, B., Gracia, V. de, Diaz, E. & Gonzalez, J. del C. (2016a) Geographic distribution. *Enulius flavitorques. Herpetological Review* 47(3), 426.

Flores-Guerrero, U.S. & Sánchez-González, J.S. (2016) Distribution notes: *Crotalus campbelli*. *Mesoamerican Herpetology* 3(2), 524–525, 2 figs.

Flores-Hernández, M.A., Fernández-Badillo, L. & Hernández-González, A.I. (2017) Distribution notes: *Hypsiglena tanzeri*. *Mesoamerican Herpetology* 4(4), 962–963, 1 fig.

Flores-Villela, O., Ríos-Muñoz, C.A., Magaña-Cota, G.E. & Quezadas-Tapia, N.L. (2016) Alfredo Dugès' type specimens of amphibians and reptiles revisited. *Zootaxa* 4092(1), 33–54, 11 figs.

Folt, B., Bauder, J., Spear, S., Stevenson, D., Hoffman, M., Oaks, J.R., Wood, P.L., Jenkins, C., Steen, D.A. & Guyer, C. (2019) Taxonomic and conservation implications of population genetic admixture, mito-nuclear discordance, and male-biased dispersal of a large endangered snake, *Drymarchon couperi*. *PLOS ONE* 14(3)e0214439, 1–21, 6 figs.

Folt, B., Laurencio, D., Goessling, J.M., Birkhead, R.D., Stiles, J., Stiles, S., Harris, A.T. & Belford, S. (2015) One hundred two new county records for amphibians and reptiles in Alabama, USA. *Herpetological Review* 46(4), 591–594.

Folt, B., Pierson, T., Goessling, J., Goetz, S.M., Laurencio, D., Thompson, D. & Graham, S.P. (2013) Amphibians and reptiles of Jasper County, Mississippi, with comments on the potentially extinct Bay Springs salamander (*Plethodon ainsworthi*). *Herpetological Review* 44(2), 283–286.

Fong G. A. & Armas, L.F. de (2011) The easternmost record for *Tropidophis spiritus* (Hedges & Garrido, 1999) (Serpentes: Tropidophiidae) in Cuba. *Herpetology Notes* 4, 111–112, 2 figs.

Fong, J.J., Grioni, A., Crow, P. & Cheung, K.-S. (2017) Morphological and genetic verification of Ovophis tonkinensis (Bourret, 1934) in Hong Kong. *Herpetology Notes* 10, 457–461, 2 figs.

Fonseca, W.L. da, Silva, J.D. da, Abegg, A.D., Rosa, C.M. da & Bernarde, P.S. (2019) Herpetofauna of Porto Walter and surrounding areas, southwest Amazonia, Brazil. *Herpetology Notes* 12, 91–107, 7 figs.

Fortes, V.B., Lucas, E.M. & Caldart, V.M. (2010) Reptilia, Serpentes, Dipsadidae, *Gomesophis brasiliensis* (Gomes, 1918): Distribution extension in state of Santa Catarina, Brazil. *Check List* 6(3), 414–415, 2 figs.

Foster, N.Y., Conway, C., Dillashaw, M., Christensen, M., Williams, M., Blackmore, M.A., Burress, S. et al. (2013) Geographic distribution of herpetofauna of middle Tennessee. *Herpetological Review* 44(3), 484–486.

Foufopoulos, J. & Richards, S. (2007) Amphibians and reptiles of New Britain Island, Papua New Guinea: Diversity and conservation status. *Hamadryad* 31(2), 176–201, 36 figs.

Fraga, R. de, Almeida, A.P. de, Moraes, L.J.C. de L., Gordo, M., Pirani, R., Zamora, R.R., Carvalho, V.T. de, Passos, P. & Werneck, F.P. (2017) Narrow endemism or insufficient sampling? Geographic range extension and morphological variation of the poorly known *Atractus riveroi* (Roze, 1961) (Serpentes: Dipsadidae). *Herpetological Review* 48(2), 281–284, 2 figs.

Fraga, R. de, Souza, E., Santos-Júnior, A.P. & Kawashita-Ribeiro, R.A. (2018) Notes on the rare *Mastigodryas moratoi* (Serpentes: Colubridae) in the Brazilian Amazon white-sand forests. *Phyllomedusa* 17(2), 299–302, 2 figs.

França, D.P.F., Barbo, F.E., Silva, N.J. da, Silva, H.L.R. & Zaher, H. (2018) A new species of *Apostolepis* (Serpentes, Dipsadidae, Elapomorphini) from the Cerrado of central Brazil. *Zootaxa* 4521(4), 539–552, 7 figs.

França, D.P.F. de, Freitas, M.A. de, Bernarde, P.S. & Uhlig, V.M. (2013) *Erythrolamprus oligolepis* (Boulenger, 1905) (Serpentes: Dipsadidae): First record for the state of Acre, Brazil. *Check List* 9(3), 668–669, 2 figs.

França, D.P.F., Freitas, M.A. de, Ramalho, W.P. & Bernarde, P.S. (2017) Diversidade local e influência da sazonalidade sobre taxocenoses de anfíbios e répteis na Reserva Extrativista Chico Mendes, Acre, Brasil. *Iheringia, Série Zoologia* 107(supl. 2017023), 1–12, 4 figs.

França, F.G.R., França, R.C. de, Germano, C.E. de S. & Filho, J.M. de O. (2012) *Bothrops neuwiedi* (Wagler, 1824) (Squamata: Serpentes: Viperidae): Distribution for the Atlantic forest, first vouchered record for Paraíba, and geographic distribution map. *Check List* 8(1), 170–171, 3 figs.

França, F.G.R., Mesquita, D.O. & Colli, G.R. (2006) A checklist of snakes from Amazonian Savannas in Brazil, housed in the Coleção Herpetológica da Universidade de Brasília, with new distribution records. *Occasional Papers Sam Noble Museum of Natural History* (17), 1–13, 1 fig.

França, R.C. de, Germano, C.E. de S. & França, F.G.R. (2012) Composition of a snake assemblage inhabiting an urbanized area in the Atlantic Forest of Paraiba state, northeast Brazil. *Biota Neotropica* 12(3), 183–195, 7 figs.

França, R.C., Morais, M.S.R., Freitas, M.A., Moura, G.J.B. & França, F.G.R. (2019) A new record of *Xenopholis scalaris* (Wucherer, 1861) (Dipsadidae) for the state of Pernambuco, Brazil. *Herpetology Notes* 12, 57–59, 2 figs.

França, R.C. de, Oitaven, L.P.C., Moura, G.J.B., Freitas, M.A. & França, F.G.R. (2018) First record of *Sibon nebulatus* (Linnaeus, 1758) (Dipsadidae) for the state of Pernambuco, Brazil. *Herpetology Notes* 11, 19–21, 2 figs.

Francisco, B.C.S., Pinto, R.R. & Fernandes, D.S. (2018) Taxonomic notes on the genus *Siagonodon* Peters, 1881, with a report on morphological variation in *Siagonodon cupinensis* (Bailey and Carvalho, 1946) (Serpentes: Leptotyphlopidae). *Copeia* 106(2), 321–328, 5 figs.

Franco, F.L., Trevine, V.C., Montingelli, G.G. & Zaher, H. (2017) A new species of *Thamnodynastes* from the open areas of central and northeastern Brazil (Serpentes: Dipsadidae: Tachymenini). *Salamandra* 53(3), 339–350, 6 figs.

Franzen, M., Hawlitscheck, O., Aßmann, O. & Bayerl, M. (2016) Würfelnatternfunde (*Natrix tessellata*) aus Bayern mit molekulargeneticher untursuchung zur herkunft der tiere. *Zeitschrift für Feldherpetologie* 23, 213–220, 3 figs.

Frazão, L., Campos, J., Oliveira, M.E., Carvalho, V.T. de & Hrbek, T. (2017) New record and revised distribution map of *Pseudoboa martinsi* (Zaher et al., 2008) (Serpentes: Colubridae: Dipsadinae) in the state of Amazonas, Brazil. *Herpetology Notes* 10, 193–195, 2 figs.

Freed, P., Hakim, J., Lester, B., Rahman, S.C., Stickley, J., Trageser, S., Wetherall, J. & Low, M.R. (2015) Geographic distribution. *Fordonia leucobalia*. *Herpetological Review* 46(4), 573.

Freitas, I., Fahd, S., Velo-Antón, G. & Martínez-Freiría, F. (2018) Chasing the phantom: Biogeography and conservation of *Vipera latastei-monticola* in the Maghreb (North Africa). *Amphibia-Reptilia* 39(2), 145–161, 3 figs.

Freitas, M.A. de (2014) Squamate reptiles of the Atlantic forest of northern Bahia, Brazil. *Check List* 10(5), 1020–1030, 5 figs.

Freitas, M.A. de, Abegg, A.D., Dias, I.R. & Moraes, E.P. de F. (2018) Herpetofauna from Serra da Jibóia, an Atlantic rainforest remnant in the state of Bahia, northeastern Brazil. *Herpetology Notes* 11, 59–72, 7 figs.

Freitas, M.A. de, Almeida, B.J.L., Almeida, M.S.M., Danin, T.S. & Moura, G.J.B. de (2014b) Rediscovery and first record of *Sibynomorphus mikanii septentrionalis* (Cunha, Nascimento & Hoge, 1980) (Squamata; Serpentes) for the state of Pará. *Check List* 10(5), 1246–1248, 6 figs.

Freitas, M.A. de, Angôlo, A.J.S., Gonner, C. & Veríssimo, D. (2014a) Biology and conservation status of Piraja's Lancehead Snake *Bothrops pirajai* (Amaral, 1923) (Serpentes: Viperidae), Brazil. *Journal of Threatened Taxa* 6(10), 6326–6334, 2 figs.

Freitas, M.A. de, Barbosa, G.G., Bernardino, K.P., Filho, J.D.P. & Abegg, A.D. (2019b) First records of the rare snake *Echinanthera cephalomaculata* Di-Bernardo, 1994 in the state of Pernambuco, Brazil (Serpentes: Dipsadidae). *Herpetology Notes* 12, 1005–1009, 2 figs.

Freitas, M.A. de, Colli, G.R., Entiauspe-Neto, O.M., Trinchão, L., Araújo, D., Lima, T. de O., França, D.P.F. de, Gaiga, R. & Dias, P. (2016) Snakes of cerrado localities in western Bahia, *Brazil. Check List* 12(1896), 1–10, 5 figs.

Freitas, M.A. de, Silva, T.F.S., Fonseca, P.M., Hamdan, B., Filadelfo, T. & Abegg, A.D. (2019a) Herpetofauna of Serra do Timbó, an Atlantic forest remnant in Bahia State, northeastern Brazil. *Herpetology Notes* 12, 245–260, 8 figs.

Freitas, M.A. de, Veríssimo, D. & Uhlig, V. (2012) Squamate reptiles of the central Chapada Diamantina, with a focus on the municipality of Mucugê, state of Bahia, Brazil. *Check List* 8(1), 16–22, 4 figs.

Freitas, M.A. de, Vieira, R.S., Entiauspe-Neto, O.M., Oliveira e Sousa, S., Farias, T., Sousa, A.G. & Moura, G.J.B. de (2017) Herpetofauna of the northwest Amazon forest in the state of Maranhão, Brazil, with remarks on the Gurupi Biological reserve. *ZooKeys* 643, 141–155, 4 figs.

Frías, J.R., Fernández-Badillo, L., López, M.A. & Escudero, J.I.A. (2015) Geographic distribution. *Crotalus totonacus. Herpetological Review* 46(2), 219.

Frick, W.F., Heady III, P.A. & Hollingsworth, B.D. (2016a) Geographic distribution. *Lichanura trivirgata. Herpetological Review* 47(1), 83–84.

Frick, W.F., Heady III, P.A. & Hollingsworth, B.D. (2016b) Geographic distribution. *Lampropeltis californiae. Herpetological Review* 47(3), 427.

Frolova, E.N. & Gapanov, S.P. (2016) "Morphology of Nikolsky's Viper in Voronezh and Lipetsk regions." *Vestnik St. Petersburg University* 3(3), 165–169. [in Russian].

Frota, J.G. da, Missassi, A.F.R., Santos-Costa, M.C. dos & Prudente, A.L. da C. (2015) New records of Imantodes lentiferus (Cope, 1894) (Squamata: Dipsadidae) from the states of Pará and Mato Grosso, Brazil. *Check List* 11(1686), 1–4, 1 fig.

Frota, J.G. da, Santos-Jr., A.P. dos, Chalkidis, H. de M. & Guedes, A.G. (2005) As serpentes da região do baixo Rio Amazonas, oeste do estado do Pará, Brasil (Squamata). *Biociências* 13(2), 211–220.

Frynta, D., Kratochvíl, L., Moravec, J., Benda, P., Dandová, R., Kaftan, M., Klosová, K., Mikulová, P., Nová, P. & Schwarzová, L. (2000) Amphibians and reptiles recently recorded in Libya. *Acta Societatis Zoologicae Bohemicae* 64, 17–26, 1 fig.

Fuentes, R.D. & Corrales, G. (2016) New distributional record and reproductive data for the Chocoan bushmaster, *Lachesis acrochorda* (Serpentes: Viperidae), in Panama. *Mesoamerican Herpetology* 3(1), 115–127, 11 figs.

Fuentes, S. de & Lizana, M. (2015) Revisión de la distribución y abundancia de la herpetofauna de los Arribes del Duero salmantinos. *Boletin de la Asociasión Herpetológica Española* 26(1), 64–71, 3 figs.

Fulton, J.N., Couch, M. & Smith, W.H. (2014) New geographic distribution records for herpetofauna in southwest Virginia, USA. *Herpetological Review* 45(1), 105–106.

Funk, A., Velechovský, M., Stěnička, J., Gračko, M., Knobloch, O., Hejduk, J. & Blažej, L. (2009) "A new herpetological observations and faunistic data from the Morocco." *Herpetologicke Informace* 8(1), 69–72. [in Czech].

Funk, A., Vrabec, V., Velechovský, M., Schwertner, J., Hlava, J. & Čadková, Z. (2007) "Herpetological observations and new data of occurrence in the Morocco." *Herpetologicke Informace* 6(1), 25–29, 3 figs. [in Czech].

Gabot-Rodríguez, E. & Marte, C. (2019) Inventario rápido de los reptiles del Refugio de Vida Silvestre El Cañón del Río Gurabo, República Dominicana. *Novitates Caribaea* 13, 117–121, 3 figs.

Galán, P. (2012) *Natrix maura* en el medio marino de las Islas Atlánticas de Galicia. *Boletin de la Asociación Herpetológica Española* 23(2), 38–43, 4 figs.

Galarza, J.A., Mappes, J. & Valkonen, J.K. (2015) Biogeography of the smooth snake (*Coronella austriaca*): Origin and conservation of the northernmost population. *Biological Journal of the Linnean Society* 114(2), 426–435, 4 figs.

Gallardo, F.B., Baldo, J.L., Vilte, A. & Scrocchi, G. (2017) *Apostolepis multicincta* (Harvey 1999) (Squamata, Dipsadidae) in Argentina. *Check List* 13(6), 913–916, 4 figs.

Gallardo, F.B., Stazzonelli, J.C. & Baldo, J. (2014) Ampliación del rango de distribución de *Tomodon orestes* (Harvey y Muñoz, 2004) (Serpentes: Dipsadidae) para el territorio argentino. *Cuadernos de Herpetología* 28(2), 161–163, 3 figs.

Gallardo, G.A., Nenda, S.J. & Scrocchi, G.J. (2019) *Tachymenis peruviana* (Wiegmann, 1834) (Serpentes, Dipsadidae) in Argentina: Geographic distribution and a new province record. *Check List* 15(1), 7–12, 4 figs.

Ganesh, S.R., Adimallaiah, D. & Prasad, K.K. (2013b) New locality records of Nagarjun Sagar racer snake, *Coluber bholanathi* Sharma, 1976. *Herpetotropicos* 9(1/2), 9–12, 2 figs.

Ganesh, S.R., Aengals, R. & Ramanujam, E. (2014b) Taxonomic reassessment of two Indian shieldtail snakes in the *Uropeltis ceylanicus* species group (Reptilia: Uropeltidae). *Journal of Threatened Taxa* 6(1): 5305–5314, 3 figs.

Ganesh, S.R. & Arumugam, M. (2016) Species richness of montane herpetofauna of southern Eastern Ghats, India: A historical resume and a descriptive checklist. *Russian Journal of Herpetology* 23(1), 7–24, 5 figs.

Ganesh, S.R. & Asokan, J.R. (2010) Catalogue of Indian herpetological specimens in the collection of the Government Museum Chennai, India. *Hamadryad* 35(1), 46–63.

Ganesh, S.R., Bhupathy, S., David, P., Sathishkumar, N. & Srinivas, G. (2014a) Snake fauna of High Wavy Mountains, Western Ghats, India: Species richness, status, and distribution pattern. *Russian Journal of Herpetology* 21(1), 53–64, 9 figs.

Ganesh, S.R., Chadramouli, S.R., Sreeker, R. & Shanker, P.G. (2013a) Reptiles of the central western Ghats, India—A reappraisal and revised checklist, with emphasis on the Agumbe Plateau. *Russian Journal of Herpetology* 20(3), 181–189, 5 figs.

Ganesh, S.R. & Chandramouli, S.R. (2018) Die verbreitung von *Trimeresurus strigatus* Gray, 1842 – eine korrigierende notiz. *Sauria* 40(1), 87–91, 3 figs.

Ganesh, S.R., Kalaimani, A., Karthik, P., Baskaran, N., Nagarajan, R. & Chandramouli, S.R. (2018) Herpetofauna of southern Eastern Ghats, India – II from Western Ghats to Coromandel Coast. *Asian Journal of Conservation Biology* 7(1), 28–45, 6 figs.

Ganesh, S.R., Nandhini, T., Deepak, V., Sreeraj, C.R., Abhilash, K.R., Purvaja, R. & Ramesh, R. (2019) Marine snakes of Indian coasts: Historical resume, systematic checklist, toxinology, status, and identification key. *Journal of Threatened Taxa* 11(1), 13132–13150, 1 fig.

Ganesh, S.R. & Ramanujam, E. (2014) Rediscovery of Beddome's coralsnake *Calliophis beddomei* Smith, 1943 from the type locality. *Journal of Threatened Taxa* 6(3), 5580–5582, 1 fig.

Ganesh, S.R., Sharma, V. & Guptha, M.B. (2017) Records of the Indian Sand Snake *Psammophis condanarus* (Merrem, 1820) (Reptilia: Lamprophiidae) in southern India. *Journal of Threatened Taxa* 9(7), 10453–10458, 3 figs.

Ganesh, S.R. & Vogel, G. (2018) Taxonomic reassessment of the common Indian wolf snakes *Lycodon aulicus* (Linnaeus, 1758) complex (Squamata: Serpentes: Colubridae) *Bonn Zoological Bulletin* 67(1), 25–36, 6 figs.

Garberoglio, F.F., Apesteguía, S., Simöes, T.R., Palci, A., Gómez, R.O., Nydam, R.L., Larsson, H.C.E., Lee, M.S.Y. & Caldwell, M.W. (2019) New skulls and skeletons of the Cretaceous legged snake *Najash*, and the evolution of the modern snake body plan. *Science Advances* 5(eaax5833), 1–8, 4 figs.

García-Cardenete, L., Jiménez-Cazalla, F., Fernández-Cardenete, J.R., Valdeón, A., Pérez-García, M.T. & Herrera-Sánchez, F.J. (2014) Contribución al conocimiento corológico de *Myriopholis algeriensis* en el suroeste de Marruecos. *Boletín de la Asociación Herpetológica Española* 25(1), 43–46, 2 figs.

García-Cobos, D. & Gómez-Sánchez, D.A. (2019) Reproductive mode and defensive behaviour of the South American aquatic snake *Helicops pastazae* (Serpentes: Dipsadidae). *Herpetology Notes* 12, 447–451, 2 figs.

García-Díez, T. & González-Fernández. J.E. (2013) The reptile type specimens preserved in the Museo Nacional de Ciencias Naturales (CSIC) of Madrid, Spain. *Zootaxa* 3619(1), 46–58, 3 figs.

García-Grajales, J., Buenrostro-Silva, A. & Mata-Silva, V. (2016) Diversidad herpetofaunística del Parque Nacional Lagunas de Chacahua y La Tuza de Monroy, Oaxaca, México. *Acta Zoológica Mexicana (New series)* 32(1), 90–100, 4 figs.

García-Morales, D., Cervantrs-Burgos, R.I. & García-Vázquez, U.O. (2017) Geographic distribution. *Tantillita lintoni. Herpetological Review* 48(3), 591.

García-Padilla, E. (2015) Predation event and a distributional record for *Atropoides occiduus* (Hoge, 1966). *Mesoamerican Herpetology* 2(3), 376–377, 1 fig.

García-Padilla, E., González-Botello, J.P., Nevárez-de los Reyes, M., Lazano, D., Mata-Silva, V., Johnson, J.D. & Wilson, L.D. (2016a). Distribution notes. *Coniophanes imperialis. Mesoamerican Herpetology* 3(3), 776, 1 fig.

García-Padilla, E., Herrera-Enríquez, G.J., Nevárez-de los Reyes, M., Lazcano, D., Mata-Silva, V., Johnson, J.D. & Wilson, L.D. (2016) Distribution notes. *Crotalus ornatus. Mesoamerican Herpetology* 3(3), 778, 1 fig.

García-Padilla, E. & Mata-Silva, V. (2013) Geographic distribution. *Tantilla bocourti. Herpetological Review* 44(4), 630.

García-Padilla, E. & Mata-Silva, V. (2014a) Noteworthy distributional records for the herpetofauna of Oaxaca, México. *Herpetological Review* 45(3), 468–469.

García-Padilla, E. & Mata-Silva, V. (2014b) Noteworthy distributional records for the herpetofauna of Chiapas, Mexico. *Mesoamerican Herpetology* 1(2), 293–295.

García-Vázquez, U.O., Pavó-Vázquez, C.J., Blancas-Hernándcz, J.C., Blancas-Calva, E. & Centenero-Alcalá, E. (2018) A new rare species of the *Rhadinaea decorata* group from the Sierra Madre del Sur of Guerrero, Mexico (Squamata, Colubridae). *ZooKeys* 780, 137–154, 7 figs.

Garda, A.A., Costa, T.B., Santos-Silva, C.R. dos, Mesquita, D.O., Faria, R.G., Conceição, B.M. da, Silva, I.R.S. da et al. (2013) Herpetofauna of protected areas in the Caatinga I: Raso da Catarina Ecological Station (Bahia, Brazil). *Check List* 9(2), 405–414, 9 figs.

Gawor, A., Pham, C.T., Nguyen, T.Q., Nguyen, T.T., Schmitz, A. & Ziegler, T. (2016) The herpetofauna of the Bai Tu Long National Park, northeastern Vietnam. *Salamandra* 52(1), 23–41, 12 figs.

Geiger, R.D., Geiger, N.D. & Ray, J.M. (2014) Geographic distribution. *Ninia sebae. Herpetological Review* 45(1), 95.

Geissler, P., Hartmann, T., Ihlow, F., Neang, T., Seng, R., Wagner, P. & Böhme, W. (2019) Herpetofauna of the Phnom Kulen National Park, northern Camvbodia – an annotated checklist. *Cambodian Journal of Natural History* 2019(1), 40–63, 6 figs.

Geniez, P. & Teynié, A. (2005) Discovery of a population of the critically endangered *Vipera darevskii* Vedmederja, Orlov & Tuniyev, 1986 in Turkey, with new elements on its identification (Reptilia: Squamata: Viperidae). *Herpetozoa* 18(1/2), 25–33, 10 figs.

Georgalis, G.L. & Scheyer, T.M. (2019) A new species of *Palaeopython* (Serpentes) and other extinct squamates from the Eocene of Dielsdorf (Zurich, Switzerland). *Swiss Journal of Geosciences* 112, 1–35, 18 figs.

Georgalis, G.L., Szyndlar, Z., Kear, B.P. & Delfino, M. (2016) New material of *Laophis crotaloides*, an enigmatic giant snake from Greece, with an overview of the largest fossil European vipers. *Swiss Journal of Geosciences* 109(1), 103–116, 4 figs.

Georgialis, G.L., Arca, M., Rook, L., Tuveri, C. & Delfino, M. (2019b) A new colubroid snake (Serpentes) from the early Pleistocene of Sardinia, Italy. *Bollettino della Società Paleontologica Italiana* 58(3), 277–294, 8 figs.

Georgialis, G.L., Rage, J.-C., Bonis, L. de & Koufos, G.D. (2018) Lizards and snakes from the late Miocene hominoid locality of Ravin de la Pluie (Axios Valley, Greece). *Swiss Journal of Geosciences* 111(1–2), 169–181, 5 figs.

Georgialis, G.L., Villa, A. & Delfino, M. (2017) Fossil lizards and snakes from Ano Metochi – a diverse squamate fauna from the latest Miocene of northern Greece. *Historical Biology* 29(6), 730–742, 7 figs.

Georgialis, G.L., Villa, A., Ivanov, M., Vasilyan, D. & Delfino, M. (2019a) Fossil amphibians and reptiles from the Neogene locality of Maramena (Greece), the most diverse European at the Miocene/Pliocene transition boundary. *Palaeontologia Electronica* 22(3), 68, 1–99, 39 figs.

Gherghel, I., Papeş, M., Brischoux, F., Sahlean, T. & Strugariu, A. (2016) A revision of the distribution of sea kraits (Reptilia, *Laticauda*) with an updated occurrence dataset for ecological and conservation research. *ZooKeys* 569, 135–148, 9 figs.

Gherghel, I., Strugariu, A., Ghiurca, D., Rosu, S. & Hutuleac-Volosciuc, M.-V. (2007) The composition and distribution of the herpetofauna from the Valea Neagra river basin (Neamţ County, Romania). *Herpetologica Romanica* 1, 70–76, 3 figs.

Ghielmi, S., Bergò, P.E. & Andreone, F. (2006) Nuove segnalazioni di *Zootoca vivipara* Jaquin e di *Vipera berus* Linnaeus, in Piemonte, Italia nord-occidentale (Novitates Herpetologicae Pedemontanae II). *Acta Herpetologica* 1(1), 29–36, 2 figs.

Ghielmi, S., Menegon, M., Marsden, S.J., Laddaga, L. & Ursenbacher, S. (2016) A new vertebrate for Europe: The discovery of a range-restricted relict viper in the western Italian Alps. *Journal of Zoological Systematics and Evolutionary Research* 54(3), 161–173, 10 figs.

Ghira, I. (2007a) Rediscovery of *Vipera ursinii rakosiensis* in Transylvania. *Herpetologica Romanica* 1, 77–81, 4 figs.

Ghira, I. (2007b) The herpetofauna of the Sighişoara area (Transylvania, Romania). *Transylvanian Review of Systematical and Ecological Research* 4, 159–168.

Ghira, I., Venczel, M., Covaciu-Marcov, S., Mara, G., Ghile, P., Hartel, T., Török, Z. et al. (2002) Mapping of Transylvanian herpetofauna. *Nymphaea* 29, 145–201, 28 figs.

Gholamhosseini, A., Schmidtler, J.F., Zareian, H. & Esmaeili, H.R. (2009) The second specimen of the Iranian *Eirenis rechingeri* Eiselt, 1971. *Herpetozoa* 22(3/4), 189–190, 3 figs.

Gholamifard, A. & Esmaeili, H.R. (2010) First record and range extension of Field's horned viper, *Pseudocerastes fieldi* (Schmidt, 1930) (Squamata: Viperidae), from Fars province, southern Iran. *Turkish Journal of Zoology* 34(4), 551–552, 2 figs.

Gholamifard, A., Rastegar-Pouyani, N. & Esmaeili, H.R. (2012) Annotated checklist of reptiles of Fars Province, southern Iran. *Iranian Journal of Animal Biosystematics* 8(2), 155–167.

Ghosh, A. & Mukherjee, N. (2019) Geographic distribution: *Boiga quincunciata. Herpetological Review* 50(1), 104.

Ghosh, A., Mukherjee, N. & Rahut, B. (2017) Geographic distribution. *Lycodon zawi. Herpetological Review* 48(1), 129.

Giambelluca, L.A. (2015) *Serpientes Bonaerenses.* Editorial de la Universidad de La Plata, Buenos Aires, 1–70, numerous color ill.

Giery, S.T. (2013) First records of Red Cornsnakes (*Pantherophis guttatus*) from Abaco Island, The Bahamas, and notes on their current distribution in the greater Caribbean region. *IRCF Reptiles & Amphibians* 20(1), 36–39, 3 figs.

Giraldo, A., Garcés-Restrepo, M.F., Quintero-Angel, A., Bolívar, W. & Velandia-Perilla, J.H. (2014) Vertebrados terrestres de Isla Palma (Bahía Málaga, Valle del Cauca, Colombia). *Boletín Científico Museo de Historia Natural* 18(2), 183–202, 3 figs.

Giraudo, A.R., Nenda, S.J., Arzamendia, V., Bellini, G.P. & Franzoy, A. (2015) Nuevos datos sobre la distribución, morfología y conservación de *Micrurus silviae* (Serpentes: Elapidae), una serpiente coral amenazada poco conocida. *Revista Mexicana de Biodiversidad* 86(4), 1041–1047, 2 figs.

Giraudo, A.R., Vidoz, F., Arzamendia, V. & Nenda, S.J. (2012) Distribution and natural history notes on *Tachymenis chilensis chilensis* (Schlegel, 1837) (Reptilia: Serpentes: Dipsadidae) in Argentina. *Check List* 8(5), 919–923, 2 figs.

Giri, V.B., Deepak, V., Captain, A., Das, A., Das, S., Rajkumar, K.P., Rathish, R.L. & Gower, D.J. (2017) A new species of *Rhabdops* (Boulenger, 1893) (Serpentes: Natricinae) from the northern Western Ghats region of India. *Zootaxa* 4319(1), 27–52, 10 figs.

Giri, V.B., Deepak, V., Captain, A., Pawar, S. & Tillack, F. (2019b) A new species of *Boiga* (Fitzinger, 1826) (Serpentes: Colubridae) from the northern Western Ghats of India. *Journal of the Bombay Natural History Society* 116, 1–11, 5 figs.

Giri, V.B., Gower, D.J., Das, A., Lalremsanga, H.T., Lalronunga, S., Captain, A. & Deepak, V. (2019a) A new genus and species of natricine snake from northeast India. *Zootaxa* 4603(2), 241–264, 7 figs.

Glavaš, O.J., Vilaj, I., Lauš, B., Dvorski, P., Koren, T., Kolarić, A., Grbac, I. & Šafarek, G. (2016) Contribution to the knowledge on amphibians and reptiles of north-western Croatia. *Acta Zoologica Bulgarica* 68(4), 519–527, 13 figs.

Glaw, F., Kucharzewski, C., Köhler, J., Vences, M. & Nagy, Z.T. (2013) Resolving an enigma by integrative taxonomy: *Madagascarophis fuchsi* (Serpentes: Lamprophiidae), a new opisthoglyphous and microendemic snake from northern Madagascar. *Zootaxa* 3630(2), 317–332, 6 figs.

Glaw, F., Kucharzewski, C., Nagy, Z.T., Hawlitschek, O. & Vences, M. (2014) New insights into the systematics and molecular phylogeny of the Malagasy snake genus *Liopholidophis* suggest at least one rapid reversal of extreme sexual dimorphism in tail length. *Organisms Diversity & Evolution* 14, 121–132, 5 figs.

Göçmen, B., Arikan, H., Yildiz, M.Z., Mermer, A. & Alpagut-Keskin, N. (2009c) Serological characterization and confirmation of the taxonomic status of *Montivipera albizona* (Serpentes, Viperidae) with an additional new locality record and some phylogenetical comments. *Animal Biology* 59(1), 87–96, 5 figs.

Göçmen, B., Atatür, M.K., Budak, A., Bahar, H., Yildiz, M.Z. & Alpagut-Keskin, N. (2009b) Taxonomic notes on the snakes of Northern Cyprus, with observations on their morphologies and ecologies. *Animal Biology* 59(1), 1–30, 10 figs.

Göçmen, B., Franzen, M., Yildiz, M.Z., Akman, B. & Yalçinkaya, D. (2009a) New locality records of eremial snake species in southeastern Turkey (Ophidia: Colubridae, Elapidae, Typhlopidae, Leptotyphlopidae). *Salamandra* 45(2), 110–114, 2 figs.

Göçmen, B., Geçít, M. & Kariş, M. (2014c) New locality records of the striped dwarf snake, *Eirenis lineomaculatus* (Schmidt, 1939) (Squamata: Ophidia: Colubridae) in Turkey. *Biharean Biologist* 8(2), 126–128, 2 figs.

Göçmen, B., Iğci, N., Akman, B. & Oğuz, M.A. (2013) New locality records of snakes (Ophidia: Colubridae: *Dolichophis, Eirenis*) in eastern Anatolia. *North-Western Journal of Zoology* 9(2), 276–283, 5 figs.

Göçmen, B., Kariş, M., Özmen, E. & Oğuz, M.A. (2018) First record of the Palestine Viper *Vipera palaestinae* (Serpentes: Viperidae) from Anatolia. *South Western Journal of Horticulture, Biology and Environment* 9(2), 87–90, 2 figs.

Göçmen, B., Mebert, K., Akman, B., Yalçinkaya, D., Kariş, M. & Erturhan, M. (2011) New locality records of snakes resembling the Big-headed Grass Snake, *Natrix megalocephala* (Orlov & Tuniyev, 1987) (Ophidia: Colubridae) in Turkey. *North-Western Journal of Zoology* 7(2), 363–367, 4 figs.

Göçmen, B., Mebert, K., Iğci, N., Akman, B., Yildiz, M.Z., Oğuz, M.A. & Altin, Ç. (2014a) New locality records for four rare species of vipers (Reptilia: Viperidae) in Turkey. *Zoology in the Middle East* 60(4), 306–313, 2 figs.

Göçmen, B., Mebert, K. & Kariş, M. (2015b) New distributional data on *Vipera* (*berus*) *barani* from western and northeastern Anatolia. *Herpetology Notes* 8, 609–615, 3 figs.

Göçmen, B., Mebert, K., Kariş, M., Oğuz, M.A. & Ursenbacher, S. (2017) A new population and subspecies of the critically endangered Anatolian meadow viper *Vipera anatolica* Eiselt and Baran, 1970 in eastern Antalya province. *Amphibia-Reptilia* 38(3), 289–305, 6 figs.

Göçmen, B., Mulder, J., Kariş, M. & Mebert, K. (2015a) New locality records of *Vipera ammodytes transcaucasiana* Boulenger, 1913 in Turkey. *South Western Journal of Horticulture, Biology and Environment* 6(2), 91–98, 2 figs.

Göçmen, B., Mulder, J., Kariş, M. & Oğuz, M. (2014b) The poorly known Anatolian meadow viper, *Vipera anatolica*: New morphological and ecological data. *Herpetologica Romanica* 8, 1–10, 3 figs.

Gojo-Cruz, P.H.P., Afuang, L.E., Gonzalez, J.C.T. & S.M.W. Gruezo (2018) Amphibians and reptiles of Luzon Island, Philippines: The herpetofauna of Pantabangan-Carranglan Watershed, Nueva Ecija Province, Caraballo Mountain Range. *Asian Herpetological Research* 9(4), 201–223, 47 figs.

Gomes de Arruda, L.A., Carvalho, M.A. de & Kawashita-Ribeiro, R.A. (2015) New records of the Amazon banded snake *Rhinobothryum lentiginosum* (Serpentes: Colubridae) from Mato Grosso state, Brazil, with natural history notes. *Salamandra* 51(2), 199–205, 2 figs.

Gómez, C. & Buitrago-González, W. (2017) *Bothriechis schlegelii* (Berthold 1846) Cabeza de candado, Víbora de pestañas. *Catálogo de Anfibios y Reptiles de Colombia* 3(1), 1–11, 3 figs.

Gómez, R.O., Garberoglio, F.F. & Rougier, G.W. (2019) A new late Cretaceous snake from Patagonia: Phylogeny and trends in body size evolution of madtsoiid snakes. *Comptes Rendus Palevol* 18, 771–781, 5 figs.

Gonçalves, D.V., Martínez-Freiría, F., Crochet, P.-A., Geniez, P., Carranza, S. & Brito, J.C. (2018) The role of climatic cycles and trans-Saharan migration corridors in species diversification: Biogeography of *Psammophis schokari* group in North Africa. *Molecular Phylogenetics and Evolution* 118, 64–74, 4 figs.

Gonçalves, F.M.P., Braine, D., Bauer, A.M., Valério, H.M., Marques, M.P. & Ceríaco, L.M.P. (2019) Rediscovery of the poorly known Angolan Adder, *Bitis heraldica* (Bocage, 1889) (Serpentes: Viperidae); new records, live photographs, and first case history of envenomation. *Herpetological Review* 50(2), 241–246, 4 figs.

Gong, S., Hitschfeld, E., Hundsdörfer, A.K., Auer, M., Wang, F., Zhou, L. & Fritz, U. (2011) Is the horned viper *Ceratrimeresurus shenlii* Liang and Liu, 2003 from China a valid *Protobothrops*? *Amphibia-Reptilia* 32(1), 132–135, 2 figs.

Gonwouo, N.L., LeBreton, M., Chirio, L., Ineich, I., Tchamba, N.M., Ngassam, P., Dzikouk, G. & Diffo, J.L. (2007) Biodiversity and conservation of the reptiles of the Mount Cameroon area. *African Journal of Herpetology* 56(2), 149–161, 2 figs.

Gonzales, L. & Montaño-F., R.R. (2005) Material tipo en las colecciones zoológicas del Museo de Historia Natural "Noel Kempff Mercado" Santa Cruz, *Bolivia. Kempffiana* 1, 6–20.

González, A. & Iturriaga, M. (2014) Geographic distribution. *Arrhyton taeniatum. Herpetological Review* 45(4), 663.

Gonzalez, R.C., Prudente, A.L. da C. & Franco, F.L. (2014a) Morphological variation of *Gomesophis brasiliensis* and *Ptychophis flavovirgatus* (Serpentes, Dipsadidae, Xenodontinae). *Salamandra* 50(2), 85–98, 8 figs.

Gonzalez, R.C., Silva-Soares, T., Castro, T.M. de & Bérnils, R.S. (2014b) Review of the geographic distribution of *Micrurus decoratus* (Jan, 1858) (Serpentes: Elapidae). *Phyllomedusa* 13(1), 29–39, 3 figs.

González-Carvajal, J.M., Medina-Rangel, G.F. & Rojas-Murcia, L.E. (2018) The first country record of the Mexican Blind Snake, *Anomalepis mexicanus* (Jan, 1860) (Serpentes, Anomalepididae), in Colombia. *Check List* 14(6), 1047–1052, 4 figs.

González-Fernández, J.E. (2006) Catálogo de los reptiles procedentes de la comunidad de Madrid (España) que se conservan en el Museo Nacional de Ciencias Naturales. *Graellsia* 62(num. extra.), 145–174, 24 figs.

González-Hernández, A., Moro-Hernández, D.M. & Cruz, J.A. (2016) Distribución y uso de habitat de *Thamnophis pulchrilatus* (Cope, 1855 [sic]) en Chignahuapan, Puebla, México. *Acta Zoológica Mexicana*, New Series 32(3), 390–392, 2 figs.

González-Sánchez, V.H., Johnson, J.D., García-Padilla, E., Mata-Silva, V., DeSantis, D.L. & Wilson, L.D. (2017) The herpetofauna of the Mexican Yucatan Peninsula: Composition, distribution, and conservation status. *Mesoamerican Herpetology* 4(2), 264–380, numerous figs.

Goodman, A.M., Esposito, L.A., Ponce, P.L., Sauer, A.L., Stiner, E.O. & Ruane, S. (2019) New record and range expansion of *Masticophis lateralis* (Hallowell, 1853) (Squamata, Colubridae) into western Baja California Sur, Mexico. *Check List* 15(2), 345–348, 3 figs.

Gosá, A. & Bergerandi, A. (1994) Atlas de distribución de los anfibios y reptiles de Navarra. *Munibe* 46, 109–189, 4 figs, numerous maps.

Gower, D.J., Giri, V., Captain, A. & Wilkinson, M. (2016a) A reassessment of *Melanophidium* (Günther, 1864) (Squamata: Serpentes: Uropeltidae) from the Western Ghats of peninsular

India, with the description of a new species. *Zootaxa* 4085(4), 481–503, 9 figs.

Gower, D.J., Wade, E.O.Z., Spawls, S., Böhme, W., Buechley, E.R., Sykes, D. & Colston, T.J. (2016b) A new large species of *Bitis* (Gray, 1842) (Serpentes: Viperidae) from the Bale Mountains of Ethiopia. *Zootaxa* 4093(1), 41–63, 12 figs.

Gower, D.J. & Wickramasinghe, L.J.M. (2016) Recharacterization of *Rhinophis dorsimaculatus* (Deraniyagala, 1941) (Serpentes: Uropeltidae), including description of new material. *Zootaxa* 4158(2), 203–212, 4 figs.

Graboski, R., Arredondo, J.C., Grazziotin, F.G., Silva, A.A.A. da, Prudente, A.L.C., Rodrigues, M.T., Bonatto, S.L. & Zaher, H. (2019) Molecular phylogeny and hemipeneal diversity of South American species of *Amerotyphlops* (Typhlopidae, Scolecophidia). *Zoologica Scripta* 48(2), 139–156, 7 figs.

Graboski, R., Filho, G.A.P., Silva, A.A.A. da, Prudente, A.L.da C. & Zaher, H. (2015) A new species of *Amerotyphlops* from northeastern Brazil, with comments on distribution of related species. *Zootaxa* 3920(3), 443–452, 4 figs.

Graitson, E. (2013) Les reptiles de Wallonie. Bilan des connaissances et évolutions récentes. *L'Echo des Rainettes* (12), 2–18, 20 figs.

Grano, M. & Cattaneo, C. (2015) First record of *Zamenis situla* (Linnaeus, 1758) (*Reptilia Serpentes*) from the Aegean island of Chalki (Dodecanese, Greece). *Naturalista Siciliano*, series 4, 39(2), 375–381, 2 figs.

Grano, M., Cattaneo, C. & Cattaneo, A. (2013) First record of *Hierophis gemonensis* (Laurenti, 1768) (Reptilia Serpentes Colubridae) in the Aegean island of Tsougriá, northern Sporades, Greece. *Biodiversity Journal* 4(4), 553–556, 2 figs.

Gray, R.J. & Hofmann, E.P. (2017) Two new herpetofaunal records for Stann Creek District, Belize. *Mesoamerican Herpetology* 4(1), 205–207, 2 figs.

Greenbaum, E., Kusamba, C., Muninga, W.M. & Aristote, M.M. (2017) Geographic distribution. *Boaedon perisilvestris. Herpetological Review* 48(1), 127.

Greenbaum, E., Portillo, F., Jackson, K. & Kusamba, C. (2015) A phylogeny of Central African *Boaedon* (Serpentes: Lamprophiidae), with the description of a new cryptic species from the Albertine Rift. *African Journal of Herpetology* 64(1), 18–38, 5 figs.

Griffin, R., Martin, G., Whitaker, R. & Lewis, T.R. (2012) Distribution, morphology, and natural history of the Medo Pit Viper, *Viridovipera medoensis* (Viperidae, Crotalinae) in Arunachal Pradesh, northeastern India. *IRCF Reptiles & Amphibians* 19(4), 237–242, 8 figs.

Griffin, R. & Powell, G. (2014) Geographic distribution. *Tropidodipsas fasciata. Herpetological Review* 45(3), 467.

Grismer, L.L., Quah, E.S.H., Anuar M.S., S., Muin, M.A., Wood, P.L. & Nor, S.A.M. (2014) A diminutive new species of cave-dwelling wolf snake (Colubridae: *Lycodon* Boie, 1826) from Peninsular Malaysia. *Zootaxa* 3815(1), 51–67, 8 figs.

Groen, J., Bok, B., Tiemann, L. & Verspui, G.J. (2019) Geographic range extension of the Rough-scaled Bush Viper, *Atheris hispida* (Serpentes: Viperidae) from Uganda, Africa. *Herpetology Notes* 12, 241–243, 2 figs.

Grosse, W.-R., Winkler, C. & Bringsøe, H. (2015) Die herpetofauna der Nordfriesischen Inseln Dänemarks und Deutschlands. *Rana* 16, 9–24, 3 figs.

Grossmann, W., Zwanzig, B.-M., Kowalski, T. & Zilger, H.-J. (2013) Weitere ergänzend herpetologische beobachtungen auf dem Saiq-Plateau und im Jebel al-Akhdar, Sultanat Oman. *Sauria* 35(3), 23–31, 12 figs.

Grünwald, C.I., Jones, J.M., Franz-Chávez, H. & Ahumada-Carrillo, I.T. (2015) A new species of *Ophryacus* (Serpentes: Viperidae:

Crotalinae) from eastern Mexico, with comments on the taxonomy of related pitvipers. *Mesoamerican Herpetology* 2(4), 388–416, 14 figs.

Grünwald, C.I., Morales-Flores, K.I., Franz-Chávez, H., Hermosillo-López, A.I. & Jones, J.M. (2016b) First report of *Porthidium ophryomegas* (Serpentes: Viperidae: Crotalinae) from Mexico, with comments on the status of an endangered biogeographical formation. *Mesoamerican Herpetology* 3(4), 1104–1107, 3 figs.

Grünwald, C.I., Pérez-Rivera, N., Ahumada-Carillo, I.T., Franz-Chávez, H. & La Forest, B.T. (2016a) New distributional records for the herpetofauna of Mexico. *Herpetological Review* 47(1), 85–90, 2 figs.

Guedes, T.B., Barbo, F.E., França, D. & Zaher, H. (2018) Morphological variation of the rare psammophilous species *Apostolepis gaboi* (Serpentes, Dipsadidae, Elapomorphini). *Zootaxa* 4418(5), 469–480, 6 figs.

Guedes, T.B., Nogueira, C. & Marques, O.A.V. (2014) Diversity, natural history, and geographic distribution of snakes in the Caatinga, northeastern Brazil. *Zootaxa* 3863(1), 1–93. 32 figs.

Guedes, T.B., Nunes, G.S.S., Prudente, A.L. da C. & Marques, O.A.V. (2011) New records and geographical distribution of the tropical banded treesnake *Siphlophis compressus* (Dipsadidae) in Brazil. *Herpetology Notes* 4, 341–346, 1 fig.

Guevara, H.J.L., Ubeda-Olivas, M.F., Castellón, J.C.G. & Sunyer, J. (2015) Distribution notes: *Chironius grandisquamis*. *Mesoamerican Herpetology* 2(4), 543–544, 1 fig.

Guillemin, M. & Martin, T.E. (2017) Geographic distribution. *Platyceps karelini*. *Herpetological Review* 48(1), 130–131.

Guiller, G., Lourdais, O. & Ursenbacher, S. (2016) Hybridization between a Euro-Siberian (*Vipera berus*) and a para-Mediterranean viper (*V. aspis*) at their contact zone in western France. *Journal of Zoology* 302, 138–147, 3 figs.

Gül, S. (2015) Potential distribution modeling and morphology of *Pelius barani* (Böhme and Joger, 1983) in Turkey. *Asian Herpetological Research* 6(3), 206–212, 4 figs.

Gül, S., Kumlutaş, Y. & Ilgaz, Ç. (2015) Climatic preferences and distribution of 6 evolutionary lineages of *Typhlops vermicularis* Merrem, 1820 in Turkey using ecological niche modeling. *Turkish Journal of Zoology* 39, 235–243, 3 figs.

Gül, S., Kumlutaş, Y. & Ilgaz, Ç. (2016a) Predicted distribution patterns of *Pelias kaznakovi* (Nikolsky, 1909) in the Caucasus hotspot with a new locality record from Turkey. *Russian Journal of Herpetology* 23(3), 224–230, 4 figs.

Gül, S., Kumlutaş, Y. & Ilgaz, Ç. (2016b) A new locality record of *Pelias barani* (Böhme et Joger, 1983) from the northeastern Anatolia. *Russian Journal of Herpetology* 23(4), 319–322, 3 figs.

Guo, P., Li, J., Chen, Y. & Wang, Y. (2012c) Designation of a neotype for *Protobothrops mangshanensis* (Zhao, 1990). *Asian Herpetological Research* 3(4), 340–341, 2 figs.

Guo, P., Liu, Q., Li, J., Cao, Y. & Wang, Y. (2016a) Catalogue of the type specimens of amphibians and reptiles in the Herpetological Museum of the Chengdu Institute of Biology, Chinese Academy of Sciences: V. Viperidae (Reptilia, Serpentes). *Asian Herpetological Research* 7(1), 59–63, 6 figs.

Guo, P., Liu, Q., Li, J., Zhing, G., Chen, Y. & Wang, Y. (2012a) Catalogue of the type specimens of amphibians and reptiles in the Herpetological Museum of the Chengdu Institute of Biology, Chinese Academy of Sciences: III. Snakes excluding viperids (Reptilia, Serpentes). *Asian Herpetological Research* 3(4), 334–339, 11 figs.

Guo, P., Liu, Q., Myers, E., Liu, S., Xu, Y., Liu, Y. & Wang, Y. (2012b) Evaluation of the validity of the ratsnake subspecies *Elaphe carinata deqenensis* (Serpent: Colubridae). *Asian Herpetological Research* 3(3), 219–226, 5 figs.

Guo, P., Liu, Q., Wen, T., Xiao, R., Fang, M., Zhong, G., Truong, N.Q., Zhu, F., Jardin, R.C. & Li, C. (2016b) Multilocus phylogeny of the Asian lance-headed pitvipers (Squamata, Viperidae, *Protobothrops*). *Zootaxa* 4093(3), 382–390, 2 figs.

Guo, P., Liu, Q., Zhong, G., Zhu, F., Yan, F., Tang, T., Xiao, R., Fang, M., Wang, P. & Fu, X. (2015) Cryptic diversity of green pitvipers in Yunnan, south-west China (Squamata, Viperidae). *Amphibia-Reptilia* 36(3), 265–276, 3 figs.

Guo, P., Liu, Q., Zhu, F., Zhong, G.H., Che, J., Wang, P., Xie, Y.L., Murphy, R.W. & Malhotra, A. (2019a) Multilocus phylogeography of the brown-spotted pitviper *Protobothrops mucrosquamatus* (Reptilia: Serpentes: Viperidae) sheds a new light on the diversification pattern in Asia. *Molecular Phylogenetics and Evolution* 133, 82–91, 5 figs.

Guo, P., Malhotra, A., Li, P.P., Pook, C.E. & Creer, S. (2007) New evidence on the phylogenetic position of the poorly known Asia pitviper *Protobothrops kaulbacki* (Serpentes: Viperidae: Crotalinae) with a redescription of the species and a revision of the genus *Protobothrops*. *Herpetological Journal, London* 17, 237–246, 4 figs.

Guo, P., Zhang, L., Liu, Q., Li, C., Pyron, R.A., Jiang, K. & Burbrink, F.T. (2013) *Lycodon* and *Dinodon*: One genus or two? Evidence from Molecular phylogenetics and morphological comparisons. *Molecular Phylogenetics and Evolution* 68, 144–149, 2 figs.

Guo, P., Zhu, F. & Liu, Q. (2019b) A new member of the genus *Sinonatrix* (Serpentes: Colubridae) from western China. *Zootaxa* 4623(3), 535–544, 7 figs.

Guo, P., Zhu, F., Liu, Q., Zhang, L., Li, J.X., Huang, Y.Y. & Pyron, R.A. (2014) A taxonomic revision of the Asian keelback snakes, genus *Amphiesma* (Serpentes: Colubridae: Natricinae), with description of a new species. *Zootaxa* 3873(4), 425–440, 3 figs.

Guptha, B., Prasad, N.V.S., Maddock, S.T. & Deepak, V. (2015) First record of *Chrysopelea taprobanica* (Smith, 1943) (Squamata: Colubridae) from India. *Check List* 11(1523), 1–3, 4 figs.

Guptha, B., Prasad, N.V.S. & Vecrappan, D. (2012) Rediscovery and range extension of *Coluber bholanathi* Sharma, 1976 from Seshachalam hills, Andhra Pradesh, India. *Herpetology Notes* 5, 447–448, 2 figs.

Gutiérrez-López, L., Martínez-Fonseca, J.G. & J. Sunyer (2015) Nature notes: *Stenorrhina freminvillii*. *Mesoamerican Herpetology* 2(4), 530–531, 1 fig.

Gutiérrez-Rodríguez, A.A. & Sunyer, J. (2016) First record of *Trimetopon pliolepis* (Cope, 1894) (Reptilia: Squamata: Dipsadidae) from Nicaragua. *Mesoamerican Herpetology* 3(2), 517–518, 1 fig.

Gutiérrez-Rodríguez, A.A. & Sunyer, J. (2017a) Distribution notes: *Senticolis triaspis*. *Mesoamerican Herpetology* 4(2), 465–466, 1 fig.

Gutiérrez-Rodríguez, A.A. & Sunyer, J. (2017b) Distribution notes: *Xenodon angustirostris*. *Mesoamerican Herpetology* 4(2), 479–480, 1 fig.

Guyer, C., Folt, B., Hoffman, M., Stevenson, D., Goetz, S.M., Miller, M.A. & Godwin, J.C. (2019) Patterns of head shape and scutellation in *Drymarchon couperi* (Squamata: Colubridae) reveal a single species. *Zootaxa* 4695(2), 168–174, 4 figs.

Haacke, W.D. (2013) Description of a new tiger snake (Colubridae, *Telescopus*) from south-western Africa. *Zootaxa* 3737(3), 280–288, 6 figs.

Habeeb, I.N., Karamiani, R. & Rastegar-Pouyani, N. (2016) Annotated checklist of semi-venomous and venomous snakes of Iraq. *Iranian Journal of Animal Biosystematics* 12(2), 239–248.

Habeeb, I.N. & Rastegar-Pouyani, N. (2016a) Geographical distribution of the snakes of Iraq. *Mesopotamia Environmental Journal* 2(3), 67–77.

Habeeb, I.N. & Rastegar-Pouyani, N. (2016b) Recent identification key to Iraqi snakes. *Mesopotamia Environmental Journal* 3(1), 60–74, numerous figs.

Halpern, B., Major, A., Péchy, T., Marinov, E.M. & Kiss, J.B. (2007) Genetic comparison of Moldavian Meadow Viper (*Vipera ursinii moldavica*) populations of the Danube-Delta. Scientific Annals of the Danube Delta Institute, *Tulcea-Romania* 13, 19–26, 5 figs.

Hamdan, B., Coelho, D.P., D'Angiolella, A.B., Dias, E.J. dos R. & Lira-da-Silva, R.M. (2013a) The reptile collection of the Museu de Zoologia, Universidade Federal da Bahia, Brazil. *Check List* 9(2), 257–262.

Hamdan, B. & Fernandes, D.S. (2015) Taxonomic revision of *Chironius flavolineatus* (Jan, 1863) with description of a new species (Serpentes: Colubridae). *Zootaxa* 4012(1), 97–119, 11 figs.

Hamdan, B., Machado, C. & Citeli, N.K. (2015) Filling gaps and a new state record of *Xenopholis scalaris* (Wucherer, 1861) (Serpentes: Dipsadidae). *Check List* 11(1746), 1–3, 2 figs.

Hamdan, B., Pereira, A.G., Loss-Oliveira, L., Rödder, D. & Schrago, C.G. (2017) Evolutionary analysis of *Chironius* snakes unveils cryptic diversity and provides clues to diversification in the Neotropics. *Molecular Phylogenetics and Evolution* 116, 108–119, 4 figs.

Hamdan, B., Scali, S. & Fernandes, D.S. (2014) On the identity of *Chironius flavolineatus* (Serpentes: Colubridae). *Zootaxa* 3794(1), 134–142, 3 figs.

Hamdan, B., Silva, Jr., N.J. da, Silva, H.L.R., Cintra, C.E.D. & Lema, T. de (2013b) Redescription of *Phalotris labiomaculatus* (Serpentes, Dipsadidae, Elapomorphini), with notes on the taxonomic boundaries within the *nasutus* group. *Zootaxa* 3693(2), 182–188, 2 figs.

Handal, E.N., Amr, Z.S. & Qumsiyeh, M.B. (2016) Some records of reptiles from the Palestinian territories. *Russian Journal of Herpetology* 23(4), 261–270, 4 figs.

Hansen, R.W. & Alamillo, J.A. (2018) Geographic distribution: *Lampropeltis greeri*. *Herpetological Review* 49(1), 77.

Hansen, R.W., Arciga, R.H. & Savage, R.F. (2015) Geographic distribution. *Lampropeltis ruthveni*. *Herpetological Review* 46(4), 574.

Hansen, R.W., Fernández-Badillo, L., Ramírez-Bautista, A. & Avalos-Torales, O. (2016) Geographic distribution. *Lampropeltis mexicana*. *Herpetological Review* 47(2), 262–263.

Hansen, R.W. & Salmon, G.T. (2017) Distribution analysis, taxonomic updates, and conservation status of the *Lampropeltis mexicana* group (Serpentes: Colubridae). *Mesoamerican Herpetology* 4(4), 699–758, 32 figs.

Hansen, R.W. & Vermilya, D.W. (2015) Geographic distribution. *Amastridium sapperi*. *Herpetological Review* 46(4), 571.

Harikrishnan, S., Choudhary, B.C. & Vasudevan, K. (2010) Recent records of snakes (Squamata: Serpentes) from Nicobar Islands, India. *Journal of Threatened Taxa* 2(11), 1297–1300, 4 figs.

Harikrishnan, S., Choudhary, B.C. & Vasudevan, K. (2012) A survey of herpetofauna on Long Island, Andaman and Nicobar Islands, India. *Herpetological Bulletin* (119), 19–28, 20 figs.

Harrington, S.M. & Reeder, T.W. (2017) Phylogenetic inference and divergence dating of snakes using molecules, morphology and fossils: New insights into convergent evolution of feeding morphology and limb reduction. *Biological Journal of the Linnean Society* 121, 379–394, 5 figs.

Hartmann, T., Handschuh, M. & Böhme, W. (2011) First record of *Psammophis indochinensis* Smith, 1943 from Cambodia, within the context of a distributional species account. *Cambodian Journal of Natural History* 2011(1), 7–10, 3 figs.

Hartmann, T., Ihlow, F., Edwards, S., Sovath, S., Handschuh, M. & Böhme, W. (2013) A preliminary annotated checklist of the amphibians and reptiles of the Kulen Promtep Wildlife Sanctuary in northern Cambodia. *Asian Herpetological Research* 4(1), 36–55, 7 figs.

Hasan, M.K., Feeroz, M.M., Ahmed, S., Ahmed, A. & Saha, S. (2013) The confirmed record of *Oligodon albocinctus* (Cantor, 1839) from Bangladesh. *Taprobanica* 5(1), 77–80, 2 figs.

Hauser, S. (2017) On the validity of *Pareas macularius* (Theobald, 1868) (Squamata: Pareidae) as a species distinct from *Pareas margaritophorus* (Jan *in* Bocourt, 1866). *Tropical Natural History* 17(1), 25–52, 14 figs.

Hauser, S. (2018) Addition of *Liopeltis frenatus* (Günther, 1858) and *Cyclophiops multicinctus* (Roux, 1907) to the herpetofauna of Thailand (Squamata: Colubridae). *Tropical Natural History* 18(1), 54–67, 4 figs.

Hauser, S. (2019) Addition of *Liopeltis stoliczkae* (Sclater, 1891) (Squamata: Colubridae) to the herpetofauna of Thailand, with notes on its distribution and natural history. *Natural History Bulletin of the Siam Society* 63(2), 127–140, 9 figs.

Head, J.J. (2015) Fossil calibration dates for molecular phylogenetic analysis of snakes 1: Serpentes, Alethinophidia, Boidae, Pythonidae. *Palaeontologia Electronica* 18(1), 6FC, 1–17, 1 fig.

Head, J.J., Bloch, J.I., Moreno-Bernal, J., Rincon-Burbano, A.F. & Bourque, J. (2013) Cranial osteology, body size, systematics, and ecology of the giant Paleocene snake *Titanoboa cerrejonensis*. *Journal of Vertebrate Paleontology* 33(suppl.), 140–141.

Head, J.J., Mahlow, K. & Müller, J. (2016) Fossil calibration dates for molecular phylogenetic analysis of snakes 2: Caenophidia, Colubroidea, Elapoidea, Colubridae. *Palaeontologia Electronica* 19(2)9FC, 1–21, 1 fig.

Hecht, V.L., Pham, C.T., Nguyen, T.T., Nguyen, T.Q., Bonkowski, M. & Ziegler, T. (2013) First report on the herpetofauna of Tay Yen Tu Nature Reserve, northeastern Vietnam. *Biodiversity Journal* 4(4), 507–552, 82 figs.

Hedges, S.B., Marion, A.B., Lipp, K.M., Marin, J. & Vidal, N. (2014) A taxonomic framework for typhlopid snakes from the Caribbean and other regions (Reptilia, Squamata). *Caribbean Herpetology* 49, 1–61, 3 figs.

Hedges, S.B., Powell, R., Henderson, R.W., Hanson, S. & Murphy, J.C. (2019) Definition of the Caribbean Islands biogeographic region, with checklist and recommendations for standardized common names of amphibians and reptiles. *Caribbean Herpetology* 67, 1–53, 3 figs.

Heimes, P. (2016) *Herpetofauna Mexicana Vol. I. Snakes of Mexico*. Edition Chimaira, Frankfurt-am-Main, 1–572, 652 figs, 198 maps.

Henderson, R.W. (2015) *Natural History of Neotropical Treeboas (genus Corallus)*. Edition Chimaira, Frankfurt-am-Main, 1–338, 308 figs.

Henderson, R.W. & Powell, R. (2009) *Natural History of West Indian Reptiles and Amphibians*. University Press of Florida, Gainesville, xxiv+495 pp, 37 figs.

Hernandez, T., Herr, M.W., Stevens, S., Cork, K., Medina-Nava, C., Vialpando, C.J., Warfel, T., Fields, N., Brodie, C. & Graham, S.P. (2019) New distribution records for amphibians and reptiles in eastern Chihuahua, *Mexico. Check List* 15(1), 79–86, 1 fig.

Hernández-Arciga, R., Hernández, C., López-Vidal, J.C., Villegas-Ruiz, J. & Elizalde-Arellano, C. (2013) Nuevo registro de *Geophis latifrontalis* (Squamata: Colubridae) para el estado Guanajuato, México. *Acta Zoológica Mexicana* (new series) 29(3), 684–687, 2 figs.

Hernández-Gallegos, O., Carmen-Cristóbal, J.M., Malvaez-Estrada, J.R., Rangel-Patiño, C.A., Granados-González, G. & Ruiz-Gómez, M. de L. (2014) Geographic distribution. *Conophis vittatus*. *Herpetological Review* 45(4), 663.

Hernández-Jiménez, C.A., Flores-Villela, O. & Campbell, J.A. (2019) A new species of patch-nosed snake (Colubridae: *Salvadora* Baird and Girard, 1853) from Oaxaca, Mexico. *Zootaxa* 4564(2), 588–600, 5 figs.

Hernández-Ordóñez, O., Arroyo-Rodríguez, V., González-Hernández, A., Russildi, G., Luna-Reyes, R., Martínez-Ramos, M. & Reynoso, V.H. (2015) Range extensions of amphibians and reptiles in the southeastern part of the Lacandona rainforest, Mexico. *Revista Mexicana de Biodiversidad* 86, 457–468, 1 fig.

Hernández-Salinas, U., Ramírez-Bautista, A. & Mata-Silva, V. (2014) Species richness of squamate reptiles from two islands in the Mexican Pacific. *Check List* 10(6), 1264–1269, 6 figs.

Hernández-Valadez, E., Hernández-Estañol, E., Barragán-Vázquez, R., Charruau, P. & López-Luna, M.A. (2016) First record and distribution extension of *Enulius flavitorques* (Cope, 1869) (Squamata: Colubridae) in Tabasco, Mexico. *Mesoamerican Herpetology* 3(2), 512–513, 1 fig.

Herr, M.W., Giovanetto, L.A., Black, M., Hernadez, J.C., Mendoza-Pérez, M.R. & Graham, S.P. (2017) Geographic distribution. *Tantilla cucullata*. *Herpetological Review* 48(4), 816.

Herrera, A. & Ray, J.M. (2016) Geographic distribution. *Erythrolamprus bizona*. *Herpetological Review* 47(1), 83.

Herrera-Montes, A., Ríos-Dróz, B., Puente-Rolón, A.R., Dávila-Casanova, D. & Ríos-López, N. (2015) Geogrphic distribution. *Xenochrophis vittatus*. *Herpetological Review* 46(1), 64.

Herrmann, H.-W. & Branch, W.R. (2013) Fifty years of herpetological research in the Namib Desert and Namibia with an updated and annotated species checklist. *Journal of Arid Environments* 93, 94–115, 3 figs.

Herse, M.R. & Ray, J.M. (2014) A review and correction of data on a poorly known leaf litter snake, *Trimetopon slevini* (Dunn, 1940), from Panama, including additional data on defensive behaviours. *Herpetology Notes* 7, 359–361, 2 figs.

Hinckley, A., Intang, R., Tuh, F. & Jaimis, B. (2017) A distribution update on the Bornean endemic *Trimeresurus (Popeia) sabahi* (David et al., 2011). *Herpetology Notes* 10, 625–626, 2 figs.

Hoaglund, E. & Smith, C.E. (2013) New and updated records of amphibians and reptiles from Minnesota, USA. *Herpetological Review* 44(3), 482–483.

Hodges, C.W., Amber, E.D. & Strine, C.T. (2018) *Boiga guangxiensis* (Wen, 1998) (Squamata: Colubridae) feeding on *Draco blanfordii* in Yunnan, China. *Herpetology Notes* 11, 981–984, 3 figs.

Hoefer, S. (2019) Range extension of the Black-banded Trinket Snake, *Oreocryptophis porphyraceus* (Cantor 1839) in Lam Dong Province, Vietnam. *IRCF Reptiles & Amphibians* 26(2), 123–124, 3 figs.

Hofmann, E.P. (2016) Geographic distribution. *Sibon dimidiatus*. *Herpetological Review* 47(1), 84.

Hofmann, E.P., Gray, R.J., Wilson, L.D. & Townsend, J.H. (2016) Discovery of the first male specimen of *Tantilla hendersoni* (Stafford, 2004) (Squamata: Colubridae), from a new locality in central Belize. *Herpetology Notes* 10, 53–57, 3 figs.

Hofmann, S., Fritzche, P. & Miehe, G. (2016) A new record of *Elaphe dione* from high altitude in western Sichuan reveals high intraspecific differention. *Salamandra* 52(3), 273–277, 2 figs.

Hofmann, S., Mebert, K., Schulz, K.-D., Helfenberger, N., Göçmen, B. & Böhme, W. (2018) A new subspecies of *Zamenis hohenackeri* (Strauch, 1873) (Serpentes: Colubridae) based on morphological and molecular data. *Zootaxa* 4471(1), 137–153, 7 figs., plus supplements.

Hofmann, S., Tillack, F. & Miehe, G. (2015) Genetic differentiation among species of the genus *Thermophis* Malnate (Serpentes, Colubridae) and comments on *T. shangrila*. *Zootaxa* 4028(1), 102–120, 2 figs.

Holbrook, J.D. (2012) *A Field Guide to the Snakes of Southern Florida*. ECO Herpetological Publishing & Distribution, Rodeo. x+179, numerous illustrations.

Holt, B.D., Barger, T.W., Peters, A.S., Taylor, C.T. & Lawrence, E. (2017) One hundred and ninety-four new county records for amphibians and reptiles in Alabama, USA. *Herpetological Review* 48(1), 138–144.

Hoogmoed, M.S., Fernandes, R., Kucharzewski, C., Moura-Leite, J.C., Bérnils, R.S., Entiauspe-Neto, O.M. & Santos, F.P.R. dos (2019) Synonymization of *Uromacer Ricardinii* Peracca, 1897 with *Dendrophis aurata* (Schlegel, 1837) (Reptilia: Squamata: Colubridae: Dipsadinae), a rare South American snake with a disjunct distribution. *South American Journal of Herpetology* 14(2), 88–102, 4 figs.

Hoogmoed, M.S. & Lima, J.D. (2018) *Epictia collaris* (Hoogmoed, 1977) (Reptilia: Squamata: Leptotyphlopidae), new record for the herpetofauna of Amapá and Brazil, with additional localities in French Guiana and a distribution map. *Boletim do Museu Paraense Emilio Goeldi, Ciências Naturais* 13(3), 461–465, 2 figs.

Hoser, R. (2013) Case 3601. *Spracklandus* (Hoser, 2009) (Reptilia, Serpentes, Elapidae): Request for confirmation of the availability of the generic name and for the nomenclatural validation of the journal in which it was published. *Bulletin of Zoological Nomenclature* 70(4), 234–237.

Hoser, R.T. (2015a) PRINO (Peer reviewed in name only) journals: When quality control in scientific publication fails. *Australasian Journal of Herpetology* 26, 3–64, illustrated.

Hoser, R.T. (2015b) Comments on *Spracklandus* (Hoser, 2009) (Reptilia, Serpentes, Elapidae): Request for confirmation of the availability of the generic name and for the nomenclatural validation of the journal in which it was published (Case 3601; see *BZN* 70: 234–237; comments *BZN* 71:30–38, 133–135). *Australasian Journal of Herpetology* 27, 37–43.

Hosseinzadeh, M.S., Ghezellou, P. & Kazemi, S.M. (2017) Predicting the potential distribution of the endemic snake *Spalerosophis microlepis* (Serpentes: Colubridae), in the Zagros Mountains, western Iran. *Salamandra* 53(2), 294–298, 3 figs.

Houssaye, A., Rage, J.-C., Bardet, N., Vincent, P., Amaghzaz, M. & Meslouh, S. (2013) New highlights about the enigmatic marine snake *Palaeophis maghrebianus* (Palaeophiidae; Palaeophiinae) from the Ypresian (lower Eocene) phosphates of Morocco. *Palaeontology* 56(3), 647–661, 10 figs.

Hsiang, A.Y., Field, D.J., Webster, T.H., Behlke, A.D.B., Davis, M.B., Racicot, R.A. & Gauthier, J.A. (2015) The origin of snakes: Revealing the ecology, behavior, and evolutionary history of early snakes using genomics, phenomics, and the fossil record. *BMC Evolutionary Biology* 15(87), 1–22, 9 figs.

Hsiou, A.S., Albino, A.M., Medeiros, M.A. & Santos, R.A.B. (2014) The oldest Brazilian snakes from the Cenomanian (early Late Cretaceous). *Acta Palaeontologica Polonica* 59(3), 635–642, 3 figs.

Hubbs, B. (2013a) *Harmless Snakes of the West*. Tricolor Books, Tempe. 1–130, numerous illustrations.

Hubbs, B. (2013b) New county records and an update for Kansas and Nebraska, USA. *Herpetological Review* 44(3), 481–482.

Hubbs, B. (2016) New county records for Iowa, Kansas, Missouri, and Nebraska, USA. *Herpetological Review* 47(1), 94–95.

Hubbs, B. & O'Connor, B. (2009) *A Guide to the Rattlesnakes of the United States*. Tricolor Books, Tempe, 1–96, numerous illustrations.

Hubbs, B. & O'Connor, B. (2012) *A Guide to the Rattlesnakes and Other Venomous Serpents of the United States*. Tricolor Books, Tempe, 1–132, numerous illustrations.

Hudson, B.D., Felix, Z.I., Oguni, J., Wilson, B., McEntire, K.D., Stratmann, T., Duff, D.D. & Seymour, Z. (2015) New geographic distributional records of amphibians and reptiles in Georgia, USA. *Herpetological Review* 46(4), 595–596.

Hughes, D.F., Behangana, M. & Wilber, L. (2017a) Geographic distribution. *Naja haje*. *Herpetological Review* 48(2), 392.

Hughes, D.F., Scharsu, C., Behangana, M. & Wilbur, L. (2017b) Geographic distribution. *Naja nigricollis*. *Herpetological Review* 48(2), 392.

Hunt, J.D. (2017) Geographic distribution. *Rena humilis humilis*. *Herpetological Review* 48(4), 814–815.

Hurtado-Gómez, J.P., Grisales-Martínez, F.A. & Rendón-Valencia, B. (2015) Starting to fill the gap: First record of *Tantilla supracincta* (Peters, 1863) (Serpentes: Colubridae) from Colombia. *Check List* 11(1713), 1–4, 3 figs.

Hurzaid, A., Bakar, M.A.A., Sharma, D., Nasir, N., Sharma, R., Aznan, A.R.Y. & Jaafar, I. (2013) An updated checklist of the herpetofauna of the Belum-Temengor forest reserves, Hulu Perak, Peninsular Malaysia. *Proceedings of the 3rd Annual Conference Syiah Kuala University (AIC Unsyiah)*, 151–158.

Hussien, N.A. & Hussein, R.M. (2013) Morphological and molecular olymorphism of snake *Psammophis schokari* (Colubridae) in the desert-mountain and coastal areas of Egypt. *World Applied Sciences Journal* 27(8), 996–1004, 8 figs.

Hutchinson, M., Swain, R. & Driessen, M. (2001) *Snakes and Lizards of Tasmania. Fauna of Tasmania Handbook 9*. Fauna of Tasmania Committee, Hobart. 1–64, numerous illustrations.

Hynková, I., Starostová, Z. & Frynta, D. (2009) Mitochondrial DNA variation reveals recent evolutionary history of main *Boa constrictor* clades. *Zoological Science* 26(9), 623–631, 3 figs.

Ibrahim, A.A. (2012) New records of the Dice Snake, *Natrix tessellata*, in the Suez Canal Zone and Sinai. *Amphibian and Reptile Conservation* 6(2), 2–4, 3 figs.

Ibrahim, A.A. (2013) The herpetology of the Suez Canal Zone, Egypt. *Vertebrate Zoology* 63(1), 87–110, 38 figs.

ICZN Opinion 2381. (2017) *Vipera latastei* (Boscá, 1878) (Reptilia, Serpentes, Viperidae): Conservation of the original spelling. *Bulletin of Zoological Nomenclature* 73(2–4), 145–147.

Iftime, A. & Iftime, O. (2008) Observations on the herpetofauna of the Giurgiu County (Romania). *Travaux du Muséum National d'Histoire Naturelle "Grigore Antipa"* 51, 209–218, 3 figs.

Iftime, A. & Iftime, O. (2010) Contributions to the knowledge of the herpetofauna of the eastern Jiu and upper Lotru drainage basins (southern Carpathians, Romania). *Travaux du Muséum National d'Histoire Naturelle "Grigore Antipa"* 53, 273–286, 6 figs.

Iftime, A. & Iftime, O. (2013) Contributions to the knowledge regarding the distribution and ecology of the herpetofauna of Ţarcu Massif (southern Carpathians, Romania). *Travaux du Muséum National d'Histoire Naturelle "Grigore Antipa"* 56(1), 81–92, 6 figs.

Iftime, A. & Iftime, O. (2014a) Notes on the herpetofauna of the Leaota Mountains, a "wildlife corridor" area. *North-Western Journal of Zoology* 10(Suppl.), S33–S37, 5 figs.

Iftime, A. & Iftime, O. (2014b) Note on the amphibians and reptiles of the "Nordul Gorjului de Est" site of community interest and adjacent areas (Southern Carpathians, Romania). *North-Western Journal of Zoology* 10(Suppl.), S44–S50, 7 figs.

Iftime, A. & Iftime, O. (2016) Contributions to the knowledge on the amphibians and reptiles of Teleorman County (Southern Romania). *Travaux du Muséum National d'Histoire Naturelle "Grigore Antipa"* 58(1–2), 63–71, 6 figs.

Iftime, A., Iftime, O. & Pop, D.A. (2009) Observations on the herpetofauna of the Iezer-Păpuşa Massif (southern Carpathians, Romania). *Herpetozoa* 22(1/2), 55–64, 4 figs.

İğci, N., Akman, B., Göçmen, B., Demirsoy, A.I. & Oğuz, M.A. (2015) Range extension of four species of snakes (Ophidia: *Eirenis, Pseudocyclophis, Platyceps*) in eastern Anatolia. *Biharean Biologist* 9(2), 166–169, 4 figs.

Ikeda, T., Otsuka, H. & Ota, H. (2016) Early Pleistocene fossil snakes (Reptilia: Squamata) Okinawajima Island in the Ryukyu Archipelago, southwestern Japan. *Herpetological Monographs* 30, 143–156, 7 figs.

Iković, V., Tomović, L. & Ljubisavljević, K. (2016) Contribution to the knowledge of the batracho- and herpetofauna of the Bjelopavlići region (Macedonia). *Bulletin of the Natural History Museum* 9, 113–125, 1 fig.

Ineich, I. (2009) The terrestrial herpetofauna of Torres and Banks groups (northern Vanuatu), with report of a new species for Vanuatu. *Zootaxa* 2198(1), 1–15, 3 figs.

Ineich, I., Chirio, L., Ascani, M., Rabeil, T. & Newby, J. (2014) Herpetofauna of Termit Massif and neighbor areas in Tenere Desert, southeastern Niger, West Africa. *Herpetology Notes* 7, 375–390, 6 figs.

Ineich, I., LeBreton, M., Lhermitte-Vallarino, N. & Chirio, L. (2015) The reptiles of the summits of Mont Oku and the Bamenda Highlands, Cameroon. *Amphibian & Reptile Conservation* 9(2), 15–38, 22 figs.

Ineich, I. & Prudent, P. (2014) Geographic distribution. *Myriopholis boueti*. *Herpetological Review* 45(3), 466.

Insacco, G., Spadola, F., Russotto, S. & Scaravelli, D. (2015) *Eryx jaculus* (Linnaeus, 1758): A new species for the Italian herpetofauna (Squamata: Erycidae). *Acta Herpetologica* 10(2), 149–153, 3 figs.

Iskenderov, T.M., Akhmedov, S.B. & Bunyatova, S.N. (2017) [Lotiev's Viper (*Pelias lotievi*, Serpentes, Viperidae), a species new to the fauna of Azerbaijan Republic.] *Zoologicheskii Zhurnal* 96(1), 121–124, 2 figs.

Itescu, Y., Jamison, S., Slavenko, A., Tamar, K., Roussos, S.A., Foufopoulos, J., Meiri, S. & Pafilis, P. (2017) The herpetofauna of Folegandros Island (Cyclades, Greece). *Herpetozoa* 29(3/4), 183–190, 5 figs.

Iturriaga, M. (2015) Geographic distribution. *Typhlops leptolepis*. *Herpetological. Review* 46(3), 387.

Ivanov, M. & Čerňanský, A. (2017) *Vipera berus* (Linnaeus, 1758) remains from the late Pleistocene of Slovakia. *Amphibia-Reptilia* 38(2), 133–144, 3 figs.

Ivanov, M., Vasilyan, D., Böhme, M. & Zazhigin, V.S. (2019) Miocene snakes from northeastern Kazakhstan: New data on the evolution of snake assemblages in Siberia. *Historical Biology* 31(10), 1284–1303, 9 figs.

Jablonski, D. (2011) K herpetofauně obce svetlice a okolí (Slovensko). *Folia Faunistica Slovaca* 16(2), 103–107, 7 figs. [in Slovak].

Jablonski, D. & Balej, P. (2015) *Xerotyphlops vermicularis* (Merrem, 1820), in the west Bulgarian Rhodope Mountains: Rediscovery after more than 100 years. *Herpetozoa* 27(3–4), 200–203, 1 fig.

Jablonski, D., Frynta, D. & Marín, G.M. del M. (2014) New records of the Awl-headed snake (*Lytorhynchus diadema*) from northeastern Morocco. *Herpetology Notes* 7, 295–297, 1 fig.

Jablonski, D., Kukushkin, O.V., Avcı, A., Bunyatova, S., Kumlutaş, Y., Ilgaz, Ç, Polyakova, E., Shiryaev, K., Tuniyev, B.S. & Jandzik, D. (2019) The biogeography of *Elaphe sauromates*

(Pallas, 1814), with a description of a new rat snake species. *PeerJ* 7(e6944), 1–44, 10 figs.

Jablonski, D. & Masroor, R. (2019) The easternmost distribution and highest elevation record of the rare Desert Cat Snake *Telescopus rhinopoma* (Reptilia: Colubridae) in Pakistan. *Journal of Threatened Taxa* 11(1), 13180–13183, 2 figs.

Jablonski, D., Musilová, R. & Zavadil, V. (2011) "A find of the Aesculapian snake (*Zamenis longissimus*) in south Bohemia." *Acta Musei Bohemiae Meridionalis in České Budějovice, Scientiae Naturales* 51, 166–169, 1 fig. [in Slovak].

Jablonski, D., Szabolcs, M., Simović, A. & Mizsei, E. (2017) Color and pattern variation of the Balkan Whip Snake, *Hierophis gemonensis* (Laurenti, 1768). *Turkish Journal of Zoology* 41(2), 363–369, 3 figs.

Jackson, K. & Blackburn, D.C. (2007) Die amphibian und reptilian des Nouabale-Ndoki Nationalparks, Republik Kongo (Brazzaville). *Salamandra* 43(3), 149–164, 2 figs.

Jacobs, S. (2018) Geographic distribution: *Boa sigma. Herpetological Review* 49(2), 287.

Jacobsen, N.H.G., Pietersen, E.W. & Pietersen, D.W. (2010) A preliminary herpetological survey of the Vilanculos Coastal Wildlife Sanctuary on the San Sebastian Peninsula, Vilankulo, Mozambique. *Herpetology Notes* 3, 181–193, 1 fig.

Jacobsen, N.H.G. & Randall, R.M. (2013) Survey of reptiles in the Wilderness section of the Garden Route National Park, South Africa. *Herpetology Notes* 6, 209–217, 1 fig.

Jadin, R.C., Blair, C., Jowers, M.J., Carmona, A. & Murphy, J.C. (2019) Hiding in the lianas of the tree of life: Molecular phylogenetics and species delimitation reveal considerable cryptic diversity of New World vine snakes. *Molecular Phylogenetics and Evolution* 134, 61–65, 2 figs.

Jadin, R.C., Burbrink, F.T., Rivas, G.A., Vitt, L.J., Barrio-Amorós, C.L. & Guralnick, R.P. (2014) Finding arboreal snakes in an evolutionary tree: Phylogenetic placement and systematic revision of the Neotropical birdsnakes. *Journal of Zoological of Systematics and Evolutionary Research* 52(3), 257–264, 3 figs.

Jairam, R. (2017) The distribution of *Taeniophallus nicagus* (Colubridae) in Suriname with some information on morphology. *IRCF Reptiles & Amphibians* 24(3), 177–179, 3 figs.

Jamal, Q., Idrees, M., Ullah, S., Adnan, M., Zaidi, F., Zaman, Q. & Rasheed, S.B. (2018) Diversity and altitudinal distribution of Squamata in two distinct ecological zones of Dir, a Himalayan sub-zone of northern Pakistan. *Pakistan Journal of Zoology* 50(5), 1835–1839.

Jandzik, D., Avci, A. & Gvoždík, V. (2013) Incongruence between taxonomy and genetics: Three divergent lineages within two subspecies of the rare Transcaucasian rat snake (*Zamenis hohenackeri*). *Amphibia-Reptilia* 34(4), 579–584, 2 figs.

Janssen, H.Y., Pham, C.T., Ngo, H.T., Le, M.D., Nguyen, T.Q. & Ziegler, T. (2019) A new species of *Lycodon* (Boie, 1826) (Serpentes, Colubridae) from northern Vietnam. *ZooKeys* 875, 1–29, 8 figs.

Jaramillo-Martínez, A.F., Valencia-Zuleta, A. & Castro-Herrera, F. (2013) *Imantodes chocoensis* Torres-Carvajal, Yánez-Muñoz, Quirola, (Smith and Almendáriz, 2012) (Squamata: Dipsadidae): First records from Colombia. *Check List* 9(5), 1070–1071, 3 figs.

Jasinski, S.E. & Moscato, D.A. (2017) Late Hemphillian colubrid snakes (Serpentes, Colubridae) from the Gray Fossil Site of northeastern Tennessee. *Journal of Herpetology* 51(2), 245–257, 4 figs.

Javed, S.M.M., Tampal, F. & Srinivasulu, C. (2010) First record of *Coelognathus radiatus* (Boie, 1827) (Reptilia: Colubridae) from the Ananthagiri Hills, Eastern Ghats, India. *Journal of Threatened Taxa* 2(9), 1172–1174, 5 figs.

Jelić, D. (2014) Checklist of Croatian amphibians and reptiles with bibliography of 250 years of research. *Natura Sloveniae* 16(2), 17–72, 2 figs.

Jesus, J., Gonçalves, R., Spínola, C. & Brehm, A. (2013) First record of *Ramphotyphlops braminus* (Daudin, 1803) on Madeira Island (Portugal). *Herpetozoa* 26(1/2), 106–109, 1 fig.

Jiménez-Robles, O., León, R., Cárdenas, S., Rebollo, B. & Martínez, G. (2017) Contributions to the natural history and distribution of *Dasypeltis sahelensis* Trape & Mané, 2006, in Morocco. *Herpetozoa* 30(1/2), 80–86, 2 figs.

Jins, V.J., Bhupathy, S. & Panigrahi, M. (2014) New record of Beddome's coral snake *Calliophis beddomei* Smith, 1943 from the southern Western Ghats, India. *Herpetology Notes* 7, 555–557, 2 figs.

Jins, V.J., Sampaio, F.L. & Gower, D.J. (2018) A new species of *Uropeltis* (Cuvier, 1829) (Serpentes: Uropeltidae) from the Anaikatty Hills of the Western Ghats of India. *Zootaxa* 4415(3), 401–422, 10 figs.

Joger, U., Bshena, I. & Essghaier, F. (2008) First record of the parthenogenetic Brahminy blind snake, *Ramphotyphlops braminus* (Daudin, 1803), from Libya (Serpentes: Typhlopidae). *Herpetology Notes* 1, 13–16, 3 figs.

Joger, U., Fritz, U., Guicking, D., Kalyabina-Hauf, S., Nagy, Z.T. & Wink, M. (2010) Relict populations and endemic clades in Palearctic reptiles: Evolutionary history and implications for conservation. In: Habel, J.C. & Assmann, T. (Eds.), *Relict Species - Phylogeography and Conservation Biology*. Springer-Verlag, Berlin, 119–132, 10 figs.

Johnson, J.D., Mata-Silva, V., García-Padilla, E. & Wilson, L.D. (2015b) The herpetofauna of Chiapas, Mexico: Composition, distribution, and conservation. *Mesoamerican Herpetology* 2(3), 272–329, some figs.

Johnson, J.D., Mata-Silva, V. & Wilson, L.D. (2014) The type locality of *Tantilla johnsoni* (Wilson, Vaughan, and Dixon, 1999) (Squamata, Colubridae) and related issues. *Mesoamerican Herpetology* 1(2), 305–308, 1 fig.

Johnson, J.D., Wilson, L.D., Mata-Silva, V., García-Padilla, E. & DeSantis, D.L. (2017) The endemic herpetofauna of Mexico: Organisms of global significance in severe peril. *Mesoamerican Herpetology* 4(3), 543–620, numerous color illus.

Jorge da Silva, N., Hamdan, B., Tonial, I.J., Silva, H.L.R. da & Cintra, C.E.D. (2012) *Hydrodynastes melanogigas* (Franco, Fernandes and Bentim, 2007) (Squamata: Serpentes: Colubridae): Range extension and new state record. *Check List* 8(4), 813–814, 2 figs.

Joshi, R. & Singh, A. (2015) Range extension and geographic distribution record for the Burmese Python, *Python bivittatus* (Kuhl, 1820) (Reptilia: Pythonidae) in northwestern India. *IRCF Reptiles & Amphibians* 22(3), 102–105, 7 figs.

Joshi, R., Singh, A. & Puri, K. (2019) The northernmost record of the colubrid snake *Dendrelaphis tristis* (Daudin, 1803) (Reptilia: Squamata: Serpentes) in India (Rajaji Tiger Reserve). *Herpetology Notes* 12, 305–308, 2 figs.

Jowers, M.J., Caut, S., Garcia-Mudarra, J.L., Alasaad, S. & Ineich, I. (2013) Molecular phylogenetics of the possibly extinct Martinique ground snake. *Herpetologica* 69(2), 227–236, 1 fig.

Jowers, M.J., Mudarra, J.L.G., Charles, S.P. & Murphy, J.C. (2019) Phylogeography of West Indies coral snakes (*Micrurus*): Island colonization and banding patterns. *Zoologica Scripta* 48(3), 263–276, 4 figs.

Juárez-Peña, C., Bartuano, A.S. & Sigüenza-Mejia, S. (2016) New herpetofaunal records for Parque Nacional Montecristo, *El Salvador*. *Mesoamerican Herpetology* 3(4), 1107–1113, 4 figs.

Kaiser, C.M., Kaiser, H. & O'Shea, M. (2018) The taxonomic history of Indo-Papuan groundsnakes, genus *Stegonotus* (Duméril et al., 1854) (Colubridae), with some taxonomic revisions and the designation of a neotype for *S. parvus* (Meyer, 1874). *Zootaxa* 4512(1), 1–73, 33 figs.

Kaiser, C.M., O'Shea, M. & Kaiser, H. (2019) A new species of Indo-Papuan groundsnake, genus *Stegonotus* (Duméril et al., 1854) (Serpentes, Colubridae), from the Bird's Head Peninsula of West Papua, Indonesia, with comments on differentiating morphological characters. *Zootaxa* 4590(2), 201–230, 7 figs.

Kaiser, H. (2013) The taxon filter, a novel mechanism designed to facilitate the relationship between taxonomy and nomenclature, vis-à-vis the utility of the *Code's* Article 81 (the Commission's plenary power). *Bulletin of Zoological Nomenclature* 70(4), 293–302.

Kaiser, H., Crother, B.I., Kelly, C.M.R., Luiselli, L., O'Shea, M., Ota, H., Passos, P., Schleip, W.D. & Wüster, W. (2013a) Best practices: In the 21st Century, taxonomic decisions in herpetology are acceptable only when supported by a body of evidence and published via peer-review. *Herpetological Review* 44(1), 8–23.

Kaiser, H., Sanchez, C., Heacox, S., Kathriner, A., Ribeiro, A.V., Soares, Z.A., Araujo, L.L. de, Mecke, S. & O'Shea, M. (2013b) First report on the herpetofauna of Ataúro Island, Timor-Leste. *Check List* 9(4), 752–762, 7 figs.

Kaito, T., Ota, H. & Toda, M. (2017) The evolutionary history and taxonomic reevaluation of the Japanese coral snake, *Sinomicrurus japonicus* (Serpentes, Elapidae), endemic to the Ryukyu Archipelago, Japan, by use of molecular and morphological analyses. *Journal of Zoological Systematics and Evolutionary Research* 55(2), 156–166, 5 figs.

Kaito, T. & Toda, M. (2016) The biogeographical history of Asian keelback snakes of the genus *Hebius* (Squamata: Colubridae: Natricinae) in the Ryukyu Archipelago, Japan. *Biological Journal of the Linnean Society* 118, 187–199, 5 figs.

Kalaentzis, K., Strachinis, I., Katsiyiannis, P., Oefinger, P. & Kazilas, C. (2018) New records and an updated list of the herpetofauna of Kastellorizo and the adjacent islet Psomi (Dodecanese, SE Greece). *Herpetology Notes* 11, 1009–1019, 6 figs.

Kalki, Y., Schramer, T.D. & Wylie, D.B. (2018) On the occurrence of *Malayopython reticulatus* (Schneider, 1801) in mainland India (Squamata: Pythonidae). *Herpetology Notes* 11, 703–708, 2 figs.

Kane, D., Goodwin, S., Verspui, G.J., Tump, A. & Marin, G.M. del M. (2019) Reptile diversity of southern Morocco: Range extensions and the role of the Djebel Ouarkziz as a biogeographical barrier. *Herpetology Notes* 12, 787–793, 2 figs.

Kapfer, J.M., Lorch, J.M., Wild, E.R., Brown, D.J., Mitchem, L., Rudolph, N., Rutzen, K. & Vogt, R.C. (2015) Distributional records for amphibians and reptiles from Wisconsin, USA. *Herpetological Review* 46(4), 587–590.

Kapfer, J.M., Sloss, B.L., Schuurman, G.W., Paloski, R.A. & Lorch, J.M. (2013) Evidence of hybridization between common gartersnakes (*Thamnophis sirtalis*) and Butler's gartersnakes (*Thamnophis butleri*) in Wisconsin, USA. *Journal of Herpetology* 47(3), 400–405, 3 figs.

Karin, B.R., Stubbs, A.L., Arifin, U., Bloch, L.M., Ramadhan, G., Iskandar, D.T., Arida, E., Reilly, S.B., Kusnadi, A. & McGuire, J.A. (2018) The herpetofauna of the Kei Islands (Maluku, Indonesia): Comprehensive report on new and historical collections, biogeographic patterns, conservation concerns, and an annotated checklist of species from Kei Kecil, Kei Besar, Tam, and Kur. *Raffles Bulletin of Zoology* 66, 704–738, 37 figs.

Kariş, M. & Göçmen, B. (2018) On a new finding of the local endemic Amanos Dwarf Snake, *Muhtarophis barani* (Serpentes: Colubridae) from southern Turkey. *South Western Journal of Horticulture, Biology and Environment* 9(1), 47–52, 2 figs.

Karunarathna, D.M.S.S. & Amarasinghe, A.A.T. (2012) Reptile diversity in Beraliya Mukalana Proposed Forest Reserve, Galle District, Sri Lanka. *Taprobanica* 4(1), 20–26, 8 figs.

Karunarathna, S., Surasinghe, T., Botejue, M. & Madawala, M. (2018) *Gerarda prevostiana* (Serpentes: Homalopsidae) in Sri Lanka: Distribution and behaviour. *The Herpetological Bulletin* 145, 8–13, 4 figs.

Kass, C., Kass, N.A., Velasco, M.A., Juri, M.D., Williams, J.D. & Kacoliris, F.P. (2018) Inventory of the herpetofauna of Talampaya National Park, a World Heritage Site in Argentina. *Neotropical Biology and Conservation* 13(3), 202–211, 3 figs.

Kawashita-Ribeiro, R.A., Ávila, R.W. & Morais, D.H. (2013) A new snake of the genus *Helicops* (Wagler, 1830) (Dipsadidae, Xenodontinae) from Brazil. *Herpetologica* 69(1), 80–90, 5 figs.

Kaya, N. & Özuluğ, O. (2017) The herpetological collection of Zoology Museum, Istanbul University. *Andolu University Journal of Science and Technology C – Life Sciences and Biotechnology* 6(2), 55–63, 3 figs.

Kazemi, E., Kaboli, M., Khosravi, R. & Khorasani, N. (2019) Evaluating the importance of environmental variables on spatial distribution of Caspian cobra, *Naja oxiana* (Eichwald, 1831) in Iran. *Asian Herpetological Research* 10(2), 129–138, 5 figs.

Kazemi, S.M., Rastegar-Pouyani, E., Shafieri Darabi, S.A., Ebrahim Tehrani, M., Hosseinzadeh, M.S., Mobaraki, A. & Mashayekhi, M. (2015) Annotated checklist of amphibians and reptiles of Qom Province, central Iran. *Iranian Journal of Animal Biosystematics* 11(1), 23–31.

Keates, C., Conradie, W., Greenbaum, E. & Edwards, S. (2019) A snake in the grass: Genetic structuring of the widespread African grass snake (*Psammophylax* Fitzinger 1843), with the description of a new genus and a new species. *Journal of Zoological Systematics and Evolutionary Research* 57(4), 1039–1066, 7 figs.

Keiser, E.D. (2010) *Snakes of the University of Mississippi Field Station*. University of Mississippi, Oxford, 1–97.

Kessler, E.J., Kuhns, A.R., Crawford, J.A., Philips, C.A., Wright, E.M., Anthonysamy, W.J.B., Esker, T.L., Gillespie, J., Jacques, L.J. & Saffer, R.S. (2013) New county records of reptiles and amphibians from state-managed properties in east-central Illinois, USA. *Herpetological Review* 44(2), 286–288.

Kessler, E.J., Kuhns, A.R., Rhoden, C.M., Philips, C.A., Cook, C.E., Keigher, A.K. & Crawford, J.A. (2015) New county records of reptiles and amphibians from state-managed properties in Illinois. *Herpetological Review* 46(4), 603–604.

Khaire, N. (2014) *Indian Snakes – A Field Guide*. Jyotsna Prakashan, Pune. 1–160, numerous illustrations.

Khan, A.A. & Sharma, V. (2018) Geographic distribution: *Lycodon jara*. *Herpetological Review* 49(3), 506.

Khan, M.S. (2012) Scientific and standard common English names of amphibians and reptiles of Pakistan transliterated in Urdu. *Pakistan Journal of Zoology Supplement Series* 11, 1–12.

Khandal, D., Rai, S. & Sharma, V. (2016a) Geographic distribution. *Boiga flaviviridis*. *Herpetological Review* 47(1), 82.

Khandal, D., Sahu, Y.K. & Sharma, V. (2016b) New record of *Elachistodon westermanni* (Reinhardt, 1863) (Serpentes, Colubridae) for Rajasthan state, India. *Russian Journal of Herpetology* 23(4), 249–253, 3 figs.

Khani, S., Kami, H.G. & Rajabizadeh, M. (2017) Geographic variation of *Gloydius halys caucasicus* (Serpentes: Viperidae) in Iran. *Zoology in the Middle East* 63(4), 303–310, 4 figs.

Kidov, A.A., Pykhov, S.G. & Dernakov, V.V. (2009) New finds of the Talysh common toad (*Bufo eichwaldi*), meadow lizard (*Darevskia praticola*) and Iranian ratsnake (*Elaphe persica*) in south-eastern Azerbaijan. *Proceedings of the Ukranian Herpetological Society* (2), 21–26, 2 figs.

Kieckbusch, M., Mader, F., Kaiser, H. & Mecke, S. (2018) A new species of *Cylindrophis* (Wagler, 1828) (Reptilia: Squamata: Cylindrophiidae) from Boano Island, northern Maluku Province, Indonesia. *Zootaxa* 4486(3), 236–250, 7 figs.

Kieckbusch, M., Mecke, S., Hartmann, L., Ehrmantraut, L., O'Shea, M. & Kaiser, H. (2016) An inconspicuous, conspicuous new species of Asian pipesnake, genus *Cylindrophis* (Reptilia: Squamata: Cylindrophiidae), from the south coast of Jawa Tengah, Java, *Indonesia, and an overview of the tangled taxonomic history of C. ruffus (Laurenti, 1768). Zootaxa* 4093(1), 1–25, 7 figs.

Kindler, C., Böhme, W.W., Corti, C., Gvoždík Jablonski, D., Jandzik, D., Metallinou, M., Široký, P. & Fritz, U. (2013) Mitochondrial phylogeography, contact zones and taxonomy of grass snakes (*Natrix natrix, N. megalocephala*). *Zoologica Scripta* 42(5), 458–472, 2 figs.

Kindler, C., Bringsøe, H. & Fritz, U. (2014) Phylogeography of grass snakes (*Natrix natrix*) all around the Baltic Sea: Implications for the Holocene colonization of Fennoscandia. *Amphibia-Reptilia* 35(4), 413–424, 2 figs.

Kindler, C., Chèvre, M., Ursenbacher, S., Böhme, W., Hille, A., Jablonski, D., Vamberger, M. & Fritz, U. (2017) Hybridization patterns in two contact zones of grass snakes reveal a new Central European snake species. *Scientific Reports* 7(7378), 1–12, 5 figs.

Kindler, C. & Fritz, U. (2014) Neue genetische erkenntnisse zur taxonomie und phylogeographie der Ringelnatter (*Natrix natrix*) sowie der Großkopfringelnatter (*N. megalocephala*). *Zeitschrift für Feldherpetologie* 21, 1–14, 6 figs.

Kindler, C. & Fritz, U. (2018) Phylogeography and taxonomy of the barred grass snake (*Natrix helvetica*), with a discussion of the subspecies category in zoology. *Vertebrate Zoology* 68(3), 269–281, 8 figs.

Kindler, C., Gracía, E. & Fritz, U. (2018b) Extra-Mediterranean glacial refuges in barred and common grass snakes (*Natrix helvetica, N. natrix*). *Scientific Reports* 8(1821), 1–13, 10 figs.

Kindler, C., Pous, P. de, Carranza, S., Beddek, M., Geniez, P. & Fritz, U. (2018a) Phylogeography of the Ibero-Maghrebian red-eyed grass snake (*Natrix astreptophora*). *Organisms Diversity & Evolution* 18(1), 143–150, 5 figs.

Kirchhof, S., Mahlow, K. & Tillack, F. (2016) The identity of *Stenorhabdium temporale* (Werner, 1909) (Serpentes: Colubroidea). *Vertebrate Zoology* 66(2), 179–190, 8 figs.

Kirchner, M. (2009) *Telescopus fallax* (Fleischmann, 1831) found in the Aegean Island of Chios, Greece. *Herpetozoa* 21(3/4), 189–190.

Kizirian, D., Nguyen, Q.T., Hanh, T.N. & Minh, D.L. (2018) *Parahelicops, Pararhabdophis*, paraphyly: Phylogenetic relationships among certain Southeast Asian natricine snakes (*Hebius*). *American Museum Novitates* (3906), 1–7, 1 fig.

Klein, C.G., Longrich, N.R., Ibrahim, N., Zouhri, S. & Martill, D.M. (2017) A new basal snake from the mid-Cretaceous of Morocco. *Cretaceous Research* 72, 134–141, 4 figs.

Knight, J.L. & Cicimurri, D.J. (2019) A late Pleistocene record of *Drymarchon* sp. (Serpentes: Colubridae: Colubrinae) from South Carolina, USA. *Paludicola* 12(2), 46–52, 2 figs.

Knight, J.L., Ruback, P., Wedow, J. & Ray, J.M. (2016) *Masticophis mentovarius* (Duméril, Bibron & Duméril, 1854) (Squamata: Serpentes: Colubridae): An update of records from Panama. *Mesoamerican Herpetology* 3(1), 185–188, 3 figs.

Koch, A., Arida, E., Riyanto, A. & Böhme, W. (2009) Islands between the realms: A revised checklist of the herpetofauna of the Talaud Archipelago, Indonesia, with a discussion about its biogeographic affinities. *Bonner Zoologische Beiträge* 56(1/2), 107–129, 16 figs.

Koch, C., Martins, A. & Schweiger, S. (2019) A century of waiting: Description of a new *Epictia* (Gray, 1845) (Serpentes: Leptotyphlopidae) based on specimens housed for more than 100 years in the collection of the Natural History Museum Vienna (NMW). *PeerJ* 7(e7411), 1–37, 16 figs.

Koch, C., Santa-Cruz, R. & Cárdenas, H. (2016) Two new endemic species of *Epictia* (Gray, 1845) (Serpentes: Leptotyphlopidae) from northern Peru. *Zootaxa* 4150(2), 101–122, 13 figs.

Koch, C. & Venegas, P.J. (2016) A large and unusually colored new snake species of the genus *Tantilla* (Squamata; Colubridae) from the Peruvian Andes. *PeerJ* 2016(2767), 1–28, 9 figs.

Koch, C., Venegas, P.J. & Böhme, W. (2015) Three new endemic species of *Epictia* (Gray, 1845) (Serpentes: Leptotyphlopidae) from the dry forest of northwestern Peru. *Zootaxa* 3964(2), 228–244, 8 figs.

Koch, C., Venegas, P.J., Santa Cruz, R. & Böhme, W. (2018) Annotated checklist and key to the species of amphibians and reptiles inhabiting the northern Peruvian dry forest along the Andean valley of the Marañón River and its tributaries. *Zootaxa* 4385(1), 1–101, 17 figs.

Köhler, G., Cedeño-Vázquez, J.R. & Beutelspacher-García, P.M. (2016a) The Chetumal snake census: Generating biological data from road-killed snakes. Part 1. Introduction and identification key to the snakes of southern Quintana Roo, Mexico. *Mesoamerican Herpetology* 3(3), 669–687, 16 figs.

Köhler, G., Cedeño-Vázquez, J.R., Kirstein, T. & Beutelspacher-García, P.M. (2016b) The Chetumal snake census: Generating biological data from road-killed snakes. Part 2. *Dipsas brevifacies, Sibon sanniolus*, and *Tropidodipsas sartorii*. *Mesoamerican Herpetology* 3(3), 688–705, 17 figs.

Köhler, G., Cedeño-Vázquez, J.R., Kraus, E.D., Beutelspacher-García, P.M. & Domínguez-Lepe, J.A. (2017b) The Chetumal snake census: Generating biological data from road-killed snakes. Part 5. *Imantodes tenuissimus, Lampropeltis triangulum*, and *Stenorrhina freminvillii*. *Mesoamerican Herpetology* 4(4), 773–789, 16 figs.

Köhler, G., Cedeño-Vázquez, J.R., Spaeth, M. & Beutelspacher-García, P.M. (2016c) The Chetumal snake census: Generating biological data from road-killed snakes. Part 3. *Leptodeira frenata, Ninia sebae*, and *Micrurus diastema*. *Mesoamerican Herpetology* 3(4), 929–947, 14 figs.

Köhler, G., Cedeño-Vázquez, J.R., Tun, A.M. & Beutelspacher-García, P.M. (2017a) The Chetumal snake census: Generating biological data from road-killed snakes. Part 4. *Coniophanes imperialis, C. meridanus*, and *C. schmidti*. *Mesoamerican Herpetology* 4(3), 527–542, 12 figs.

Köhler, G. & Kieckbusch, M. (2014) Two new species of *Atractus* from Colombia (Reptilia, Squamata, Dipsadidae). *Zootaxa* 3872(3), 291–300, 6 figs.

Köhler, G., Vargas, J., Köhler, J.J. & Vesely, M. (2013) Noteworthy distributional records of amphibians and reptiles from Costa Rica. *Herpetological Review* 44(2), 280–283, 8 figs.

Koirala, B.K., Gurung, D.B., Lhendup, P. & Phuntsho, S. (2016) Species diversity and spatial distribution of snakes in Jogma Dorji National Park and adjoining areas, western Bhutan. *Journal of Threatened Taxa* 8(12), 9461–9466, 9 figs.

Kornilios, P. (2017) Polytomies, signal and noise: Revisiting the mitochondrial phylogeny and phylogeography of the Eurasian blindsnake species complex (Typhlopidae, Squamata). *Zoologica Scripta* 46(6), 665–674, 2 figs.

Kornilios, P., Giokas, S., Lymberakis, P. & Sindaco, R. (2013a) Phylogenetic position, origin and biogeography of Palearctic and Socotran blind-snakes (Serpentes: Typhlopidae). *Molecular Phylogenetics and Evolution* 68, 35–41, 2 figs.

Kornilios, P., Thanou, E., Lymberakis, P., Sindaco, R., Liuzzi, C. & Giokas, S. (2013b) Mitochondrial phylogeography, intraspecific diversity and phenotypic convergence in the four-lined snake (Reptilia, *Squamata*). *Zoologica Scripta* 43(2), 149–160, 4 figs.

Korsós, Z., Barina, Z. & Pifkó, D. (2008) First record of *Vipera ursinii graeca* in Albania (Reptilia: Serpentes, Viperidae). *Acta Herpetologica* 3(2), 167–173, 3 figs.

Kraus, F. (2009 [2010]) New species of *Toxicocalamus* (Squamata: Elapidae) from Papua New Guinea. *Herpetologica* 65(4), 460–467, 4 figs.

Kraus, F. (2013) Further range extensions for reptiles and amphibians from Papua New Guinea. *Herpetological Review* 44(2), 277–280.

Kraus, F. (2017a) New species of blindsnakes (Squamata: Gerrhopilidae) from offshore islands of Papua New Guinea. *Zootaxa* 4299(1), 75–94, 5 figs.

Kraus, F. (2017b) Two new species of *Toxicocalamus* (Squamata: Elapidae) from Papua New Guinea. *Journal of Herpetology* 51(4), 574–581, 4 figs.

Kraus, F. & Cameron, H.D. (2016) A note on the proper nomenclature for the snake currently known as *Thamnophis sauritus*. *Herpetological Review* 47(1), 74–75.

Krecsák, L. & Zamfirescu, S. (2008) *Vipera (Acridophaga) ursinii* in Romania: Historical and present distribution. *North-Western Journal of Zoology* 4(2), 339–359, 3 figs.

Kropachev, I.I. & Orlov, N.L. (2017) A new subspecies of the Halys Pit Viper *Gloydius halys* (Pallas, 1776) (Viperidae, Crotalinae) from Tuva and western Mongolia. *Proceedings of the Zoological Institute, St. Petersburg* 321(2), 129–179, 14 figs.

Kropachev, I.I., Orlov, N.L. & Orlova, V.F. (2016) *Gloydius ussuriensis* (Emelianov, 1929) [Serpentes: Viperidae: Crotalinae] – a new snake species for the herpetofauna of Mongolia. *Russian Journal of Herpetology* 23(2), 108–114, 9 figs.

Kropachev, I.I., Shiryaev, K.A., Nguyen, T.T. & Orlov, N.L. (2015) New record of *Protobothrops* cf. *maolanensis* in northeastern Vietnam, with data on its morphology and biology. *Russian Journal of Herpetology* 22(2), 93–102, 11 figs.

Krysko, K.L., Granatosky, M.C., Nuñez, L.P. & Smith, D.J. (2016b) A cryptic new species of Indigo Snake (genus *Drymarchon*) from the Florida Platform of the United States. *Zootaxa* 4138(3), 549–569, 11 figs.

Krysko, K.L., Nuñez, L.P., Lippi, C.A., Smith, D.J. & Granatosky, M.C. (2016a) Pliocene-Pleistocene lineage diversifications in the Eastern Indigo Snake (*Drymarchon couperi*) in the southeastern United States. *Molecular Phylogenetics and Evolution* 98, 111–222, 4 figs.

Krysko, K.L., Steadman, D.W., Mead, J.I., Albury, N.A., MacKenzie-Krysko, C.A. & Swift, S.L. (2013) New island records for amphibians and reptiles on the Little Bahama Bank, Commonwealth of the Bahamas. *IRCF Reptiles & Amphibians* 20(3), 152–154, 3 figs.

Krysko, K.L., Nuñez, L.P., Newman, C.E. & Bowen, B.W. (2017) Phylogenetics of Kingsnakes, *Lampropeltis getula* Complex (Serpentes: Colubridae), in eastern North America. *Journal of Heredity* 108(3), 226–238, 6 figs.

Krysko, K.L., Steadman, D.W., Nuñez, L.P. & Lee, D.S. (2015) Molecular phylogeny of Caribbean dipsadid (Xenodontinae: Alsophiini) snakes, including identification of the first record from the Cay Sal Bank, The Bahamas. *Zootaxa* 4028(3), 441–450, 4 figs.

Kucharzewski, C. (2015) Herpetologische reiseeindrücke aus der Südwest-Türkei. *Sauria* 37(1), 3–15, 14 figs.

Kucharzewski, C., Raselimanana, A.P., Wang, C. & Glaw, F. (2014) A taxonomic mystery for more than 150 years: Identity, systematic position and Malagasy origin of the snake *Elapotinus picteti* Jan, 1862, and synonymy of *Exallodontophis* (Cadle, 1999) (Serpentes: Lamprophiidae). *Zootaxa* 3852(2), 179–202, 5 figs.

Kühnis, J. (2006) Die Reptilien des Fürstentums Liechtenstein. *Naturkundliche Forschung im Fürstentum Liechtenstein* 23, 1–52, 82 figs.

Kukushkin, O. (2009) *Vipera renardi puzanovi* ssp. nov. (Reptilia: Serpentes: Viperidae) as a new subspecies of steppe viper from mountains of Crimea. *Contemporary Herpetology* 9(1/2), 18–40. (In Russian).

Kukushkin, O., Iskenderov, T., Axmedov, S., Bunyatov, S. & Zinenko, O. (2012) Additions to the distribution of *Vipera eriwanensis* (Serpentes: Viperidae) in Transcaucasia, with comments on the identity of vipers in northeastern Azerbaijan. *Herpetology Notes* 5, 423–427, 4 figs.

Kulenkampff, K., Van Zyl, F., Klaus, S. & Daniels, S.R. (2019) Molecular evidence for cryptic species in the common slug eating snake *Duberria lutrix lutrix* (Squamata, Lamprophiidae) from South Africa. *ZooKeys* 838, 133–154, 4 figs.

Kumar, G.C., Srinivasulu, C. & Prasad, K.K. (2017) First records of the Dumeril's Black-headed Snake *Sibynophis subpunctatus* (Duméril, Bibron & Duméril, 1854) (Reptilia, Colubridae) from Telangana state, India. *Check List* 13(5), 577–580, 4 figs.

Kumlutaş, Y., Ilgaz, Ç. & Candan, K. (2015a) Westernmost record of *Montivipera wagneri* (Nilson & Andrén, 1984). *Herpetozoa* 28(1/2), 98–101, 2 figs.

Kumlutaş, Y., Ilgaz, Ç. & Candan, K. (2015b) [The herpetofaunic diversity of the Fethiye-Göcek (Muğla) Specially Protected Area]. *Anadolu Doğa Bilimleri Dergisi* 6, 155–162, 1 fig.

Kumlutaş, Y., Ilgaz, Ç. & Yakar, O. (2017) Herpetofauna of Karabük Province. *Acta Biologica Turcica* 30(4), 102–107, 2 figs.

Kumlutaş, Y., Sözen, M. & Ilgaz, Ç. (2013) New record of the rare *Vipera barani* Böhme & Joger, 1983. *Herpetozoa* 25(3/4), 183–188, 3 figs.

Kuriyama, T., Brandley, M.C., Katayama, A., Mori, A., Honda, M. & Hasegawa, M. (2011) A time-calibrated phylogenetic approach to assessing the phylogeography, colonization history and phenotypic evolution of snakes in the Japanese Izu Islands. *Journal of Biogeography* 38(2), 259–271, 3 figs.

Kurniawan, N., Putri, M.M., Kadafi, A.M., Chrestella, D.J., Fauzi, M.A. & Kurnianto, A.S. (2017) Phylogenetics and biogeography of Cobra (Squamata: *Naja*) in Java, Sumatra, and other Asian region. *Journal of Experimental Life Science* 7(2), 94–101, 5 figs.

Kutrup, B., Bülbül, U. & Yilmaz, N. (2005) On the distribution and morphology of the Steppe Viper, *Vipera eriwanensis* (Reuss, 1933), from Gavur Mountain (Gümüşhane). *Turkish Journal of Zoology* 29(4), 321–325, 4 figs.

Kyriazi, P., Kornilios, P., Nagy, Z.T., Poulakakis, N., Kumlutaş, Y., Ilgaz, Ç., Avci, A., Göçmen, B. & Lymberakis, P. (2012) Comparative phylogeography reveals distinct colonization patterns of Cretan snakes. *Journal of Biogeography* 40(6), 1143–1155, 5 figs.

Labanowski, R.J. & Lowin, A.J. (2011) A reptile survey in a dry deciduous forest fragment in northern Madagascar showing new records for the little-known snake *Pararhadinaea melanogaster* and a range extension for the skink *Amphiglossus tanysoma*. *Herpetology Notes* 4, 113–121, 8 figs.

Lác, J. & Lechovič, A. (1964) A historical review of research of the snakes on the territory of Slovakia up to 1963. *Acta Rerum Naturalium Musei Nationalis Slovaci, Bratislava* 10, 124–154, 11 maps.

LaDuc, T.J. (1996) A taxonomic revision of the Adelphicos quadrivirgatum *species group* (Serpentes: Colubridae). *Masters Thesis*, xiv+94, 14 figs.

Lalbiakzuala & Lalremsanga, H.T. (2017) Geographic distribution. *Lycodon fasciatus. Herpetological Review* 48(1), 19.

Lalbiakzuala & Lalremsanga, H.T. (2019a) Geographic distribution: *Hebius venningi. Herpetological Review* 50(2), 330.

Lalbiakzuala & Lalremsanga, H.T. (2019b) Geographic distribution: *Pareas margaritophorus. Herpetological Review* 50(2), 332.

Lalremsanga, H.T., Lalbiakzuala & Lalrinsanga (2017) Geographic distribution. *Protobothrops mucrosquamatus. Herpetological Review* 48(1), 131.

Lalremsanga, H.T. & Zothansiama (2015) Morphological variations in *Sinomicrurus macclellandi macclellandi* (Serpentes: Elapidae), the only coral snake species in northeast India. *Science Vision* 15(4), 212–221, 5 figs.

Landestoy, T.M.A. (2017) New localities and prey records for the Hispaniolan Brown Racer (*Haitiophis anomalus*), with comments on dorsal pattern variation. *IRCF Reptiles & Amphibians* 24(2), 106–111, 7 figs.

Lang, R. de (2013) *The snakes of the Moluccas (Maluku), Indonesia*. Edition Chimaira, Frankfurt-am-Main, 1–417, 397 figs.

Lanza, B. & Broadley, D.G. (2014) A review of the genus *Gonionotophis* in north-eastern Africa (Squamata: Lamprophiidae). *Acta Herpetologica* 9(1), 89–97, 4 figs.

Lara, L.C.E., Sosa-Bartuano, Á. & Ray, J.M. (2015) Range extension and natural history observations of a rare Panamanian snake, *Geophis bellus* (Myers, 2003) (Colubridae: Dipsadinae). *Check List* 11(1675), 1–5, 2 figs.

Lara-Resendiz, R.A., Wyman, J. & Rosen, P.C. (2016) A new record, distributional range extension, and notes on *Coniophanes lateritius* (Squamata: Colubridae) in Sonora, Mexico. *Mesoamerican Herpetology* 3(3), 801–804, 1 fig.

Lares, R.V., Martínez, R.M., Gadsden, H., León, G.A., Gaytán, G.C. & Trápaga, R.G. (2013) Checklist of amphibians and reptiles of the state of Durango, México. *Check List* 9(4), 714–724, 1 fig.

Largen, M. & Spawls, S. (2011) Amphibians and reptiles recorded from the Bale Mountains. In: Randall, D., Thirgood, S. & Kinahan, A. (Eds.), *Walia-Special Edition on the Bale Mountains*, Addis Ababa, 89–91.

Laspiur, A. & Nenda, J.J. (2018) Geographic distribution: *Boiruna maculata. Herpetological Review* 49(3), 505.

Lassiter, E., Akre, T., Ruther, E., Dragon, J., Fink, M., Krichbaum, S., Morse, B. & Rutherford, E. (2017) New county records and additional records for amphibians and reptiles in Virginia, USA. *Herpetological Review* 48(4), 822–825.

Lazăr, V., Covaciu-Markav, S.-D., Sas, I., Pusta, C. & Kovács, É.-H. (2005) The herpetofauna of the district of Dolj (Romania). *Analele Stiintifice ale Universității "Al.I. Cuza" Iaşi, s. Biologie Animală* 51, 151–158.

Le, D.T., Dao, A.N., Pham, D.T., Ziegler, T. & Nguyen, T.Q. (2018) New records and an updated list of snakes from Yen Bai Province, *Vietnam. Herpetology Notes* 11, 101–108, fig. 3.

Le, D.T., Pham, A.V., Pham, C.T., Nguyen, S.H.L., Ziegler, T. & Nguyen, T.Q. (2015) Review of the genus *Sinonatrix* in Vietnam with a new country record of *Sinonatrix yunnanensis* Rao et Yang, 1998. *Russian Journal of Herpetology* 22(2), 84–88, 4 figs.

Leão, S.M., Pelegrin, N., Nogueira, C. de C. & Brandão, R.A. (2014) Natural history of *Bothrops itapetiningae* (Boulenger, 1907) (Serpentes: Viperidae: Crotalinae), an endemic species of the Brazilian Cerrado. *Journal of Herpetology* 48(3), 324–331, 2 figs.

LeClere & Jeffrey B. (2013) *A Field Guide to the Amphibians and Reptiles of Iowa*. ECO Herpetological Publishing & Distribution, Rodeo. viii+349, numerous illustrations.

Lee, J.L., Miller, A.H., Connette, G.M., Oo, K.S., Zug, G.R. & Mulcahy, D.G. (2018a) First record of the Malaysia Bridle Snake, *Dryocalamus subannulatus* (Duméril, Bibron & Duméril, 1854), in Myanmar (Reptilia, Serpentes, Colubridae). *Check List* 14(2), 341–345, 3 figs.

Lee, J.L., Thompson, A. & Mulcahy, D.G. (2016) Relationships between numbers of vertebrae, scale counts, and body size, with implications for taxonomy in nightsnakes (Genus: *Hypsiglena*). *Journal of Herpetology* 50(4), 616–620, 2 figs.

Lee, J.L., Thura, M.K., Mulcahy, D.G. & Zug, G.R. (2015) Three colubrid snakes new to Myanmar. *Herpetology Notes* 8, 217–220, 4 figs.

Lee, J.L., Vogel, G., Miller, A.H. & Zug, G.R. (2018b) Rediscovery of *Xenochrophis bellulus* (Stoliczka, 1871) from Myanmar (Serpentes: Natricinae) with comments on its taxonomic status. *Proceedings of the Biological Society of Washington* 131, 19–35, 7 figs.

Lee, M.S.Y., Palci, A., Jones, M.E.H., Caldwell, M.W., Holmes, J.D. & Reisz, R.R. (2016b) Aquatic adaptations in the four limbs of the snake-like reptile *Tetrapodophis* from the lower Cretaceous of Brazil. *Cretaceous Research* 66, 194–199, 3 figs.

Lee, M.S.Y., Sanders, K.L., King, B. & Palci, A. (2016a) Diversification rates and phenotypic evolution in venomous snakes (Elapidae). *Royal Society Open Science* 3(150277), 1–11, 2 figs.

Lei, J., Sun, X., Jiang, K., Vogel, G., Booth, D.T. & Ding, L. (2014) Multilocus phylogeny of *Lycodon* and the taxonomic revision of *Oligodon multizonatum. Asian Herpetological Research* 5(1), 26–37, 5 figs.

Lele, Y., Kiba, V., Sethi, P. & Edake, S. (2018) A not-so-rare species: Sightings of Mandarin Ratsnakes, *Euprepiophis mandarinus* (Cantor 1842), in the Zunheboto District of Nagaland, India. *IRCF Reptiles & Amphibians* 25(3), 197–198, 1 fig.

Lema, T. de (2004) Description of a new species of *Apostolepis* (Cope, 1861) (Serpentes, Elapomorphinae) from Brazilian cerrado. *Acta Biológica Leopoldensia* 26(1), 155–160.

Lema, T. de (2006) Redescrição do holótipo de *Elapomorphus coronatus* Sauvage 1877, com a observação de Gymnophiona no estômago (Serpentes, Colubridae, *Elapomorphinae*). *Neotropical Biology and Conservation* 1(1), 39–41, 1 fig.

Lema, T. de (2015) Remarks on *Apostolepis goiasensis* (Serpentes, Xenodontinae), with presentation of the holotype. *Caderno de Pesquisa, Série Biologia, Santa Cruz do Sul* 27(2), 20–27, 6 figs.

Lema, T. de (2016) Description of new species of *Apostolepis* (Serpentes: Dipsadidae: Xenodontinae: Elapomorphini) from Serra do Roncador, central Brazil. *Caderno de Pesquisa, Biologia* 28(1), 1–12, 5 figs.

Lema, T. de & Campbell, P. (2017) New species of *Apostolepis* (Serpentes, Dipsadinae, Elapomorphini) from Bolivia, from the *Apostolepis borellii* group. *Research & Reviews: Journal of Zoological Sciences* 5(1), 19–28, 14 figs.

Lema, T. de, Queiroz, A.N. de & Martins, L.A. (2017) Color variation in *Apostolepis nigrolineata* (Serpentes, *Colubridae: Dipsadinae: Elapomorphini*), and a contribution to the knowledge of the nigrolineata group. *Cuadernos de Herpetología* 31(2), 93–101, 5 figs.

Lema, T. de & Renner, M.F. (2004) New specimens of *Apostolepis vittata* (Cope, 1887) (Serpentes, Elapomorphinae). *Caderno de Pesquisa, Série Biologia, Santa Cruz do Sul* 16(1), 51–56, 1 fig.

Lema, T. de & Renner, M.F. (2005) Discovery of new specimens of *Apostolepis freitasi* Lema (Serpentes, Elapomorphinae), with redescription of the species. *Caderno de Pesquisa, Série Biologia, Santa Cruz do Sul* 17(2), 151–158, 1 fig.

Lema, T. de & Renner, M.F. (2007) On the status of *Apostolepis freitasi* (Serpentes, Elapomorphinae) by examination of new specimens. *Neotropical Biology and Conservation* 2(2), 90–93, 1 fig.

Lema, T. de & Renner, M.F. (2011) A new species of *Apostolepis* (Serpentes, Colubridae, Elapomorphini), belonging to *assimilis* group, found in Brazilian Cerrado. *Ciência em Movimento* 13(27), 71–76, 8 figs.

Lema, T. de & Renner, M.F. (2012) *Apostolepis* specimens observed in collections from Goiás region, central Brazil (Serpentes, Xenodontinae, Elapomorphini). *Neotropical Biology and Conservation* 7(2), 144–147, 4 figs.

Lema, T. de & Renner, M.F. (2015) Status of *Apostolepis borellii* (Peracca, 1904) (Serpentes, Xenodontinae), with restriction of the *A. nigroterminata* concept. *Neotropical Biology and Conservation* 11(2), 62–71, 8 figs.

Lemberk, V. (2013) To the knowledge of Dice Snake (*Natrix tessellata*) in eastern Bohemia. *Východočeské Sborník Přírody – Práce a Studie* 20, 175–180, 2 figs.

Lemos-Espinal, J.A. & Dixon, J.R. (2013) *Amphibians and reptiles of San Luis Potosí*. Eagle Mountain Publishing, Eagle Mountain. xii+300, numerous illustrations.

Lemos-Espinal, J.A. & Smith, G.R. (2015) Amphibians and reptiles of the state of Hidalgo, Mexico. *Check List* 11(1642), 1–11, 1 fig.

Lemos-Espinal, J.A. & Smith, G.R. (2016) Amphibians and reptiles of the state of Coahuila, Mexico, with comparison with adjoining states. *ZooKeys* 593, 117–137, 6 figs.

Lemos-Espinal, J.A., Smith, H.M. & Cruz, A. (2013) *Amphibians & reptiles of the Sierra Tarahumara of Chihuahua, Mexico*. ECO Herpetological Publishing & Distribution, Rodeo. viii+405, 92 figs, 258 photos, 71 maps.

Lemos-Espinal, J.A., Smith, G.R. & Cruz, A. (2016) Amphibians and reptiles of the state of Nuevo León, Mexico. *ZooKeys* 594, 123–141, 1 fig.

Lemos-Espinal, J.A., Smith, G.R. & Cruz, A. (2018c) *Amphibians & Reptiles Nuevo León*. ECO Herpetological Publishing & Distribution, Rodeo. x+370, 98 figs., 260 photos, 83 maps.

Lemos-Espinal, J.A., Smith, G.R., Gadsden-Esparza, H., Valdez-Lares, R. & Woolrich-Piña, G.A. (2018a) Amphibians and reptiles of the state of Durango, Mexico, with comparisons with adjoining states. *ZooKeys* 748, 65–87, 5 figs.

Lemos-Espinal, J.A., Smith, G.R. & Rorabaugh, J.C. (2019) A conservation checklist of the amphibians and reptiles of Sonora, Mexico, with updated species lists. *ZooKeys* 829, 131–160, 6 figs.

Lemos-Espinal, J.A., Smith, G.R. & Woolrich-Piña, G.A. (2018b) Amphibians and reptiles of the state of San Luis Potosí, Mexico, with comparisons with adjoining states. *ZooKeys* 753, 83–106, 6 figs.

Lemos-Espinal, J.A., Smith, G.R., Woolrich-Piña, G.A. & Cruz, A. (2017) Amphibians and reptiles of the state of Chihuahua, Mexico, with comparisons with adjoining states. *ZooKeys* 658, 105–130, 5 figs.

Leviton, A.E., Brown, R.M. & Siler, C.D. (2014) The dangerously venomous snakes of the Philippine Archipelago with identification keys and species accounts. In: Williams, G.C. &

Gosliner, T.M. (Eds.), *The Coral triangle: The 2011 Hearst Philippine Biodiversity Expedition*. California Academy of Sciences, San Francisco, 473–530, 52 figs.

Leviton, A.E., Siler, C.D., Weinell, J.L. & Brown, R.M. (2018) Synopsis of the snakes of the Philippines. A synthesis of data from biodiversity repositories, field studies, and literature. *Proceedings of the California Academy of Sciences*, Series 4, 64(14), 399–568, 119 figs, 148 maps.

Leyte-Manrique, A., Berriozabal-Islas, C., Mata-Silva, V. & Morales-Castorena, J.P. (2018) Herpetofaunal diversity in Área Natural Protegida Las Musas, Guanajuato, Mexico. *Mesoamerican Herpetology* 5(1), 121–136, 5 figs.

Leyte-Manrique, A., Mata-Silva, V. & Hansen, R.W. (2017) New records for *Lampropeltis polyzona* (Cope, 1860) (Reptilia: Squamata: Colubridae), from Guanajuato, Mexico. *Mesoamerican Herpetology* 4(2), 463–464, 1 fig.

Leyte-Manrique, A., Navarro, E.M.H. & Escobedo-Morales, L.A. (2015) Herpetofauna de Guanajuato: un análisis histórico y contemporáneo de su conocimento. *Revista Mexicana de Herpetología* 1(1), 1–14, 3 figs.

Li, Y.-X., Hu, S.-M., Wang, S.-J., Zhang, Y.-X., Li, J., Xie, K. & Chen, Y. (2019) Fossil lizards and snakes from the middle Pleistocene of China. *Journal of Vertebrate Paleontology* 39(3)(e1631175), 1–5, 2 figs.

Lilley, R. (2013) Geographic distribution. *Oligodon bitorquatus*. *Herpetological Review* 44(1), 110.

Lim, B.L., Noor Alif Wira, O., Chan, K.O., Daicus, B. & Norhayati, A. (2010) An updated checklist of the herpetofauna of Pulau Singa Besar, Langkawi, peninsular Malaysia. *Malaysian Applied Biology* 39(1), 13–23, 1 fig.

Lim, K.K.P. & Lim, L.J. (1999) The terrestrial herpetofauna of Pulau Tioman, peninsular Malaysia. *The Raffles Bulletin of Zoology*, (Supplement 6), 131–155, 3 figs.

Litvinchuk, S.N., Kuranova, V.N., Kazakov, V.I. & Schepina, N.A. (2013) A northernmost record of the grass snake (*Natrix natrix*) in the Baikal Lake region, Siberia. *Russian Journal of Herpetology* 20(1), 43–50, 4 figs.

Liu, Q., Zhong, G.-H., Wang, P., Liu, Y. & Guo, P. (2018) A new species of the genus *Hebius* (Squamata: Colubridae) from Sichuan, China. *Zootaxa* 4483(2), 385–304, 7 figs.

Liuzzi, C. & Scillitani, G. (2010) L'Erpetofauna della Puglia; aggiornamenti e integrazioni. In: Di Tizio, L., Di Cerbo, A.R., Di Francesco, N. & Cameli, A. (Eds.), *Atti VIII Congresso Nazionale Societas Herpetologica Italica (Chieti, 22–26 Settembre 2010)*, Ianieri Edizioni, Pescara, 31–36, 2 figs.

Llanqui, I.B., Salas, C.Y. & Oblitas, M.P. (2019) A preliminary checklist of amphibians and reptiles from the vicinity of La Nube Biological Station, Bahuaja-Sonene National Park, Peru. *Check List* 15(5), 773–796, 7 figs.

Lo Cascio, P. & Rivière, V. (2014) Herpetofaunal inventory of Kuriat and Jbel islets (Tunisia). *Biodiversity Journal* 5(2): 391–396, 3 figs.

Loc-Barragán, J.A. & Ahumada-Carrillo, I.T. (2016) Distribution notes: *Pituophis deppei*. *Mesoamerican Herpetology* 3(1), 190, 1 fig.

Loebmann, D. & Lema, T. de (2012) New data on the distribution of the rare and poorly known *Apostolepis goiasensis* (Prado, 1943) (Serpentes, Xenodontinae, Elapomorphini) with remarks on morphology and colouration. *Herpetology Notes* 5, 523–525, 2 figs.

Loredo, A.I., Wood, P.L., Quah, E.S.H., Anuar, S., Greer, L.F., Ahmad, N. & Grismer, L.L. (2013) Cryptic speciation within *Asthenodipsas vertebralis* (Boulenger, 1900) (Squamata: Pareatidae), the description of a new species from Peninsular Malaysia, and the resuurection of *A. tropidonotus* (Lidth de

Jude [*sic*], 1923) from Sumatra: An integrative taxonomic analysis. *Zootaxa* 3664(4), 505–524, 7 figs.

Lorvelec, O., Berchel, J. & Barré, N. (2016) First report of the Flowerpot Blindsnake, *Indotyphlops braminus* (Daudin, 1803), from La Désirade (Guadeloupe Archipelago, the French West Indies). *Caribbean Herpetology* 55, 1–2, 1 fig.

Loza, J.C., Gutiérrez, L., Salazar-Saavedra, M., Martínez-Fonseca, J.G., Fernández, M. & Sunyer, J. (2017) First record of *Rhadinella godmani* (Reptilia: Squamata: Dipsadidae) from Nicaragua. *Mesoamerican Herpetology* 4(2), 476–478, 1 fig.

Lozano, S.A. & Sierra, T.A. (2018) *Pseudoboa neuwiedi* (Duméril, Bibron y Duméril, 1854). *Catálogo de Anfibios y Reptiles de Colombia* 4(1), 60–67, 3 figs.

Luja, V.H. & Grünwald, C.L. (2015) New distributional records of amphibians and reptiles from Nayarit, México. *Herpetological Review* 46(2), 223–225, 1 fig.

Luque-Fernández, C.R. & Paredes, L.N.V. (2017) First record of the Fitzinger's False Coral snake, *Oxyrhopus fitzingeri* (Tschudi, 1845) (Reptilia: Dipsadidae) in Atiquipa, southwestern Peru. *Check List* 13(2085), 1–3, 3 figs.

Luría-Manzano, R., Ramírez-Bautista, A. & Canseco-Márquez, L. (2014) Rediscovery of the rare snake *Rhadinaea cuneata* (Myers, 1974) (Serpentes: Colubridae: Dipsadinae). *Journal of Herpetology* 48(1), 122–124, 1 fig.

Luu, V.Q., Bonkowski, M., Nguyen, T.Q., Le, M.D., Calame, T. & Ziegler, T. (2018) A new species of *Lycodon* (Boie, 1826) (Serpentes: Colubridae) from central Laos. *Revue Suisse de Zoologie* 125(2), 263–276, 7 figs.

Luu, V.Q., Nguyen, T.Q., Lehmann, T., Bonkowski, M. & Ziegler, T. (2015) New records of the horned pitviper, *Protobothrops cornutus* (Smith, 1930) (Serpentes: Viperidae), from Vietnam with comments on morphological variation. *Herpetology Notes* 8, 149–152, 4 figs.

Luu, V.Q., Ziegler, T., Ha, N.V., Le, M.D. & Hoang, T.T. (2019) A new species of *Lycodon* (Boie, 1826) (Serpentes: Colubridae) from Thanh Hoa Province, Vietnam. *Zootaxa* 4586(2), 261–277, 7 figs.

Luz, N.M.C. de la, Lemos-Espinal, J. & Smith, G.R. (2016) A diversity and conservation inventory of the herpetofauna of the Cuautlapan Valley, Veracruz, Mexico. *Zootaxa* 4205(2), 127–142, 7 figs.

Lyakurwa, J.V. (2017) The reptiles of the Uzungwa Scarp Forest Reserve (USFR): An updated checklist with notes on Dagger-tooth Vine Snake *Xyelodontophis uluguruensis*. *Journal of East African Natural History* 106(2), 57–65, 2 figs.

Lyakurwa, J.V., Howell, K.M., Munishi, L.K. & Treydte, A.C. (2019) Uzungwa Scarp Nature Forest Reserve: A unique hotspot for reptiles in Tanzania. *Acta Herpetologica* 14(1), 3–14, 5 figs.

MacCulloch, R.D. & Reynolds, R.P. (2013) Baseline inventory of amphibians and reptiles in the vicinity of Kurupukari, Guyana. *Check List* 9(6), 1378–1382, 2 figs.

Maddock, S.T., Childerstone, A., Fry, B.G., Williams, D.J., Barlow, A. & Wüster, W. (2017) Multi-locus phylogeny and species delimitation of Australo-Papuan blacksnakes (*Pseudechis* Wagler, 1830: Elapidae: Serpentes). *Molecular Phylogenetics and Evolution* 107, 48–55, 2 figs.

Maddock, S.T., Ellis, R.J., Doughty, P., Smith, L.A. & Wüster, W. (2015) A new species of death adder (*Acanthophis*: Serpentes: Elapidae) from north-western Australia. *Zootaxa* 4007(3), 301–326, 9 figs.

Madella-Auricchio, C.R., Auricchio, P. & Soares, E.S. (2017) Reptile species composition in the middle Gurguéia and comparison with inventories in the eastern Parnaíba River Basin, state of Piauí, Brazil. *Papéis Avulsos de Zoologia* 57(28), 375–386, 5 figs.

Madl, R. (2017) First record of *Elaphe quatuorlineata* (Bonnaterre, 1790), from the Island of Dugi Otok (Croatia). *Herpetozoa* 30(1/2), 96–100, 7 figs.

Magalhães, F. de M., Laranjeiras, D.O., Costa, T.B., Juncá, F.A., Mesquita, D.O., Röhr, D.L., Silva, W.P. da, Vieira, G.H.C. & Garda, A.A. (2015) Herpetofauna of protected areas in the Caatinga IV: Chapada Diamantina National Park, Bahia, Brazil. *Herpetology Notes* 8, 243–261, 12 figs.

Magalhães, J.J., Assis, C.L. de, Silva, D.H. da & Feio, R.N. (2017) New records and notes on defensive behavior of *Thamnophis rutilus* (Prado 1942). *Neotropical Biology and Conservation* 12(2), 154–158, 2 figs.

Malakova, N., Sakelarieva, L. & Pulev, A. (2018) Species composition of the amphibians and reptiles in the Natura 2000 site "Oranovski Prolom – Leshko," Bulgaria. *ZooNotes* (124), 1–4.

Malhotra, A. & Thorpe, R.S. (1999) *Reptiles & Amphibians of the Eastern Caribbean*. MacMillan Education, London. ix+134, 91 figs.

Malkmus, R. (2005) Die herpetofauna eines mittelportugiesischen karstgebietes. *Zeitschrift für Feldherpetologie* 12, 211–236, numerous ill.

Malkmus, R. (2011) Die herpetofauna der portugiesischen küstenregion. *Zeitschrift für Feldherpetologie* 18, 221–254, 9 figs.

Malkmus, R. & Loureiro, A. (2010) Die herpetofauna der südöstlichen Beira Alta (Portugal). *Zeitschrift für Feldherpetologie* 17, 201–230, numerous ill.

Malkmus, R. & Loureiro, A. (2012) Die herpetofauna der Barroso-Region (Portugal). *Zeitschrift für Feldherpetologie* 19, 255–252, numerous ill.

Mallik, A.K., Achyuthan, N.S., Ganesh, S.R., Pal, S.P., Vijayakumar, S.P. & Shanker, K. (2019) Discovery of a deeply divergent new lineage of vine snake (Colubridae: Ahaetuliinae: *Proahaetulla* gen. nov.) from the southern Western Ghats of peninsular India with a revised key for Ahaetuliinae. *PLOS ONE* 145(7), e0218851, 1–21 figs.

Mallik, S., Parida, S.P., Mohanty, A.K., Mallik, A., Purohit, K.L., Mohanty, S., Nanda, S. et al. (2014) A new species of wolf snake (Serpentes: Colubridae: *Lycodon*) from Berhampur, Ganjam, Odisha, India. *Russian Journal of Herpetology* 21(3), 205–216, 14 figs.

Malonza, P.K., Bauer, A.M. & Ngwava, J.M. (2016) A new species of *Letheobia* (Serpentes: Typhlopidae) from central Kenya. *Zootaxa* 4093(1), 143–150, 4 figs.

Mamou, R., Boissinot, A., Bensidehoum, M., Amroun, M. & Marniche, F. (2014) Inventaire de l'herpétofaune du sud de la Kabylie (Bouira et Bordj Bou Arreridj). *Algérie. La Revue Ivoirienne des Sciences et Technologie* 23, 259–273.

Mané, Y. & Trape, J.-F. (2017) Le régime alimentaire des serpents de la famille des Elapidae Boie, 1827, au Sénégal. *Bulletin de la Société Herpétologique de France* 164, 15–28.

Mangiacotti, M., Limongi, L., Sannolo, M., Sacchi, R., Zuffi, M.A.L. & Scali, S. (2014) Head shape variation in eastern and western Montpellier snakes. *Acta Herpetologica* 9(2), 167–177, 6 figs.

Manhas, A., Raina, R. & Wanganeo, A. (2018a) Reptilian diversity of the Bhopal Region in the state of Madhya Pradesh in central India. *IRCF Reptiles & Amphibians* 25(2), 104–114, 18 figs.

Manhas, A., Raina, R. & Wanganeo, A. (2018b) Reptilian diversity and distributions in the Doda District of Jammu and Kashmir, India. *IRCF Reptiles & Amphibians* 25(3), 164–169, numerous figs.

Mano-Cuellar, K., Pinto-Viveros, M.A., Escalante, R.S., Villarroel, D. & Pinto-Ledezma, J.N. (2015) Reptile fauna of the Mutun region (Santa Cruz Department, Bolivia): Species list and conservation status. *Kempffiana* 11(1), 66–69.

Marcos, G.G. de, Garin-Barrio, I. & Collazos, E.P. (2018) Inventario herpetológico en habitats mediterráneos de un municipio del sur de Álava (País Vasco). *Munibe, Ciencias Naturales* 66, 59–69, 2 figs.

Marín, C.M., Toro, F.A. & Daza, J.M. (2017) First trans-Andean record of *Atractus occipitoalbus* (Jan, 1862), (Squamata: Dipsadidae), from Colombia. *Herpetology Notes* 10, 49–51, 2 figs.

Marin, J., Donnellan, S.C., Hedges, S.B., Puillandre, N., Aplin, K.P., Doughty, P., Huchinson, M.N., Couloux, A. & N. Vidal. (2013) Hidden species diversity of Australian burrowing snakes (*Ramphotyphlops*). *Biological Journal of the Linnean Society* 110(2), 427–441, 4 figs.

Marmol-Marin, G.M. del & Fernandez, B.R. (2012) An important new record of *Echis leucogaster* Roman, 1972 from Morocco. *Herpetology Notes* 5, 229–231, 2 figs.

Marosi, B., Zinenko, O.I., Ghira, I.V., Crnobrnja-Isailović Lymberakis, P., Sos, T. & Popescu, O. (2012) Molecular data confirm recent fluctuations of northern border of dice snake (*Natrix tessellata*) range in eastern Europe. *North-Western Journal of Zoology* 8(2), 374–377, 2 figs.

Marques, M.P., Ceríaco, L.M.P., Blackburn, D.C. & Bauer, A.M. (2018) Diversity and distribution of the amphibians and terrestrial reptiles of Angola atlas of historical and bibliographic records (1840–2017). *Proceedings of the California Academy of Sciences*, series 4, 65(Suppl. 2), 1–50, 413 figs.

Marques, R., Mebert, K., Fonseca, É., Rödder, D., Solé, M. & Tinôco, M.S. (2016) Composition and natural history notes of the coastal snake assemblage from northern Bahia, Brazil. *ZooKeys* 611, 93–142, 7 figs.

Marques, R., Tinôco, M.S., Browne-Ribeiro, H.C. & Fazolato, C.P. (2012) *Phimophis guerini* (Duméril, Bibron and Duméril, 1854) (Squamata, Colubridae): Distribution extension in the northeast coast of the state of Bahia, Brazil. *Check List* 8(5), 963–965, fig. 2.

Marques, R., Tinôco, M.S., Couto-Ferreira, D., Fazolato, C.P., Browne-Ribeiro, H.C., Travassos, M.L.O., Dias, M.A. & Mota, J.V.L. (2011) Reserva Imbassaí Restinga: Inventory of snakes on the northern coast of Bahia, Brazil. *Journal of Threatened Taxa* 3(11), 2184–2191, 5 figs.

Marques, R., Tinôco, M.S., Rödder, D. & Browne-Ribeiro, H.C. (2013) Distribution extension of *Thamnodynastes pallidus* and new records within the distribution of *Erythrolamprus reginae*, *Imantodes cenchoa* and *Siphlophis compressus* (Serpentes, Dipsadidae) for the north coast of Bahia, Brazil. *Herpetology Notes* 6, 529–532, 2 figs.

Martens, H. (1996) The rediscovery of the Grass Snake *Natrix natrix* (L.) in the Levant. *Zoology in the Middle East* 12, 59–64, 1 fig.

Martill, D.M., Tischlinger, H. & Longrich, N.R. (2015) A four-legged snake from the early Cretaceous of Gondwana. *Science* 349(6246), 416–419, 5 figs.

Martin, T.E., Guillemin, M., Nivet-Mazerolles, V., Landsmann, C., Dubos, J., Eudeline, R. & Stroud, J.T. (2017) The herpetofauna of central Uzbekistan. *Amphibian & Reptile Conservation* 11(1), 93–107, 32 figs.

Martínez, N., Espínola, V., Bauer, F. & Ortiz, B. (2019) Ampliación del área de distribución de *Oxyrhopus guibei* Hoge & Romano, 1977 en Paraguay. *Cuadernos de Herpetología* 33(2), 87–89, 2 figs.

Martínez, O. (2017) Acerca de la culebra endémica *Philodryas boliviana* (Boulenger 1896) (Dipsadidae): una nueva especie en el valle de la ciudad de La Paz (Bolivia) y estado de conservación de los valles secos en la region. *Kempffiana* 13(1), 132–145, 9 figs.

Martínez, T.A., Muñoz, M.J.R., Galdeano, A.P. & Acosta, J.C. (2015) New record of *Boa constrictor occidentalis* (Philippi, 1873) (Serpentes: Boidae) in San Juan province, Argentina. *Check List* 11(1775), 1-1-4, 2 figs.

Martínez-Fonseca, J.G., Loza, J., Fernández, M., Salazar-Saavedra, M. & Sunyer, J. (2019) First country record of *Rhinobothryum bovallii* (Andersson, 1916) (Squamata: Colubridae) from Nicaragua. *Check List* 15(4), 555–563, 3 figs.

Martínez-Fonseca, J.G., Reid, F.A. & Sunyer, J. (2016a) Nature notes: *Senticolis triaspis*. *Mesoamerican Herpetology* 3(2), 505, 1 fig.

Martínez-Fonseca, J.G., Yasuda, K. & Sunyer, Y. (2016b) Distribution notes: *Bothrops asper*. *Mesoamerican Herpetology* 3(3), 777, 1 fig.

Martínez-Freiría, F. & Brito, J.C. (2013) Integrating classical and spatial multivariate analyses for assessing morphological variability in the endemic Iberian viper *Vipera seoanei*. *Journal of Zoological Systematics and Evolutionary Research* 51(2), 122–131, 4 figs.

Martínez-Freiría, F., Crochet, P.-A., Fahd, S., Geniez, P., Brito, J.C. & Velo-Antón, G. (2017a) Integrative phylogeographic and ecological analyses reveal multiple Pleistocene refugia for Mediterranean *Daboia* vipers in north-west Africa. *Biological Journal of the Linnean Society* 122(2), 366–384, 5 figs.

Martínez-Freiría, F., García-Cardenete, L., Alaminos, E., Fahd, S., Feriche, M., Stols, V.F., Jiménez-Cazalla, F. et al. (2017b) Contribution to the knowledge on the reptile fauna of Jebel Sirwa (Morocco), with some insights into the conservation status of *Vipera latastei-monticola*. *Boletín de la Asociación Herpetológica Española* 28(1), 54–60, 2 figs.

Martínez-Freiría, F., Velo-Antón, G. & Brito, J.C. (2015) Trapped by climate: Interglacial refuge and recent population expansion in the endemic Iberian adder *Vipera seoanei*. *Diversity and Distributions* 21(3), 331–344, 5 figs.

Martínez-Fuentes, R.G., Cervantes-Burgos, R.I., Sánchez-García, J.C., Valdenegro-Brito, A.E. & García-Vázquez, U.O. (2017) First verified locality for *Drymobius margaritiferus* (Schlegel, 1837) from Estado de México, Mexico. *Mesoamerican Herpetology* 4(2), 461–462, 1 fig.

Martínez-Silvestre, A. & Soler, J. (2018) Caso de albinismo en *Malpolon monspessulanus* (Hermann, 1804). *Boletín de la Asociación Herpetológica Española* 29(2), 22–24, 1 fig.

Martínez-Vaca, O.I., Bello-Sánchez, E.A. & Morales-Mávil, J.E. (2016) Nuevos registros para la distribución geográfica de la serpiente cornuda Mexicana esmeralda *Ophryacus smaragdinus*, en la zona centro del Estado de Veracruz. *Acta Zoológica Mexicana, New Series*, 32(3), 393–397, 2 figs.

Martín-Regalado, C.N., Gómez-Ugalde, R.M. & Cisneros-Palacios, M.E. (2011) Herpetofauna del Cerro Guiengola, Istmo de Tehuantepec, *Oaxaca*. *Acta Zoológica Mexicana (nueva serie)* 27(2), 359–376, 6 figs.

Martins, A., Koch, C., Pinto, R., Folly, M., Fouquet, A. & Passos, P. (2019) From the inside out: Discovery of a new genus of threadsnakes based on anatomical and molecular data, with discussion of the leptotyphlopid hemipenial morphology. *Journal of Zoological Systematics and Evolutionary Research* 57(4), 840–863, 14 figs.

Martins, L.A. & Lema, T. de (2015) Elapomorphini (Serpentes, Xenodontinae) do Brasil sudoeste. *Neotropical Biology & Conservation* 10(2), 93–102, 12 figs.

Martins, L.A. & Lema, T. de (2017) The distribution of *Phalotris tricolor* group in Argentina and Paraguay, with notes on its taxonomy (Serpentes, Elapomorphini). *Neotropical Biology & Conservation* 12(2), 100–108, 11 figs.

Mason, A.J., Grazziotin, F.G., Zaher, H., Lemmon, A.R., Lemmon, E.M. & Parkinson, C.L. (2019) Reticulate evolution in nuclear

Middle America causes discordance in the phylogeny of palm-pitvipers (Viperidae: *Bothriechis*). *Journal of Biogeography* 46(5), 833–844, 4 figs.

Masseti, M. (2014) Herpetological enigmas from the Arabian seas, with particular reference to the Sarso island racer, Platyceps insularis (Mertens, 1965) (Farasan archipelago, Saudi Arabia). In Capula, M. & Corti, C. (Eds.), *Scripta Herpetologica. Studies on Amphibians and Reptiles in Honour of Benedetto Lanza.* Edizioni Belvedere, Latina, 99–116, 8 figs.

Masseti, M. & Böhme, W. (2014) Vipern, Mönche und Arzneien: Die Aspisviper (*Vipera aspis*) auf der Insel Montechristo im nördlichen Tyrrhenischen Meer. *Sekretär* 14(1), 34–42, 7 figs.

Masseti, M. & Zuffi, M.A.L. (2011) On the origin of the asp viper *Vipera aspis hugyi* Schinz, 1833, on the island of Montecristo, northern Tyrrhenian Sea (Tuscan archipelago, Italy). *Herpetological Bulletin* 117, 1–9, 2 figs.

Mata-Silva, V., DeSantis, D., García-Padilla, E. & Wilson, L.D. (2015b) Distribution notes: *Thamnophis proximus. Mesoamerican Herpetology* 2(3), 360, 1 fig.

Mata-Silva, V., García-Padilla, E., DeSantis, D.L., Rocha, A., Wilson, L.D., Simón-Salvador, P.R., Mayoral-Halla, C., Montiel-Altamirano, B.F. & Ramírez-Bautista, A. (2017) New herpetofaunal distribution records for the state of Oaxaca, Mexico. *Mesoamerican Herpetology* 4(3), 679–683, 1 fig.

Mata-Silva, V., Johnson, J.D., Wilson, L.D. & García-Padilla, E. (2015a) The herpetofauna of Oaxaca, Mexico: Composition, physiographic distribution, and conservation status. *Mesoamerican Herpetology* 2(1), 6–62, numerous figs.

Mata-Silva, V., Rocha, A., Ramírez-Bautista, A., Berriozabal-Islas, C. & Wilson, L.D. (2019) A new species of forest snake of the genus *Rhadinaea* from tropical montane rainforest in the Sierra Madre del Sur of Oaxaca, Mexico (Squamata, Dipsadidae). *ZooKeys* 813, 55–65, 5 figs.

Mata-Silva, V. & Wilson, L.D. (2016) The taxonomic status of *Tantilla marcovani* (Lema 2004) (Squamata: Colubridae). *Zootaxa* 4092(3), 421–425, 1 fig.

Mata-Silva, V., Wilson, L.D., Johnson, J.D., Mata-González, S., Ramírez-Bautista, A. & García-Grajáles, J. (2013) New distribution and elevation records for the snake *Pseudelaphe flavirufa* (Cope, 1867) (Squamata: Colubridae) in Oaxaca, Mexico. *Check List* 9(4), 790–792. [Retracted on 21 February 2014 by the authors: Specimen was *Senticolis triaspis*, not *Pseudelaphe flavirufa* as reported.].

Mateo, J.A. (2015) Los anfibios y los reptiles introducidos en Baleares: un repaso a loque sabemos y un ejemplo de puerta de entrada. *Libre Verd de Proteccío d'Espècies a les Balears* 2015:447–454, 1 fig.

Matos, S.A. de & Melo-Sampaio, P.R. (2013) Geographic distribution. *Siphlophis worontzowi. Herpetological Review* 44(3), 477–478.

Mays, J.D., Enge, K.M., Emerick, A. & Hill, E.P. (2017) New island records for reptiles in the Florida Keys, Monroe County, Florida, USA. *Herpetological Review* 48(1), 145–146.

Maza, E., Feldman, A., Fishelson, L. & Meiri, S. (2015) *Platyceps largeni* (Schätti, 2001) – sixth specimen and a distribution extension. *Check List* 11(1517), 1–3, 2 figs.

Mazuch, T., Šmíd, J., Price, T., Frýdlová, P., Awale, A.I., Elmi, H.S.A. & Frynta, D. (2018) New records of one of the least known snakes, *Telescopus pulcher* (Squamata: Colubridae) from the Horn of Africa. *Zootaxa* 4462(4), 483–496, 6 figs.

McCartney, J. & Seiffert, E.R. (2016) A late Eocene snake fauna from the Fayum Depression, Egypt. *Journal of Vertebrate Paleontology* 36(1), 1–20, 7 figs.

McCartney, J.A., Roberts, E.M., Tapanila, L. & O'Leary, M. (2018) Large palaeophiid and nigerophiid snakes from Paleogene trans-Saharan Seaway deposits of Mali. *Acta Palaeontologica Polonica* 63(2), 207–220, 8 figs.

McCartney, J.A., Stevens, N.J. & O'Connor, P.M. (2014) The earliest colubroid-dominated snake fauna from Africa: Perspectives from the late Oligocene Nsungwe Formation of southwestern Tanzania. *PLOS ONE* 9(e90415), 1–17, 9 figs.

McCartney-Melstad, E., Waller, T., Micucci, P.A., Barros, M., Draque, J., Amato, G. & Mendez, M. (2012) Population structure and gene flow of the Yellow Anaconda (*Eunectes notaeus*) in northern Argentina. *PLoS One* 7(e37473), 1–9, 2 figs.

McCluskey, E.M. & Bender, D. (2015) Genetic structure of western massasauga rattlesnakes (*Sistrurus catenatus tergeminus*). *Journal of Herpetology* 49(3), 343–348, 2 figs.

McCranie, J.R. (2014a) Geographic distribution. *Trimorphodon quadruplex. Herpetological Review* 45(2), 286.

McCranie, J.R. (2014b) First departmental records of amphibians and reptiles from Intibucá, Lempira, and Ocotepeque in southwestern Honduras. *Herpetological Review* 45(2), 291–293.

McCranie, J.R. (2017) A new species of *Rhadinella* (Serpentes: Dipsadidae) from the Sierra de Agalta, Honduras. *Mesoamerican Herpetology* 4(2), 244–253, 6 figs.

McCranie, J.R., Bedrossian, P.R. & Valdés-Orellana, L. (2014a) Geographic distribution. *Indotyphlops braminus. Herpetological Review* 45(2), 285.

McCranie, J.R., Centeno, R.D., Ramos, J., Valdés-Orellana, L., Mérida, J.E. & Cruz, G.A. (2014b) Eight new records of lizards and snakes (Reptilia: Squamata) from subhumid areas in El Paraíso, Honduras, and morphometry of the poorly-known pitviper *Agkistrodon howardgloydi. Cuadernos de Investigación UNED* 6(1), 99–104, 4 figs.

McCranie, J.R. & Gutsche, A. (2013a) Geographic distribution. *Coniophanes piceivittus. Herpetological Review* 44(3), 475.

McCranie, J.R. & Gutsche, A. (2013b) Geographic distribution. *Leptophis mexicanus. Herpetological Review* 44(3), 477.

McCranie, J.R. & Gutsche, A. (2016) The herpetofauna of islands in the Golfo de Fonseca and adjacent waters, Honduras. *Mesoamerican Herpetology* 3(4), 842–899, 9 figs.

McCranie, J.R., Harrison, A. & Valés-Orellana, L. (2017) Updated population and habitat comments about the reptiles of the Swan Islands, Honduras. *Bulletin of the Museum of Comparative Zoology* 161(7), 265–284, numerous figs.

McCranie, J.R. & Hedges, S.B. (2016) Molecular phylogeny and taxonomy of the *Epictia goudotii* species complex (Serpentes: Leptotyphlopidae: Epictinae) in Middle America and northern South America. *PeerJ* 4(e1551), 1–27, 5 figs.

McCranie, J.R. & Smith, E.N. (2017) A review of the *Tantilla taeniata* species group (Reptilia: Squamata: Colubridae: Colubrinae) in Honduras, with the description of three new species. *Herpetologica* 73(4), 338–348, 9 figs.

McCranie, J.R. & Valdés-Orellana, L. (2012) *Typhlops tycherus* (Townsend, Wilson, Ketzler and Luque-Montes, 2008) (Squamata: Serpentes: Typhlopidae): Significant range extension for this Honduran endemic. *Check List* 8(6), 1308–1309, 2 figs.

McCranie, J.R. & Valdés-Orellana, L. (2014) New island records and updated nomenclature of amphibians and reptiles from the Islas de la Bahía, Honduras. *Herpetology Notes* 7, 41–49, 2 figs.

McCranie, J.R. & Valdés-Orellana, L. (2015) Geographic distribution. *Indotyphlops braminus. Herpetological Review* 46(2), 220.

McCranie, J.R., Valdés-Orellana, L. & Gutsche, A. (2013b) New departmental records for amphibians and reptiles in Honduras. *Herpetological Review* 44(2), 288–289, 1 fig.

McCranie, J.R., Valdés-Orellana, L. & Sheehy III, C.M. (2013a) Morphological and molecular variation in the endemic and poorly known Honduran jumping pitviper *Atropoides indomitus* (Serpentes: Viperidae), with notes on distribution. *Herpetological Review* 44(1), 37–40, 2 figs.

McKelvy, A.D. & Burbrink, F.T. (2017) Ecological divergence in the yellow-bellied kingsnake (*Lampropeltis calligaster*) at two North American biodiversity hotspots. *Molecular Phylogenetics and Evolution* 106(1), 61–72, 4 figs.

McKelvy, A.D., Ozelski-McKelvy, A. & Figueroa, A. (2016) A new non-coastal record for the Pine Woods Littersnake, *Rhadinaea flavilata* (Cope, 1871) (Squamata: Colubridae), in Russell County, Alabama, USA. *Check List* 12(1913), 1–8, 3 figs.

McLellan, F. (2013) New snake records for the island of Nosy Be, northwest Madagascar: *Mimophis mahfalensis* (Grandidier, 1867) and *Pseudoxyrhopus quinquelineatus* (Günther, 1881). *Herpetology Notes* 6, 295–297, 3 figs.

McNab, A., Sanders, M. & Vanderduys, E. (2014) New records of blind snakes resembling the robust blind snake *Anilios ligatus* (Peters 1879), on Cape York Peninsula. *Memoirs of the Queensland Museum* 59, 8, 1 fig.

McVay, J.D. & Carstens, B. (2013) Testing monophyly without well-supported gene trees: Evidence from multi-locus nuclear data conflicts with existing taxonomy in the snake tribe Thamnophiini. *Molecular Phylogenetics and Evolution* 68, 425–431, 2 figs.

Mead, J.I. & Schubert, B.W. (2013) Extinct *Pterygoboa* (Boidae, Erycinae) from the latest Oligocene and early Miocene of Florida. *Southeastern Naturalist* 12(2), 427–438, 2 figs.

Mebert, K., Göçmen, B., İğcı, N., Kariş, M. & Ursenbacher, S. (2015) New records and search for contact zones among parapatric vipers in the genus *Vipera* (*barani, kaznakovi, darevskii, eriwanensis*), *Montivipera* (*wagneri, raddei*), and *Macrovipera* (*lebetina*) in northeastern Anatolia. *The Herpetological Bulletin* 133, 13–22, 9 figs.

Mebert, K., Göçmen, B. & Kariş, M. (2017) Range extension of the critically endangered Anatolian Meadow Viper *Vipera anatolica senliki* in eastern Antalya Province. *South Western Journal of Horticulture, Biology and Environment* 8(2), 65–77, 8 figs.

Mebert, K., Göçmen, B., Kariş, M., İğcı, N. & Ursenbacher, S. (2016) The valley of four viper species and a highland of dwarfs: Fieldwork on threatened vipers in northeastern Turkey. *IRCF Reptiles & Amphibians* 23(1), 1–9, 9 figs.

Mebert, K., Masroor, R. & Chaudhry, M.J.I. (2013) The Dice Snake, *Natrix tessellata* (Serpentes: Colubridae) in Pakistan: Analysis of its range limited to few valleys in western Karakoram. *Pakistan Journal of Zoology* 45(2), 395–410, 2 figs.

Mediani, M., Brito, J.C. & Fahd, S. (2015b) Atlas of the amphibians and reptiles of northern Morocco: Updated distribution and patterns of habitat selection. *Basic and Applied Herpetology* 29, 81–107, 2 figs, 58 maps.

Mediani, M., Fahd, S., Chevalier, F. & Brito, J.C. (2015a) Another record of *Lytorhynchus diadema* (Duméril, Bibron & Duméril, 1854) from Moroccan Atlantic Sahara. *Herpetozoa* 27(3/4), 197–200, 2 figs.

Mediani, M., Fahd, S., Chevalier, F., Qninba, A. & Samlali, M.L. (2013) New distribution limit of Clifford's diadem snake *Spalerosophis diadema* (Serpentes: Colubridae) in southern Morocco. *Herpetology Notes* 6, 453–456, 2 figs.

Medina-Flores, M., Murillo, J.L. & Townsend, J.H. (2016) Geographic distribution. *Atropoides indomitus. Herpetological Review* 47(2), 261.

Medina-Flores, M. & Townsend, J.H. (2016) Geographic distribution. *Tropidodipsas sartorii. Herpetological Review* 47(2), 266–267.

Medina-Rangel, G.F. (2011) Diversidad alfa y beta de la comindad de reptiles en el complejo cenagoso de Zapatosa, Colombia. *Revista de Biología Tropical* 59(2), 935–968, 8 figs.

Medina-Rangel, G.F. (2015) Geographic distribution. *Ninia atrata. Herpetological Review* 46(4), 574–575.

Medina-Rangel, G.F., Cárdenas-Arévalo, G. & Rentaría-M., L.E. (2018) Rediscovery and first record of the Phantasma Tree Snake, *Imantodes phantasma* (Myers, 1982) (Serpentes, Colubridae), in Colombia. *Check List* 14(1), 237–242, 4 figs.

Medina-Rangel, G.F. & López-Perilla, Y.R. (2015) Geographic distribution. *Porthidium lansbergii. Herpetological Review* 46(4), 575–576.

Meetei, A.B., Das, S., Campbell, P.D., Raha, S. & Bag, P. (2018) A study on *Ptyas doriae* (Boulenger, 1888) with comments on the status of *Ptyas hamptoni* (Boulenger, 1900) (Squamata: Colubridae: Colubrinae). *Zootaxa* 4457(4), 537–548, 5 figs.

Meewattana, P. (2010) Note on the variability of the Orange-necked Keelback *Macropisthodon flaviceps* (Duméril, Bibron and Duméril, 1854) in southern Thailand. *The Thailand Natural History Museum Journal* 4(2), 97–99, 2 figs.

Meijide, M.W. (1985) Localidades nuevas o poco conocidas de anfibios y reptiles de la España continental. *Doñana, Acta Vertebrata* 12(2), 318–323.

Meik, J.M., Schaack, S., Flores-Villela, O. & Streicher, J.W. (2018) Integrative taxonomy at the nexus of population divergence and speciation in insular speckled rattlesnakes. *Journal of Natural History* 52(13–16): 989–1016, 5 figs.

Meik, J.M., Streicher, J.W., Lawing, A.M., Flores-Villela, O. & Fujita, M.K. (2015) Limitations of climatic data for inferring species boundaries: Insights from Speckled Rattlesnakes. *PLOS ONE* 10(6)(e0131435), 1–19, 6 figs.

Mekinić, S., Ževrnja, N., Boban, J., Piasevoli, G. & Vladović, D. (2015) Snakes in the herpetological collection of the Natural History Museum in Split (Croatia) collected from 1924 until 2015. *Hyla* 2, 1–8, 2 figs.

Melo-Sampaio, P.R., Passos, P., Fouquet, A., Prudente, A.L. da C. & Torres-Carvajal, O. (2019) Systematic review of *Atractus schach* (Serpentes: Dipsadidae) species complex from the Guiana Shield with description of three new species. *Systematics and Biodiversity* 17(3), 207–229, 12 figs.

Melvinselvan, G., Narayanan, R.S. & Sharma, V. (2016) Geographic distribution. *Lycodon flavomaculatus. Herpetological Review* 47(3), 427.

Melvinselvan, G. & Nibedita, D. (2016) An observation on fish predation by Ornate Flying Snake, *Chrysopelea ornata* (Shaw, 1802) (Serpentes: Colubridae), from southern Western Ghats, Tamil Nadu, India. *Russian Journal of Herpetology* 23(4), 311–314, 2 figs.

Mendes-Pinto, T.J. & Souza, S.M. de (2011) Preliminary assessment of amphibians and reptiles from Floresta Nacional do Trairão, with a new snake record for the Pará state, Brazilian Amazon. *Salamandra* 47(4), 199–206, 4 figs.

Mendoza-Miranda, P., Callapa, G. & Muñoz-S., A. (2017) Variación y primer registro de *Apostolepis multicincta* (Harvey, 1999) (Squamata: Dipsadidae) para el departamento de Cochabamba, Bolivia. *Cuadernos de Herpetología* 31(1), 59–63, 2 figs.

Menegon, M., Doggart, N. & Owen, N. (2008) The Nguru mountains of Tanzania, an outstanding hotspot of herpetofaunal diversity. *Acta Herpetologica* 3(2), 107–127, 3 figs.

Menegon, M., Loader, S.P., Marsden, S.J., Branch, W.R., Davenport, T.R.B. & Ursenbacher, S. (2014) The genus *Atheris* (Serpentes: Viperidae) in East Africa: Phylogeny and the role of rifting and climate in shaping the current pattern of species diversity. *Molecular Phylogenetics and Evolution* 79, 12–22, 4 figs.

Meneses-Pelayo, E. & Caballero, D. (2019) New records and an updated map of distribution of *Micrurus camilae* (Renjifo & Lundberg, 2003) (Elapidae) for Colombia. *Check List* 15(3), 465–469, 3 figs.

Meneses-Pelayo, E., Echavarría-Rentería, J.D., Bayona-Serrano, J.D., Caicedo-Portilla, J.R. & Rengifo-Mosquera, J.T. (2016) New records and an update of the distribution of *Sibon annulatus* (Colubridae: Dipsadinae: Dipsadini) for Colombia. *Check List* 12(1931), 1–5, 2 figs.

Meneses-Pelayo, E., Echavarría-Rentería, J.D., Bayona-Serrano, J.D., Caicedo-Portilla, J.R. & Rengifo-Mosquera, J.T. (2018) *Sibon annulatus* (Günther, 1872). *Catálogo de anfibios y reptiles de Colombia* 4(3), 70–77, 4 figs.

Meneses-Pelayo, E. & Passos, P. (2019) New polychromatic species of *Atractus* (Serpentes: Dipsadidae) from the eastern portion of the Colombian Andes. *Copeia* 107(2), 250–261, 8 figs.

Menezes, F. de A., Abegg, A.D., Silva, B.R. da, Franco, F.L. & Feio, R.N. (2018) Composition and natural history of the snakes from the Parque Estadual da Serra do Papagaio, southern Minas Gerais, Serra da Mantiqueira, Brazil. *ZooKeys* 797, 117–160, 4 figs.

Mesquita, P.C.M.D., Passos, D.C., Borges-Nojosa, D.M. & Cechin, S.Z. (2013) Ecologia e história natural das serpentes de uma área de Caatinga no nordeste Brasileiro. *Papéis Avulsos de Zoologia* 53(8), 99–113, 1 fig.

Messenger, K.R. & Wang, Y. (2015) Notes on the natural history and morphology of the Ning-shan Lined Snake (*Stichophanes ningshaanensis* Yuen, 1983; Ophidia: Colubridae) and its distribution in the Shennongjia National Nature Reserve, China. *Amphibian & Reptile Conservation* 9(2), 111–119, 17 figs.

Meza-Joya, F.L. (2015) First record of *Ninia atrata* (Hallowell, 1845) (Squamata: Colubridae) from Sierra Nevada de Santa Marta, northern Colombia. *Check List* 11(1584), 1–3, 7 figs.

Mezzasalma, M., Andreone, F., Glaw, F., Guarino, F.M., Odierna, G., Petraccioli, A. & Picariello, O. (2019) Changes in heterochromatin content and ancient chromosome fusion in the endemic Malagasy boid snakes *Sanzinia* and *Acrantophis* (Squamata: Serpentes). *Salamandra* 55(2), 140–144, 3 figs.

Mezzasalma, M., Dall'Asta, A., Loy, A., Cheylan, M., Lymberakis, P., Zuffi, M.A.L., Tomović, L., Odierna, G. & Guarino, F.M. (2015) A sisters' story: Comparative phylogeography and taxonomy of *Hierophis viridiflavus* and *H. gemonensis* (Serpentes, Colubridae). *Zoologica Scripta* 44(5), 495–508, 7 figs.

Michael, D.R. & Lindenmayer, D.B. (2011) *Diplodactylus tessellatus* (Gunther, 1875) (Squamata: Diplodactylidae), *Parasuta dwyeri* Greer, 2006 and *Suta suta* (Peters, 1863) (Squamata: Elapidae): Distribution extension in the Murray catchment of New South Wales, South-eastern Australia. *Check List* 7(5), 578–580, 4 figs.

Michels, J.P. & Bauer, A.M. (2004) Some corrections to the scientific names of amphibians and reptiles. *Bonner Zoologische Beiträge* 52(1/2), 83–94.

Miinala, M. (2011) New location record for the recently described *Liophidium pattoni* Vieites, Ratsoavina, Randrianiaina, Nagy, Glaw & Vences 2010. *Herpetology Notes* 4, 181, 2 figs.

Miller, A.H. & Zug, G.R. (2016) Morphology and biology of the Asian Common Mockviper, *Psammodynastes pulverulentus* (Boie, 1827) (Serpentes: Lamprophiidae): A focus on Burmese populations. *Proceedings of the Biological Society of Washington* 129, 173–194, 6 figs.

Milto, K.D. (2003) "The distribution of the Grass Snake (*Natrix natrix*) in the northern European Russia." *Modern Herpetology* 2, 100–123, 4 figs. [in Russian].

Milto, K.D. (2014) First report of the herpetofauna of Phi Phi Archipelago, Andaman Sea, Thailand. *Russian Journal of Herpetology* 21(4), 269–273, 8 figs.

Milto, K.D. (2017) New records of reptiles on the Red Sea Coast, Egypt, with notes on Zoogeography. *Russian Journal of Herpetology* 24(1), 11–21, 4 figs.

Milto, K.D. & Zinenko, O.I. (2005) Distribution and morphological variability of Vipera berus in Eastern Europe. In: Ananjeva, N. & Tsinenko, O. (Eds.) *Herpetologia Petropolitana – Proceedings of the 12th Ordinary General Meeting of the Societas Europeae Herpetologica.* St. Petersburg, 64–73, 5 figs.

Minoli, I., Morando, M. & Avila, L.J. (2015) Reptiles of Chubut province, Argentina: Richness, diversity, conservation status and geographic distribution maps. *ZooKeys* 498, 103–126, 3 figs.

Miralles, A., Marin, J., Markus, D., Herrel, A., Hedges, S.B. & Vidal, N. (2018) Molecular evidence for the paraphyly of Scolecophidia and its evolutionary implications. *Journal of Evolutionary Biology* 31(12), 1782–1793, 4 figs.

Miranda, D.B., Venâncio, N.M. & Albuquerque, S. de (2014) Rapid survey of the herpetofauna in an area of forest management in eastern Acre, Brazil. *Check List* 10(4), 893–899, 3 figs.

Miranda, J.P., Costa, J.C.L. & Rocha, C.F.D. (2012) Reptiles from Lençóis Maranhenses National Park, Maranhão, northeastern Brazil. *ZooKeys* 246, 51–68, 7 figs.

Miranda-Calle, A.B. & Aguilar-Kirigin, A.J. (2011) *Bothrops sanctaecrucis* (Hoge 1966) (Squamata: Viperidae). *Cuadernos de Herpetología* 25(1), 29–31.

Mirza, Z.A. & Pal, S. (2008) A checklist of the reptiles and amphibians of Sanjay Gandhi National Park, Mumbai, *Maharashtra. Cobra* 2(4), 14–19.

Mirza, Z.A. & Patel, H. (2018) Back from the dead! Resurrection and revalidation of the Indian endemic snake genus *Wallophis* (Werner, 1929) (Squamata: Colubridae) insights from molecular data. *Mitochondrial DNA Part A*, 29(3), 331–334, fig. 1.

Mirza, Z.A., Vyas, R., Patel, H., Maheta, J. & Sanap, R.V. (2016) A new Miocene-divergent lineage of Old World racer snake from India. *PLOS ONE*, 11(0148380), 1–17, 10 figs.

Missassi, A.F.R., Costa, J.C.L. & Prudente, A.L.C. (2015) Range extension of the Chocoan blunt-headed vine snake: *Imantodes chocoensis* (Serpentes: Dipsadidae) in northwestern Colombia. *Salamandra* 51(3), 269–272, 2 figs.

Missassi, A.F.R. & Prudente, A.L.C. (2015) A new species of *Imantodes* (Duméril, 1853) (Serpentes, Dipsadidae) from the Eastern Cordillera of Colombia. *Zootaxa* 3980(4), 562–574, 7 figs.

Mizsei, E., Jablonski, D., Roussos, S.A., Dimali, M., Ioannidis, Y., Nilson, G. & Nagy, Z.T. (2017a) Nuclear markers support the mitochondrial phylogeny of *Vipera ursinii-renardi* complex (Squamata: Viperidae) and species status for the Greek meadow viper. *Zootaxa* 4227(1), 75–88, 4 figs.

Mizsei, E., Jablonski, D., Végvári, Z., Lengyel, S. & Szabolcs, M. (2017b) Distribution and diversity of reptiles in Albania: A novel database from a Mediterranean hotspot. *Amphibia-Reptilia* 38(2), 157–173, 5 figs, plus 40 maps in supplement.

Mizsei, E., Üveges, B., Vági, B., Szabolcs, M., Lengyel, S., Pfliegler, W.P., Nagy, Z.T. & Tóth, J.P. (2016) Species distribution modelling leads to the discovery of new populations of one of the least known European snakes, *Vipera ursinii graeca*, in Albania. *Amphibia-Reptilia* 37(1), 55–68, 3 figs.

Mizsei, E., Zinenko, O., Sillero, N., Ferri, V., Roussos, S.A. & Szabolcs, M. (2018) The distribution of meadow and steppe vipers (Vipera graeca, V. renardi and V. ursinii): A revision of the New Atlas of Amphibians and Reptiles of Europe. *Basic and Applied Herpetology* 32, 77–83, 1 fig.

Mlíkovský, J., Benda, P., Moravec, J. & Šanda, R. (2011) Type specimens of recent vertebrates in the collections of the National Museum, Prague, Czech Republic. *Journal of the National Museum (Prague), Natural History Series* 180(10), 133–164.

Moadab, M., Zargan, J., Rastegar-Pouyani, E. & Hajinourmohammadi, A. (2018) Modelling the potential distribution of *Spalerosophis diadema* (Schlegel, 1837) (Serpents [sic]: Colubridae) in Iran. *Herpetology Notes* 11, 805–808, 1 fig.

Mohamad, S.I. & Afrasiab, S.R. (2015) Two new records of dwarf snakes of the genus *Eirenis* Jan (Repttilia [sic], Colubridae) in Iraqi Kurdistan (north and northeastern Iraq) with annotated Checklist, for the genus *Eirenis* in Iraq. *Bulletin of the Iraq Natural History Museum* 13(3), 77–83, 5 figs.

Mohammad, M.K., Ali, H.H., Ali, B.A.A. & Hadi, A.M. (2013) The biodiversity of Bahr Al-Najaf depression, Al-Najaf Al-Ashraf Province. *Bulletin of the Iraq Natural History Museum* 12(3), 21–30.

Mohammed, R.S., Manickchan, S.A., Charles, S.P. & Murphy, J.C. (2014) The herpetofauna of southeast Trinidad, Trinidad and Tobago. *Living World, Journal of the Trinidad and Tobago Field Naturalists' Club* 2014, 12–20, 3 figs.

Mohan, A.V., Visvanathan, A.C. & Vasudevan, K. (2018) Phylogeny and conservation status of the Indian egg-eating snake, *Elachistodon westermanni* (Reinhardt, 1863) (Serpentes, Colubridae). *Amphibia-Reptilia* 39(3), 317–324, 3 figs.

Mohapatra, P.P., Dutta, S.K., Kar, N.B., Das, A., Murthy, B.H.C.K. & Deepak, V. (2017) *Ahaetulla nasuta anomala* (Annandale, 1906) (Squamata: Colubridae), resurrected as a valid species with marked sexual dichromatism. *Zootaxa* 4263(2), 318–332, 7 figs.

Mohapatra, P.P., Schulz, K.-D., Helfenberger, N., Hofmann, S. & Dutta, S.K. (2016) A contribution to the Indian trinket Snake, *Coelognathus helena (Daudin*, 1803), with the description of a new subspecies. *Russian Journal of Herpetology* 23(2), 115–144, 30 figs.

Monasterio, C., Álvarez, P., Trape, J.-F. & Rödel, M.-O. (2016) The herpetofauna of the Dindefelo Natural Community Reserve, Senegal. *Herpetology Notes* 9, 1–6, 2 figs.

Montes-Correa, A.C., Arévalo-Páez, M., Rada-Vargas, E., Portillo-Mozo, A. del, Granda-Rodríguez, H.D. & Rivero-Blanco, C. (2017) First record of *Atractus turikensis* (Squamata: Colubridae: Dipsadinae) from the Colombian Perijá highlands. *The Herpetological Bulletin* 141, 35–39, 3 figs.

Montingelli, G.G., Grazziotin, F.G., Battilana, J., Murphy, R.W., Zhang, Y.-P. & Zaher, H. (2019) Higher-level phylogenetic affinities of the Neotropical genus *Mastigodryas* (Amaral, 1934) (Serpentes: Colubridae), species-group definition and description of a new genus for *Mastigodryas bifossatus*. *Journal of Zoological Systematics and Evolutionary Research* 57(2), 205–239, 9 figs.

Montingelli, G.G., Valencia, J.H., Benavides, M.A. & Zaher, H. (2011) Revalidation of *Herpetodryas reticulata* (Peters, 1863) (Serpentes: Colubridae) from Ecuador. *South American Journal of Zoology* 6(3), 189–197, 5 figs.

Moore, B. & Slone, T. (2002) *Kentucky Snakes*. Kentucky Department of Wildlife and Fisheries Resources, Frankfort. 1–32, numerous illustrations.

Moradi, N., Rastegar-Pouyani, N. & Rastegar-Pouyani, E. (2014) Geographic variation in the morphology of *Macrovipera lebetina* (Linnaeus, 1758) (Ophidia: Viperidae) in Iran. *Acta Herpetologica* 9(2), 187–202, 9 figs.

Moradi, N., Shafiei, S. & Sehhatisabet, M.E. (2013) The snake fauna of Khabr National Park, southeast of Iran. *Iranian Journal of Animal Biosystematics* 9(1), 41–55, 5 figs.

Moraes, L.J.C.L., Almeida, A.P. de, Fraga, R. de, Rojas, R.R., Pirani, R.M., Silva, A.A.A., Carvalho, V.T. de, Gordo, M. & Werneck, F.P. (2017) Integrative overview of the herpetofauna from Serra de Mocidade, a granitic mountain range in northern Brazil. *ZooKeys* 715, 103–159, 18 figs.

Moraes-da-Silva, A., Amaro, R.C., Nunes, P.M.S., Strüssmann, C., Junior, M.T., Andrade-Jr., A., Sudré, V., Recoder, R., Rodrigues, M.T. & Curcio, F.F. (2019) Chance, luck and a fortunate finding: A new species of watersnake of the genus *Helicops* (Wagler, 1828) (Serpentes: Xenodontinae), from the Brazilian Pantanal wetlands. *Zootaxa* 4651(3), 445–470, 7 figs.

Morais, D.H., Ávila, R.W., Kaqwashita-Ribeiro, R.A. & Carvalho, M.A. de (2011) Squamata, Elapidae, *Micrurus surinamensis* (Cuvier, 1817): New records and distribution map in the state of Mato Grosso, Brazil, with notes on diet and activity period. *Check List* 7(3), 350–351, 2 figs.

Morais, D.H., Mott, T., Kawashita-Ribeiro, R.A. & Santos-Jr., A.P. (2010) Reptilia, Squamata, Dipsadidae, Xenodontinae, *Taeniophallus brevirostris* (Peters, 1863): Distribution extension and new state record. *Check List* 6(3), 456–457, 2 figs.

Morales-Capellán, N., Fernández-Badillo, L., López-Mejía, A., Sánchez-Martínez, G. & Goyenechea, I. (2016) Conformation of the night snake *Hypsiglena tanzeri* in Hidalgo, Mexico, and a new record for Reserva de la Biósfera de la Barranca de Metztitlán. *Mesoamerican Herpetology* 3(4), 1097–1100, 3 figs.

Morán, E.S., Henríquez, V., Greenbaum, E., López, J.G.C. & Rivera, A.M. (2015) Geographic distribution. Sibon dimidiatus. *Herpetological Review* 46(4), 576.

Moreno-Arias, R. & Quintero-Corza, S. (2015) Reptiles del valle seco del Río Magdalena (Huila, Colombia). *Caldasia* 37(1), 183–195, 2 figs.

Moreno-Rodríguez, J.D. (1995) Nuevas localidades para Aragón y primeras citas en Huesca de *Coluber hippocrepis*. *Lucas Mallada* 7, 279–280.

Mori, A. & Nagata, E. (2016) Relying on a single anuran species: Feeding ecology of a snake community on Kinkasan Island, Miyagi Prefecture, Japan. *Current Herpetology* 35(2), 106–114, 2 figs.

Moser, C.F., Avila, F.R. de, Oliveira, R.B. de, Oliveira, J.M. de, Borges-Martins, M. & Tozetti, A.M. (2018) Reptile diversity of Sinos River Basin. *Biota Neotropica* 18(20183 figs.0530), 1–8.

Moskvitin, S.V. & Kuranova, V.N. (2006) Amphibians and reptiles in the collection of the Zoological Museum of the Tomsk State University (Western Siberia, Russia). In: Vences, M., Köhler, J., Ziegler, T. & Böhme, W. (Eds.), *Herpetologia Bonnensis II. Proceedings of the 13th Congress of the Societas Europaea Herpetologica*, Bonn, 99–102, 2 figs.

Mossman, A., Culhane, K., Miller, Z., Brock, K.M., Pafilis, P. & Donihue, C.M. (2016) *Natrix natrix* (Linnaeus, 1758), found on the small islet of Tigani (Central Cyclades, Greece). *Herpetozoa* 29(1/2), 107–109, 1 fig.

Mouane, A., Si Bachir, A., Ghennoum, I. & Harrouchi, A.K. (2014) Premières données sur la diversité de l'herpétofaune de l'Erg oriental (Région du Souf - Algérie). *Bulletin de la Société Herpétologique de France* (148), 491–502, 6 figs.

Moura, M.R., Costa, H.C. & Pirani, R.M. (2013a) Rediscovery of *Phalotris concolor* (Serpentes: Dipsadidae: Elapomorphini). *Zoologia* 30(4), 430–436, 13 figs.

Moura, M.R., Motta, A.P., Fernandes, V.D. & Feio, R.N. (2012) Herpetofauna da Serra do Brigadeiro, um remenescente de Mata Atlântica em Minas Gerais, Sudeste do Brasil. *Biota Neotropica* 12(1), 209–235, 12 figs.

Moura, M.R., Pirani, R.M. & Silva, V.X. (2013b) New records of snakes (Reptilia: squamata) in Minas Gerais, Brazil. *Check List* 9(1), 99–103, 6 figs.

Mrinalini, Thorpe, R.S., Creer, S., Lallias, D., Dawnay, L., Stuart, B.L. & Malotra, A. (2015) Convergence of multiple markers and analysis methods defines the genetic distinctiveness of cryptic pitvipers. *Molecular Phylogenetics and Evolution* 92, 266–279, 5 figs.

Mulcahy, D.G., Lee, J.L., Miller, A.H., Chand, M., Thura, M.K. & Zug, G.R. (2018) Filling in the BINs of life: Report of an amphibian and reptile survey of the Tanintharyi (Tenasserim) region of Myanmar, with DNA barcode data. *ZooKeys* 757, 85–152, 5 figs.

Mulcahy, D.G., Lee, J.L., Miller, A.H. & Zug, G.R. (2017) Troublesome times: Potential cryptic speciation of the *Trimeresurus* (*Popeia*) *popeiorum* complex (Serpentes: Crotalidae) around the Isthmus of Kra (Myanmar and Thailand). *Zootaxa* 4347(2), 301–315, 3 figs.

Mulcahy, D.G., Martínez-Gómez, J.E., Aguirre-León, G., Cervantes-Pasqualli, J.A. & Zug, G.R. (2014) Rediscovery of an endemic vertebrate from the remote Islas Revillagigedo in the eastern Pacific Ocean: The Clarión Nightsnake lost and found. *PLOS ONE* 9(e97682), 1–8, 2 figs.

Mulder, J. (2017) A review of the distribution of *Vipera ammodytes transcaucasiana* (Boulenger, 1913) (Serpentes: Viperidae) in Turkey. *Biharean Biologist* 11(1), 23–26, 2 figs.

Muliya, S.K., Nath, A. & Das, A. (2018) First report of death-feigning behaviour in the yellow collared wolf snake (*Lycodon flavicollis*). *The Herpetological Bulletin* 143, 41–42, 2 figs.

Muñoz-Nolasco, F.J., Cruz-Sáenz, D., Rodríguez-Ruvalcaba, O.J. & Terrones-Ferreiro, I.E. (2015) Notes on the herpetofauna of western Mexico 12: Herpetofauna of a temperate forest in Mazamitla, southeastern Jalisco, Mexico. *Bulletin of the Chicago Herpetological Society* 50(4), 45–50, 6 figs.

Murphy, J.C., Braswell, A.L., Charles, S.P., Auguste, R.J., Rivas, G.A., Borzée, A., Lehtinen, R.M. & Jowers, M.J. (2019a) A new species of *Erythrolamprus* from the oceanic island of Tobago (Squamata, Dipsadidae). *ZooKeys* 817, 131–157, 6 figs.

Murphy, J.C., Charles, S.P., Lehtinen, R.M. & Koeller, K.L. (2013) A molecular and morphological characterization of Oliver's parrot snake, *Leptophis coeruleodorsus* (Squmata: Serpentes: Colubridae) with the description of a new species from Tobago. *Zootaxa* 3718(6), 561–574, 5 figs.

Murphy, J.C. & Rutherford, M.G. (2014) The snail-eating snake *Dipsas variegata* (Duméril, Bibron and Duméril) on Trinidad, and its relationship to the microcephalic *Dipsas trinitatis* Parker (Squamata, Dipsadidae). *Herpetology Notes* 7, 757–760, 2 figs.

Murphy, J.C., Rutherford, M.G. & Jowers, M.J. (2016) The threadsnake tangle: Lack of genetic divergence in *Epictia tenella* (Squamata, Leptotyphlopidae): Evidence for introductions or recent rafting to the West Indies. *Studies on Neotropical Fauna and Environment* 51(3), 197–205, 4 figs.

Murphy, J.C., Salvi, D., Braswell, A.L. & Jowers, M.J. (2019b) Phylogenetic position and biogeography of three-lined snakes (*Atractus trilineatus*: Squamata, Dipsadidae) in the eastern Caribbean. *Herpetologica* 75(3), 247–253, 3 figs.

Murphy, J.C. & Voris, H.K. (2013) An unusual, fangless short-tailed snake (Squamata, Serpentes, Homalopsidae) from Sumatra, Indonesia. *Asian Herpetological Research* 4(2), 140–146, 5 figs.

Murphy, J.C. & Voris, H.K. (2014) A checklist and key to the homalopsid snakes (Reptilia, Squamata, Serpentes), with the description of new genera. *Fieldiana Life and Earth Sciences* (8), iv+43, 53 figs.

Murray, S., Shedd, J.D. & Dugan, E.A. (2015) Geographic distribution. *Hypsiglena sleveni*. *Herpetological Review* 46(1), 62.

Murray-Dickson, G., Ghazali, M., Ogden, R., Brown, R. & Auliya, M. (2017) Phylogeography of the reticulated python (*Malayopython reticulatus* ssp): Conservation implications for the worlds' most traded snake species. *PLOS ONE* 12(8:e0182049), 1–25, 4 figs.

Murta-Fonseca, R.A., Franco, F.L. & Fernandes, D.S. (2015) Taxonomic status and morphological variation of *Hydrodynastes bicinctus* (Hermann, 1804) (Serpentes: Dipsadidae). *Zootaxa* 4007(1), 63–81, 10 figs.

Musilova, R., Zavadil, V. & Kotlík, P. (2007) Isolated population of *Zamenis longissimus* (Reptilia: Squamata) above the northern limit of the continuous range in Europe: Origin and conservation status. *Acta Societatis Zoologicae Bohemicae* 71, 197–208, 3 figs.

Myers, C.W. & McDowell, S.B. (2014) New taxa and cryptic species of Neotropical snakes (Xenodontinae), with commentary on hemipenes as generic and specific characters. *Bulletin of the American Museum of Natural History* (385), 1–112, 40 figs.

Myers, E.A., Bryson, R.W., Hansen, R.W., Aardema, M.L., Lazcano, D. & Burbrink, F. (2019) Exploring Chihuahuan Desert diversification in the gray-banded kingsnake, *Lampropeltis alterna* (Serpentes: Colubridae). *Molecular Phylogenetics and Evolution* 131, 211–218, 2 figs.

Myers, E.A., Burgoon, J.L., Ray, J.M., Martínez-Gómez, J.E., Matías-Ferrer, N., Mulcahy, D.G. & Burbrink, F.T. (2017b) Coalescent species tree inference of *Coluber* and *Masticophis*. *Copeia* 105(4), 640–648, 2 figs.

Myers, E.A., Hickerson, M.J. & Burbrink, F.T. (2017a) Asynchronous diversification of snakes in the North American warm deserts. *Journal of Biogeography* 44(2), 461–474, 3 figs.

Myers, E.A., Rodríguez-Robles, J.A., Denardo, D.F., Staub, R.E., Stropoli, A., Ruane, S. & Burbrink, F. (2013b) Multilocus phylogeographic assessment of the California Mountain Kingsnake (*Lampropeltis zonata*) suggests alternative patterns of diversification for the California Floristic Province. *Molecular Ecology* 22, 5418–5429, 3 figs.

Myers, E.A., Ruane, S., Knight, K., Knight, J.L. & Ray, J.M. (2013c) Distribution record of *Tantilla alticola* (Boulenger, 1903) (Squamata: Colubridae) in Coclé Province, Republic of Panama. *Check List* 9(1), 151–152, 1 fig.

Myers, E.A., Weaver, R.E. & Alamillo, H. (2013a) Population stability of the northern desert nightsnake (*Hypsiglena chlorophaea deserticola*) during the Pleistocene. *Journal of Herpetology* 47(3), 432–439, 5 figs.

Nagy, Z.T., Glaw, F. & Vences, M. (2010) Systematics of the snake genera *Stenophis* and *Lycodryas* from Madagascar and the Comoros. *Zoologica Scripta* 39(5), 426–435, 1 fig.

Nagy, Z.T., Gvoždík, V., Meirte, D., Collet, M. & Pauwels, O.S.G.. (2014) New data on the morphology and distribution of the enigmatic Schouteden's sun snake, *Helophis schoutedeni* (de Witte, 1922) from the Congo Basin. *Zootaxa* 3755(1), 96–100, 3 figs.

Nagy, Z.T., Kusamba, C., Collet, M. & Gvoždík, V. (2013) Notes on the herpetofauna of western Bas-Congo, Democratic Republic of the Congo. *Herpetology Notes* 6, 413–419, 3 figs.

Nagy, Z.T., Marion, A.B., Glaw, F., Miralles, A., Nopper, J., Vences, M. & Hedges, S.B. (2015) Molecular systematics and undescibed diversity of Madagascan scolecophidian snakes (Squamata: Serpentes). *Zootaxa* 4040(1), 31–47, 1 fig.

Nappi, A., Cipolla, R.M., Gabtiele, R., Masseti, M., Corti, C. & Arcidiacono, G. (2007) Anfibi, retilli e mammiferi delle isole del Golfo di Napoli: Check-list commentate. *Studi Trentini di Scienze Naturali, Acta Biologica* 83, 93–97.

Narayana, B.L., Sandeep, M. & Dogra, S. (2018) A new locality record for the Yellow-collared Wolfsnake, *Lycodon flavicollis* Mukherjee and Bhupathy 2007, from Hyderabad, Telangana, India. *IRCF Reptiles & Amphibians* 25(1), 55–56, 3 figs.

Narayanan, A. & Satyanarayan, K. (2012) Glossy-bellied racer snake *Platyceps ventromaculatus* (Gray, 1843) (Squamata: Serpentes: Colubridae): New locality record in Delhi National Capital Region (NCR), India. *Check List* 8(6), 1356–1358, 5 figs.

Narayanan, S., Joseph, N., Kumar, R. & Vengatesan, A. (2017a) Occurrence of the Sri Lankan Flying Snake, *Chrysopelea* cf. *taprobanica* (Smith 1943) in Tamil Nadu, India. *IRCF Reptiles & Amphibians* 24(1), 58–60, 3 figs.

Narayanan, S., Joseph, N., Selvan, M. & Jerith, A. (2017b) Notes on the southernmost distributional record for the Yellow-spotted Wolfsnake, *Lycodon* cf. *flavomaculatus* (Wall 1907), from Tamil Nadu, India. *IRCF Reptiles & Amphibians* 24(2), 115–117, 2 figs.

Nascimento, B.T.M. do., Maffei, F., Moya, G.M. & Donatelli, R.J. (2017) *Atractus albuquerquei* (Cunha & Nascimento, 1983) (Serpentes, Dipsadidae): First record for the state of Minas Gerais, southeastern Brazil. *Check List* 13(2078), 1–4, 3 figs.

Nascimento, L.P., Siqueira, D.M. & Santos-Costa, M.C. dos (2013) Diet, reproduction, and sexual dimorphism in the Vine Snake, *Chironius fuscus* (Serpentes: Colubridae), from Brazilian Amazonia. *South American Journal of Herpetology* 8(3), 168–174, 1 fig.

Nasrabadi, R., Rastegar-Pouyani, E., Hosseinian-Yousefkhani, S.S. & Khani, A. (2016) A checklist of herpetofauna from Sabzevar, Northeastern Iran. *Iranian Journal of Animal Biosystematics* 12(2), 255–259, 1 fig.

Natera-Mumaw, M., Esqueda-González, L.F. & Castelaín-Fernández, M. (2015) *Atlas serpientes Venezuela – una vision actual de su diversidad.* Dimacofi Negocios Avanzados S.A., Santiago. xii+441, 419 figs.

Natusch, D.J.D. & Lyons, J.A. (2014) Geographic and sexual variations in body size, morphology, and diet among five populations of green pythons (*Morelia viridis*). *Journal of Herpetology* 48(3), 317–323, 5 figs.

Navarro-Cornejo, G. & Gonzales, L. (2013) Geographic distribution of *Philodryas laticeps* (Werner, 1900) (Serpentes: Dipsadidae) in Bolivia: New distribution records. *Kempffiana* 9(1), 34–37, 2 figs.

Neang, T., Chan, S., Chhin, S., Samorn, V., Poyarkov, N.A., Stephens, J., Daltry, J.C. & Stuart, B.L. (2017) First records of three snake species from Cambodia. *Cambodian Journal of Natural History* 2017(2), 142–146, 5 figs.

Neang, T., Chhin, S., Meanrith, K. & Hun, S. (2011) First records of two reptile species (Gekkonidae: *Hemidactylus garnotii* Duméril & Bibron, 1836; Viperidae: *Ovophis convictus* Stoliczka, 1870) from Cambodia. *Cambodian Journal of Natural History* 2011(2), 86–92, 3 figs.

Neang, T., Grismer, L.L., Hun, S. & Phan, C. (2015) New herpetofauna records and range extensions for *Daboia siamensis* (Smith, 1917) and *Gekko petricolus* Taylor, 1962 from Cambodia. *Cambodian Journal of Natural History* 2015(2), 172–182, 8 figs.

Neang, T., Hartmann, T., Hun, S., Souter, N.J. & Furey, N.M. (2014) A new species of wolf snake (Colubridae: *Lycodon* Fitzinger, 1826) from Phnom Samkos Wildlife Sanctuary, Cardamom Mountains, southwest Cambodia. *Zootaxa* 3814(1), 68–80, 6 figs.

Neang, T. & Hun, S. (2013) First record of *Oligodon annamensis* (Leviton, 1953) (Squamata: Colubridae) from the Cardamom Mountains of southwest Cambodia. *Herpetology Notes* 6, 271–273, 2 figs.

Nekrasova, O.D., Gavris, G.G. & Kuybida, V.V. (2013) Changes in the northern border of the home range of the dice snake, *Natrix tessellata* (Reptilia, Colubridae), in the Dnipro Basin (Ukraine). *Vesnik Zoologii* 47(5), 67–71, 1 fig.

Nemes, L., Babb, R., Van Devender, R.W., Nguyen, K.V., Le, Q.K., Vu, T.N., Rauhaus, A., Nguyen, T.Q. & Ziegler, T. (2013) First contribution to the reptile fauna of Quang Ngai Province, central Vietnam. *Biodiversity Journal* 4(2), 301–326, 40 figs.

Neri-Castro, E.E., Montalbán-Huidobro, C. & Ortiz-Medina, J.A. (2017) First record of *Tantilla moesta* (Squamata: Colubridae) from the state of Campeche, Mexico. *Mesoamerican Herpetology* 4(3), 673–674, 1 fig.

Neumann-Denzau, G. & Denzau, H. (2010) The Brown vine snake *Ahaetulla pulverulenta* (Duméril, Bibron & Duméril, 1854) in the Sundarbans, Bangladesh – first record from the eastern part of the Indian subcontinent. *Herpetology Notes* 3, 271–272, 3 figs.

Nevárez-de los Reyes, M., Banda-Leal, J., Lazcano, D., Bryson, R.W. & Hansen, R.W. (2016c) Noteworthy records of snakes of the *Lampropeltis mexicana* complex from northeastern Mexico. *Mesoamerican Herpetology* 3(4), 1055–1058, 1 fig.

Nevárez-de los Reyes, M., Lazcano, D. & Banda-Leal, J. (2016a) Geographic distribution. *Crotalus ornatus. Herpetological Review* 47(2), 261.

Nevárez-de los Reyes, M., Lazcano, D. & Banda-Leal, J. (2017b) Geographic distribution. *Tantilla nigriceps. Herpetological Review* 48(3), 591.

Nevárez-de los Reyes, M., Lazcano, D., Banda-Leal, J. & Recchio, I. (2014) Notes on Mexican herpetofauna 22: Herpetofauna of the continental portion of the Municipality of Hermosillo, Sonora, Mexico. *Bulletin of the Chicago Herpetological Society* 59(8), 105–115, 8 figs.

Nevárez-de los Reyes, M., Lazcano, D. & Bonilla-Vega, A. (2017a) Geographic distribution. *Leptophis mexicanus. Herpetological Review* 48(3), 590.

Nevárez-de los Reyes, M., Lazcano, D., García-Padilla, E., Mata-Silva, V., Johnson, J.D. & Wilson, L.D. (2016b) The herpetofauna of Nuevo León, Mexico: Composition, distribution, and conservation. *Mesoamerican Herpetology* 3(3), 557–638, numerous color illus.

Nguyen, H.N., Tran, B.V., Nguyen, L.H., Neang, T., Yushchenko, P.V. & Poyarkov, N.A. (2020) A new species of *Oligodon* Fitzinger, 1826 from the Langbian Plateau, southern Vietnam, with additional information on *Oligodon annamensis* (Leviton, 1953) (Squamata: Colubridae). *PeerJ* 8(e8332), 1–36, 11 figs.

Nguyen, S.N., Nguyen, V.D.H., Le, S.H. & Murphy, R.W. (2016) A new species of kukri snake (Squamata: Colubridae: *Oligodon* Fitzinger, 1826) from Con Dao Islands, southern Vietnam. *Zootaxa* 4139(2), 261–273, 3 figs.

Nguyen, S.N., Nguyen, L.T., Nguyen, V.D.H., Phan, H.T., Jiang, K. & Murphy, R.W. (2017b) A new species of the genus *Oligodon* (Fitzinger, 1826) (Squamata: Colubridae) from Cu Lao Cham Islands, central Vietnam. *Zootaxa* 4286(3), 333–346, 6 figs.

Nguyen, S.N., Nguyen, V.D.H., Nguyen, T.Q., Le, N.T.T., Nguyen, L.T., Vo, B.D., Vindum, J.V., Murphy, R.W., Che, J. & Zhang, Y.-P. (2017a) A new color pattern of the *Bungarus candidus* complex (Squamata: Elapidae) from Vietnam based on morphological and molecular data. *Zootaxa* 4268(4), 563–572, 5 figs.

Nguyen, T.Q., Nguyen, T.V., Pham, C.T., Ong, A.V. & Ziegler, T. (2018) New records of snakes (Squamata: Serpentes) from Hoa Binh Province, northwestern Vietnam. *Bonn Zoological Bulletin* 67(1), 15–24, 3 figs.

Nguyen, T.Q., Pham, A.V., Nguyen, S.L.H., Le, M.D. & Ziegler, T. (2015) First country record of *Parafimbrios lao* (Teynié, David, Lottier, Le, Vidal et Nguyen, 2015) (Squamata: Xenodermatidae) for Vietnam. *Russian Journal of Herpetology* 22(4), 297–300, 3 figs.

Nguyen, T.Q., Phung, T.M., Schneider, N., Botov, A., Tran, D.T.A. & Ziegler, T. (2014) New records of amphibians and reptiles from southern Vietnam. *Bonn Zoological Bulletin* 63(2), 148–156, 6 figs.

Nguyen, T.Q., Stenke, R., Nguyen, H.X. & Ziegler, T. (2011) The terrestrial reptile fauna of the Biosphere Reserve Cat Ba Archipelago, Hai Phong, Vietnam. In Schuchman, K.-L. (Ed.), *Tropical Vertebrates in a Changing World*. Bonner Zoologische Monographien 57, 99–15, 5 figs.

Nilson, G. & Rastegar-Pouyani, N. (2013) The occurrence of *Telescopus nigriceps* (Ahl, 1924) in western Iran, with comments on the genus *Telescopus* (Serpentes: Colubridae). *Zoology in the Middle East* 59(2), 131–135, 4 figs.

Nistri, A. (2010) La collezione erpetologica della sezione di Zoologia "La Specola" del Museo di Storia Naturale dell'Università di Firenze. *Museologia Scientifica Memorie* 5, 118–128, 2 figs.

Nneji, L.M., Adeola, A.C., Okeyoyin, A., Oladipo, O.C., Saidu, Y., Samuel, D., Usongo, J.Y. et al. (2019) Diversity and distribution of amphibians and reptiles in Gashaka Gumti National Park, *Nigeria. Herpetology Notes* 12, 543–559, 3 igs.

Nogueira, C.C., Argôlo, A.J.S., Arzamendia, V., Azevedo, J.A., Barbo, F.E., Bérnils, R.S., Bolochio, B.E. et al. (2019) Atlas of Brazilian snakes: Verified point-locality maps to mitigate the Wallacean shortfall in a megadiverse snake fauna. *South American Journal of Herpetology* 14(Special Issue 1), 1–274, 411 plates.

Nogueira, C.C., Barbo, F.E. & Ferrarezzi, H. (2012) Redescription of *Apostolepis albicollaris* Lema, 2002, with a key for the species groups of the genus *Apostolepis* (Serpentes: Dipsadidae: Elapomorphini). *South American Journal of Herpetology* 7(3), 213–225, 3 figs.

Noguez, A.A. & Ramírez-Bautista, A. (2008) A checklist of reptiles from the Parque Nacional Isla Contoy, Mexico. *Boletin de la Sociedad Herpetológica Mexicana* 16(2), 36–40.

Nulchis, V., Biaggini, M., Carretero, M.A. & Harris, D.J. (2008) Unexpectedly low mitochondrial DNA variation within the ladder snake *Rhinechis scalaris*. *North-Western Journal of Zoology* 4(1), 119–124, 1 fig.

Nurngsomsri, P., Chuaynkern, Y., Chuaynkern, C. & Thongpun, P. (2014) Geographic distribution. *Chrysopelea ornata*. *Herpetological Review* 45(2), 284–285.

Obando, L.A. & Sunyer, J. (2016a) Distribution notes. *Porthidium nasutum*. *Mesoamerican Herpetology* 3(3), 779, 1 fig.

Obando, L.A. & Sunyer, J. (2016b) Distribution notes. *Bothriechis schlegelii*. *Mesoamerican Herpetology* 3(4), 1064–1065, 1 fig.

Obregón-Esparza, M., Balderas-Valdivia, C.J., Cabirol, N., Rojas-Oropeza, M. & Gonzalez-Hernández, A. (2018) Four herpetofaunal records from Municipio de Ixmiquilpan, Hidalgo, Mexico. *Mesoamerican Herpetology* 5(1), 191–192, 1 fig.

Ocampo, M. & Fernandez, G.P. (2014) *Bothrops diporus* (Cope, 1862). Nuevo registro para Bolivia y ampliación en su distribución norteña. *Cuadernos de Herpetología* 28(1), 47–48, 2 figs.

O'Connell, K.A. & Smith, E.N. (2018) The effect of missing data on coalescent species delimitation and a taxonomic revision of whipsnakes (Colubridae: *Masticophis*). *Molecular Phylogenetics and Evolution* 127, 356–366, 4 figs.

O'Leary, M.A., Bouaré, M.L., Claeson, K.M., Heilbronn, K., Hill, R.V., McCartney, J., Seesa, J.A. et al. (2019) Stratigraphy and paleobiology of the Upper Cretaceous-Lower Paleogene sediments from the Trans-Saharan Seaway in Mali. *Bulletin of the American Museum of Natural History* (436), 1–177, 86 figs, 2 plates.

Oliveira, D.S., Silva, A.V., Santos, R.C., Barriga, J.V.M., Gabriel, R.C. & Costa-Campos, C.E. (2015) Geographic distribution. *Helicops trivittatus*. *Herpetological Review* 46(2), 219.

Oliveira, L.S. de, Araújo, I.S., Prudente, A.L. da C., Fraga, R. de, Almeida, A.P. de & Ascenso, A.C. (2018) New distributional records of the Toad-headed Pitviper *Bothrocophias hyoprora* (Amaral, 1935) in Brazil. *Amphibian & Reptile Conservation* 12(1), 1–4, 3 figs.

Oliveira, P.S. de, Rocha, M.T., Castro, A.G., Betancourt, I.R., Wen, F.H., Neto, A.P., Bastos, M.L., Tambourgi, T.V. & Sant'Anna, S.S. (2016) New records of Gaboon Viper (*Bitis gabonica*) in Angola. *The Herpetological Bulletin* 136, 42–43, 2 figs.

Olvera, C.R.O. & Badillo, L.F. (2016a) Geographic distribution. *Crotalus molossus*. *Herpetological Review* 47(1), 83.

Olvera, C.R.O. & Badillo, L.F. (2016b) Geographic distribution. *Thamnophis scalaris*. *Herpetological Review* 47(1), 85.

Onary, S.Y., Fachini, T.S. & Hsiou, A.S. (2017) The snake fossil record from Brazil. *Journal of Herpetology* 51(3), 365–374, 5 figs.

Onary, S.Y. & Hsiou, A.S. (2018) Systematic revision of the early Miocene fossil *Pseudoepicrates* (Serpentes: Boidae): Implications for the evolution and historical biogeography of the West Indian boid snakes (*Chilabothrus*). *Zoological Journal of the Linnean Society* 184(2), 453–470, 10 figs.

Onn, C.K., Grismer, L.L., Matsui, M., Nishikawa, K., Wood, P.L., Grismer, J.L., Belabut, D. & Ahmad, N. (2010) Herpetofauna of Gunung Panti Forest Reserve, Johor, Peninsular Malaysia. *Tropical Life Sciences Research* 21(1), 71–82, 1 fig.

Onn, C.K., Grismer, L.L., Sharma, D.S., Belabut, D. & Ahmad, N. (2009) New herpetofaunal records for Perlis State Park and adjacent areas. *Malayan Nature Journal* 61(4), 277–284.

Oraie, H., Rastegar-Pouyani, E., Khosravani, A., Moradi, N., Akbari, A., Sehhatisabet, M.E., Shafiei, S., Stümpel, N. & Joger, U. (2018) Molecular and morphological analyses have revealed a new species of blunt-nosed viper of the genus *Macrovipera* in Iran. *Salamandra* 54(4), 233–248, 10 figs.

Orcés, G. & Almendáriz, A. (1994) Presencia de *Rhinobotryum* [sic] *lentiginosum* (Scopoli, 1785) en el Ecuador. *Politecnica* 19(2), 155–163, 2 figs.

Orlov, N.L., Ryabov, S.A. & Nguyen, T.T. (2013) On the taxonomy and the distribution of snakes of the genus *Azemiops* Boulenger, 1888: Description of a new species. *Russian Journal of Herpetology* 20(2), 110–128, 32 figs.

Orlov, N.L., Sundukov, Y.N. & Kropachev, I.I. (2014) Distribution of pitvipers of "*Gloydius blomhoffii*" complex in Russia with the first records of *Gloydius blomhoffii blomhoffii* at Kunashir Island (Kuril Archipelago, Russian Far East). *Russian Journal of Herpetology* 21(3), 169–178, 22 figs.

Orozco-Terwengel, P., Nagy, Z.T., Vieites, D.R., Vences, M. & Louis, E. (2008) Phylogeography and phylogenetic relationships of Malagasy tree and ground boas. *Biological Journal of the Linnean Society* 95(3), 640–652, 4 figs.

Orriols, F.A. & Fernàndez, J.M.R. (2003) Distribució dels amfibis I rèptils del Principat d'Andorra. *Butlletí de la Societat Catalana d'Herpetologia* 16, 42–72, 10 figs.

Orriols, N.T. (2014) Aproximació a l'origen peninsular de les *Hemorrhois hippocrepis* recentment arribades a les Illes

Balears a partir de l'extracció I seqüenciació del gen citocrom B de diversos exemplars. *Butlletí de la Societat Catalana d'Herpetologia* 21, 174–183, 2 figs.

Ortiz, F.R., Freitas, H.S. de, Rodrigues, A.P., Abegg, A.D. & Franco, F.L. (2017) Snakes from the municipality of São José do Barreiro, State of São Paulo, *Brazil*. *Herpetology Notes* 10, 479–486, 2 figs.

Ortiz-Medina, J.A. (2017) Distribution notes: *Indotyphlops braminus. Mesoamerican Herpetology* 4(2), 480–481, 1 fig.

O'Shea, M., Allison, A. & Kaiser, H. (2018b) The taxonomic history of the enigmatic Papuan snake genus *Toxicocalamus* (Elapidae: Hydrophiinae), with the description of a new species from the Managalas Plateau of Oro Province, Papua New Guinea, with a revised dichotomous key. *Amphibia-Reptilia* 39(4), 403–433, 9 figs.

O'Shea, M., Herlihy, B., Paivu, B., Parker, F., Richards, S.J. & Kaiser, H. (2018a) Rediscovery of the rare Star Mountains Worm-eating Snake, *Toxicocalamus ernstmayri* (O'Shea et al., 2015) (Serpentes: Elapidae: Hydrophiinae) with the description of its coloration in life. *Amphibian & Reptile Conservation* 12(1), 27–33, 5 figs.

O'Shea, M. & Kaiser, H. (2016) The first female specimen of the poorly known Arfak Stout-tailed Snake, *Calamophis sharonbrooksae* (Murphy, 2012) (Serpentes: Colubroidea: Homalopsidae), from the Vogelkop Peninsula of Indonesian West New Guinea, with comments on the taxonomic history of primitive homalopsids. *Amphibian & Reptile Conservation* 10(2), 1010, 3 figs.

O'Shea, M. & Kaiser, H. (2018) Erroneous environs or aberrant activities? Reconciling unexpected collection localities for three New Guinea worm-eating snakes (*Toxicocalamus*, Serpentes, Elapidae) using historical accounts. *Herpetological Review* 49(2), 189–207, 12 figs.

O'Shea, M., Kusuma, K.I. & Kaiser, H. (2018c) First record of the Island Wolfsnake, *Lycodon capucinus* (H. Boie in F. Boie 1827), from New Guinea, with comments on its widespread distribution and confused taxonomy, and a new record for the Common Sun Skink, *Eutropis multifasciata* (Kuhl 1820). *IRCF Reptiles & Amphibians* 25(1), 70–84, 7 figs.

O'Shea, M., Parker, F. & Kaiser, H. (2015) A new species of New Guinea worm-eating snake, genus *Toxicocalamus* (Serpentes: Elapidae), from the Star Mountains of Western Province, Papua New Guinea, with a revised dichotomous key to the genus. *Bulletin of the Museum of Comparative Zoology* 161(6), 241–264, 9 figs.

Ospina-L., A.M. (2017) *Bothrops punctatus* (García 1896). *Catálogo de Anfibios y Reptiles de Colombia* 3(1), 25–30, 3 figs.

Ostrovskikh, S.V., Pestov, M.V. & Shaposhnikov, A.V. (2010) On a question of the Caspian Racer, Hierophis caspius (Gmelin, 1789), distribution in between of Volga and Ural Rivers. In: *Researches in Kazakhstan and Adjacent Countries*. Almati, 252–254, 2 figs. [in Russian].

Outeiral, A.B., Balestrin, R.L., Cappellari, L.H., Lema, T. de & Ferreira, V.L. (2018) Snake assemblage from Serra do Sudeste, *Pampas Biome, southern Brazil*. *Herpetology Notes* 11, 733–745, 5 figs.

Özdemir, A. & Baran, İ. (2002) Research on the herpetofauna of Murat Mountain (Kütahya-Uşak). *Turkish Journal of Zoology* 26(2), 189–195, 1 fig.

Padial, J.M. (2006) Commented distributional list of the reptiles of Mauritania (West Africa). *Graellsia* 62(2), 159–178, 1 fig.

Padial, J.M., Castroviejo-Fisher, S., Merchan, M., Cabot, J. & Castrioviejo, J. (2003) The herpetological collection from Bolivia in the "Estación Biológica de Doñana" (Spain). *Graellsia* 59(1), 5–13.

Padilla-Pérez, D.J., Murillo-Monsalve, J.D., Rincon-Barón, E.J. & Daza, J.M. (2015) Non-specialized caudal pseudoautotomy in the Emerald Racer snake *Drymobius rhombifer* (Günther, 1860). *Herpetology Notes* 8, 567–569, 1 fig.

Padrón, D.F., De Freitas, M. & Camargo, E. (2016) Primer reporte al norte del Río Orinoco para *Siphlophis cervinus* (Laurenti, 1768) (Serpentes: Dipsadidae: Xenodontinae) en la Península de Paria, Estado Sucre, Venezuela. *Saber* 28(1), 171–176, 4 figs.

Palacios-Aguilar, R. & Flores-Villela, O. (2018) An updated checklist of the herpetofauna from Guerrero, Mexico. *Zootaxa* 4422(1), 1–24, 3 figs.

Palacios-Aguilar, R., Santos-Bibiano, R. & Beltrán-Sánchez, E. (2016) Notable distributional records of amphibians and reptiles from Guerrero, Mexico. *Mesoamerican Herpetology* 3(2), 527–531, 3 figs.

Palacios-Aguilar, R., Santos-Bibiano, R., Smith, E.N. & Campbell, J.A. (2018) First records of the snake *Coniophanes lateritius* (Cope, 1862) (Squamata: Dipsadidae) from Guerrero, Mexico with notes on its natural history. *Herpetology Notes* 11, 651–653, 2 figs.

Palci, A. & Caldwell, M.W. (2014) The upper Cretaceous snake *Dinilysia patagonica* Smith-Woodward, 1901, and the crista circumfenestralis of snakes. *Journal of Morphology* 275, 1187–1200, 6 figs.

Palci, A., Caldwell, M.W. & Albino, A.M. (2013a) Emended diagnosis and phylogenetic relationships of the upper Cretaceous fossil snake *Najash rionegrina* Apesteguía and Zaher, 2006. *Journal of Vertebrate Paleontology* 33(1), 131–140, 5 figs.

Palci, A., Caldwell, M.W. & Nydam, R.L. (2013b) Reevaluation of the anatomy of the Cenomanian (Upper Cretaceous) hindlimbed marine fossil snakes *Pachyrhachis*, *Haasiophis*, and *Eupodophis*. *Journal of Vertebrate Paleontology* 33(6), 1328–1342, 8 figs.

Palci, A., Caldwell, M.W. & Scanlon, J.D. (2014) First report of a pelvic girdle in the fossil snake *Wonambi naracoortensis* Smith, 1976, and a revised diagnosis for the genus. *Journal of Vertebrate Paleontology* 34(4), 965–969, 1 fig.

Palot, M.J. (2015) A checklist of reptiles of Kerala, India. *Journal of Threatened Taxa* 7(13), 8010–8022, 12 figs.

Palot, M.J. & Radhakrishnan, C. (2010) First record of Yellow-bellied Sea Snake *Pelamis platurus* (Linnaeus, 1766) (Reptilia: Hydrophiidae) from a riverine tract in northern Kerala, India. *Journal of Threatened Taxa* 2(9), 1175–1176, 1 fig.

Pan, H., Chettri, B., Yang, D., Jiang, K., Wang, K., Zhang, L. & Vogel, G. (2013) A new species of the genus *Protobothrops* (Squamata: Viperidae) from southern Tibet, China and Sikkim, India. *Asian Herpetological Research* 4(2), 109–115, 7 figs.

Pančišin, L. & Klembara, J. (2003) [Amphibians and reptiles of the spring area of the Cirocha River in the National Park Poloniny]. *Folia Faunistica Slovaka* 8, 83–86, 1 fig. [In Slovak].

Pandey, D.P. (2012) Snakes in the vicinity of Chitwan National Park, Nepal. *Herpetological Conservation and Biology* 7(1), 46–57, 6 figs.

Pandey, D.P., Jelic, D., Sapkota, S., Lama, H.M., Lama, B., Pokharel, K., Goode, M & Kuch, U. (2018) New records of snakes from Chitwan National Park and vicinity, central Nepal. *Herpetology Notes* 11, 679–696, 9 figs.

Pantoja, D.L. & Fraga, R. de (2012) Herpetofauna of the Reserva Extrativista do Rio Gregório, Juruá Basin, southwest Amazonia, *Brazil*. *Check List* 8(3), 360–374, numerous ill.

Panzera, A., Guerrero, J.C. & Maneyro, R. (2017) Delimiting the geographic distribution of *Lygophis anomalus* (Günther, 1858) (Squamata, Dipsadidae) from natural history and ecological niche modeling. *South American Journal of Herpetology* 12(1), 24–33, 1 fig.

Panzera, A. & Maneyro, R. (2014) Feeding biology of *Lygophis anomalus* (Dipsadidae, Xenodontinae). *South American Journal of Herpetology* 9(2), 75–82, 3 figs.

Paradiz-Dominguez, M. (2016) Geographic distribution. *Indotyphlops braminus. Herpetological Review* 47(4), 630.

Park, J., Kim, I.-H., Fong, J.J., Koo, K-S., Choi, W.-J., Tsai, T.-S. & Park, D. (2017b) Northward dispersal of sea kraits (*Laticauda semifasciata*) beyond their typical range. *PLOS ONE* 12(0179871), 1–9, 1 fig.

Park, J., Koo, K.-S., Kim, I.-H., Choi, W.-J. & Park, D. (2017a) First record of the Blue-banded Sea Krait (*Laticauda laticaudata*, Reptilia: Squamata: Elapidae: Laticaudinae) on Jeju Island, South Korea. *Asian Herpetological Research* 8(2), 131–136, 3 figs.

Parker, J.M. & Brito, S. (2013) *Reptiles and Amphibians of the Mojave Desert – A Field Guide*. Snall Press, Las Vegas. 1–184, numerous illustrations.

Parmar, D.S. (2018) First record of a Yellow-bellied Sea Snake, *Hydrophis platurus* (Linnaeus 1766), from Gujarat, India. *IRCF Reptiles & Amphibians* 25(2), 137–138, 2 figs.

Parmar, D.S. (2019) First record of an Annulated Seasnake, *Hydrophis cyanocinctus* (Daudin 1803), from the Surat District, South Gujarat, India. *IRCF Reptiles & Amphibians* 26(1), 56–57, 2 figs.

Parmar, D.S. & Tank, S.K. (2019) Herpetofauna of Veer Narmad South Gujarat University, Surat, India. *IRCF Reptiles & Amphibians* 26(1), 21–34, 32 figs.

Parnazio, T. & Vrcibradic, D. (2018) Sexual dimorphism in two species of *Sibynomorphus* (Squamata, Dipsadidae) from Brazil. *Herpetology Notes* 11, 329–335.

Parra-Hernández, R.M., Zambrano, D.F. & Bernal, M.H. (2019) New record of *Tantilla alticola* (Boulenger, 1903) (Serpentes, Colubridae) for the Central Cordillera in the department of Tolima, Colombia. *Check List* 15(2), 485–488, 3 figs.

Passos, P., Azevedo, J.A.R., Nogueira, C.C., Fernandes, R. & Sawaya, R.J. (2019b) An integrated approach to delimit species in the puzzling *Atractus emmeli* complex (Serpentes: Dipsadidae). *Herpetological Monographs* 33, 1–25, 14 figs.

Passos, P., Echevarría, L.Y. & Venegas, P.J. (2013b) Morphological variation of *Atractus carrioni* (Serpentes: Dipsadidae). *South American Journal of Herpetology* 8(2), 109–120, 8 figs.

Passos, P., Junior, M.T., Recoder, R.S., Sena, M.A. de, Dal Vechio, F., Pinto, H.B. de A., Mendonça, S.H.S.T., Cassimiro, J. & Rodrigues, M.T. (2013a) A new species of *Atractus* (Serpentes: Dipsadidae) from Serra do Cipó, Espinhaço Range, southeastern Brazil, with proposition of a new species group to the genus. *Papéis Avulsos de Zoologia* 53(6), 75–85, 7 figs.

Passos, P., Kok, P.J.R., Albuquerque, N.R. de & Rivas, G.A. (2013c) Groundsnakes of the Lost World: A review of *Atractus* (Serpentes: Dipsadidae) from the Pantepui region, northern South America. *Herpetological Monographs* 27, 52–86, 21 figs.

Passos, P., Martins, A. & Pinto-Coelho, D. (2016b) Population morphological variation and natural history of *Atractus potschi* (serpents: Dipsadidae) in northeastern Brazil. *South American Journal of Herpetology* 11(3), 188–211, 10 figs.

Passos, P., Prudente, A.L.C. & Lynch, J.D. (2016a) Redescription of *Atractus punctiventris* and description of two new *Atractus* (Serpentes: Dipsadidae) from Brazilian Amazonia. *Herpetological Monographs* 30, 1–20, 14 figs.

Passos, P., Prudente, A.L.C. de, Ramos, L.O., Caicedo-Portilla, J.R. & Lynch, J.D. (2018a) Species delimitations in the *Atractus collaris* complex (Serpentes: Dipsadidae). *Zootaxa* 4392(3), 491–520, 13 figs.

Passos, P., Ramos, L.O., Fouquet, A. & Prudente, A.L.C. de (2017) Taxonomy, morphology, and distribution of *Atractus flammigerus* (Boie 1827) (Serpentes: Dipsadidae). *Herpetologica* 73(4), 349–363, 9 figs.

Passos, P., Ramos, L.O., Pinna, P.H. & Prudente, A.L.C. (2013d) Morphological variation and affinities of the poorly known snake *Atractus caxiuana* (Serpentes: Dipsadidae). *Zootaxa* 3745(1), 35–48, 6 figs.

Passos, P., Scanferla, A., Melo-Sampaio, P.R., Brito, J. & Almendariz, A. (2019a) A giant on the ground: Another large-bodied *Atractus* (Serpentes: Dipsadinae) from Ecuadorian Andes, with comments on the dietary specializations of the goo-eaters snakes. *Anais da Academia Brasileira de Ciéncias* 91(supl. 1)(e20170976), 1–14, 7 figs.

Passos, P., Sudré, V., Doria, G. & Campbell, P.D. (2018b) The taxonomic status of the "forgotten" Bolivian snakes, *Atractus balzani* Boulenger 1898 and *Atractus maculatus sensu* (Boulenger 1896) (Serpentes: Dipsadidae). *Zootaxa* 4438(1), 176–182, 3 figs.

Patel, H. & Vyas, R. (2019) Reptiles of Gujarat, India: Updated checklist, distribution, and conservation status. *Herpetology Notes* 12, 765–777, 8 figs.

Patel, H., Vyas, R. & Dudhatra, B. (2019) Might *Dendrelaphis caudolineatus* (Gray, 1834) (Squamata: Colubridae) present in India? *Zootaxa* 4571(2), 278–280, 1 fig.

Patel, H., Vyas, R. & Tank, S.K. (2015) On the distribution, taxonomy, and natural history of the Indian Smooth Snake, *Coronella brachyura* (Günther, 1866). *Amphibian & Reptile Conservation* 9(2), 120–125, 4 figs.

Paternina, R.F. & Capera-M., V.H. (2017) *Atractus crassicaudatus* (Duméril, Bibron y Duméril 1854). *Catálogo de Anfibios y Reptiles de Colombia* 3(2), 7–13, 4 figs.

Pauwels, O.S.G., Albert, J.-L., Arrowood, H., Mvele, C., Casanova, M., Dodane, J.-B., Morgan, J., Primault, L., Thepenier, L. & Fenner, J.N. (2017d) Miscellanea herpetological Gabonica X. *Bulletin of the Chicago Herpetological Society* 52(8), 133–138, 13 figs.

Pauwels, O.S.G., Carlino, P., Chirio, L. & Albert, J.-L. (2016a) Miscellanea herpetologica Gabonica IV. *Bulletin of the Chicago Herpetological Society* 51(5), 73–79, 7 figs.

Pauwels, O.S.G., Carlino, P., Chirio, L., Meunier, Q., Okouyi Okouyi, J.V., Orbell, C., Rousseaux, D. & Testa, O. (2017c) Miscellanea herpetological Gabonica IX. *Bulletin of the Chicago Herpetological Society* 52(6), 97–102, 18 figs.

Pauwels, O.S.G. & Chan-ard, T. (2016) Geographic distribution. *Calamaria pavimentata. Herpetological Review* 47(1), 82–83.

Pauwels, O.S.G., Chirio, L., Neil, E.J., Berry, S., Texier, N. & Rosin, C. (2017b) Miscellanea herpetological Gabonica VIII. *Bulletin of the Chicago Herpetological Society* 52(3), 41–46, 14 figs.

Pauwels, O.S.G., David, P. & Chan-ard, T. (2015) First confirmed record of the stream-dwelling snake *Amphiesma leucomystax* (Squamata: Natricidae) in Thailand. *Russian Journal of Herpetology* 22(2), 136–138, 2 figs.

Pauwels, O.S.G., Essono, T.B.B., Carlino, P., Chirio, L., Huijbregts, B., Leuteritz, T.E.J., Rousseaux, D., Tobi, E., Vigna, C. & Van neer, W. (2017a) Miscellanea herpetologica Gabonica VII. *Bulletin of the Chicago Herpetological Society* 52(1), 1–7, 11 figs.

Pauwels, O.S.G. & Grismer, L.L. (2016) First documented record of the long-tailed ringneck *Gongylosoma longicauda* (Squamata: Colubridae) in Thailand. *Russian Journal of Herpetology* 23(3), 239–242, 2 figs.

Pauwels, O.S.G., Larsen, H., Suthanthangjai, W., David, P. & Sumontha, M. (2017e) A new kukri snake (Colubridae: *Oligodon*) from Hua Hin District, and the first record of *O. deuvei* from Thailand. *Zootaxa* 4291(3), 531–548, 13 figs.

Pauwels, O.S.G., Le Garff, B., Ineich, I., Carlino, P., Melcore, I., Boundenga, L., Vigna, C. et al. (2016b) Miscellanea herpetologica Gabonica V & VI. *Bulletin of the Chicago Herpetological Society* 51(11), 177–185, 10 figs.

Pauwels, O.S.G. & Sumontha, M. (2016) Taxonomic identity of two enigmatic aquatic snake populations (Squamata: Homalopsidae: *Cerberus* and *Homalopsis*) from southern Thailand. *Zootaxa* 4107(2), 293–300, 6 figs.

Pavan, F.A., Mainardi, Á.A., Rocha, M.C. da, Cechin, S.Z. & Lipinski, V.M. (2018) New locality record and distribution extension for *Pseudoboa haasi* (Boettger, 1905) (Serpentes: Dipsadidae). *Herpetology Notes* 11, 1027–1028, 1 fig.

Pavlov, A.V., Svinin, A.O., Litvinchuk, S.N. & Zabiyakin, V.A. (2013) The annotated list of amphibians and reptiles being registered in the reserve in 2009–2012. In: *Scientific works of the State Natural Reserve 'The Great Kokshaga.'* Yoshkar-Ola, Vol. 6, 216–232, 1 fig.

Pavón-Vázquez, C.J., Canseco-Márquez, L. & Nieto-Montes de Oca, A. (2013) A new species in the *Geophis dubius* group (Squamata: Colubridae) from northern Puebla, México. *Herpetologica* 69(3), 358–370, 4 figs.

Pavón-Vázquez, C.J., García-Vázquez, U.O., Meza-Lázaro, R.N. & Nieto-Montes de Oca, A. (2015) First record of the coralsnake, *Micrurus nebularis* Roze 1989, from the state of Puebla, Mexico. *Mesoamerican Herpetology* 2(1), 131–133, 1 fig.

Pavón-Vázquez, C.J., Gray, L.N., White, B.A., García-Vázquez, U.O. & Harrison, A.S. (2016a) First records for Cozumel Island, Quintana Roo, Mexico: *Eleutherodactylus planirostris* (Anura: Eleutherodactylidae), *Trachycephalus typhonius* (Anura: Hylidae), and *Indotyphlops braminus* (Squamata: Typhlopidae). *Mesoamerican Herpetology* 3(2), 531–533, 1 fig.

Pavón-Vázquez, C.J., Maayan, I.P., White, B.A. & Harrison, A.S. (2016b) Three noteworthy herpetofaunal records from Belize. *Mesoamerican Herpetology* 3(3), 780–782, 1 fig.

Paysant, F. (1999) Nouvelles données sur l'herpétofaune de Castellorizzo (sud-est de l'archipel Egéen, Grèce). *Bulletin de la Société Herpétologique de France* 91, 5–12, 3 figs.

Peabotuwage, I., Bandara, I.N., Samarasinghe, D., Perera, N., Madawala, M., Amarasinghe, C., Kandambi, H.K.D. & Karunarathna, D.M.S.S. (2012) Range extension for *Duttaphrynus kotagamai* (Amphibia: Bufonidae) and a preliminary checklist of herpetofauna from the Uda Mäliboda Trail in Samanala Nature Reserve, Sri Lanka. *Amphibian and Reptile Conservation* 5(2), 52–64, 24 figs.

Pedrocchi, V. & Pedrocchi-Rius, C. (1994) Adición de *Coronella girondica* a la herpetofauna de las Islas Medes (Girona). *Boletín de la Asociación Herpetológia Española* 5, 16–17.

Pedrosa, I.M.M. de C., Costa, T.B., Faria, R.G., França, F.G.R., Laranjeiras, D.O., Oliveira, T.C.S.P. de, Palmeira, C.N.S. et al. (2014) Herpetofauna of protected areas in the Caatinga III: The Catimbau National Park, Pernambuco, Brazil. *Biota Neotropica* 14(0046), 1–12, 9 figs.

Pedroso-Santos, F., Sanches, P.R. & Costa-Campos, C.E. (2019) Anurans and reptiles of the Reserva Extrativista Beija-Flor Brilho de Fogo, Amapá state, eastern Amazon. *Herpetology Notes* 12, 799–807, 2 figs.

Peng, L., Lu, C., Huang, S., Guo, P. & Zhang, Y. (2014) A new species of the genus *Thermophis* (Serpentes: Colubridae) from Shangri-La, northern Yunnan, China, with a proposal for an eclectic rule for species delimitation. *Asian Herpetological Research* 5(4), 228–239, 5 figs.

Peng, L., Wang, L., Ding, L., Zhu, Y., Luo, J., Yang, D., Huang, R., Lu, S. & Huang, S. (2018) A new species of the genus

Sinomicrurus (Slowinski, Boundy and Lawson, 2001) (Squamata: Elapidae) from Hainan Province, China. *Asian Herpetological Research* 9(2), 65–73, 5 figs.

Penner, J., Gonwouo, L.N. & Rödel, M.-O. (2013) Second record of the West African hairy bush viper *Atheris hirsuta* (Ernst & Rödel, 2002) (Serpentes: Viperidae). *Zootaxa* 3694(2), 196–200, 1 fig.

Peralta-Fonseca, Z.A. & García-Padilla, E. (2015) Distribution notes: *Crotalus culminatus*. *Mesoamerican Herpetology* 2(2), 208.

Perez, C.H.F., Pérez, D.R. & Ávila, L.J. (2010) New records of *Leptotyphlops borrichianus* Degerbøl, 1923 in northwestern Patagonia. *Herpetology Notes* 3, 65–67, 2 figs.

Pérez de Ana, J.M. (2014) Nuevos datos de anfibios y reptiles en el País Vasco. *Munibe (Ciencias Naturales-Natur Zientziak)* 62, 135–144, 1 fig.

Pérez-Hernández, R.J., Raya-García, E. & Chaves, A. (2015) Geographic distribution. *Clelia scytalina*. *Herpetological Review* 46(1), 61.

Pérez-Z., J. & Lleellish, M. (2015) Reptiles terrestres de la isla San Lorenzo, Lima, Perú. *Revista Peruana de Biología* 22(1), 119–122, 3 figs.

Perry, G. (2012) On the appropriate names for snakes usually identified as *Coluber rhodorachis* (Jan, 1865) *or* why ecologists should approach the forest of taxonomy with great care. *IRCF Reptiles & Amphibians* 19(2), 90–100, 14 figs.

Persons, T. & Mays, J.D. (2017) Past and present distribution of the North American Racer (Coluber constrictor) in Maine, USA. *Herpetological Review* 48(1), 147–151.

Petrov, B.P. (2000) The herpetofauna (Amphibia and Reptilia) of the eastern Rhodopes (Bulgaria and Greece). In: Beron, P. & Popov, A. (Eds.), *Biodiversity of Bulgaria. 2. Biodiversity of Eastern Rhodopes (Bulgaria and Greece)*. Pensoft, Sofia, 863–879, 2 figs.

Pham, A.V., Nguyen, S.L.H. & Nguyen, T.Q. (2014) New records of snakes (Squamata: Serpentes) from Son La Province, Vietnam. *Herpetology Notes* 7:771–777, 8 figs.

Pham, A.V., Pham, C.T., Hoang, N.V., Ziegler, T. & Nguyen, T.Q. (2017) New records of amphibians and reptiles from Ha Giang Province, *Vietnam. Herpetology Notes* 10, 183–191, 8 figs.

Piccoli, A.P., De Lorenzis, A. & Fortuna, F. (2017) Osservazioni preliminari sui rettili dell'Oasi Lipu Castel di Guido (Lazio Settentrionale, Italia). *Naturalista Siciliano*, series 4, 41(2), 147–159, 2 figs.

Pierson, T., Stratmann, T., White, E.C., Clause, A.G., Carter, C., Herr, M.W., Jenkins, A.J., Vogel, H., Knoerr, M. & Folt, B. (2014) New county records of amphibians and reptiles resulting from a bioblitz competition in north-central Georgia, USA. *Herpetological Review* 45(2), 296–297.

Pierson, T.W. & Sollenberger, D. (2014) New county records of amphibians and reptiles resulting from a bioblitz competition in southwestern Georgia, USA. *Herpetological Review* 45(2), 295–296.

Pietersen, D.W., Pietersen, E.W. & Conradie, W. (2017) Preliminary herpetological survey of Ngonye Falls and surrounding regions in south-western Zambia. *Amphibian & Reptile Conservation* 11(1), 24–43, 3 figs.

Pinto, R.R. & Fernandes, R. (2017) Morphological variation of *Trilepida macrolepis* (Peters, 1857), with reappraisal of the taxonomic status of *Rena affinis* (Boulenger, 1884) (Serpentes: Leptotyphlopidae: Epictinae). *Zootaxa* 4244(2), 246–260, 4 figs.

Pinto, R.R., Franco, F.L. & Hoogmoed, M.S. (2018) *Stenostoma albifrons* (Wagler, 1824) (Squamata: Leptotyphlopidae): A name with two neotypes? *Salamandra* 54(4), 291–296, 1 fig.

Pinto-Erazo, M.A. & Medina-Rangel, G.F. (2018) First record of *Corallus blombergi* (Rendahl & Vestergren, 1941) (Serpentes: Boidae) from Colombia. *Check List* 14(1), 183–188, 3 figs.

Pinto-Viveros, M.A., Mano-Cuellar, K., Escalante, R.S., Villaroel, D. & Pinto-Ledezma, J.N. (2017) Historia Natural del Cerro Mutún: IV. La herpetofauna. *Kempffiana* 13(1), 106–118, 2 figs.

Pires, M.G., Feitosa, D.T., Prudente, A.L. da C. & Silva, N.J. da (2013) First record of *Micrurus diana* (Roze, 1983) (Serpentes: Elapidae) fror Brazil and extension of its distribution in Bolivia, with notes on morphological variation. *Check List* 9(6), 1556–1560, 3 figs.

Pires, M.G., Silva, N.J. da, Feitosa, D.T., Prudente, A.L. da C., Filho, G.A.P. & Zaher, M. (2014) A new species of triadal coral snake of the genus *Micrurus* (Wagler, 1824) (Serpentes: Elapidae) from northeastern Brazil. *Zootaxa* 3811(4), 569–584, 10 figs.

Platt, S.G., Zug, G.R., Platt, K., Ko, W.K., Myo, K.M., Soe, M.M., Lwin, T. et al. (2018) Field records of turtles, snakes and lizards in Myanmar (2009–2017) with natural history observations and notes on folk herpetological knowledge. *Natural History Bulletin of the Siam Society* 63(1), 67–114, 25 figs.

Pleguezuelos, J.M. (1989) Distribución de los reptiles en la provincial de Granada (SE. Península Ibérica). *Doñana* 16(1), 15–44, 4 figs.

Poiteu-Charentes Nature (Eds.). (2002) *Amphibiens et reptiles du Poiteu-Charentes – Atlas preliminaire.* Cahiers Techniques du Poiteu-Charentes, Poiteu-Charentes Nature, Poitiers. 1–112, numerous illustrations.

Pokrant, F., Kindler, C., Ivanov, M., Cheylan, M., Geniez, P., Böhme, W. & Fritz, U. (2016) Integrative taxonomy provides evidence for the species status of the Ibero-Maghrebian grass snake *Natrix astreptophora*. *Biological Journal of the Linnean Society* 118(4), 873–888, 8 figs.

Polović, L. & Čađenović, N. (2013) The herpetofauna of Krnovo (Montenegro). *Natura Montenegrina* 12(1), 109–115, 5 figs.

Popgeorglev, G.S., Tzankov, N.D., Kornilev, Y.V., Naumov, B.Y. & Stoyanov, A.Y. (2014) Amphibians and reptiles in Ponor Special Protection Area (Natura 2000), western Bulgaria: Species diversity, distribution and conservation. *Acta Zoologica Bulgarica, Supplementum* (5), 85–96, 4 figs.

Porras, L.W., Wilson, L.D., Schuett, G.W. & Reiserer, R.S. (2013) A taxonomic reevaluation and conservation assessment of the common cantil, *Agkistrodon bilineatus* (Squamata: Viperidae): A race against time. *Amphibian & Reptile Conservation* 7(1), 48–73, 16 figs.

Portik, D.M., Jongsma, G.F.M., Kouete, M.T., Scheinberg, L.A., Freiermuth, B., Tapondjou, W.P. & Blackburn, D.C. (2016) A survey of amphibians and reptiles in the foothills of Mount Kupe, Cameroon. *Amphibian & Reptile Conservation* 10(2), 37–67, 11 figs.

Portik, D.M., Mulungu, E.A., Sequeira, D. & McEntee, J.P. (2013) Herpetological surveys of the Serra Jeci and Namuli Massifs, Mozambique, and an annotated checklist of the southern Afromontane Archipelago. *Herpetological Review* 44(1), 394–406, 5 figs.

Portillo, F., Branch, W.R., Conradie, W., Rödel, M.-O., Penner, J., Barej, M.F., Kusamba, C. et al. (2018) Phylogeny and biogeography of the African burrowing snake subfamily Aparallactinae (Squamata: Lamprophiidae). *Molecular Phylogenetics & Evolution* 127, 288–303, 6 figs.

Portillo, F., Branch, W.R., Tilbury, C.R., Nagy, Z.T., Hughes, D.F., Kusamba, C., Muninga, W.M., Aristote, M.M., Behangana, M. & Greenbaum, E. (2019b) A cryptic new species of *Polemon* (Squamata: Lamprophiidae, Aparallactinae) from the Miombo woodlands of central and east Africa. *Copeia* 107(1), 22–35, 7 figs.

Portillo, F., Stanley, E.L., Branch, W.R., Conradie, W., Rödel, M.-O., Penner, J., Barej, M.F. et al. (2019a) Evolutionary history of burrowing asps (Lamprophiidae: Atractaspidinae) with emphasis on fang evolution and prey selection. *PLOS ONE* 14(4:e0214889), 1–32, 6 figs.

Poulakakis, N., Kapli, P., Kardamaki, A., Skourtanioti, E., Göcmen, B., Ilgaz, Ç., Kumlutaş, Y., Avci, A. & Limnerakis, P. (2013) Comparative phylogeography of six herpetofauna species in Cyprus: Late Miocene to Pleistocene colonization routes. *Biological Journal of the Linnean Society* 108(3), 619–635, 4 figs.

Pour, F.E., Rastegar-Pouyani, E. & Ghorbani, B. (2016) A preliminary study of the reptile's fauna in northwestern Yazd Province, Iran. *Russian Journal of Herpetology* 23(4), 243–248, 2 figs.

Pous, P. de, Simó-Riudalbas, M., Els, J., Jayasinghe, S., Amat, F. & Carranza, S. (2016) Phylogeny and biogeography of Arabian populations of the Persian horned viper *Pseudocerastes persicus* (Duméril, Bibron & Duméril, 1854). *Zoology in the Middle East* 62(3), 231–238, 2 figs.

Pous, P. de, Speybroeck, J., Bogaerts, S., Pasmans, F. & Beukema, W. (2012) A contribution to the atlas of the terrestrial herpetofauna of Sardinia. *Herpetology Notes* 5, 391–405, 11 figs.

Powell, R., Conant, R. & Collins, J.T. (2016) *Peterson field guide to reptiles and amphibians of eastern and central North America.* 4th edn. Houghton Mifflin Harcourt, Boston. xiii + 494, 207 figs., 47 plates.

Powell, R.L., Eversole, C.B., Crocker, A.V., Lizarro, D. & Bravo, R.C. (2016) *Xenopholis werdingorum,* (Jansen, Álvarez & Köhler, 2009) (Squamata: Dipsadidae): Range extension with comments on distribution. *Check List* 12(1985), 1–3, 2 figs.

Poyarkov, N.A., Nguyen, T.V. & Vogel, G. (2019) A new species of the genus *Liopeltis* Fitzinger, 1843 from Vietnam (Squamata: Colubridae). *Journal of Natural History* 53(27–28), 1647–1672, 8 figs.

Prairie, A., Chandler, K., Ruback, P. & Ray, J.M. (2015) Dumeril's Coralsnake (*Micrurus dumerilii* Jan, 1858) in Panama. *Mesoamerican Herpetology* 2(3), 253–259, 3 figs.

Premkumar, S. & Sharma, V. (2017) Geographic distribution. *Psammophis longifrons. Herpetological Review* 48(1), 131.

Price, M.S. & Dimler, T.M. (2015) New distributional records for the herpetofauna of Texas. *Herpetological Review* 46(4), 605–607.

Prigioni, C., Borteiro, C. & Kolenc, F. (2011) Amphibia and Reptilia, Quebrada de los Cuervos, Departamento de Treinta y Tres, Uruguay. *Check List* 7(6), 763–767, 3 figs.

Prigioni, C., Borteiro, C., Kolenc, F., Colina, M. & Gonzáles, E.M. (2013) Geographic distribution and apparent decline of *Crotalus durissus terrificus* (Laurenti 1768; Serpentes, Viperidae) in Uruguay. *Cuadernos de Herpetología* 27(2), 163–165, 1 fig.

Pritchard, A.C., McCartney, J.A., Krause, D.W. & Kley, N.J. (2014) New snakes from the upper Cretaceous (Maastrichtian) Maevarano Formation, Mahajanga Basin, Madagascar. *Journal of Vertebrate Paleontology* 34(5), 1080–1093, 4 figs.

Prudente, A.L. da C., Menks, A.C., Silva, F.M. da & Maschio, G.F. (2014) Diet and reproduction of the Western Indigo Snake *Drymarchon corais* (Serpentes: Colubridae) from the Brazilian Amazon. *Herpetology Notes* 7, 99–108, 3 figs.

Prudente, A.L. da C., Ramos, L.A. da C., Silva, T.M. da, Sarmento, J.F. de M., Dourado, A.C.M., Silva, F.M., Almeida, P.C.R. de, Santos, C.R.M. dos & Sousa, M.P.A. (2019) Dataset from the snakes (Serpentes, Reptiles) collection of the Museu Paraense Emílio Goeldi, Pará, Brazil. *Biodiversity Data Journal* 7(e34013), 1–12, 6 figs.

Prudente, A.L. da C., Silva, F.M. da, Meireles, M. dos S. & Puorto, G. (2017) Morphological variation in *Siphlophis worontzowi* (Squamata: Serpentes: Dipsadidae) from the Brazilian Amazon. *Salamandra* 53(2), 245–256, 6 figs.

Pulev, A.N., Domozetski, L.D., Sakelarieva, L.G. & Manolev, G.N. (2018a) Distribution of the Eurasian Blind Snake *Xerotyphlops vermicularis* (Merrem, 1820) (Reptilia: Typhlopidae) in south-western Bulgaria and its zoological significance. *Acta Zoologica Bulgarica*, (Supplement 12), 41–49, 2 figs.

Pulev, A.N., Naumov, B.Y., Sakelarieva, L.G., Manolev, G.N. & Domozetski, L.D. (2018b) Distribution and seasonal activity of Eastern Montpellier Snake *Malpolon insignitus* (Geoffroy Saint-Hilaire, 1827) (Reptilia: Psammophiidae) in south-western Bulgaria. *Acta Zoologica Bulgarica*, (Supplement 12), 51–58, 2 figs.

Pulev, A.N. & Sakelarieva, L. (2011) Serpentes (Reptilia) in the territory of the Blagoevgrad Municipality. *Faculty of Mathematics & Natural Science – FMNS* 2011, 618–626, 3 figs.

Purkayastha, J. (2018) Urban biodiversity: An insight into the terrestrial vertebrate diversity of Guwahati, India. *Journal of Threatened Taxa* 10(10), 12299–12316, 63 figs.

Purkayastha, J., Das, M., Sengupta, S. & Dutta, S.K. (2010) Notes on *Xenochrophis schnurrenbergeri* (Kramer, 1977) (Serpentes: Colubridae) fcrom Assam, India with some comments on its morphology and distribution. *Herpetology Notes* 3, 175–180, 5 figs.

Purkayastha, J., Das, M., Vogel, G., Bhattacharjee, P.C. & Sengupta, S. (2013) Comments on *Xenochrophis cerasogaster* (Cantor, 1839) (Serpentes: Natricidae) with remarks on its natural history and distribution. *Hamadryad* 36(2), 149–156, 6 figs.

Purkayastha, J. & David, P. (2019) A new species of the snake genus *Hebius* Thompson from northeast India (Squamata: Natricidae). *Zootaxa* 4555(1), 79–90, 4 figs.

Purkayastha, J., Kalita, J., Brahma, R.K., Doley, R. & Das, M. (2018) A review of the relationships of *Xenochrophis cerasogaster* (Cantor, 1839) (Serpentes: Colubridae) to its congeners. *Zootaxa* 4514(1), 126–136, 5 figs.

Pyron, R.A. (2017) Comment (Case 3688) – On the proposed suppression of *Charinidae* (Gray, 1849) (Reptilia, Squamata, Serpentes): A counter-application. *Bulletin of Zoological Nomenclature* 73(2–4), 124–126.

Pyron, R.A., Arteaga, A., Echevarría, L.Y. & Torres-Carvajal, O. (2016a) A revision and key for the tribe Diaphorolepidini (Serpentes: Dipsadidae) and checklist for the genus *Synophis*. *Zootaxa* 4171(2), 293–320, 4 figs.

Pyron, R.A., Burbrink, F.T. & Wiens, J.J. (2013) A phylogeny and revised classification of Squamata, including 4161 species of lizards and snakes. *BMC Evolutionary Biology* 13(93), 1–53, 28 figs.

Pyron, R.A., Ganesh, S.R., Sayyed, A., Sharma, V., Wallach, V. & Somaweera, R. (2016c) A catalogue and systematic overview of the shield-tailed snakes (Serpentes: Uropeltidae). *Zoosystema* 38(4), 453–506, 9 figs.

Pyron, R.A., Guayasamin, J.M., Peñafiel, N., Bustamante, L. & Arteaga, A. (2015) Systematics of Nothopsini (Serpentes, Dipsadidae), with a new species of *Synophis* from the Pacific Andean slopes of southwestern Ecuador. *ZooKeys* 541, 109–147, 9 figs.

Pyron, R.A., Hendry, C.R., Chou, V.M., Lemmon, E.M., Lemmon, A.R. & Burbrink, F.T. (2014a) Effectiveness of phylogenomic data and coalescent species-tree methods for resolving difficult nodes in the phylogeny of advanced snakes (Serpentes: Caenophidia). *Molecular Phylogenetics and Evolution* 81, 221–231, 4 figs.

Pyron, R.A., Hsieh, F.W., Lemmon, A.R., Lemmon, E.M. & Hendry, C.R. (2016b) Integrating phylogenomic and morphological data to assess candidate species-delimitation models in brown and red-bellied snakes (*Storeria*). *Zoological Journal of the Linnean Society* 177(4), 937–949, 2 figs.

Pyron, R.A., Reynolds, R.G. & Burbrink, F.T. (2014b) A taxonomic revision of Boas (Serpentes: Boidae). *Zootaxa* 3846(2), 249–260, 1 fig.

Pyron, R.A. & Somaweera, R. (2019) Further notes on the Sri Lankan uropeltid snakes *Rhinophis saffragamus* (Kelaart, 1853) and *Uropeltis ruhunae* Deraniyagala, 1954. *Zootaxa* 4560(3), 592–600, 5 figs.

Pyron, R.A. & Wallach, V. (2014) Systematics of the blindsnakes (Serpentes: Scolecophidia: Typhlopoidea) based on molecular and morphological evidence. *Zootaxa* 3829(1), 1–81, 3 figs.

Qing, N., Xiao, Z., Watkins-Colwell, G.J., Hou, M., Lu, W.-H., Lazell, J. & Sun, Z.-W. (2015) Additions to the reptile and amphibian fauna of Nan Ao Island: A Chinese treasure trove of biogeographic patterns. *Bulletin of the Peabody Museum of Natural History* 56(1), 107–124, 2 figs.

Quah, E.S.H., Anuar, S., Grismer, L.L., Wood, P.L. & Azizah, S. (2019b) Systematics and natural history of mountain reed snakes (genus *Macrocalamus*; Calamariinae). *Zoological Journal of the Linnean Society* 188(prepublication), 1–41, 18 figs.

Quah, E.S.H., Grismer, L.L., Jetten, T., Wood, P.L., Miralles, A., Shahrul Anuar M.S., Guek, K.H.P. & Brady, M.T. (2018b) The rediscovery of Schaefer's Spine-jawed Snake *Xenophidion schaeferi* (Günther & Manthey, 1995) (Serpentes, Xenophidiidae) from Peninsular Malaysia, with notes on its variation and the first record of the genus from Sumatra, Indonesia. *Zootaxa* 4441(2), 366–378, 5 figs.

Quah, E.S.H., Grismer, L.L., Lim, K.K.P., Anuar, M.S.S. & Imbun, P.Y. (2019a) A taxonomic reappraisal of the Smooth Slug Snake *Asthenodipsas laevis* (Boie, 1827) (Squamata: Pareidae) in Borneo with the description of two new species. *Zootaxa* 4646(3), 501–526, 8 figs.

Quah, E.S.H., Grismer, L.L., Wood, P.L., Thura, M.K., Zin, T., Kyaw, H., Lwin, N., Grismer, M.S. & Murdoch, M.L. (2017) A new species of Mud Snake *Serpentes, Homalopsidae, Gyiophis* (Murphy & Voris, 2014) from Myanmar with a first molecular assessment of the genus. *Zootaxa* 4238(4), 571–582, 5 figs.

Quah, E.S.H., Lim, K.K.P., Leong, E.H.H. & Shahrul Anuar M.S. (2018a) Identification and a new record from Penang Island of the rare red-bellied reed snake (*Calamaria albiventer*) (Gray, 1835) (Serpentes: Calamariinae). *Raffles Bulletin of Zoology* 66, 486–493, 3 figs.

Quah, E.S.H., Shahrul Anuar M.S., Grismer, L.L., Mohd Abdul Muin, M.A., Chan, K.G. & Grismer, J.L. (2011) Preliminary checklist of the herpetofauna of Jerejak Island, Penang, Malaysia. *Malayan Nature Journal* 63(3), 595–600.

Quah, E.S.H., Shahrul Anuar M.S., Mohd Abdul Muin, M.A., Rahman, N.A.A., Mustafa, F.S. & Grismer, L.L. (2013) Species diversity of herpetofauna of Bukit Panchor State Park, Penang, Peninsular Malaysia. *Malayan Nature Journal* 64(4), 193–211.

Quah, E.S.H., Wood, P.L., Grismer, L.L. & Shahrul Anuar M.S. (2018c) On the taxonomy and phylogeny of the rare Selangor Mud Snake (*Raclitia indica*) Gray (Serpentes, Homalopsidae) from Peninsular Malaysia. *Zootaxa* 4514(1), 53–64, 3 figs.

Questel, K. (2011) Alsophis rijgersmaei (Anguilla Bank Racer). Distribution. *Caribbean Herpetology* 23, 1, 1 fig.

Quick, J.S., Reinert, H.K., Cuba, E.R. & de Odum, R.A. (2005) Recent occurrence and dietary habits of *Boa constrictor* on

Aruba, Dutch West Indies. *Journal of Herpetology* 39, 304–307, 1 fig.

Quiñones, J., Burneo, K.G. & Barragan, C. (2014) Rediscovery of the Yellow-bellied Sea Snake, *Hydrophis platurus* (Linnaeus, 1766) in Máncora, northern Perú. *Check List* 10(6), 1563–1564, 2 figs.

Quiñones-Betancourt, E., Díaz-Ricaurte, J.C., Angarita-Sierra, T., Guevara-Molina, E.C. & Díaz-Morales, R.D. (2018) *Bothrops atrox* (Linnaeus, 1758). *Catálogo de anfibios y reptiles de Colombia* 4(3), 7–13, 7 figs.

Quintela, F.M., Loebmann, D. & Gianuca, N.M. (2006) Répteis continentais do Município de Rio Grande, Rio Grande do Sul, Brasil. *Biociências* 14(2), 180–188, 1 fig.

Quintela, F.M., Medvedovsky, I.G., Ibarra, C., Neves, L.F. de M. & Figueiredo, M.R.C. (2011) Reptiles recorded in Marinheiros Island, Patos Lagoon estuary, southern Brazil. *Herpetology Notes* 4, 57–62, 3 figs.

Quintero-A., D. & Shear, W.A. (2016) Case 3688 – CHARINIDAE (Gray 1849) (Reptilia, Squamata, Serpentes): Proposed suppression. *Bulletin of Zoological Nomenclature* 73(1), 25–29.

Quintero-Ángel, A., Osorio-Dominguez, D., Vargas-Salinas, F. & Saavedra-Rodríguez, C.A. (2012) Roadkill rate of snakes in a disturbed landscape of Central Andes of Colombia. *Herpetology Notes* 5, 99–105, 2 figs.

Quintero-Díaz, G.E. & Carbajal-Márquez, R.A. (2017a) Nature notes: *Oxybelis aeneus. Mesoamerican Herpetology* 4(1), 181–182.

Quintero-Díaz, G.E. & Carbajal-Márquez, R.A. (2017b) Nature notes: *Trimorphodon tau. Mesoamerican Herpetology* 4(1), 183–184, 1 fig.

Quintero-Díaz, G.E. & Carbajal-Márquez, R.A. (2017c) The Western Diamond-backed Rattlesnake, *Crotalus atrox* (Baird & Girard, 1853) (Squamata: Viperidae): A new state record for Aguascalientes, México. *Herpetology Notes* 10, 251–253, 2 figs.

Quintero-Díaz, G.E. & Carbajal-Márquez, R.A. (2019) New state record of *Lampropeltis mexicana* (Squamata: Colubridae) for Jalisco and second record for Aguascalientes, Mexico. *Herpetology Notes* 12, 995–998, 3 figs.

Quintero-Díaz, G.E., Cardona-Arceo, A. & Carbajal-Márquez, R.A. (2016) The Great Plains Ratsnake, *Pantherophis emoryi* (Baird & Girard, 1853), (Squamata: Colubridae), a new state record from Aguascalientes, México. *Check List* 12(1903), 1–16, 1 fig.

Quintero-Díaz, G.E., Rojas-Quezada, S. & Carbajal-Márquez, R.A. (2014a) Geographic distribution. *Lampropeltis polyzona. Herpetological Review* 45(4), 664.

Quintero-Díaz, G.E., Sigala-Rodríguez, J.J. & Carbajal-Márquez, R.A. (2014b) Geographic distribution. *Storeria storerioides. Herpetological Review* 45(4), 666.

Quinteros-Muñoz, O. (2013) Serpentes, Dipsadidae, *Atractus occipitoalbus*: Second record and distribution extension in Bolivia. *Check List* 9(1), 76–77, 3 figs.

Quinteros-Muñoz, O. (2015) A new prey item for the snake *Boiruna maculata* (Serpentes: Dipsadidae) in the yungas of Bolivia. *Phyllomedusa* 14(1), 79–81, 1 fig.

Quiroga, L. & Ferrer, D. (2016) *Phalotris cuyanus* (Cei, 1984) (Serpentes, Dipsadidae). Primer registro documentado para la Reserva de Biósfera Ñacuñán, Mendoza (Argentina). *Cuadernos de Herpetología* 30(1), 43–44, 2 figs.

Rage, J.-C. & Bailon, S. (2011) Amphibia and Squamata. In: Harrison, T. (Ed.), *Paleontology and Geology of Laetoli: Human Evolution in Context. Volume 2: Fossil Hominins and the Associated Fauna.* Springer, Dordrecht, 467–478, 4 figs.

Rage, J.-C., Métais, G., Bartolini, A., Brohi, I.A., Lashari, R.A., Marivaux, L., Merle, D. & Solangi, S.H. (2014) First report of the giant snake *Gigantophis* (Madtsoiidae) from the Paleocene of Pakistan: Paleobiogeographic implications. *Geobios* 47(3), 147–153, 3 figs.

Rage, J.-C., Vullo, R. & Néraudeau, D. (2016) The mid-Cretaceous snake *Simoliophis rochebrunei* (Sauvage, 1880) (Squamata: Ophidia) from its type area (Charentes, southwestern France): Redescription, distribution, and palaeoecology. *Cretaceous Research* 58, 234–253, 9 figs.

Raha, S., Das, S., Bag, P., Debnath, S. & Pramanick, K. (2018) Description of a new species of genus *Trachischium* with a redescription of *Trachischium fuscum* (Serpentes: Colubridae: Natricinae). *Zootaxa* 4370(5), 549–561, 5 figs.

Rahadian, R. & Das, I. (2013) A new record of *Pseudoxenodon inornatus* (Boie in: Boie, 1827) from Gunung Gedeh National Park, West Java, Indonesia (Squamata: Pseudoxenodontidae). *Hamadryad* 36(2), 174–177, 3 figs.

Rais, M., Baloch, S., Rehman, J., Anwar, M., Hussain, I. & Mahmood, T. (2012) Diversity and conservation of amphibians and reptiles in North Punjab, Pakistan. *Herpetological Bulletin* 122, 16–21, 4 figs.

Rajabizadeh, M., Adriaens, D., Kaboli, M., Sarafraz, J. & Ahmadi, M. (2015a) Dorsal colour pattern variation in Eurasian mountain vipers (genus *Montivipera*): A trade-off between thermoregulation and crypsis. *Zoologischer Anzeiger* 257, 1–9, 5 figs.

Rajabizadeh, M., Nagy, Z.T., Adriaens, D., Avci, A., Masroor, R., Schmidtler, J., Nazarov, R., Esmaeili, H.R. & Christiaens, J. (2015b) Alpine-Himalayan orogeny drove correlated morphological, molecular, and ecological diversification in the Persian dwarf snake (Squamata: Serpentes: *Eirenis persicus*). *Zoological Journal of the Linnean Society* 176(4), 878–913, 14 figs.

Rajabizadeh, M., Nilson, G., Kami, H.G. & Naderi, A.R. (2011) Distribution of the subgenus *Acridophaga* (Reuss, 1927) (Serpentes: Viperidae) in Iran. *Iranian Journal of Animal Biosystematics* 7(1), 83–87, 3 figs.

Rajabizadeh, M., Yazdanpanah, A. & Ursenbacher, S. (2012) Preliminary analysis of dorsal pattern variation and sexual dimorphism in *Montivipera latifi* (Mertens, Darevsky and Klemmer, 1967) (Ophidia: Viperidae). *Acta Herpetologica* 7(1), 13–21, 2 figs.

Ramalho, W.P., França, D.P.F., Guerra, V., Marciano, R., Vale, N.C. do & Silva, H.L.R. (2018b) Herpetofauna of Parque Estadual Altamiro de Moura Pacheco: One of the last remnants of seasonal forest in the core region of the Brazilian Cerrado. *Papéis Avulsos de Zoologia* 58(e20185851), 1–12, 6 figs.

Ramalho, W.P., Silva, J.R., Soares, P.T., Ferraz, D., Arruda, F.V. & Prado, V.H.M. (2018a) The anurans and squamates of a peri-urban cerrado remnant in the state of Goiás, central Brazil. *Herpetology Notes* 11, 573–583, 4 figs.

Ramírez-Bautista, A., Hernández-Salinas, U., García-Vázquez, U.O., Leyte-Manrique, A. & Canseco-Márquez, L. (2009) *Herpetofauna del Valle de México: Diversidad y conservación.* D.R. Comisión Nacional Para el Conocimiento y Uso de la La Biodiversidad, Tlalpan. xxiv+213, numerous illustrations, 69 maps.

Ramírez-Bautista, A., Wilson, L.D. & Berriozabal-Islas, C. (2014) Morphological variation in a population of *Tantilla calamarina* (Cope, 1866) (Squamata: Colubridae) from Guerrero, Mexico, and comments on fossoriality in the *calamarina* group and *Geagras redimitus. Herpetology Notes* 7, 797–805, 1 fig.

Ramírez-Chaves, H.E. & Solari, S. (2014) *Bothrops ayerbei* Folleco-Fernández, 2010 y *Bothrops rhombeatus* (García,

1896) (Serpentes: Viperidae) son un nombre no disponible y un *nomen dubium*, respectivamente. *Boletín Científico de Museo de Historia Natural* 18(1), 138–141.

Ramos, B., Gallardo, F.B. & Baldo, J.L. (2013) *Tantilla melanocephala* (Linnaeus, 1758) – (Serpentes: Colubridae). Primeros registros para la Provincia de Jujuy y confirmación de su prescencia en el noroeste Agentino. *Cuadernos de Herpetología* 27(1), 81–83, 2 figs.

Ramos, E. & Meza-Joya, F.L. (2018) Reptile road mortality in a fragmented landscape of the middle Magdalena Valley, Colombia. *Herpetology Notes* 11, 81–91, 4 figs.

Rangel-Patiño, C.A., Hernández-Gallegos, O., Ruiz-Gómez, M. de L. & Cristobal, J.M.C. (2015) Geographic distribution. *Coniophanes imperialis. Herpetological Review* 46(4), 572.

Rao, K.T., Ghate, H.V., Sudhakar, M., Javed, S.M.M. & Krishna, I.S.R. (2005) Fauna of protected areas 19 – Herpetofauna of Nallamalai Hills with eleven new records from the region including ten new records from Andhra Pradesh. *Zoos' Print Journal* 20(1), 1737–1740, 1 fig.

Rasmussen, A.R., Elmberg, J., Gravlund, P. & Ineich, I. (2011a) Sea snakes (Serpentes: subfamilies Hydrophiinae and Laticaudinae) in Vietnam: A comprehensive checklist and an updated identification key. *Zootaxa* 2894(1), 1–20, 12 figs.

Rasmussen, A.R., Ineich, I., Elmberg, J. & McCarthy, C. (2011b) Status of the Asiatic sea snakes of the *Hydrophis nigrocinctus* group (*H. nigrocinctus, H. hendersoni*, and *H. walli*; Elapidae, Hydrophiinae). *Amphibia-Reptilia* 32(4), 459–464, 2 figs.

Rasmussen, A.R., Sanders, K.L., Guinea, M.L. & Amey, A.P. (2014) Sea snakes in Australian waters (Serpentes: subfamilies Hydrophiinae and Laticaudinae)—a review with an updated identification key. *Zootaxa* 3869(4), 351–371, 17 figs.

Rassati, G. (2012) Contributo alla conoscenza della distribuzione di alcune specie di Amphibia e di Reptilia in Friuli Venezia Giulia e in Veneto. *Atti del Museo Civico di Storia Naturale di Trieste* 55, 91–135, numerous figs.

Rastegar-Pouyani, E., Ebrahimipour, F. & Hosseinian, S. (2017) Genetic variability and differentiation among the populations of dice snake, *Natrix tessellata* (Serpentes, Colubridae) in the Iranian Plateau. *Biochemical Systematics and Ecology* 72, 23–28, 2 figs.

Rastegar-Pouyani, E., Eskandarzadeh, N. & Darvish, J. (2014) Re-evaluation of the taxonomic status of sand boas of the genus *Eryx* (Daudin, 1803) (Serpentes: Boidae) in northeastern Iran using sequences of the mitochondrial genome. *Zoology in the Middle East* 60(4), 320–326, 2 figs.

Rastegar-Pouyani, E., Oraie, H., Khosravani, A., Kaboli, M., Mobaraki, A., Yousefi, M., Behrooz, R., Fakharmenesh, Z. & Wink, M. (2014a) A re-evaluation of taxonomic status of *Montivipera* (Squamata: Viperidae) from Iran using a DNA barcoding approach. *Biochemical Systematics and Ecology* 57, 350–356, 3 figs.

Rastegar-Pouyani, N., Afroosheh, M., Kami, H.G., Mashayekhi, M., Motesharei, A. & Rajabizadeh, M. (2010) New records of Typhlopidae (Reptilia: Ophidia) from the Iranian Plateau. In: DiTizio, L., DiCerbo, A.R., Di Francesco, N. & Camel, A. (Eds.), *Atti VIII Congresso Nazionale Societas Herpetologica Italica*. Ianieri Edizioni, Pescara, 171–175, 2 figs.

Rastegar-Pouyani, N., Kami, H.G., Rajabizadeh, M., Shafiei, S. & Anderson, S.C. (2008) Annotated checklist of amphibians and reptiles of Iran. *Iranian Journal of Animal Biosystematics* 4(1), 7–30.

Ratnarathorn, N., Harnyuttanakorn, P., Chanhome, L., Evans, S.E. & Day, J.J. (2019) Geographical differentiation and cryptic diversity in the monacled cobra, *Naja kaouthia* (Elapidae), from Thailand. *Zoologica Scripta* 48(6), 711–726, 3 figs.

Ratnikov, V.Y. (2002) Muchkapian (early Neopleistocene) amphibians and reptiles of the east-European Plain. *Russian Journal of Herpetology* 9(3), 229–236, 5 figs.

Ratnikov, V.Y. (2005) Likhvinian (middle Neopleistocene) amphibians and reptiles of the east-European Plain. *Russian Journal of Herpetology* 12(1), 7–12, 4 figs.

Rato, C., Brito, J.C., Carretero, M.A., Larbes, S., Shacham, B. & Harris, D.J. (2007) Phylogeography and genetic diversity of *Psammophis schokari* (Serpentes) in North Africa based on mitochondrial DNA sequences. *African Zoology* 42(1), 112–117, 2 figs.

Rato, C., Silva-Rocha, I., González-Miras, E., Rodríguez-Luque, F., Fariña, B. & Carretero, M.A. (2015) A molecular assessment of European populations of *Indotyphlops braminus* (Daudin, 1803). *Herpetozoa* 27(3/4), 179–182.

Rautsaw, R.M., Holding, M.L., Strickland, J.L., Gaytán, J.J.C., González, F.C.G., Gaytán, J.G.C., Jiménez, J.M.B. & Parkinson, C.L. (2018) Geographic distribution: *Hypsiglena tanzeri. Herpetological Review* 49(2), 287.

Ray, J.M. (2015) Geographic distribution. *Ungaliophis panamensis. Herpetological Review* 46(1), 64.

Ray, J.M., DeCero, K., Ruback, P., Wedow, J.D. & Knight, J.L. (2013) Geographic distribution notes on *Trimetopon barbouri* Dunn 1930 from western Panama. *Check List* 9(6), 1573–1575, 2 figs.

Ray, J.M. & P. Ruback (2015) Updated checklists of snakes for the provinces of Panamá and Panamá Oeste, *Republic of Panama. Mesoamerican Herpetology* 2(2), 168–188, 30 figs.

Ray, J.M. & Santana, P. (2014) Geographic distribution. *Urotheca fulviceps. Herpetological Review* 45(1), 97.

Reaño-Hernández, I., Wilson, L.D., Ramírez-Bautista, A., Cruz-Elizalde, R. & Hernández-Melo, J.A. (2015) Distribution notes: *Masticophis flagellum. Mesoamerican Herpetology* 2(3), 357–358, 2 figs.

Reeder, T.W., Townsend, T.M., Mulcahy, D.G., Noonan, B.P., Wood, P.L., Sites, J.W. & Wiens, J.J. (2015) Integrated analyses resolve conflicts over squamate reptile phylogeny and reveal unexpected placements for fossil taxa. *PLoS One* 2015 (0118199), 1–22, 1 fig.

Ren, J.-L., Wang, K., Guo, P., Wang, Y.-Y., Nguyen, T.T. & Li, J.-T. (2019) On the generic taxonomy of *Opisthotropis balteata* (Cope, 1895) (Squamata: Colubridae: Natricinae): Taxonomic revision of two natricine genera. *Asian Herpetological Research* 10(2), 105–128, 7 figs.

Ren, J.-L., Wang, K., Jiang, K., Guo, P. & Li, J.-T. (2017) A new species of the Southeast Asian genus *Opisthotropis* (Serpentes: Colubridae: Natricinae) from western Hunan, China. *Zoological Research* 38(5), 251–263, 7 figs.

Ren, J.-L., Wang, K., Nguyen, T.T., Hoang, C.V., Zhong, G.-H., Jiang, K., Guo, P. & Li, J.-T. (2018) Taxonomic re-evaluation of the monotypic genus *Pararhabdophis* (Bourret, 1934) (Squamata: Colubridae: Natricinae) with discovery of its type species, *P. chapaensis*, from China. *Zootaxa* 4486(1), 31–56, 9 figs.

Restrepo, A., Molina-Zuluaga, C., Hurtado, J.P., Marín, C.M. & Daza, J.M. (2017) Amphibians and reptiles from two localities in the northern Andes of Colombia. *Check List* 13(4), 203–237, 14 figs.

Reyes-Vera, A.M., Torres-Ángeles, F. & Iturbe-Morgado, J.C. (2017) Distribution notes: *Rena dulcis. Mesoamerican Herpetology* 4(1), 200–201.

Reynolds, R.G. & Henderson, R.W. (2018) Boas of the world (Superfamily Booidae [sic]): A checklist with systematic, taxonomic, and conservation assessments. *Bulletin of the Museum of Comparative Zoology* 162(1), 1–58, 16 figs.

Reynolds, R.G. & Niemiller, M.L. (2010a) *Typhlops platycephalus* (Puerto Rican White-tailed Blindsnake). Distribution. *Caribbean Herpetology* 15, 1.

Reynolds, R.G. & Niemiller, M.L. (2010b) *Epicrates chryso-gaster* (Southern Bahamas Boa). Distribution. *Caribbean Herpetology* 14, 1.

Reynolds, R.G., Niemiller, M.L., Hedges, S.B., Dornburg, A., Puente-Rolón, A.R. & Revell, L.J. (2013) Molecular phylogeny and historical biogeography of West Indian boid snakes (*Chilabothrus*). *Molecular Phylogenetics and Evolution* 68, 461–470, 4 figs.

Reynolds, R.G., Niemiller, M.L. & Revell, L.J. (2014b) Toward a tree-of-life for the boas and the pythons: Multilocus species-level phylogeny with unprecedented taxon sampling. *Molecular Phylogenetics and Evolution* 71, 201–213, 3 figs.

Reynolds, R.G. & Puente-Rólon, A.R. (2016) Geographic distribution. *Chilabothrus strigilatus. Herpetological Review* 47(3), 425.

Reynolds, R.G., Puente-Rolón, A.R., Barandiaran, M. & Revell, L.J. (2014a) Hispaniolan Boa (*Chilabothrus striatus*) on Vieques Island, Puerto Rico. *Herpetology Notes* 7, 121–122, 1 fig.

Reynolds, R.G., Puente-Rolón, A.R., Burgess, J.P. & Baker, B.O. (2018a) Rediscovery and redescription of the Crooked-Acklins Boa, Chilabothrus schwartzi (Buden, 1975), comb. nov. *Breviora* (558), 1–16, 6 figs.

Reynolds, R.G., Puente-Rolón, A.R., Castle, A.L., Van De Schoot, M. & Geneva, A.J. (2018b) Herpetofauna of Cay Sal Bank, Bahamas and phylogenetic relationships of Anolis fairchildi, Anolis sagrei, and Tropidophis curtus from the region. *Breviora* (560), 1–19, 8 figs.

Reynolds, R.G., Puente-Rolón, A.R., Geneva, A.J., Aviles-Rodriguez, K.J. & Herrmann, N.C. (2016) Discovery of a remarkable new boa from the Conception Island Bank, Bahamas. *Breviora* (549), 1–19, 9 figs.

Reynosa, V.H., González-Hernández, A.J. & Cruz-Silva, J.A. (2014) Geographic distribution. *Enulius oligostichus. Herpetological Review* 45(2), 285.

Rezaie-Atagholipour, M., Ghezellou, P., Hesni, M.A., Dakhteh, S.M.H., Almadian, H. & Vidal, N. (2016) Sea snakes (Elapidae, Hydrophiinae) in their westernmost extent: An updated and illustrated checklist and key to the species in the Persian Gulf and Gulf of Oman. *ZooKeys* 622, 129–164, 26 figs.

Rhodin, A.G.J. & 69 co-authors (2015) Comment on *Spracklandus* (Hoser, 2009) (Reptilia, Serpentes, Elapidae): Request for confirmation of the availability of the generic name and for the nomenclatural validation of the journal in which it was published. *Bulletin of Zoological Nomenclature* 72(1), 65–78.

Rio, J.P. & Mannion, P.D. (2017) The osteology of the giant snake *Gigantophis garstini* from the upper Eocene of North Africa and its bearing on the phylogenetic relationships and biogeography of Madtsoiidae. *Journal of Vertebrate Paleontology* 34(4)(e1347179), 1–20, 12 figs.

Rios-Soto, J.A., Arango-Lozano, J. & Rivera-Molina, F.A. (2018) *Micrurus mipartitus* (Duméril, Bibron y Duméril, 1854). *Catálogo de Anfibios y Reptiles de Colombia* 4(1), 37–44, 4 figs.

Rivallin, P., Barrioz, M., Massary, J.-C. de & Lescure, J. (2017) Présence de la couleuvre verte et jaune, *Hierophis viridiflavus* (Lacepède, 1789) (Squamata: Colubridae), en île-de-France et an Normandie: des données nouvelles. *Bulletin de la Société Herpetogique de France* 161, 75–84, 4 figs.

Rivas, L.R., Mendoza-Miranda, P. & Saravia, A.M. (2018) Lista preliminar de la herpetofauna del Parque Nacional Torotoro, Potosí, *Bolivia. Cuadernos de Herpetología* 32(1), 41–46, 2 figs.

Roberto, I.J. & Loebmann, D. (2016) Composition, distribution patterns, and conservation priority areas for the herpetofauna of the state of Ceará, northeastern Brazil. *Salamandra* 52(2), 134–152.

Roberto, I.J., Oliveira, C.R. de, Filho, J.A. de A. & Ávila, R.W. (2014) *Dipsas sazimai* (Fernandes, Marques & Argolo, 2010) (Squamata: Dipsadidae): Distribution extension and new state record. *Check List* 10(1), 209–210, 2 figs.

Roberto, I.J., Oliveira, C.R. de, Filho, J.A. de A., Oliveira, H.F. de & Ávila, R.W. (2017) The herpetofauna of the Serra do Urubu mountain range: A key biodiversity area for conservation in the Brazilian Atlantic Forest. *Papéis Avulsos de Zoologia* 57(27), 347–373, 15 figs.

Roberts, S.H. & Daly, C. (2014) A rapid herpetofaunal assessment of Nosy Komba Island, northwestern Madagascar, with new locality records for seventeen species. *Salamandra* 50(1), 18–26, 3 figs.

Rocha, A., Mata-Silva, V., García-Padilla, E., DeSantis, D.E. & Wilson, L.D. (2016) Third known specimen and first locality record in Oaxaca, Mexico, for *Tantilla sertula* (Wilson & Campbell, 2000) (Squamata: Colubridae). *Mesoamerican Herpetology* 3(3), 771–774, 4 figs.

Rocha, C.F.D., Telles, F.B. da S., Vrcibradic, D. & Nogueira-Costa, P. (2018) The herpetofauna from Ilha Grande (Angra dos Reis, Rio de Janeiro, Brazil): Updating species composition, richness, distribution and endemisms. *Papéis Avulsos de Zoologia* 58(e20185825), 1–12, 4 figs.

Rocha, S., Harris, D.J., Perera, A., Silva, A., Vasconcelos, R. & Carretero, M.A. (2009) Recent data on the distribution of lizards and snakes of the Seychelles. *The Herpetological Bulletin* 110, 20–32, 18 figs.

Rödel, M.-O. & Glos, J. (2019) Herpetological surveys in two proposed protected areas in Liberia, West Africa. *Zoosystematics and Evolution* 95(1), 15–35, 7 figs.

Rödel, M.-O., Kucharzewski, C., Mahlow, K., Chirio, L., Pauwels, O.S.G., Carlino, P., Sambolah, G. & Glos, J. (2019) A new stiletto snake (Lamprophiidae, Atractaspidinae, *Atractaspis*) from Liberia and Guinea, West Africa. *Zoosystematics and Evolution* 95(1), 107–123, 8 figs.

Rodrigues, G.M., Maschio, G.F. & Prudente, A.L. da C. (2015) Snake assemblages of Marajó Island, Pará state, Brazil. *Zoologia* 33(1), 1–13, 7 figs.

Rodrigues, R., Albuquerque, R.L. de, Santana, D.J., Laranjeiras, D.O., Protázio, A.S., Rodrigues França, F.G. & Mesquita, D.O. (2013) Record of the occurrence of *Lachesis muta* (Serpentes, Viperidae) in an Atlantic Forest fragment in Paraíba, Brazil, with comments on the species' preservation status. *Revista Biotemas* 26(2), 283–286, 2 figs.

Rodríguez-Canseco, J.M., González-Estupiñán, K.L. & López-Rodríguez, L.E. (2013) Geographic distribution. *Drymarchon melanurus. Herpetological Review* 44(3), 476.

Rodríguez-Robles, J.A., Jezkova, T., Fujita, M.K., Tolson, P.J. & García, M.A. (2015) Genetic divergence and diversity in the Mona and Virgin Islands Boas, *Chilabothrus monensis* (*Epicrates monensis*) (Serpentes: Boidae), West Indian snakes of special conservation concern. *Molecular Phylogenetics and Evolution* 88, 144–153, 4 figs.

Rodríguez-Rodríguez, E.J., Carmona-González, R. & García-Cardenete, L. (2018) Actualización de la distribución de los reptiles en la provincial de Sevilla. *Boletín de la Asociación Herpetológica de Española* 29(2), 46–53, 2 figs.

Roelke, C. & Smith, E.N. (2010) Herpetofauna, Parc National des Volcans, North Province, Republic of Rwanda. *Check List* 6(4), 525–531, 12 figs.

Rojas-Morales, J.A., Cabrera-Vargas, F.A. & Ruiz-Valdarrama, D.H. (2018a) *Ninia hudsoni* (Serpentes: Dipsadidae) as prey of the coral snake *Micrurus hemprichii ortonii* (Serpentes: Elapidae) in northwestern Amazonia. *Boletín Científico Museo de Historia Natural* 22(1), 102–105, 1 fig.

Rojas-Morales, J.A., González-Durán, G.A. & Basto-Riascos, M.C. (2017) *Atractus manizalesensis* Prado 1940. *Catálogo de Anfibios y Reptiles de Colombia* 3(1), 37–42, 4 figs.

Rojas-Morales, J.A., Marín-Martínez, M. & Zuluaga-Isaza, J.C. (2018b) Aspectos taxonómicos y ecogeográficos de algunas serpientes (Reptilia: Colubridae) del área de influencia de la Central Hidroeléctrica Miel I, Caldas, Colombia. *Biota Colombiana* 19(2), 73–91, 9 figs.

Rojas-Murcia, L.E., Carvajal-Cogollo, J.E. & Cabrejo-Bello, J.A. (2016) Reptiles del bosque seco estacional en el Caribe Colombiano: Distribución de los habitats y del recurso alimentario. *Acta Biológica Colombiana* 21(2), 365–377, 1 fig.

Rooijen, J. van & Vogel, G. (2016) On the status of three nominal species in the synonymy of *Dendrelaphis calligaster* (Günther, 1867) (Serpentes: Colubridae). *Zootaxa* 4093(2), 293–300, 5 figs.

Rooijen, J. van, Vogel, G. & Somaweera, R. (2015) A revised taxonomy of the Australo-Papuan species of the colubrid genus *Dendrelaphis* (Serpentes: Colubridae). *Salamandra* 51(1), 33–56, 13 figs.

Rorabaugh, J.C. & Lemos-Espinal, J.A. (2016) *A field guide to the amphibians and reptiles of Sonora, Mexico.* ECO Herpetological Publishing and Distribution, Rodeo, 1–688, numerous color photos.

Rorabaugh, J.C., Turner, D., Van Devender, T.R., Hugo-Cabrera, V., Maynard, R.J., Van Devender, R.W., Villa, R.A. et al. (2019) Herpetofauna of the Mesa Tres Ríos area in the northern Sierra Madre Occidental of Sonora, Mexico. *Herpetological Review* 50(2), 251–259, 4 figs.

Rosa, G.M., Mercurio, V., Crottini, A. & Andreone, F. (2010) Predation of the snake *Leioheterodon modestus* (Günther, 1863) upon the rainbow frog *Scaphiophryne gottlebei* Busse & Böhme, 1992 at Isalo, southern Madagascar. *Herpetology Notes* 3, 259–261, 1 fig.

Rosa, G.M., Noël, J. & Andreone, F. (2012) Updated distribution map and additional record for the cryptic leaf-nosed snake, *Langaha madagascariensis* (Bonnaterre, 1790) (Serpentes: Lamprophiidae) from Madagascar. *Herpetology Notes* 5, 435–436, 2 figs.

Rosado, D., Brito, J.C. & Harris, D.J. (2015) Molecular screening of Hepatozoon (Apicomplexa: Adeleorina) infections in *Python sebae* from West Africa using 18S rRNA gene sequences. *Herpetology Notes* 8, 461–463, 1 fig.

Rosado, D., Harris, D.J., Perera, A., Jorge, F., Tomé, B., Damas-Moreira, I., Tavares, I. et al. (2016) Moroccan herpetofauna distribution updates including a DNA barcoding approach. *Herpetozoa* 28(3/4), 171–178, 2 figs.

Roth-Monzón, A.J., Mendoza-Hernández, A.A. & Flores-Villela, O. (2018) Amphibian and reptile biodiversity in the semi-arid region of the municipality of Nopala de Villagrán, Hidalgo, Mexico. *PeerJ* 6(e4202), 1–21, 5 figs.

Rouag, R. & S. Benyacoub (2006) Inventaire et écologie des reptiles du Parc National d'El Kala (Algérie). *Bulletin de la Société Herpétologique de France* 117, 25–40, 1 fig.

Roussos, S.A. (2013) Geographic distribution. *Vipera ammodytes*. *Herpetological Review* 44(4), 631.

Roussos, S.A. (2016) Geographic distribution. *Eryx jaculus*. *Herpetological Review* 47(2), 262.

Roux, P. & Slimani, T. (1992) Nouvelles données sur la repartition et l'écologie des reptiles du Maroc (la region de Marrakech : Haouz et Jebilet). *Bulletin de l'Institut Scientifique, Rabat* 16, 122–131, 3 figs.

Row, J.R., Brooks, R.J., Mackinnon, C.A., Lawson, A., Crother, B.I., White, M. & Lougheed, S.C. (2011) Approximate Bayesian computation reveals the factors that influence genetic diversity and population structure of foxsnakes. *Journal of Evolutionary Biology* 24, 2364–2377, 4 figs.

Rowntree, N., Griffiths, J. & Rowntree, P. (2017) Geographic distribution. *Xenophidion acanthognathus*. *Herpetological Review* 48(1), 132–133.

Ruane, S., Bryson, R.W., Pyron, R.A. & Burbrink, F.T. (2014) Coalescent species delimitation in Milksnakes (genus *Lampropeltis*) and impacts on phylogenetic comparative analyses. *Systematic Biology* 63(2), 231–250, 4 figs.

Ruane, S., Burbrink, F.T., Randriamahatantsoa, B. & Raxworthy, C.J. (2016) The cat-eyed snakes of Madagascar: Phylogenyand description of a new species of *Madagascarophis* (Serpentes: Lamprophiidae) from the Tsingy of Ankarana. *Copeia* 2016(3), 712–721, 5 figs.

Ruane, S., Myers, E.A., Lo, K., Yuen, S., Welt, R.S., Juman, M., Futterman, I. et al. (2018a) Unrecognized species diversity and new insights into colour pattern polymorphism within the widespread Malagasy snake *Mimophis* (Serpentes: Lamprophiidae). *Systematics and Biodiversity* 16(3), 229–244, 4 figs.

Ruane, S., Raxworthy, C.J., Lemmon, A.R., Lemmon, E.M. & Burbrink, F.T. (2015) Comparing species tree estimation with large anchored phylogenomic and small Sanger-sequenced molecular datasets: An empirical study on Malagasy pseudoxyrhophiine snakes. *BMC Evolutionary Biology* 15(221), 1–14, 3 figs.

Ruane, S., Richards, S.J., McVay, J.D., Tjaturadi, B., Krey, K. & Austin, C.C. (2018b) Cryptic and non-cryptic diversity in New Guinea ground snakes of the genus *Stegonotus* Duméril, Bibron and Duméril, 1854: A description of four new species (Squamata: Colubridae). *Journal of Natural History* 52(13–16), 917–944, 8 figs.

Rubio, M. & Keyler, D.E. (2013) *Venomous Snakebite in the Western United States.* ECO Herpetological Publishing & Distribution, Rodeo, 1–170, many figs.

Ruchin, A.B. & Rhyzhov, M.K. (2006) *Amphibians and Reptiles of Mordovia: Species Diversity, Distribution, and Numbers.* Mordovian State University, Saransk, 1–159, 42 figs. [in Russian].

Rueda-Solano, L.A. & Castellanos-Barliza, J. (2010) Herpetofauna de Neguanje, Parque Nacional Natural Tayrona, Caribe Colombiano. *Acta Biológica Colombiana* 15(1), 195–206, 2 figs.

Ryabov, S.A. & Orlov, N.L. (2010) Reproductive biology of *Boiga guangxiensis* (Wen, 1998) (Serpents: Colubridae). *Asian Herpetological Research* 1(1), 44–47, 4 figs.

Ryan, M.J., Latella, I.M., Willink, B., García-Rodríguez, A. & Gilman, C.A. (2015) Notes on the breeding habits and new distribution records of seven species of snakes from southwest Costa Rica. *Herpetology Notes* 8, 669–671.

Sabbaghzadeh, A. & Mashayekhi, M. (2015) Survey of reptiles fauna of Nazmabad of Arak, Markazi Province, Iran. *African Journal of Basic & Applied Sciences* 7(2), 101–108, 2 figs.

Sadeghi, N., Hosseinian-Yousefkhani, S.S., Rastegar-Pouyani, N. & Rajabizadeh, M. (2014b) Skull comparison between *Eirenis collaris* and *Dolichophis jugularis* (Serpentes: Colubridae) from Iran. *Iranian Journal of Animal Biosystematics* 10(2), 87–100, 7 figs.

Sadeghi, N., Rajaizadeh, M., Rastegar-Pouyani, N. & Hosseinian-Yousefkhani, S.S. (2014a) Updated distribution of *Eirenis collaris* (Ménétriés, 1832) (Serpentes: Colubridae) in Iran. *Herpetological Notes* 7, 245–246, 2 figs.

Saenger, M.V. & Ray, J.M. (2016) Geographic distribution. *Chironius flavopictus*. *Herpetological Review* 47(1), 83.

Saenz, D., Podlipny, H.V., Tasi, P.-Y., Burt, D.B. & Yuan, H.-W. (2009) A survey of reptiles and amphibians on Kinmen Island, Taiwan. *The Herpetological Bulletin* 108, 3–9, 6 figs.

Safaei-Mahroo, B. & Ghaffari, H. (2015) New data on presence of the smooth snake *Coronella austriaca* (Laurenti, 1768) (Serpentes: Colubridae) in Iran with notes on habitat. *Herpetology Notes* 8, 235–238, 3 figs.

Safaei-Mahroo, B., Ghaffari, H., Fahimi, H., Broomand, S., Yazdanian, M., Najafi-Majd, E., Hosseinian-Yousefkhani, S.S. et al. (2015) The herpetofauna of Iran: Checklist of taxonomy, distribution and conservation status. *Asian Herpetological Research* 6(4), 257–290, 5 figs.

Safaei-Mahroo, B., Ghaffari, H., Salmabadi, S., Kamangar, A., Almasi, S., Kazemi, S.M. & Ghafoor, A. (2017) Eastern Montpellier Snake (*Malpolon insignitus fuscus*) ophiophagy behavior from Zagros Mountains. *Russian Journal of Herpetology* 24(1), 69–71, 2 figs.

Safdarian, P., Todehdehghan, F., Hojati, V. & Shiravi, A. (2016) Seasonal changes in the testicular activity of the Iranian Mountain Viper, *Montivipera albicornuta* (Nilson & Andrén, 1985) (Reptilia: Viperdae). *Zoology in the Middle East* 62(1), 39–45, 5 figs.

Sagadevan, J., Ganesh, S.R., Anandan, N. & Rajasingh, R. (2019) Recent records of the Banded Racer *Argyrogena fasciolata* (Shaw, 1802) (Reptilia: Squamata: Colubridae) from southern Coromandel Coast, peninsular India. *Journal of Threatened Taxa* 11(5), 13567–13572, 2 figs.

Sahlean, T.C., Gavril, V.D., Gherghel, I. & Strugariu, A. (2015) Back in 30 years: A new record for the rare and highly elusive sand boa, *Eryx jaculus turcicus* (Reptilia: Boidae) in Romanian Dobruja. *North-Western Journal of Zoology* 11(2), 366–368, 3 figs.

Sahlean, T.C., Gherghel, I., Papeş, M., Strugariu, A. & Zamfirescu, Ş.R. (2014) Refining climate change projections for organisms with low dispersal abilities: A case study of the Caspian Whip Snake. *PLOS ONE* 9(3)(e91994), 1–12, 3 figs.

Salazar-Valenzuela, D., Martins, A., Amador-Oyola, L. & Torres-Carvajal, O. (2015) A new species and country record of threadsnakes (Serpentes: Leptotyphlopidae: Epictinae) from northern Ecuador. *Amphibian & Reptile Conservation* 8(1), 107–119, 7 figs.

Salazar-Valenzuela, D., Torres-Carvajal, O. & Passos, P. (2014) A new species of *Atractus* (Serpentes: Dipsadidae) from the Andes of Ecuador. *Herpetologica* 70(3), 350–363, 5 figs.

Saldarriaga-Córdoba, M.M., Sasa, M., Pardo, R. & Méndez, M.A. (2009) Phenotypic differences in a cryptic predator: Factors influencing morphological variation in the terciopelo *Bothrops asper* (Garman, 1884; Serpentes: Viperidae). *Toxicon* 54(7), 923–937, 3 figs.

Saleh, M. & Sarhan, M. (2016) The egg-eating snake (Colubridae: *Dasypeltis*) of Faiyum. Egypt, with the description of a new species. *Bulletin de la Société Herpétolgique de France* 160, 25–48, 7 figs.

Salemi, A., Heydari, N. & Mahin, M.J. (2018) A new distribution record for the rare Maynard's Longnose Sand Snake, *Lytorhynchus maynardi* Alcock and Finn, 1896 from Nikshahr, southeastern Iran. *Herpetology Notes* 11, 617–619, 3 figs.

Sales, R.F.D., Lima, M.L.S. & França, B.R.A. (2019) Dead but delicious: An unusual feeding event by the Sertão Muçurana snake (*Boiruna sertaneja*) on a bird carcass. *Herpetology Notes* 12, 941–943, 1 fig.

Sales, R.F.D. de, Lisdboa, C.M.C.A. & Freire, E.M.X. (2009) Répteis Squamata de remanescentes florestais do Campus da Universidade Federal do Rio Grande do Norte, Natal-RN, Brasil. *Cuadernos de Herpetología* 23(2), 77–88, 3 figs.

Sallaberry-Pincheira, N., Garin, C.F., González-Acuña, D., Sallaberry, M.A. & Vianna, J.A. (2011) Genetic divergence of Chilean long-tailed snake (*Philodryas chamissonis*) across latitudes: Conservation threats for different lineages. *Diversity and Distributions* 17, 152–162, 4 figs.

Salvador, A., Busack, S.D., McDiarmid, R., Ineich, I. & Brito, J.C. (2014) *Vipera latastei* (Boscá, 1878) (Reptilia, Serpentes, Viperidae): Request for conservation of the original spelling. *Bulletin of Zoological Nomenclature* 71(1), 22–25.

Salvi, D. & Bombi, P. (2010) Reptiles of Sardinia: Updating the knowledge on their distribution. *Acta Herpetologica* 5(2), 161–177, 4 figs.

Salvi, D., Lucente, D., Mendes, J., Liuzzi, C., Harris, D.J. & Bologna, M.A. (2017) Diversity and distribution of the Italian Aesculapian snake *Zamenis lineatus*: A phylogeographic assessment with implications for conservation. *Journal of Biological Systematics and Evolutionary Research* 55(3), 222–237, 5 figs.

Salvi, D., Mendes, J., Carranza, S. & Harris, D.J. (2018) Evolution, biogeography and systematics of the western Palaearctic *Zamenis* ratsnakes. *Zoologica Scripta* 47(4), 441–461, 5 figs.

Sampaio, I.L.R., Santos, C.P., França, R.C., Pedrosa, I.M.M.C., Solé, M. & França, F.G.R. (2018) Ecological diversity of a snake assemblage from the Atlantic Forest at the south coast of Paraíba, northeast Brazil. *ZooKeys* 787, 107–125, 6 figs.

Samson, A., Santhoshkumar, P., Ramakrishnan, B., Karthick, S. & Gnaneswar, C. (2017) New distribution record of Nagarjunasagar Racer *Platyceps bholanathi* (Reptilia: Squamata: Colubridae) in Sigur, Nilgiris landscape, India. *Journal of Threatened Taxa* 9(3), 10014–10017, 3 figs.

Sanches, P.R., Santos, F.P. dos, Gama, C. de S. & Costa-Campos, C.E. (2018) Predation on *Iguana iguana* (Squamata: Iguanidae) by *Boa constrictor* (Squamata: Boidae) in a fluvial island in the Amazonas river, Brazil, including a list of saurophagy events with *Boa constrictor* as predator. *Cuadernos de Herpetología* 32(2), 129–132, 1 fig.

Sánchez-García, J.C., Canseco-Márquez, L., Pavón-Vázquez, C.J., Cruzado-Cortés, J. & García-Vázquez, U.O. (2019) New records and morphological variation of *Rhadinaea marcellae* (Taylor, 1949) (Squamata, Colubridae) from Sierra Madre Oriental, México. *Check List* 15(5), 729–733, 2 figs.

Sánchez-Martínez, P.M. & Rojas-Runjaic, F.J.M. (2018) *Plesiodipsas perijanensis* (Alemán, 1953). *Catálogo de Anfibios y Reptiles de Colombia* 4(3), 59–64, 5 figs.

Sanders, K.L., Lee, M.S.Y., Mumpuni Bertozzi, T. & Rasmussen, A.R. (2013) Multilocus phylogeny and recent rapid radiation of the viviparous sea snakes (Elapidae: Hydrophiinae). *Molecular Phylogenetics and Evolution* 66, 575–591, 4 figs.

Sanders, K.L., Mumpuni Hamidy, A., Head, J.J. & Gower, D.J. (2010) Phylogeny and divergence times of filesnakes (*Acrochordus*): Inferences from morphology, fossils and three molecular loci. *Molecular Phylogenetics & Evolution* 56, 857–867, 3 figs.

Sanders, K.L., Schroeder, T., Guinea, M.L. & Rasmussen, A.R. (2015) Molecules and morphology reveal overlooked populations of two presumed extinct Australian sea snakes (*Aipysurus*: Hydrophiinae). *PLOS ONE* 10(e0115679), 1–13, 3 figs.

Sanguila, M.B., Cobb, K.A., Siler, C.D., Diesmos, A.C., Alcala, A.C. & Brown, R.M. (2016) The amphibians and reptiles of Mindanao Island, southern Philippines, II: The herpetofauna of northeast Mindanao and adjacent islands. *ZooKeys* 624, 1–132, 81 figs.

Santana, D.O., Caldas, F.L.S., Matos, D.S., Machado, C.M.S., Vilanova-Júnior, J.L. & Faria, R.G. (2017) Morphometry of hatchlings of *Thamnodynastes pallidus* (Linnaeus, 1758) (Serpentes: Dipsadidae: Xendodontinae: Tachmenini). *Herpetology Notes* 10, 589–591, 1 fig.

Santos, D.L., Andrade, S.P. de, Victor-Jr., E.P. & Vaz-Silva, W. (2014) Amphibians and reptiles from southeastern Goiás, central Brazil. *Check List* 10(1), 131–148, 11 figs.

Santos, F.J.M. dos, Entiauspe-Neto, O.M., Araújo, J. da S., Souza, M.B. de, Lema, T. de, Strüssmann, C. & Albuquerque, N.R. de (2018) A new species of burrowing snake (Serpentes: Dipsadidae: *Apostolepis*) from the state of Mato Grosso, central-west region of Brazil. *Zoologia* 35(e26742), 1–10, 10 figs.

Santos, F.J.M. dos & Reis, R.E. (2018) Two new blind snake species of the genus *Liotyphlops* (Peters, 1881) (Serpentes: Anomalepididae), from central and south Brazil. *Copeia* 106(3), 507–514, 7 figs.

Santos, F.J.M. dos & Reis, R.E. (2019) Redescription of the blind snake *Anomalepis colombia* (Serpentes: Anomalepididae) using high-resolution X-Ray computed tomography. *Copeia* 107(2), 239–243, 6 figs.

Santos, F.J.M. dos, Repenning, M., Beier, C. & Pontes, G.M.F. (2015) First record of *Chironius maculoventris* (Dixon, Wiest & Cei, 1993) (Squamata: Serpentes: Colubridae) in Brazil. *Herpetology Notes* 8, 169–171, 2 figs.

Santos, M.B. dos, Oliveira, M.C.L.M. de & Tozetti, A.M. (2012) Diversity and habitat use by snakes and lizards in coastal environments of southernmost Brazil. *Biota Neotropica* 12(3), 78–87, 6 figs.

Santos, M.M. dos, Ávila, R.W. & Kawashita-Ribeiro, R.A. (2011) Checklist of the amphibians and reptiles in Nobres municipality, Mato Grosso state, central Brazil. *Herpetology Notes* 4, 455–461, 4 figs.

Santos, X., Rato, C., Carranza, S., Carretero, M.A. & Pleguezuelos, J.M. (2012) Complex phylogeography in the Southern Smooth Snake (*Coronella girondica*) supported by mtDNA sequences. *Journal of Zoological Systematics and Evolutionary Research* 50(3), 210–219, 2 figs.

Santos, X., Roca, J., Pleguezuelos, J.M., Donaire, D. & Carranza, S. (2008) Biogeography and evolution of the smooth snake *Coronella austriaca* (Serpentes: Colubridae) in the Iberian Peninsula: Evidence for Messinian refuges and Pleistocene range expansions. *Amphibia-Reptilia* 29(1), 35–47, 2 figs.

Santos, X., Vidal-García, M., Brito, J.C., Fahd, S., Llorente, G.A., Martínez-Freiría, F., Parellada, X., Pleguezuelos, J.M. & Sillero, N. (2014) Phylogeographic and environmental correlates support the cryptic function of the zigzag pattern in a European viper. *Evolutionary Ecology* 28(4), 611–626, 4 figs.

Santos-Costa, M.C. dos, Maschio, G.F. & Prudente, A.L. da C. (2015) Natural history of snakes from Floresta Nacional de Caxiuanã, eastern Amazonia, Brazil. *Herpetology Notes* 8, 69–98, 8 figs.

Santos-Jr, A.P., Adams, G.B., Buhler, D., Ribeiro, S. & Carvalho, T.S. (2017) Distribution extension for *Hydrodynastes melanogigas* (Franco, Fernandes & Bentim, 2007) (Serpentes: Dipsadidae: Xenodontinae) in the Araguaia-Tocantins basin, Brazilian cerrado. *Check List* 13(2135), 1–3, 2 figs.

Sapkota, S. & Sharma, V. (2017) Geographic distribution. *Oligodon juglandifer. Herpetological Review* 48(3), 390.

Sarikaya, B., Yildiz, M.Z. & Sezen, G. (2017) The herpetofauna of Adana Province (Turkey). *Commagene Journal of Biology* 1(1), 1–11, 1 fig.

Savage, J.M. (2015) What are the correct family names for the taxa that include the snake genera *Xenodermus, Pareas,* and *Calamaria? Herpetological Review* 46(4), 664–665.

Savage, J.M. & Crother, B.I. (2017) Comment (Case 3688) – On the proposed suppression of Charinidae (Gray, 1849) (Reptilia, Squamata, Serpentes): A counter proposal. *Bulletin of Zoological Nomenclature* 73(2–4), 122–123.

Savage, J.M. & McDiarmid, R.W. (2017) *The herpetological contributions of Giorgio Jan (1791–1866) – with an introduction, annotated bibliography, synopsis of herpetological taxa, and a comprehensive guide to the* Iconographie générale des Ophidiens. *Society for the Study of Amphibians and Reptiles,* viii+926, numerous illustrations.

Sawant, N.S., Jadhav, T.D. & Shyama, S.K. (2010) Habitat suitability, threats and conservation strategies of Hump-nosed Pit Viper *Hypnale hypnale* Merrem (Reptilia: Viperidae) found in Western Ghats, Goa, India. *Journal of Threatened Taxa* 2(11), 1261–1267, 4 figs.

Sawaya, R.J., Marques, O.A.V. & Martins, M. (2008) Composição e história natural das serpents de Cerrado de Itirapina, São Paulo, sudeste do Brasil. *Biota Neotropica* 8(2), 127–149, 45 figs.

Sayyed, A. (2016) Faunal diversity of Satara District, Maharashtra, India. *Journal of Threatened Taxa* 8(13), 9537–9552, 4 figs.

Scalera, R., Venchi, A., Carafa, M., Pellegrini, M., Capula, M. & Bologna, M.A. (2006) Amphibians and reptiles of the Majella National Park (Central Italy). *Aldrovandia* 2, 31–47, 2 figs.

Scali, S. (2010) History and scientific importance of the herpetology collection of the Museo Civico di Storia Naturale of Milan. *Museologia Scientifica Memorie* (5), 69–77, 4 figs.

Scanferla, A. & Agnolín, F.L. (2015) Nuevos aportes al conocimiento de la herpetofauna de la formación Cerro Azul (Mioceno superior), provincia de La Pampa, Argentina. *Papéis Avulsos de Zoologia* 55(23), 323–333, 4 figs.

Scanferla, A. & Canale, J.I. (2007) The youngest record of the Cretaceous snake genus *Dinilysia* (Squamata, Serpentes). *South American Journal of Herpetology* 2(1), 76–81, 2 figs.

Scanferla, A., Zaher, H., Novas, F.E., Muizon, C. de & Céspedes, R. (2013) A new snake skull from the Paleocene of Bolivia sheds light on the evolution of macrostomatans. *PLOS ONE* 8(e57583), 1–9, 5 figs.

Scartozzoni, R.R., Salamão, M. da G. & Almeida-Santos, S.M. de (2009) Natural history of the vine snake *Oxybelis fulgidus* (Serpentes, Colubridae) from Brazil. *South American Journal of Herpetology* 4(1), 81–89, 5 figs.

Schargel, W.E., Lamar, W.W., Passos, P., Valencia, J.H., Cisneros-Heredia, D.F. & Campbell, J.A. (2013) A new giant *Atractus* (Serpentes: Dipsadidae) from Ecuador, with notes on some other large Amazonian congeners. *Zootaxa* 3721(5), 455–474, 8 figs.

Schätti, B. & Kucharzewski, C. (2017) Identity, origin, and distribution of Auguste Ghiesbreght's Mexican amphibians and reptiles. *Mesoamerican Herpetology* 4(4), 84–110, 2 plates, 2 figs.

Schätti, B. & Stutz, A. (2016) *A Short Account of the Snakes of Southern Oaxaca, Mexico.* Privately Published, Oaxaca de Juárez, 1–40, 20 color illus.

Schätti, B., Tillack, F. & Kucharzewski, C. (2014) *Platyceps rhodorachis* (Jan, 1863) - a study of the racer genus *Platyceps* Blyth, 1860 east of the Tigris (Reptilia: Squamata: Colubridae). *Vertebrate Zoology* 64(3), 297–405, 25 figs.

Schembri, B. & Jolly, C.J. (2017) A significant range extension of the unbanded shovel-nosed snake (*Brachyurophis incinctus* Storr, 1968) in the Einasleigh Uplands. *Memoirs of the Queensland Museum* 60, 113 117, 4 figs.

Schiavo, R.M. & Ferri, V. (1996) Anfibi e rettili di alcune aree di rilevanza ambientale della provincia di Cremona. *Pianura* 8, 69–94, 7 figs.

Schield, D.R., Adams, R.H., Card, D.C., Corbin, A.B., Jezkova, T., Hales, N.R., Meik, J.M. et al. (2018) Cryptic genetic diversity, population structure, and gene flow in the Mojave rattlesnake

(*Crotalus scutulatus*). *Molecular Phylogenetics and Evolution* 127, 669–681, 5 figs.

Schield, D.R., Card, D.C., Adams, R.H., Jezkova, T., Reyes-Velasco, J., Proctor, F.N., Spencer, C.L., Herrmann, H.-W., Mackessy, S.P. & Castoe, T.A. (2015) Incipient speciation with biased gene flow between two lineages of the Western Diamondback Rattlesnake (*Crotalus atrox*). *Molecular Phylogenetics and Evolution* 83, 213–223, 5 figs.

Schleip, W.D. (2014) Two new species of *Leiopython* (Hubecht, 1879) (Pythonidae: Serpentes): Non compliance with the International Code of Zoological Nomenclature leads to unavailable names in zoological nomenclature. *Journal of Herpetology* 48(2), 272–275, 1 fig.

Schlüter, A., Icochea, J. & Perez, J.M. (2004) Amphibians and reptiles of the lower Río Llullapichis, Amazonian Peru: Updated species list with ecological and biogeographical notes. *Salamandra* 40(2), 141–160, 6 figs.

Schoenig, E.J. (2017) Geographic distribution. *Lichanura trivirgata*. *Herpetological Review* 48(1), 128–129.

Schrey, A.W., Evans, R.K., Netherland, M., Ashton, K.G., Mushinsky, H.R. & McCoy, E.D. (2015) Phylogeography of the Peninsula crowned snake (*Tantilla relicta relicta*) on the Lake Wales Ridge in central Florida. *Journal of Herpetology* 49(3), 415–419, 2 figs.

Schuett, G.W., Feldner, M.J., Smith, C.F. & Reiserer, R.S. (2016a) *Rattlesnakes of Arizona Species Accounts and Natural History*. Volume 1. ECO Publishing, Rodeo, 1–736, many figs.

Schuett, G.W., Feldner, M.J., Smith, C.F. & Reiserer, R.S. (2016b) *Rattlesnakes of Arizona Conservation, Behavior, Venom and Evolution*. Volume 2. ECO Publishing, Rodeo, 1–488, many figs.

Schulz, K.-D. (2010) Übersicht der variationen des *Orthriophis taeniurus* unterarten-komplexes mit anmerkungen zum status von *Coluber taeniurus pallidus* Rendahl, 1937 under der beschreibung einer neuen unterart (Reptilia: Squamata: Serpentes. Colubridae). *Sauria* 32(2), 3–26, 36 figs.

Schulz, K.-D., Tillack, F., Das, A. & Helfenberger, N. (2015) On the identity and taxonomic status of *Coluber nuthalli* Theobald, 1868, with redescription of the type specimens of *Coluber nuthalli* and *Elaphis yunnanensis* (Anderson, 1879) (Reptilia, Squamata, Colubridae). *Asian Herpetological Research* 6(1), 1–10, 3 figs.

Scrocchi, G.J. & Giraudo, A.R. (2012) First records of *Phalotris sansebastiani* (Jansen & Köhler, 2008) (Serpentes: Dipsadidae) from Argentina. *Check List* 8(5), 900–902, 4 figs.

Scrocchi, G.J. & Kretzchmar, S. (2017) Catálogo de los especímenes tipo de la Colección Herpetológica de la Fundación Miguel Lillo, Tucumán, Argentina. *Acta Zoológica Lilloana* 61(2), 87–135.

Scrocchi, G.J., Stazzonelli, J.C. & Cabrera, P. (2019) Nuevas citas de Squamata (Gekkonidae, Phyllodactylidae y Dipsadidae) para la provincia de Tucumán, Argentina. *Cuadernos de Herpetología* 33(2), 75–78, 1 fig.

Seetharamaraju, M. & Srinivasulu, C. (2013) Discovery and description of male specimen of *Coluber bholanathi* (Sharma, 1976) (Reptilia: Colubridae) from Hyderabad, India. *Taprobanica* 5(1), 32–35, 4 figs.

Seetharamaraju, M., Srinivasulu, C. & Srinivasulu, B. (2011) New records of *Oligodon taeniolatus* (Jerdon, 1853) (Reptilia: Colubridae) in Andhra Pradesh, India. *Herpetology Notes* 4, 421–423, 2 figs.

Seglie, D. & Sindaco, R. (2013) Segnalazioni faunistiche Piemontesi e Valdostane, VI. (Reptilia, Colubridae). *Rivista Piemontese di Storia Naturale* 34, 439–452, 4 figs.

Seijas, A.E., Araujo-Quintero, A. & Velásquez, N. (2013) Mortalidad de vertebrados en la carretera Guanare-Guanarito, estado Portuguesa, Venezuela. *Revista de Biología Tropical* 61(4), 1619–1636, 4 figs.

Sen, M.K. & Nama, K.S. (2013) Documentation of ophiofauna of Mukundara Hills National Park. Kota, Rajasthan (India). *International Journal of Recent Biotechnology* 1(1), 21–24, 1 fig.

Sengupta, D., Borah, C.G. & Phukon, J. (2019) Assessment of the reptilian fauna in the Brahmaputra Plains of two districts in Assam, *India. IRCF Reptiles & Amphibians* 26(1), 65–67, 1 fig.

Serna-Botero, V. & Ramírez-Castaño, V.A. (2017) Curaduría y potencial de investigación de la Colección Herpetológica del Museo de Historia Natural de la Universidad de Caldas, Manizales, Colombia. *Boletín Científico Museo de Historia Natural* 21(1), 138–153, 7 figs.

Serrano, F. & Díaz-Ricaurte, J.C. (2018) *Erythrolamprus aesculapii* (Linnaeus, 1758). *Catálogo de Anfibios y Reptiles de Colombia* 4(3), 48–53, 3 figs.

Setser, K., Meik, J.J. & Mulcahy, D.G. (2002) Herpetofauna of the southern Snake Range of Nevada and surrounding valleys. *Western North American Naturalist* 62(2), 234–239, 1 fig.

Shafiei, S., Fahimi, H., Sehhatisabet, M.E. & Moradi, N. (2015) Rediscovery of Maynard's Longnose Sand Snake, *Lytorhynchus maynardi*, with the geographic distribution of the genus *Lytorhynchus* Peters, 1863 in Iran. *Zoology in the Middle East* 61(1), 32–37, 4 figs.

Shahriza, S., Jaafar, I., Shahrul-Anuar, M.S., Nur, H.I., Amiruddin, I., Amirah, H. & Zalina, A. (2013) An addition of reptiles of Gunung Inas, Kedah, Malaysia. *Russian Journal of Herpetology* 20(3), 171–180, 21 figs.

Shankar, P.G. & Ganesh, S.R. (2009) Sighting record and range extension of *Calliophis* (=*Callophis*) *bibroni* (Jan, 1858) (Reptilia, Squamata, Serpentes, Elapidae). *The Herpetological Bulletin* 108, 10–13, 6 figs.

Sharma, V. (2014) On the distribution of *Elachistodon westermanni* (Reinhardt, 1863) (Serpentes, Colubridae). *Russian Journal of Herpetology* 21(3), 161–165, 3 figs.

Sharma, V. & Jani, M. (2015) Geographic distribution. *Lycodon travancoricus. Herpetological Review* 46(4), 574.

Sharma, V., Jain, A. & Bhandari, R. (2015) A new locality for the elusive and endemic Yellow-spotted Wolf Snake (*Lycodon flavomaculatus* Wall 1907), with notes on distribution and habitat. *IRCF Reptiles & Amphibians* 22(4), 164–167, 2 figs.

Sharma, V., Karoo, R. & Zire, U. (2016) Geographic distribution. *Boiga flaviviridis. Herpetological Review* 47(1), 82.

Sharma, V., Louies, J. & Vattam, A. (2013) A contribution to *Coluber bholanathi* (Sharma, 1976) (Serpentes: Colubridae). *Russian Journal of Herpetology* 20(4), 259–263, 3 figs.

Shea, G.M. (2015) A new species of *Anilios* (Scolecophidia: Typhlopidae) from Central Australia. *Zootaxa* 4033(1). 103–116, 5 figs.

Sheehy III, C.M., Yánez-Muñoz, M.H., Valencia, J.H. & Smith, E.N. (2014) A new species of *Siphlophis* (Serpentes: Dipsadidae: Xenodontinae) from the eastern Andean slopes of Ecuador. *South American Journal of Herpetology* 9(1), 30–45, 8 figs.

Shi, J., Wang, G., Chen, X., Fang, Y., Ding, L., Huang, S., Mou, M., Liu, J. & Li, P. (2017) A new moth-preying alpine pit viper species from Qinghai-Tibetan Plateau (Viperidae, Crotalinae). *Amphibia-Reptilia* 38(4), 517–532, 6 figs.

Shi, J., Yang, D., Zhang, W.Y., Peng, L., Orlov, N.L., Jiang, F., Ding, L. et al. (2018) A new species of the *Gloydius strauchi* complex (Crotalinae: Viperidae: Serpentes) from Qinghai,

Sichuan, and Gansu, China. *Russian Journal of Herpetology* 25(2), 126–138, 8 figs.

Shi, J., Yang, D.W., Zhang, W.Y., Qi, S., Li, P.P. & Ding, L. (2016) Distribution and infraspecific taxonomy of *Gloydius halys-Gloydius intermedius* complex in China (Serpentes: Crotalinae). *Chinese Journal of Zoology* 51(5), 777–798, 6 figs. (in Chinese).

Shibata, H., Chijiwa, T., Hattori, S., Terada, K., Ohno, M. & Fukumaki, Y. (2016) The taxonomic position and the unexpected divergence of the Habu viper, *Protobothrops* among Japanese subtropical islands. *Molecular Phylogenetics and Evolution* 101, 91–100, 6 figs.

Shupe, S. (2013) *Venomous snakes of the World – A manual for use by U.S. Amphibious Forces.* Skyhorse Publishing, New York. ix+320, 297 figs.

Sievert, G. & Sievert, L. (2011) *A Field Guide to Oklahoma's Amphibians and Reptiles.* 3rd edn. Oklahoma Department of Wildlife Conservation, Oklahoma City. vi+211, numerous illustrations.

Sigé, B., Hugueney, M., Crochet, J.-Y., Legendre, S., Mourer-Chauviré, C., Rage, J.-C. & Simon-Coinçon, R. (1998) Baraval, nouvelle faune de l'Oligocène inférieur (MP22) des phosphorites du Quercy. Apport à la signification chronologique des remplissages karstiques. *Bulletin de la Société d'Histoire Naturelle de Toulouse* 134, 85–90, 2 figs.

Siler, C., Oliveros, C.H., Santanen, A. & Brown, R.M. (2013) Multilocus phylogeny reveals unexpected diversification patterns in Asian wolf snakes (genus *Lycodon*). *Zoologica Scripta* 42(3), 262–277, 3 figs.

Siler, C.D., Swab, J.C., Oliveros, C.H., Diesmos, A.C., Averia, L., Alcala, A.C. & Brown, R.M. (2012) Amphibians and reptiles, Romblon Island group, central Philippines: Comprehensive herpetofaunal inventory. *Check List* 8(3), 443–462, 24 figs.

Silva, M.B. da, Rocha, W.A. da & Nogueira-Paranhos, J.D. (2016) Checklist of reptiles of the Amazonia-Caatinga-Cerrado ecotonal zone in eastern Maranhão, *Brazil. Herpetology Notes* 9, 7–14, 3 figs.

Silva, M.C. da, Oliveira, R.H. de, Morais, D.H., Kawashita-Ribeiro, R.A., Brito, E.S. de & Ávila, R.W. (2015) Amphibians and reptiles of a Cerrado area in Primavera do Leste Municipality, Mato Grosso State, central Brazil. *Salamandra* 51(2), 187–194, 5 figs.

Silva-Rocha, I., Salvi, D., Sillero, N., Mateo, J.A. & Carretero, M.A. (2015) Snakes on the Balearic Islands: An invasion tale with implications for native biodiversity conservation. *PLOS ONE* 2015(0121026), 1–18, 3 figs.

Silva-Soares, T., Ferreira, R.B., Salles, R. de O.L. & Rocha, C.F.D. (2011) Continental, insular and coastal marine reptiles from the municipality of Vitória, state of Espírito Santo, southeastern Brazil. *Check List* 7(3), 290–298, 7 figs.

Silveira, A.L. (2014a) Geographic distribution. *Apostolepis flavotorquata. Herpetological Review* 45(1), 94.

Silveira, A.L. (2014b) Geographic distribution. *Micrurus brasiliensis. Herpetological Review* 45(1), 95.

Silveira, A.L. (2014c) Geographic distribution. *Paraphimophis rusticus. Herpetological Review* 45(1), 96.

Silveira, A.L. (2014d) Geographic distribution. *Rodriguesophis iglesiasi. Herpetological Review* 45(1), 96.

Silveira, A.L., Ribeiro, L.S.V.B., Dornas, T.T. & Fernandes, T.N. (2018) New records of *Dispas* [sic] *albifrons* (Serpentes, Dipsadidae) in the Atlantic Forest of Minas Gerais, Brazil, with morphological data. *Herpetology Notes* 11, 809–815, 5 figs.

Simonov, E., Lisachov, A., Oreshkova, N. & Krutovsky, K.V. (2018) The mitogenome of *Elaphe bimaculata* (Reptilia: Colubridae) has never been published: A case with the complete

mitochondrial genome of *E. dione. Acta Herpetologica* 13(2), 185–189, 1 fig.

Simonov, E. & Wink, M. (2012) Population genetics of the Halys pit viper (*Gloydius halys*) at the northern distribution limit in Siberia. *Amphibia-Reptilia* 33(2), 273–283, 3 figs.

Sinaiko, G., Magoty-Cohen, T., Meiri, S. & Dor, R. (2018) Taxonomic revision of Israeli snakes belonging to the *Platyceps rhodorachis* species complex (Reptilia: Squamata: Colubridae). *Zootaxa* 4379(3), 301–346, 9 figs.

Sindaco, R., Nincheri, R. & Lanza, B. (2014) Catalogue of Arabian reptiles in the collections of the "La Specola" Museum, Florence. In: Corti, C. & Capula, M. (Eds.), *Scripta Herpetologica. Studies on Amphibians and Reptiles in Honour of Benedetto Lanza.* Edizioni Belvedere, Latina, 137–164.

Sindaco, R., Venchi, A. & Grieco, C. (2013) *The Reptiles of the Western Palearctic 2. Annotated Checklist and Distributional Atlas of the Snakes of Europe, North Africa, Middle East and Central Asia, with an Update of the* Vol. 1. Edizioni Belvedere, Latina, 1–543, 342 figs.

Sindaco, R., Ziliani, U., Razzetti, E., Pupin, F., Carugati, C., Fasola, M., Grieco, C. & Pella, F. (2009) The status of knowledge of the herpetofauna of Socotra Island (Yemen). In: Corti, C. (Ed.), *Herpetologia Sardiniae.* Edizioni Belvedere, Latina, 454–459, 2 figs.

Singh, A. & Joshi, R. (2018) Sighting of the Himalayan Trinket Snake, *Orthriophis hodgsonii* (Günther, 1860) (Reptilia: Colubridae), in Sahastra Dhara, Uttarakhand: A new elevational record. *Amphibian & Reptile Conservation* 12(1), 15–17, 3 figs.

Singh, A., Puri, K. & Joshi, R. (2017) New distributional records for the Himalayan White-lipped Pitviper, *Trimersurus septentrionalis* (Kramer, 1977) (Reptilia: Viperidae) from the Garhwal Himalaya in northwestern India. *IRCF Reptiles & Amphibians* 24(3), 197–200, 3 figs.

Singh, A., Puri, K. & Joshi, R. (2018) New locality record for the Common Cliff Racer, *Platyceps rhodorachis rhodorachis* (Jan, 1863) (Reptilia: Squamata: Colubridae), from Garhwal Himalaya, northwestern India. *IRCF Reptiles & Amphibians* 25(2), 153–154, 3 figs.

Siqueira, D.M., Nascimento, L.P., Montingelli, G.G. & Santos-Costa, M.C. dos (2013) Geographic variation in the reproduction and sexual dimorphism of the Boddaert's tropical racer, *Mastigodryas boddaerti* (Serpentes: Colubridae). *Zoologia* 30(5), 475–481, 1 fig.

Smaga, C.R., Ttito, A. & Catenazzi, A. (2019) *Arcanumophis*, a new genus and generic allocation for *Erythrolamprus problematicus* (Myers 1986), Xenodontinae (Colubridae) from the Cordillera de Carabaya, southern Peru. *Zootaxa* 4671(1), 129–138, 6 figs.

Smart, U., Smith, E.N., Murthy, B.H.C.K. & Mohanty, A. (2014) Report of Nagarjunasagar Racer *Coluber bholanathi* (Sharma, 1976) (Squamata: Serpentes: Colubridae) from the Gingee Hills, Tamil Nadu, India. *Journal of Threatened Taxa* 6(4), 5671–5674, 5 figs.

Šmíd, J., Göçmen, B., Crochet, P.-A., Trape, J.-F., Mazuch, T., Uvizl, M. & Nagy, Z.T. (2019) Ancient diversification, biogeography, and the role of climatic niche evolution in the Old World cat snakes (Colubridae, *Telescopus*). *Molecular Phylogenetics and Evolution* 134, 35–49, 6 figs.

Šmíd, J., Martínez, G., Gebhart, J., Aznar, J., Gállego, J., Göçmen, B., Pous, P. de, Tamar, K. & Carranza, S. (2015) Phylogeny of the genus *Rhynchocalamus* (Reptilia; Colubridae) with a first record from the Sultanate of Oman. *Zootaxa* 4033(3), 380–392, 4 figs.

Smith, K.T. (2013) New constraints on the evolution of the snake clades Ungaliophiinae, Loxocemidae and Colubridae (Serpentes), with comments on the fossil history of erycine boids in North America. *Zoologischer Anzeiger* 252, 157–182, 13 figs.

Smith, P. & Atkinson, K. (2017) Observations of two predation events involving herps and birds. *Herpetology Notes* 10, 635–637, 2 figs.

Smith, P., Atkinson, K., Brouard, J.-P. & Pheasey, H. (2016) Reserva Natural Laguna Blanca, Departamento San Pedro: Paraguay's first important area for the conservation of amphibians and reptiles? *Russian Journal of Herpetology* 23(1), 25–34, 1 fig.

Smith, P., Cacciali, P., Scott, N., Castillo, H. del, Pheasey, H. & Atkinson, K. (2014) First record of the globally-threatened Cerrado endemic snake *Philodryas livida* (Amaral, 1923) (Serpentes, Dipsadidae) from Paraguay, and the importance of the Reserva Natural Laguna Blanca to its conservation. *Cuadernos de Herpetología* 28(2), 169–171, 1 fig.

Smith, P., Scott, N., Cacciali, P. & Atkinson, K. (2013) *Rhachidelus brazili* (Squamata: Serpentes): First records from Paraguay and clarification of the correct spelling of the generic name. *Salamandra* 49(1), 56–58, 2 figs.

Smith, T., Kumar, K., Rana, R.S., Folie, A., Solé, F., Noiret, C., Steeman, T., Sahni, A. & Rose, K.D. (2016) New early Eocene vertebrate assemblage from western India reveals a mixed fauna of European and Gondwana affinities. *Geoscience Frontiers* 7(6), 969–1001, 24 figs.

Smits, T. & Hauser, S. (2019) First record of the krait *Bungarus slowinskii* Kuch, Kizirian, Nguyen, Lawson, Donelly [sic] and (Mebs, 2005) (Squamata: Elapidae) from Thailand. *Tropical Natural History* 19(2), 43–50, 6 figs.

Snyder, S.J., Schmidt, R.E. & Tirard, N. (2019) First report of the Brahminy Blindsnake, *Indotyphlops braminus* (Daudin), from the Caribbean island of Montserrat. *Caribbean Herpetology* 65, 1–2, 1 fig.

Sokolov, A.S. (2005) On the taxonomic status of the Common Adder of the partially wooded steppe of the Oka – Don Plain. In: Ananjeva, N. & Tsinenko, O. (Eds.), *Herpetologia Petropolitana – Proceedings of the 12th Ordinary General Meeting of the Societas Europeae Herpetologica*. St. Petersburg, 96–99.

Solano-Zavaleta, I., Pavón-Vázquez, C.J., Campillo-García, G., Arenas-Monroy, J.C., Pérez-Ramos, E., Centenero-Alcalá, E., Avendaño-Pazos, J.J. & Nieto-Montes de Oca, A. (2014) New record and comments on the distribution of the Mexican colubrid snake *Coniophanes melanocephalus* (Peters, 1869). *Mesoamerican Herpetology* 1(2), 295–298, 2 figs.

Solís, J.M., Ayala-Rojas, C.G. & O'Reilly, C.M. (2017a) New records for two reptiles from the Bay Islands, Honduras. *Mesoamerican Herpetology* 4(3), 669–671, 2 figs.

Solís, J.M., Espinal, M. & O'Reilly, C.M. (2014b) Distribution notes. *Enulius flavitorques* (Cope, 1869). *Mesoamerican Herpetology* 1(2), 292–293.

Solís, J.M., Espinal, M.R., Wostl, E., Mora, J.M., Zuñiga, L.G. & Bonilla, J. (2017b) New distribution and habitat records for *Atropoides indomitus* (Serpentes: Viperidae), a Honduran endemic. *Mesoamerican Herpetology* 4(4), 988–992, 3 figs.

Solís, J.M., Valle, R.E., Herrera, L.A., O'Reilly, C.M. & Downing, R. (2015) Range extensions and new departmental records for amphibians and reptiles in Honduras. *Mesoamerican Herpetology* 2(4), 557–561, 3 figs.

Solís, J.M., Wilson, L.D. & Townsend, J.H. (2014a) An updated list of the amphibians and reptiles of Honduras, with comments on their nomenclature. *Mesoamerican Herpetology* 1(1), 123–144.

Solórzano, A. (2011) Variación de color de la serpiente marina *Pelamis platura* (Serpentes: Elapidae) en el Golfo Dulce, Puntarenas, Costa Rica. *Cuadernos de Investigación UNED* 3(1), 89–96, 4 figs.

Solovyov, V.A., Scopin, A.E. & Solovyov, A.N. (2016) Distribution and status of adder *Vipera* (*Pelias*) *berus* populations in the Vladimir Region. *Modern Herpetology* 16(3/4), 142–150, 4 figs. [in Russian].

Sos, T. (2007) Notes on the distribution and current status of herpetofauna in the northern area of Braşov County (Romania). *North-Western Journal of Zoology* 3(1), 34–52, 13 figs.

Sosa-Bartuano, A. (2017) Distribution notes: *Urotheca fulviceps*. *Mesoamerican Herpetology* 4(4), 965–966, 1 fig.

Sousa, B.M. de, Nascimento, A.E.R. do, Gomides, S.C., Rios, C.H.V., Hudson, A. de A. & Novelli, I.A. (2010) Répteis em fragmentos de cerrado e mata Atlântica no Campo das Vertentes, Estado de Minas Gerais, Sudeste do Brasil. *Biota Neotropica* 10(2), 129–138, 8 figs.

Souto, N.M., Pinna, P.H., Machado, A.S. & Lopes, R.T. (2017) New records, morphological variation, and description of the skull of *Liophis dorsocorallinus* (Esqueda, Natera, La Marca and Ilija-Fistar, 2005) (Serpentes: Dipsadidae). *Herpetological Review* 48(3), 532–537, 5 figs.

Souza, J.R.D. de, Venância, N.M., Freitas, M.A. de, Souza, M.B. de & Veríssimo, D. (2013) First record of *Bothrops taeniatus* (Wagler, 1824) (Reptilia: Viperidae) for the state of Acre, Brazil. *Check List* 9(2), 430–431, 3 figs.

Souza, S.M., Junqueira, A.B., Jakovac, A.C.C., Assunção, P.A. & Correia, J.A. (2011) Feeding behavior and ophiophagous habits of two poorly known Amazonian coral snakes, *Micrurus albicinctus* Amaral 1926 and *Micrurus paraensis* (Cunha and Nascimento 1973) (Squamata, Elapidae). *Herpetology Notes* 4, 369–372, 1 fig.

Souza-Filho, G.A. de, Plombon, L.L. & Capola, D.J.V. (2015) Reptiles of the Complexo Energético Fundão-Santa Clara, central south region of Paraná state, southern Brazil. *Check List* 11(1655), 1–6, 3 figs.

Sow, A.S., Martínez-Freiría, F., Crochet, P.-A., Geniez, P., Ineich, I., Dieng, H., Fahd, S. & Brito, J.C. (2014) Atlas of the distribution of reptiles in the Parc National du Banc d'Arguin, Mauritania. *Basic and Applied Herpetology* 28, 99–111, 6 figs.

Speybroek, J., Bohle, D., Razzetti, E., Dimaki, M., Kirchner, M.K. & Beukema, W. (2014) The distribution of amphibians and reptiles on Samos Island (Greece) (Amphibia: Reptilia). *Herpetozoa* 27(1/2), 39–63, 10 figs.

Srinivasulu, C., Srinivasulu, B. & Molur, S. (2014) *The Status and Distribution of Reptiles in the Western Ghats, India*. Wildlife Information Liaison Development Society, Coimbatore, 1–148, numerous figs. and maps.

Srinivasulu, C., Venkateshwarlu, D. & Seetharamaraju, M. (2009) Rediscovery of the Banded Krait *Bungarus fasciatus* (Schneider, 1801) (Serpentes: Elapidae) from Warangal District, Andhra Pradesh, India. *Journal of Threatened Taxa* 1(6), 353–354, 3 figs.

Stark, T., Laurijssens, C. & Weterings, M. (2014) Distributional and natural history notes on five species of amphibians and reptiles from Isla Ometepe, Nicaragua. *Mesoamerican Herpetology* 1(2), 308–312, 2 figs.

Stein, A.C. & Kalinina, V. (2016) Confirmation of the Red-backed Rat Snake *Oocatochus rufodorsatus* (Cantor, 1842) (Squamata: Colubridae) in Amur Oblast', Russian Federation. *Russian Journal of Herpetology* 23(1), 81–82, 2 figs.

Sterijovski, B. (2014) Systematic survey of amphibian and reptile fauna in the Bosilegrad region of southern Serbia. *Biologia Serbica* 36(1/2), 65–68.

Sterijovski, B., Tomović, L. & Ajtić, R. (2014) Contribution to the knowledge of the reptile fauna and diversity in FYR of Macedonia. *North-Western Journal of Zoology* 10(1), 83–92, 3 figs.

Stevenson, D.J., Jenkins, C.L., Stohlgren, K.M., Jensen, J.B., Bechler, D.L., Deery, I., Graham, S.P. et al. (2015) Significant new records of amphibians and reptiles from Georgia, USA. *Herpetological Review* 46(4), 597–601.

Stevenson, D.J., Jensen, J.B., Frank, P.G., Bond, W., Duff, D., Moore, M., Nelson, K. et al. (2014) New county records for amphibians and reptiles of Georgia, USA. *Herpetological Review* 45(1), 102–104.

Stickel, A.L., Abarca, J.G. & Pounds, J.A. (2017) Geographic distribution. *Anomalepis mexicanus. Herpetological Review* 48(3), 589.

Stille, B. & Stille, M. (2016) The herpetofauna of Mathraki, Othonoi and Erikoussa, Diapontia Islets, Greece. *Herpetozoa* 28(3/4), 193–197, 9 figs.

Strachinis, I. & Artavanis, D. (2017) Additions to the known herpetofauna of the Island of Ithaki, Ionian Sea, Greece. *Herpetozoa* 30(1/2), 64–66, 2 figs.

Strachinis, I. & Lymberakis, P. (2013) New records of *Hemorrhois nummifer* (Reuss, 1834) from two Greek islands, Chios and Samothrace and consequent biogeographic implications. *Herpetology Notes* 6, 513–515, 2 figs.

Strachinis, I. & Roussos, S.A. (2016) Terrestrial herpetofauna of Limnos and Agios Efstratios (northern Aegean, Greece), including new species records for *Malpolon insignitus* (Geoffroy Saint-Hilaire, 1827) and *Pelobates syriacus* Boettger, 1889. *Herpetology Notes* 9, 237–248, 7 figs.

Strachinis, I. & Wilson, M. (2014) New record of *Typhlops vermicularis* Merrem, 1820 from Symi Island, Greece. *Herpetology Notes* 7, 9–10, 2 figs.

Streicher, J.W. & Wiens, J.J. (2016) Phylogenomic analyses reaveal novel relationships among snake families. *Molecular Phylogenetics and Evolution* 100, 160–169, 4 figs.

Strickland, J.L., Carter, S., Kraus, F. & Parkinson, C.L. (2016) Snake evolution in Melanesia: Origin of the Hydrophiinae (Serpentes, Elapidae), and the evolutionary history of the enigmatic New Guinean elapid *Toxicocalamus. Zoological Journal of the Linnean Society* 178(3), 663–678, 4 figs.

Strickland, J.L., Parkinson, C.L., McCoy, J.K. & Ammerman, L.K. (2014) Phylogeography of *Agkistrodon piscivorus* with emphasis on the western limit of its range. *Copeia* 2014(4), 639–649, 6 figs.

Strickland, J.M. & Hartman, G.D. (2015) New distribution records of amphibians and reptiles in west-central Georgia, USA. *Herpetological Review* 46(4), 601–602.

Strödicke, M. & Gerish, B. (1999) Morphologische Merkmalsvariabilität bei *Elaphe longissima* (Laurenti, 1768), unter besonderer Berücksichtigung zweier isolierter Populationen an der Nordgrenze des Artareals (Squamata: Serpentes: Colubridae). *Herpetozoa* 11(3/4), 121–139, 7 figs.

Strugariu, A., Sahlean, T.C., Gherghel, I., Dincă, P.C. & Zamfirescu, Ş.R. (2016) First record of the Dice Snake, *Natrix tessellata* (Reptilia: Colubridae) from north-eastern Romania. *Russian Journal of Herpetology* 23(4), 323–326, 5 figs.

Strugariu, A., Sahlean, T.C., Huţuleac-Volosciuc, M.H. & Puşcaşu, C.M. (2006) Preliminary data regarding the distribution of reptilian fauna in Suceava County (Romania). *North-Western Journal of Zoology* 2(1), 39–43.

Strugariu, A., Sos, T., Gherghel, I., Ghira, I., Sahlean, T.C., Puşcaşu, C.M. & Huţuleac-Volosciuc, M.V. (2008) Distribution and current status of the herpetofauna from the northern Măcin Mountains area (Tulcea County, Romania). *Analele Ştiinţifice*

ale Universităţi "Al. I. Cuza" Iaşi, s. Biologie Animală 54, 191–206, 11 figs.

Strugariu, A., Sos, T., Sotek, A., Gherghel, I. & Hegyeli, Z. (2009) New locality records for the adder (*Vipera berus*) in the Carpathian Corner, Romania. *AES Bioflux* 1(2), 99–103, 3 figs.

Stuebing, R.B., Inger, R.F. & Lardner, B. (2014) *A Field Guide to the Snakes of Borneo*. 2nd edn. Natural History Publications (Borneo), Kota Kinabalu, viii+310, numerous color photos.

Stümpel, N. & Joger, U. (2009) Recent advances in phylogeny and taxonomy of Near and Middle Eastern vipers – an update. *ZooKeys* 31, 179–191, 9 figs.

Stümpel, N., Rajabizadeh, M., Avci, A., Wüster, W. & Joger, U. (2016) Phylogeny and diversification of mountain vipers (*Montivipera*, Nilson et al., 2001) triggered by multiple Plio-Pleistocene refugia and high-mountain topography in the Near and Middle East. *Molecular Phylogenetics and Evolution* 101, 336–351, 8 figs.

Suárez-Atilano, M., Burbrink, F.T. & Vázquez-Domínguez, E. (2014) Phylogeographical structure within *Boa constrictor imperator* across the lowlands and mountains of Central America and Mexico. *Journal of Biogeography* 41(12), 2371–2384, 4 figs.

Suazo-Ortuño, I., Gutiérrez, J.G.P., Medina-Aguilar, O., Bucio, L.E., González-Hernández, A.J. & Reynoso, V.H. (2014) Geographic distribution. *Coniophanes lateritus. Herpetological Review* 45(2), 284.

Sudré, V., Curcio, F.F., Nunes, P.M.S., Pellegrino, K.C.M. & Rodrigues, M.T. (2017) Who is the red-bearded snake, anyway? Clarifying the taxonomic status of *Chironius pyrrhopogon* (Wied, 1824) (Serpentes: Colubridae). *Zootaxa* 4319(1), 143–156, 6 figs.

Sumarli, A.X., Grismer, L.L., Anuar, S., Muin, M.A. & Quah, E.S.H. (2015) First report on the amphibians and reptiles of a remote mountain, Gunung Tebu in northeastern Peninsular Malaysia. *Check List* 11(1679), 1–32, 41 figs.

Sumontha, M., Kunya, K., Dangsri, S. & Pauwels, O.S.G. (2017) *Oligodon saiyok*, a new limestone-dwelling kukri snake (Serpentes: Colubridae) from Kanchanaburi Province, western Thailand. *Zootaxa* 4294(3), 316–328, 9 figs.

Suntrarachun, S., Chanhome, L. & Sumontha, M. (2014) Phylogenetic analysis of the king cobra, *Ophiophagus hannah* in Thailand based on mitochondrial DNA sequences. *Asian Biomedicine* 8(2), 269–274, 1 fig.

Sunyer, J. (2014) An updated checklist of the amphibians and reptiles of Nicaragua. *Mesoamerican Herpetology* 1(2), 185–202, 2 fig.

Sunyer, J., Fonseca, J.G.M., Fernández, M.A., Olivas, M.F.U. & Obando, L.A. (2014) Noteworthy snake records from Nicaragua (Reptilia: Serpentes). *Check List* (10(5), 1134–1147, 5 figs.

Sunyer, J., Jirón, C., Antón, A.A.A. & Rodríguez, A.A.G. (2017) Distribution notes: *Cerrophidion wilsoni. Mesoamerican Herpetology* 4(4), 967–969, 2 figs.

Sunyer, J. & Köhler, G. (2007) New country and departmental records of herpetofauna in Nicaragua. *Salamandra* 43(1), 57–62, 5 figs.

Sunyer, J. & Leonardi, R. (2015) Nature notes: *Leptodrymus pulcherrimus. Mesoamerican Herpetology* 2(4), 523, 1 fig.

Supsup, C.E. (2016) Geographic distribution. *Oligodon ancorus. Herpetological Review* 47(3), 428.

Supsup, C.E., Guinto, F.M., Redoblado, B.R. & Gomez, R.S. (2017) Amphibians and reptiles from the Mt. Hamiguitin Range of eastern Mindanao Island, Philippines: New distribution records. *Check List* 13(2121), 1–14, 6 figs.

Supsup, C.E., Puna, N.M., Asis, A.A., Redoblado, B.R., Panaguinit, M.F.G., Guinto, F.M., Rico, E.B., Diesmos, A.C., Brown,

R.M. & Mallari, N.A.D. (2016) Amphibians and reptiles of Cebu, Philippines: The poorly understood herpetofauna of an island with very little remaining natural habitat. *Asian Herpetological Research* 7(1), 151–179, 36 figs.

Suraprasit, K., Jaeger, J.-J., Chaimanee, Y., Chavasseau, O., Yamee, C., Tian, P. & Panha, S. (2016) The middle Pleistocene vertebrate fauna from Khok Sung (Nakhon Ratchasima, Thailand): Biochronological and paleobiogeographical implications. *ZooKeys* 613, 1–157, 48 figs.

Sutradhar, S. & Nath, A. (2013) An account on poorly known corral [sic] red snake *Oligodon kheriensis* Acharji et Ray, 1936 from Assam, India. *Russian Journal of Herpetology* 20(4), 247–252, 5 figs.

Sutton, W.B., Czech, H.A., Wang, Y. & Schweiter, C.J. (2014) New records of amphibians and reptiles from Alabama, USA. *Herpetological Review* 45(2), 293–294.

Sy, E.Y. (2016a). Geographic distribution. *Ophiophagus hannah.* *Herpetological Review* 46(2), 263.

Sy, E.Y. (2016b). Geographic distribution. *Ophiophagus hannah.* *Herpetological Review* 46(2), 264.

Sy, E.Y. (2017). Geographic distribution. *Ophiophagus hannah.* *Herpetological Review* 48(1), 130.

Sy, E.Y. & Balete, D.S. (2017) Geographic distribution. *Naja philippinensis.* *Herpetological Review* 48(3), 590.

Sy, E.Y., Baniqued, R.D. & Diesmos, A.C. (2016a). Geographic distribution. *Ophiophagus hannah.* *Herpetological Review* 46(2), 263–264.

Sy, E.Y. & Binaday, J.W.B. (2016). Geographic distribution. *Aplopeltura boa.* *Herpetological Review* 47(2), 260–261.

Sy, E.Y. & Boos, R. (2015a) Geographic distribution. *Ophiophagus hannah.* *Herpetological Review* 46(2), 220.

Sy, E.Y. & Boos, R. (2015b) Geographic distribution. *Trimeresurus flavomaculatus.* *Herpetological Review* 46(2), 222.

Sy, E.Y. & Dichaves, J.A.L. (2017) Geographic distribution. *Ophiophagus hannah.* *Herpetological Review* 48(3), 591.

Sy, E.Y., Dycoco, Jr., Q.S., Rodriguez, A.Z.A., Du, S.G., Wisco, M.A.S., Manuel M., Pangilinan, M. & Banaguas, G.S. (2016b) Geographic distribution. *Naja sumatrana.* *Herpetological Review* 47(3), 427–428.

Sy, E.Y. & Gerard, T. (2014) Geographic distribution. *Naja philippinensis.* *Herpetological Review* 45(4), 664–665.

Sy, E.Y. & Labatos, B.V. (2017) Geographic distribution. *Ophiophagus hannah.* *Herpetological Review* 48(1), 130.

Sy, E.Y., Layola, L.A. de, Yu, C.M.T. & Diesmos, A.C. (2015) Geographic distribution. *Ophiophagus hannah.* *Herpetological Review* 46(2), 220–221.

Sy, E.Y. & Letana, S.D. (2017) Geographic distribution. *Ophiophagus hannah.* *Herpetological Review* 48(1), 130.

Sy, E.Y. & Tan, E.K. (2014) Geographic distribution. *Gonyosoma oxycephalum.* *Herpetological Review* 44(2), 275.

Sy, E.Y. & Tan, E.K. (2015a) Geographic distribution. *Coelognathus erythrurus psephenourus.* *Herpetological Review* 46(2), 218.

Sy, E.Y. & Tan, E.K. (2015b) Geographic distribution. *Malayopython reticulatus.* *Herpetological Review* 46(2), 220.

Sy, E.Y. & Vargas, M.G. (2017) Geographic distribution. *Naja philippinensis.* *Herpetological Review* 48(3), 590.

Sy, E.Y. & Wallbank, M. (2013) Geographic distribution. *Ophiophagus hannah.* *Herpetological Review* 44(1), 110.

Sztencel-Jabłonka, A., Mazgajski, T.D., Bury, S., Najbar, B., Rybacki, M., Bogdanowicz, W. & Mazgajska, J. (2015) Phylogeography of the smooth snake *Coronella austriaca* (Serpentes: Colubridae): Evidence for a reduced gene pool and a genetic discontinuity in central Europe. *Biological Journal of the Linnean Society* 115(1), 195–210, 6 figs.

Takeuchi, H., Savitsky, A.H., Ding, L., Silva, A. de, Das, I., Nguyen, T.T., Tsai, T.-S. et al. (2018) Evolution of nuchal glands, unusual defensive organs of Asian natricine snakes (Serpentes: Colubridae), inferred from a molecular phylogeny. *Ecology and Evolution* 8(20), 10219–10232, 3 figs.

Takeuchi, H., Zhu, G.-X., Ding, L., Tang, Y., Ota, H., Mori, A., Oh, H.-S. & Hikida, T. (2014) Taxonomic validity and phylogeography of the east Eurasian natricine snake, *Rhabdophis lateralis* (Berthold, 1859) (Serpentes: Colubridae), as inferred from mitochondrial DNA sequence data. *Current Herpetology* 33(2), 148–153, 2 figs.

Tamar, K., Šmid, J., Göçmen, B., Meiri, S. & Carranza, S. (2016) An integrative systematic revision and biogeography of *Rhynchocalamus* snakes (Reptilia, Colubridae) with a description of a new species from Israel. *PeerJ* 2016(2769), 1–37, 6 figs.

Tanaka, R.M., Rotenberg, E.L. & Muscat, E. (2018) *Tropidophis paucisquamis* (Müller in Schenkel, 1901) (Reptilia, Squamata, Tropidophiidae): Notes on natural history and gap-filling record for lowland Atlantic Forest in Ubatuba, state of São Paulo, Brazil. *Herpetology Notes* 11, 243–244, 2 figs.

Tandavanitj, N., Mitani, S. & Toda, M. (2013b) Origins of *Laticauda laticaudata* and *Laticauda semifasciata* (Elapidae: Laticaudinae) individuals collected from the main islands of Japan as inferred from molecular data. *Current Herpetology* 32(2), 135–141, 2 figs.

Tandavanitj, N., Ota, H., Cheng, Y.-C. & Toda, M. (2013a) Geographic genetic structure in two laticaudine sea kraits, *Laticauda laticaudata* and *Laticauda semifasciata* (Serpentes: Elapidae), in the Ryukyu-Taiwan region as inferred from mitochondrial cytochrome *b* sequences. *Zoological Science* 30, 633–641, 3 figs.

Tapley, B. & Muurmans, M. (2011) Herpetofaunal records from Pulau Bangkaru, Sumatra. *Herpetology Notes* 4, 413–417, 3 figs.

Tarn, N., Tobi, E. & Dixon-MacCallum, G.P. (2018a) Geographic distribution: *Dipsadoboa underwoodi.* *Herpetological Review* 49(1), 76.

Tarn, N., Tobi, E., Dixon-MacCallum, G.P. & Bamba-Kaya, A. (2018b) Geographic distribution: *Dipsadoboa viridis.* *Herpetological Review* 49(1), 76.

Tarroso, P., Pereira, R.J., Martínez-Freiría, F.M., Godinho, R. & Brito, J.C. (2014) Hybridization at an ecotone: Ecological and genetic barriers between three Iberian vipers. *Molecular Ecology* 23(5), 1108–1123, 6 figs.

Tavares, J.R., Melo, C.E. de, Campos, V.A., Oda, F.H. & Strüssmann, C. (2012) Snakes from Canoa Quebrada Hydroelectric Power Plant, state of Mato Grosso, Brazil. *Herpetology Notes* 5, 543–546, 1 fig.

Terán-Juárez, S.A., García-Padilla, E., Leyto-Delgado, F.E. & García-Morales, L.J. (2015) New records and distributional range extensions for amphibians and reptiles from Tamaulipas, Mexico. *Mesoamerican Herpetology* 2(2), 208–214.

Terán-Juárez, S.A., García-Padilla, E., Mata-Silva, V., Johnson, J.D. & Wilson, L.D. (2016) The herpetofauna of Tamaulipas, Mexico: Composition, distribution, and conservation status. *Mesoamerican Herpetology* 3(1), 43–113, numerous figs.

Teynié, A., David, P., Lottier, A., Le, M.D., Vidal, N. & Nguyen, T.Q. (2015) A new genus and species of xenodermatid snake (Squamata: Caenophidia: Xenodermatidae) from northern Laos People's Democratic Republic. *Zootaxa* 3926(4): 523–540, 8 figs.

Teynié, A., David, P., Ohler, A. & Luanglath, K. (2004) Notes on a collection of amphibians and reptiles from southern Laos, with a discussion of the occurrence of Indo-Malayan species. *Hamadryad* 29(1), 33–62, 8 figs.

Teynié, A. & Hauser, S. (2017) First record of *Parafimbrios lao* (Teynié, David, Lottier, Le, Vidal et Nguyen, 2015) (Squamata: Caenophidia: Xenodermatidae) for Thailand. *Russian Journal of Herpetology* 24(1), 41–48, 9 figs.

Teynié, A., Lottier, A., David, P., Nouyen, T.Q. & Vogel, G. (2014) A new species of the genus *Opisthotropis* Günther, 1872 from northern Laos (Squamata: Natricidae). *Zootaxa* 3774(2), 165–182, 6 figs.

Thaler, R., Folly, H., Galvão, C. & Silva, L.A. da (2018) First predation report of *Leptodactylus chaquensis* (Anura, Leptodactylidae) by *Helicops infrataeniatus* (Squamata, Dipsadidae) and distribution extension for this water snake in the Brazilian Cerrado ecoregion. *Herpetology Notes* 11, 539–541, 2 figs.

Thesing, B.J. & Clause, A.G. (2018) Filling gaps with specimens: New herpetofaunal county records from Georgia, USA. *Herpetological Review* 49(1), 80–84.

Thomas, T.M., Suarez, E. & Enge, K.M. (2015) New county records of reptiles from the Suwannee River, Florida, USA. *Herpetological Review* 46(4), 608.

Thomassen, H., Costa, H.C., Silveira, A.L., Garcia, P.C.A. & Bérnils, R.S. (2015) First records of the snake *Siphlophis leucocephalus* (Günther, 1863) in Minas Gerais, Brazil, and a review of the geographic distribution of *S. longicaudatus* (Andersson, 1901) (Squamata: Dipsadidae). *Check List* 11(1637), 1–10, 5 figs.

Tiburcio, I.C.S., Lisbia, B.S. & Araujo-Vieira, K. (2016) Geographic distribution. *Xenopholis undulatus. Herpetological Review* 47(2), 267.

Timms, J., Chaparro, J.C., Venegas, P.J., Salazar-Valenzuela, D., Scrocchi, G., Cuevas, J., Leynaud, G. & Carrasco, P.A. (2019) A new species of pitviper of the genus *Bothrops* (Serpentes: Viperidae: Crotalinae) from the central Andes of South America. *Zootaxa* 4656(1), 99–120, 12 figs.

Tiutenko, A. (2018) A new record of *Pseudoboodon gascae* (Peracca, 1897) (Squamata: Serpentes: Lamprophiidae) from the southern edge of the Ethiopian Highlands. *Herpetology Notes* 11, 201–203, 5 figs.

Tok, C.V., Özkan, B., Gürkan, M. & Yakin, B.Y. (2014) Çanakkale'nin tetrapodları (Amphibia, Reptilia, Aves, Mammalia) ve korunma statüleri. *Anadolu Doğa Bilimleri Dergisi* 5(2), 36–53.

Tok, C.V., Tosunoğlu, M., Gül, C., Yiğini, B., Türkakın, M., Saruhan, G. & Kaya, S. (2006) Erythrocyte count and size in some colubrids (Reptilia: Ophidia) from Turkey. *Russian Journal of Herpetology* 13(2), 97–100, 1 fig.

Tomović, L., Ajtić, R., Ljubisavljević, K., Urošević, A., Jović, D., Krizmanić, I., Labus, N. et al. (2014) Reptiles in Serbia -- Distribution and diversity patterns. *Bulletin of the Natural History Museum* 7, 129–158, 12 figs.

Tomović, L., Urošević, A., Ajtić, R., Krizmanić, I., Simović, A., Labus, N., Jović, D. et al. (2015) Contribution to the knowledge of distribution of colubrid snakes in Serbia. *Ecologica Montenegrina* 2(3), 162–186, 4 figs.

Torki, F. (2017a) Description of a new species of *Lytorhynchus* (Squamata: Colubridae) from Iran. *Zoology in the Middle East* 63(2), 109–116, 5 figs.

Torki, F. (2017b) A new species of blind snake, *Xerotyphlops*, from Iran. *The Herpetological Bulletin* 140, 1–5. 4 figs.

Torres, J., Martínez-Muñoz, C.A. & Gandia, A.C. (2019) Geographic distribution: *Typhlops leptolepis. Herpetological Review* 50(1), 107.

Torres, J., Rodríguez-Cabrera, T.M. & Martínez-Muñoz, C.A. (2016) Geographic distribution. *Tropidophis galacelidus. Herpetological Review* 47(1), 85.

Torres-Carvajal, O., Echevarría, L.Y., Lobos, S.E., Venegas, P.J. & Kok, P.J.R. (2019) Phylogeny, diversity and biogeography of Neotropical sipo snakes (Serpentes: Colubrinae: *Chironius*). *Molecular Phylogenetics and Evolution* 130, 315–329, 4 figs.

Torres-Carvajal, O., Echevarría, L.Y., Venegas, P.J., Chávez, G. & Camper, J.D. (2015) Description and phylogeny of three new species of *Synophis* (Colubridae, Dipsadinae) from the tropical Andes in Ecuador and Peru. *ZooKeys* 546, 153–179, 13 figs.

Torres-Orriols, N. (2014) Aproximació a l'origen peninsular de les *Hemorrhois hippocrepis* recentment arribades a les Illes Balears a partir de l'extracció i seqüenciació del gen citocrom B de diversos exemplars. *Butlletí de la Societat Catalana d'Herpetologia* 21, 174–183, 2 figs.

Torres-Pérez-Coeto, J., Alvarado-Díaz, J., Suazo-Ortuño, I. & Wilson, L.D. (2016) *Ficimia publia* (Cope, 1866) (Squamata: Colubridae): First record for the herpetofauna of Michoacán, México. *Acta Zoológica Mexicana* 32(1), 123–125, 1 fig.

Torres-Santana, C. (2010) Borikenophis portoricensis (Puerto Rican Racer). *Caribbean Herpetology* (13), 1, 1 fig.

Torres-Solís, M.A., Nahuat-Cervera, P.E. & Ortiz-Medina, J.A. (2017) First record of *Leptophis ahaetulla* (Linnaeus, 1758) (Squamata: Colubridae) from the state of Yucatán, Mexico. *Mesoamerican Herpetology* 4(1), 199–200, 1 fig.

Tosunoğlu, M., Gül, C. & Uysal, I. (2009) The herpetofauna of Tenedos (Bozcaada, Turkey). *Herpetozoa* 22(1/2), 75–78, 2 figs.

Tóth, T., Heltai, M., Keszi, A., Sušić, G., Moharos, L., Farkas, B., Géczy, C., Torda, O. & Gál, J. (2017) Herpetofauna inventory of the small islands of the Cres-Lošinj Archipelago (North Adriatic Sea, Croatia). *Herpetozoa* 30(1/2), 21–28, 3 figs.

Townsend, J.H. (2014) Characterizing the Chortis Block Biogeographic Province: Geological, physiographic, and ecological associations and herpetofaunal diversity. *Mesoamerican Herpetology* 1(2), 203–252, 11 figs.

Townsend, J.H., Herrera-B., L.A., Medina-Flores, M., Gray, L.N., Stubbs, A.L. & Wilson, L.D. (2010) Notes on the second male specimen of the cryptozoic snake *Geophis damiani* (Wilson, McCranie & Williams, 1998). *Herpetology Notes* 3, 305–308, 2 figs.

Townsend, J.H., Medina-Flores, M., Wilson, L.D., Jardin, R.C. & Austin, J.D. (2013b) A relict lineage and new species of green palm-pitviper (Squamata, Viperidae, *Bothriechis*) from the Chortís Highlands of Mesoamerica. *ZooKeys* 29, 77–105, 7 figs.

Townsend, J.H., Wilson, L.D., Medina-Flores, M., Aguilar-Urbina, E., Atkinson, B.K., Cerrato-Mendoza, C.A., Contreras-Castro, A. et al. (2012) A premontane hotspot for herpetological endemism on the windward side of Refugio de Vida Silvestre Texíguat, Honduras. *Salamandra* 48(2), 92–114, 9 figs.

Townsend, J.H., Wilson, L.D., Medina-Flores, M. & Herrera-B., L.A. (2013a) A new species of centipede snake in the *Tantilla taeniata* group (Squamata: Colubridae) from Premontane rainforest in Refugio De Vida Silvestre Texíguat, Honduras. *Journal of Herpetology* 47(1), 191–200, 4 figs.

Trape, J.-F. (2014) Une espèce et un genre nouveaux de Rhinoleptini Hedges, Adalsteinsson & Branch, 2009, du Mali (Reptilia, Squamata, Leptotyphlopidae). *Bulletin de la Société Herpétologique de France* 152, 45–56, 8 figs.

Trape, J.-F. (2018) Partition d'*Echis ocellatus* (Stemmler, 1970) (Squamata, Viperidae), avec la description d'une espèce nouvelle. *Bulletin de la Société Herpétologique de France* 167, 13–34, 15 figs.

Trape, J.-F. (2019) Scolecophidiens (Squamata: Ophidia) nouveaux d'Afrique centrale. *Bulletin de la Société Herpétologique de France* 169, 27–44, 13 figs.

Trape, J.-F. & Baldé, C. (2014) A checklist of the snake fauna of Guinea, with taxonomic changes in the genera *Philothamnus* and *Dipsadoboa* (Colubridae) and a comparison with the snake fauna of some other West African countries. *Zootaxa* 3900(3), 301–338, 2 figs.

Trape, J.-F. & Chirio, L. (2019) Une nouvelle espèce de Leptotyphlopidae (Squamata: Ophidia) d'Afrique centrale. *Bulletin de la Société Herpétologique de France* 169, 45–52, 6 figs.

Trape, J.-F., Crochet, P.-A., Broadley, D.G., Sourouille, P., Mané, Y., Burger, M., Böhme, W. et al. (2019) On the *Psammophis sibilans* group (Serpentes, Lamprophiidae, Psammophiinae) north of 12°S, with the description of a new species from West Africa. *Bonn Zoological Bulletin* 68(1), 61–91, 23 figs.

Trape, J.-F. & Mané, Y. (2015) The snakes of Niger. *Amphibian & Reptile Conservation* 9(2), 39–55, 7 figs.

Trape, J.-F. & Mané, Y. (2017) The snakes of Mali. *Bonn Zoological Bulletin* 66(2), 107–133, 12 figs.

Trape, J.-F. & Mediannikov, O. (2016) Cinq serpents nouveaux du genre *Boaedon* (Duméril, Bibron & Duméril, 1854) (Serpentes: Lamprophiidae) en Afrique centrale. *Bulletin de la Société Herpétologique de France* 159, 61–111, 32 figs.

Trevine, V., Forlani, M.C., Haddad, C.F.B. & Zaher, H. (2014) Herpetofauna of Paranapiacaba: Expanding our knowledge on a historical region in the Atlantic forest of southeastern Brazil. *Zoologia* 31(2), 126–146, 86 figs.

Trivedi, K. & Desai, S. (2019) Unusual morphs of Russell's Viper, *Daboia russelii* (Shaw and Nodder, 1797), from Goa, India. *IRCF Reptiles & Amphibians* 26(1), 70–72, 3 figs.

Trutnau, L., Monzel, M. & Consul, A. (2016) *Amphibien und reptilian des Paraguayischen Chaco.* Edition Chimaira, Frankfurt-am-Main, 1–543, 836 figs.

Tshewang, S. & Letro, L. (2018) The herpetofauna of Jigme Singye Wangchuck National Park in central Bhutan: Status, distribution and new records. *Journal of Threatened Taxa* 10(11), 12489–12498, 34 figs.

Tuniyev, B.S. (2016) Rare species of shield-head vipers in the Caucasus. *Nature Conservation Research* 1(3), 11–15, 9 figs.

Tuniyev, B.S., Avcı, A., Ilgaz, Ç., Olgun, K., Petrova, T.V., Bodrov, S.Y., Geniez, P. & Teynié, A. (2018a) On taxonomic status of shield-headed vipers from Turkish Lesser Caucasus and East Anatolia. *Proceedings of the Zoological Institute Russian Academy of Science* 322(1), 3–44, 21 figs.

Tuniyev, B.S., Iremashvili, G.N., Petrova, T.V. & Kravchenko, M.V. (2018b) Rediscovery of the Steppe Viper in Georgia. *Proceedings of the Zoological Institute Russian Academy of Science* 322(2), 87–107, 17 figs.

Tuniyev, B.S., Lotiev, K.Y., Tuniyev, S.B., Gabaev, V.N. & Kidov, A.A. (2017) "Amphibians and reptiles of South Ossetia." *Nature Conservation Research* 2(2), 1–23, 10 figs. (in Russian).

Tuniyev, S.B., Iremashvili, G.N., Heras, B. de las & Tuniyev, B.S. (2014) About type locality and finds of Darevsky's Viper [*Pelias darevskii* (Vedmederja, Orlov et Tuniyev, 1986), Reptilia: Viperinae] in Georgia. *Russian Journal of Herpetology* 21(4), 281–290, 12 figs.

Tuniyev, S.B., Kidov, A.A. & Tuniyev, B.S. (2016) Additions to the description and rapid assessment of the current staus of a population of the relic viper (*Pelias magnifica* (Tuniyev et Ostrovskikh, 2001)), (Ophidia: Viperinae) at a type locality. *Modern Herpetology* 16(1/2), 43–50, 10 figs.

Tuniyev, S.B., Orlov, N.L., Tuniyev, B.S. & Kidov, A.A. (2013) On the taxonomical status of Steppe viper from foothills of the south macroslope of the East Caucasus. *Russian Journal of Herpetology* 20(2), 129–146, 16 figs.

Tuniyev, S.B., Tuniyev, B.S. & Mazanaeva, L.F. (2011) Distribution and variation of Lotiev's Viper Pelias lotievi (Nilson, Tuniyev, Orlov, Hoggren et Andren, 1995) (Serpentes: Viperinae). In: *Proceedings of IV Meeting of A.M. Nikolsky Herpetological Society (Kazan, October 12–17, 2009).* St. Petersburg, 250–266, 4 figs.

Tupikov, A.I. & Zinenko, O.I. (2015a) Distribution of the Dione snake *Elaphe dione* (Reptilia, Colubridae) in Ukraine: Historical aspect and the current state. *Visnik Dnipropetrovs'kogo Universitetu. Series Biology, Ecology* 23(2), 91–99, 3 figs.

Tupikov, A.I. & Zinenko, O.I. (2015b) Distribution of the steppe viper *Vipera renardi* (Reptilia, Viperidae) in Kharkiv region. *Visnik Dnipropetrovs'kogo Universitetu. Series Biology, Ecology* 23(2), 172–176, 1 fig.

Turcios-Casco, M.A., Galdámez, J.R., Salazar-Saavedra, M. & McCranie, J.R. (2018) A second locality for *Rhinobothryum bovallii* Andersson (Colubridae) in nuclear Central America, with comments on its habitat. *Mesamerican Herpetology* 5(1), 137–144, 1 fig.

Turner, D.S., Holm, P.A., Wirt, E.B. & Schwalbe, C.R. (2003) Amphibians and reptiles of the Whetstone Mountains, Arizona. *The Southwestern Naturalist* 48(3), 347–355, 2 figs.

Turrisi, G.F. (1996) Gli anfibi e i rettili. In: *Atti del Convegno su La Fauna degle Iblei tenuto dall'Ente Fauna Siciliana a Noto il 13 e 14 maggio* 1995, 103–116.

Turrisi, G.F., Lo Cascio, P. & Vaccaro, A. (2008) Anfibi e rettili (Amphibia et Reptilia). In: *Atlante della biodiversità della Sicilia; Vertebrati terrestri. Collana Studi e Ricerche dell'ARPA Sicilia*, Palermo, pp. 251–372, numerous figs. and maps.

Turrisi, G.F. & Vaccaro, A. (1998) Contributo alla conoscenza degli anfibi e dei rettili di Sicilia. *Bollettino dell'Accademia Gioenia di Scienze Naturali* 30(353), 5–88, 36 figs.

Ubeda-Olivas, M.F. & Sunyer, J. (2015a) Distribution notes: *Senticolis triaspis. Mesoamerican Herpetology* 2(4), 546–547, 1 fig.

Ubeda-Olivas, M.F. & Sunyer, J. (2015b) Distribution notes: *Conophis lineatus. Mesoamerican Herpetology* 2(4), 548, 1 fig.

Uetz, P., Cherikh, S., Shea, G., Ineich, I., Campbell, P.D., Doronin, I.V., Rosado, J. et al. (2019) A global catalog of primary reptile type specimens. *Zootaxa* 4695(5), 438–450, 4 figs.

Uetz, P., Davison, R. & Ellis, J. (2014) [Review] Snakes of the World: Catalogue of living and extinct species. *Herpetological Review* 45(4), 721–722.

Uğurtaş, I.H., Durmuş, S.H. & Kete, R. (2000a) [Some poisonous animals found in Bursa Uludağ]. *Ocak-Şubat-Mart* 2000 34, 3–8, 6 figs. [in Turkish].

Uğurtaş, I.H., Kaya, R.S. & Akkaya, A. (2007) The herpetofauna of the islands in Uluabat Lake (Bursa). *Ekoloji* 17(65), 7–10, 2 figs.

Uğurtaş, I.H., Sevinç, M., Öz, M. & Kaya, R.S. (2006) New localities for *Leptotyphlops macrorhynchus* (Jan, 1862) (Reptilia: Leptotyphlopidae) in Turkey. *Turkish Journal of Zoology* 30(4), 373–376, 1 fig.

Uğurtaş, I.H., Yildirimhan, H.S. & Öz, M. (2000b) Herpetofauna of the eastern region of the Amanos Mountains (Nur). *Turkish Journal of Zoology* 24(3), 257–261, 1 fig.

Uhrin, M., Havaš, P., Minařík, M., Kodejš, I., Danko, S., Husák, T., Koleska, D. & Jablonski, D. (2016) Distribution updates to amphibian and reptile fauna for the Republic of Macedonia. *Herpetology Notes* 9, 201–220, 5 figs.

Ukuwela, K.D.B., Lee, M.S.Y., Rasmussen, A.R., Silva, A de, Mumpuni, Fry, B.G., Ghezellou, P., Rezaie-Atagholipour, M. & Sanders, K.L. (2016) Evaluating the drivers of Indo-Pacific biodiversity: Speciation and dispersal of sea snakes (Elapidae: Hydrophiinae). *Journal of Biogeography* 43(2), 243–255, 3 figs.

Ukuwela, K.D.B., Silva, A. de & Sanders, K.L. (2017) Further specimens of the mud snake, *Gerarda prevostiana* (Homalopsidae) from Sri Lanka with insights from molecular phylogenetics. *Raffles Bulletin of Zoology* 65(1), 29–34, 3 figs.

Ullenbruch, K. & Böhme, W. (2017) Six new records of Afrotropical lizard and snake species (Reptilia: Squamata) from the Republic of South Sudan. *Bonn Zoological Bulletin* 66(2), 139–144, 7 figs.

Urbina-Cardona, J.N., Londoño-Murcia, M.C. & García-Ávila, D.G. (2008) Dinámica espacio-temporal en la diversidad de serpientes en cuatro habitats con diferente grado de alteración antropogénica en el Parque Nacional Natural Isla Gorgona, Pacífico Colombiano. *Caldasia* 30(2), 479–493, 3 figs.

Uriarte-Garzón, P. & García-Vázquez, U.O. (2014) Primer registro de *Nerodia erythrogaster bogerti* (Conant, 1953) (Serpentes: Colubridae) para el Estado de Chihuahua, México. *Acta Zoológica Mexicana (n.s.)* 30(1), 221–225, 4 figs.

Urioste, J.A. de & Mateo, J.A. (2011) Nuevos datos acerca de la culebrilla ciega de las macetas, *Ramphotyphlops braminus*, en Canarias. *Boletin de la Asociación Herpetológica Española* 22, 135–137, 2 figs.

Urriza, R.C., Elec, J.A. & Palacpac, A.S. (2017) Geographic distribution. *Ophiophagus hannah*. *Herpetological Review* 48(3), 591.

Ursenbacher, S., Guillon, M., Cubizolle, H., Dupoué, A., Blouin-Demers, G. & Lourdais, O. (2015) Postglacial recolonization in a cold climate specialist in western Europe: Patterns of genetic diversity in the adder (*Vipera berus*) support the central-marginal hypothesis. *Molecular Ecology* 24(14), 3639–3651, 4 figs.

Uysal, I. & Tosunoğlu, M. (2012) [Herpetofaunal richness of the Kavak Delta (Saroz Bay)]. *Anadolu Doğa Bilimleri Dergisi* 3(2), 52–58, 2 figs. [in Turkish].

Vaccaro, A. & Turrisi, G.F. (2007) Ritrovamento di *Zamenis longissimus* (Laurenti, 1768) (Reptilia, Colubridae) sull'Isola d'Elba (Toscana, Italia). *Acta Herpetologica* 2(1), 59–63, 3 figs.

Valdez-Rentería, S.Y. & Fernández-Badillo, L. (2016) Nuevos registros herpetofaunísticos para el Municipio de Tezontepec de Aldama, Hidalgo, México. *Acta Zoológica Mexicana (n.s.)* 32(2), 199–201, 1 fig.

Valdez-Villavicencio, J.H., Peralta-García, A. & Yee-Pérez, H. (2016) Distribution notes: *Indotyphlops braminus. Mesoamerican Herpetology* 3(1), 205, 1 fig.

Valencia, J.H. & Garzon-Tello, K. (2018) Reproductive behavior and development in *Spilotes sulphureus* (Serpentes: Colubridae) from Ecuador. *Phyllomedusa* 17(1), 113–126, 5 figs.

Valencia, J.H., Garzón-Tello, K. & Barragán-Paladines, M.E. (2016) *Serpientes venenosas del Ecuador.* Fundación Herpetológica Gustavo Orcés, Quito, 1–652, 280 figs., 137 plates.

Valencia-Herverth, J. & Fernández-Badillo, L. (2013) Geographic distribution. *Micrurus bernadi. Herpetologcal Review* 44(1), 109–110.

Vamdev, Y.H., Varma, V. & Parashar, A. (2019) Range extension of Forsten's Catsnake, *Boiga forsteni* (Duméril, Bibron, and Duméril 1854) in the Damoh District of Madhya Pradesh, India. *IRCF Reptiles & Amphibians* 26(1), 73–74, 2 figs.

Vanderduys, E.P. (2013) Additional information on *Ramphotyphlops aspina* (Couper, Covacevich & Wilson 1998) (Reptilia: Typhlopidae), a poorly known blind snake from the Mitchell Grass Downs of Queensland. *Memoirs of the Queensland Museum* 56(2), 615–619, 4 figs.

Van Devender, T.R., Hernández-Jimenez, R. & Silva-Kurumiya, H. (2014) Geographic distribution. *Sonora aemula. Herpetological Review* 45(1), 96–97.

Van Devender, T.R. & Holm, P. (2014) Geographic distribution. *Sonora aemula. Herpetological Review* 45(1), 96.

Vanegas-Guerrero, J., Batista, A., Medina, I. & Vargas-Salinas, F. (2015) *Tantilla alticola* (Boulenger, 1903) (Squamata: Colubridae): Filling a geographical distribution gap in western Colombia. *Check List* 11(1555), 1–3, 2 figs.

Vanegas-Guerrero, J., Mantilla-Castaño, J.C. & Passos, P. (2014) *Atractus titanicus* (Passos, Arredondo, Fernandes & Lynch, 2009) (Serpentes: Dipsadidae): Filling gaps in its geographical distribution. *Check List* 10(3), 672–673, 2 figs.

Vanegas-Guerrero, J., Martins, A., Quiñones-Betancurt, E. & Lynch, J.D. (2019) Rediscovery of the rare Andean blindsnake *Anomalepis colombia* (Marx, 1953) (Serpentes: Anomalepididae) in the wild. *Zootaxa* 4623(3), 595–600, 4 figs.

Vanzolini, P.E. & Calleffo, M.E.V. (2002) On some aspects of the reproductive biology of Brasilian *Crotalus* (Serpentes, Viperidae). *Biologia Geral e Experimental* 3(1), 3–37, 6 figs.

Vanzolini, P.E. & Myers, C.W. (2015) The herpetological collection of Maximilian, Prince of Wied (1782–1867), with special reference to Brazilian materials. *Bulletin of the American Museum of Natural History* (395), 1–155, 29 figs., 56 plates.

Vasaruchapong, T., Laoungbua, P., Tangrattanapibul, K., Tawan, T. & Chanhome, L. (2017) *Protobothrops muscrosquamatus* (Cantor, 1839), a highly venomous species added to the snake fauna of Thailand (Squamata: Viperidae). *Tropical Natural History* 17(2), 111–115, 4 figs.

Vasconcelos, R., Montero-Mendieta, S., Simó-Riudalbas, M., Sindaco, R., Santos, X., Fasola, M., Llorente, G., Razzatti, E. & Carranza, S. (2016) Unexpectedly high levels of cryptic diversity uncovered by a complete DNA barcoding of reptiles of the Socotra Archipelago. *PLOS ONE* 11(e0149985), 1–19, 5 figs.

Vasile, Ş., Csiki-Sava, Z. & Venczel, M. (2013) A new madtsoiid snake from the upper Cretaceous of the Hațeg Basin, western Romania. *Journal of Vertebrate Paleontology* 33(5), 1100–1119, 6 figs.

Vásquez-Restrepo, J.D. & Toro-Cardona, F.A. (2019) *Rhadinaea decorata* (Günther, 1858). *Catálogo de Anfibios y Reptiles de Colombia* 5(1), 56–63, 6 figs.

Vassilieva, A.B. (2015) A new species of the genus *Oligodon* (Fitzinger, 1826) (Squamata: Colubridae) from coastal southern Vietnam. *Zootaxa* 4058(2), 211–226, 7 figs.

Vassilieva, A.B., Galoyan, E.A., Poyarkov, N.A. & Geissler, P. (2016) *A Photographic Field Guide to the Amphibians and Reptiles of the Lowland Monsoon Forests of Southern Vietnam.* Edition Chimaira, Frankfurt-am-Main, 1–324, 396 figs.

Vassilieva, A.B., Geissler, P., Galoyan, E.A., Poyarkov, N.A., Van Devender, R.W. & Böhme, W. (2013) A new species of Kukri Snake *Oligodon* (Fitzinger, 1826) (Squamata: Colubridae) from the Cat Tien National Park, southern Vietnam. *Zootaxa* 3702(3), 233–246, 6 figs.

Vaz, R.I. & Ortega-Chinchilla, J.E. (2019) Thermographic record of predation of *Rhinella ornata* (Spix, 1824) (Anura: Bufonidae) by *Xenodon neuwiedii* (Günther, 1863) (Squamata: Dipsadidae) with feeding behaviour notes. *Herpetology Notes* 12, 235–239, 1 fig.

Vázquez-Vega, L.F., Canseco-Márquez, L., Acosta-Sánchez, M.M. & Flores-Villela, O. (2016) Distribution notes: *Geophis semidoliatus. Mesoamerican Herpetology* 3(2), 514–515, 1 fig.

Vaz-Silva, W., Guedes, A.G., Azevedo-Silva, P.L. de, Gontijo, F.F., Barbosa, R.S., Aloísio, G.R. & Oliveira, F.C.G. de (2007)

Herpetofauna, Espora Hydroelectric Power Plant, state of Goiás, *Brazil. Check List* 3(4), 338–345, 2 figs.

Vecchiet, J.A., Ray, J.M., Knight, J.L. & Wedow, J. (2014) Geographic distribution *Dipsas articulata. Herpetological Review* 45(1), 94.

Vences, M., Glaw, F., Mercurio, V. & Andreone, F. (2004) Review of the Malagasy tree snakes of the genus *Stenophis* (Colubridae). *Salamandra* 40(2), 161–179, 3 figs.

Venchi, A., Wilson, S.K. & Borsboom, A.C. (2015) A new blind snake (Serpentes: Typhlopidae) from an endangered habitat in south-eastern Queensland, Australia. *Zootaxa* 3990(2), 272–278, 4 figs.

Venczel, M. (2000) *Quaternary snakes from Bihor (Romania).* Publishing House of the Tǎrii Crişurilor Museum, Oradea, 1–142, 72 figs.

Venczel, M. (2011) Middle-Late Miocene snakes from the Pannonian Basin. *Acta Palaeontologica Romaniae* 7, 343–349, 2 plates.

Venczel, M. & Sanchíz, B. (2006) Lower Miocene amphibians and reptiles from Oschiri (Sardinia, Italy). *Hantkeniana* 5, 72–75, fig. 1.

Venegas, P.J. (2005) Herpetofauna del bosque seco ecuatorial de Peru: Taxonomía, ecología y biogeografía. *Zonas Áridas* 9, 9–26, 1 fig.

Venegas, P.J., Chávez-Arribasplata, J.C., Almora, A., Girilli, P. & Duran, V. (2019) New observations on diet of the South American two-striped forest-pitviper *Bothrops bilineatus smaragdinus* (Hoge, 1966). *Cuadernos de Herpetología* 33(1), 29–31, 1 fig.

Vera-Pérez, L.E. (2019) A new species of *Sibon* (Fitzinger, 1826) (Squamata: Colubridae) from southwestern Colombia. *Zootaxa* 4701(5), 443–453, 6 figs.

Vera-Pérez, L.E. & Zúñiga-Baos, J.A. (2015) First record of *Urotheca dumerilli* [sic] (Bibron, 1840) (Squamata: Dipsadidae) in Cauca state, Colombia and notes on natural history. *Check List* 11(1794), 1–5, 7 figs.

Vera-Pérez, L.E., Zúñiga-Baos, J.A. & Alzate-Basto, E. (2019) *Rhinobothryum bovallii* (Andersson, 1916) (Serpentes: Colubridae): Nuevas localidades para Colombia, descripción hemipeneal y comentarios sobre su historia natural. *Revista Novedades Colombianas* 14, 57–68, 5 figs.

Vera-Pérez, L.E., Zúñiga-Baos, J.A. & González, S.A. (2018) Reptiles del Parque Nacional Natural Munchique, Colombia. *Revista Novedades Colombianas* 13(1), 97–131, numerous figs.

Vera-Pérez, L.E., Zúñiga-Baos, J.A. & Montingelli, G.G. (2015) First record of *Tantilla alticola* (Boulenger, 1903) (Serpentes: Colubridae) in Cauca state, Colombia, filling distribution gap and notes on natural history. *Check List* 11(1578), 1–4, 4 figs.

Verducci, D. & Zuffi, M.A.L. (2015) Anfibi e rettili del Padule di Bientina (Toscano centro-settentrionale): Note ecologiche e distributive. *Quadernos Museo di Storia Naturale del Livorno* 26, 1–105 figs.

Vianey-Liaud, M., Comte, B., Marandat, V., Peigné, S., Rage, J-G. & Sudre, J. (2014) A new early Late Oligocene (MP 26) continental vertebrate fauna from Saint-Privat-des-Vieux (Alès Basin, Gard, southern France). *Geodiversitas* 36(4), 565–622, 17 figs.

Vignaud, P., Duringer, P., Mackaye, H.T., Likius, A., Blondel, C., Boisserie, J.-R., Bonis, L. de et al. (2002) Geology and palaeontology of the Upper Miocene Toros-Menalla hominid locality, Chad. *Nature* 418(6894), 152–155, 2 figs.

Vignoli, L., Segniagbeto, G.H., Eniang, E.A., Hema, E., Petrozzi, F., Akani, G.C. & Luiselli, L. (2015) Aspects of natural history in a sand boa, *Eryx muelleri* (Erycidae) from arid savannahs in Burkina Faso, Togo, and Nigeria (West Africa). *Journal of Natural History* 50(11/12), 749–758, 4 figs.

Vilela, B., De Lima, M.G., Gonçalves, U. & Skuk, G.O. (2011) *Siphlophis compressus* (Daudin, 1803) (Squamata: Dipsadidae): First records for the Atlantic forest north of the São Francisco river, northeastern Brazil. *Cuadernos de Herpetología* 25(1), 23–24.

Villa, R.A., Van Devender, T.R., Valdéz-Colonel, C.M. & Burkhardt, T.R. (2015) Peripheral and elevational distribution, and a novel prey item for *Drymarchon melanurus* in Sonora, Mexico. *Mesoamerican Herpetology* 2(3), 378–380, 3 figs.

Villalobos-Juárez, I. & Sigala-Rodríguez, J. (2019) Crotalus atrox. *Herpetological Review* 50(2), 330.

Villatoro-Castañeda, M. & Ariano-Sánchez, D. (2017) Rediscovery of the Guatemalan yellow-lipped snake, *Chapinophis xanthocheilus* (Serpentes: Dipsadidae), with comments on its distribution and ecology. *Herpetological Review* 48(1), 25–28, 3 figs.

Vink, J. & Shonleben, S. (2015) Geographic distribution. *Boiga nigriceps. Herpetological Review* 46(2), 218.

Visvanathan, A.C., Anne, S. & Kolli, A.K. (2017) New locality records of the Stout Sand Snake *Psammophis longifrons* (Boulenger, 1890) (Reptilia: Squamata: Lamprophiidae) in Telangana, India. *Journal of Threatened Taxa* 9(11), 10968–10970, 23 figs.

Vite-Silva, V.D., Ramírez-Bautista, A. & Hernández-Salinas, U. (2010) Diversidad de anfibios y reptiles de la Reserva de la Biosfera Barranca de Metztitlán, Hidalgo, México. *Revista Mexicana de Biodiversidad* 81, 473–485, 5 figs.

Vlček, P., Najbar, B. & Jablonski, D. (2010) First records of the Dice Snake (*Natrix tessellata*) from the north-eastern part of the Czech Reupblic and Poland. *Herpetology Notes* 3, 23–26, 5 figs.

Vogel, G. (2015) A new montane species of the genus *Pareas* (Wagler, 1830) (Squamata: Pareatidae) from northern Myanmar. *Taprobanica* 7(1), 1–7, 5 figs.

Vogel, G., Amarasinghe, A.A.T. & Ineich, I. (2016) Resurrection of *Pseudorabdion torquatum* (A.M.C. Duméril, Bibron & A.H.A. Duméril, 1854), a former synonym of *P. longiceps* (Cantor, 1847) (Colubridae: Calamariinae) from Sulawesi, Indonesia. *Zootaxa* 4121(3), 337–345, 2 figs.

Vogel, G. & David, P. (2019) A new species of the *Lycodon fasciatus* complex from the Khorat Plateau, eastern Thailand (Reptilia, Squamata, Colubridae). *Zootaxa* 4577(3), 515–528, 4 figs.

Vogel, G., David, P. & Chandramouli, S.R. (2014a) On the systematics of *Trimeresurus labialis* Fitzinger *in* Steindachner, 1867, a pitviper from the Nicobar Islands (India), with revalidation of *Trimeresurus mutabilis* (Stoliczka, 1870) (Squamata, Viperidae, Crotalinae). *Zootaxa* 3786(5), 557–573, 5 figs.

Vogel, G., David, P. & Reza, F. (2018) A contribution to the systematics of the Indonesian snake genus *Elapoidis* (Boie, 1827) (Squamata: Colubridae). *Russian Journal of Herpetology* 25(2), 113–120, 5 figs.

Vogel, G., David, P. & Sidik, I. (2014b) On *Trimeresurus sumatranus* (Raffles, 1822), with the designation of a neotype and the description of a new species of pitviper from Sumatra (Squmata: Viperidae: Crotalinae). *Amphibian & Reptile Conservation* 8(2), 1–29, 16 figs.

Vogel, G. & Ganesh, S.R. (2013) A new species of cat snake (Reptilia: Serpentes: Colubridae: *Boiga*) from dry forests of eastern Peninsular India. *Zootaxa* 3637(2), 158–168, 3 figs.

Vogel, G. & Han-Yuen, H.K. (2010) Death feigning behavior in three colubrid species of Tropical Asia. *Russian Journal of Herpetology* 17(1), 15–21, 5 figs.

Vogel, G. & Harikrishnan, S. (2013) Revalidation of *Lycodon hypsirhinoides* (Theobald, 1868) from Andaman Islands (Squamata: Serpentes: Colubridae). *Taprobanica* 5(1), 19–31, 9 figs.

Vogel, G. & Hauser, S. (2013) Addition of *Ptyas nigromarginata* (Blyth, 1854) (Squamata: Colubridae) to the snake fauna of Thailand with preliminary remarks on its distribution. *Asian Herpetological Research* 4(3), 166–181, 6 figs.

Vogel, G., Lalremsanga, H.T. & Vanlalhrima (2017) A second species of the genus *Blythia* (Theobald, 1868) (Squamata: Colubridae) from Mizoram, India. *Zootaxa* 4276(4), 569–581, 8 figs.

Volynchik, S. (2012) Morphological variability in *Vipera palaestinae* along an environmental gradient. *Asian Herpetological Research* 3(3), 227–239, 4 figs.

Voris, H.K. (2015) Marine snake diversity in the mouth of the Muar River, Malaysia. *Tropical Natural History* 15(1), 1–21, 10 figs.

Voris, H.K. (2017) Diversity of marine snakes on trawling grounds in the Straits of Malacca and the South China Sea. *Tropical Natural History* 17(2), 65–87, 8 figs.

Voris, H.K., Murphy, J.C., Karns, D.R., Kremer, E. & O'Connell, K. (2012) Differences among populations of the Mekong Mud Snake (*Enhydris subtaeniata*: Serpentes: Homalopsidae) in Indochina. *Tropical Natural History* 12(2), 175–188, 5 figs.

Vullo, R. (2019) A new species of *Lapparentophis* from the mid-Cretaceous Kem Kem beds, Morocco, wiith remarks on the distribution of lapparentophiid snakes. *Comptes Rendus Palevol* 18(7), 765–770, 2 figs.

Vyas, R. (2013) Notes and comments on distribution of a snake: Indian egg eater (*Elachistodon westermanni*). *Russian Journal of Herpetology* 20(1), 39–42, 3 figs.

Vyas, R. (2014) Report on some remarkable specimens with unusual color morph recorded from two species of snakes (Reptilia: Serpentes) from Gujarat, India. *Russian Journal of Herpetology* 21(1), 47–52, 6 figs.

Vyas, R. & Patel, H. (2013) Notes on distribution and natural history of *Psammophis longifrons* (Boulenger 1896) (Serpentes: Psammophiidae: Psammophiinae) in Gujarat, India. *Russian Journal of Herpetology* 20(3), 217–222, 6 figs.

Vyas, R.V., Murphy, J.C. & Voris, H.K. (2013) The dog-faced water snake (*Cerberus rynchops*) and Gerard's mud snake (*Gerarda prevostiana*) at the western edge of their distribution. *Herpetological Review* 44(1), 34–36, 2 figs.

Wagner, P., Bauer, A.M., Leviton, A.E., Wilms, T.M. & Böhme, W. (2016b) A checklist of the amphibians and reptiles of Afghanistan exploring herpetodiversity using biodiversity archives. *Proceedings of the California Academy of Sciences*, series 4, 63(13), 457–565, 26 figs, 14 plates.

Wagner, P., Tiutenko, A., Mazepa, G., Borkin, L.J. & Simonov, E. (2016a) Alai! Alai! – a new species of the *Gloydius halys* (Pallas, 1776) complex (Viperidae, Crotalinae), including a brief review of the complex. *Amphibia-Reptilia* 37(1), 15–31, 6 figs.

Waldez, F., Menin, M. & Vogt, R.C. (2013) Diversidade de anfíbios e répteis Squamata na região do baixo rio Purus, Amazônia Central, *Brasil. Biota Neotropica* 13(1), 300–316, 6 figs.

Wallach, V. (2016) Morphological review and taxonomic status of the *Epictia phenops* species group of Mesoamerica, with description of six new species and discussion of South American *Epictia albifrons*, *E. goudotii*, and *E. tenella* (Serpentes: Leptotyphlopidae: Epictinae). *Mesoamerican Herpetology* 3(2), 216–374, 16 figs.

Wallach, V. & Gemel, R. (2018) *Typhlops weidholzi* n. inedit.. a new species of Letheobia from the Republic of Cameroon, and a synopsis of the genus (Squamata: Serpentes: Scolecophidia: Typhlopidae). *Herpetozoa* 31(1/2), 27–46, 17 figs.

Wallach, V., Williams, K.L. & Boundy, J. (2014) *Snakes of the World: A Catalogue of Living and Extinct Species*. CRC Press, Boca Raton, xxvii + 1209 pp.

Waller, T., Micucci, P.A., Barros, M., Draque, J. & Estavillo, C. (2012) Conservation status of the Argentine Boa Constrictor (*Boa constrictor occidentalis*) 20 years after being listed in CITES Appendix I. *IRCF Reptiles & Amphibians* 19(1), 1–10, 17 figs.

Wang, H., Wang, H., Xiao, Y., Wang, X., Sun, L., Shi, J., Lin, A., Feng, J. & Wu, Y. (2015) Low genetic diversity and moderate interbreeding risk of an insular endemic pit viper (*Gloydius shedaoensis*): Implication for conservation. *Journal of Herpetology* 49(2), 190–199, 5 figs.

Wang, J., Li, Y., Zeng, Z.-C., Lyu, Z.-T., Sung, Y.-H., Li, Y.-Y., Lin, C.-Y. & Wang, Y.-Y. (2019) A new species of the genus *Achalinus* from southwestern Guangdong Province, China (Squamata: Xenodermatidae). *Zootaxa* 4674(4), 471–481, 4 figs.

Wang, J., Lyu, Z.-T., Zeng, Z., Liu, Z.-Y. & Wang, Y.-Y. (2017) Re-description of *Opisthotropis laui* (Yang, Sung and Chan, 2013) (Squamata: Natricidae). *Asian Herpetological Research* 8(1), 70–74, 3 figs.

Wang, J.-H., Huang, X.-Y., Ye, J.-F., Yang, S.-P., Zhang, X.-C. & Chan, B.P.-L. (2019) A report on the herpetofauna of Tengchong section of Gaoligongshan National Nature Reserve, China. *Journal of Threatened Taxa* 11(11), 14434–14451, 7 figs.

Wang, K., Jiang, K., Jin, J.-Q., Liu, X. & Che, J. (2019a) Confirmation of *Trachischium guentheri* (Serpentes: Colubridae) from Tibet, China, with description of Tibetan *T. monticola*. *Zootaxa* 4688(1), 101–110, 4 figs.

Wang, K., Ren, J.-L., Dong, W., Jiang, K., Shi, J.-S., Siler, C.D. & Che, J. (2019b) A new species of plateau pit viper (Reptilia: Serpentes: *Gloydius*) from the Upper Lancang (=Mekong) Valley in the Hengduan Mountain region, Tibet, China. *Journal of Herpetology* 53(3), 224–236, 7 figs.

Wang, P., Shi, L. & Guo, P. (2019) Morphology-based intraspecific taxonomy of *Oreocryptophis porphyraceus* (Cantor, 1839) in mainland China (Serpentes: Colubridae). *Zoological Research* 40(4), 324–330, 3 figs.

Wang, X., Messenger, K., Zhao, E. & Zhu, C. (2014) Reclassification of *Oligodon ningshaanensis* (Yuan, 1983) (Ophidia: Colubridae) into a new genus, *Stichophanes* gen. nov. with description on its malacophagous behavior. *Asian Herpetological Research* 5(3), 137–149, 7 figs.

Wang, Y.-T., Guo, Q., Liu, Z.-Y., Lyu, Z.-T., Wang, J., Luo, L., Sun, Y.-J. & Zhang, Y.-W. (2017) Revision of two poorly known species of *Opisthotropis* (Günther, 1872) (Squamata: Colubridae: Natricinae) with description of a new species from China. *Zootaxa* 4247(4), 391–412, 7 figs.

Wanger, T.C., Motzke, I., Saleh, S. & Iskandar, D.T. (2011) The amphibians and reptiles of the Lore Lindu National Park area, central Sulawesi, Indonesia. *Salamandra* 47(1), 17–29, numerous figs.

Watson, J.A., Spencer, C.L., Schield, D.R., Butler, B.O., Smith, L.L., Flores-Villela, O., Campbell, J.A., Mackessy, S.P., Castoe, T.A. & Meik, J.M. (2019) Geographic variation in morphology in the Mohave Rattlesnake (*Crotalus scutulatus* Kennicott 1861) (Serpentes: Viperidae): Implications for species boundaries. *Zootaxa* 4683(1), 129–143, 7 figs.

Wegener, J.E., Swoboda, S., Hawlitschek, O., Franzen, M., Wallach, V., Vences, M., Nagy, Z.T., Hedges, S.B., Kohler, J. & Glaw, F. (2013) Morphological variation and taxonomic reassessment of the endemic Malagasy blind snake family Xenotyphlopidae. *Spixiana* 36(2), 269–282, 7 figs.

Weiler, A. & Ortega, A. (2016) Primer registro de *Chironius quadricarinatus* (Serpentes, Colubridae) para el departamento de Cordillera, Paraguay. *Kempffiana* 12(2), 54–56, 2 figs.

Weinell, J.L. & Austin, C.C. (2017) Refugia and speciation in North American scarlet snakes (*Cemophora*). *Journal of Herpetology* 51(1), 161–171, 4 figs.

Weinell, J.L. & Brown, R.M. (2018) Discovery of an old, archipelago-wide, endemic radiation of Philippine snakes. *Molecular Phylogenetics and Evolution* 119, 144–150, 3 figs.

Weinell, J.L., Hooper, E., Leviton, A.E. & Brown, R.M. (2019) Illustrated key to the snakes of the Philippines. *California Academy of Sciences*, Series 4, 66(1), 1–49, 47 figs.

Weller, W.F., McCarter, J., Gagnon, L., Brinker, S., Copeland, C.E., Oldham, M.J., Shueler, F.W. & Thompson, S. (2017) New records of amphibians and reptiles from Cockburn Island, Manitoulin District, Ontario, Canada. *Herpetological Review* 48(3), 593–595.

Werner, Y.L. & Shacham, B. (2010) The herpetological collection (Section of the Amphibians and Reptiles), report of the Section. *Haasiana* 5, 15–33, 3 figs.

Werner, Y.L. & Shapira, T. (2011) A brief review of morphological variation in *Natrix tessellata* in Israel: Between sides, among individuals, between sexes, and among regions. *Turkish Journal of Zoology* 35(4), 451–466, 11 figs.

Westerström, A. (2005) Some notes on the herpetofauna in western Bulgaria. In: Ananjeva, N. & Tsinenko, O. (Eds.) *Herpetologia Petropolitana – Proceedings of the 12th Ordinary General Meeting of the Societas Europeae Herpetologica.* St. Petersburg, 241–244.

Wickramasinghe, L.J.M. (2016) A new canopy-dwelling species of *Dendrelaphis* (Serpentes: Colubridae) from Sinharaja, World Heritage Site, Sri Lanka. *Zootaxa* 4162(3), 504–518, 14 figs.

Wickramasinghe, L.J.M., Bandara, I.N., Vidanapathirana, D.R. & Wickramasinghe, N. (2019) A new species of *Aspidura* (Wagler, 1830) (Squamata: Colubridae: Natricinae) from Knuckles, World Heritage Site, Sri Lanka. *Zootaxa* 4559(2), 265–280, 12 figs.

Wickramasinghe, L.J.M., Vidanapathirana, D.R., Kandambi, H.K.D., Pyron, R.A. & Wickramasinghe, N. (2017b) A new species of *Aspidura* (Wagler, 1830) (Squamata: Colubridae: Natricinae) from Sri Pada sanctuary (Peak Wilderness), Sri Lanka. *Zootaxa* 4347(2), 275–292, 11 figs.

Wickramasinghe, L.J.M., Vidanapathirana, D.R., Rajeev, M.D.G. & Gower, D.J. (2017a) A new species of *Rhinophis* (Hemprich, 1820) (Serpentes: Uropeltidae) from the central hills of Sri Lanka. *Zootaxa* 4263(1), 153–164, 5 figs.

Wiens, J.J., Hutter, C.R., Mulcahy, D.G., Noonan, B.P., Townsend, T.M., Sites, J.W. & Reeder, T.W. (2012) Resolving the phylogeny of lizards and snakes (Squamata) with extensive sampling of genes and species. *Biology Letters* 2012(8), 1043–1046, 1 fig.

Williams, R.J., Ross, T.N., Morton, M.N., Daltry, J.C. & Isidore, L. (2016) Update on the natural history and conservation status of the Saint Lucia racer, *Erythrolamprus ornatus* (Garman, 1887) (Squamata: Dipsadidae). *Herpetology Notes* 9, 157–162, 3 figs.

Wilson, L.D. & Mata-Silva, V. (2014) Snakes of the genus *Tantilla* (Squamata: Colubridae) in Mexico: Taxonomy, distribution, and conservation. *Mesoamerican Herpetology* 1(1), 5–95, 31 figs.

Wilson, L.D. & Mata-Silva, V. (2015) A checklist and key to the snakes of the *Tantilla* clade (Squamata: Colubridae), with comments on taxonomy, distribution, and conservation. *Mesoamerican Herpetology* 2(4), 418–498, numerous figures.

Wilson, L.D., Mata-Silva, V. & Johnson, J.D. (2013) A conservation assessment of the reptiles of Mexico based on the EVS measure. *Amphibian & Reptile Conservation* 7(1), 1–47, numerous illustrations.

Wilson, L.D. & McCranie, J.R. (1992) Bothriechis marchi. *Catalogue of American Amphibians and Reptiles* 554.1–554.2, 3 figs.

Wilson, L.D. & Townsend, J.H. (2007) Biogeography and conservation of the herpetofauna of the upland pine-oak forests of Honduras. *Biota Neotropica* 7(1), 131–142, 2 figs.

Wilson, M.J. (2006) Herpetological observations on the Greek islands of Kefallinia and Zakynthos. *The Herpetological Bulletin* 97, 19–28, 8 figs.

Wilson, M.J. & Stille, B. (2014) Herpetofauna of Paxos, Ionian Islands, Greece, including two species new to the island. *Herpetozoa* 27(1/2), 108–112, 4 figs.

Winden, J. van der, Bogaerts, S. & Strijbosch, H. (1997) Herpetofauna des Göksu Deltas und des umliegenden gebirges, Türkei. *Salamandra* 33(1), 9–24, 6 figs.

Winkler, F.J.M., Waltenberg, L.M., Almeida-Santos, P., Nascimento, D.S. do, Vrcibradic, D. & Van Sluys, M. (2011) New records of anuran prey for *Thamnodynastes strigatus* (Günther, 1858) (Serpentes: Colubridae) in a high-elevation area of southeast Brazil. *Herpetology Notes* 4, 123–124, 3 figs.

Wood, D.A., Fisher, R.N. & Vandergast, A.G. (2014) Fuzzy boundaries: Color and gene flow patterns among parapatric lineages of the Western Shovel-nosed Snake and taxonomic implication. *PLOS ONE* 9(97494), 1–13, 5 figs.

Wood, D.A., Halstead, B.J., Casazza, M.L., Hansen, E.C., Wylie, G.D. & Vandergast, A.G. (2015) Defining population structure and genetic signatures of decline in the giant gartersnake (*Thamnophis gigas*): Implications for conserving threatened species within highly altered landscapes. *Conservation Genetics* 16(5), 1025–1039, 3 figs.

Woolrich-Piña, G.A., García-Padilla, E., DeSantis, D.L., Johnson, J.D., Mata-Silva, V. & Wilson, L.D. (2017) The herpetofauna of Puebla, Mexico: Composition, distribution, and conservation status. *Mesoamerican Herpetology* 4(4), 790–884, numerous color illus.

Woolrich-Piña, G.A., Ponce-Campos, P., Loc-Barragán, J., Ramírez-Silva, J.P., Mata-Silva, V., Johnson, J.D., García-Padilla, E. & Wilson, L.D. (2016) The herpetofauna of Nayarit, Mexico: Composition, distribution, and conservation status. *Mesoamerican Herpetology* 3(2), 376–448, numerous figs.

Wostl, E., Hamidy, A., Kurniawan, N. & Smith, E.N. (2017) A new species of wolf snake of the genus *Lycodon* H. Boie in Fitzinger (Squamata: Colubridae) from the Aceh Province of northern Sumatra, Indonesia. *Zootaxa* 4276(4), 539–553, 4 figs.

Wostl, E., Sidik, I., Trilaksono, W., Shaney, K.J., Kurniawan, N. & Smith, E.N. (2016) Taxonomic status of the Sumatran pitviper *Trimeresurus* (*Popeia*) *toba* (David, Petri, Vogel & Doria, 2009) (Squamata: Viperidae) and other Sunda Shelf species of the subgenus *Popeia*. *Journal of Herpetology* 50(4), 633–641, 2 figs.

Wüster, W. & Bérnils, R.S. (2011) On the generic classification of the rattlesnakes, with special reference to the Neotropical *Crotalus durissus* complex (Squamata: Viperidae). *Zoologia* 28(4), 417–419.

Wüster, W., Chirio, L., Trape, J.-F., Ineich, I., Jackson, K., Greenbaum, E., Barron, C. et al. (2018) Integration of nuclear and mitochondrial gene sequences and morphology reveals unexpected diversity in the forest cobra (*Naja melanoleuca*) species complex in Central and West Africa (Serpentes: Elapidae). *Zootaxa* 4455(1), 68–98, 9 figs.

Wylie, D.B. & Grünwald, C.I. (2016) First report of *Bothriechis schlegelii* (Serpentes: Viperidae: Crotalinae) from the state of Oaxaca, Mexico. *Mesoamerican Herpetology* 3(4), 1066–1067, 1 fig.

Wynn, A.H., Diesmos, A.C. & Brown, R.M. (2016) Two new species of *Malayotyphlops* from the northern Philippines, with redescriptions of *Malayotyphlops luzonensis* (Taylor) and *Malayotyphlops ruber* (Boettger). *Journal of Herpetology* 50(1), 157–168, 5 figs.

Xing, L., Caldwell, M.W., Chen, R., Nydam, R.L., Palci, A., Simões, T.R., McKellar, R.C. et al. (2018) A mid-Cretaceous embryonic-to-neonate snake in amber from Myanmar. *Science Advances* 2018(4)(eaat5042), 1–8, 4 figs.

Yaakob, N.S. (2003) A record of *Anomochilus leonardi* (Smith, 1940) (Anomochilidae) from Peninsular Malaysia. *Hamadryad* 27(2), 285–286, 1 fig.

Yadav, O. & Yankanchi, S. (2015) Occurrence of *Ophiophagus hannah* (Cantor, 1836) (Squamata, Elapidae) in Tillari, Maharashtra, India. *Herpetology Notes* 8, 493–494, 2 figs.

Yakin, B.Y., Şahın, U., Günay, U.K. & Tol, C.V. (2018) New records and rediscovery of some snakes from Gökçeada (Imbros), Turkey. *Biharean Biologist* 12(1), 17–20, 2 figs.

Yánez-Muñoz, M.H., Pozo-Zamora, G.M., Sornoza-Molina, F. & Brito-M., J. (2017) Dos nuevos registros de vertenrados en la dieta de *Corallus hortulanus* (Squamata: Boidae) en el noroeste de la Amazonía. *Cuadernos de Herpetología* 31(1), 41–47, 3 figs.

Yang, J.-H., Sung, Y.-H. & Chan, B.P.-L. (2013) A new species of the genus *Opisthotropis* (Günther, 1872) (Squamata: Colubridae: Natricidae) from Guangdong Province, China. *Zootaxa* 3646(3), 289–296, 4 figs.

Yang, J.-H. & Zheng, X. (2018) A new species of the genus *Calamaria* (Squamata: Colubridae) from Yunnan Province, China. *Copeia* 106(3), 485–491, 5 figs.

Yanthungbeni, M., Kwabamli, Hemsu, H., Phom, L., Phom, M.M., Yanthan, R., Vijila, V. & Nandakumar, R. (2018) Reptiles diversity in Dimapur of north east India. *World Scientific News* 114, 164–176, 11 figs.

Yao, C.-Y. (2004) Reptilian fauna and zoogeographic division of Gansu Province. *Sichuan Journal of Zoology* 23(3), 217–221, 1 fig. (in Chinese).

Yildiz, M.Z. (2011) Distribution and morphology of *Platyceps ventromaculatus* (Gray, 1834) (Serpentes: Colubridae) in southeastern Anatolia, Turkey. *North-Western Journal of Zoology* 7(2), 291–296, 3 figs.

You, C.-W., Poyarkov, N.A. & Lin, S.-M. (2015) Diversity of the snail-eating snakes *Pareas* (Serpentes, Pareatidae) from Taiwan. *Zoologica Scripta* 44(4), 349–361, 4 fgs.

Yousefkhani, S.S.H., Yousefi, M., Khani, A. & Rastegar-Pouyani, E. (2014) Snake fauna of Shirahmad wildlife refuge and Parvand protected area, Khorasan Razavi province, *Iran. Herpetology Notes* 7, 75–82, 4 figs.

Zadhoush, B., Van den Brink, M. & Rajabizadeh, M. (2016) Geographic distribution. *Eirenis rechingeri. Herpetological Review* 47(2), 262.

Zaher, H., Apesteguía, S. & Scanferla, C.A. (2009) The anatomy of the upper Cretaceous snake *Najash rionegrina* Apesteguía & Zaher, 2006, and the evolution of limblessness in snakes. *Zoological Journal of the Linnean Society* 156(4), 801–826, 14 figs.

Zaher, H., Arredondo, J.C., Valencia, J.H., Arbeláez, E., Rodrigues, M.T. & Altamirano-Benevides, M. (2014) A new Andean species of *Philodryas* (Dipsadidae: Xenodontinae) from Ecuador. *Zootaxa* 3785(3), 469–480, 6 figs.

Zaher, H., Murphy, R.W., Arredondo, J.C., Graboski, R., Machado-Filho, P.R., Mahlow, K., Montingelli, G.G. et al. (2019) Large-scale molecular phylogeny, morphology, divergence-time estimation, and the fossil record of advanced caenophidian snakes (Squamata: Serpentes). *PLOS ONE* 14(5)(e0216148), 1–82, 30 figs.

Zaher, H., Yánez-Muñoz, M.H., Rodrigues, M.T., Graboski, R., Machado, F.A., Altamirano-Benevides, M., Bonatto, S.L. & Grazziotin, F.G. (2018) Origin and hidden diversity within the poorly known Galápagos snake radiation (Serpentes: Dipsadidae). *Systematics and Biodiversity* 16(7), 614–642, 9 figs.

Zakaria, N., Senawi, J., Musa, F.H., Belabut, D., Onn, C.K., Nor, S.M. & Ahmad, N. (2014) Species composition of amphibians and reptiles in Krau Wildlife Reserve, Pahang, Peninsular Malaysia. *Check List* 10(2), 335–343, 4 figs.

Zamprogno, C., Zamprogno, M. das G.F. & Lema, T. de (1998) Contribuição ao conhecimento de *Apostolepis cearensis* Gomes, 1915, serpent fossorial do Brasil (Colubridae: Elapomorphinae). *Acta Biologica Leopoldensia* 20(2), 207–216, 4 figs.

Zaracho, V.H., Ingaramo, M. del R., Semhan, R.V., Etchepare, E.G., Acosta, J.L., Falcione, A.C. & Álvarez, B.B. (2014) Herpetofauna de la Reserva Natural Provincial Isla Apipé Grande (Corrientes, Argentina). *Cuadernos de Herpetología* 28(2), 153–160, 1 fig.

Zarrintab, M., Milto, K.D., Eskandarzadeh, N., Zangi, B., Jahan, M., Kami, H.G., Rastegar-Pouyani, N., Rastegar-Pouyani, E. & Rajabizadeh, M. (2017) Taxonomy and distribution of sand boas of the genus *Eryx* (Daudin, 1803) (Serpentes: Erycidae) in Iran. *Zoology in the Middle East* 63(2), 117–129, 9 figs.

Zhang, B.-L. & Huang, S. (2013) Relationship of Old World *Pseudoxenodon* and New World Dipsadinae, with comments on underestimation of species of Chinese *Pseudoxenodon*. *Asian Herpetological Research* 4(3), 155–165, 3 figs.

Zhang, B.-W., Huang, X., Pan, T., Zhang, L., Zhou, W., Song, T. & Han, D. (2013) Systematics and species validity of the Dabieshan Pit Viper *Protobothrops dabieshanensis* Huang et al., 2012: Evidence from a mitochondrial gene sequence analysis. *Asian Herpetological Research* 4(4), 282–287, 2 figs.

Zheng, Y. & Wiens, J.J. (2016) Combining phylogenomic and supermatrix approaches, and a time-calibrated phylogeny for squamate reptiles (lizards and snakes) based on 52 genes and 4162 species. *Molecular Phylogenetics and Evolution* 94, 537–547, 2 figs.

Zhong, C.-F. (2004) Reptilian fauna and zoogeographic division of Jiangxi Province. *Sichuan Journal of Zoology* 23(3), 222–229, 1 fig.

Zhong, G., Liu, Q., Li, C., Peng, P. & Guo, P. (2017) Sexual dimorphism and geographic variation in the Asian Lance-headed Viper *Protobothrops mucrosquamatus* in the mainland China. *Asian Herpetological Research* 8(2), 118–122, 3 figs.

Zhong, G.-H., Chen, W.-D., Liu, Q., Zhu, F., Peng, P.-H. & Guo, P. (2015) Valid or not? Yunnan mountain snake *Plagiopholis unipostocularis* (Serpentes: Colubridae: Pseudoxenodontinae). *Zootaxa* 4020(2), 390–396, 3 figs.

Zhou, Z.-Y., Sun, Z.-Y., Qi, S., Lu, Y.-Y., Lyu, Z.-T., Wang, Y.-Y., Li, P. & Ma, J.-Z. (2019) A new species of the genus *Hebius* (Squamata: Colubridae: Natricinae) from Hunan Province, China. *Zootaxa* 4674(1), 68–82, 6 figs.

Zhu, G.-X., Wang, Y.-Y., Takeuchi, H. & Zhao, E.-M. (2014) A new species of the genus *Rhabdophis* (Fitzinger, 1843) (Squamata:

Colubridae) from Guangdong Province, southern China. *Zootaxa* 3765(5), 469–480, 5 figs.

Ziegler, T., David, P., Ziegler, T.N., Pham, C.T., Nguyen, T.Q. & Le, M.D. (2018b) Morphological and molecular review of Jacob's Mountain Stream Keelback *Opisthotropis jacobi* (Angel & Bourret, 1933) (Squamata: Natricidae) with description of a sibling species from northern Vietnam. *Zootaxa* 4374(4), 476–496, 12 figs.

Ziegler, T., Luu, V.Q., Nguyen, T.T., Ha, N.V., Ngo, H.T., Le, M.D. & Nguyen, T.Q. (2019a) Rediscovery of Andrea's keelback, *Hebius andreae* (Ziegler & Le, 2006): First country record for Laos and phylogenetic placement. *Revue Suisse de Zoologie* 126(1), 61–71, 6 figs.

Ziegler, T., Ngo, H.N., Pham, A.V., Nguyen, T.T., Le, M.D. & Nguyen, T.Q. (2018a) A new species of *Parafimbrios* from northern Vietnam (Squamata: Xenodermatidae). *Zootaxa* 4527(2), 269–276, 4 figs.

Ziegler, T., Nguyen, T.Q., Pham, C.T., Nguyen, T.T., Pham, A.V., Schingen, M. van, Nguyen, T.T. & Le, M.D. (2019b) Three new species of the snake genus *Achalinus* from Vietnam (Squamata: Xenodermatidae). *Zootaxa* 4590(2), 249–269, 12 figs.

Ziegler, T., Pham, C.T., Nguyen, T.V., Nguyen, T.Q., Wang, J., Wang, Y.-Y., Stuart, B.L. & Le, M.D. (2019c) A new species of *Opisthotropis* from northern Vietnam previously misidentified as the Yellow-spotted Mountain Stream Keelback *O. maculosa* (Stuart & Chuaynkern, 2007) (Squamata: Natricidae). *Zootaxa* 4613(3), 579–586, 3 figs.

Ziegler, T., Tran, V.A., Babb, R.D., Jones, T.R., Molcr, P.E., Van Devender, R.W. & Nguyen, T.Q. (2019d) A new species of reed snake, *Calamaria* Boie, 1827 from the central highlands of Vietnam (Squamata: Colubridae). *Revue Suisse de Zoologie* 126(1), 17–26, 7 figs.

Ziegler, T., Tran, D.T.A., Nguyen, T.Q., Perl, R.G.B., Wirk, L., Kulisch, M., Lehmann, T. et al. (2014) New amphibian and reptile records from Ha Giang Province, northern Vietnam. *Herpetology Notes* 7, 185–201, 6 figs.

Ziegler, T., Ziegler, T.N., Peusquens, J., David, P., Vu, T.N., Pham, C.T., Nguyen, T.Q. & Le, M.D. (2017) Expanded morphological definition and molecular phylogenetic position of the Tam Dao mountain stream keelback *Opisthotropis tamdaoensis* (Squamata: Natricidae) from Vietnam. *Revue Suisse de Zoologie* 124(2), 377–389, 4 figs.

Zinenko, O., Avcı, A., Spitzenberger, F., Tupikov, A., Shiryaev, K., Bozkurt, E., Ilgaz, Ç. & Stümpel, N. (2016b) Rediscovered and critically endangered: *Vipera anatolica* Eiselt & Baran, 1970, of the western Taurus Mountains (Turkey), with remarks on its ecology (Squamata: Serpentes: Viperidae). *Herpetozoa* 28(3/4), 141–148, 2 figs.

Zinenko, O., Sovic, M., Joger, U. & Gibbs, H.L. (2016a) Hybrid origin of European vipers (*Vipera magnifica* and *Vipera orlovi*) from the Caucasus determined using genomic scale DNA markers. *BMC Evolutionary Biology* 16(76), 1–13, 7 figs.

Zinenko, O., Stümpel, N., Mazanaeva, L., Bakiev, A., Shiryaev, K., Pavlov, A., Kotenko, T. et al. (2015) Mitochondrial phylogeny shows multiple independent ecological transitions and northern dispersion despite of Pleistocene glaciations in meadow and steppe vipers (*Vipera ursinii* and *Vipera renardi*). *Molecular Phylogenetics and Evolution* 84, 85–100, 6 figs.

Zuazo, Ó., Freitas, I., Zaldívar, R. & Martínez-Freiría, F. (2019) Coexistence and intermediate morphological forms between *Vipera aspis* and *V. latastei* in the intensive agriculture fields of north-western Iberian system. *Boletín de la Asociación Herpetológica Española* 30(1), 35–41, 4 figs.

Zuffi, M.A.L. (2008) Colour pattern variation in populations of the European whip snake, *Hierophis viridiflavus*: Does geography explain everything? *Amphibia-Reptilia* 29(2), 229–233, 1 fig.

Zug, G.R. (2013) *Reptiles and Amphibians of the Pacific Islands – A Comprehensive Guide.* University of California Press, Berkeley, x+306, 29 figs., 35 plates.

Index

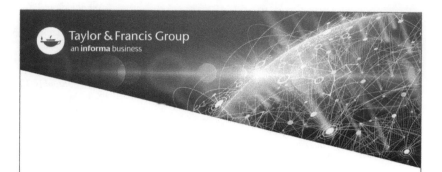